ECOLOGY IN A CHANGING WORLD

ECOLOGY IN A CHANGING WORLD

Michael P. Marchetti
St. Mary's College of California

Julie L. Lockwood
Rutgers University

Martha F. Hoopes
Mount Holyoke College

W. W. NORTON & COMPANY
Celebrating a Century of Independent Publishing

W. W. Norton & Company has been independent since its founding in 1923, when William Warder Norton and Mary D. Herter Norton first published lectures delivered at the People's Institute, the adult education division of New York City's Cooper Union. The firm soon expanded its program beyond the Institute, publishing books by celebrated academics from America and abroad. By midcentury, the two major pillars of Norton's publishing program—trade books and college texts—were firmly established. In the 1950s, the Norton family transferred control of the company to its employees, and today—with a staff of five hundred and hundreds of trade, college, and professional titles published each year—W. W. Norton & Company stands as the largest and oldest publishing house owned wholly by its employees.

Editors: Betsy Twitchell and Jake Schindel
Developmental Editor: Sunny Hwang
Senior Associate Managing Editor, College: Carla L. Talmadge
Editorial Assistants: Danny Vargo and Clare Lewis
Managing Editor, College: Marian Johnson
Associate Director of Production, College: Benjamin Reynolds
Media Editor: Kate Brayton
Associate Media Editor: Jasmine N. Ribeaux
Media Project Editor: Jesse Newkirk
Media Assistant Editor: Kara Zaborowsky
Managing Editor, College Digital Media: Kim Yi
Ebook Production Manager: Kate Barnes
Marketing Manager, Biology: Lib Triplett
Design Director: Rubina Yeh
Designer: DeMarinis Design LLC
Director of College Permissions: Megan Schindel
Permissions Associate: Patricia Wong
Senior Photo Editor: Thomas Persano
Composition: Graphic World, Inc./Project Manager: Gary Clark
Illustrations: DeMarinis Design LLC; Graphic World, Inc.; Spark Life Science Visuals
Manufacturing: Transcontinental—Beauceville QC

Permission to use copyrighted material is included at the back of the book.

Library of Congress Cataloging-in-Publication Data

Names: Marchetti, Michael P., author. | Lockwood, Julie L., author. |
  Hoopes, Martha F., author.
Title: Ecology in a changing world / Michael Marchetti, St. Mary's College
  of California, Julie Lockwood, Rutgers University, Martha Hoopes, Mount
  Holyoke College.
Description: First edition. | New York, NY : W.W. Norton & Company, Inc.,
  [2023] | Includes bibliographical references and index.
Identifiers: LCCN 2022030836 | **ISBN 9780393892307 (paperback)** | ISBN
  9780393892338 (epub)
Subjects: LCSH: Ecology--Research.
Classification: LCC QH541.2 .M36 2023 | DDC 577.072--dc23/eng/20220713
LC record available at https://lccn.loc.gov/2022030836

W. W. Norton & Company, Inc., 500 Fifth Avenue, New York, NY 10110
wwnorton.com
W. W. Norton & Company Ltd., 15 Carlisle Street, London W1D 3BS

1 2 3 4 5 6 7 8 9 0

CONTENTS

chapter 5   Individuals: Physiology and Behavior 165

chapter 6   Populations 209

chapter 7 Competition 259

chapter 8 Exploitation 307

chapter 9 — Mutualism and Facilitation 353

chapter 10 — Multispecies Interactions and Food Webs 387

chapter 11   Biodiversity 433

chapter 12   Spatial Dynamics 483

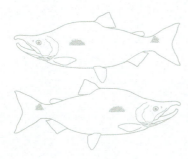

chapter 15 — Anthropocene Ecology 621

PREFACE

Narrative Themes

When the three of us considered the scope of an ecology textbook that would enhance our teaching, we decided that it *must* clearly articulate the ecological consequences of the largest single challenge facing the planet today: global anthropogenic change. Evidence that the entire Earth system is changing under the weight of human influences is now overwhelming. The climate is warming, seas are rising and becoming more acidic, fires and hurricanes are more common, and urban and agricultural lands are supplanting wildlands at high rates. Humans appropriate more than 20% of the Earth's primary productivity each year to supply us with food, fuel, wood, and fiber. Given these effects, we felt that the science and teaching of ecology must address the reality of a changed and changing planet. We, therefore, made discussion of anthropogenic influences on ecological systems the foundational element of every chapter in the book.

This approach allowed us to weave information on the historical, cultural, and social context of ecological interactions into the fabric of the book. For example, in Chapter 3 (Global Climate and Biomes), we directly address the effects of greenhouse gases, the evidence for their anthropogenic sources, and the resultant shifts in average surface temperatures and sea levels. Similarly, in Chapter 6 (Populations), we introduce students to how population models inform efforts to recover populations that are at risk of extinction due to human actions. In Chapter 14 (Nutrient Cycling and Ecosystem Services), we explicitly consider how human activities alter the carbon, nitrogen, and phosphorus cycles and how those alterations tie to loss of different habitat types. It seemed essential to roll the human component into explanations of ecological principles, and we feel doing so highlights the value of the science and makes the material more accessible for students.

Pedagogical Approach

We have taught for a mind-boggling combined total of almost six decades. All those lectures, discussions, labs, and hours of grading convinced us that we wanted our text to reflect the realities of modern ecology and science in two fundamental ways. First, learning and teaching need to mirror the way ecologists actually conduct science, which means we felt a strong need to present observations, use them to create hypotheses and conceptual models, formulate those conceptual models into quantitative models where appropriate, and then confront those models with real-world evidence. We wanted students to see the feedbacks within this sequence, appreciate the utility and limitations of experiments, and see how models help to extend our ability to generalize and explore the boundaries of our current understanding. We strongly feel that this knowledge is as important as the facts and theories we seek to share with students.

Traversing these steps requires the mastery of quantitative concepts and tools, which are common to several branches of science and even to everyday decision-making. Thus, our second and likely more revolutionary goal was to build the text from the ground up as a way to teach basic quantitative thinking and data literacy skills to students. Many readers will interpret that statement as signifying that we put numbers to every concept, but we mean something slightly different. All scientific concepts involve a model of some kind. Not all of these models need quantitative approaches, but they all demand a clarity around relationships, interactions, and expectations. We tried to make those kinds of models (conceptual models) the backbone of even the least quantitative chapters so that students could get used to thinking about factors and outcomes, cause and effect, evidence, and ways to alter models as information changes. With those pieces in place, it was easy to draw on our experiences teaching quantitative aspects of ecology to pace the content and flow of quantitative information and make it accessible to even the more math-phobic students.

We have integrated the nuts and bolts of ecology with a deliberate dive into quantitative thinking for two reasons. First, ecology is by its nature a quantitative science often focused on change, and quantitative reasoning gives us ways to frame and explore such change. Second, we firmly believe quantitative reasoning is not a distraction or a frill but instead a central tenet of ecology (and modern science) that deepens student understanding and appreciation of the natural world.

We define quantitative thinking as cognitively working with and utilizing measured or numerical information. Quantitative thinking is sometimes thought to be synonymous with mathematics, and quite often the two are clearly linked. For example, both involve learnable skills with practical applications. Yet there are strong differences between mathematics and quantitative thinking. Although a mathematician might get excited over the formal process of proving a theorem or deriving an equation, a reasonably well-educated student can apply quantitative thinking skills to a variety of real-world issues. In the context of ecology, for example, understanding the inherent uncertainty involved with global climate change predictions or the differences between two US National Park Service plans to restore mountain lion populations are highly relevant and practical uses of quantitative thinking.

In a similar manner, we define data literacy as the ability to ask, answer, and communicate meaningful questions by collecting, analyzing, and making sense of data. Data literacy fosters an understanding of what numbers and data actually mean, and thankfully, it is also a skill that one can learn. Data literacy and quantitative thinking, therefore, go hand in hand. A student who becomes data literate has acquired such skills as how to read graphs and charts appropriately, draw reasonable conclusions from data, and recognize when data are being used in misleading or inappropriate ways.

We begin each chapter with an easy-to-hold, logical conceptual anchor, adding increasingly complex and quantitative content as the chapter progresses while preserving the narrative thread in an accessible and informal manner throughout. We believe that this approach provides a roadmap for students to learn the foundational concepts that underlie the discipline while offering them tools to forge their own pathways to deeper insight. The ability to break an idea into its constituent parts and to think analytically about how those components fit together is essential to writing equations, but it is also essential to general problem-solving. Acquainting students with these approaches gives them more tools for communicating about ecological phenomena, and it also prepares them to interact with the many forms of quantitative news, graphs, and potential insights that will continue to infuse our modern lives. The rise of the "information age" has created an increasingly quantitative world. An understanding of data and quantitative reasoning helps students navigate complex news cycles, job markets, and everyday decision-making.

Organization and Coverage

Traditionally, textbooks struggle to cover an enormous body of content and often provide the reader with an overstuffed text. Instructors are then given the task of picking and choosing among topics because it is generally impossible to teach all the material in these texts during a single-semester course. Although authors work hard to tie this dense material together, the narrative threads and intellectual scaffolding binding the concepts and ideas are often lost in the tangled web of details that confront students or in the instructor's picking and choosing.

We feel that a text should pace the material in much the same way that instructors move through content when addressing a classroom of students, with each chapter providing material for approximately one week of lectures. We therefore have designed this text for use in single-semester, introductory ecology classes that are taught across a variety of institutional settings. We expect the text to be useful for students with a wide array of backgrounds and future plans. We have grouped the material into 15 chapters with the hope of fitting the textbook into a traditional semester.

Our commitment to this pedagogical approach made it necessary to narrow our focus, homing in on fewer theories, models, and examples in each chapter. We recognize that by doing so we made the choice not to dive too deeply into several ecological principles that may appear in other texts. For instructors who desire more complexity in their courses than we provide, this text can serve as a primary pedagogical tool, allowing them greater latitude to add primary literature and to control their content. Our text provides an opportunity for instructors to "flip" their classrooms; chapters can serve as highly accessible outside-of-class "lectures," freeing in-class time for meaningful discussions and experiential exercises.

In that vein, although we use and teach all the appropriate terms, we avoid unnecessary jargon and overly academic styling. Each chapter has some challenging content, but the words and explanations themselves should not be part of the challenge. To help students assess their understanding of the material before exiting the chapter, we also include both Conceptual and Quantitative Questions at the end of each chapter; answers to select questions are provided at the end of the book to allow students to evaluate their responses.

Resources

Simulations

A set of simulations based on a subset of classic ecological models allows students to adjust parameters and see the effects in a visual representation. These simulations are assignable in Smartwork and are also housed on the book's digital landing page. They are accompanied by suggestions in the Norton Teaching Tools site for use within in-person or virtual classrooms.

Smartwork
digital.wwnorton.com/ecology

Smartwork is an online homework platform that allows the instructor to assign questions with answer-specific feedback, hints, and solutions. Smartwork questions and activities vary in type—including ranking, sorting, labeling, and multiple choice—to engage students and help them refine their understanding of the course concepts. Written by and for ecology instructors and aligned to the textbook's learning objectives, questions are customizable by instructors, and a premade assignment is also available for each chapter. Quantitative questions support quantitative reasoning and data interpretation skills. Application-level questions emphasize the relevance of ecology. Questions also incorporate the book's vibrant art as well as the interactive simulations based on ecological models that allow students to practice interpreting ecological data.

Norton Ebook
digital.wwnorton.com/ecology

The Norton Ebook for *Ecology in a Changing World* can be viewed on all computers and mobile devices, and it allows students to add notes, view vocabulary definitions, search, bookmark, and read offline. All the art in the text has been specifically optimized for a digital reading experience and can be expanded for a closer look. Students can also engage with simulations of important ecological models directly within the ebook as they read.

Norton Teaching Tools

The Norton Teaching Tools site is a searchable database of active learning–focused course materials. Filter the library by chapter and content type to find class activities, case studies with discussion prompts and questions, information on the simulations in your course, and more.

Norton Testmaker

The Test Bank features 600 questions, including both multiple-choice and short-answer question types. Expert accuracy checkers have ensured that every question in the Test Bank is scientifically reliable and truly tests students' understanding of the most important topics in each chapter so that the questions can be assigned with confidence.

The Test Bank utilizes Norton Testmaker, which allows instructors to create assessments for their courses from anywhere with an internet connection, without the need to download files or install specialized software. The format makes it easy to search and filter Test Bank questions by chapter, type, difficulty, learning objectives, and other criteria. Instructors can also customize Test Bank questions to fit each course and easily export tests or Norton's ready-to-use quizzes to Microsoft Word or Common Cartridge files for the school's LMS.

PowerPoint Lecture Slides

Designed for classroom use in both live and online settings, these slides present section headings, photographs, and line art from the book in a form that has been optimized for use in the PowerPoint environment.

Art PowerPoint Slides and Files

Images and figures from the book are available to download in PowerPoint and JPG formats.

Resources for your LMS

Easily add Norton's digital resources to your online, hybrid, or lecture course.

ACKNOWLEDGMENTS

First, we give a hearty thank you to Ben Roberts for first suggesting this idea to us many years ago. His fine publishing company would have been a great place for this text. We always enjoyed our discussions and interactions with Ben, and we thank him for his fantastic introduction to the wonderful Norton team. At Norton, we have been lucky to work with the phenomenal Betsy Twitchell, our amazing editor and chief. You took a chance with us after another publisher wanted to go in a very different direction, and you convinced the folks at Norton that we had something worth their time. Your strong but gentle guidance with us over the many years helped shape the nature and accessibility of this book more than you may realize. All three of us always perked up when we had meetings with you (sadly, way too many of them were virtual due to circumstances), and we greatly respect your ideas, direction, and vision. We couldn't have asked for a better editor.

We also have to thank Sunny Hwang as our developmental editor and the first/second/third reader of the text. Sunny's dedication to this project was legendary despite all the unbelieveable challenges. In our wildest dreams we couldn't have asked for a better, more knowledgeable reader. Your comments were always spot on, showing that you not only understood the nuances of what we were saying but also had excellent and insightful suggestions that always made the text better. If it were possible, we would have you read and comment on all our scientific publications before sending them out. Thanks a ton.

The artwork that is such a fundamental and integrated part of a text such as this would not be possible without the spectacular abilities of artist/illustrator extraordinaire Anne DeMarinis. Our weekly Friday meetings were often epic in their length and scope but always thoroughly enjoyable. Anne took our admittedly sketchy sketches and made them into educational artworks, always with a smile on her face and seeming joy in her heart. We wish we had our own personal DeMarinis to brighten and illustrate our other teaching duties because all our students would benefit greatly.

We also wish to thank Gwen Burda, Carla Talmadge, and Jake Schindel. Your help with the final rounds of the editing, construction, and polish was invaluable. There is no way we could have found all the incorrect en dashes and missed punctuation. At least one of us still falls asleep thinking about side discussions of why compound adjectives are hyphenated while compound nouns are not. Yes, we are nerds, but the lovely people at Norton are even more accomplished nerds, and that has been wonderful for us. Also, a project like this would never have been finished without the tireless assistance of people like Danny Vargo, Clare Lewis, and Tommy Persano. Thanks for all your hard work.

The real thanks for the kind of herculean effort it requires to create such a tome as this lies firmly with our friends and family. MPM would like to thank St. Mary's College of California and the Fletcher Jones Foundation for giving him a job where efforts like this are both encouraged and valued. Not every institution of higher learning makes room for this type of nonresearch activity. MPM is grateful that St. Mary's and its Biology Department offered an endowed chair that he just couldn't refuse, thereby luring him from his more northern Northern California home. He also thanks Dean Roy Wensley, who consistently made accommodations for the amount of effort a task like this requires. MFH is thankful to Mount Holyoke College, her colleagues, and the many students who helped hone lectures that found their way into this text. JLL is appreciative of Rutgers University and her colleagues and graduate students for their consistent support.

Many ecological colleagues helped with this work over the years. This includes Sudeep Chandra (University of Nevada, Reno), Seth Riley (US National Park Service), Pete Trenham (herpetologist extreme), Peter Moyle (University of California, Davis), Rich VanBuskirk (Pacific University), Peter Hodum (University of Puget Sound), Jim Pesevento (St. Mary's College of California), Christine Parisek (University of California, Davis), Gordon Wolfe (California State University, Chico), Tag Engstrom (California State University, Chico), and Don Miller (California State University, Chico). Thank you all for your help, as little or big as it was.

We also want to extend thanks to the people who helped mentor us as young ecologists. Many of the ideas, concepts, and understanding in basic and applied ecology discussed in this book are a direct result of their influence. MPM thanks Peter Moyle and the late Cathy Toft, whose guidance made him a much better scientist. He hopes that Dr. Toft is somewhere nice, smiling alongside her border collies. JLL thanks Michael Moulton and Dan Simberloff for their vision, kindness, and guidance over the years. MFH thanks Susan Harrison for her gentle guidance (and many discussion questions), Alan Hastings for his focus on the important things in life and his love of math (even while counting on his fingers), and Kevin Rice for his real-world practicality.

But, of course, the real inspiration for the day-to-day efforts needed to construct a book like this all come from home.

MPM: Al Marchetti was the first scientist I ever met and one who set the standard for professionalism and honesty in his work. Thanks also go to David Marchetti (Western Colorado University) whose good nature, geologic acumen, and spectacular fly-fishing skills set a high bar to emulate. It is difficult to express sufficient gratitude to Sarah Zaner and Matteo Marchetti. Marrying someone like Sarah, who is so intimately familiar with the publishing industry and education, was a genius move on my part. I have to this day never met anyone with a more blazing intellect, open heart, and giving nature than my partner, Sarah. Walking through life's challenges with you has made me a better person, scientist, and human being. Thank you from the bottom of my heart! The many late nights, long weeks, sabbatical leaves, photography excursions, bike rides, epic hikes, and constant search for salamanders and mushrooms would be more than enough to break anyone. But both you and Matteo stuck with me and are so clearly my foundation, my strength, and 100% of my inspiration. I hope that this small bit of scientific education will help us as a planet lean into a more sustainable future for Matteo and all the other young people living today. You all deserve our best efforts, and this is mine.

JLL: Thanks go to my spouse, Tabby, and our two bright, curious, and funny sons: Henry and Tanner. Looking at page proofs and engaging in long talks about various critters with Tanner at night, curled up in his bed, was a highlight of this process. I also thank my parents, Dianne and Bob Lockwood, for their encouragement and support even when they were not really sure what I was doing with my life. Thank you to my brothers, David and Robert, for teaching me resilience and the value of unconditional love. Finally, thank you to the gigantic Fenn clan for making me one of the family, including providing a space in the Adirondacks to unwind from textbook writing and ride out stretches of the pandemic.

MFH: I would never have written a book like this without the lifelong focus on questions and discovery from Abby Hoopes and the constant refining struggle of discussion with four annoying, brilliant, and fabulous sisters. The day-to-day work of this book, though, would not have been possible without the patience of Crystal, who was my inspiration throughout. C, this book is for you. May it help to make the world a slightly better place so that there is still some beauty and some wildness left for you when you are my age.

Reviewers

Finally, we are extraordinarily grateful for the many instructors whose feedback helped shape the development of this book over the years, enabling it to become what we hope is a useful and impactful text for ecology students and teachers alike. These reviewers are:

Henry Adams, Washington State University
David Allard, Texas A&M University, Texarkana
Peter Alpert, University of Massachusetts, Amherst
David Argent, California University of Pennsylvania
Bradley Bergstrom, Valdosta State University, Georgia
Joydeep Bhattacharjee, University of Louisiana, Monroe
Brent Blair, Xavier University, Ohio
Christopher Bloch, Bridgewater State University, Massachusetts
Jere Boudell, Clayton State University, Georgia
Jennifer Boyd, University of Tennessee, Chattanooga
Justin Boyles, Southern Illinois University, Carbondale
Rebekah Chapman, Georgia State University
D. Liane Cochran-Stafira, Saint Xavier University, Illinois
Phyllis Coley, University of Utah
Scott Connelly, University of Georgia
Heather Cutway, Mercer University, Georgia
Joseph D'Silva, Norfolk State University, Virginia
Matthew Dugas, Illinois State University
Mara Evans, University of North Carolina, Chapel Hill
John Fauth, University of Central Florida

Caitlin Fisher-Reid, Bridgewater State University, Massachusetts
Thomas Gehring, Central Michigan University
Frank Gilliam, University of West Florida
Laura Gonzalez, University of Texas, Austin
Kirsten Grorud-Colvert, Oregon State University
Thayer Hallidayschult, University of Oklahoma
Floyd Hayes, Pacific Union College, California
Christiane Healey, University of Massachusetts, Amherst
Siti Hidayati, Middle Tennessee State University
Dagne Hill, Grambling State University, Louisiana
Tara Jo Holmberg, Northwestern Connecticut Community College
Traci Hudson, University of Arkansas, Pine Bluff
Jodee Hunt, Grand Valley State University, Michigan
Peter Jenkins, Charleston Southern University, South Carolina
Rupesh Kariyat, University of Texas, Rio Grande Valley
Eric Keeling, State University of New York, New Paltz
Zion Klos, Marist College, New York
Ned Knight, Linfield University, Oregon

Marguerite Koch-Rose, *Florida Atlantic University*
William Kroll, *Loyola University, Chicago*
Eric Long, *Seattle Pacific University*
Zachary Long, *University of North Carolina, Wilmington*
Sheila Lyons-Sobaski, *Albion College, Michigan*
Silvia Maciá, *Barry University, Florida*
Jay Mager, *Ohio Northern University*
Juan Luis Mata, *University of Southern Alabama*
Jennie McLaren, *University of Texas, El Paso*
Joseph Milanovich, *Loyola University, Chicago*
Maynard Moe, *California State University, Bakersfield*
Cy Mott, *Eastern Kentucky University*
Crima Pogge, *City College of San Francisco*
Mark Pyron, *Ball State University, Indiana*
Molly Redmond, *University of North Carolina, Charlotte*
Aaron Reedy, *University of Virginia*
Ulrich Reinhardt-Segawa, *Eastern Michigan University*
Peter Sakaris, *Georgia Gwinnett College*
Michael Shaughnessy, *Northeastern State University, Oklahoma*
Dennis Shiozawa, *Brigham Young University, Utah*

Ann Showalter, *Clayton State University, Georgia*
Matthew Simmons, *University of Minnesota, Crookston*
Joseph Staples, *University of Southern Maine*
Donald Strong, *University of California, Davis*
Stephen Sumithran, *Eastern Kentucky University*
Thilina Surasinghe, *Bridgewater State University, Massachusetts*
Cara Thompson, *Arizona State University*
Ryan Vazquez, *Texas Tech University*
Robert Wallace, *Ripon College, Wisconsin*
Lixin Wang, *Indiana University–Purdue University Indianapolis*
Gideon Wasserberg, *University of North Carolina, Greensboro*
Sara Weaver, *Texas A&M University, San Antonio*
Peter White, *University of North Carolina, Chapel Hill*
Susan Whitehead, *Virginia Tech*
Christopher Widmaier, *Rochester Institute of Technology, New York*
Marsha Williams, *University of Texas, Tyler*

ABOUT THE AUTHORS

 **Michael P. Marchetti** is the Fletcher Jones Endowed Chair of Biology at St. Mary's College of California. Michael holds an MSc and PhD in ecology from the University of California, Davis, and a BSc in biology and chemistry from Bucknell University. He has over 30 years of experience working on aquatic ecosystems, conservation, invasion biology, and food web ecology. He has worked in California, Oregon, Nevada, Mexico, Hawai'i, and New Zealand and has authored numerous publications. Michael also coauthored *Invasion Ecology* and *Protecting Life on Earth*.

 **Julie L. Lockwood** is professor of ecology, evolution, and natural resources at Rutgers University, where she currently serves as the director of the Institute of Earth, Ocean and Atmospheric Sciences. Julie holds a PhD in zoology from the University of Tennessee and an MSc and BSc in biology from Georgia Southern University. She oversees a research program that encompasses biodiversity conservation, invasion science, and climate change. She is a fellow of the Ecological Society of America and of the American Association for the Advancement of Science. Beyond publishing research papers, she coauthored *Avian Invasions* and *Invasion Ecology* and coedited *Biotic Homogenization* and *Coastal Conservation*.

 **Martha F. Hoopes** is professor of biology at Mount Holyoke College, where she teaches ecology, conservation biology, and biostatistics. Martha holds a PhD in ecology from the University of California, Davis, and earned a BA in English and physics from Williams College. Her research focuses on spatial interactions in invasions and conservation biology. She has authored numerous scholarly papers and book chapters and coauthored *Invasion Ecology* and several chapters in *Metacommunities: Spatial Dynamics and Ecological Communities* (Holyoak, Leibold, and Holt, eds).

ECOLOGY IN A CHANGING WORLD

In the Anthropocene, the philosophical division between nature-dominated landscapes and human-dominated landscapes is dissolving. This view of a tropical rainforest just outside the major urban environment of Panama City, Panama, illustrates this point well. Increasingly, the urban is encroaching on the nonurban, and vice versa.

ECOLOGY IN THE ANTHROPOCENE

LEARNING OBJECTIVES

- Explain what the term *ecology* means, and describe what ecologists study.

- Define the term *Anthropocene*.

- Distinguish among the four approaches ecologists use in their science.

- Differentiate between conceptual and mathematical models.

- Explain the scientific method and its iterative nature.

1.1 The Ecological "House"

Many readers of this book likely have some idea about what ecology is and what ecologists do, but we, the authors, have found that even our own families are sometimes a little confused about where ecology sits in the realm of the sciences. At the simplest level, ecology is a branch of biology. Academic programs around the world often split biology into two subdisciplines, one that focuses on processes occurring inside organisms (e.g., intra- and intercellular) and another that focuses on processes occurring between whole organisms and their environments. Ecology rests firmly in the latter category. As you will discover in this book, though, these realms of biology often intertwine, and there are many ways to specialize within the field of ecology.

The word *ecology* derives from the German word *Oekologie*, meaning roughly the "doctrine of the household." The German scientist Ernst Haeckel first used the term in 1866. If ecology is the study of a household, then the "house" is the place or location where all the organisms live, which also includes the nonbiological aspects of the place, such as the air, water, soil, and rocks (**Figure 1.1**). The term **organism** refers to a single individual of any type of living creature on the planet, including trees, birds, insects, worms, microbes, and all the other species that inhabit the Earth. A more complete definition of **ecology** is *the study of the interactions that determine the distribution and abundance of organisms*. These interactions can be among individuals that are members of the same species, among individuals of two or more species, or among individuals and their physical environment. In this

Figure 1.1 Ecologists focus on Ernst Haeckel's *Oekologie* and seek to understand the diverse, complex interactions among the living and nonliving parts of the planet that form our "ecological house."

context an **ecosystem** can be defined as a group of interacting organisms and their physical environment.

If we return to Haeckel's definition of ecology as the study of our "house" and the interactions within it, then it is also by its very nature the study of the mundane or commonplace. Most people see trees, birds, grass, insects, and worms nearly every day of their lives, and for that reason alone they often pay little attention to them and their interactions with us, in the process underestimating their importance to global health and human society.

However, ecologists resist the urge to overlook the mundane aspects of the world. Instead, they look closely and deeply at the things that compose our ecological house. For many budding ecologists, this initial act of careful observation is like taking off a blindfold to behold a vibrant spectrum of interactions and fascinating questions sitting before them. We all dwell in a truly remarkable house. It sometimes appears quite simple but wonderfully surprising; at other times, it appears so complex that its very existence is bewildering. The primary goal of this textbook is to help you remove the metaphorical blindfold and begin to see and understand your ecological home.

1.2 Ecology in the Anthropocene

Undertaking this deep inquiry into the planet where we live leads quickly to noticing signs of damage. Much like a house that has been battered by a bad storm, our planet has ecological damage. Sitting in the living room, we may not at first notice the extent of the damage; but once we begin walking around and closely inspecting the house, the deterioration becomes apparent. And many of these ecological wounds are caused by humans. As Aldo Leopold wrote in his classic work, *A Sand County Almanac,*

> One of the penalties of an ecological education is that one lives alone in a world of wounds. Much of the damage inflicted on the land is quite invisible to laymen. An ecologist must either harden his shell and make believe that the consequences of science are none of his business, or he must be the doctor who sees the marks of death in a community that believes itself well and does not want to be told otherwise. (Leopold 1949)

Leopold published his seminal book in 1949. Over the ensuing decades, the metaphorical wounds that he described have escalated in both frequency and intensity. Incontrovertible scientific evidence indicates that human actions dominate the planet and have led to a biological world that is rapidly shifting toward an unknown future state. The ubiquitous influence of people on the biosphere means that the ecologists of today must use scientific principles and evidence to help ameliorate human effects and perhaps heal a rapidly changing Earth.

In 2000, Eugene F. Stoermer, a freshwater biologist, and Paul J. Crutzen, a Nobel Prize–winning atmospheric chemist, suggested that the pervasive planetary damage by humans warrants calling the current geological epoch the **Anthropocene** (Crutzen and Stoermer 2000). An epoch is a subdivision of the geological timescale, and Crutzen and Stoermer were emphasizing the central role that humans (*anthropo-* from the Greek for "human") play in the current era (*-cene* from the Greek for "new or recent"). This was not really a new term. Starting in the 1960s, Russian scientists began using a version of the term to refer to the Quaternary (the geological period from about 2.6 million years ago to the present), and Stoermer used the term in the 1980s; but it was after the year 2000 that it graduated to more popular use. An interdisciplinary group of scientists considering evidence for the Anthropocene in 2017 concluded that "human impact has now grown to the point that it has changed the course of Earth history by at least many millennia, in terms of the anticipated long-term climate effects, and in terms of the extensive and ongoing transformation of the biota, including a geologically unprecedented phase of human-mediated species invasions, and by species extinctions which are accelerating" (Zalasiewicz et al. 2017).

Perhaps there is no better way to intuitively understand how the planet has changed since Leopold's time than to consider what are known as the Great Acceleration graphs (**Figure 1.2**), produced by Will Steffen and his colleagues

Figure 1.2 The Great Acceleration refers to the sudden, intense increase across a range of planetary metrics due to human influence. In each graph, the dashed vertical line at 1950 helps highlight the acceleration in change after 1950; notice the way the curves get steeper to the right of the dashed lines. In their 2015 paper, Will Steffen and colleagues showed that these effects are both sociological (orange graphs) and ecological (blue graphs). Although only nine relationships are presented here, the paper includes 24 separate graphs, all revealing the same general trends and collectively suggesting a pervasive pattern of human-influenced change since 1950 (Steffen et al. 2015).

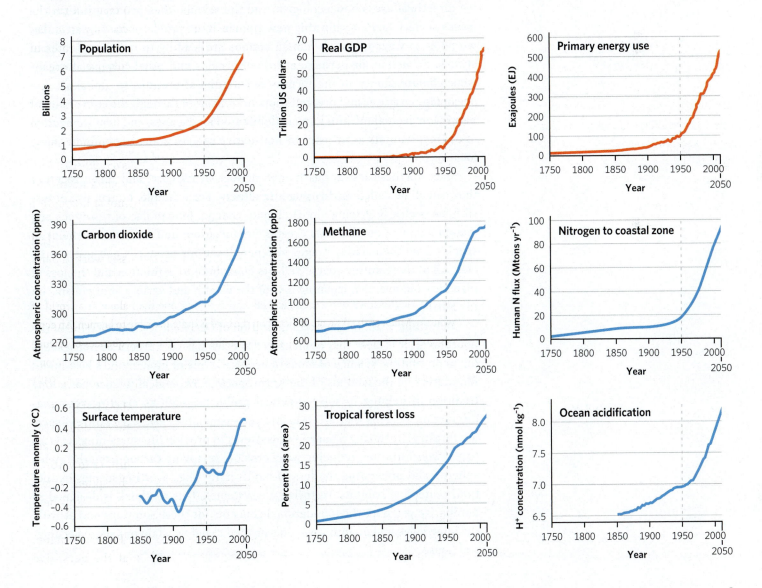

(Steffen et al. 2015). In this series of graphs, data from a variety of sources, standardized for time frame (1750–2010), show the temporal trajectories of socioeconomic and ecological metrics. Taken together, these metrics, or measurements, demonstrate massive increases in the human population and our energy use; huge additions of carbon and methane into the atmosphere; a rise in air surface temperatures; the loss of tropical forests; and the global acidification and nitrification of the oceans. What the graphs do not show as clearly is that these trends are occurring worldwide. Across all parts of the Earth, most of these trends are now apparent.

So, how should ecologists proceed? Richard Corlett (2015) suggests that recognizing the accelerating nature of human influence creates a kind of moral imperative that we conserve and restore biodiversity. We need to consider how the rest of the world's species can coexist with humans, and whether the novel assemblages of species that we have created have any social or ecological value.

Corlett suggests that most ecologists recognize the global trends outlined in the Great Acceleration graphs and work toward integrating these new rates of change into their theories, models, and applied practice. In this book, we directly address how ecologists and the science they produce inform the issues society faces within this new human-dominated epoch. In particular, we point to ways in which human actions are leading to increasing rates of species *extinction*, or permanent loss of species and other groups of organisms; *climate change* in the form of substantial alterations to global air and ocean temperatures and precipitation; *urbanization* through the conversion of natural environments into human-dominated landscapes; and *homogenization* of distinct habitats as invasive or non-native species come to dominate many locations.

Each of these issues has human actions at its core, and taken together they can all be called **anthropogenic effects**. For example, human use of fossil fuels is clearly driving global climate change; human use of resources and manipulation of our habitats creates urbanization; and human movement of species from their native ranges into new locations produces non-native species. All of these anthropogenic effects contribute to extinction and the loss of planetary biodiversity, meaning loss of the wealth and variety of organisms on the Earth.

Returning again to Haeckel's definition of *Oekologie* as the "doctrine of the household," it is clear that humans as a species have some serious housekeeping to do, and the science of modern ecology is one of humankind's best hopes for addressing the wounds of the Anthropocene. Ecological science has a long tradition of helping to solve practical problems, such as creating resource-production systems that are valuable yet sustainable (e.g., agriculture, fisheries, forestry), restoring damaged ecosystems to provide life-sustaining services (e.g., water filtration, reducing wave erosion, removing carbon from the atmosphere), and preventing the loss of biodiversity (e.g., averting species extinction, protecting habitats, restoring ecosystems). This textbook is designed to familiarize you with the foundational concepts for these solutions and provide the knowledge and quantitative tools needed to help sustain our shared ecological house.

1.3 What Do Ecologists Do?

If ecology is the study of the interactions that determine the distribution and abundance of organisms, then how does one go about actually doing this science? Scientists know there are millions of different species on Earth, all of them interacting in some way with each other. From what you can easily observe in your surroundings, you can see that individuals of these species respond to a variety of conditions. These range from microscale gradients, such as soil moisture, to global-scale processes, such as currents that circulate across entire oceanic basins. Given this complexity, how does ecological science even begin to comprehend these patterns and processes and understand what produces them?

There is no single answer to that question, so ecological scientists have embraced a variety of approaches. In what follows, we briefly review four avenues for doing ecological science and show how the approaches complement one another and collectively offer incredible insight into the workings of the natural world. These four approaches are observation and natural history, experimental ecology and null hypothesis testing, multiple hypothesis testing with best fit comparisons, and ecological modeling. In Chapter 2, we follow this introduction with a short primer on the quantitative foundations for these approaches, which may be useful as a reference as you move through later chapters.

Observation and Natural History

The first ecologists were called **naturalists**, or natural historians (1600s–1950s). These early scientists were skilled and patient observers of nature who inferred patterns and processes from their observations. Charles Darwin was an early pioneer in the use of observation to study ecology (**Figure 1.3A**), providing an

A

B

Figure 1.3 We can see from illustrations that early naturalists were making very detailed observations of the natural world around them, such as in drawings of Ⓐ goose barnacles by Charles Darwin and Ⓑ interactions between arachnids and ants by Maria Sibylla Merian.

evolutionary foundation for our understanding of ecological patterns and processes in his book *On the Origin of Species*, published in 1859. In fact, Darwin's role as a young man aboard HMS *Beagle*, the British survey ship, was officially that of "naturalist" during its second voyage (1831–1836). Even earlier was Maria Sibylla Merian, a seventeenth-century German woman who raised two daughters and managed a household while also building a successful career as an artist, botanist, naturalist, and entomologist (**Figure 1.3B**). Although Merian was well respected in her lifetime, her work was seen as outside the scope of a woman's duties, and it consequently faded from the scientific literature around the 1800s. Nonetheless, Merian's careful observations and illustrations of insects helped scientists and the general public understand not only insect metamorphosis but also the fact that insects did not emerge spontaneously out of mud (a common belief in her time).

A lesser-known early ecologist who thought about ways to categorize his observations was Edward Forbes, a British naturalist and oceanographer who studied the distribution and aggregation of benthic marine invertebrate species (those living in the mud and sediment at the bottom of an aquatic environment) in the 1840s. In lieu of writing lengthy descriptions, he developed a systematic survey approach that recorded the depth at which his benthic marine samples were taken and the numbers and types of organisms collected in each depth range. From these data, he defined a series of benthic marine depth zones that were inhabited by characteristic groups of invertebrate species, effectively producing early descriptions of patterns in ecological communities (**Figure 1.4**).

Figure 1.4 The careful observations of the Aegean Sea by Edward Forbes led to our first understanding of marine biological zones. Ⓐ He catalogued that different species were found at different depths and offered a first basic sketch of this zonation (Forbes 1843). Ⓑ Modern illustrations of these zones are more colorful but owe a debt to his early observations.

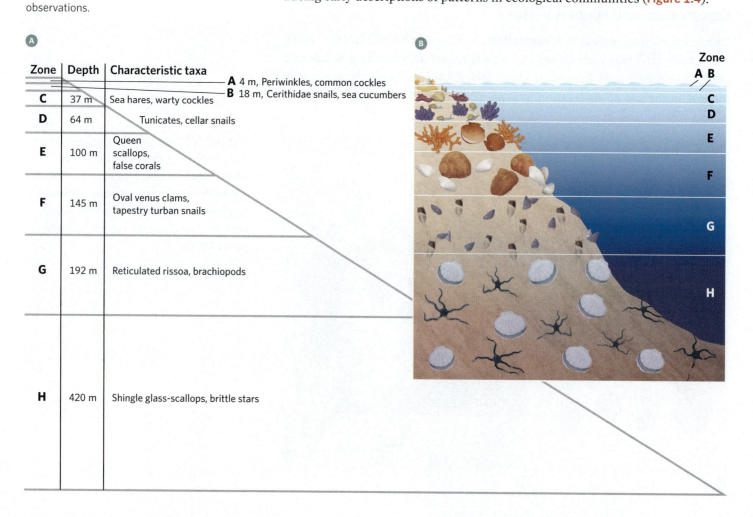

Zone	Depth	Characteristic taxa
		A 4 m, Periwinkles, common cockles
		B 18 m, Cerithidae snails, sea cucumbers
C	37 m	Sea hares, warty cockles
D	64 m	Tunicates, cellar snails
E	100 m	Queen scallops, false corals
F	145 m	Oval venus clams, tapestry turban snails
G	192 m	Reticulated rissoa, brachiopods
H	420 m	Shingle glass-scallops, brittle stars

Many other such naturalists have played an important role in the history of ecological science. Some held formal academic positions, but many did not. Most of them had keen observational skills, and the insights they generated were due to enthusiastic documentation of the species that lived around them. Their work laid the foundation for a variety of subjects covered in this text, and much of the terminology they developed is still in use today. Observational skills were, therefore, the historical precursor for much of what is currently called ecology.

To this day, the importance of careful observation has not diminished and remains the starting point for almost all ecological investigations. Yet producing descriptive observations of nature and the natural world without some level of quantification is much rarer in modern ecology than it was in the days of Darwin, Merian, and Forbes. By the early 1950s, more and more ecologists started to combine observations with other approaches, and the pure natural history approach to ecological science began to fall out of favor.

This shift has many advantages (as we will discuss shortly), but observations are still critical to ecological insights. An excellent modern example is that, as climatic alterations combine with other anthropogenic effects to produce large-scale global change, descriptions of the effects of these changes have become ever more important. Although we have fewer people devoted to the intense observation necessary to describe new species, modern ecology has put significant resources into making consistent observations on a large scale.

For example, in the United States, the National Science Foundation (NSF) has allocated substantial funding for the National Ecological Observatory Network (NEON), which is a series of data collection stations across the country. At these stations, sensors collect data on abiotic (nonliving) factors automatically, and field ecologists collect additional data following standardized protocols at 81 locations spread across 20 different ecological domains around the country (**Figure 1.5**). The NEON concept began at the turn of the last century, and scientists spent about five years in workshops discussing how to collect the most critical data. Construction and installation of the sensors began in 2006, and NEON data went online in 2019. Experienced scientists with significant expertise in natural history are

Figure 1.5 Each picture in this photo array from the National Ecological Observatory Network (NEON) website represents a subset of the data collected by NEON. The network offers more than 175 types of open-access data, ranging from atmospheric data collected by eddy flux towers, remote sensing data taken from airplanes, many types of data on aquatic and terrestrial organisms, information on plant and wildlife diseases, and data on the timing of specific events each year (phenological information).

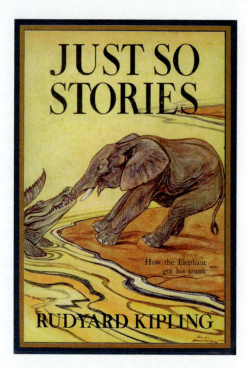

Figure 1.6 The cover of the 1912 edition of Rudyard Kipling's *Just So Stories* shows an elephant being tugged toward the water by a crocodile. The story suggests that the elephant got its long trunk from this tug-of-war with the crocodile. Modern approaches to ecology try to avoid "just-so stories" by scientifically testing hypotheses about the way the world works.

still needed, but NEON is an enormous resource moving forward and represents a sophisticated type of observational ecology.

Experimental Ecology and Null Hypothesis Testing

Driven in part by the successes of other scientific disciplines, ecologists in the 1950s increasingly began to approach their research through **manipulative experiments** and statistical **hypothesis testing**. At the core of this approach is the idea that researchers can clearly state **hypotheses**—unproven ideas—and then produce evidence to support or reject these ideas by conducting carefully designed experiments (more on manipulative experiments in a moment).

One of the key aspects of the experimental approach to ecology is its focus on converting observations into falsifiable hypotheses. This effort requires scientists to identify specific mechanisms that influence an ecological system and thus avoid **ad hoc fallacies**, sometimes referred to as "just-so stories." An ad hoc fallacy is a story or narrative that explains an observed pattern or process, and the "neatness" of the explanation or story is presented as evidence of the correctness of the explanation. The term *just-so story* comes from Rudyard Kipling, who published a series of amusing children's stories in 1902 with fanciful explanations for a variety of animal traits, such as explaining that elephants got their long trunks when a crocodile pulled on an elephant's nose (**Figure 1.6**). Kipling's explanations make for entertaining stories, but they are fanciful leaps of imagination that make no attempt to support or falsify their claims. Scientists staunchly try to avoid this type of explanation. Instead, they look for evidence in the natural world and allow their "explanations" to change based on the evidence. Experiments provide a way to gather such evidence.

When designing an ecological experiment, the goal is to alter or manipulate various key aspects, or **focal factors**, of an ecological system to see how the changes affect ecological outcomes or responses. This is what we mean by a *manipulative experiment*. The manipulated aspects are considered **treatments**, and researchers collect data on how the ecological system responds to the treatments. In general, scientists compare outcomes in systems with treatments to outcomes in systems serving as **controls**, in which the focal factors are not manipulated. For example, if we grew tomato plants using three different types of fertilizer, the types of fertilizer would be the treatments. Typically, in an experiment like this, we would also grow some of the plants without any of the three fertilizers; these plants would be the controls.

In reality, there may be many factors acting at once in a system, so ecologists rarely try to isolate or completely control all possible **extraneous factors**, meaning the nonfocal elements. Instead, ecologists generally set up experiments that try to manipulate a single factor even if the rain falls, the wind blows, or organisms move around. By allowing the rest of the system to operate as it always does, scientists can see the effect of the manipulated factor in the context of the real world. Of course, this context changes with location, time frame, season, and scale. This variability makes it impossible to generalize the insights from experiments—and from any scientific inquiry—to all situations.

Our inability to include and consider all potential factors that may influence a system is why scientists generally refrain from saying that they can *prove* anything. Science can only *disprove*. We can demonstrate that a mechanism does not act in a particular way, but it is difficult to say that a particular mechanism is the only explanation for an observed pattern. Instead, ecologists conduct experiments in many different locations under a suite of different conditions, and then, over time, the larger discipline accumulates support for one or more hypotheses.

Hypothesis testing, which is commonly associated with the scientific method, is generally taught as a step-wise process (**Figure 1.7**), with the first step being to make observations about the natural world that raise questions in the observer's mind. These questions can lead to more observations but eventually result in the researcher's formulating a concrete idea that might explain the pattern or process observed. The next step is to propose this idea as both a **null hypothesis** and an **alternative hypothesis**. The null hypothesis is a statement proposing that the focal explanatory factors do not have an effect. The alternative hypothesis is a statement proposing that the focal explanatory factors *do* have an effect, and scientists will often specify an effect that accounts for the observed pattern or process. The null and alternative hypotheses are refined until they are specific enough that a critical experiment can be designed that could potentially falsify the null hypothesis by demonstrating that the focal factors do, in fact, have an effect. In the third step, the scientist conducts the experiment and collects **data**. Ecologists can then analyze the collected data to determine whether the results support or refute the null hypothesis. It is important to note that null and alternative hypotheses are statistical concepts, but that scientists generally focus on the explanatory ideas in alternative hypotheses when using the term "hypotheses."

Typically, interpreting the results of an experiment leads to more questions, which are then formulated into new hypotheses. At times, the results refute the null hypothesis but not in the way that scientists expect. It is essential that scientists use such information to redirect their hypotheses and look for new answers. Clinging to explanations despite evidence to the contrary is the opposite of good science. Scientists try to remain flexible in their thinking, open to new ideas and greater understanding. Often experiments highlight the importance of additional contributing factors, which may prompt additional observations and generate new experiments. One of the very useful aspects of experiments is that we can use them to test the effects of more than one factor.

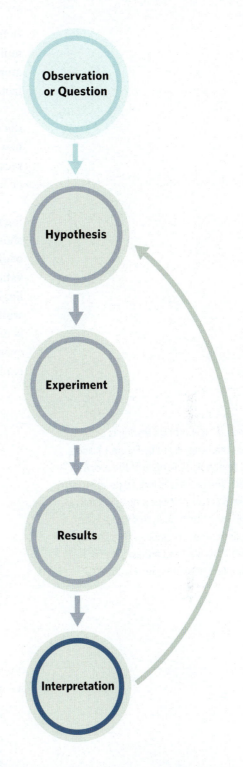

Figure 1.7 This is a common representation of hypothesis testing and the scientific method. Starting with observations, scientists develop a question, which they hone into a focused hypothesis. The hypothesis is tested with an experiment, and the data and results are used to arrive at an interpretation. The interpretation often leads to new questions, and the process starts over again (in other words, it is an iterative process). This simplistic representation does not really convey the many ways that the scientific process encourages questioning and reassessment, but it is a good starting point. We expand on this basic representation throughout this chapter.

In fact, experiments are often most useful for exploring the interacting effects of multiple factors. There is a limit, though. Additional factors make experiments more complicated, so scientists generally try to include only a handful of focal factors in any one experiment.

But all experiments lead to more questions and more investigations. Of course, the investigations do not all have to be experimental. The investigator may collect new observational data or may use one of the other methods described in the next sections. The scientific process continues to the limits of the researcher's interest or funding and often grows to involve multiple researchers in multiple places.

The activity of refining and revising ideas actually makes the process of testing hypotheses closer to that depicted in **Figure 1.8**, and quite different from that shown in Figure 1.7. Perhaps most importantly, the process can start from a variety of different places. It is true that ecologists often devise new ideas based on observations, but the process is equally likely to start from combining insights from earlier investigations, from reading the work of other ecologists, or from continuing work from a lifetime of study. All of these sources of inspiration mean that there is no single starting point for this iterative process. In fact, Figure 1.8 could have even more loops and arrows, which we demonstrate later in the chapter after we explore more of the approaches used by the modern ecologist.

Figure 1.8 Hypothesis testing is actually less linear than suggested by Figure 1.7 and by many textbook depictions of the process. Notice that the process is indeed iterative, as suggested by Figure 1.7, but it does not follow a straight-line path. In fact, this figure is also too simplistic. The remainder of the chapter expands on this idea and adds detail. For a more complete depiction, see Figure 1.16.

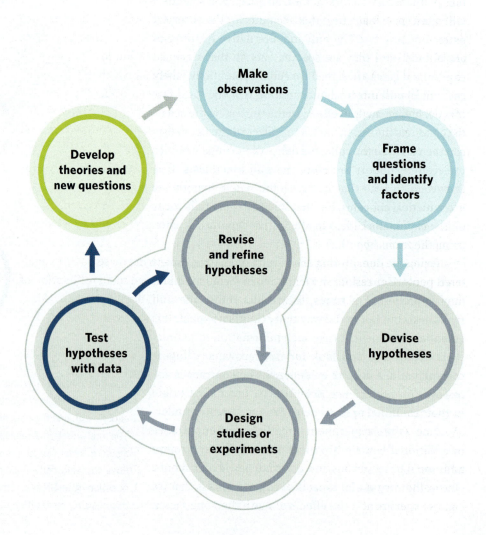

The advantages of an experimental approach and hypothesis testing for understanding the natural world are fourfold. First, the approach requires researchers to carefully consider their observations and use them to craft specific hypotheses that are falsifiable. Second, it encourages scientists to identify critical factors relevant to the hypothesis. Third, the manipulation of various factors in experiments is a good way to demonstrate the effects of focal factors and disentangle the influence of these factors from all other extraneous factors. Fourth, the methods should be repeatable, so different researchers can look for similar results in different places or situations.

The appeal of this approach is clear, in that a well-designed experimental test can point to specific effects and produce verifiable results that offer the scientific community a measure of trust in the new findings. Unlike long descriptions of natural history patterns, ecological experiments drill into the hows and whys behind ecological patterns.

For a quick example of this approach, we can examine the research of Jane Lubchenco, an eminent ecologist who has been both the president of the Ecological Society of America and the administrator of the National Oceanic and Atmospheric Administration (NOAA), the entity that provides weather and marine information of all kinds across the United States. As a graduate student, Lubchenco was interested in the factors affecting the ecological communities in New England tide pools (**Figure 1.9A**). She noticed that pools with *Enteromorpha* spp. (**Figure 1.9B**), fast-growing, lettuce-like algae, tended to have very few common periwinkle snails (*Littorina littorea*; **Figure 1.9C**), which consume algae. On the other hand, pools with many periwinkles had very little *Enteromorpha* but lots of *Chondrus* spp. (**Figure 1.9D**), much thicker red algae that grow more slowly than *Enteromorpha*. Pools with intermediate amounts of snails had many more algal species.

Lubchenco hypothesized that a combination of competition for space between the algae and consumption by the snails determined the eventual number of algal species. She devised an elegant series of experiments to identify the snails' eating preferences by putting them in tanks with food choices. She found that *Enteromorpha* was the preferred food, whereas *Chondrus* got barely a nibble from the snails. After identifying the snails' eating preferences, Lubchenco suspected that pools with lots of snails had no *Enteromorpha* because the snails quickly gobbled all the young *Enteromorpha* that settled and started to grow. She tested this hypothesis by altering the density of snails in some pools and comparing these pools to unaltered pools. Her results showed that removing the snails led to a proliferation of the fast-growing *Enteromorpha* and loss of *Chondrus*. On the other hand, adding snails to a pool with predominantly *Enteromorpha* led to its decline and the slow growth of *Chondrus*.

This series of experiments helped explain a tide pool pattern observed in the natural world. The work also shows the iterative nature of the scientific method and hypothesis testing. Observations led to questions and hypotheses, and experimental results led to new ideas and new experiments.

Successful scientific experiments require unbiased data collection and well-designed experimental treatments, which can sometimes be expensive and time-consuming. These types of practical requirements often place significant physical limits on experimental designs. For example, most species used in ecological experiments tend to be small because it is easier to manage them in natural

Figure 1.9 Jane Lubchenco explored the factors that affect the algal community in New England tide pools, like the one pictured here located in Gloucester, Massachusetts Ⓐ. She found that the faster-growing and less-defended green algal species, such as *Enteromorpha* spp. Ⓑ, were more likely to be eaten by the common periwinkle snail, *Littorina littorea* Ⓒ. Periwinkles tended not to eat the slower-growing and more-defended red algal species, such as *Chondrus* spp. Ⓓ. Tide pools without the snail and with little wave action tended to lose *Chondrus* and become dominated by *Enteromorpha*. Tide pools with snails often held a wider range of algal species or lost *Enteromorpha* when there were lots of snails.

field sites or laboratory settings. Experimental species also tend to reproduce on short timescales, such as days or months, allowing for meaningful data collection over a reasonable time frame.

Exceptions exist, of course. Government programs in the United States, Austria, Germany, New Zealand, and Israel fund decades-long experiments often replicated across multiple habitats. There are also a few excellent examples of experiments that were run by individual laboratories for decades, such as Charles Krebs's experiments in Canada on the effects of large-bodied predators on their prey (Krebs et al. 1995; **Figure 1.10**). These types of long-term and spatially replicated experiments are critical in our quest to understand ecological variability and track long-term patterns in how species respond to large-scale factors such as climate variation.

Nevertheless, outcomes of even well-designed and well-executed ecological experiments have their limitations. Local climatic conditions or the evolutionary

Figure 1.10 Charles Krebs (pictured here) and his colleagues set up large-scale experiments involving predator exclusion and food manipulation in the Canadian Yukon in the 1970s and then proceeded to follow lynx and hare populations for decades in an attempt to understand the causes of the population cycles.

history of the species under study make it difficult (if not impossible) to produce universally applicable ecological insights from the results of experiments alone. This problem is particularly evident in ecology, where context is often more critical than in some of the other experimental sciences. For example, electricity works the same way in Canada as it does in Brazil, but the effect of a predator on its prey may differ substantially between these two locations.

Additionally, simple experiments offer limited guidance on some of the Anthropocene's more pressing ecological problems, such as the global effects of climate change, species extinctions, species invasions, and large-scale land use change. For example, there is no way to accurately determine how a species near extinction will respond to a warmer and drier future climate through experiments alone. Even if we could manipulate temperature and water availability across a variety of locations, we do not know if all the other environmental variables adequately represent the conditions of the future. In addition, there are ethical considerations to manipulating natural systems on a large scale. If we were studying climate change, for example, would we really want to limit water availability across a large geographic area if the species living there were already struggling to get water? One might argue that such a large-scale experiment could provide invaluable information on water conservation efforts. But is this experiment an ethically or morally reasonable thing to do? Even if we were able to manipulate factors like this over enormous areas, there could be massive unintended consequences to the research.

Multiple Hypothesis Testing with Best-Fit Comparisons

In situations where we cannot logistically or ethically conduct an experiment, scientists can collect quantitative observations of large-scale phenomena and use these data to understand an ecological process. In fact, with technological advances, it is possible for scientists (and others) to collect reams of quantitative observational data on large spatial or temporal scales without sacrificing the precision of those measurements. Thanks to low-cost sensors and massive computing power, programs such as NEON are almost like having 1,000 individual researchers working 24 hours a day collecting data synchronously across all of North America.

Making huge numbers of quantitative observations in real time and across large geographic scales is revolutionary and often allows scientists to craft many simultaneous hypotheses. As a result, ecologists need a way to determine when data support or refute particular hypotheses. In this context, modern ecologists have adopted a variety of approaches to assess how well multiple hypotheses fit real-world data.

Ray Hilborn and Marc Mangel coined the term *ecological detective* to describe an ecologist who gleans data from a complex and often hard-to-observe ecological world, and then uses the data to assess the strength of evidence for a suite of hypotheses (Hilborn and Mangel 1997). This approach to ecological science replaces a single falsifiable hypothesis (i.e., the null hypothesis) and alternative hypothesis with multiple more realistic (or alternative) hypotheses that "compete" with one another in their relative fit to the data. The set of competing hypotheses represents an informed understanding of how an ecological system works. In this multiple hypothesis approach, the role of data shifts from falsifying a hypothesis suggesting a particular factor (or set of factors) has no effect (i.e., a null hypothesis) to evaluating which of several ecologically reasonable hypotheses is most likely to be true (**Figure 1.11**).

This may appear similar to what Lubchenco did in the tide pools, but her experiments allowed her to reject hypotheses. The philosophical shift in using data to judge among competing hypotheses is substantial and involves changing from absolute yes-or-no answers to weighing hypotheses by their levels of support. The statistical differences are more complicated than we can explain here, but it may be useful to envision an ecological Sherlock Holmes ferreting out an answer from a suite of competing hypotheses. By comparison, the standard null hypothesis testing would involve checking a single alibi.

For example, a detective such as Sherlock Holmes investigates a murder by collecting evidence at the scene of the crime and coming up with initial explanations. Learning more about the victim and the victim's interactions with others, as well as interviewing people the victim knew, are all steps that offer bits of information, like the data that ecologists collect in relation to their questions and hypotheses. The evidence the detective collects is weighed against the initial conjectures. The evidence may strongly suggest that one or more of the potential explanations are very likely to be true, while not supporting other explanations. Despite this, a good detective keeps collecting information and continually evaluating how these data fit with the hypotheses. Eventually, the fit between the evidence and the explanations eliminates several potential explanations and offers strong enough support for one of the explanations for the detective to hand the case over to a prosecutor for legal action.

An ecologist can approach scientific research questions the same way, by formulating a number of initial hypotheses about an ecological phenomenon. For example, suppose an ecologist were interested in why a species went extinct from a local wetland habitat. The ecologist would propose a series of competing hypotheses for the loss—hypotheses informed by prior observations or experiments, as well as by the published work of others. The next step would be to collect quantitative data on the wetland. As a detective, the ecologist would not try to falsify a single null hypothesis but instead use the data to evaluate the strength of evidence supporting each of the competing hypotheses. Some hypotheses would

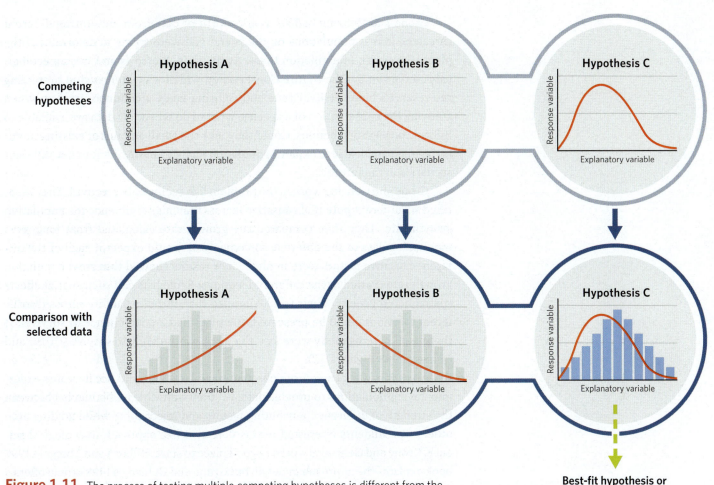

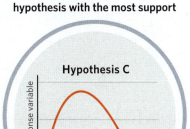

Best-fit hypothesis or hypothesis with the most support

Figure 1.11 The process of testing multiple competing hypotheses is different from the usual hypothesis testing in that researchers do not simply look to see if a factor or combination of factors has a significant effect. Instead, each hypothesis is presented as a prediction that focuses on the expected response of the variable of interest. Researchers compare the actual data to predictions from the competing hypotheses and arrive at a *best-fit hypothesis*, or the hypothesis with the most support. In the figure, we can see that hypothesis C offers a much better fit to the data (blue histogram) than do hypotheses A and B, even though the fit is not perfect. In real life, the fit is rarely perfect, and a slight mismatch between the best-fit hypothesis and the data often drives further investigation.

have considerable support, others marginal support, and still others essentially no support. In the example of the wetland, if the data strongly supported one (or more) hypotheses, the ecologist would feel comfortable using this hypothesis to provide recommendations to the local wetland manager to help recover the lost species.

An example of this approach can be found in the work of Elizabeth E. Crone and Janet L. Gehring (1998), who explored the potential for extinction of the threatened plant species Columbia yellow cress (*Rorippa columbiae*) on Pierce Island off the coast of Washington State. Species in danger of extinction receive different designations depending on their level of endangerment. Many nations, including the United States, have lists of rare, threatened, or endangered species, with the risk of extinction increasing as you progress from rare to threatened to endangered.

Crone and Gehring had six years' worth of data from monitoring different threatened cress populations on the island and were trying to determine if the plant was at risk of extinction there. At the time, Pierce Island was an ecological reserve run by The Nature Conservancy, which was interested in protecting the reserve's biodiversity. Crone and Gehring faced a common problem when studying the conservation of a species: what to do when few data are available to inform conservation actions. Could they take the small amount of existing monitoring data and use it to explore population changes in the plant under different scenarios in the future?

Crone and Gehring worked the problem like ecological detectives. They identified six different potential causative factors that might influence the population growth rate. They then compared the growth rate calculated from long-term monitoring data to the different outcomes they would expect if each of the different causative factors were in play. Their results showed that cress population growth rates varied at the different locations. Ecologists call such spatial effects spatial heterogeneity. Their results implied that, if the variability across the different locations could be preserved, the plant had a good chance of survival. If the spatial heterogeneity were lost, the plant was unlikely to survive (Crone and Gehring 1998; **Figure 1.12**).

All of this work came from observational data. They did not have any experimental manipulations to provide data, yet they were able to distinguish between the strength of different contributing factors and reject some while holding onto others. For students interested in this detective-like approach to ecological science, Crone and Gehring's work is a good place to start. Hilborn and Mangel's 1997 book explains the approach in detail, but Crone and Gehring's 1998 article offers a concise short primer, with each step explained carefully.

The approach of multiple hypothesis testing is most often used in the context of large-scale research projects, but the logic and philosophy can be applied equally well to small-scale experiments or sets of observations. One of the key

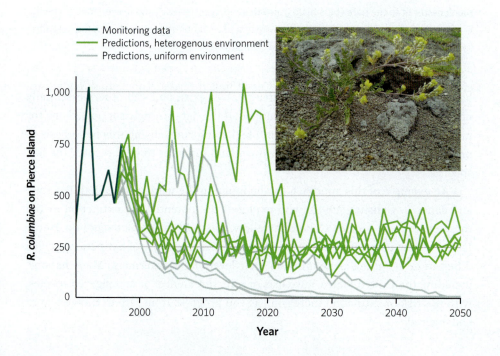

Figure 1.12 Crone and Gehring ran simulations based on previously collected monitoring data to predict the future population sizes of Columbia yellow cress (*Rorippa columbiae*; inset) on Pierce Island, Washington. The dark green line on the left side of the graph represents population fluctuations found during their six years of monitoring. To the right of that, gray lines show different potential outcomes in a uniform environment. The light green lines indicate potential population changes in a heterogenous environment. Individual lines are only representative "runs," or projections, of the model. With thousands of projections, the simulated runs in a heterogenous environment never went extinct, and the ones in a uniform environment almost always did (Crone and Gehring 1998).

advantages to being an "ecological detective" is that it allows scientists to acknowledge uncertainty and explore multiple potential explanations from the beginning. This approach also mimics the way humans accumulate and evaluate evidence more closely than does null hypothesis testing. It can also easily accommodate real-world complexities without requiring ecologists to set up large-scale experiments or to sacrifice the benefits of careful hypothesis generation followed by statistical evaluation.

Ecological Modeling

Ecological modeling provides another way of exploring how various factors affect ecological dynamics without having to manipulate real ecosystems. For this reason, models are used in theory and practice by nearly all modern ecologists. Students learning ecology need familiarity with modeling for two important reasons. First, models are incredibly useful for understanding the mechanisms that produce ecological patterns and for predicting how complex ecological systems will respond to environmental factors. Second, many fundamental tenets in the field are based on models, and understanding the logic that led to these principles requires understanding the models.

Given this importance, it is critical for students to think about what models are and how ecologists use them. The multitude of types of ecological models makes it difficult to produce a comprehensive taxonomy for them, but a good starting place is defining the relationship between conceptual and mathematical models.

A **conceptual model** can take many forms but is essentially a theoretical construct that puts various components in relationship to each other. It is a formalized idea about how things work. In ecology the components may be energy, water, species, groups of species, resources, or a host of other factors that affect the environment. For example, Crone and Gehring constructed conceptual models to examine how different factors affected the population growth of the cress plant.

In a conceptual model, the components are linked together in a logical fashion to explore the connections among these components and their combined impacts on ecological outcomes. In some cases, the model may represent the changing states of these components through time or space. For example, conceptual models can track individuals as they grow from one developmental state to another, describe how resource pools change in response to season, or illustrate how species drift toward extinction via a series of linked forces. Conceptual models are extremely useful and help us organize and formalize abstract thinking on a subject in much the same way that a hypothesis does. In fact, in the case of Crone and Gehring's work, each hypothesis was treated as a separate conceptual model.

Scientists use conceptual models to frame and analyze a set of questions about how an ecological system performs under various conditions. Many of us make what amount to conceptual models in our heads all the time. For example, the game of rock-paper-scissors is a conceptual model. Rock beats scissors; scissors cut paper; and paper covers rock (**Figure 1.13A**). No one actually plays the game with real implements (we hope?!), but the game (i.e., model) illustrates the idea that no weapon or strategy is perfect in all situations and that each approach can win or fail, depending on the circumstances. The model may seem overly simplistic, so you may be surprised to learn that a variation of the rock-paper-scissors scenario does a pretty good job of describing the mating strategies of male side-blotched

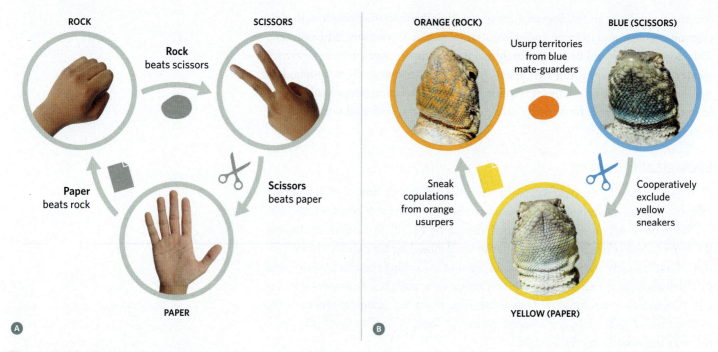

ROCK   SCISSORS

Rock beats scissors

Paper beats rock

Scissors beats paper

PAPER

A

ORANGE (ROCK)   BLUE (SCISSORS)

Usurp territories from blue mate-guarders

Sneak copulations from orange usurpers

Cooperatively exclude yellow sneakers

YELLOW (PAPER)

B

Figure 1.13 **A** The popular game rock-paper-scissors is an example of a conceptual model in that it suggests *relationships* between the different "weapons" and even dictates that no single weapon wins in all situations. **B** The rock-paper-scissors model fits the general way that male side-blotched lizards (*Uta stansburiana*) compete for mates (Sinervo and Lively 1996). Yellow-throated males (known as "sneakers") beat aggressive orange-throated males (known as "usurpers"), usurpers beat blue-throated males (known as "mate-guarders"), and mate-guarders beat sneakers (Dickinson and Koenig 2003).

lizards (*Uta stansburiana*) that inhabit much of the western United States (Sinervo and Lively 1996; **Figure 1.13B**).

Sometimes the relationships depicted in conceptual models can be formalized into the language of mathematics, thus creating a **mathematical model**, our second model type. Mathematical representations require researchers to refine their thinking about what parts of a system are most important to include in a model and how these parts relate to one another. By formalizing relationships among elements using mathematics, ecologists can set a mathematical model "in motion" and explore how the system responds when one or more variables change.

Sometimes an ecological process can be simple enough that the response of the entire system can be demonstrated with fairly modest techniques, as is possible with a basic model of population growth:

$$\text{Change in Population Size} = \text{Births} - \text{Deaths}$$

We will expand on such a model in Chapter 2 and much more in Chapter 6 when we talk about population growth. Here, it is important to know that we call these mathematical models **analytical models** because we can solve the equations or produce important insights from analyzing the relationships among variables. Frequently, though, ecological systems are complex enough that we cannot solve model equations in a straightforward manner but instead need to explore the model's behavior with simulations. A **simulation model**, therefore, is a version of a conceptual or mathematical model that is solved or run on a computer in order to predict the model's performance. Simulation modeling has become much easier with the vast amounts of computing power available in the modern world.

Computer simulations have become so commonplace in our lives that it is hard to imagine what life was like without them. How did people ever play baseball effectively without having the benefit of playing simulated baseball on their computer or television screens? Or drive a race car, or fly an airplane, or help defeat the incoming zombie hordes? The games we play on our computer screens are so visually sophisticated that it is hard to imagine that each element of the game has to be specified in excruciating detail by a series of functions, most of which have mathematical forms, such as the velocity and angle of the baseball as it leaves a virtual bat.

Like game simulations, ecological simulations use a series of functions generally represented by mathematical equations to connect the different components of a system. The goal of ecological simulations is to provide a user with the ability to alter one or more components, such as species traits or nutrient levels, and track how the other components of the ecosystem respond. If that goal sounds familiar, that is because it is almost identical to the goal of experiments. Ecologists use simulations most frequently when their models incorporate real-world variations in factors of interest (Chapter 2). Ecological simulation models are handy for exploring how altering small elements or adding some uncertainty might change an ecological outcome the same way that a gamer might explore how changing the type of baseball pitch or range of pitch speeds influences the opponent's number of base hits, number of strikeouts, or which team wins the game.

The level of realism needed for a simulation to capture the ecological dynamics of interest varies substantially depending on the purpose of the model. In general, the more realism we put into a simulation model, (1) the more the results become system-specific, (2) the more complex the model becomes, and (3) the more we need field data to reflect the real system dynamics. If we design a simulation model for use in the day-to-day management of a specific ecosystem, achieving this level of realism is quite important. In other cases, a broader level of abstraction offers more relevance across systems.

You can think of this trade-off as being similar to how much realism you need if you are constructing a model airplane. If your goal is to create a model airplane that accurately represents a Boeing 777 to be used for special effects in a Hollywood movie, realism is important, even if this visual realism means the model cannot possibly fly or even glide (**Figure 1.14A**). However, if your goal is to produce a model airplane that can fly, a simpler model may be much better (**Figure 1.14B**). It only takes a few creases in a sheet of paper to create a very efficient, gliding model airplane. Paper airplanes do not look much like real airplanes, but they reproduce the key trait of flying quite well.

Figure 1.14 Detailed and simple models each have their advantages and disadvantages. **A** For example, a detailed model can offer realism for specific features but may sacrifice some ability to generalize. This detailed version of a Boeing 777 looks beautiful and may be painted to resemble a tiny version of an actual plane, but it lacks the most general feature of all planes—it cannot fly. **B** A basic model will never be mistaken for the real thing, but it may capture general aspects of the process it is meant to explain. In the model airplane analogy, the simple paper airplane will never be mistaken for a real plane, but it captures the general capability of flight much better than the intricate Boeing 777 model. Similarly, simple ecological models often allow us to see more general outcomes and effects than complex models do.

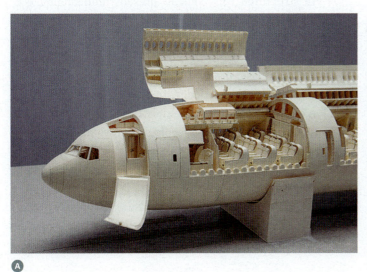

A

B

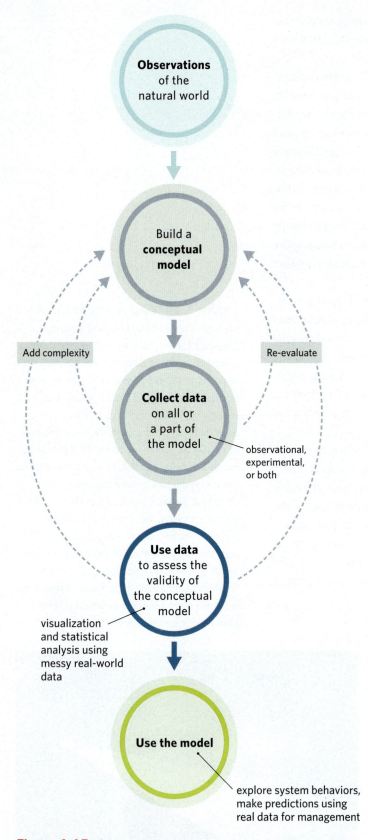

Figure 1.15 This schematic of the scientific process is not complete, but it fits well with the way we try to present each chapter.

Modern Ecological Use of the Four Approaches

All four approaches—observation, experimentation plus null hypothesis testing, multiple hypothesis testing, and modeling—are important in modern ecological investigations. The process of observing, coming up with an idea about how things work, testing the ideas, and learning from mistakes is central to science, ecology, and this textbook. Ecology, like all other branches of science, only moves forward if we keep gathering new information. Some of that information will support our current ideas, but some of it will not. All scientists have to be willing to drop an idea when data demonstrate that the model or hypothesis does not fit the real world. The good news is that, as a scientific community, ecologists learn from this iterative experience. We expect our readers are willing to go on that same journey: to observe, to try to see patterns, to construct explanations, to make mistakes, to challenge themselves, and to be willing to try new things. That is what ecological scientists do.

1.4 How This Book Is Organized

Textbooks on any subject try to organize and present fundamental principles for their readers. Authors organize this information into logical groupings of chapters and paragraphs and often present information in a somewhat simplified manner. These approaches make the material more accessible, but they cannot bring the material to life. If you ask most scientists why they like their jobs, their first response will often convey their love of discovering the unknown. It is not what we already know that is so exciting about science; it is what we do not yet know. And the knowledge and skills necessary for us to discover those unknowns are right at our fingertips.

We structured this book to present fundamental ecological knowledge in a way that mirrors how people actually conduct the science. As just discussed, ecologists can take a variety of approaches to their work; however, all ecologists must follow a basic process for the advancement and application of ecological knowledge. In this process, observations lead to ideas and conceptual models, observations and experiments yield data that allow ecologists to validate (or invalidate) those concepts, and analytical or simulation models (based on the conceptual model) explore and predict the behavior of the system being studied (**Figure 1.15**). Of course, there are more complicated ways to think about this process in a larger context (**Figure 1.16**): feedback is received from

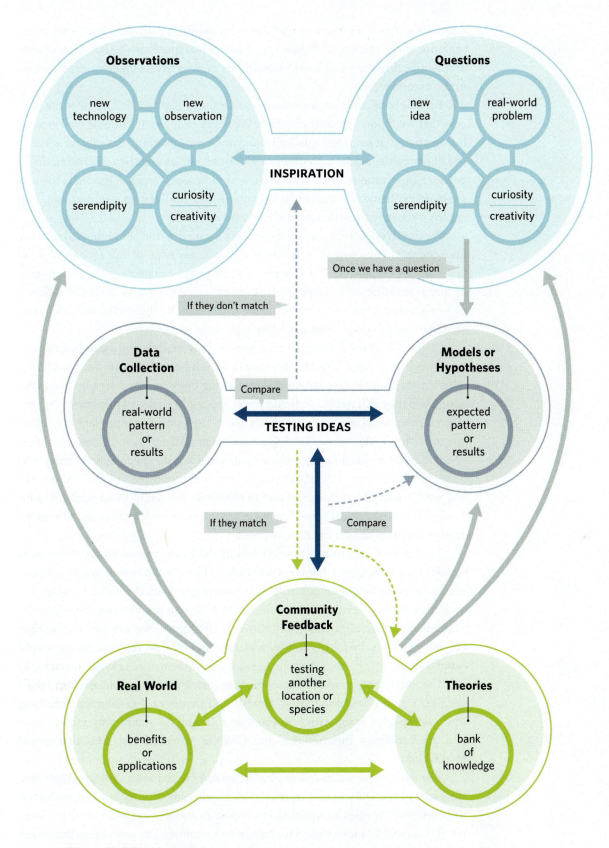

Figure 1.16 This model is a far cry from the simple model of the scientific method shown in Figure 1.7 or even Figure 1.8, but it is still perhaps too simplistic.

various sources along the way, and the exact research pathway varies with every scientific investigation. In Figure 1.16, you can imagine how the iterative aspects of scientific discovery can lead to a multitude of ways to link together explorations of the natural world.

For this book, the simpler scheme presented in Figure 1.15 illustrates the process that we follow for most chapters. We begin chapters with some background information that leads to a qualitative or conceptual model of the subject at hand. We illustrate the model with observations or examples and then use the model to produce expectations about how an ecological process "works." In some chapters we build mathematical models to help clarify our thinking on a subject and then confront the models with data or observations.

We feel strongly that, for anyone embarking on the study of ecological science, knowledge of the process is as important as knowledge of the ecological facts and theories themselves. Traversing these steps in ecology requires the mastery of many quantitative concepts and tools that are common to several branches of science and even to everyday decision-making. By the end of this book, you should be able to move more easily through these steps.

Chapter 2 offers a more detailed look at how models and quantitative concepts work in ecology. This chapter may be a useful primer for anyone wondering whether the models throughout the rest of the book will be overwhelming. We assure you they will not be. Although some models in ecology are complex, we try to present material in a way that makes sense, the same way that planning before you take a trip makes sense. The logical and quantitative reasoning "muscles" that you will build while stepping from a conceptual model to ecological theory will be useful if you continue in the ecological sciences, but these skills will also serve you well in everyday life. By the end of the book, you should find yourself more comfortable with logical formulations in a wide variety of settings and also more prepared to *do* ecological science yourself.

To that end, we group the chapters of the book around some of the subspecialties of ecology. Chapter 3 offers insight into the way ecology and climate interact to form recognizable environments around the globe. Chapter 4 is an introduction to evolutionary ecology, to help you understand the way populations and species change over time. In Chapter 5, we explore physiological and behavioral ecology and look at how individuals respond to the world around them.

The next four chapters focus on the basics of population ecology, with Chapter 6 exploring population growth and demography, Chapter 7 delving into two-species competitive interactions, Chapter 8 examining exploitative and predator-prey interactions, and Chapter 9 examining mutualisms and facilitation.

The five chapters that follow explore larger, multispecies ecological systems. Chapters 10 and 11 focus on community and landscape ecology, starting with food webs and multispecies interactions in Chapter 10 and moving to diversity in Chapter 11. Chapter 12 tackles spatial ecology and explores the way interactions and outcomes change when we consider that species and individuals move around.

This movement is essential to thinking about the way communities change through time, the topic of Chapter 13. Chapter 14 examines ecosystem ecology by adding the flow of nutrients through communities.

Finally, we end the book in Chapter 15 back where we started, with a focus on the Anthropocene and human effects on ecological systems. This chapter introduces you to some of the most pressing ecological issues of our time. We hope that by the time you reach this final chapter, you will have an appreciation for a variety of ecological interactions, an interest in the biggest problems facing us, and the tools to help find their solutions.

SUMMARY

1.1 The Ecological "House"

- The word *ecology* derives from the German word *Oekologie*, meaning "doctrine of the household."

- A more modern definition of *ecology* is "the study of the interactions that determine the distribution and abundance of organisms."

- Ecologists have a long history of tackling practical questions and studying the damage that humans have created.

1.2 Ecology in the Anthropocene

- Humans have damaged the "house" that we all live in, and this issue must be a part of any current study of ecology.

- The Anthropocene is the modern geological epoch characterized by human influence around the globe.

- Topics that arise in studying ecology in the Anthropocene include extinction, climate change, urbanization, homogenization, invasive species, and anthropogenic effects.

1.3 What Do Ecologists Do?

- Ecologists typically use one or more of the following approaches when doing ecological science:

Observation and Natural History

- This involves carefully watching nature and natural phenomena and constructing narrative explanations.

- Observation is the main way early ecologists studied nature, and it is still important today.

Experimental Ecology and Null Hypothesis Testing

- Hypothesis testing through the use of manipulative experiments became common among ecologists during the 1950s.

- Constructing a falsifiable hypothesis—a null hypothesis—that could be evaluated using treatments, controls, and collected data became the main avenue for ecological research.

- This approach imposed some limits on the types of organisms studied (generally smaller) and the timescales under consideration (generally shorter).

- An example of this approach is Jane Lubchenco's work on a snail-and-algae interaction in New England tide pools.

Multiple Hypothesis Testing with Best-Fit Comparisons

- More recently an "ecological detective" approach to the science has been suggested in which multiple hypotheses are simultaneously considered.

- Results and conclusions are based on the strength of evidence supporting one or more of the competing hypotheses.

- An example of this approach is Elizabeth Crone and Janet Gehring's research on the conservation of the threatened Columbia yellow cress in Washington State.

- This approach is often used in the context of large-scale studies in which multiple causative factors are at play.

Ecological Modeling

- Building a model of an ecological situation allows researchers to examine the mechanisms behind effects without the need to physically manipulate a system.

- Two general types of models are examined:

  - Conceptual models: These are theoretical constructs that build various interacting components into a logical framework.

  - Mathematical models: These involve refining conceptual models by including mathematical representations for parts of the model. They often involve computer simulations.

Modern Ecological Use of the Four Approaches

- All four approaches to conducting ecological science are included in a modern ecologist's tool bag, although individual researchers may not always have skills in all four.

1.4 How This Book Is Organized

- The text is structured to reflect how ecologists generally conduct their science: that is, first conceptual models are formulated and then observations are made and experiments performed to evaluate these concepts.

- Each chapter, therefore, begins with a conceptual model, illustrated by observations and examples. The model is then used to produce expectations about how the process works and is sometimes made mathematically explicit. Finally, the model is confronted with data or observations to evaluate the scientific fit.

KEY TERMS

ad hoc fallacy	data	manipulative experiment
alternative hypothesis	ecology	mathematical model
analytical model	ecosystem	naturalist
Anthropocene	extraneous factor	null hypothesis
anthropogenic effect	focal factor	organism
conceptual model	hypothesis	simulation model
control	hypothesis testing	treatment

CONCEPTUAL QUESTIONS

1. Does the field of ecology differ from other biological fields with which you are familiar? In what ways is it similar to other fields?

2. General observation of the natural world was the main approach of early ecologists. Discuss how observation still plays an important role in modern studies.

3. Describe the scientific method as practiced by ecologists, as shown in Figure 1.15.

4. Generate a concrete ecological question, and show how you might use the scientific method as outlined in Figure 1.15 to determine an answer.

5. Compare and contrast the various ways that ecologists use the word *model*. In what ways is the word used in fairly conventional terms? What ways are new to you?

SUGGESTED READING

Crutzen, P. J. and E. F. Stoermer. 2000. The "anthropocene." *Global Change Newsletter* 41: 17–18.

Hilborn, R. and M. Mangel. 1997. *The Ecological Detective: Confronting Models with Data*. Monographs in Population Biology 28. Princeton, NJ: Princeton University Press.

Lubchenco, J. 1978. Plant species diversity in a marine intertidal community: Importance of herbivore food preference and algal competitive abilities. *American Naturalist* 112(983): 23–39.

Steffen, W., W. Broadgate, L. Deutsch, O. Gaffney, and C. Ludwig. 2015. The trajectory of the Anthropocene: The Great Acceleration. *Anthropocene Review* 2(1): 81–98.

A scientist is collecting quantitative data from a coral reef ecosystem in the Maldives. A tremendous amount of ecological research involves measuring aspects of the natural world and then using these numerical data to gain a deeper understanding of how species interact with each other and the environment.

chapter 2

QUANTITATIVE REASONING

LEARNING OBJECTIVES

• Define quantitative reasoning and data literacy.

• Explain the process of building a conceptual model.

• Compare and contrast the features of conceptual and mathematical models.

• Define analytical and simulation approaches to model building.

• Characterize the different types of data.

• Explain how probability is used in ecological research.

2.1 What Is Quantitative Reasoning?

Although observations and descriptions of natural history are important in ecology, many critical ecological insights come from the quantitative side of the discipline. Understanding the interactions that determine the distribution and abundance of organisms requires understanding how various factors combine, interact, and change through time. The study of ecology, therefore, requires students to master fundamental skills for presenting and analyzing quantitative data and to understand how insights from such data can forecast future states of nature. Although most schooling from kindergarten onward highlights the distinction between math skills and reading skills, both words and equations can be used to tell ecological stories and explore relationships. We hope that, by the end of this book, you will be able to accomplish this and do so with both pleasure and creativity.

We start by explaining what we mean by quantitative reasoning. In a simple sense, a *quantitative approach* to a topic requires using information related to quantities rather than qualities. In order to get quantitative information, we need to measure or count things (**Figure 2.1**). **Quantitative reasoning** involves applying math to this measured or counted information to produce a deeper understanding of the topic being investigated.

Most people use quantitative reasoning more frequently than they realize. If we think of quantitative reasoning as using numbers to make a decision, reach a conclusion, or make a comparison, then most people use quantitative reasoning many times a day. These uses may not involve calculus, but they often involve

Figure 2.1 Quantitative reasoning requires numbers, and in ecology we often get those numbers by measuring aspects of nature with a range of tools. **A** For example, we can use calipers to measure the length of dragonfly larvae. Such calipers also work nicely to measure the diameter of small branches or trees. **B** For larger trees, we can use a special tape measure that converts circumference to diameter by incorporating π into its scale. We can also get numbers by counting. Counting individuals is slightly different from measuring in that it produces whole numbers. For example, we can count **C** sea turtle eggs on a beach, **D** salamander larvae collected from a pond, or **E** acorns scattered across a field.

algebraic relationships. For example, suppose you have to calculate whether you have enough cash in your pocket to buy ice cream for two friends and yourself. Solving this pleasant problem involves using some quick algebra.

In ecology, quantitative reasoning is used to explore how species interact with each other and their environment. Sometimes this involves simple skills, on par with figuring out if you have enough money to buy ice cream. At other times, it involves thinking about how quantities change over time or in relation to other quantities. In daily life, we use such skills informally when thinking about whether to get our ice cream at noon or 3 pm based on how we think the line will change. Formal analyses of such problems sometimes use calculus or linear algebra. Although both approaches are frequently used in ecology, in this book we try to distill most of the math down to simple algebra. At times, we point to calculus or even present equations based on calculus, but you should be able to understand the central points or do any necessary manipulations of the equations with basic algebra. These tools offer insight into a host of fascinating ecological questions, and we hope that exploring these questions will lead you to develop a broad interest in ecology and more comfort with an array of quantitative approaches in life and science.

An important aspect of quantitative reasoning is **data literacy** (**Figure 2.2**). Literacy is defined by the United Nations Educational, Scientific, and Cultural Organization (UNESCO) as the "ability to identify, understand, interpret, create, communicate and compute, using printed and written materials" (UNESCO Institute for Statistics 2021). By broadening the context of literacy beyond written text, we can define data literacy as the ability to collect, analyze, make sense of, and communicate the meaning of quantitative and qualitative data. Just as print literacy requires the skills of decoding text and writing coherent sentences,

Figure 2.2 Data literacy in ecology involves being able to work with numerical data, make graphs, interpret graphs, and communicate with quantitative information.

data literacy requires such skills as reading graphs and charts, drawing appropriate conclusions from data, and recognizing misleading or inappropriate uses of data. Data literacy and quantitative reasoning frequently go hand in hand.

In our experience, many incoming students fear the quantitative aspects of an ecology course. Like many subjects encountered for the first time, ecology requires learning new terms. Explaining those new terms by way of mathematical equations may seem like expecting an English speaker to learn Russian by explaining it to them in Korean. But that is not the case. Remember from Chapter 1 that models in ecology are often used to simplify a concept and break it down to its component parts. Rather than teaching in a language you do not understand, we use equations to create simple generalizations: it is more like teaching with cartoons. Additionally, we use "cartoons" (i.e., equations) that you can understand. Those of you who have always enjoyed math will find these generalizations clarifying. Others who may have feared math will too. Seriously.

If you have "math fear," try to leave it behind. We have two good reasons for saying this. First, there is no scientific evidence whatsoever that people come in one of two "flavors": those who are good at math and those who are not. There is, however, good evidence demonstrating that people who tell themselves (or are told by others) they are not good at math consistently do worse on math tests (Burkley et al. 2010). Math prodigies do exist, just as musical prodigies exist, but most people can enjoy singing or appreciate a beautiful song without having to be a prodigy. For non–math prodigies (and we definitely count ourselves among them), quantitative methods require some practice, and those with better preparation or who have practiced more can sometimes use the methods more easily. Think of math and quantitative methods as skills. No one would expect you to speak a new language fluently just by opening a book, or to play a sport at a professional or varsity level without a lot of practice. To use the quantitative methods employed in ecology, you will need to practice.

The second reason for opening your mind to quantitative methods is that, throughout the book, these methods are tightly linked to ecological concepts and ideas. Unfortunately, math and quantitative approaches are often taught in a way that divorces the methods from the question or application. There is beauty in pure math, just as there is beauty in pure musical notes, but most of us are more motivated to work hard or use a skill when we can follow a tune or see a purpose or application. We try throughout this text to tie the quantitative approaches to ecological ideas, concepts, and applications, so you can see why these approaches are useful and how they improve ecological understanding.

To make sure all readers start at the same level and are prepared for the quantitative methods used in this book, we introduce in this chapter some of the essential elements of quantitative reasoning. These include the basic skills of building models and collecting, using, and interpreting data. This effort takes us partway into the world of statistics and linked equations, but the material should be fairly accessible. You may find yourself returning to these pages for a refresher as you encounter theories, experiments, and models in other chapters.

2.2 Building Models

The British statistician George Box famously said, "All models are wrong, but some are useful" (Box 1979). He was referring mostly to models in statistics, but his words so succinctly capture any attempt to build a realistic model that they have become a mantra for all model builders. Models by their very nature are abstractions of reality, and because they are not "the real thing," model builders must pick and choose the elements to include that will best represent the important or interesting aspects of what they are trying to model. The resulting model is, therefore, useful only if it captures some essential insight or generalization about reality.

In this section, we describe the process of building ecological models and pay particular attention to the differences between qualitative and quantitative models. As detailed in Chapter 1, ecological modelers generally move through a series of qualitative steps to build a conceptual model that encompasses the known and unknown elements of the central question or concept at hand. Once they feel comfortable with the conceptual model, they slowly add in quantitative (or numerical) elements that translate the conceptual model into a quantitative or mathematical model. Although there are logical steps to the process, modeling is essentially a creative exercise aimed at expressing known features of a system or problem and using them to explore unknown features of that system or problem. Following the logical structure of environmental model building described by Andrew Ford (2009), we have separated this section into two parts, the qualitative and quantitative steps of model building.

Conceptual Models: Qualitative Steps

The first step in building a model is to become familiar with the question or problem at hand. For example, suppose you want to know what factors affect the rate of population growth for a particular species. Or maybe you are tasked with solving a very specific problem, such as how to deliver sufficient freshwater flows to streams to protect salmon while also providing needed water for agricultural use. No matter the reason for building a model, this contemplative step is critical because it is when you gather the information to make the model relevant. Good places to begin this process include reading pertinent published and unpublished literature, brainstorming with peers and colleagues, talking to people familiar with the issue, and creatively imagining possible outcomes for the question or problem. For larger applied questions, this stage may require observing the organism or system closely and talking with individuals who control key environmental resources or have a stake in resolving the problem.

It is almost inevitable that the first simple question will morph into a more complex mosaic of subquestions during this initial step. Part of the nature of learning about a subject is becoming aware of how much you do not know and recognizing when your initial impressions and questions are too simplistic. The more you learn, the more you realize how complex things are.

The next step, then, is to identify the specific aspects of the question that are the most important to include in the model. This step is like forming a hypothesis, as described in Chapter 1, and it requires identifying the key elements that produce the pattern or effect that the model will explore. Note that, thus far, this process is remarkably similar to the steps you would take to understand a problem in any discipline, and that many disciplines use conceptual or qualitative models.

Many ecological models track something that changes through time or space, which makes sense if you remember that ecology is defined as the study of the interactions that determine the distribution and abundance of organisms. So, a good starting point for a conceptual model is to ask, What factors am I going to track? If there is more than one answer, the next question is, Which of these factors is the most critical to track, given the end goal? In ecology, you can track a huge range of factors, such as population abundance (number of individuals in the population), nutrient availability, and gene frequency, to name just a few. What-

Figure 2.3 The signal crayfish (*Pacifastacus leniusculus*) is an invasive species in Lake Tahoe, California.

ever factor is determined to be the most important, you will likely measure it in multiple places or at multiple times (or both).

The final qualitative step to building a conceptual model often involves drawing. To start, you could draw a box that represents the one factor determined to be the most critical to track through time, then label this box with an informative name. For example, suppose the goal is to track the size of the population of the signal crayfish (*Pacifastacus leniusculus*) in California's Lake Tahoe (**Figure 2.3**). This crayfish species naturally occurs in northern parts of western North America but has been introduced by humans in several parts of California and Europe, where it has led to declining population abundances of native species. The variable of interest would be the number of signal crayfish in the lake, otherwise known as the *population abundance*, perhaps scaled by area (number of signal crayfish per square meter, for example). An appropriate label for the box would be "signal crayfish population abundance" (**Figure 2.4A**).

Next, think about how this box is "filled" or "emptied." You could give the box an inflow arrow (for how it is filled) and an outflow arrow (for how it is emptied), and label them with informative names. In the signal crayfish example, these labels could be "births" for the inflow and "deaths" for the outflow. These arrows represent specific *numbers* (number of births and number of deaths).

Now, let's think about *rates* of inflow and outflow—that is, the increase or decrease in these numbers *per unit of time*. In our example, the rate of inflow would be the birth rate (number of signal crayfish births per unit of time), and the rate of outflow would be the death rate (number of signal crayfish deaths per unit of time). Obviously, birth rates and death rates can have an influence on population abundance—that is, they can influence the number of births and number of deaths, respectively. We can illustrate rates in our conceptual drawing by sketching a "valve" on each flow arrow to "control" the flow.

Other features that may be incorporated in a conceptual diagram are additional changes in the critical or focal factor due to different causes. For example, Lake Tahoe's signal crayfish population may also increase when people move boats between bodies of water and unintentionally transport tiny juvenile crayfish attached to their boats (**Figure 2.4B**). This pool of transported individuals is influenced by several additional factors, including seasonal differences in the number of people fishing (e.g., more in summer, fewer in winter), the way that the crayfish young cling to boats, and the length of time a boat is out of the water. Is it important to include some, or all, of these additional factors in the model?

It is useful to remember that a model is an approximation. It is a perfectly good modeling practice to reach a point of complexity and say, "That's enough for now." No model can re-create a real situation exactly. If it could, it would cease being a model; it would be the thing itself. Models are not attempts to re-create reality. Instead, they are representations and explorations of a particular facet, process, or characteristic of reality. Focusing on the elements you think may be most important and ignoring other factors is perfectly fine.

Think back to the paper airplane example from Chapter 1. You don't need a highly intricate and complex design to create a paper airplane that models flight. Just as a paper airplane design focuses on key aspects of flight, so an ecological model should focus on the most important elements of the system that the model is meant to capture. This is not to say that later in the process you won't need to go

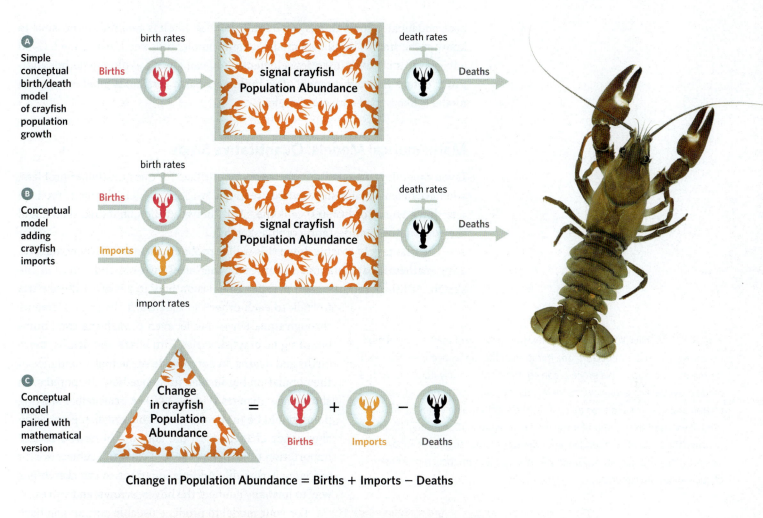

A Simple conceptual birth/death model of crayfish population growth

birth rates

Births

signal crayfish Population Abundance

death rates

Deaths

B Conceptual model adding crayfish imports

birth rates

Births

Imports

signal crayfish Population Abundance

death rates

Deaths

import rates

C Conceptual model paired with mathematical version

Change in crayfish Population Abundance = Births + Imports − Deaths

Change in Population Abundance = Births + Imports − Deaths

D Conceptual model with iterative mathematical version

Population Abundance **Next Year** = Population Abundance **This Year** + Change in crayfish Population Abundance

Population Abundance Next Year = Population Abundance This Year + Change in Population Abundance

Figure 2.4 Four increasingly detailed versions of a model of signal crayfish population growth in Lake Tahoe, California. The first version **A** shows a simple conceptual-model sketch indicating how crayfish population abundance can grow or decline based on the number of births and deaths (indicated by arrows) as well as the rates at which births and deaths happen (indicated by valves). The second version **B** adds a factor that can influence population abundance—namely, the import of additional crayfish from elsewhere via recreational boating traffic. The third version **C** pulls the factors from the conceptual model into a word equation that focuses on change in population abundance. The fourth version **D** illustrates how the model can be run in an iterative fashion so that the final population abundance from one iteration becomes the starting population abundance for the next iteration, allowing the model to track population abundance through time.

back and include some things you initially left out. But it is generally impossible to know at the start of building a model which complexities need to be added. Good advice is to start with just a few elements. You can always explore whether additional details improve a model after building a basic version. A good conceptual model is like a good concept: simple.

Mathematical Models: Quantitative Steps

If you can follow the qualitative process just outlined for one question or problem, you have taken a major step toward creating a working model. The next step is to tear down what Ford (2009) calls the "wall" between the qualitative sketch and the quantitative manifestation of this sketch.

The first task in tearing down this wall is probably the most challenging: creating mathematical equations to represent the boxes, arrows, and valves in the sketch. At this point, it is good practice to envision how to get all of these parts to "talk to each other" so they track the same elements through time. For example, even though the total number of signal crayfish varies with births and deaths, those births and deaths do not provide the actual abundance of the population but instead influence how the population abundance changes (**Figure 2.4C**). A reasonable starting point would be to say that the signal crayfish population abundance changes with the addition of newborns and importation of juveniles via boats and the subtraction of individuals that die. Adding equations to the sketch is a way to formally connect the boxes, arrows, and valves.

For your model to produce useable output, you have to figure out what numbers to plug into the mathematical model. This is the next quantitative task. But first, there is a terminological issue here that deserves some attention. In mathematics, a changeable value within an equation is represented as a **variable**, and variables are generally expressed as capital letters, such as X, Y, or N. Note that the variables do not have to be represented by letters, as they could just as easily be called "Population Size" or even "Greta," but capital letters are the norm. The letter is a placeholder that identifies that variable (element of the model) in shorthand in the equation. Each of the elements in the word equation in Figure 2.4C is a variable that could be represented by a capital letter.

Let's return to the changeability of variables. How and when do they change? In general, a model will have at least two variables, and scientists are often interested in how one variable affects or alters the other variable. For example, all the variables on the right-hand side of the equation (that is, to the right of the equal sign) in Figure 2.4C change the value of the single variable on

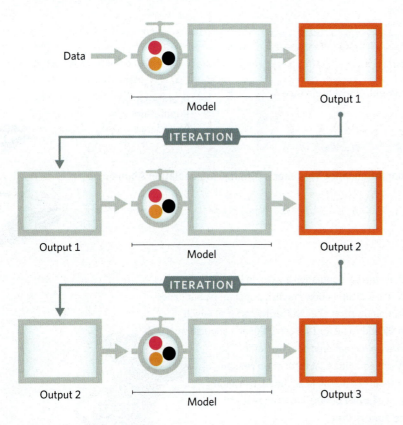

Figure 2.5 This illustration of the iterative process of mathematical models shows how the output from one iteration of the model becomes the input for the next iteration. This process can continue for as many iterations as the modeler would like and allows modelers to set initial conditions (e.g., population abundance) and then follow how the focal variable changes through time. If we used this method for the model in Figure 2.4D, we could iteratively produce the population abundance for the signal crayfish in Lake Tahoe for several years into the future. Figure 2.7 offers a mathematical representation of such an iterative model.

the left-hand side. The model, therefore, represents the relationship envisioned between the variables. Although ecologists may include more than one variable in a simple model, one of the variables in the model is the focal factor, as identified in the qualitative stage of model building. Thus, this is *the* variable of interest across model iterations, like population abundance for the signal crayfish.

If you are interested in the abundance of signal crayfish over successive years, you may build the model a little differently from the one in Figure 2.4C and focus not on the change in population abundance but on the actual population abundance (**Figure 2.4D**). In this new formulation, you can obtain the population abundance next year (the output) by looking at how births, imports, and deaths alter this year's population abundance. To project forward in time, you would then take that output and use it as the new starting point, add more births and imports, and subtract deaths to get the population abundance two years into the future; and so on, iteratively, for the population abundance moving forward.

This example helps answer the question of how variables change. In order to see the change that comes from a modeled relationship, ecologists often look at the output of the model multiple times and call each new pass through the equation an **iteration**. With some models, ecologists may use the "output" from one iteration as the "input" for the next iteration (**Figure 2.5**), as described for the signal crayfish population abundance in Lake Tahoe. In general, this output and input come from the focal variable.

Other elements of the equations may have fixed values (or a specific range of values) that make ecological sense. These values will *not* change across iterations of the model. These unchanging model elements are called **parameters**, and they are generally represented by lowercase letters of the Roman or Greek alphabet (e.g., a, b, c, or α, β, γ). In our crayfish example, you could track quantities other than population abundance by using parameters. For example, instead of following total births as a variable, you could use a parameter based on average egg production per individual (**Figure 2.6**) and multiply that parameter by the number of individuals in the population (**Figure 2.7**). You could do the same thing with deaths by multiplying population abundance by the average rate of death. By using average parameters for birth and death rates, the only thing that makes reproduction change from year to year is the number of individuals in the population—in other words, population abundance, the focal variable.

Notice that the research question determines the focal factor in a model and, therefore, determines which elements are variables and which are parameters. For example, if you were interested in how egg production changes with weather or temperature, then egg production would need to be a variable, not a parameter.

An important aspect to model building is identifying realistic value ranges for parameters. Parameters based on reality offer insights based on reality. Luckily, ecological modelers often have access to previously published

Figure 2.6 A signal crayfish with a load of eggs attached to its tail. Quantifying the number of eggs per individual for a large population allows us to estimate the average egg production for crayfish in a particular lake.

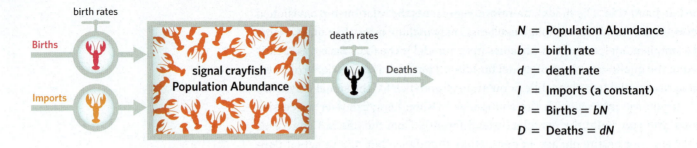

Population Abundance Next Year = Population Abundance This Year + Imports + Births − Deaths

$$N_{\text{next year}} = N_{\text{this year}} + I + bN_{\text{this year}} - dN_{\text{this year}}$$

Figure 2.7 If we want to explore what makes the Lake Tahoe signal crayfish population increase or decrease, we could focus on the parameters for birth rates and death rates. In this version of our model, we have turned imports into a constant (notice the loss of the valve over the imports arrow), and we have connected the number of births and deaths to both the rate and the population size. Our rates represent what happens per individual crayfish, so we can multiply that rate by the number of individuals to replace the B and D variables with a parameter multiplied by the population abundance variable (N). We explore population modeling in much more detail in Chapter 6.

experimental or observational studies of their focal ecosystem, which can provide these numbers. If parameter estimates are obtained from solid ecological sources, a modeler can be confident that the range of model outcomes will be realistic.

In some instances, ecologists do not have detailed information to inform parameter values, but this lack of information does not have to be a catastrophe. It is a fairly common practice to start with parameter values based on informed opinions, best guesses made by scientific experts, or even just a few observations. These preliminary parameter values allow ecologists to explore the behavior of the model and the potential outcomes such a model might produce. Once additional observations are acquired, researchers can return to the model, input the new, more realistic parameter values, and adjust the model to produce more realistic outcomes.

Variation in Models

It is useful to think about how ecologists use these models, how specific or general they want them to be, and whether a model should include variations like those found in the real world. Such variation may make model outcomes differ from location to location, just as experiments often differ between locations. The way ecologists introduce variation comes from the way they treat parameters, so we need to discuss parameters a bit more.

Ecologists can look at model outcomes across different time frames and different parameter values. The modeler may have information on appropriate parameters for a particular location, but using only those values provides insight only into the likely outcomes for that single location. If, instead, the modeler is more interested in understanding the general behavior of the model, then looking at the relationship between parameters and seeing the outcomes across a broad range of parameter values may be a better approach. Both ways of understanding the general behavior of a model are *analytical approaches*. As mentioned in Chapter 1, an analytical approach requires solving equations or looking at relationships between parameters to identify the range of potential outcomes, which is sometimes also called exploring parameter space. Let's return again to the Lake Tahoe signal crayfish to illustrate.

Suppose you are interested in whether the population of signal crayfish in Lake Tahoe is increasing or decreasing. Population abundance is still the factor of interest, but you can look at the other elements in the equation to find the range of conditions that produce population increases. From Figure 2.4, you can see that the population will grow when births and boat imports together are greater than deaths. The birth rate and death rate could be parameters that represent the average egg production and death probability per individual crayfish (Figure 2.7). We will discuss the specific effects of birth and death rates on population growth much more extensively in Chapter 6. If you let the number of boat imports be a constant (because you decide not to worry about the way fishing changes with the seasons), then an analytical approach provides a general sense of the birth rates and death rates necessary for population growth. Sometimes, this sort of insight is more important than running an iterative model with just one set of parameters.

So far, the models we have outlined are all **deterministic models**, meaning they always produce the same result as long as we begin with the same parameter values and starting values for our variables. The way things change in these models is completely "determined" by the way the model links elements together. This approach means that a particular set of initial values or conditions (for both variables and parameters) will lead to the same model outcome every time the model is run. Deterministic models lend themselves well to analytical approaches.

But what if you want the model to represent the actual way ecological factors change across space or time? In such a situation, a *simulation approach* might be more useful. A simulation approach is one in which a model is run several different times, and each run represents a version, or simulation, of the potential real-world conditions.

For example, previously we suggested using the average number of eggs per individual signal crayfish as a parameter in the Lake Tahoe model. This effectively means that each crayfish in the model produces the same number of eggs regardless of size, health, or age. Even if the parameter somehow represented an average across all crayfish of the entire species, such reproductive consistency would be unrealistic. Signal crayfish biology suggests that factors such as temperature, pH, and nutrient availability are all likely to fluctuate through time and thereby alter the average number of eggs per individual.

To incorporate this type of natural-world variation in the model, instead of choosing a single value for the number of eggs per individual, you can allow the egg production parameter to assume a range of values. This range should reflect the variability seen in the real world. During each iteration of the model, you (or a computer program) can randomly choose a value from within this range of egg production and plug that number into the model to simulate a single potential set of circumstances. In every subsequent iteration, you can put another random value from within the range of egg production values into the model, until all iterative runs are complete.

This type of model not only lets the parameter values change between iterations but also allows them to change randomly based on an arbitrary pick from the range of specified parameter values (**Figure 2.8**). We call such models **stochastic models**, because they include an element of randomness (*stochastic* means

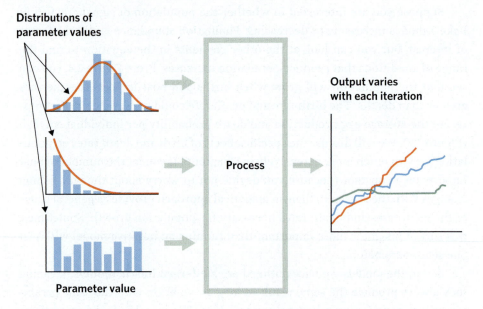

Figure 2.8 Parameters do not have to be fixed and remain the same through an entire model run the way they do in a deterministic model. Using a stochastic modeling approach, a researcher or computer program can instead change parameters with each iteration to reflect the variability one might see in those factors in the real world. With stochastic models, researchers choose a distribution of parameter values and pick the value "plugged into" the model randomly at each iteration. The range of parameter values may follow a normal distribution, a uniform distribution (see Figure 2.18), or a more complex range of values. Each run leads to a different outcome, represented by different colored lines in the outcome graph. In Chapter 1, we saw such runs in Crone and Gehring's (1998) exploration of different models of population persistence for *Rorippa columbiae* (Figure 1.12).

"random"). Stochastic models lend themselves best to simulation approaches because they have no single analytical solution. By running the model many times, the researcher sees the range of potential outcomes, as well as the frequency of particular outcomes.

Stochastic models can provide a more accurate sense of how real-world variation alters the ecological outcomes scientists are interested in exploring. Of course, how well a stochastic model reflects actual ecological dynamics depends on how accurately the selection of parameter values represents real-world variation. In the next section, we explore how ecologists identify and include real-world variation and how observations produce information (data) that helps to parameterize and test models.

2.3 The Lowdown on Data

We started the chapter by pointing out that every ecological investigation is motivated by a question. Data are the pieces of information that help us to both frame the question and arrive at the answer. The word *data* is confusing because it is generally plural in the sciences (singular, *datum*) but singular in everyday language. You may notice that we slip back and forth a bit depending on how we are using the word.

oregonensis

platensis

picta

xanthoptica

croceater

eschscholtzii

klauberi

We could have started this chapter with a discussion of data, but we wanted first to address the process of framing an ecological question and identifying essential elements and factors. Nonetheless, it is important to realize that every ecological question is generally informed by some type of data or data collection. The data at times will show a pattern, perhaps something unexpected or interesting, and that pattern can lead to further questions. In assembling information to build a conceptual model, ecologists often rely on previously collected data or on their own observations. Once they have built the model, conceptual or mathematical, they can collect additional data to help answer the question being investigated and parameterize the model.

Figure 2.9 Ecologists can distinguish subspecies of the *Ensatina* salamander (*Ensatina eschscholtzii*) across the Pacific United States and Canada based on color differences. Each color type is a unique subspecies, as designated by the subspecific epithet (an addition to the species name) under each picture.

Data Types

Categorical Data Sometimes ecological data have no inherent numerical value but occur instead as categories. For example, in various Pacific North American populations of the common *Ensatina* salamander (*Ensatina eschscholtzii*), individuals may have spots that are red, orange, or yellow, or they may have no spots at all, depending on where they live (**Figure 2.9**). Notice that the category of spot color is not only nonnumerical, but the individual "items" in the category, or

Nominal categorical variables

Ordinal categorical variables

Figure 2.10 Nominal categorical variables have no inherent numerical value associated with them; examples include color, sex, and status. Ordinal categorical variables have an inherent order to them; examples include life stage and size.

pieces of categorical data (i.e., the various spot colors), have no inherent order to them: yellow is not "better" or "higher" or more important than orange. Categorical data of this type are called **nominal categorical variables**, meaning they are defined by names. Common examples (**Figure 2.10**) include color (as in our salamander example), sex (male, female), status (alive, dead), and species (a list of names).

Yet sometimes categorical data do have a natural order to them. Examples include life stages (egg, larva, pupa, adult; arranged from youngest to oldest), calendar months, or sizes (small, medium, large). Categorical data that have a natural order are called **ordinal categorical variables** (Figure 2.10).

It may be unclear how categorical data fit into a quantitative approach, or why they are important, but scientists can do a great deal quantitatively with categorical variables. First, these variables are essential when constructing models. If you think back to how ecologists often use models to study changes through time or space, you can see that each distinct time period or location can be a separate piece of categorical data. In fact, sampling period, location, and species are perhaps the most common and important categorical variables in ecological models. Second, ecologists frequently collect numerical data that are closely associated with, or tied to, categorical variables, such as measuring the mass of different sexes of salamanders. Ecologists cannot look for differences in mass between sexes unless they also record the categorical data on sex from each individual and pair each mass data point with a sex category.

Numerical Data   **Numerical variables** also come in two types, continuous and discrete. As a general rule, **continuous numerical variables** come from measurements. A measured factor can take any value, including fractions and decimals. Examples of continuous numerical variables in ecology include mass, leaf area, lake size, tree height, body weight (**Figure 2.11**), and nitrogen concentra-

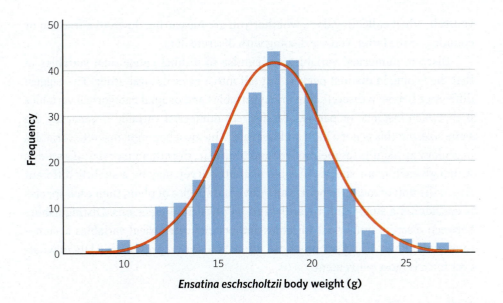

tion. Continuous numerical variables can be measured to whatever degree of precision the measuring instrument allows; this may include a few digits to the right of the decimal point.

Not all numerical variables can be measured to a tiny fraction, though. Many numerical factors of interest only come in whole numbers, in which case they are referred to as **discrete numerical variables**. The most common in ecology are count data, such as the number of insects on a plant (**Figure 2.12**), the number of bones in a leg or arm, the number of lakes in a given area, the number of wins or losses in encounters between fighting individuals, and so on. Such factors only occur as whole or discrete units. For example, there is no such thing as 26.3 insects. An insect may be small or large, but it is still one single insect. There may, however, be an *average* of 26.3 insects per square meter. The distinction between discrete and continuous numerical variables can be confusing. If you are unsure, ask

Figure 2.12 The number of insects on a plant is a discrete numerical value because it is not possible to find 26.3 insects. Live insects only come in whole numbers.

yourself, when collecting data, whether you are measuring down to a fraction or counting. If the latter, you are dealing with discrete data.

Discrete numerical variables are similar to ordinal categorical variables in that they come in distinct packets but fall into a clear natural order. The biggest difference between discrete numerical variables and ordinal categorical variables is that each instance or packet of a discrete numerical variable is generally the same size, but this is not a hard and fast rule. The most common discrete variables in ecology are counts (as mentioned) and time (e.g., the nth hour or day of a study). Although each plant species in a biodiversity survey may be a slightly different size, if the unit of measurement is an individual species of plant, then each species is considered a single unit regardless of its physical dimensions. Distinguishing between discrete numerical variables and ordinal categorical variables is sometimes unimportant, and scientists can often choose to represent the data in whatever form is most convenient.

Collecting Data

As mentioned in Chapter 1, scientists are a lot like detectives, and information (data) provides them with the clues that lead to answers. As is true in detective novels, the information at hand is almost never complete. In most scientific research, it is impossible to observe or measure every single member of a population. For example, if you wanted to know the average surface area of a sugar maple leaf (*Acer saccharum*) in order to increase maple syrup production, waffle consumption, and, therefore, general world happiness, you would not be able to measure every single leaf over the whole geographic range of sugar maple trees (**Figure 2.13**). Leaves on these trees grow in the spring, drop in the fall, and might be eaten by insects or hit by hail in between. It would be impossible to physically reach, let alone measure, every single leaf. One solution would be to visit a number of places with sugar maple trees and randomly select a sample of leaves from each place at peak growth in that location. If you then measured the leaves in this subset of all sugar maple leaves, these measurements could represent the variation in leaf size across all locations where sugar maples grow.

Generally, when ecologists collect data, they are forced to collect information on only a subset of a whole group. The data gleaned from the subset are then used to characterize and make inferences about the larger group. In the context of sampling and statistics, this bigger group is called the **population**, and the smaller measured group is called a **sample**. In the leaf example, the population would be all sugar maple leaves across a region, and the sample would be the subset of leaves that were actually measured.

Notice that we now have two different meanings of the word *population*. In the signal crayfish example, the population was all of the signal crayfish in Lake Tahoe. That is the biological and ecological use of the term. In other words, a biological or ecological population consists of all the individuals of a species living and reproducing in a given area (see Chapters 4 and 6 for more detailed treatment). In contrast, the statistical meaning of *population* refers to *any* real-world group of subjects, be they individuals of a species, lakes spread out across Minnesota, or leaves on a tree.

Figure 2.13 (A) The spatial area of a single leaf can be quantified with a portable leaf-area meter. (B) The geographic range of sugar maple trees (*Acer saccharum*) is quite large, making it physically impossible to collect and measure all sugar maple leaves across the entire geographic range of the species. Instead, a researcher could take a sample of leaves, measure those, and use the mean and standard deviation of the sampled leaves to estimate leaf size for the whole population. (Adapted from Peters et al. 2020.)

It is important to get a good, representative sample of the true statistical population. A good sample will be *unbiased*, meaning it will be a random subset of the statistical population. *Biased* samples are nonrandom (meaning individuals in the population have unequal probabilities of being sampled) and, unfortunately, lead to biased answers, which no scientist wants. For example, you should not generalize about egg production across all female signal crayfish if you only sampled females from one location in Lake Tahoe. Similarly, you would not want to generalize about the surface area of sugar maple leaves across all trees in North America if you only sampled one cluster of trees outside of Montpelier, Vermont.

A good statistical textbook offers guidelines for ways to avoid bias, but a rule of thumb is to generalize about the population (or populations) under study and avoid extending the generalizations to other groups. To avoid biased data collection, consider whether any particular types of individuals might be left out of a sample because of the sampling process itself (e.g., selecting for small body size, only one sex, only hungry individuals, or only individuals found in the spring). This type of thought exercise can help reveal biases in data and ways those biases could skew answers.

Categorical and numerical

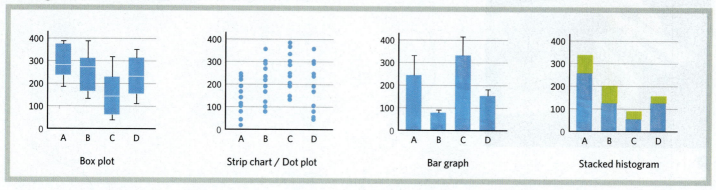

| Box plot | Strip chart / Dot plot | Bar graph | Stacked histogram |

Numerical and numerical

Categorical and categorical

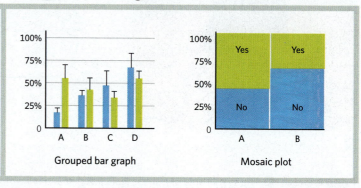

| Scatterplot | Line plot | Grouped bar graph | Mosaic plot |

Figure 2.14 Common graph forms in ecology. In the top row, the categorical variable is always on the x-axis and the numerical variable is on the y-axis. Note that the examples are not comprehensive, as there is a wide variety of graph types and an even wider variety of ways they can be presented.

Using and Summarizing Data

Ecologists use data in many ways, but three stand out as the most common. First, data can provide generalizations, such as averages, which can be useful for estimating parameters for models. For example, one could calculate an average value from collected data to serve as a fixed parameter in a deterministic model. Second, ecological data can provide a sense of the natural variability that exists in a focal factor (e.g., variation in the number of eggs a female signal crayfish will produce or variation in the surface area of sugar maple leaves). This second application uses the frequency distribution (or some other measure of variability in the data) to provide a range of values from which to draw a random parameter for each iteration in a stochastic model. Third, ecologists can use data to look for relationships among variables. Examples of this application are often found in news stories that discuss scientific results, differences between voters in one area and another, changes in household income, and a range of other everyday topics.

Creating graphs or summaries to represent averages and the range of variation in data requires creativity and considerable care. **Figure 2.14** shows some common graphs for different types of data. When examining or creating graphs, it helps to focus first on the point you are trying to convey with the graph. Next, check that the graph is "honest," meaning all the data are presented fairly and equally, and differences are not somehow magnified. For example, textbooks and

Figure 2.15 Examples of how a graph can mislead a reader. Each graph represents the mean number of insect species collected from two sites, a hillside and near a stream. Notice that the mean values for the two sites look very different from each other in graph **Ⓐ** but less so in graph **Ⓑ**. Starting the y-axis at zero forces graph B not to overemphasize the differences. Graph **Ⓒ** adds error bars that represent one standard error, a common measure of variability in an estimate of a mean. These error bars suggest that our estimates of the means overlap and that there is no significant difference between the mean number of insect species collected in the two locations. Error bars do not always represent standard error, but they do tend to represent something about the range or variation in values in the sampled data.

even newspaper articles often use bar charts with error bars to show differences among groups. The top of each bar usually indicates a mean (average), and error bars typically represent some measure of variability present in the sample that leads to uncertainty in the estimate of the mean. Differences between bars look much bigger when the graph only shows part of the y-axis (i.e., the tops of bars rather than entire bars; **Figure 2.15**). There are many other ways that graphs or representations of data can mislead readers, so look at graphs, figures, results, and summaries with a critical eye to make sure that the data offer useful information without bias.

These considerations are also important for exploring relationships among variables. Generally, data are a critical part of tackling ecological questions, so it makes sense to follow some of the same steps with data that ecologists follow when creating a model. As a first step, it is helpful to outline your initial ideas by creating a conceptual model and identifying where data fit in this model. What relationship does the hypothesis or conceptual model suggest is likely to emerge?

If ecologists take any two variables they think may be associated, they can often use a conceptual model to identify one variable as an **explanatory variable** and the other as a **response variable**. An explanatory variable affects, changes, or explains why the response variable varies as it does. Returning to the signal crayfish example, suppose you are interested in determining whether a relationship exists between boat traffic and crayfish population abundance. The model in Figure 2.4B suggests that boats might carry crayfish or crayfish eggs from other lakes and deposit them into Lake Tahoe. In this case, boat traffic (number of boats) would be the explanatory variable because it contributes to, or helps

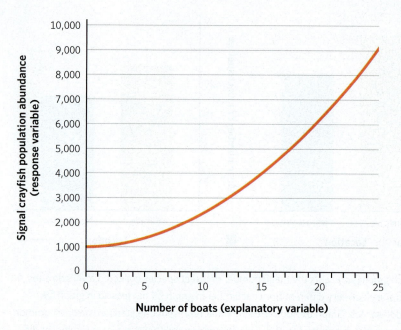

Signal crayfish population abundance (response variable) — y-axis

Number of boats (explanatory variable) — x-axis

explain, crayfish population abundance, which is the response variable. After collecting data, you may see that an increase in boat traffic (explanatory variable) leads to an increase in the signal crayfish population abundance (response variable; **Figure 2.16**).

Often a response variable will be the critical factor that is tracked in a conceptual, or qualitative, model. Note that some texts refer to explanatory and response variables as independent and dependent variables, respectively. When creating a graph, generally the explanatory variable is on the x-axis and the response variable on the y-axis. In this book, we use many graphs to show generalizations, trends, and relationships between variables. We always keep explanatory variables on the x-axis and response variables on the y-axis, and we make sure that the graphs and figures do not distort the data to make a point.

Figure 2.16 In exploring the relationship between two variables, it is important to think about which variable affects or changes the other. The variable that explains the change is the *explanatory variable* and goes on the x-axis. The variable that changes in response to the explanatory variable is the *response variable* and goes on the y-axis. Using the signal crayfish example, we can envision that the amount of boat traffic (represented here by number of boats) might alter the signal crayfish population abundance. If we visited a series of lakes of approximately equal size, we might see a change in signal crayfish population abundance as the number of boats increases.

Probability

How do we know whether generalizations seen in data or relationships between variables are important? The answer hinges on probability. The study of probability is complex and significantly more involved than we can cover in a few paragraphs, but most people have some intuitive grasp of probability.

People make statements of probability all the time. For example, you might say, "I will probably need to get more gas before I arrive in Cedar Rapids" or "I'm afraid we will run out of onion dip because so many people showed up at the party." In ecology, statements of probability are also quite common. One type of probability-related statement you might hear ecologists and other scientists make is that something (an experimental result) is "statistically significant." In fact, you may have come across the concept of *statistical significance* already in your studies. We will come back to this shortly.

As discussed in Chapter 1, one of the common experimental approaches used in ecology and other sciences involves forming null and alternative hypotheses. A null hypothesis is an expectation of results from an experiment that assumes that variation in the sampled data is due to random chance and not to a particular focal factor (the explanatory variable). In the Lake Tahoe signal crayfish example, suppose you wanted to know whether, in general, heavier boat traffic leads to more crayfish imports and, therefore, to a faster-growing signal crayfish population. It would be hard to answer this question with data from Lake Tahoe alone, so instead you could compare the growth of signal crayfish populations across multiple lakes, some with more boat traffic and others with less. A null hypothesis for this investigation might be: *There is no difference between the average signal crayfish population growth rate in lakes with high levels of boat traffic and the average signal crayfish population growth rate in lakes with low levels of boat traffic.* Null hypotheses generally take the form of negative statements—that is, statements that the explanatory factor has "no effect."

An alternative hypothesis proposes that variation in the sampled data *is* related to the explanatory variable (e.g., the import of crayfish by boats) and often takes the form of a positive statement. For example, an alternative hypothesis for the crayfish investigation might be: *There is a difference between the average signal crayfish population growth rate in lakes with high levels of boat traffic and the average signal crayfish population growth rate in lakes with low levels of boat traffic.* Having formed your null and alternative hypotheses, you would go about gathering your data. Next, using standard statistical methods for testing experimental outcomes, you would evaluate the pair of hypotheses and calculate what is known as a **p-value**, usually represented by *P*. You can interpret the p-value as the probability of collecting your observed data if the null hypothesis were true. A very low p-value (i.e., a very small number) suggests that the probability of getting your results by random chance—that is, the probability that the null hypothesis is true—is very low.

Ecologists (and most scientists) use the phrase **statistically significant** to describe when this probability is less than a previously chosen significance level, usually 5%. Convention suggests rejecting a null hypothesis if the p-value is below the chosen significance level ($P < 0.05$). In our example, you would reject the idea that there is no difference between the crayfish population growth rate in lakes with high boat traffic versus lakes with low boat traffic. Strangely, though, what you cannot do based on p-values is ever accept the alternative hypothesis, which in the crayfish example is that there *is* a difference in the population growth rate between lakes with high and low boat traffic. The results may support the alternative hypothesis, but they can never prove an alternative hypothesis because scientists can never rule out the possibility that some unusual event has transpired to produce these results. This is the convoluted and confusing part of null hypothesis testing and *inferential frequentist statistics*. We leave it to your current or future statistics professor to help explain these conventions in more detail and provide alternative ways to think about the probability of particular ecological outcomes.

An important distinction to keep in mind is the difference between statistical significance and biological significance. In the hypothesis-testing framework just described, the statistical significance is entirely dependent on the p-value generated from the statistical test. It is generally good advice not to interpret p-values as the sole indicator of importance. Ecologists are much more interested in *biological significance*, which is the importance of a factor in a real-world biological or ecological sense.

A factor is biologically significant if it is important to the organism or situation in question. That importance is determined by the size of its effect on the organism or situation, not by the size of the p-value. For example, a drug company may explore a new allergy medication that reduces hives in 99% of its test subjects. Such a pervasive effect is likely to yield a significant p-value. Yet, if the drug only reduced the severity of hives by 1%, the result is not very biologically significant, nor would it persuade anyone to take the medication.

An additional issue with p-values is that many ecological approaches do not lend themselves well to single null and alternative hypotheses, as discussed in Chapter 1. Ecologists are often interested in exploring a suite of hypotheses and

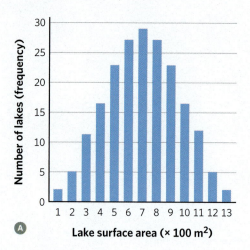

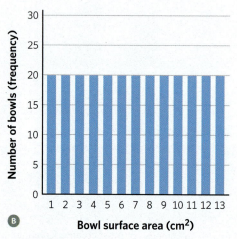

Figure 2.17 Here we have two hypothetical frequency distributions of objects with particular surface areas. **Ⓐ** The hypothetical distribution of lake surface areas shows a bell-shaped curve with the most lakes having 700 square meters (m²) of surface area and decreasing numbers of lakes having greater or smaller surface areas. We see very few, if any, lakes smaller than 100 m² or larger than 1,300 m². **Ⓑ** If we, instead, looked at the frequency of surface areas among some human-designed objects, such as kitchen bowls in a department store, we might see an even, or uniform, distribution of data if the store always stocks 20 packages of each size bowl. Uniform distributions are rare in nature, so this was a fairly contrived example (but bowls are similar to tiny lakes).

assessing which hypotheses best fit the data and offer the most insight into the way the world works. These multimodel, ecological-detective approaches rely on probabilities and likelihood but not generally on p-values. Although p-values and discussions of statistical significance appear in many ecological research articles, we recommend that students and readers not blindly accept p-values as indicators of biological significance.

Distributions

In order to make statements about probability, scientists also need to know something about the way data points are "distributed." A data distribution represents all the possible values observed in the data and indicates how often each value occurs. For example, one may ask whether the data are clustered around some central value or whether they are evenly spread out across all values (**Figure 2.17**). Imagine that someone offers to give you a bag of candy-coated chocolates if you can guess the color of the one piece in their closed fist. If you really want that bag of chocolates, you may also want more information about whether some colors are more common than others. The data distribution would give you the proportion of chocolates of each color in an average bag. Data distributions provide scientists with critical information on variability in their data and help scientists identify probabilities.

One way to determine the distribution of data is to make a graph. For example, if you were to collect data on the size (surface area) of 200 lakes in the Rocky Mountains, you would find a variety of lake sizes. You could divide these observed sizes into categories (small, medium, large) or "place" them in numerical bins (0–100 m², 100–200 m², and so on to the largest surface area) and then plot the frequency of each category or bin on a graph (Figure 2.17A). This sort of graph is called a **frequency distribution**. The graph in Figure 2.11 is an example of a frequency distribution, as is the stacked histogram in Figure 2.14, although the latter shows two frequencies at once.

When data points are evenly distributed across a range of values, this is called a *uniform* distribution (Figure 2.17B). Uniform distributions are fairly uncommon in nature. A more likely distribution in nature is one that produces a bell-shape on a graph, with more observational data points clustered at the center and fewer in the data bins that are farther away from the center (Figure 2.17A). If we could take the bins and make them super-tiny so that the curve became smooth and continuous, we might have something that approximates a **normal distribution** (**Figure 2.18**), the bell-shaped curve presented in every statistics textbook.

Normal distributions have several useful properties that help us understand an entire population. These are: (1) the mean value (average) is the most common and sits at the center of the distribution; (2) the distribution is symmetric around the mean so that one side of the distribution is a mirror image of the other; (3) the area under the line, or curve, is equal to 1.00, or 100% (true of all distributions); and (4) the proportion of the distribution that lies at a particular distance from the center, or mean, is known (Figure 2.18). The distances away from the mean in Figure 2.18 are measured in standard deviations. The **standard deviation** has a specific statistical definition, but is essentially a measure of the average distance that all the points lie from the mean. In other words, if we added up the distance from the mean of every point in the distribution and then took the average of all those points, we would have something extremely close to the standard deviation. Standard deviations can vary for different data sets with normal distributions, but the proportion of the data that lies within one standard deviation or two standard deviations will always remain the same, as shown in Figure 2.18.

This last property is the key to many statistical concepts you will learn, or have already learned, in introductory statistics classes. For data graphed as a frequency distribution, calculating the area under the curve produces a statement of probability, even if the graph does not perfectly fit a normal distribution. Thinking about the area under the curve also lets us see why biology and ecology majors are often encouraged to take calculus. If you have never studied calculus, it is still possible to visualize what is meant by the "area under the curve" by thinking of the shape made from a line along the top of the humped frequency distribution and extending down to the x-axis (Figure 2.18). Calculus provides a way of formally calculating an area under a curve without having to visually identify and measure that area, as shown in Figure 2.18.

2.4 Where Do We Go from Here?

With so much to learn about the intersection of models, data, and quantitative aspects in ecology, many students feel overwhelmed and at times underprepared. If you feel that way after reading this chapter, you are not alone, but you also should not worry. The effective use of quantitative approaches requires time and practice. In reality, ecologists tend to specialize their use of quantitative approaches in terms of their particular research focus. As you work through this book, we do not expect you to be an expert; we only ask that you be open to learning and using the basic elements of quantitative approaches in the study of ecology. From this base, you will be able to build and logically extrapolate your own ideas from the natural world.

In the chapters that follow, we aim to present basic elements of a topic in a way that allows you to conceptualize relationships, consider how to quantify those relationships, and understand how to use data to test and refine models. Our ultimate goal with this book is to produce thinkers who can deduce answers to questions based on their intuitive understanding of data and the relationship between variables—that is, thinkers who have quantitative reasoning and data literacy skills.

You can grow your quantitative skills in a variety of ways beyond those provided in this book. If you choose to pursue ecology as a career, you will almost certainly be asked to take a series of formal courses that develop your quantitative skills, including one or more courses in mathematics and statis-

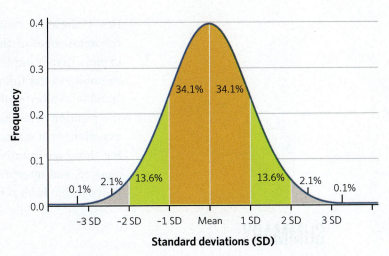

Figure 2.18 This is the standardized normal distribution, showing the frequency on the y-axis and the mean value and standard deviations away from the mean on the x-axis. The percentages indicate how much of the total area under the curve is represented by each outlined colored section. The percentages do not add up to 100% because we have rounded to the nearest tenth for simplicity, but the proportion in the orange and green zones encompasses 95.45% of the distribution. Notice that the distribution is symmetrical around the highest, or most common, point in the middle.

tics. If these are taught in a way that divorces content from application, use the approaches of this book to identify core concepts, find an example, and explore that example, even if you have to do it on your own. You will find many examples in the following pages of how to apply quantitative tools to understand or explore ecological systems. Most importantly, though, put yourself in the position to discuss concepts in quantitative terms, share your hard-won experiences in mastering quantitative tools with your peers, and keep exercising those quantitative muscles. Seek out experiences where you can collect, analyze, and interpret data. Quantitative reasoning and data literacy require practice, but both skills are attainable and will be useful throughout your life.

SUMMARY

2.1 What Is Quantitative Reasoning?

- A quantitative approach to research uses information related to quantities rather than qualities of objects.

- Quantitative reasoning applies mathematics to measured or numerical quantities to produce a deeper understanding of a topic.

- People use quantitative reasoning every day, as when they determine whether they have enough money to do an activity.

- Data literacy is the ability to ask and answer questions and to communicate meaningfully using quantitative data.

- Many people suffer from "math fear" because they were once told, or they told themselves, that they were no good at math. Fortunately, this is not true; there is no evidence to support the notion of "good" versus "bad" math brains.

- Becoming good at quantitative reasoning takes practice.

- Linking quantitative methods to ecological concepts helps us learn both the methods and the concepts.

2.2 Building Models

- All models are, by their very nature, approximations of reality and should capture some essential insight or generalization about reality.

Conceptual Models: Qualitative Steps

- The first step in building a conceptual model is to become familiar with the question, problem, or system of interest by researching and employing the best ideas and information available.

- The next step is to develop a specific question of interest and outline the key factors that play a role.

- The final step is to sketch out the idea on paper, a chalkboard, or a tablet computer. Sketches of conceptual models can involve arrows, valves, and flows.

Mathematical Models: Quantitative Steps

- The first, and sometimes most challenging, step in moving from the qualitative to the quantitative—from a conceptual model to a mathematical model—is to create mathematical equations that represent the elements in the conceptual model.

- With a mathematical model in place, the equations can be populated with numbers (i.e., values for variables, parameters, and so on) that represent real-world values or approximations.

- With mathematical models, we sometimes use iteration to explore how the focal factor changes through time.

Variation in Models

- Analytical approaches to modeling look for generalizations based on specific relationships. They can give expected outcomes for specific parameter values. They can also identify the ranges of parameters that lead to each possible type of outcome.

- Simulation approaches are useful when analytical models are too complicated for algebraic solutions or when scientists want to include real-world variation in a model. Simulations require running a model many times to examine the range of outcomes.

- Simulation approaches may involve stochastic modeling, a method in which parameters vary randomly within a range for each run of the model. This element of randomness allows simulation models to capture the range of outcomes that we may see when factors vary randomly in the real world.

2.3 The Lowdown on Data

- All ecological questions are informed by some type of data or data collection.

- Data collection, manipulation, and interpretation are fundamental aspects of ecological science.

Data Types

- Categorical data have no inherent numerical value associated with them but instead are presented as categories that may be either ordered or nominal (not ordered).

- Numerical data have inherent numerical values associated with them in either continuous or discrete forms.

Collecting Data

- It is not feasible to collect all the possible data on a subject, so instead one must collect a sample of all the available data.

- It is important when collecting sample data to be as unbiased in the data collection as possible.

Using and Summarizing Data

- It is often necessary to summarize ecological data using graphs and figures, which allow us to see patterns and make generalizations.

Probability

- Some data generalizations involve statements of probability and statistics.

- Two key concepts in understanding and determining probability are p-value and statistical significance.

- Biological significance is different from statistical significance, and the two should not be conflated.

Distributions

- Statements of probability depend on variation in data and the way data points are distributed.

- It is common for data to be normally distributed, forming a bell-shaped curve on a graph.

2.4 Where Do We Go from Here?

- In the chapters ahead, you will learn how to conceptualize and quantify relationships and to use data to test and refine models.

- You will build your quantitative reasoning and data literacy skills.

- We will not expect you to be an expert, only to be open to learning.

KEY TERMS

continuous numerical variable
data literacy
deterministic model
discrete numerical variable
explanatory variable
frequency distribution
iteration

nominal categorical variable
normal distribution
numerical variable
ordinal categorical variable
parameter
population
p-value

quantitative reasoning
response variable
sample
standard deviation
statistical significance
stochastic model
variable

CONCEPTUAL QUESTIONS

1. Describe what quantitative reasoning and data literacy mean to you after reading this chapter. Would you have come up with the same definitions for these terms before reading this chapter? Explain your answer.

2. In what ways are quantitative reasoning and data literacy skills useful or transferable outside of the ecological sciences? Do you think these skills are important? Why or why not?

3. Do you currently, or have you ever, personally felt math phobia or math anxiety? Where do you think it originated from? Have you ever had a triumphant or intellectually satisfying "math moment" in your life? What were the circumstances?

4. Consider the statement, "There are two flavors of people: those who are good at math and those who are not." Do you agree or disagree with this statement? Explain your answer.

5. Sketch out a conceptual model for your favorite organism, ecological system, or phenomenon. Can you generate a testable hypothesis from your model?

QUANTITATIVE QUESTIONS

1. Try to populate your conceptual model from Conceptual Question 5 with mathematical terms. Is this easy or difficult? Discuss your answer.

2. Find three examples of graphs from the news or elsewhere that you think may be misleading, and explain why.

3. For each of the following, state whether the variable is categorical or numerical. If you think the variable could fall into either type, explain why and explain the circumstances under which you would describe it as a categorical variable.

   a. The names of species found in a reserve

   b. The length of amphibians from the tips of their noses to the ends of their bodies (before the tail)

   c. The height of trees

   d. The biomass of baby newts

   e. The days of the week

   f. The months of the year

   g. Each year of a 20-year study

   h. The age of an organism

   i. The life stage of an organism

   j. The number of pollinators that visit a particular flower

   k. The color of the flower visited by the pollinators

4. For each of the following questions, identify the explanatory variable and the response variable.

   a. Does plant biomass increase as water availability increases?

   b. Does pollination by honeybees lead to higher seed production than pollination by bumblebees?

   c. Is there an association between the color of a flower and the species of its dominant pollinator?

   d. Does water depth affect the amount of photosynthesis in 100 liters of ocean water?

   e. Does birthweight affect how long juvenile narwhals stay with their pod?

5. For each question in Quantitative Question 4, identify the explanatory and response variables as categorical or numerical.

6. For each question in Quantitative Question 4, try to draw a graph that could potentially show the relationship between the variables. There is more than one correct answer for most questions, but do not worry about devising more than one graph for each.

7. Look at Figure 2.17 and try to come up with a natural example from your own experience that fits a normal distribution curve and another that fits a uniform distribution curve.

SUGGESTED READING

Box, G. E. P. 1979. Robustness in the strategy of scientific model building. Pp. 201–236 in R. L. Launer and G. N. Wilkinson, eds. *Robustness in Statistics*. New York: Academic Press.

Burkley, M., J. Parker, S. P. Stermer, and E. Burkley. 2010. Trait beliefs that make women vulnerable to math disengagement. *Personality and Individual Differences* 48(2): 234–238.

Ford, A. 2009. *Modeling the Environment*, 2nd ed. Washington, DC: Island Press.

A view of the Earth from the Moon, as seen in 1969 by the astronauts of the Apollo 11 spacecraft. When humans first saw our planet from a distance, something changed for the whole species. We realized our home was a tiny "blue marble" floating in the whole of the universe.

chapter 3

GLOBAL CLIMATE AND BIOMES

- Distinguish between climate and weather.

- Explain the basis of the conceptual climate model.

- Summarize ways the conceptual climate model should be modified in order to better fit the observed climate.

- Explain how changes in elevation and latitude produce similar outcomes in the conceptual climate model.

- Summarize how terrestrial biomes are formed as the result of climatic features.

- Identify and explain the features we use to differentiate terrestrial biomes.

- Identify the nine terrestrial biomes and explain the physical features that characterize them.

- Summarize three distinguishing features of aquatic environments.

- Identify the nearshore aquatic biological zones and explain the biological and physical features for each.

- Describe some of the human effects on the global climate.

3.1 What Is the Biosphere?

When we look at the world around us, our view is often limited. We gaze out the window of our homes and see the few meters contained by our yard, apartment complex, or dorm environs. At the top of a mountain or flying in an airplane, our view may be extended, so we can see for tens of miles. But even with such grand views, we observe only a teeny-tiny fraction of 1% of the Earth's surface. That is why, when the first space explorers returned to the Earth from the Moon, the most significant thing they remembered was not setting foot on the lunar surface but the startling view of the Earth from the Moon.

Oddly enough the overriding sensation I got looking at the Earth was, my god that little thing is so fragile out there. —Michael Collins, *Apollo 11* astronaut

This planet is not *terra firma*; it is a delicate flower. . . . It's lonely. It's small. It's isolated. . . . This is our home, and this is all we've got. —Scott Carpenter, *Mercury 7* astronaut

As a result of humans' limited ability to experience the entire Earth except at local scales, most of us have an incomplete understanding of how the planet functions as a dynamic "whole." Therefore, as a first step in the study of ecology, it is helpful to gain an appreciation for the incredible physical and biological variability seen around the planet. In this chapter, we focus on global climatic patterns and how they produce biomes. These global patterns, measured across both space and time, provide the background for the diversity found among the

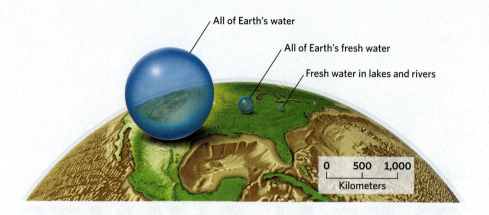

Figure 3.1 If all the water on the planet were collected into a large sphere, the diameter of that sphere would be 1,300 km (808 mi.). If only the fresh water were collected, the sphere would be much smaller (275 km, or 171 mi., in diameter) (Marshak and Rauber 2020).

living organisms of Earth. We examine biological variation and its causes and consequences in much greater detail in later chapters (Chapters 4 and 11). The world has many component systems, including the *atmosphere* (envelope of gases around the Earth), the *hydrosphere* (water system), the *geosphere* (solid parts of the earth), and the *cryosphere* (ice system). Ecologists are primarily concerned with the component called the **biosphere**. The biosphere is all the living organisms on the Earth, plus the environments in which they live.

When viewed from space, the Earth is very blue, covered mostly by water and ice. But when we consider the mass of the entire Earth down to its core, the vast majority of the planet is rock of one form or another. If we represented the Earth with all the water removed and pooled together in a sphere (the hydrosphere) and set it next to the rocky part of the planet (the geosphere) to show its relatively small volume, we would get **Figure 3.1**. The volume of water in Figure 3.1 may seem small relative to the rock mass of the Earth, but the biosphere is even tinier. Even though scientists have found microbes as far above the Earth as 15 km (9 mi.) and as deep into the depths of the sea as we have gone (roughly 11 km, or 7 mi.), the vast majority of the Earth's living biomass exists between 6 km in elevation and 5 km in depth (between 3.7 and 3.1 mi.) (DeLeon-Rodriguez et al. 2013; Nunoura et al. 2015). If we could spread all the living biomass on the Earth out into a thin, consistent layer over the whole Earth and look at it from space, the biosphere would be a very thin film, much like a soap bubble, covering the planet. Yet the biosphere is the stage upon which all life as we know it, past and present, carries out its fantastically complicated evolutionary and ecological play. It is by definition the realm that occupies the attention of ecologists.

We focus on climate and the resulting planetary biomes in this chapter because they provide the foundation for all life on Earth. Understanding climate helps us account for the distribution and abundance of organisms, the formation and ecology of soils, and the global circulation of water, energy, and nutrients. In short, global climatic patterns form the backbone for the science of ecology, and the Earth's biomes are a critical outcome of climatic forces.

Climate or Weather?

A common point of confusion for anyone initially learning about climate is the distinction between **climate** and **weather**. A handy rule of thumb is, "Climate is what you expect, and weather is what you get." But a deeper understanding requires using some quantitative reasoning. The physical factors that make up the world (e.g., precipitation, heat, wind) change on a daily, if not hourly, basis. However, if you were to measure and track these factors near your home over a long enough period of time, you would begin to see obvious patterns.

No matter where you live, a basic pattern repeats itself every year, and this repeating pattern is so consistent that you reflexively structure your daily and annual schedules around it. In fact, this pattern is so consistent over time that you experience patterns similar to those that your grandparents experienced and even to those experienced by people hundreds of years in the past who lived in the same location. This expected pattern is called climate. This pattern, though, does not tell you exactly what will happen on a particular day. Even if you know that snow will fall in the winter months where you live, it will not be so easy to predict that 1 m (3.3 ft.) of snow will fall on a particular day at your house. You expect snow in winter, but it is hard to predict which days will have snow. What happens on any given day is called weather.

Weather is dictated by climate, but it is much harder to predict. A woman walking her dog on a leash along the beach can offer a useful metaphor (**Figure 3.2**). Envision the beach as the space within a two-dimensional figure, with the x-axis as time (hours, days, or months) and the y-axis as air temperature. The woman walks a straight line across the beach but in an upward direction, so that her footsteps trace a slow, steady increase in air temperature through time (think winter to summer). The dog, on the other hand, traces a more erratic path as he explores everywhere he can within the reach allowed by his leash. The dog's tracks represent much more variability in air temperature across the same time frame. In this metaphor, the dog is weather and the woman is climate. Note that the dog always has to be in the general area of the woman because of the leash, and for this reason he generally follows along with the woman. Weather is something like this; it may not be so predictable from one day to the next, but it will generally always follow the longer-term trend set by the climate.

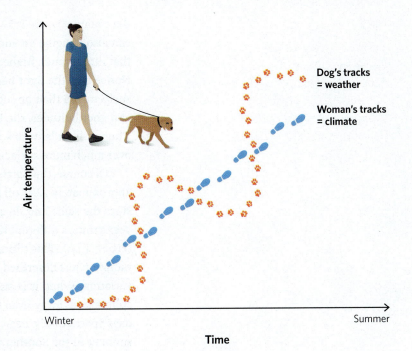

Figure 3.2 This graphic uses the metaphor of a woman walking a dog on a beach to explain the relationship between climate and weather. The woman's tracks represent climate, in that they trace a steady increase in air temperature over time from winter to summer months. The dog's tracks represent weather, in that their path varies more widely. The dog may stray from the woman but never further than the limit of the leash. Climate is the expectation, whereas weather is what you get on any given day (Brown 2013).

3.2 Building a Conceptual Climate Model

As a first step toward understanding climate and the biomes that result, we will create a conceptual model that describes some of the key phenomena that produce global climatic patterns. Our model will be simplistic at first, but as you learn more in this chapter, its complexity and explanatory power will increase.

Solar Radiation

One of the simplest observations serves as our foundation: the Earth is a sphere, and it is located very, very, very far away from the Sun. If we picture the Earth as a smooth, uniform ball with the poles at the top and bottom and the equator around the middle, we can also imagine a circular beam of sunlight hitting the Earth at the equator (**Figure 3.3A**). The Sun is big, and its light comes in many beams, so we can also imagine an additional beam from the same directional source (the Sun) that hits at much higher latitudes, further north or south, away from the equator. Notice that the light beam hitting the Earth at higher latitudes illuminates more surface area than an identical light beam hitting at the equator. This is because, at higher latitudes, the Earth's surface bends away from the light source, the Sun. Thus, at high latitudes, the same amount of sunlight, or *solar radiation*, is spread over much more surface area.

Of course, the Earth does not sit straight up and down on its axis but is tilted at approximately 23.4° off perpendicular (**Figure 3.3B**). Although the axial tilt always stays the same and does not flip-flop, as the Earth travels around the Sun over a year's time, a different hemisphere is oriented toward the Sun in different months (**Figure 3.3C**). This tilting accentuates the equatorial-versus-polar pattern in solar radiation just described and gives the planet its seasonal temperature and daylight variation. When it is summer in the Northern Hemisphere, the Southern Hemisphere faces away from the Sun and is significantly colder and darker with shorter days (particularly near the South Pole); in other words, it is winter. When it is summer in the Southern Hemisphere, the Northern Hemisphere is pointed away from the Sun and is colder and darker and experiencing winter.

The greatest amount of direct sunlight hits the equator at two times during the year, the March and September equinoxes (around March 20 and September 22,

Figure 3.3 Ⓐ The light from the Sun hits the Earth and spreads over the land differently depending on latitude. Spreading the same amount of light over a larger area decreases the amount of solar energy per unit area, leaving it colder than at the equator. Ⓑ Because of the Earth's tilt on its axis (23.4°), the planet's surface is heated even more unevenly as it travels around the Sun. For part of the year the Northern Hemisphere is pointed at the Sun and, thus, receives more direct heat. At the same time, the Southern Hemisphere is pointed away from the Sun and receives much less heat. This spatial pattern switches during the second half of the year. Ⓒ The annual cycle of the Earth's motion around the Sun dictates the equinoxes and solstices. Shown here are the seasons for the Northern Hemisphere.

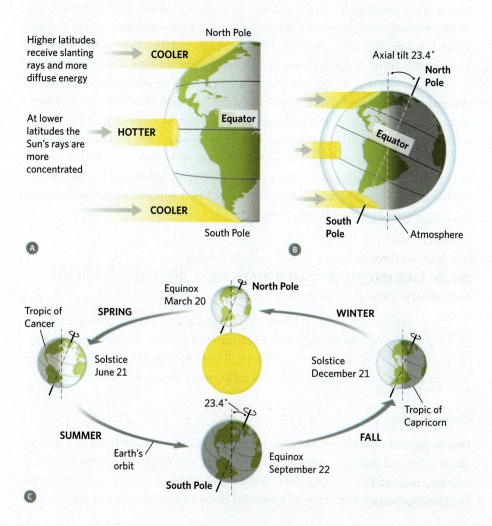

respectively), when the Earth is not tilted toward the Sun. The longest and shortest days of the year are the solstices, which occur around June 21 and December 21. The shift in the length of the days is most notable at high latitudes, which either receive lots of sunlight (summer solstice) or very little sunlight (winter solstice). At each solstice, the tilt of the Earth points directly toward the Sun. Peak sunlight hits the Tropic of Cancer (23°26′ north latitude) in June—summer in the Northern Hemisphere and winter in the Southern Hemisphere—or the Tropic of Capricorn (23°26′ south latitude) in December—summer in the Southern Hemisphere and winter in the Northern Hemisphere (Figure 3.3C). Notice that the latitudes for the Tropics of Cancer and Capricorn match the tilt of the Earth—23°26′ latitude can also be expressed as 23.4° latitude—and that the names of these latitudes match the zodiac sign for the month when full solar radiation hits them (June for Cancer and December for Capricorn). These latitudes were so named because of this seasonal pattern in solar radiation.

Sunlight is the main source of energy for life on Earth, providing heat, light, and radiant energy for processes such as photosynthesis (Chapter 5). More sunlight per unit area contributes more energy per unit area. If we examine a graph showing the estimated total amount of solar radiation by latitude over a whole year (**Figure 3.4**), we notice three things. First, the seasonal variation in solar energy produces a radical shift between the two poles, with one in almost complete darkness while the other is receiving near constant (although indirect) sunlight. Second, at certain seasons, peak solar energy is not at the equator but is at the Tropic of Capricorn or the Tropic of Cancer. And, third, even with these seasonal shifts in the amount of solar radiation, the equator receives the largest amount of total annual solar energy. This last observation may suggest that temperatures at the equator are the hottest on the planet, but this is not generally the case. Desert environments and the centers of continents are often significantly hotter (as will be discussed shortly). Instead, the equatorial regions of the world are notable for their consistently warm temperatures throughout the year.

Atmospheric Circulation

Now that we have a foundational understanding of how solar energy is distributed across the planet and how it varies through a year, we can explore how thermal (or heat) energy is distributed across the Earth. Atmospheric circulation is the large-scale movement of air and the mechanisms that create differences in air temperature across the Earth. To build a conceptual model for atmospheric circulation, we use five simple observations, one of which we just made (that there is more solar energy at the equator) and four that we are going to generate now. Before we get there, we want to show off a bit and suggest that, with this simple model, we can explain why rainforests are found in particular places on the planet and deserts are found in others, and why regions such as Southern France, Spain, and California produce such wonderful wines.

Figure 3.4 This graph depicts Earth's relative solar energy input by latitude. The blue line represents the total yearly solar energy input; the red and green lines indicate partial solar inputs for the two halves of the year. The sum of the values for the red and green lines equals the annual total (blue line) (Haurwitz and Austin 1944).

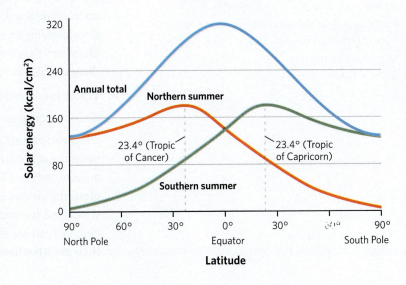

We start by asking you to answer a series of simple and somewhat intuitive questions. We also ask that you back up your answers using observational evidence from the world around you. Try to answer each question and identify some evidence before you read the answer below the question.

1. Which weighs more, a volume of hot air or the same volume of cold air?

   Answer: The cold air weighs more than the hot air.

   Observational evidence: Evidence of this is all around. For example, when you turn on a heater in a house, the room gets warm near the ceiling before it gets warm near the floor. Alternatively, if you turn on a stove or burner it gets very hot above it, but not so much to the sides and below. Hot-air balloons are also evidence, as they rise off the ground and carry people into the air.

2. Where is there more atmospheric pressure, at the ocean's surface (or on a sea-level beach) or high up a mountain at an elevation of 3,000 m (9,843 ft.)?

   Answer: There is more atmospheric pressure at the ocean's surface than atop a mountain.

   Observational evidence: You may have observed this pattern if you have walked, driven, or flown to a high altitude with a sealed bag of snacks and noticed that the bag expands. The bag still contains the same number of air particles (and the same number of snacks), but the air takes up more space because there is less pressure compressing the air particles together. The snacks inside do not change detectably because they are a solid and cannot spread out the same way that a gas does.

3. When air is compressed, does it heat up or cool down? When air expands, does it heat up or cool down?

   Answer: Compressed air heats up; expanding air cools down.

   Observational evidence: If you have ever sprayed a can of compressed air (or any other gas) for more than a few seconds, you may have noticed that the can gets very cold. By spraying, you let the compressed air inside the can expand outward, and the process cooled the can. Alternatively, if you have pumped air into a bike or car tire and felt the pump nozzle afterward, you may have noticed that it was warm or even hot. Compressing air into the tire generates heat. Or you may remember from chemistry that compressing air creates more molecular collisions among the gas molecules, and those collisions produce heat.

4. Which holds more water, warm air or cold air?

   Answer: Warm air holds more water than cold air.

   Observational evidence: One way to observe this phenomenon is to breathe against a cool pane of glass. Your breath is moist with water in a warm, gaseous form. When your breath makes contact with the cold glass, the air you exhaled rapidly cools and can no longer hold as much water, and the water vapor condenses onto the glass. People living in cold climates can see this every time they go outside in the winter and can see their breath when they exhale.

	Question:	Which part of the Earth gets the most solar radiation annually?	Which weighs more, a volume of hot air or the same volume of cold air?	How does atmospheric pressure change with altitude?	Does compression change air temperature?	Which holds more water, warm air or cold air?
Answer:		The equator gets the most total solar radiation.	Cold air weighs more.	Pressure decreases as altitude increases.	Compression heats air up. Expansion cools it down.	Warm air holds more water.
Observational evidence:		There is no cold season at the equator.	Hot-air balloons rise.	A bag of chips expands as you ascend in altitude.	Tire pump nozzles get hot. Spray cans get cold.	Water vapor is visible in breath on cold days.

These four observations, plus the observation that the equator receives more solar energy than elsewhere on the planet, are summarized in **Figure 3.5**.

Figure 3.5 The questions used to build our conceptual climate model, with answers and general observations to demonstrate the principles.

Putting It All Together

With these five observations, we can build a simple global atmospheric circulation model that helps explain large-scale climate patterns. Sunlight strikes the equator, and the energy from the Sun's rays acts to heat up the equatorial air. As this air heats up, it rises; and as it rises, there is less pressure on it, so it expands. Expansion is a cooling process, so the warm equatorial air cools significantly as it rises. We know that cool air holds less moisture than warm air, so as it rises and cools, the water vapor held by the warm equatorial air condenses, forming clouds and significant rainfall. High solar radiation at the equator throughout the year generally leads to plentiful rainfall all year. Therefore, near the equator, we see vegetation that can thrive in warm and moist conditions. These kinds of vegetation assemblages are called tropical rainforests, which we will soon describe in more detail.

This cycle of warming, rising, and cooling air creates what meteorologists call an area of *low atmospheric pressure*, which leads to a rising air column. Low-pressure zones around the planet are generally associated with increased rainfall, as the air in these areas is continually rising, cooling, and losing its ability to hold water in a gaseous state. In the high, cool air, water vapor condenses and forms rain (**Figure 3.6A**). The change in the air temperature due to the expansion (or contraction) of air is called an *adiabatic process*. More precisely, an adiabatic process is one in which the temperature of a gas (in our case, air) changes solely in

Figure 3.6 Ⓐ The Hadley circulation cell starts as warm, moist air rises at the equator due to solar radiation. As this air ascends, it expands and cools, releasing its moisture as rainfall. This dynamic produces tropical rainforests near the equator. Cold air is denser and drops back toward the Earth. As it descends, it warms. The warm, dry air can absorb water from the Earth's surface, producing deserts at around 30° north and south latitudes. Ⓑ This idealized diagram shows the atmospheric circulation cells across the planet by latitude. Atmospheric low pressures result from rising columns of air, and high pressures result from descending columns of air. A rising column of air cannot rise without limit, so it also moves north or south, away from the equator. This movement toward higher latitudes, combined with the rise and fall of warmed and cooled air, creates interconnected cells of circulating air between the equator and the poles. This circulation pattern produces alternating high- and low-pressure zones, as well as alternating bands of relatively wet and dry habitats at fairly predictable latitudes across the Earth's surface.

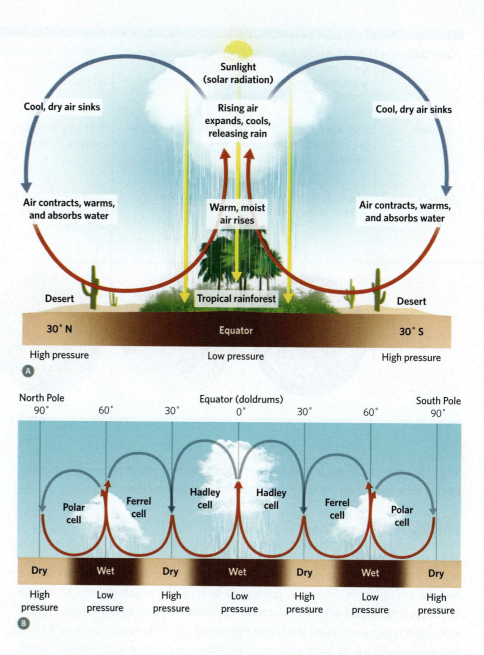

response to a change in pressure and not as a result of the addition or removal of an outside heat source. As the air warmed by the Sun rises and expands, the cooling and eventual condensation of the water vapor is almost solely due to adiabatic processes.

Luckily for us, the Earth is a relatively closed system, so this rising, cooling air must go somewhere and is forced either north or south from the equator (Figure 3.6A). Because cold air is heavier than warm air, and gravity pulls matter toward the Earth's surface, this cold, dry air descends as it moves north or south. This descending air mass produces what meteorologists call an area of *high pressure*, which is generally associated with sunny, dry weather. The descending air generally "lands" around the 30° north and south latitudes. As the cold air descends, it experiences more atmospheric pressure and warms up adiabatically. By the time it nears the Earth's surface, the air has become warm and especially dry. At that point, it has a great capacity to hold water and sucks in the moisture from its surrounding environment. Perhaps not surprisingly, we tend to find the warm and very dry environments of the world's deserts at 30° north and south latitudes (Figure 3.6A).

This airflow from the equator to 30° north and 30° south creates two circular "cells" of moving air that were first described by the amateur meteorologist George Hadley in 1735 and now bear his name: the **Hadley cells** (**Figure 3.6B**). This circular movement of air around the globe is responsible for the formation of the huge swaths of tropical rainforest, many of the world's major deserts, and the so-called trade winds and doldrums. *Trade winds* are generated as the air mass descending at 30° is forced back toward the equator along the Earth's surface, creating wind. The *doldrums* are maritime equatorial regions around the globe where, due to a rising (low-pressure) air mass, there are often no significant winds at the Earth's surface.

The Hadley cells are not the only atmospheric circulation pattern. As you can see from Figure 3.6B, air masses continue to circulate as you go north and south in latitude. The air mass that circulates approximately between latitudes of 30° and 60° north and south is called a **Ferrel cell** and works generally the same way as the Hadley cell. The dry, warm air that sank down and sucked up moisture in the desert band at 30° latitude continues moving away from the equator. At 60° latitude, another band of low pressure forms, as the air has now warmed and taken on water from the Earth's surface as it moved away from the deserts. At 60° latitude, the air is so warm that it rises, which leads to expansion and cooling, thus forming condensation and rain. The high-latitude limit of the Ferrel cell produces two additional bands of fairly heavy rainfall that strongly influence the local vegetation. In parts of the Pacific Northwest of North America, for example, there are habitats called temperate rainforests, such as the Tongass rainforest of southern Alaska, which receives more than 5,700 mm (224 in.) of rainfall per year.

The final atmospheric circulation cell as we approach the poles is called a **Polar cell** and produces another high-pressure area at each of the polar regions (Figure 3.6B). This high-pressure area is the result of cold, dry air returning to the Earth's surface, and it has the same effect on the amount of precipitation as it does in the deserts at 30° latitude. This observation tells us that the poles (both North and South) can be considered deserts. This may seem confusing. There is clearly a lot of water in the form of ice and snow at both poles. How can the poles be considered deserts if they have clear evidence of so much moisture? The answer is that the total amount of new precipitation that falls each year at the poles is very low, but the snow that does fall fails to melt and, therefore, has accumulated bit by bit over eons, forming the polar ice caps.

We have described each of these circulation cells as separate, but notice that they do interact (**Figure 3.7A**). The air in each cell cycles, but it also feeds into the adjoining cell. Instead of three separate loops, we could think of these three interacting cells as sort of a "figure eight and a half." This simple model explains how air, particles, pollutants, and even microscopic organisms can move from warm parts of the globe to cold parts of the globe. It illustrates how the factors that affect one part of the Earth are likely to affect other parts of the globe. It also helps scientists generate a map with predictions of the latitudinal climate patterns that we would see across the globe based on this model (**Figure 3.7B**), which we will compare with actual climate patterns throughout the rest of this chapter.

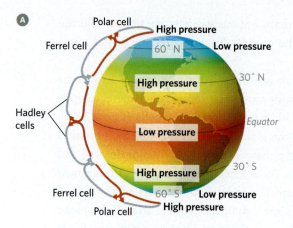

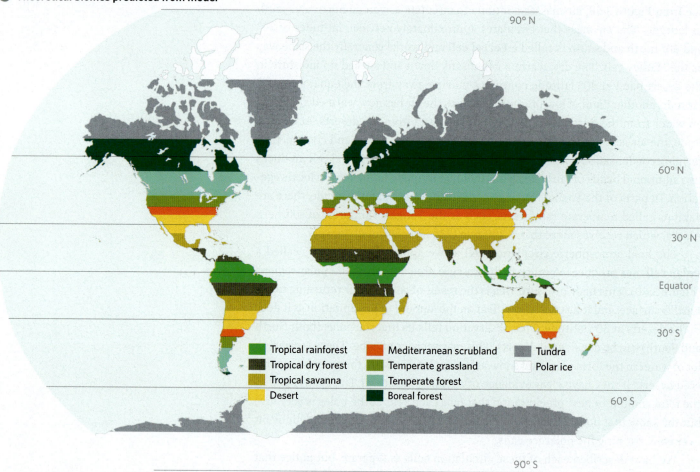

B Theoretical biomes predicted from model

Tropical rainforest
Tropical dry forest
Tropical savanna
Desert
Mediterranean scrubland
Temperate grassland
Temperate forest
Boreal forest
Tundra
Polar ice

Figure 3.7 **A** The three major atmospheric circulation cells interact with each other, making it possible to move air and the things it holds from the equator to the poles. **B** Putting together all of our insights about decreasing sunlight with latitude, seasonality, and the effects of high and low pressure on precipitation, we can predict climatic patterns that move from the equator toward the poles. Those climatic patterns translate into the terrestrial biomes. If these latitudinal effects were the only factors that determined Earth's biomes, we would see global bands for each of these biomes, as depicted here.

3.3 Additional Complexity

Looking at an actual map of the world's major ecological groupings (or biomes; **Figure 3.8**), we can see that the conceptual model does a fairly impressive job of predicting where habitats such as tropical rainforests and deserts exist. But does the model fit the full complexity of reality very well?

One way to answer this question is to look at one of the predictions from the model to see if it fits what we see in nature through direct observations. For example, do we actually see alternating bands of high and low precipitation as we go from the equator to the poles? If we look at average annual precipitation by latitude, the patterns support the model quite well (**Figure 3.9**).

First, there is a definite peak in precipitation at or near the equator. Second, there is also a significant dip in precipitation at 30°, a smaller peak near 60°

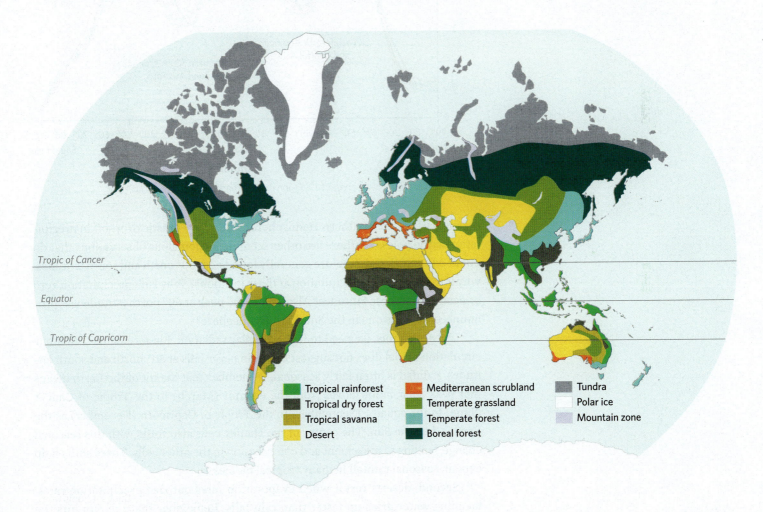

Figure 3.8 The location of the terrestrial biomes and polar ice across the planet's surface.

Figure 3.9 Average precipitation variation (y-axis) by latitude (x-axis) based on NOAA data, 1950–2009. As expected, based on the precipitation portion of the conceptual model (see Figure 3.6), precipitation varies with latitude. Higher average precipitation in all decades occurs at latitudes with low pressure (near the equator and near 60° north and south latitudes), and lower average precipitation occurs at latitudes with higher pressure (at the poles and near 30° north and south latitudes) (Roper 2016).

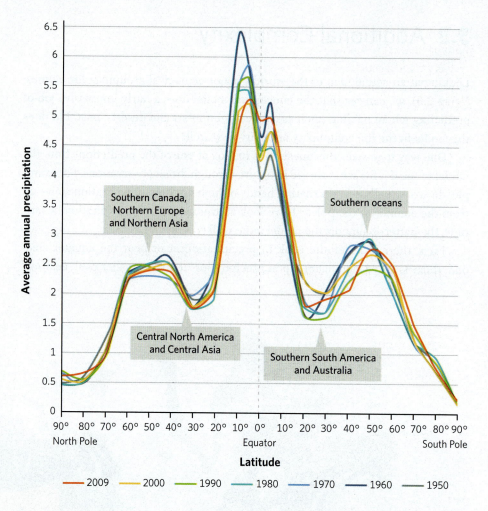

(particularly in the Southern Hemisphere), and a very drastic drop-off in precipitation at the poles. Yet there are other seemingly anomalous observations that do not fit our model. If deserts are generally found near 30° north and south latitudes, why is there so much precipitation at these locations? Why isn't precipitation zero or near zero at these latitudes? Also why is the peak at 60° north latitude less pronounced than the one in the Southern Hemisphere?

To explain this, we need to focus on solar radiation. First, remember that our circulation model does not suggest that rain *never* falls at 30° north and south latitudes. Rainfall is often fairly seasonal. Remember that the tilt of the Earth means that peak solar input moves from the equator (March) to the Tropic of Cancer (June) to the equator (September) to the Tropic of Capricorn (December) as the Earth orbits the Sun. The center of the Hadley cells also shifts with this seasonal change, leading to movement and compression in the other cells. These shifts help explain seasonal rainfall in many tropical biomes.

Second, deserts result when evaporation rates outpace precipitation rates, meaning water dries up faster than rain falls. Remember that the Sun hits the Earth with much higher heating power closer to the equator than it does at higher latitudes. High solar radiation has the potential to evaporate water, making it shift from a liquid to a gas and thereby carrying it away from the Earth's surface. At 30° north or south latitude, we have enough solar radiation to evaporate quite a bit of water. Even though the decrease in precipitation at 30° latitudes may not be as big a dip as might be expected, the combination of lower precipitation, high solar radiation, and dry air is a recipe for a very dry environment at the Earth's surface.

As we will discuss, the climate patterns at any specific location reflect interactions between rainfall and solar radiation. There is no doubt, though, that even after we add the effects of solar radiation at different latitudes, our model does not completely explain global average rainfall patterns, which are quite complex (**Figure 3.10**). A number of additional factors can be added to the model to help explain the differences between our simple model (Figure 3.7) and the maps in Figures 3.8 and 3.10.

Coriolis Effect

One factor we can add to our model is the **Coriolis effect**, which alters wind patterns and ocean circulation, with resulting effects on temperature and precipitation. It will take a few steps to explain why, but let's start with the simple fact that

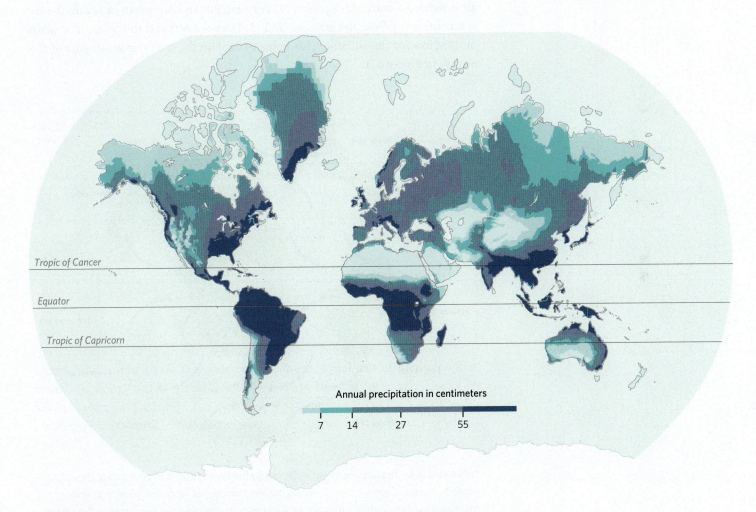

Figure 3.10 Annual global precipitation map, 1960–1990 (SAGE and CRU, n.d.). Note that the idealized patterns from our conceptual climate model do not universally hold.

the Earth spins on its axis on a daily basis, producing a 24-hour day length. Visualize the Earth spinning on its axis, and imagine one person standing on the equator and another one at the North Pole. In a 24-hour period, the person at the equator spins (or travels) a distance equal to the Earth's widest circumference, approximately 40,075 km (24,901 mi.). Meanwhile, the person standing on the North Pole spins around a really tiny circle traveling almost no distance at all. This means that the air at the equator is moving east (the direction of the Earth's spin) at a much faster speed, measured as distance per unit time, than the air at the poles. This difference produces a rotational shift in the air circulation, known as the Coriolis effect, and causes air masses moving north or south over the surface of the Earth to be deflected to the east or west (**Figure 3.11**).

Let's break this down. From Figure 3.7A, we can see that the Hadley, Ferrel, and Polar cells have air moving in one direction high in the atmosphere and in the opposite direction at the Earth's surface. We can get more specific and notice that, at the Earth's surface: (1) between the equator and 30° north or south, the air is moving toward the equator; (2) between 30° and 60° north or south, the air is moving away from the equator; and (3) between 60° and the poles, it is again moving toward the equator. This helps to explain the north or south part of the arrows in Figure 3.11.

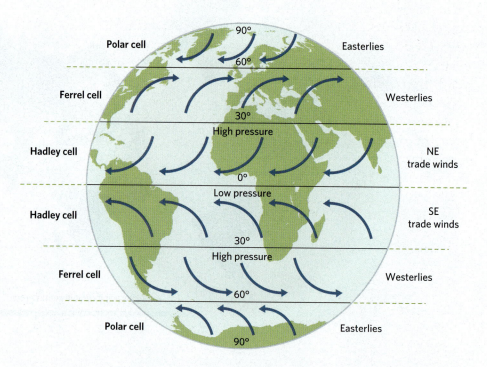

Figure 3.11 The Coriolis effect is caused by the Earth's spin and adds an easterly or westerly aspect to surface winds. Winds moving from higher latitudes to lower latitudes move eastward more slowly than the ground beneath them moves, which means they move westward relative to the ground. Winds that move away from the equator come from a part of the Earth with a large circumference, so they spin east faster than the ground beneath them and are perceived by us as moving east. The northward or southward direction of the winds comes from the Hadley, Ferrel, and Polar cells. To add one more level of complication, the convention is to name winds based on the direction from which they come. Therefore, the winds between 30° and 60° latitudes are *westerlies*, and the others are all *easterlies*.

Now let's focus on the deflection. If we recognize that the air near the planet's surface has the same tendency to spin eastward as the ground, then we can explain the deflection to the right that we see in the Northern Hemisphere and the deflection to the left that we see in the Southern Hemisphere. The air that moves away from the equator starts at a wide spot on the Earth, so it spins fast toward the east. As this air moves away from the equator, it approaches narrower parts of the Earth, where the ground beneath it spins eastward a little more slowly. This means the air moves to the east faster than the ground it approaches. For those of us standing on the ground, this creates a wind that moves from the west toward the east, or an apparent deflection eastward. Air that moves toward the equator experiences the opposite effect. It comes from a section of the Earth that spins more slowly and thus moves east more slowly. As it passes over the Earth, the Earth's surface slides to the east under it, making the air appear to move west.

It may not come as a surprise to realize that these deflected atmospheric circulation cells also drive the movement and circulation of the waters in the world's oceans, especially when you consider that the main force behind the movement of the ocean's water is wind. In the Northern Hemisphere, ocean waters generally move in a clockwise fashion, whereas in the Southern Hemisphere, they rotate counterclockwise.

Even a casual look at the patterns of oceanic circulation in **Figure 3.12** yields interesting observations. For example, incorporating the water movement into our conceptual model of atmospheric circulation helps to explain why the coast and beaches off Oregon and Northern California are generally cold and foggy in the summer, while the beaches at almost the exact same latitude on the east coast of North America are warm and sunny. The ocean currents that run along the Pacific coastline come down from the cold, arctic waters of the north, and that cold water cools down the coastline. The relatively cold wind coming off the ocean often produces copious fog in the summer and, therefore, a fairly uninviting summer beach experience for those seeking a tan. Over on the eastern seaboard of North America, the current brings water up from the warm, tropical Caribbean region. The winds that blow ashore create a wonderfully pleasant summer beach experience. The fact that ocean waters carry tropical heat and act as a sort of heat trap

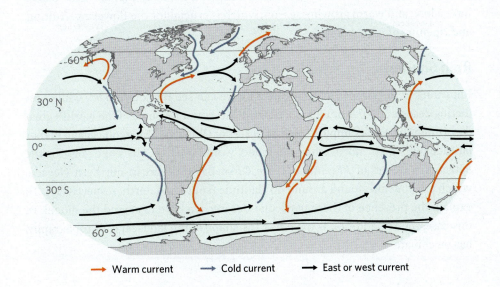

Figure 3.12 Circulation of oceanic waters around the globe. Red arrows indicate warm-water currents, and blue arrows indicate cold-water currents. The spinning of the Earth leads to circular patterns of movement in the world's ocean waters. Notice that the circulation is generally clockwise in the Northern Hemisphere and counterclockwise in the Southern Hemisphere.

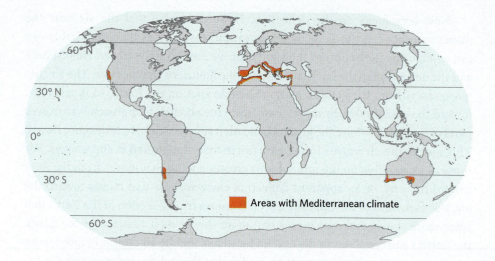

Figure 3.13 Locations around the globe with a Mediterranean climate.

Areas with Mediterranean climate

is also responsible for the relatively warm temperatures enjoyed in Great Britain. For example, Thurso, a town at the northern tip of Scotland, is at almost the same latitude as Juneau, Alaska, but Thurso has a much milder climate than Juneau, thanks to the buffering effects of warm winds coming ashore as a result of northward oceanic circulation.

These oceanic currents and circulation patterns also produce a particular type of climate found at approximately 30°–45° north and south latitudes, but only on the west coasts of continents. This is where you find all the major wine-growing regions of the world—Spain, Southern France, Chile, South Africa, California, and Western Australia (**Figure 3.13**). All of these places have what is known as a Mediterranean climate. We discuss the specifics of the Mediterranean climate later in the chapter, but for now it is enough to know that this type of climate has hot, dry summers and wet winters that rarely see freezing temperatures. These conditions are a result of the proximity of these areas to a cold ocean current. Most of these locations receive cold Arctic or Antarctic waters through ocean circulation patterns (see Figure 3.12). The ocean, in these cases, has a fairly strong moderating effect on the nearby landmass. Cold oceanic currents tend to produce rain in the winter in these regions, but in the summer high-pressure cells prevent water-soaked winds from penetrating onshore. The end result is that all these places have a long and warm growing season, which is crucial for the flowering, fruiting, and ripening of wine grapes.

Rain Shadows

Returning to the biomes map (Figure 3.8), we can see that our model, while predicting deserts at 30° latitudes, does not fully account for some of the world's great desert regions. For example, the Gobi desert of China and Mongolia is at a higher latitude (above 40° north) than the model would predict. The same goes for the Patagonian desert of Argentina or the Great Basin desert in northern Nevada, which are both also found around 40° latitude (south and north, respectively). To explain the presence of these particular deserts, we have to amend the model. In this case, the key addition is the effect the shape of the landscape, or topography, has on climate.

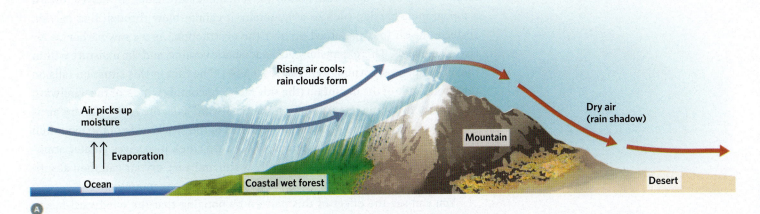

Figure 3.14 Ⓐ Formation of a rain shadow. As warm, moist air is forced to higher elevations over a mountain, it expands, cools, and loses its water as rain. On the back side of the mountain, air that has become cold and dry descends, warms, and absorbs water from the land surface, producing dry conditions. Ⓑ Satellite view of a rain shadow formed by the Sierra Nevada mountains between California and Nevada, with ground-level photographs in inserts. Wind from the west creates moist forests and lakes on the windward side of the mountains, while the leeward east side is dry and parched.

High-elevation mountain ranges block air circulation and cause climatic patterns of their own called **rain shadows**. When moisture-laden air moves toward a topographic barrier, such as a mountain, it cannot blow through that barrier. Instead, the air mass is forced up over the obstruction. As we saw earlier, as air rises, it expands. This expansion causes adiabatic cooling, and the moisture within the air condenses, forming rain (**Figure 3.14A**). The rain in this situation falls on the "front" side of the mountain (the side facing oncoming winds, or the windward side). After dropping its moisture on the windward side, the now cold, dry air is forced over the mountain. As the air descends to the "back" side of the mountain (the side facing away from the winds, or leeward side), it receives greater atmospheric pressure and warms. The newly warm and very dry air can then absorb moisture from the environment on the leeward side of the mountain.

You can see the effect of this pattern by noticing that the windward-facing slopes of large mountains have dense vegetation, whereas the leeward slopes are sparsely vegetated (**Figure 3.14B**). This leeward part of the effect is the actual rain shadow and often produces a desert. We normally think of a shadow as blocked light, but this is a shadow of blocked rain. The Gobi, Patagonian, and Great Basin deserts are all produced largely from the effects of rain shadows along the Himalayan, Andean, and Sierran mountain ranges, respectively.

Heat Capacity and Continental Effects

Another factor to include in our conceptual climate model is *heat capacity*. Heat capacity quantifies how much heat energy has to be added to a substance in order to raise its temperature by 1°C. In other words, heat capacity is the ability of a substance to store (and release) thermal energy. Water has an especially high heat capacity because H_2O molecules form hydrogen bonds among themselves that require energy to break. These bonds provide water with five times the heat capacity of soil. This difference in heat capacity means it takes a lot of thermal energy to raise the temperature of large bodies of water, such as oceans and lakes. Once these large bodies are warm, though, they release this stored heat energy very slowly.

Heat capacity plays a role in temperature differences at the interior of landmasses or along coasts or near large bodies of water. Inland locations on large landmasses tend to see big temperature swings as the Earth's surface heats up and cools down rapidly. Yet for coastal areas, the ocean buffers the temperature changes and creates a moderating, stabilizing climatic effect, leading to cooler summers and warmer winters. Scientists call this difference between the centers of continents and the coasts the **continental effect**.

The continental effect accounts for the large differences in yearly seasonal temperature extremes between coastal or interior locations at the same latitude on large land areas such as North America, Eastern Europe, Australia, and Asia. For example, if we look at the annual range of temperatures for Wichita, Kansas, which is right in the middle of the United States, and compare it to that of a coastal city at the same latitude, such as San Francisco, California (**Figure 3.15A**), we see a huge disparity. Wichita experiences a roughly 28°C (51°F) difference between its average highest temperatures in July and its average lowest temperatures in January (**Figure 3.15B**). San Francisco, on the other hand, has only a 9°C (16°F) difference between its high and low average temperatures (Fig-

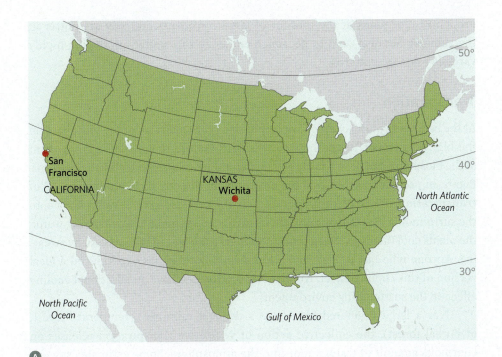

Figure 3.15 Proximity to an ocean has a distinct moderating effect on local climate. For example, cities in the interior of large continents, such as North America, often show considerable annual temperature variations. **A** San Francisco, California, and Wichita, Kansas, lie at about the same latitude, but San Francisco is on the West Coast, while Wichita lies in the interior of the continent. **B** Wichita has a much larger variation in temperature than San Francisco because the latter enjoys air temperatures that are moderated by the Pacific Ocean (National Drought Mitigation Center 2006). Pictured here are the two locations on a typical day in January.

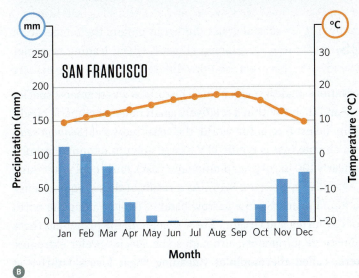

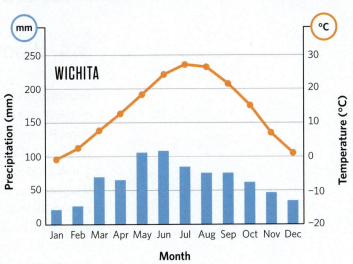

ure 3.15B). Locations at the interior of large continents not only get hotter in the summer but also tend to get much colder in the winter, producing highly variable seasonal patterns.

Transpiration

At the beginning of the chapter, we emphasized the small size of the biosphere on this big, rocky planet. It is perhaps surprising, therefore, that the living component of the biosphere can have a fairly significant influence on global climatic patterns. Plants, in particular, act to cool the environment directly through the process of **transpiration**. When plant tissues heat up, they release excess water vapor into the surrounding air from pores in their leaves called stomata. By releasing water, the plants cool themselves in a process similar to the way mammals cool by sweating. Anyone who has walked into a forest on a hot day has experienced a local version of this cooling effect. All this evaporation can have an additional cooling effect on the surrounding environment.

Over a large vegetated area (e.g., a tropical rainforest, temperate forest, or boreal forest), the collective action of vegetative transpiration releases an enormous amount of water vapor into the atmosphere. Some estimates suggest that a single tree in the Amazonian rainforest can transpire more than 1,000 L (264 gal.) of water per day, whereas a square meter of the ocean surface evaporates only approximately 1 L (0.3 gal.) of water per day. Scaling this up to include every tree in a rainforest, we can clearly see how this large amount of transpiration can have a marked impact on the total water vapor in the atmosphere. This massive volume of atmospheric water produces increased precipitation and cloud cover, both of which are key components of a region's climate. Thus, the world's forests produce transpired "aerial rivers," huge volumes of water vapor that move with winds to other parts of the Earth. As humans clear or burn these forests and allow no substantial tree regrowth, we essentially "turn off the tap" to these aerial rivers, which alters vegetation growth and climatic conditions all over the world.

Now that we have looked at how sunlight, atmospheric and oceanic circulation, topography, and water vapor affect climate, let's explore how climate affects the biosphere in different locations.

3.4 Biomes

One of the first observations made by early natural historians was that most species on the planet survive and thrive within a relatively small range of climatic conditions. For example, the tallest tree in the world, the coast redwood (*Sequoia sempervirens*), requires a climate that is typified by a large amount of annual rainfall (up to 2,500 mm, or nearly 100 in., of precipitation per year), fairly cool year-round temperatures, and extensive fog (**Figure 3.16**). As a result of these requirements, coast redwoods are found primarily in a narrow band of habitat along the north and central coasts of California and Oregon. Every species on the planet has these types of limiting climatic requirements, although some species prefer extremes. For example, bacteria called thermophiles (meaning "heat lovers") thrive at

Figure 3.16 The tallest trees in the world, coast redwood trees (*Sequoia sempervirens*), require copious fog in order to thrive.

extremely high temperatures of 41°C–122°C (106°F–252°F) (**Figure 3.17**), whereas other species, such as those belonging to the family Channichthyidae (white-blooded icefish), survive in oceanic waters that regularly reach temperatures below freezing (**Figure 3.18**).

Biomes can be defined as large geographic areas affected by similar climatic and physical factors, leading to distinctive formations of plants and animals. What constitutes a biome is scalable, meaning that at smaller (or larger) spatial scales of observation, different numbers and types of biomes can be defined. As a result, there is no scientific agreement on the number of different biomes present on the planet. Some scientists define a few larger biomes, while others parse the world into more, smaller biomes. Although we also make an idiosyncratic choice about which biomes we delineate in this chapter, we recommend strongly that students focus more on the factors that produce the biomes than on the specific definitions and locations for each one.

Terrestrial biomes are generally determined by the climatic factors we have discussed (water, sunlight, and temperature), plus soils, but they are usually described by the characteristics of the dominant plant community. The focus on plant characteristics involves describing variation in traits, such as plant

Figure 3.17 In Morning Glory Pool (hot spring) in Yellowstone National Park, different species of thermophilic bacteria thrive at different temperatures. Because each species is a different color, their temperature preferences lead to bands of different colors.

growth form (trees, shrubs, grasses), variation in size or morphology (tall, short, shrubby), leaf characteristics (broadleaf, needleleaf), and plant spacing (dense forest, open woodland, savanna). In general, when defining terrestrial biomes, we focus on factors that affect **primary productivity**—the synthesis of organic material by plants through the process of photosynthesis—because this has a direct effect on the organismal composition of the biome. You will notice that, in the pages that follow, we present the biomes in the order that one would expect to encounter them if starting a trip at the equator and moving toward the poles. Aquatic biomes, on the other hand, cannot be defined in terms of primary productivity because these systems generally have few complex plants. Instead, aquatic biomes are defined by variation in physical and chemical characteristics, such as water temperature, water depth, substrate (underlying surface), light availability, salinity, oxygen concentration, and water movement.

3.5 Terrestrial Biomes

Recall that Figure 3.8 shows the distribution of the Earth's nine major terrestrial biomes, the locations of which are generally explained by climatic factors. Our conceptual climate model used air circulation patterns and the Earth's tilt and relationship to the Sun to understand latitudinal patterns of precipitation and temperature.

Figure 3.18 This blackfin icefish (*Chaenocephalus aceratus*) is a species in the family Channichthyidae, which is unique in that it is the only vertebrate family on the planet that lacks hemoglobin. This trait allows it to live at temperatures below the freezing point of water.

Available sunlight and water lead to more plant growth and more structurally complex biomes. But because heat causes water to evaporate, and a lack of heat makes water unavailable for use by plants (because it is frozen), these two factors also interact biologically, altering the plant formations that we find at different latitudes. For example, we would not expect abundant plant growth in extremely hot, dry areas or in extremely cold, iced-over areas, even though one of our focal factors is high in each situation (radiation and precipitation, respectively).

As a way to visualize water and heat together, we use a **climate diagram**. Climate diagrams were originally developed by Heinrich Walter and Helmut Lieth (1960–1967). A climate diagram is a graph that depicts two aspects of a region's climate, in this case average monthly temperature and precipitation across a year. Our climate diagram has temperature and precipitation as two independent y-axes, with temperature on the left and precipitation on the right (**Figure 3.19**). For each biome, we provide a theoretical climate diagram as well as two actual climate diagrams for comparison. We encourage you to use the theoretical one to visualize how broad climatic patterns dictate terrestrial biomes rather than focusing on the specifics of a single geographic location.

Before diving into each biome, we need to consider how soils influence biomes and vice versa. Soils play a considerable role in determining the vegetation that grows in any one location, mostly because water availability for plants is controlled by the soil's capacity to hold water. How well a soil holds water (and makes it available to plants) is largely dictated by the depth of the soil, the composition and texture, and the amount of organic material in the soil. Soil is basically made up of rocks and minerals in differently sized particles (sand, silt, and clay from biggest to smallest) along with small pieces of organic matter. Although the minerals can affect how water moves, it is the organic material that creates the sponge-like qualities of soil. Think of filling a bucket with different-sized rocks and water. If you poke a hole in the bucket's bottom, the water will quickly run out. If instead you add a bunch of ground-up leaves among your rocks, the bucket will hold the water for much longer.

In nature, the mineral (nonorganic) portion of soils develops slowly from rock that breaks down over eons due to weathering. The organic material in soils is determined by the vegetation growing on the soil (i.e., the biome) and the rate at which the vegetation decomposes. Decomposition rates are determined jointly by the climatic conditions in a location and the type of vegetation present. Thus, soils simultaneously help define biomes through their ability to dictate what vegetation grows at a location, while also being a product of the vegetation itself. So, do soils determine a biome, or do biomes determine the soil? Many times, the direction of the relationship is not clear. Because of this "chicken-and-egg" problem, we characterize the soils in each terrestrial biome to provide a sense of how they contribute to water availability.

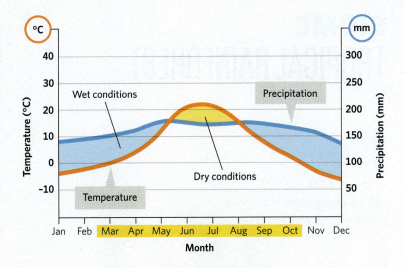

Figure 3.19 In this generalized climate diagram, the orange line indicates average monthly temperature (°C), and the blue line indicates average monthly precipitation (mm) over a year. Months of the year highlighted in yellow along the x-axis indicate time periods when the temperature is generally above freezing. Water availability for plant growth is indicated by the shading above and below the temperature curve. Blue shading above the temperature line but below the precipitation line indicates precipitation levels that are likely to be sufficient for growth. Yellow shading below the temperature line but above the precipitation line indicates times when evaporation outpaces precipitation, leading to dry conditions that may limit plant growth.

BIOME

TROPICAL RAINFOREST

Figure 3.20

Ⓐ Map showing the locations of tropical rainforests around the world.

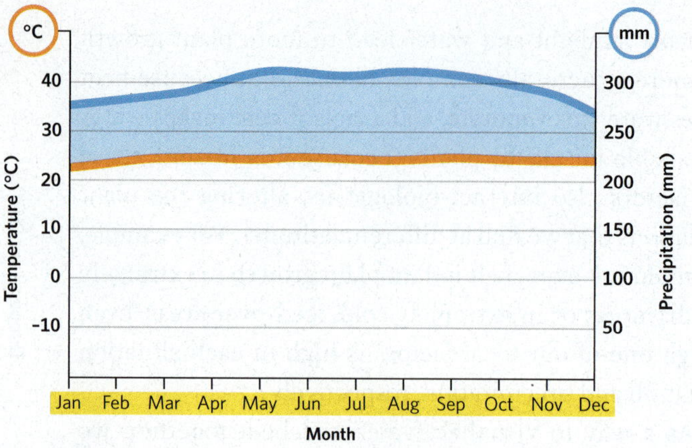
°C
mm
40
30
20
10
0
-10
300
250
200
150
100
50
Temperature (°C)
Precipitation (mm)
Jan Feb Mar Apr May Jun Jul Aug Sep Oct Nov Dec
Month

Ⓑ Theoretical climate diagram for tropical rainforests.

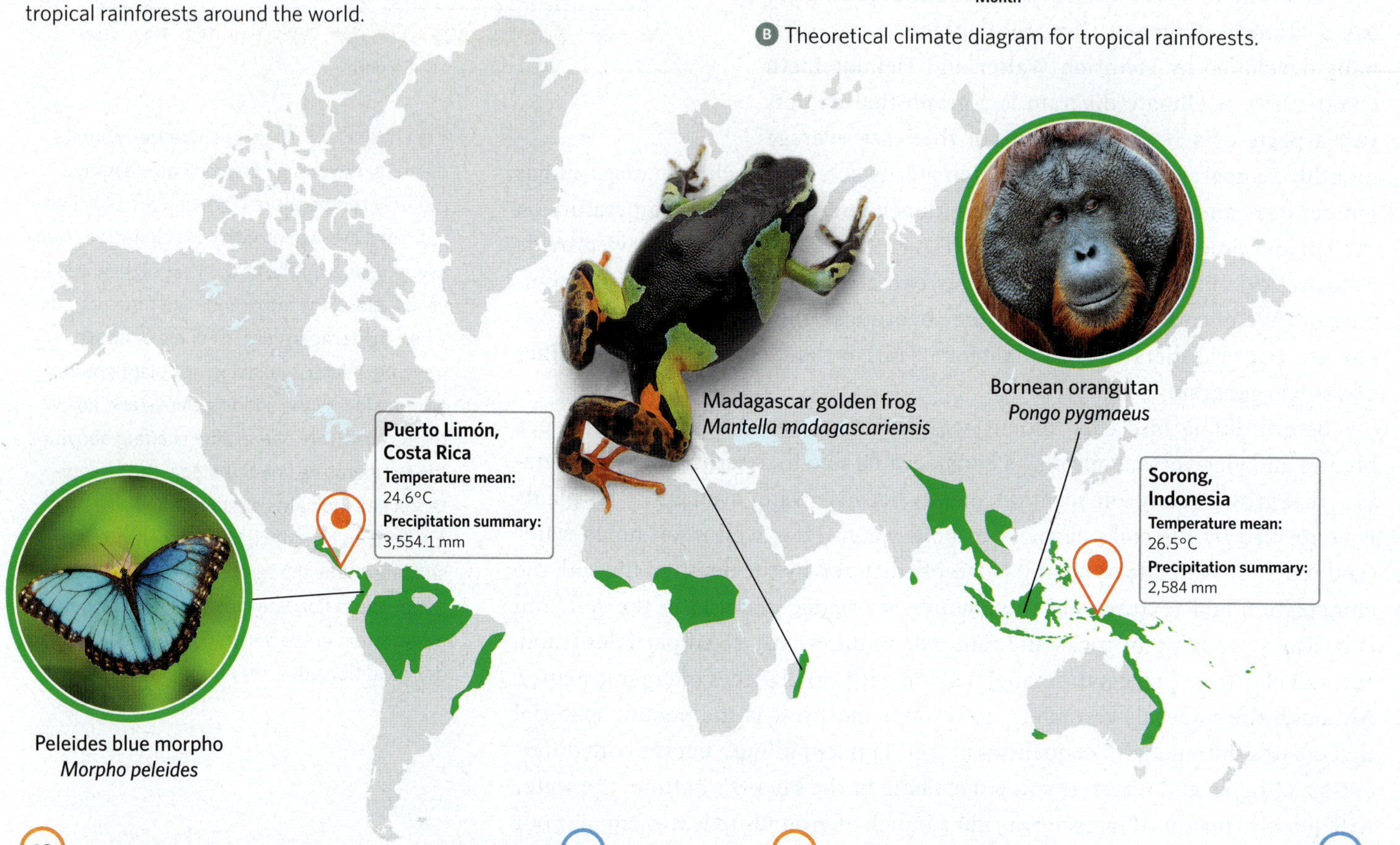
Madagascar golden frog
Mantella madagascariensis
Bornean orangutan
Pongo pygmaeus
Puerto Limón, Costa Rica
Temperature mean: 24.6°C
Precipitation summary: 3,554.1 mm
Sorong, Indonesia
Temperature mean: 26.5°C
Precipitation summary: 2,584 mm
Peleides blue morpho
Morpho peleides

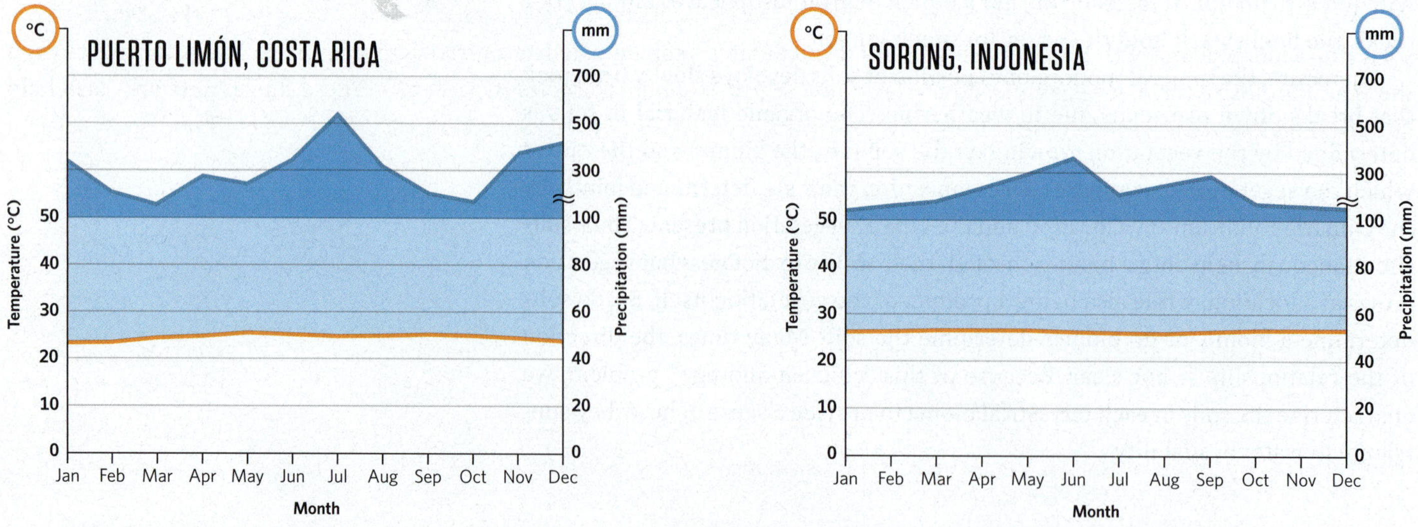
°C
mm
PUERTO LIMÓN, COSTA RICA
700
500
300
100
80
60
40
20
0
50
40
30
20
10
0
Temperature (°C)
Precipitation (mm)
Jan Feb Mar Apr May Jun Jul Aug Sep Oct Nov Dec
Month
°C
mm
SORONG, INDONESIA
700
500
300
100
80
60
40
20
0
50
40
30
20
10
0
Temperature (°C)
Precipitation (mm)
Jan Feb Mar Apr May Jun Jul Aug Sep Oct Nov Dec
Month

Ⓒ Actual climate diagrams for Costa Rica and Indonesia; dark-blue shading indicates extremely high precipitation and a change in the scale on the precipitation axis (Zepner et al. 2021).

D Typical rainforest vegetation with tall trees, dense understory, and lots of species growing on other species in the canopy. **E** Tropical rainforests are home to a huge diversity and abundance of animals demonstrating a wide variety of sizes, shapes, and colors.

The **tropical rainforest** biome is found around the world at or near the equator (between approximately 20° north latitude and approximately 20° south latitude; **Figure 3.20A**), where the temperature is uniformly warm year-round (average temperature, 25°C–27°C [77°F–80°F]; **Figure 3.20B**). As the name implies, tropical rainforests have very high precipitation, with an average of more than 2,000 mm (79 in.) annually. Although water availability tends to be high enough to allow year-round plant growth, the actual amounts of rainfall can vary from seasonal deluges to rainfall that occurs almost every single day (**Figure 3.20C**). Regardless, tropical rainforests typically do not experience seasons like the rest of the world. These temperature and precipitation patterns make sense, given the high solar input the tropics experience.

Because of the fairly constant availability of water and solar radiation, plant growth is prolific and occurs throughout the year. Rainforest trees, for example, do not lay down yearly growth rings like trees at higher latitudes but instead produce wood year-round. In addition, rainforest soils are surprisingly low in nutrients; although decomposition is fast in the warm, wet conditions, the living plants grab all the available nutrients as quickly as they can. With so many organisms growing, nutrients disappear from the soil into new plant tissues quite rapidly. With so much plant growth, it is not surprising that the physical structure of tropical rainforests includes a dense canopy of tall trees reaching 30–45 m (98–148 ft.) in height. The leaves of the canopy trees intercept up to 95% of the incoming sunlight, and competition to capture that sunlight is part of what drives the trees to grow so tall. These canopy trees are generally evergreen, meaning they retain their leaves all year (**Figure 3.20D**).

Below the canopy are complex layers of plant growth, with vines climbing up the trees and with epiphytes (plants growing on other plants) taking advantage of slightly elevated light levels and trapped water to grow on tree branches. The plants in this rainforest understory (underlying layer of vegetation) have evolved strategies for living in low-light conditions, such as large leaves and the ability to grow and lengthen rapidly when a light gap appears. Because almost no sunlight reaches the forest floor, tropical rainforests are relatively free of small ground-hugging plants. Instead, the forest floor is a repository of dead and decaying plant and animal matter and generally harbors dense mats of saprophytic fungi (fungi that consume dead organic matter). Huge networks of fungi in the soil help deliver nutrients to trees in exchange for sugars from photosynthesis (Chapter 9). These fungal networks also help to drive the aforementioned low levels of soil nutrients. In tropical rainforests the vast amount of the biome's nutrients reside inside the living plants, fungi, and animals.

Tropical rainforests are home to an astonishing wealth of planetary biodiversity and contain 40%–75% of all (described) species on the planet (see Chapter 11). A single hectare of tropical rainforest may contain up to 42,000 different species of insects, more than 300 species of trees, and upwards of 1,500 species of other plants (Newman 2002). Many of the iconic rainforest animals are brightly colored (e.g., scarlet macaws, poison arrow frogs, and morpho butterflies; **Figure 3.20E**), yet most animals in the rainforest are camouflaged and hard to see.

BIOME
TROPICAL DRY FOREST

Figure 3.21

A Map showing the locations of tropical dry forests around the world.

B Theoretical climate diagram for tropical dry forests in the Northern Hemisphere; peak rainfall occurs in the summer months, which depends on the hemisphere of the specific location.

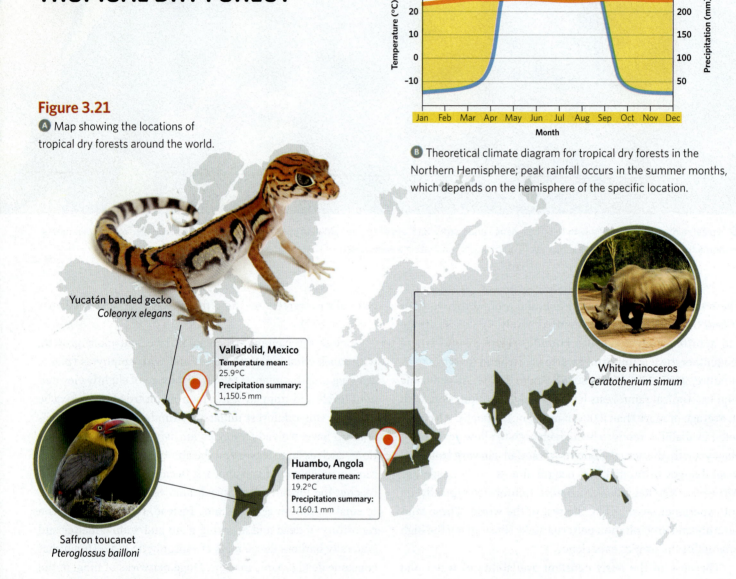

Yucatán banded gecko
Coleonyx elegans

Valladolid, Mexico
Temperature mean:
25.9°C
Precipitation summary:
1,150.5 mm

Huambo, Angola
Temperature mean:
19.2°C
Precipitation summary:
1,160.1 mm

White rhinoceros
Ceratotherium simum

Saffron toucanet
Pteroglossus bailloni

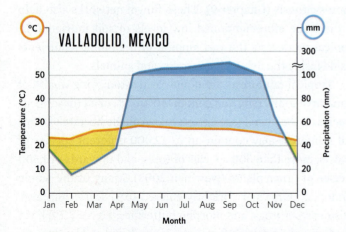
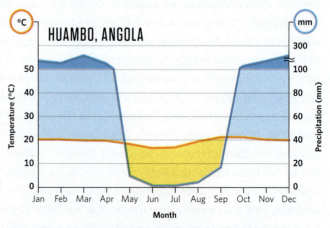

C Actual climate diagrams for Valladolid, Mexico (Northern Hemisphere), and Huambo, Angola (Southern Hemisphere); notice that extremely high rainfall leads to a change in the scale on the right y-axis and is again indicated by dark-blue shading (Zepner et al. 2021).

D Typical vegetation in the dry season (left) and the early part of the wet season (right) in the Yucatán, Mexico.

Tropical dry forests are found at slightly higher latitudes (10°–20° north and south) than tropical rainforests (**Figure 3.21A**). Notice that this range overlaps somewhat with the range for tropical rainforests because biomes do not come in perfect bands like those shown in Figure 3.7B. Instead, the patterns are more complex, as seen in Figure 3.8. This biome is warm to hot year-round, with an average temperature of 25°C–39°C (77°F–102°F), and it receives a substantial annual rainfall of 1,300–2,800 mm (51–110 in.) concentrated into just three to six months (**Figure 3.21B**). As a result, there is a very rainy *wet season* and a notable and sometimes extended *dry season* (**Figure 3.21C**).

In the wet season, a tropical dry forest becomes an emerald tangle of life, rivaling that of tropical rainforests; in the dry season, though, the majority of trees drop their leaves, and the landscape can appear brown and parched (**Figure 3.21D**). This parched appearance is sometimes increased by an understory of herbs and grasses that flourish briefly after rains but eventually turn yellow or tan. These understory plants are possible in a tropical dry forest because the deciduous trees allow sunlight to reach the ground during the dry season. In addition, the physical limitations posed by the protracted dry season lead to a lower canopy height in tropical dry forests than in tropical rainforests.

The cyclical precipitation timing also allows organic matter to build up in the soils of tropical dry forests because decomposition and nutrient uptake slow when water availability is low, leaving the soils richer in organic matter and containing more nutrients than soils in nearby tropical rain-forests. These richer soils allow rapid plant growth when the rains return. The complex plant growth in this biome supports a diversity of animal life, including species such as coatis, iguanas, wild pigs, tigers, monkeys, and parrots (**Figure 3.21E**).

E The biome supports a rich diversity of plant and animal life.

BIOME

TROPICAL SAVANNA

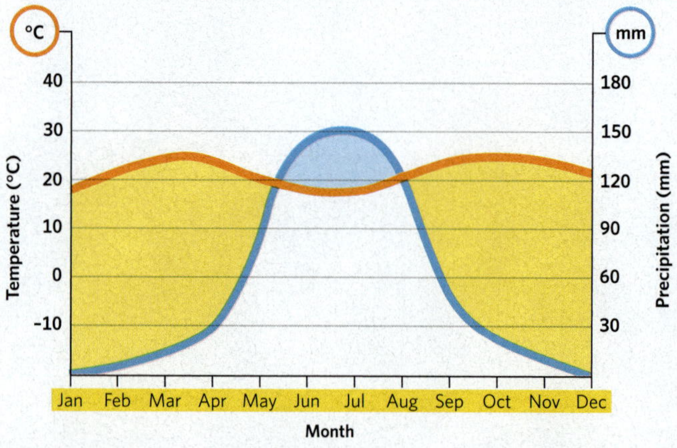

°C
mm
Temperature (°C)
Precipitation (mm)
40
30
20
10
0
-10
180
150
120
90
60
30
Jan Feb Mar Apr May Jun Jul Aug Sep Oct Nov Dec
Month

Figure 3.22

Ⓐ Map showing the locations of tropical savannas around the world.

Ⓑ Theoretical climate diagram for tropical savannas.

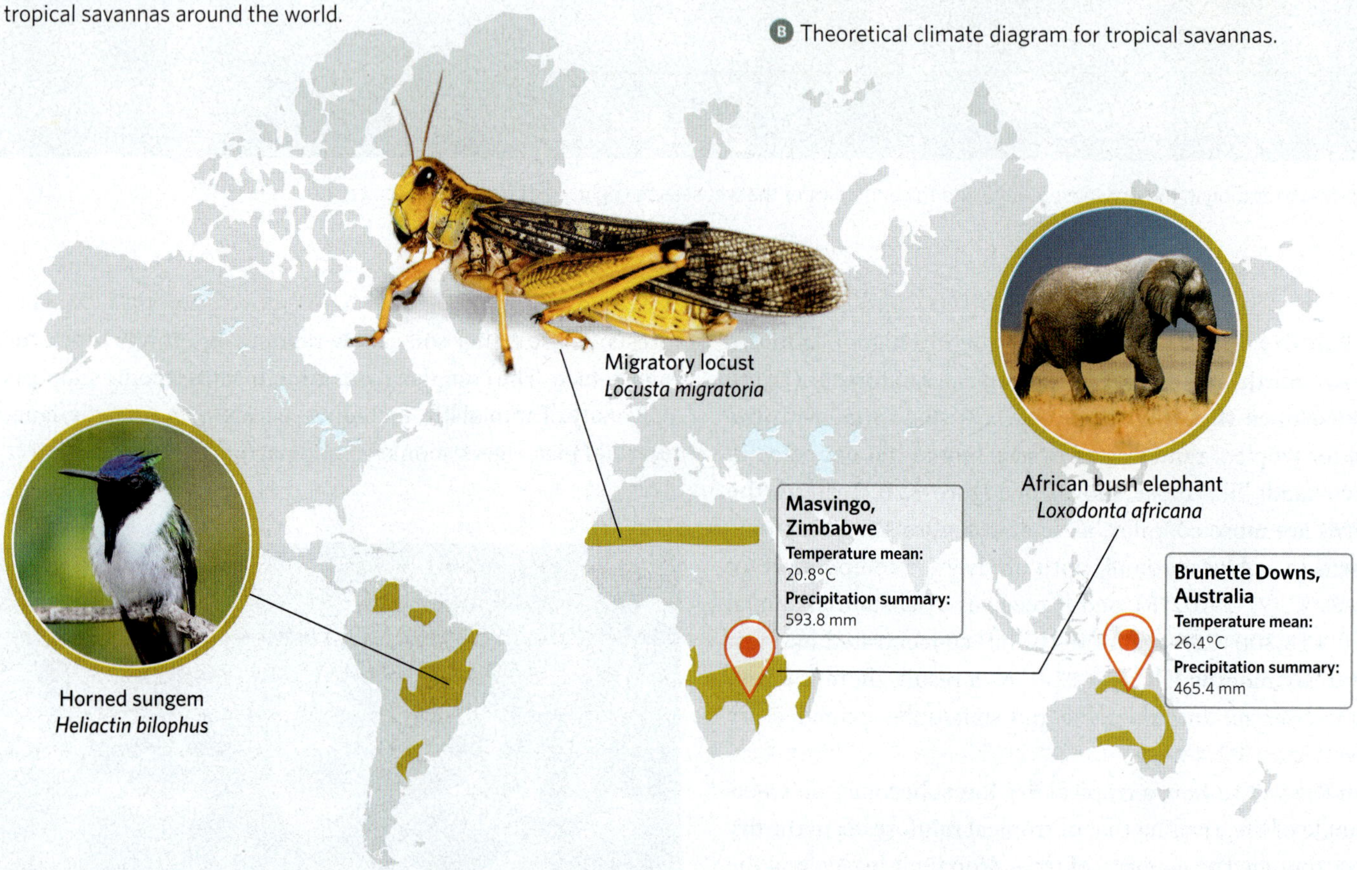

Migratory locust
Locusta migratoria
African bush elephant
Loxodonta africana
Horned sungem
Heliactin bilophus
Masvingo, Zimbabwe
Temperature mean:
20.8°C
Precipitation summary:
593.8 mm
Brunette Downs, Australia
Temperature mean:
26.4°C
Precipitation summary:
465.4 mm

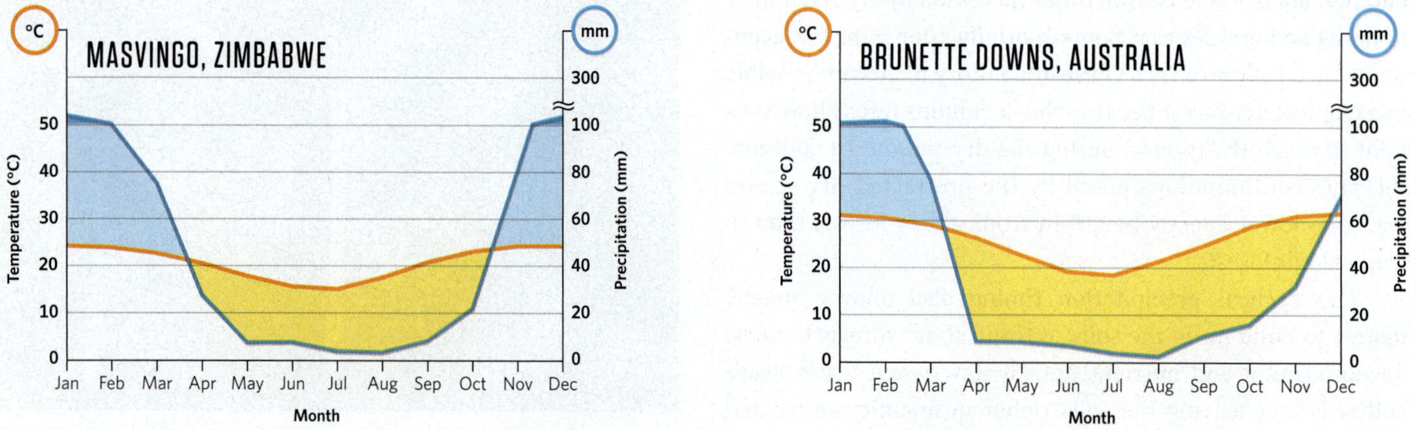

MASVINGO, ZIMBABWE
°C
mm
Temperature (°C)
Precipitation (mm)
50
40
30
20
10
0
300
100
80
60
40
20
0
Jan Feb Mar Apr May Jun Jul Aug Sep Oct Nov Dec
Month
BRUNETTE DOWNS, AUSTRALIA
°C
mm
Temperature (°C)
Precipitation (mm)
50
40
30
20
10
0
300
100
80
60
40
20
0
Jan Feb Mar Apr May Jun Jul Aug Sep Oct Nov Dec
Month

Ⓒ Actual climate diagrams for Masvingo, Zimbabwe, and Brunette Downs, Australia (Zepner et al. 2021).

 Typical open grassland vegetation, dotted with the occasional woody shrub or tree.

Tropical savannas occur in bands around the globe at approximately 10°–25° north and south latitudes (**Figure 3.22A**)—at slightly higher latitudes and in slightly drier climates than tropical dry forests. This biome is characterized by large, open grasslands with only a few isolated trees. The largest and most extensive tropical savannas are in Africa and South America. Tropical savannas are believed to be the evolutionary "birthplace" of our species, *Homo sapiens*, because the first primate ancestral fossils for modern humans were found in the African savanna.

The climate of a tropical savanna has a pronounced periodicity to it. Although the temperature remains fairly consistent throughout the year (average temperature, 24°C–29°C [75°F–84°F]), rain comes in a short wet season, followed by a long, hot dry season (**Figure 3.22B**). The wet season sometimes lasts only a few weeks (**Figure 3.22C**), but the rainfall in this time period can be intense, producing somewhere between 400 and 1,200 mm (16–47 in.) of rain per year in those weeks. The periods of rain arrive with large thunderstorms and frequent lightning strikes, which can lead to recurrent and extensive fires. The period of plant growth in the brief wet season is also short, and the following extended dry period leads to plants that can either survive with little water or be dormant during the dry season.

Soils in tropical savannas tend to have a thin organic layer that does not hold water or nutrients very well. The hot, dry periods can bake the underlying soil and create an impermeable layer that at times, paradoxically, leads to standing water. The combination of potential fires from lightning, low water availability, and thin soils with limited nutrients restricts tree growth in tropical savannas, producing the iconic picture of a golden grassland with herbivores and trees dotted across the landscape (**Figure 3.22D**). In general, grasses are well suited for such environments because many can grow in low-moisture environments and can regrow quickly after fires.

The large open spaces of tropical savannas support enormous numbers and varied assemblages of animals, particularly herbivores. There are vertebrate grazers and browsers, such as springbok, elephants, and giraffes, as well as literally tons of invertebrate grazers (e.g., locusts, grasshoppers, beetles) present in these grasslands (**Figure 3.22E**). Most of the vertebrate animals found in tropical savannas have life-history patterns tuned to the wet-dry cycles, especially as this cycle dictates food and water availability. For many animals, this means a wandering or migratory life cycle in which they follow the rains, which determine the pattern of greening grass that they eat.

 Grasslands have many grazing animals and enormous numbers of insects.

BIOME
DESERT

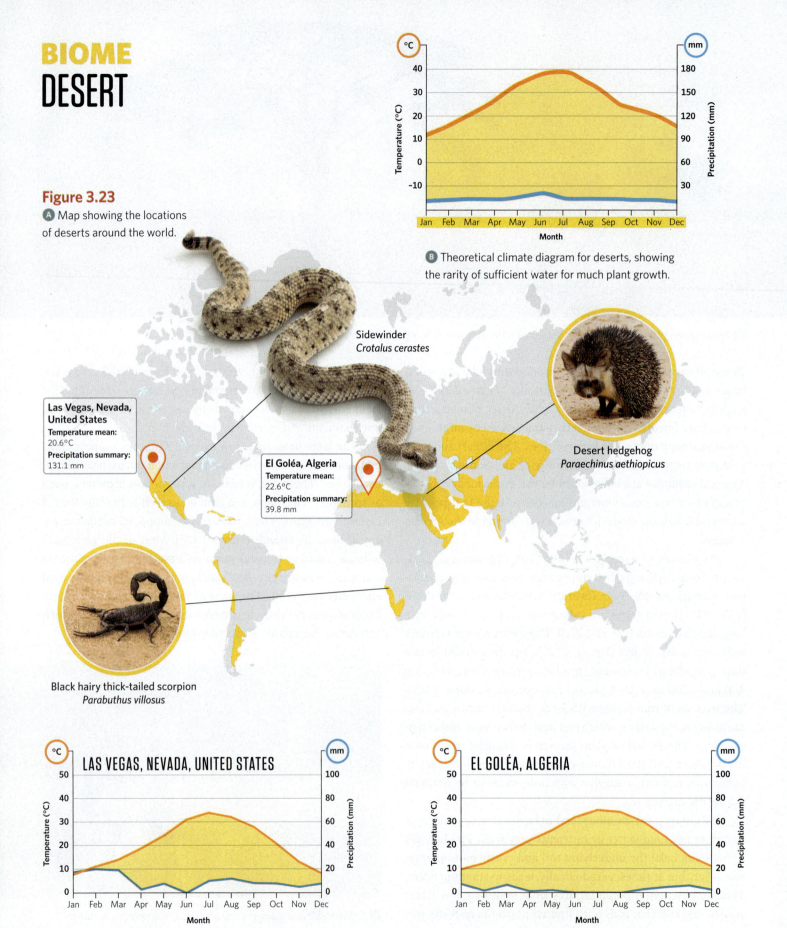

Figure 3.23

Ⓐ Map showing the locations of deserts around the world.

Ⓑ Theoretical climate diagram for deserts, showing the rarity of sufficient water for much plant growth.

Sidewinder
Crotalus cerastes

Desert hedgehog
Paraechinus aethiopicus

Las Vegas, Nevada, United States
Temperature mean:
20.6°C
Precipitation summary:
131.1 mm

El Goléa, Algeria
Temperature mean:
22.6°C
Precipitation summary:
39.8 mm

Black hairy thick-tailed scorpion
Parabuthus villosus

LAS VEGAS, NEVADA, UNITED STATES

EL GOLÉA, ALGERIA

Ⓒ Actual climate diagrams for Las Vegas, Nevada, and El Goléa, Algeria (in the Sahara) (Zepner et al. 2021).

D The typical desert vegetation can be very sparse; even in areas of more abundant vegetation, the plants tend to have characteristics that allow them to reflect the Sun's rays (such as spines) and conserve water.

Deserts are present on almost every continent (**Figure 3.23A**), and are defined as areas where evaporation generally exceeds precipitation. In general, they receive less than 250 mm (10 in.) of precipitation per year (**Figures 3.23B** and **3.23C**), yet not all deserts are created alike. There are hot deserts (the Sahara in North Africa) and cold deserts (Antarctica, Greenland), as well as hyperarid deserts (the Namib in Namibia) and wet deserts (the Sonoran Desert in Arizona). Earlier, we discussed why hot deserts are mostly found near 30° (between 25° and 35°) north and south latitudes, and there are similar but colder polar deserts resulting from the dry Polar cells. These areas of high pressure tend to inhibit precipitation and lead to high rates of evapotranspiration when coupled with warm temperatures.

Unlike other terrestrial biomes that are defined by the dominant vegetation type, deserts are generally defined by evaporation that exceeds precipitation and have low primary productivity because photosynthesis is quite limited. The scarcity of plants in deserts exposes the soils and other geologic features. As a result, deserts are great places to study geology, and they lend themselves to amazingly dramatic landscape photography (**Figure 3.23D**). The plant life that does exist in deserts tends to be sparsely distributed and often displays growth forms that are rarely seen in other biomes (e.g., cacti and succulents) or have unique adaptations to the dry climate. Life in the desert seems incredibly harsh, but many plants and animals (e.g., bighorn sheep, snakes, and scorpions; **Figure 3.23E**) are well adapted to the environmental conditions (see also Chapter 4). Animal abundance is often low, and individual animals generally spread themselves out across the landscape. However, because of their unique traits, desert animal species are often found only in this biome, leading to high levels of *endemic biodiversity*, meaning high numbers of species that are found nowhere else (Chapter 11).

E Animal population densities also tend to be low in this biome, but desert species display a wide range of characteristics not found in inhabitants of other biomes.

BIOME
MEDITERRANEAN SCRUBLAND

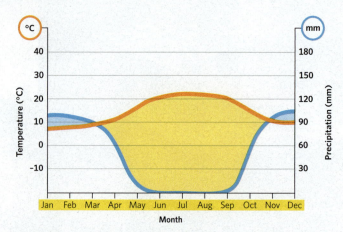

Figure 3.24

A Map showing the locations of Mediterranean scrublands around the world.

B Theoretical climate diagram showing the mismatch between peak sunlight and peak water availability in Mediterranean scrublands.

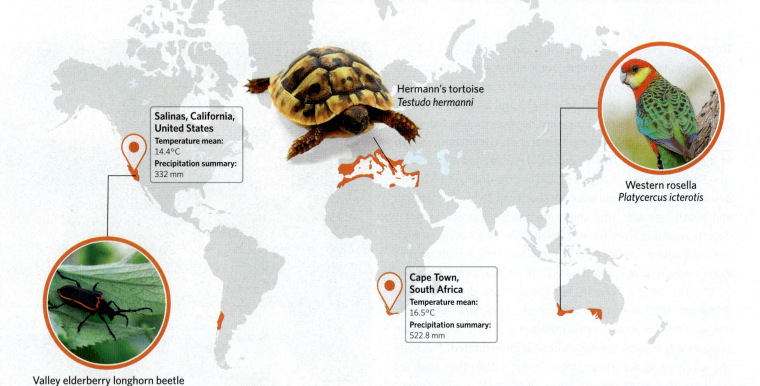

Hermann's tortoise
Testudo hermanni

Western rosella
Platycercus icterotis

Salinas, California, United States
Temperature mean:
14.4°C
Precipitation summary:
332 mm

Cape Town, South Africa
Temperature mean:
16.5°C
Precipitation summary:
522.8 mm

Valley elderberry longhorn beetle
Desmocerus californicus

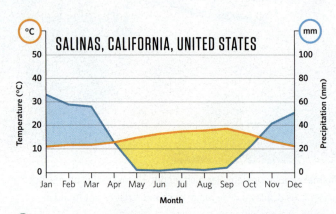

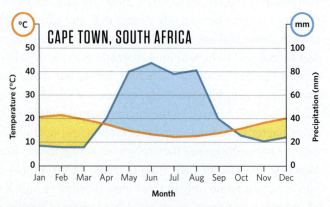

C Actual climate diagrams for Salinas, California, and Cape Town, South Africa (Zepner et al. 2021).

D The typical vegetation is often well suited to frequent fires.

There are five **Mediterranean scrubland** regions on the planet, and all are located between 30° and 45° north or south latitude, on the west coast of a continent, and next to a cool ocean current (except for those on the Mediterranean Sea itself; **Figure 3.24A**). As a result of this suite of conditions, each of these areas has cool, wet winters and hot, dry summers (**Figure 3.24B**). The annual temperature fluctuations in this biome are moderate in comparison to other temperate biomes. We generally characterize temperate zones as being in the zone of influence of the Ferrel cells (meaning between 30° and 60° latitude, north and south). Because the vast majority of the rainfall occurs during a few months in the winter, moderate to severe drought conditions characterize summers in Mediterranean scrublands (**Figure 3.24C**).

Each of the five Mediterranean scrubland regions has a unique local name for the vegetation found in the biome. It is called *fynbos* in South Africa, *matorral* in Chile, *kwongan* in Australia, *chaparral* in California, and *maquis* in the areas around the Mediterranean Sea. No matter the name, shrubs, small trees, and grasses dominate the vegetation and generally have adaptations to withstand drought conditions (**Figure 3.24D**). Vegetation falls into three groups: (1) grasses and herbaceous plants that condense all their growth into the cooler, wetter, low-light conditions of the winter; (2) trees and shrubs that lose their leaves in the summer drought; and (3) evergreen trees, shrubs, and grasses with longer roots and thick, leathery leaves to reduce water loss. All of these adaptations help the plants retain needed water during the long, dry months of summer. In addition, fire plays a significant role in this biome, where dry summer conditions can lead to some locations burning on average every 10–25 years. The fires tend to burn fairly hot and leave large open gaps within formerly dense vegetation. Many Mediterranean plants fare well with frequent fire, because of fire-resistant bark, highly flammable leaves, and seeds that require either the heat or scarring from a hot fire before they will germinate.

Due to the seasonal nature of the rainfall, animals in this biome must be able to survive periods of little to no water. In many locations, there are herbivorous species, such as goats and sheep, that consume a wide variety of plants compared to other grazing herbivores that rely mostly on grasses. Although unexpected, based on the fire frequency and the dry summers, Mediterranean scrublands are home to an array of amphibians and reptiles (**Figure 3.24E**). The wet, cool winters create moist environments that are perfect for amphibians in terms of their physiology, and the long, hot summers are excellent for reptiles. The wet but not frozen winters also create critical stopover spots for many migratory bird species. Adaptation to the separation of rainfall and high temperatures leads to unique organisms, so Mediterranean scrublands, like deserts, often contain endemic species.

E Animals here can survive periods of very low water availability, as well as periods with inundated or soaked soils.

BIOME
TEMPERATE GRASSLAND

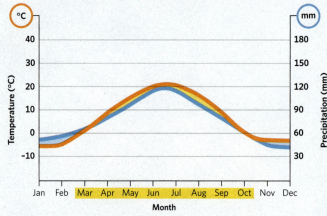

Figure 3.25

A Map showing the locations of temperate grasslands around the world.

B Theoretical climate diagram for temperate grasslands.

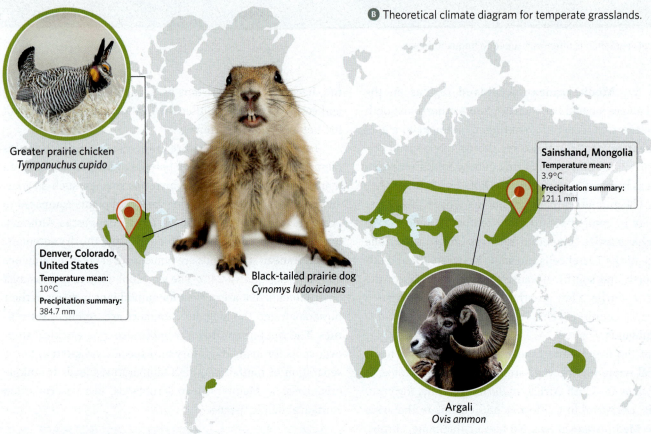

Greater prairie chicken
Tympanuchus cupido

Sainshand, Mongolia
Temperature mean:
3.9°C
Precipitation summary:
121.1 mm

Black-tailed prairie dog
Cynomys ludovicianus

Denver, Colorado, United States
Temperature mean:
10°C
Precipitation summary:
384.7 mm

Argali
Ovis ammon

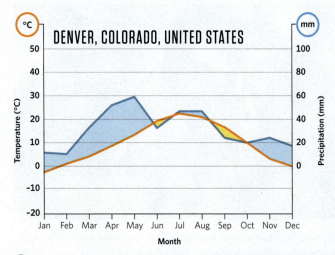

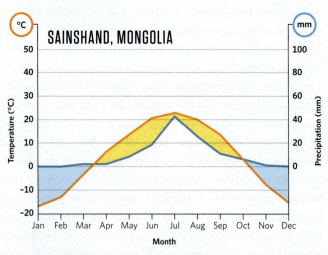

C Actual climate diagrams for Denver, Colorado, and Sainshand, Mongolia (Zepner et al. 2021).

D Grasses are the typical vegetation, with some growing quite tall.

Temperate grasslands, also called prairies, are regions where the predominant vegetation is grasses or small shrubs. These are found at 30°–55° north and south latitudes and are often located in the interior of continents (**Figure 3.25A**). The temperature varies strongly between seasons, with a cold, frozen winter and a hot summer (**Figure 3.25B**). The rainfall pattern generally follows this cycle as well, with the majority of the precipitation falling during the hot summer months (**Figure 3.25C**). Thunderstorms associated with rainfall in the summer produce lightning strikes causing frequent fires. As in the tropical savannas, the presence of periodic fires acts to both produce and maintain the grassland, as the fire eliminates all but a few trees. Temperate grasslands receive approximately 300–1,000 mm (12–39 in.) of precipitation per year, which is generally wetter than deserts but not as wet as other biomes.

Some temperate grasslands are dominated by annual plants that grow anew from seed every year, and in other locations the plants are deep-rooted perennial grasses that regrow from the underground portions of the plant (**Figure 3.25D**). In the Great Plains of North America, grasses, such as Indiangrass (*Sorghastrum nutans*), big bluestem (*Andropogon gerardii*), and switchgrass (*Panicum virgatum*), grow incredibly tall, reaching an average height of 1.5–2.0 m (4.9–6.6 ft.) and occasionally growing as tall as 2.5–3.0 m (8.2–9.8 ft.). It is not hard to imagine the difficulty of trying to walk across these areas or to understand why early European horse-drawn wagons were sometimes referred to as "prairie schooners," because they "sailed" through the vast acres of grasslands.

The annual dieback of plants in the grasslands produces a substantial layer of organic mulch that, over time, forms deep rich soils that are attractive for agriculture. It is interesting to think about the similarities and differences between temperate grasslands and tropical savannas. In temperate grasslands, the heat is less intense and less consistent than in tropical savannas. The lack of intense heat combined with a long, cold winter slows decomposition in temperate grasslands, allowing a buildup of organic material. This deep organic layer holds nutrients and water better than soils in tropical savannas. For these reasons, temperate grasslands are considered the world's breadbaskets and are now home to vast stretches of commercially grown grain crops.

A characteristic feature of temperate grasslands is the presence of large herbivorous grazing mammals. Huge herds of bison and antelope in North America, pampas deer in South America, yaks in Mongolia, and kangaroos in Australia range far and wide over these biomes (**Figure 3.25E**). The grasses evolved with constant grazing, and the act of grazing may even stimulate their growth. The tall, herbaceous plants that dominate the grasslands follow a distinct seasonal pattern, rapid growth in the springtime and dieback in autumn. In addition to herds of herbivores, this biome harbors an array of burrowing and ground-nesting vertebrates.

E Animals include many insects and herd animals, as well as large numbers of burrowing organisms and ground-nesting birds.

BIOME
TEMPERATE FOREST

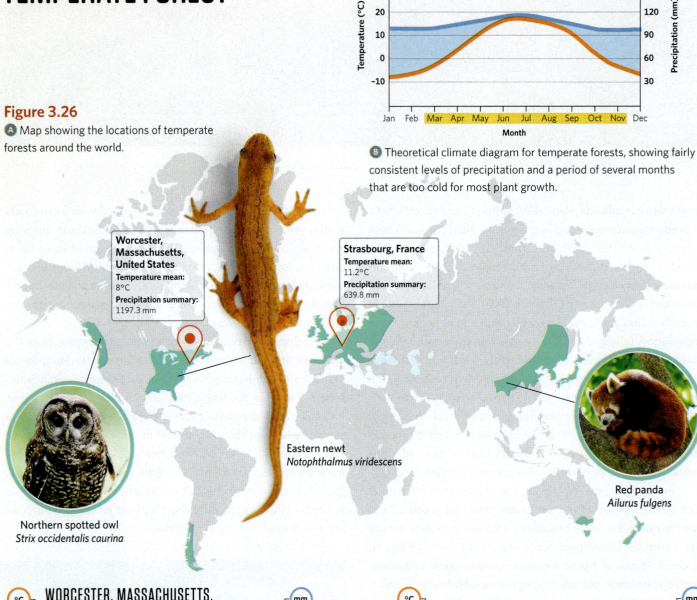

Figure 3.26

Ⓐ Map showing the locations of temperate forests around the world.

Ⓑ Theoretical climate diagram for temperate forests, showing fairly consistent levels of precipitation and a period of several months that are too cold for most plant growth.

Worcester, Massachusetts, United States
Temperature mean:
8°C
Precipitation summary:
1197.3 mm

Strasbourg, France
Temperature mean:
11.2°C
Precipitation summary:
639.8 mm

Eastern newt
Notophthalmus viridescens

Northern spotted owl
Strix occidentalis caurina

Red panda
Ailurus fulgens

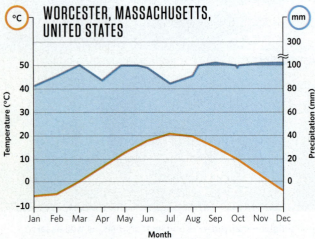

WORCESTER, MASSACHUSETTS, UNITED STATES

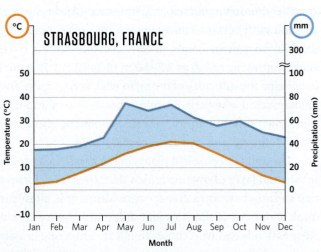

STRASBOURG, FRANCE

Ⓒ Actual climate diagrams for Worcester, Massachusetts, and Strasbourg, France (Zepner et al. 2021).

D The vegetation is dominated by trees that lose their leaves in the fall, often producing fabulous color changes.

Temperate forests are dominated by deciduous trees, or ones that lose their leaves each year. These areas are found between 35° and 60° latitude north and south, predominantly in the Northern Hemisphere on the eastern edges of both North America and Asia and across much of Europe (**Figure 3.26A**). At these latitudes, there is a distinct seasonality to the temperatures: the summers are warm and humid, while the winters are often below freezing, with snow possible during several months of the year (**Figure 3.26B**). Precipitation patterns vary around the globe for this biome, but in general there is little seasonal variation (**Figure 3.26C**).

The most striking feature of the temperate forest is a shift from dense vegetation in the spring and summer months to leafless forms in the fall and winter months. This shift often comes with incredible color displays in autumn (**Figure 3.26D**). Prior to *leaf abscission* (leaf dropping), the tree begins a systematic shutdown of vital fluid transport to the leaves. This physiological reverse course means that the abundant deep-green chlorophyll used to capture sunlight degrades and eventually disappears. Hiding "beneath" the chlorophyll in the leaf tissue are other color-bearing molecules and compounds (e.g., carotenoids, anthocyanins), which in the summer are masked by the abundant presence of chlorophyll. With the chlorophyll gone, these other colors come to the fore, and the leaves take on a spectacular but short-lived fall color display. At the lowest and highest latitudes, temperate forests can become less deciduous, with conifers mixing into the forest at higher latitudes and evergreen, broad-leaved trees at lower latitudes. These non-deciduous broad-leaved trees are primarily in the warmest and wettest portions of the temperate forest.

The temperate forest biome has a distinguishable vertical structure to it, similar to that found in tropical rainforests. There is a dense canopy of dominant trees that intercepts a large majority of the incoming sunlight. Below the canopy is a *secondary canopy* (or understory) of shorter trees, followed by a distinct shrub layer, and finally a ground layer consisting of various ferns, herbs, and mosses. Due to the annual influx of dead plant material (leaves), the soils are high in carbon and nutrients. The decay of this organic matter plays a large role in the ecology of this biome in that it not only adds significant carbon and nutrients to the soil but also allows the soil to house an enormous portion of the biome's biodiversity. Because many of the world's largest human population centers are in temperate forest biomes, many animals familiar to city dwellers are found in these temperate forests, including white-tailed deer, cardinals, and squirrels (**Figure 3.26E**).

E Many of the animals are familiar to city dwellers because many of the world's largest cities are in temperate forest biomes.

BIOME
BOREAL FOREST

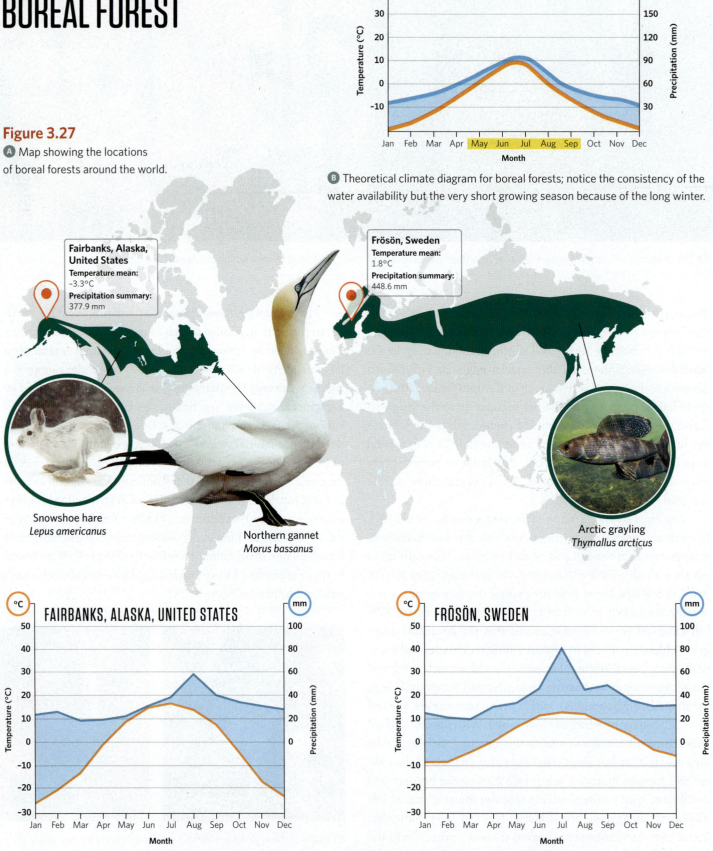

Figure 3.27

A Map showing the locations of boreal forests around the world.

B Theoretical climate diagram for boreal forests; notice the consistency of the water availability but the very short growing season because of the long winter.

Fairbanks, Alaska, United States
Temperature mean:
−3.3°C
Precipitation summary:
377.9 mm

Frösön, Sweden
Temperature mean:
1.8°C
Precipitation summary:
448.6 mm

Snowshoe hare
Lepus americanus

Northern gannet
Morus bassanus

Arctic grayling
Thymallus arcticus

C Actual climate diagrams for Fairbanks, Alaska, and Frösön, Sweden (Zepner et al. 2021).

The typical boreal forest vegetation is conifer forests of moderately tall trees with evergreen needles. The fauna includes many insects but also some of our largest terrestrial mammal species.

The **boreal forest** biome is characterized by coniferous forests. Also referred to as **taiga**, from a Russian word meaning "land of little sticks," this is a large terrestrial biome on the planet, accounting for approximately 11% of the Earth's total surface and covering one-third of the planet's forested land (**Figure 3.27A**). Boreal forests are not visually showy or particularly biologically diverse, but they do contain an assortment of physical habitats, such as streams, lakes, ponds, and wetlands.

Boreal forests are largely found between 50° and 65° north latitude. The Southern Hemisphere has little, if any, continental landmass in this same latitudinal belt, and the lack of land in this region restricts the boreal forest biome primarily to the Northern Hemisphere. At these higher northern latitudes, there is a strong seasonal pattern to the temperature (**Figure 3.27B**). Winters are quite long, and below-freezing temperatures can extend for six or more months of the year. Some of the more northern areas experience bitterly intense cold in winter. For example, in Fairbanks, Alaska, winter temperatures can often be as low as −51°C to −59°C (−60°F to −74°F; **Figure 3.27C**). Yet, because of the Earth's tilt on its axis, the long hours of daylight in the Alaskan summer can routinely bring the temperature up to 10°C–21°C (50°F–70°F) in June and July. In general, across this biome, the summers are short, lasting only two to three months. Precipitation, on the other hand, is fairly constant throughout the year (although on the lower side compared with other biomes), with totals of 200–600 mm (8–24 in.) per year, the majority of which comes in the form of snow.

The extreme weather patterns largely determine the ecological and vegetative structure of these forests. Not only must the plants deal with below-freezing temperatures, but the protracted nature of the winters means that the soils in which they grow also freeze. The nearly constant below-freezing air temperatures produce permafrost at the higher latitudes of this biome. Permafrost is a subsurface layer of soil that remains frozen year-round, limiting deep-rooted plants and preventing rainfall and snowmelt from draining away through the soil. This pooling of precipitation creates very wet and waterlogged ground, which limits the decomposition of organic detritus during the brief summer period. The narrow, needlelike leaves on evergreen coniferous trees have denser tissues than broad, deciduous leaves and are acidic and slow to decompose, which keeps soil nutrients in a thick, slowly decomposing, organic layer. Because of the interactions of the soil, water, and temperature, the dominant vegetation is shallow-rooted, evergreen conifers, such as spruce (*Picea* spp.) and fir (*Abies* spp.) that grow to 15–30 m (49–98 ft.) in height. There is often a dense canopy of these cone-bearing trees that produces a relatively open understory (**Figure 3.27D**).

Boreal forests are home to some iconic animals, such as moose, lynx, snowshoe hare, wolves, loons, ptarmigan, and grizzly bears (**Figure 3.27E**). Overall, this biome has low biodiversity, though. Boreal forests are dominated by a few tree species. In comparison, the tropical rainforest can have more than 300 tree species in a single hectare.

BIOME
TUNDRA

Figure 3.28

(A) Map showing the locations of tundra around the world.

(B) Theoretical climate diagram showing that the tundra is frozen for most of the year.

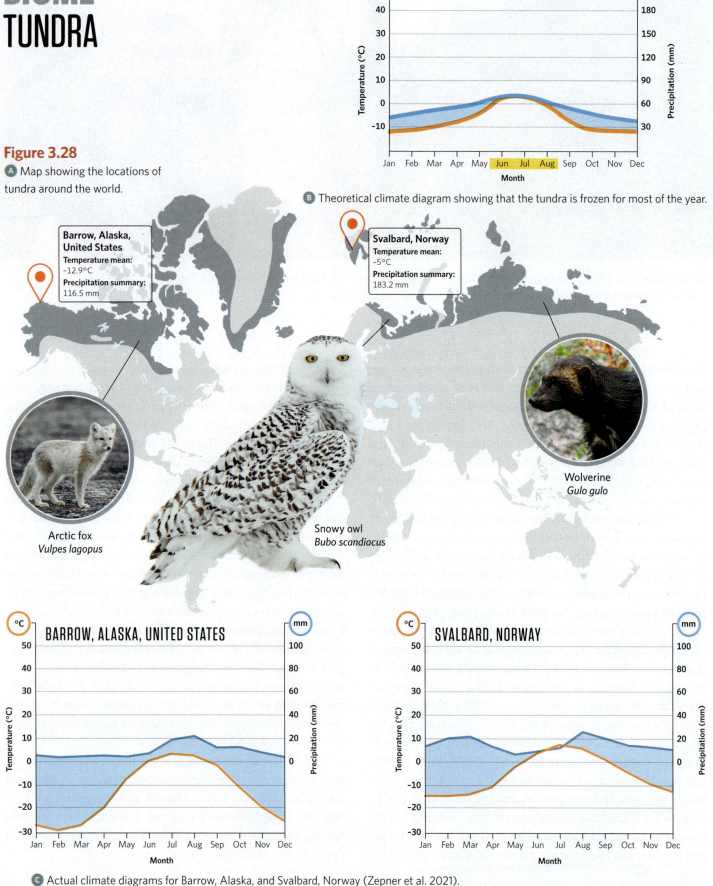

Barrow, Alaska, United States
Temperature mean:
–12.9°C
Precipitation summary:
116.5 mm

Svalbard, Norway
Temperature mean:
–5°C
Precipitation summary:
183.2 mm

Arctic fox
Vulpes lagopus

Snowy owl
Bubo scandiacus

Wolverine
Gulo gulo

(C) Actual climate diagrams for Barrow, Alaska, and Svalbard, Norway (Zepner et al. 2021).

D The typical vegetation is low growing and consists of many dwarf plants and almost no plants taller than a few feet.

At latitudes above 65° north or south, because of the low temperatures (average temperature, –10°C to –16°C [14°F–3°F]), the land ceases to be able to sustain plant growth in the form of trees and is called **tundra** (Figure 3.28A). The tree line, which marks the boundary beyond which trees can no longer grow, also marks the boundary between the boreal forest and tundra biomes. The transition between boreal forest and tundra is only visible in the Northern Hemisphere because all the Southern Hemisphere tundra is in Antarctica. The summer growing season in the tundra is quite short, sometimes lasting only one to two months (Figure 3.28B), although the biome may receive 20–24 hours of daylight during that time. The plants survive the long harsh winters (8–10 months long) by becoming dormant and maintaining their tissue integrity under an insulating layer of snow or even soil. Precipitation falls throughout the year, mainly as snow, and generally has little if any periodicity (Figure 3.28C).

The open tundra landscape is covered with only a few species of slow-growing lichens, mosses, and shrubs (e.g., *Salix* spp.), interspersed with small streams, rivers, and ponds (Figure 3.28D). No plants rise more than a few feet above the ground, but the landscape, colors, and broad open views create areas of stunning natural grandeur and beauty. In the summer, these areas are almost uniformly wet and soggy, as a layer of permafrost underlies the entire region. The ubiquity of still water across the landscape also creates the perfect habitat for mosquitoes, blackflies, deerflies, and other biting insects. The drone of thousands of mosquitoes looking for a blood meal can instantly kill a tundra visitor's ecstatic communal interaction with nature.

In terms of animal life, this biome is one of the last areas on Earth to support large populations of wild mammals, such as reindeer, musk oxen, caribou, bears, Arctic foxes, and wolves, as well as numerous small mammals, including lemmings, weasels, and squirrels (Figure 3.28E). In addition, this far-northern biome is home to a host of seasonal migratory bird species that nest and roost in the long daylight hours (Chapter 12). This list includes tundra swans, Canada geese, eiders, Arctic terns, and many more. This biome contains some of the last large tracts of true wilderness on the planet, as humans have largely avoided these cold northern climates except for oil extraction.

E The fauna includes many migratory birds, large populations of mammals, and copious biting insects.

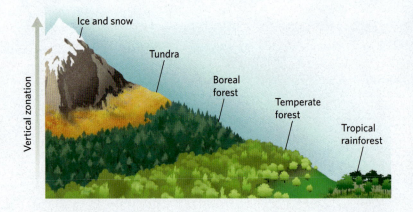

Figure 3.29 Vegetation zonation as altitude increases up a tropical mountain is similar to how vegetation changes with latitude.

Mountain Zones

Approximately one-fourth of the Earth's terrestrial surface is mountainous. As discussed earlier, large mountain ranges can change the climate from what we might expect at particular latitudes, and you can see these effects by comparing Figures 3.7B and 3.8. Although mountains are not biomes in and of themselves, as you go up in altitude, the landscape echoes several facets of the various terrestrial biomes. On a typical ascent up a mountain, the temperature will decrease approximately 6.4°C for every 1,000 m increase in elevation (or 3.6°F per 1,000 ft.) due to adiabatic cooling. This change in air temperature is equivalent to moving north by 13° of latitude (or approximately 1,400 km [870 miles]). As a result of this temperature change, mountains show distinct biome bands as elevation increases (**Figure 3.29**). When mountaintops are viewed across a larger landscape, it is easy to think of them as "sky islands" surrounded by "seas" of lower-elevation habitat, such as a forest surrounded by desert. Some sky islands have been found to serve as refuges for boreal species that have been stranded there since the last ice age.

Biomes Tied to Climate

We started this section on terrestrial biomes by suggesting that the availability of water and intensity of sunlight (or heat) dictate the major vegetation groupings that define biomes. We can put both factors together on a single graph, with temperature (a result of sunlight intensity) along the x-axis and precipitation (indicating water availability) along the y-axis, to look for patterns in the biome distribution. Using this scheme, the biomes fit together but overlap some (**Figure 3.30**).

Figure 3.30 Graph showing average annual temperature (°C) versus average annual precipitation (mm) and the relative locations of the terrestrial biomes. Note that the shape representing each biome is approximate and represents the average range of values rather than the extremes.

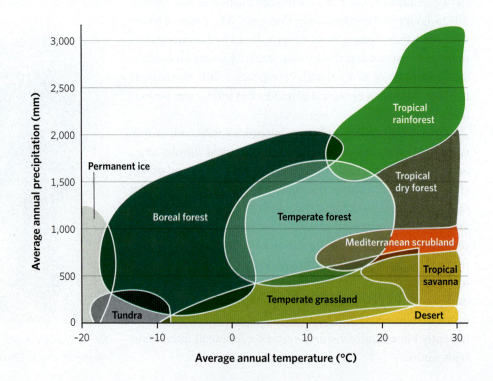

Remember, though, that other physical factors are missing, such as the seasonality of water availability and the role of soils in dictating vegetation. Polar regions, for example, are often quite dry in terms of precipitation and water availability, but they have lots of stored precipitation. Similarly, although boreal forests rarely receive as much water as temperate forests or tropical rainforests, the permafrost can create inundated soil conditions that make for wet roots. This sort of diagram is always going to be cartoonish, but it is useful for organizing and examining ecological patterns particularly in terms of understanding plant distribution, and it makes an interesting comparison with Figures 3.7B and 3.8. The overall patterns differ when we examine the aquatic realms, but there, too, we can think about the way environmental factors interact to create zones where one would expect to find certain types of organisms.

3.6 Aquatic Biological Zones

More than 70% of the planet is covered by water containing a rich diversity of plant and animal life. Although solar radiation still plays a big role in determining where this aquatic diversity is found, precipitation and soils cannot help define aquatic biological zones. One of the biggest challenges faced by ecologists in trying to define aquatic, and particularly oceanic, biomes is that most of the water on Earth sits in large, fairly homogeneous water bodies. In an ocean, the boundaries between ecological regions are rather fluid (pun firmly intended) and messy. We cannot rely on plant distributions or soil types to help organize the oceans into broad ecological units.

The factors that do help define **aquatic biological zones** are light and nutrient availability, temperature, salinity, the physical movement of water, and the structure of the benthic (bottom) surface. All of these factors change dramatically between oceans and freshwater environments and with increasing distance from coastal shorelines. Nutrients running off from the terrestrial biomes are in higher concentrations near the shore. The ocean gets deeper (generally) farther from shore. Water depth affects temperature and light availability, which in turn affects the organisms that can produce three-dimensional structures, such as kelp forests or coral reefs. Salinity also fundamentally affects the organisms that can survive in a particular environment. We will expand on each of these factors for the aquatic biological zones, but the special relationship between sunlight and water warrants extra attention before we dive into the zones.

One of the most important environmental factors we use to describe aquatic environments is the depth to which sunlight penetrates. On average, sunlight will penetrate only the top 100–200 m (328–656 ft.) of the ocean or a lake, and this top "layer" is called the **photic zone** (**Figure 3.31**). Almost all photosynthesis within the aquatic realms occurs in the photic zone, because phytoplankton

Figure 3.31 Marine zones are characterized based on four metrics: (1) proximity to the shoreline, marked with horizontal arrows; (2) physical location (i.e., the bottom surface [benthic] versus the water column [pelagic]); (3) depth of the pelagic zones, demarcated in shades of blue; and (4) light penetration, marked with vertical arrows.

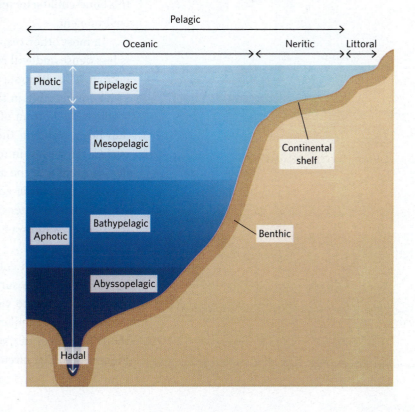

(small, unicellular algae), seaweeds, and kelp cannot photosynthesize without sunlight. Below the photic zone is the **aphotic zone**, where light levels are too low for photosynthesis, limiting the numbers of organisms present at these depths (Figure 3.31). Recent discoveries have highlighted some chemosynthesizing organisms (which do not require sunlight to capture energy; see Chapter 5) at those deeper depths. Nonetheless, approximately 90% of the ocean's life exists in the well-lit photic zone, despite the fact that the vast majority of the volume of the ocean is at the deeper, darker depths.

Another important feature of the aquatic realm results from the curious chemical nature of the water molecule itself. As mentioned earlier, water molecules (H_2O) can form loose hydrogen bonds with other water molecules, allowing them to "stick" to each other. In itself, this is not so special. However, this bond produces an unusual effect when water freezes: the molecules solidify into a crystal in which they are actually farther apart than they are in the liquid phase. Most substances become denser as they cool, with liquids being generally denser than gases, and solids being denser than liquids. But water is an exception. The greater intermolecular distance in ice than in liquid water means ice has a lower density than water. You already know this from observing ice cubes floating in a glass of water. This observation is so ordinary that we often overlook it, but from a chemistry point of view, it is fairly strange.

In terms of ecology, though, this bizarre fact of water chemistry has an enormous impact on aquatic environments and life on Earth. If solid ice were denser than liquid water, then lakes and oceans would freeze from the bottom up and would eventually become solid blocks of ice. Cold liquid water is denser than warm liquid water, and water has its highest density at 4°C (39°F). This means that at the bottom of large bodies of water, such as the oceans, it is pitch-black and very cold. In fact, the deep ocean has always been and will always remain at this bone-chilling temperature, making it a very stable (although not so appealing) environment.

In most other respects, water behaves similarly to air. When it is warm, it is less dense and will rise to the surface; when it is cold, it will sink toward the bottom. This dynamic works in all water bodies and accounts for why you can swim comfortably in the surface waters of a lake during the summer months, while at the bottom of that same lake the water may be unbearably cold. The Sun heats water in the photic zone, and warmer water rises or floats to the surface. However, in temperate regions of the planet, eventually the length of day shortens and the amount of heat energy transferred from the Sun into the water of the photic zone decreases, causing the surface water to cool down. As this surface water cools, it becomes denser and begins to sink, which shifts the deeper waters and pushes water up from the cold aphotic zone toward the surface.

This change in energy input from the Sun creates thermally driven circulation between the surface layers and deep layers of water. This circulation is sometimes easier to visualize in a lake system because of its smaller size (**Figure 3.32**), but a similar process happens in the oceans over a single year, with the added effect of global ocean currents that move water long distances (see Figure 3.12). The circulation of oceanic water drives aspects of the world's cli-

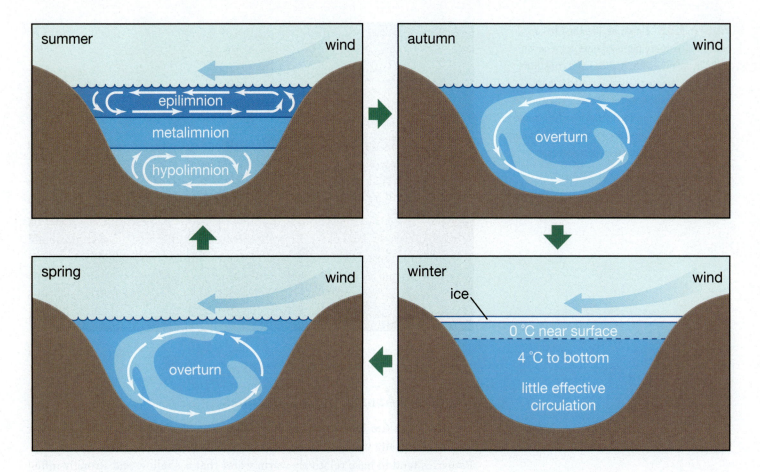

Figure 3.32 These four panels show thermally driven water-circulation patterns in a temperate-latitude lake in each of the four seasons. In the winter, the lake is frozen, and the densest, 4°C (39°F) water is at the bottom. When all the water reaches 4°C in the spring due to warming from the Sun, the lake mixes from top to bottom with the help of wind across the lake's surface. In the warm summer months, the lake becomes stratified, with cold water deep in the lake (hypolimnion) and a layer of warm surface water (epilimnion) heated by the Sun. In the fall, when the surface and middle layers of water approach the same temperature as a result of less solar warming, the lake again mixes from top to bottom with the help of wind.

mates, as you learned earlier. Remember that the tropical oceans get more direct sunlight than the polar waters, so the warm equatorial ocean waters essentially "float" over colder polar waters.

Marine ecologists and oceanographers have developed names for different locations within the ocean that roughly correspond to what we define as biomes in the terrestrial world (Figure 3.30). We will first look at the aquatic world in terms of the oceans, dividing them into nearshore biological zones and open-water zones, and then we will discuss some of the physical aspects of fresh waters. Note that the oceans and aquatic realms can be almost infinitely divided into different categories or types. We chose to present some of the major ones here, but we could easily have highlighted many others as well.

Figure 3.33 Estuaries, like the Chesapeake Bay estuary pictured here, experience shifts in salinity with the tides and constitute a rare zone where marine, terrestrial, and freshwater environments interact. They are one of three marine-terrestrial interface zones where complex plants are found. Estuaries can be extremely productive environments.

Nearshore Aquatic Biological Zones

Estuaries   At the juncture of rivers and oceans sit the extremely productive and economically valuable marine environments called **estuaries** (**Figure 3.33**). Estuaries tend to have relatively warm water that is shallow and strongly influenced by ocean tides. They also have a pronounced salinity gradient, with fresher waters inland grading into salty water closer to the ocean. The fresh water is produced by river outflow, which gradually combines with ocean water in the estuarian zone.

A unique environment is created within the zone where salt water and fresh water mix, which is due in large part to the dissolved salts in the marine water forcing the organic debris in the fresh water to *flocculate* (clump). This clumping makes the water of estuaries very turbid, or muddy. This turbid zone moves toward and away from the ocean on daily and seasonal cycles following tides and the amount of freshwater river outflow, respectively. In estuaries that abut terrestrial biomes with highly seasonal precipitation, the amount of water running from rivers into estuaries will follow these seasonal rain pulses. The turbid zone in estuaries often contains a nutrient-rich mixture of terrestrial runoff, contributing to the high plant productivity of estuaries.

Estuaries are one of the few aquatic biological zones in which some sections are dominated by plants, such as cordgrass (*Spartina* spp.), rather than algae. These vegetated zones at the interface of the terrestrial and aquatic environments are sometimes called marshes, and they line the edges of the most seaward reaches of rivers, creating large, complex arrays of creeks and tidally flooded grasslands. In terms of biological diversity, estuaries have comparatively few species. They are, however, one of the most productive ecosystems on the planet, with estimates of primary productivity easily approaching 8,000 g of carbon produced per square meter per year.

Figure 3.34 Mangrove forests can serve as important zones for moderating influences between marine and terrestrial environments. They catch much of the runoff from terrestrial environments and reduce tidal surges into terrestrial environments. The mangrove trees shown here are in Belize.

Mangrove Forests **Mangrove forests** are a second marine-terrestrial interface that is dominated by complex plants, in this case by woody vegetation (**Figure 3.34**). Sixteen different species of evergreen trees and shrubs are collectively called mangroves because they all share some fundamental characteristics. They all thrive in salt water, using a suite of adaptations, including salt exudation from leaf surfaces, aerial roots that lift the roots out of the anoxic (oxygen-free) marine sediments, and specialized structures called *pneumatophores* that facilitate gas exchange. Mangrove forests are exclusively found in the tropics and subtropics between latitudes of 25° north and 25° south and can slow the movement of nutrients from the terrestrial to marine environment. The physical structure of a dense tangle of ocean-dwelling roots, typical of mangrove forests, provides a vital nursery and rearing area for many marine fish and invertebrate species. In addition, because their complex root systems dissipate wave energy, mangrove forests are able to moderate tidal and storm effects on terrestrial environments.

Figure 3.35 A hawksbill turtle (*Eretmochelys imbricata*) grazes on a seagrass bed in the Caribbean. Seagrass beds are diverse and productive systems that can harbor many species.

Seagrass Beds **Seagrass beds** are the third marine-terrestrial interface dominated by complex plants (**Figure 3.35**). *Seagrass* is a collective term for a suite of submerged flowering marine plants (e.g., *Zostera* spp.) that, despite the common name, are not actual grasses but are related to herbaceous aquatic plants. Seagrasses share morphological characteristics with terrestrial grasses in that they have distinct roots, stems, leaves, and flowers, which are pollinated under the water via fish visitations. Seagrass beds form a link between shallow mangrove systems and nearby coral reefs. They are generally found at an intermediate depth between the two and are fully submerged at all but the lowest tides, unlike mangroves

Figure 3.36 Warm-water coral reefs, such as this one in Cozumel, Mexico, cover only one-tenth of a percent of the ocean, but they harbor approximately 30% of the world's fish species.

and the highest reaches of estuaries. Seagrass beds are diverse and productive systems that can harbor many species, including juvenile and adult fish, small and large algae, mollusks, marine worms, and nematodes. They also provide food for a variety of larger herbivorous (plant-eating) species, including sea turtles, dugongs, manatees, fish, geese, swans, sea urchins, and crabs.

Coral Reefs   In overall biodiversity, warm-water **coral reef** systems rival tropical rainforests. Visually striking in terms of the variety of colors, shapes, and forms they display (**Figure 3.36**), these systems are generally found between the latitudes of 30° north and 30° south. Coral animals, including anemones, sponges, corals, and many other taxa, need warm water to thrive, with the optimal temperature being between 26°C (79°F) and 27°C (81°F). Very few warm-water coral reefs exist in waters that dip below 18°C (64°F). The three-dimensional reef structure is formed from calcium carbonate laid down by the growth of the coral, so in a sense the coral organisms build the framework that houses the entire ecosystem.

Paradoxically, these diverse systems exist in extremely nutrient-poor waters; corals themselves get their nutrients by way of a symbiotic relationship with *zooxanthellae* (Chapter 9), which are single-celled photosynthetic protists. Because of this photosynthetic relationship, warm-water corals require high light availability and are not found in the turbid, cloudy water near estuaries or other zones where rivers move silt into the marine environment. Warm-water coral reefs cover approximately 0.1% of the ocean yet harbor more than 30% of the world's fish species.

Rocky-Substrate Zones   As you move to higher latitudes, water temperature gets substantially cooler, and the shorelines of the world become dominated by one of two biological habitats: rocky substrate and sandy bottom. **Rocky-substrate zones** are present when the shoreline consists of a hard substrate (or seafloor) (**Figure 3.37**). This hard substrate provides a stable anchoring point for numerous species of marine invertebrates and algae. Many species of kelp (a type of macroalgae) are able to attach securely to the rocks in these zones and form dense kelp "forests" (**Figure 3.38**). These marine environments are analogous to terrestrial forests in that they have a three-dimensional structure with a distinct "canopy" and an "understory." The underwater towers of kelp that grow here provide food and shelter for thousands of fish, invertebrates, and marine mammal species. Some marine organisms aggregate and spawn in kelp forests, whereas others only use these areas as nursery habitat. In addition, large predatory sharks and marine mammals hunt in the long corridors that form between rows of individual kelp plants. Kelp forests are believed to be one of the more productive marine ecosystems on the planet.

At the marine-terrestrial interface of these rocky-substrate zones, tide pools form a habitat for a fascinating collection of organisms (Figure 3.37). This habitat experiences challenging physical forces, including daily tidal action and wave energy. For stationary marine organisms living along the shoreline, a typical 24 hours includes two periods where they may be out of the water for more than two to three hours. Exposure to air for such a long period of time is not ideal for most marine creatures. Those species that survive and thrive exhibit interesting adaptations that combat these periods of drying and warming. In addition, and as any surfer will tell you, dealing with a constant barrage of waves all day, regardless of the size of those waves, can leave an organism battered and beaten. Yet rocky-substrate marine animals have adapted to this challenge as well. When you really observe a rocky-substrate area for the first time, you may be struck by the significant zonation to the organisms present on the rocks. It sometimes looks as if the rocky shoreline has stripes, or "zones." These zones reflect aspects of individual species' physiological niches, which we discuss in Chapter 5.

Figure 3.37 Tidal pools frequently found along the rocky-substrate zone are home to many species of marine invertebrate and fish species.

Figure 3.38 Kelp forests found along some of the world's rocky-substrate zones create an environment that is rich in marine life.

Figure 3.39 Sandy-bottom marine habitats are common. In these locations most of the organisms, like mole crabs (*Emerita* sp.), live beneath the shifting sands.

Sandy-Bottom Zones   The long, sandy beaches popular with swimmers, surfers, beachcombers, and bird-watchers appear at first glance to be devoid of life (**Figure 3.39**). **Sandy-bottom zones** are extremely dynamic, as the sand grains are constantly changing position in response to wind, waves, and tides. This shifting substrate makes it difficult for most marine invertebrates to settle and complete their life cycle because there is no spot for them to establish sturdy or permanent attachments. Instead, most of the life found in sandy areas either lives buried in the sand (*infauna*, such as mole crabs, clams, and polychaetes) or constantly clambering above the sand to avoid being buried (mobile organisms such as crabs and shrimp).

Despite the relatively low diversity of life forms, those that live on sandy areas are often found in large numbers because of the sheer extent of this habitat. Sandy marine habitat is very common, covering approximately two-thirds of the planet's ice-free coastlines. These areas serve as critical buffer zones that protect the terrestrial coastline, sea cliffs, and dunes from direct wave action.

Open-Water Aquatic Biological Zones

Open Ocean   It is hard to comprehend how large, extensive, and featureless the open ocean really is. People who have sailed across the Pacific or other large stretches of ocean describe weeks under sail where there is nothing to break up the endless views of water and sky. In addition to their sheer size, oceans are also relatively featureless (**Figure 3.40**), as most of the open ocean is just uninterrupted vistas of salt water.

The open ocean is sometimes called the **oceanic zone**, with the watery part (as opposed to the ocean floor, or **benthic zone**) defined as the **pelagic zone** (see Figure 3.31). There are many quadrillions (i.e., 10^{15}) of organisms living in the pelagic zones of the world's oceans, but they are dispersed over great distances. This is particularly true for larger animals such as fish, marine mammals, and cephalopods (e.g., octopuses and squids). Microscopic organisms, on the other hand, can sometimes form enormous concentrations that can even be seen from space. For example, **Figure 3.41** is an image taken by a satellite of a coccolithophore bloom (coccolithophores are unicellular marine phytoplankton) in the Atlantic Ocean off the coast of France in 2004. Most of the life forms in the open ocean are planktonic, meaning they float or have limited swimming abilities throughout their entire lives and are relatively small (from 2 µm to 20 mm in size). Planktonic organisms—or plankton, collectively—can be plants (phytoplankton), such as coccolithophores, or animals (zooplankton), such as copepods.

Deep Ocean   Not only are the world's oceans big, but they are also very deep. **Figure 3.42** shows the relative depths of the large lakes of the world compared to the depths found in the ocean. The deepest part of the ocean, in the Mariana Trench near Guam, goes more than 2,000 m (6,562 ft.) deeper than the heights of the tallest mountains on Earth. The photic zone is indicated in the figure by the tiny band of light blue at the top of the figure. Notice that the rest of the depth of the ocean is shown as dark blue. As previously discussed, essentially no light penetrates that far down into the water column. The photic zone sits in the epipelagic zone, or the top 200 m (656 ft.) of water (see Figure 3.31). The depth of the photic zone varies because suspended particles and turbidity can affect light penetration, but it is rare for the photic zone to extend below 100 m (328 ft.). Thus, even part of the epipelagic, or top layer of the pelagic, is without sunlight; and the vast majority of the ocean's water is dark and very cold (around 4°C, or 39°F). The lower depths of the pelagic zone are the meso-, bathy-, and abyssopelagic zones (Figure 3.31), which extend downward as follows: mesopelagic, 200–1,000 m (656–3,281 ft.); bathypelagic, 1,000–4,000 m (3,281–13,123 ft.); and abyssopelagic, 4,000–6,000 m (13,123–19,685 ft.). In deep trenches, below the abyssopelagic zone, we find the hadal zone, an area that scientists did not know existed when the other zones were named.

In terms of sheer size, the deep pelagic ocean is clearly the largest single environment on the planet, and yet we have a better understanding of the surface of the Moon than we do of the deep ocean. Part of this is due to the extreme inaccessibility of this aquatic biological zone to humans. For example, the pressures in the deepest zones are staggering and can reach more than 1,000 times the atmospheric pressure, making it nearly impossible for humans to enter this zone without serious technology. Nevertheless, it appears that the deep ocean, and particularly the bottom sediments at these depths, may house large numbers of marine species.

The Census of Marine Life was a 10-year, multinational scientific expedition to explore the world's oceans. Affiliated scientists tried to estimate the diversity of this deep ocean biome, and they are only partly done identifying the thousands of samples collected. Nevertheless, the

Figure 3.40   On the surface, the open ocean is a wide, open, almost featureless habitat that stretches for thousands of kilometers.

Figure 3.41   This photograph captures a coccolithophore bloom off the coast of France as seen from space. Although most organisms in the open ocean are so dispersed that they are hard for us to encounter or see, algal blooms can become dense and enormous.

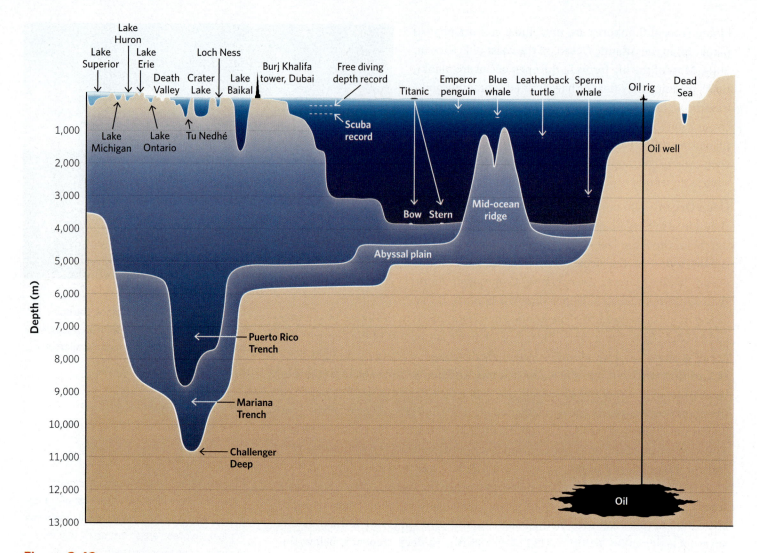

Figure 3.42 Just how deep is the "deep ocean"? This sketch shows the relative depths of the ocean versus lakes and other aspects of the planet. The deepest parts of our oceans go far deeper than the heights of our tallest mountains (Munroe, n.d.).

current estimates of deep-ocean diversity from this effort range from 0.5 million to 10 million species. Most of these organisms have never before been described by scientists, and many species are quite bizarre looking because of the severe adaptations required for life in the pitch-dark and under severe atmospheric pressure (**Figure 3.43**).

Freshwater Biological Zones

Fresh water is defined as water having less than 500 parts per million (ppm) of dissolved salts, whereas salt water has between 500 and 35,000 ppm of dissolved salt. Freshwater systems have many of the same defining features of marine systems, but the spatial extent of freshwater habitats on the planet is dwarfed by those of marine and terrestrial habitats (recall Figure 3.1). Freshwater habitats cover roughly 0.01% of the Earth's surface, whereas terrestrial systems account for about 30% and marine systems about 70%. The vast majority of fresh water on the planet is locked up in glaciers and the ice caps, which is much of the reason why there are so few freshwater habitats on Earth.

Many freshwater habitats, particularly streams and rivers, are not structured the same way as terrestrial or marine systems, in part because of their limited extent and linear nature. As a result, few areas can be labeled as discrete freshwater biomes. If others exist, they are very small and quite specific to a particular geographic location or latitude. Freshwater systems can be categorized as *lotic* (flowing streams and rivers), *lentic* (stationary ponds and lakes), or *aquifer* (underground or subsurface water). Lentic systems cover significantly more area than lotic ones and behave similarly to marine systems, as previously noted. Freshwater environments are home to myriad species of fish, aquatic invertebrates, zooplankton, and phytoplankton.

Figure 3.43 These deep-ocean organisms were found by the Census of Marine Life.

3.7 Human Effects on Global Climate and Biomes

Thus far, we have focused on the way that precipitation, sunlight, and the Earth's physical features determine global climate and biomes. As mentioned earlier, there are also some very powerful biological feedbacks to global climate. Our example was that forest transpiration moves massive amounts of water vapor into the air, altering global precipitation and temperature patterns. It is also critical to discuss some of the ways that humans are changing our planet and our climate.

Overwhelming scientific evidence from a variety of sources points to accelerating changes in the Earth's climate from human production and release of greenhouse gases (gases that absorb and emit thermal energy). Political bickering may affect policies in response to these changes, but it cannot obscure the fact that alterations to the global climate are plainly visible today. Although the biggest changes lie ahead of us, climate change is not just a thing of the future. It is happening right now. Let's start with a basic explanation of how greenhouse gases increase global temperatures, then briefly explore two effects of rising temperatures on biomes.

Greenhouse Gases and Increasing Average Global Temperatures

Climate patterns can change for many reasons. There is good evidence that the Earth's climate has changed several times over its long history, giving us such major phenomena as the ice ages (**Figure 3.44**). These changes come from distinctly nonhuman factors such as the Earth's tilt (which actually changes through time; at the moment, the tilt is slowly decreasing) and position relative to the Sun, plate tectonics, volcanic activity, and meteor impacts.

The shifts in climate we are seeing today are mostly a product of human activities, particularly the release of energy-absorbing molecules, such as carbon dioxide, methane, and other "greenhouse gases" (**Figure 3.45**). These molecules

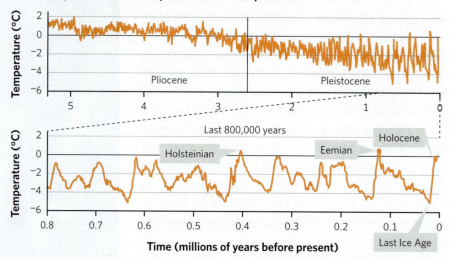

Figure 3.44 In this figure, all temperatures shown are relative to peak temperatures in the Holocene (12,000 years ago to the present). Temperatures above the horizontal line at 0°C (32°F) are warmer than today, while those below are colder than today. On the top are global surface temperatures for the past 5.3 million years as inferred from cores of ocean sediments taken from all around the globe. The last 800,000 years are expanded in the lower half of the figure. There have been two periods in the last 800,000 years when the average temperature has been higher than it is now. In those periods, the world's oceans were also as much as 6 m (almost 20 ft.) higher than they are now. In the very early Pliocene, when the Earth's average temperature was also warmer, global oceans were 15–25 m (49–82 ft.) higher than they are today (Hansen and Sato 2011).

allow solar radiation to penetrate into the Earth's atmosphere but block some of that energy from escaping the atmosphere when reflected from the Earth's surface. Such heat retention is important for the functioning of our planet under normal circumstances, just as heat retention is important in a greenhouse. But too much heat retention causes problems. This effect is equivalent to the air inside your car heating up on a sunny day. The windows in the car allow solar radiation into the interior. Although light can reflect back out, much of the heat stays inside as the glass and steel hold the warmed air molecules in place. The air in a car can heat so far above the outside air temperature that pets or small children can overheat in the car even when outside temperatures are not particularly high. Because of this **greenhouse effect**, the global mean surface temperature of the Earth is rising—it is already more than 1°C (1.8°F) hotter on average than a century ago (**Figure 3.46**)—and scientists are quite confident that the change is due to human influences (**Figure 3.47**).

Should we be concerned about this change in mean global air temperature when we know that the Earth's climate has changed in the past? The answer is an emphatic yes. First, the change is already altering some of the patterns discussed in this chapter, and more changes are expected. The world is dynamic, though, and change is the bread and butter of the study of ecology. The critical factor with global climate change is not just the change itself but the *rate* at which that change is occurring. As Figures 3.46 and 3.47 reveal, the size of the deviations, or the rate of change, has been steadily increasing over the past 50 years.

Figure 3.45 One of the reasons for the increase in global warming comes from the increase in the amount of carbon dioxide (CO_2) in the Earth's atmosphere from human use of fossil fuels. At present, the amount of CO_2 in the Earth's atmosphere is higher than at any other period in the Earth's history (NASA 2021a).

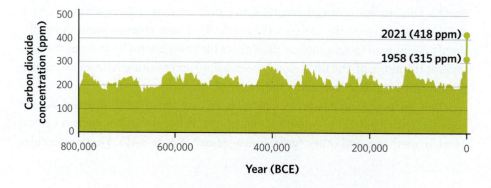

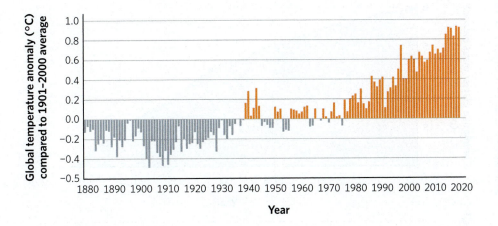

Figure 3.46 By calculating the average global air temperature during the twentieth century (1901–2000), we can see how each year from 1880 to 2020 deviates from that long-term average. Gray bars show how much colder cold years are and orange bars show how much warmer warm years are than the average. The clustering of gray bars on the left and orange bars on the right indicates a general warming trend over time, and it looks as though the temperature is increasing more rapidly in more recent years (NOAA National Centers for Environmental Information 2020).

What happens to the biosphere under this new rate of change? Can species alter their geographical distributions or evolve fast enough to keep up with these climate changes? Or will they simply go extinct because they cannot relocate or evolve quickly enough? Will whole biomes disappear, while other biomes expand? Or will brand new biomes develop? A way to begin answering these questions is to ask what aspects of our climate have already changed and then find ways to demonstrate how ecological systems have responded to these changes. There are numerous documented changes in climate that accompany rising surface temperatures, but we discuss only two of them here—sea-level rise and extreme weather events—as illustrations. Other critical issues involving global climate change will be discussed at length in Chapters 14 and 15.

Sea-Level Rise

One of the main impacts of the Earth's increased surface temperature is a worldwide rise in mean sea level. Mean sea level is measured as the long-term average level of the ocean after averaging out physical dynamics such as tides. This level has changed fairly substantially through geologic time in response to fluctuations in global climate. For example, geologists estimate that about 21,000 years

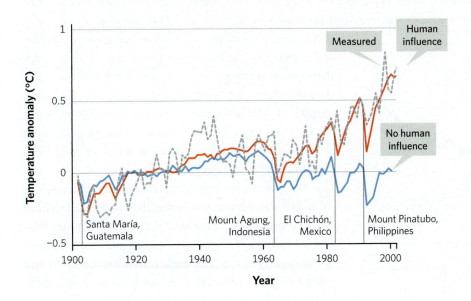

Figure 3.47 The dashed gray line shows deviation from the average global temperature of the twentieth century. We can break down the portion of the warming shown in Figure 3.46 that comes from natural effects and the portion that is anthropogenic, meaning it comes from human effects. There are natural cooling effects, which we can see from the decreases after volcanic eruptions (marked with gray vertical lines and the volcano names and locations). Nonetheless, the human effects are stronger (red line). Without any human influence, we would currently be in a cooling cycle (blue line) (NASA 2010).

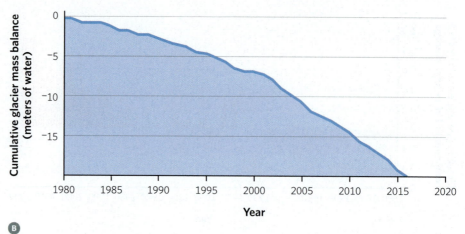

A **B**

Figure 3.48 Glaciers and ice sheets are melting, and the rate at which they are melting appears to be accelerating. **A** This image shows patches of melting ice in Greenland. **B** Data from the National Aeronautics and Space Administration indicate that the Greenland ice sheet lost almost eight times as much ice from 2012 to 2016 as the total amount lost in the 10-year period from 1991 to 2000 (NASA 2021b). That acceleration is evident in the curved line in this graph showing cumulative global glacial melt from 1980 to 2016 (Lindsey 2020).

Figure 3.49 This graph shows how sea level has been lower or higher than the average level between 1993 and 2008. Not only is sea level increasing, but the slight upward curve suggests that the rate of increase may be accelerating. The dark blue line is from three-season data, and the light blue line is from a newer single-measurement approach (Lindsey 2021).

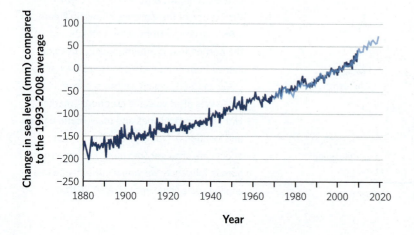

ago, mean sea level was about 125 m (about 410 ft.) lower than it is today. About 125,000 years ago, it was probably about 6 m (20 ft.) higher than it is today.

The mean sea level changes primarily because of two interrelated forces, glacial melt and rising sea temperatures. We mentioned previously that polar regions are not warm enough to melt the small amounts of frozen precipitation that fall each year, so ice from one year is layered on top of ice from previous years. Over thousands of years, polar ice caps and glaciers have accumulated literally tons of frozen water. The current increase in air temperatures is melting glaciers and ice caps (**Figure 3.48**), taking that frozen precipitation "out of storage" and adding it to the world's oceans, thereby forcing mean sea levels upward. Although the oceans are huge, this input of water is not trivial. Glaciers cover more than 10% of the Earth's terrestrial surface. The complete melting of the ice sheet (essentially a really big glacier) that extends across Greenland is expected—on its own—to raise the mean sea level by 6 m (20 ft.). If we add in the water contained in the rest of the Earth's glaciers, the mean sea level would rise about 60 m (197 ft.) or more, which would put most of our important coastal cities under water.

Even without adding water from melting glaciers, the sea level will rise substantially just from applying additional heat to the existing ocean volume. Just like air, water expands as it warms. Because water holds onto heat much better than air does, it may heat up more slowly, but it also does not cool down easily. From long-term shipboard records and modern satellite measurements, we know that the ocean's surface waters have been warming at an accelerating rate over the last 100 years, just as the air has. In addition, the ocean depths at which we are observing increases in water temperature have been increasing, indicating that heat is now being captured well below what we might call the "surface" of the ocean.

So, what are the actual effects of the combined forces of added water and expansion with heat? Sea levels have been rising steadily for at least the past half century (**Figure 3.49**). In New York City, for example, mean sea level is now nearly 30 cm (1 ft.) higher than it was in 1900, and entire countries that exist on low-lying oceanic islands

Figure 3.50 Sea-level rise is leading to loss of terrestrial habitat in many parts of the world. This picture of the Marshall Islands in the central Pacific shows just one example of the many places where land is being lost to rising oceans.

are witnessing the loss of their land from rising oceans (**Figure 3.50**). Salt water is entering environments that have not previously experienced salinity, including groundwater aquifers. These *saltwater intrusions* can make fresh water scarce for humans, animals, and plants alike.

Extreme Weather Events

In addition to rising sea level, we also need to worry about extreme weather events. Although everyone from radio hosts to schoolkids now commonly blames each unexpected weather event on climate change, that connection is a bit too simplistic. Remember that it is hard to predict any given day's weather even if there is a longer-term pattern in climate. What we can do is look for trends in the long-term records of weather events and essentially ignore the short-term changes. With this approach, we can see if extreme events have become more common.

As an example, consider the number of single-day records for extreme heat or cold in the United States from 1931 to 2016 (**Figure 3.51**). You might expect that somewhere in the United States on any given day, a single location will break a

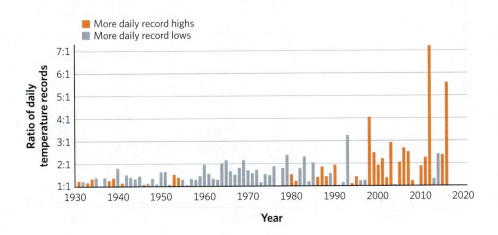

Figure 3.51 In this graph of the ratio of record-setting daily temperatures in the contiguous United States for 1931–2016, orange bars indicate a year with more daily record highs than daily record lows, and gray bars indicate a year with more record lows than highs. For example, a gray bar indicating a ratio of 2:1 means that there were twice as many daily record lows as daily record highs that year (Vose et al. 2017).

record for being the hottest or coldest version of that calendar date recorded at that location. All of these records across the United States from 1931 to 2016 were tallied and separated into record highs and record lows, and then the ratio of extreme-heat records to extreme-cold records for each year was plotted as shown in Figure 3.51. If there were more record high temperatures than record low temperatures in a year, the bar for that year is orange; if there were more record low temperatures, the bar is gray. If the world was experiencing no directional changes in temperature or in extreme events, we might expect about the same number of new heat records and new cold records, and we might expect a random distribution of orange and gray bars. However, in Figure 3.51 we see more gray bars on the left and more orange bars on the right, which is consistent with the Earth's average temperature rising over the past few decades. In fact, there is a clear shift toward more daily heat records starting in the 1980s and accelerating into the 2000s.

It is possible to visualize why this is happening by revisiting the metaphor of a woman walking her dog along the beach (Figure 3.2). However, a more nuanced understanding comes from recalling the concept of normal distribution from Chapter 2. If we recorded the daily temperature for, say, the month of July in Oklahoma City during the 1950s, we could graph these values in a histogram. The result would be similar to **Figure 3.52**, with an average July temperature, but days just above (hotter than) and below (colder than) the average would be fairly common. It is, however, rare to have an extremely hot or cold day in July. These temperatures in the "tails" of the distribution are the extreme weather events. Extreme-temperature days make you sit up and pay attention.

Because the average surface temperature of the Earth has increased by 1°C–2°C (1.8°F–3.6°F) since the 2000s, the average temperature for Oklahoma City for that decade has also likely increased. This would shift the distribution of July daily temperatures to the right in Figure 3.52, toward the hotter end of the scale. Thus, the expectations have changed in that an average day in 2020 is likely to be a few degrees hotter than an average day in 1955. But the location of the

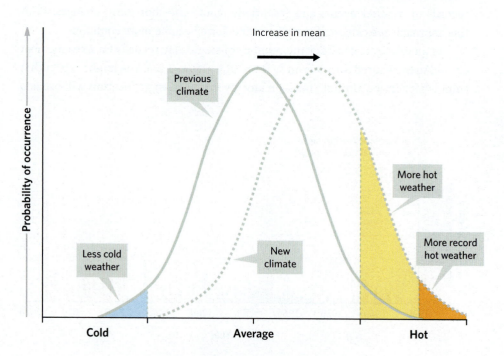

Figure 3.52 Schematic showing the effect on extreme temperatures when the mean temperature increases, for a normal temperature distribution (IPCC 2007).

"tails" of that distribution has also shifted. The really cold days from 1955 happen less frequently, but the frequency of really hot days has increased.

This is a simplified explanation of why we see more extremes. Even with the shift toward more heat, some places may also experience more extreme cold events because of shifts in global wind patterns or changes in ocean current flows. As was mentioned, atmospheric flows and currents can have a big effect on climate. Many places are not only seeing a shift toward warmer temperatures but an overall increase in variability. This increase means that the new temperature curve in Figure 3.52 may widen toward the right (hotter), while still keeping some of the cooler temperatures on the left. These types of considerations are what make climate models so complicated. If you are interested, reports prepared by the United Nations' Intergovernmental Panel on Climate Change (available online at ipcc.ch/reports/) have illuminating discussions of all these considerations.

Human Effects on Biomes

Clearly, these human effects on the climate are going to reverberate through the biosphere (**Figure 3.53**). Species have evolved in the context of climatic conditions that have prevailed for millions of years. They have evolved to survive and sometimes thrive after extreme weather events, such as floods and tropical cyclones, because these events have been regular enough in their occurrence to act as a common agent of natural selection. Climate fundamentally shapes the way species respond to and modify their environment. Although we have barely touched the complexity of the climate system here, and certainly have not fully explored how climate is likely to change in the future, we hope to have ignited your interest in the climate and its importance to the science of ecology and life on the planet.

A full recognition of the structure of biomes can inform ecological approaches to combating climate change. For example, as we try to reduce greenhouse gases, we need to focus on how biomes hold or release carbon (Chapter 14). Despite their

Figure 3.53 As a result of climate change, desert biomes are increasing in size and extent. Here, we see encroaching desert along the Yellow River in Shapotou, Ningxia, China.

lack of diversity, boreal forests are vitally important for global carbon because of the vast amount of land covered by these forests and the sheer volume of carbon locked into the permafrost and other low decomposition soils. Humans have not often chosen to settle in the boreal forest biome, but recently the extent of logging and clear-cutting of the forest has intensified. As the global climate continues to change, the vast stores of carbon locked up in boreal forest soils may be released through rapid decomposition, making this area critical in our efforts to slow increases in global temperatures (Chapter 14).

In addition, each biome experiences a fairly unique human impact that is driven in part by the biological and physical patterns presented in this chapter. This impact will dictate the fate of each biome as we move further into the Anthropocene (Chapter 15). The biomes and climate diagrams that you learn about today are likely to be fundamentally altered in the future. Estuaries, for example, are one of the most degraded marine biological zones. Because they straddle the ocean-freshwater interface, humans have tended to situate ports and large cities within them, such as Los Angeles, Honolulu, London, Melbourne, and Buenos Aires. Similarly, many of the major cultural and political centers of the world are located within the temperate forest biome, including New York City, Boston, London, Paris, Berlin, Beijing, and Tokyo. Finally, more than 88% of the world's temperate grasslands have been converted to agriculture or grazing lands. Agriculture around the world has altered the water retention of soil and converted many areas from dry forest or grassland to desert (Figure 3.53).

Within each of the remaining chapters, we ground the conceptual and quantitative information presented in real-world applications within the context of these sometimes-massive alterations to biomes. In Chapter 15, we explore in detail how the human influence on global patterns in plants and animals is now larger than the effect of the climate.

SUMMARY

3.1 What Is the Biosphere?

- The biosphere is the portion of the Earth that houses all living organisms, plus the environments in which they live.

Climate or Weather?

- Climate is the expected pattern in weather involving statistical trends in temperature, humidity, atmospheric pressure, wind, and precipitation.

- Weather is the day-to-day variation in the same parameters.

- The climate is the primary factor that influences the distribution and abundance of organisms throughout the biosphere.

3.2 Building a Conceptual Climate Model

Solar Radiation

- Because the Earth is a sphere, the equator receives more solar radiation than the poles.

- Sunlight is the main source of energy for life on the planet.

- Because the Earth is tilted, the Northern and Southern Hemispheres experience contrasting seasons.

Atmospheric Circulation

- We can make four key observations about atmospheric circulation:

  - Cold air weighs more than hot air (therefore, hot air rises and cold air sinks).

- Atmospheric pressure decreases as you go up in elevation.
- Compressed air heats up, whereas expanding air cools down.
- Warm air holds more water than cold air.

Putting It All Together

- Combining one of our observations about solar radiation—namely, that the equator receives more solar radiation than the poles—with the four observations about atmospheric circulation, we can construct a conceptual model of how climate works around the globe.
- Hot air rises at the equator because the equator receives more sunlight.
- This hot air carries a large amount of water vapor.
- As this air rises, it expands and cools due to less atmospheric pressure at higher altitudes.
- This cool air loses its moisture in the form of rain at the equator, producing tropical rainforests.
- The cool, dry air is pulled by gravity back to the Earth, compressing and heating up as it descends.
- The descending air is warm and dry and produces deserts at 30° latitude north and south.
- This cyclical movement of air over the Earth from the equator to 30° latitude north (or south) is called the Hadley cell.
- Similar patterns continue both north and south and produce two more air circulation cells in each hemisphere, the Ferrel cell and the Polar cell.
- These patterns can help us arrive at a conceptual model of the effects of water and sunlight on patterns of vegetation.

3.3 Additional Complexity

- The climatic circulation cells go a long way toward explaining major climatic patterns across the planet, but many features are not explained by this simple model.

Coriolis Effect

- The Earth's daily rotation on its axis produces shifts in the atmospheric and oceanic circulation patterns.

Rain Shadows

- Large mountain ranges interrupt the prevailing wind patterns, forcing air to ascend to higher altitudes, where the air expands, cools, and loses moisture, producing rainfall.

- As a result, the windward sides of mountains tend to be much wetter than the leeward sides.

Heat Capacity and Continental Effects

- Interior portions of large landmasses tend to exhibit much wider variations in temperature than coastal regions because of the moderating effects of large bodies of water (oceans and lakes).

Transpiration

- Plants and forests "breathe out" water vapor from their leaves and affect rainfall patterns.

3.4 Biomes

- Biomes are large geographic areas affected by similar climatic and physical factors.
- Biomes are somewhat arbitrary designations.
- Physical aspects of the environment, such as water availability, sunlight, temperature, and soils, help delineate terrestrial biomes.
- Aquatic biomes are defined by variations in chemical and physical features (water temperature, depth, substrate, light availability, salinity, oxygen concentration, and water movements.)

3.5 Terrestrial Biomes

- The conceptual climate model points out the importance of temperature and precipitation.
- Combined, these can form climate diagrams that help depict an area's climate in a two-dimensional graphic form.
- Soils are a critical aspect of biome formation and persistence.
- Tropical rainforests are found near the equator; they have a complex and dense canopy and very high biodiversity despite soils with low nutrient availability.
- Tropical dry forests have distinct wet and dry periods, and the vegetation responds by dropping its leaves.
- Tropical savannas are open grasslands with few trees. Lightning from thunderstorms and resultant fires limit the number and density of trees in this biome, and migrations by large mammal herbivores (e.g., elephants, zebras, giraffes) happen here.
- Deserts are defined by evaporation that exceeds precipitation. Organisms that live here have extreme adaptations to the environment.

- Mediterranean scrublands are located between 30° and 45° north or south latitude and on the west coasts of continents; they usually develop next to a cool, or cold ocean current and have hot, dry summers and cool, wet winters. They tend to have unique and locally adapted organisms.

- Temperate grasslands are found at temperate latitudes, with thunderstorms and lightning acting to limit the number and density of trees. Very deep, nutrient-rich soils make them an attractive location for agriculture. Much of the world's temperate grasslands have been converted to farmland.

- Temperate forests are characterized by the dramatic annual fall color change in the trees, which is precipitated by cold winter temperatures. Many of the world's big cities are located in this biome.

- Boreal forests are one of the largest biomes; they are dominated by coniferous trees and experience extremely cold winters that produce soil permafrost.

- Tundra is generally too cold to sustain trees and is, therefore, dominated by low shrubs and mosses. Summer periods are very short in the tundra, and it is one of the last areas on the planet to contain herds of wild mammals.

- Mountain zones themselves are not biomes, but as the altitude increases, the changes in the mountain landscape mirror the changes in terrestrial biomes with increasing latitude.

- The biomes are tied to climate. By graphing average annual precipitation against average annual temperature, we can see how the terrestrial biomes are distributed around the world as a result of water availability and sunlight intensity.

3.6 Aquatic Biological Zones

- Aquatic biological zones are less clearly delimited than biomes.

- The unique properties of water help to define the aquatic zones.

  — Sunlight only penetrates the upper surface of water.

  — Cold water is denser than warm water.

  — Solid ice floats on liquid water.

  — We use water temperature, light availability, depth, salinity, proximity to shorelines, and substrate to define aquatic areas.

Nearshore Aquatic Biological Zones

- Estuaries sit at the juncture of rivers and the ocean and are extremely productive.

- Mangrove forests are dominated by woody plants and are found at tropical and subtropical latitudes.

- Seagrass beds are dominated by complex plants (i.e., seagrasses), providing an ecological link between mangrove forests and coral reefs.

- Coral reefs are the second most biodiverse system on Earth; they are found in warm, nutrient-poor water and built by coral animals.

- Rocky-substrate zones are found on hard (rock) substrate near the shore and include kelp forests.

- Sandy-bottom zones are characterized by large expanses of sandy-bottom habitat with many organisms that burrow in the substrate.

Open-Water Aquatic Biological Zones

- Open ocean is a huge expanse of featureless low-biodiversity blue water that is important to the world's fisheries.

- Deep ocean is the largest single habitat on the planet, characterized by dark, cold water where sunlight never penetrates and that is largely inaccessible to humans.

Freshwater Biological Zones

- Freshwater zones are a small part of the planet's aquatic realm; they are divided into flowing (lotic), stationary (lentic), and underground (aquifer) waters.

3.7 Human Effects on Global Climate and Biomes

- Scientific evidence suggests that humans are changing the climate.

Greenhouse Gases and Increasing Average Global Temperatures

- Greenhouse gases warm the planet and are at their highest levels ever.

- Scientific evidence suggests that the increase in temperature is due to human influences.

Sea-Level Rise

- The warming planet is leading to rising sea levels.

- Glaciers and ice sheets are melting at increasing rates.

- Seawater expands as it warms and holds heat more than air.

- The rising seas affect habitation and can lead to saltwater intrusion.

Extreme Weather Events

- Increasing temperatures lead to more extreme hot weather.
- Although individual weather events cannot be tied to climate warming, there is a clear upward trend in the number of warming events.
- This trend is likely to accelerate as previously extreme temperatures become commonplace.

Human Effects on Biomes

- Human influences on climate are likely to affect our biomes. The best way to combat these effects is to understand biome ecology.
- We need to focus on biomes that store or release carbon.

- We can already see human effects on several different biomes:
  - Boreal forests are threatened by human expansion and tree removal, and their permafrost contains a huge pool of carbon trapped in the soil that will be released as the planet warms.
  - Estuaries are one of the most degraded marine biological zones because humans tend to locate ports and large cities within them.
  - More than 88% of the world's temperate grasslands have been converted to agriculture or grazing lands.

KEY TERMS

aphotic zone	estuary	rocky-substrate zone
aquatic biological zone	Ferrel cell	sandy-bottom zone
benthic zone	greenhouse effect	seagrass bed
biome	Hadley cell	taiga
biosphere	mangrove forest	temperate forest
boreal forest	Mediterranean scrubland	temperate grassland
climate	oceanic zone	transpiration
climate diagram	pelagic zone	tropical dry forest
continental effect	photic zone	tropical rainforest
coral reef	Polar cell	tropical savanna
Coriolis effect	primary productivity	tundra
desert	rain shadow	weather

CONCEPTUAL QUESTIONS

1. Explain with pictures and words how our generalized atmospheric circulation model produces variations in climate by latitude.

2. Using the generalized climate model from this chapter, explain (with pictures and words) why deserts are often found at 30° north and south latitudes.

3. Explain three of the four ways the generalized atmospheric circulation model that we presented in this chapter can be modified or added to in order to make it more realistic.

4. How might the Earth's tilt help to explain seasonal rainfall at the Tropic of Cancer or the Tropic of Capricorn? Hint: Think about air movement in cells between the equator and the poles.

5. The Atacama Desert in Chile lies between about 18° and 30° south latitude, just west of the Andes mountains. Although the top of the Andes (east of the Atacama) gets rain and snow, the Atacama is often described as the driest place on Earth, with no recorded rainfall at some rain stations. Using what you have learned in this chapter, can you explain why such a difference in precipitation occurs between these two areas?

6. How is a Mediterranean climate defined? Explain why Western Australia has one.

7. Which biomes do you expect to be most affected by rising sea levels? Explain your answer.

QUANTITATIVE QUESTIONS

1. In this chapter we built a global climate model from some simple principles and observations of the natural world. This climate model is not a quantitative or numerical model but is conceptual in nature. Conceptual models have many of the same characteristics as numerical or quantitative models and can be manipulated and tweaked to better understand their behavior. One thing we can do with conceptual models is to alter one of the factors and examine how the outcome of the model would change. For our global climate model, this would entail a thought experiment, because it is not possible to build an alternative or experimental Earth on which to examine our ideas. Let's try imagining what happens if we change our "world"—that is, if we change our model.

a. What would happen if the Earth were shaped like a cylinder (e.g., a can) with its north and south "poles" occupying the flat circular areas? If all the other physical factors remained the same (heat still rises, expanding air cools, and so on), how would the global climate model be altered? Would we still see the same distribution of climatic features such as rainforests and deserts over the surface of the Earth?

b. If we go back to a spherical Earth, how would the output of the climatic model change if cold air held more water than warm air?

c. How would rain shadows be affected if mountains were permeable to wind?

d. Can you think of other theoretical tweaks we could make to our global climate model and then explain how they would affect the climatic patterns we observe on the Earth?

2. Explain how you can tell from Figures 3.46 and 3.47 that the planet is warming.

3. Revisit Figures 3.48B and 3.49, and explain why the slight curve to their lines indicates an accelerating rate of glacier melt (Figure 3.48B) and an accelerating rise in sea level (Figure 3.49).

4. Revisit Figure 3.52.

a. In your own words, re-explain why Figure 3.52 suggests that we will see more days with extreme heat in the future.

b. If that graph only showed an increase in variation, would we still experience more extreme heat? Draw such a graph to illustrate your answer.

SUGGESTED READING

Haurwitz, B. and J. M. Austin. 1944. *Climatology*. New York: McGraw-Hill.

Lieth, H., J. Berlekamp, S. Fuest, and S. Riediger. 1999. *Climate Diagram World Atlas*. Leiden, Netherlands: Backhuys.

Vose, R. S., D. R. Easterling, K. E. Kunkel, A. N. LeGrande, and M. F. Wehner. 2017. Temperature changes in the United States. Pp. 185–206 in D. J. Wuebbles, D. W. Fahey, K. A. Hibbard, D. J. Dokken, B. C. Stewart, and T. K. Maycock, eds. *Climate Science Special Report: Fourth National Climate Assessment*, vol. 1. Washington, DC: U.S. Global Change Research Program. DOI: 10.7930/J0N29V45.

The harlequin lady beetle (*Harmonia axyridis*) is one of the most variable species in the world, exhibiting a huge range of color and spot patterns on its wing coverings. The species is native to eastern Asia but was originally introduced to North America and Europe to control aphids in greenhouses, crop fields, and gardens. Currently, it is considered an invasive species and a significant pest in much of Europe and North America.

chapter 4
EVOLUTION AND SPECIATION

4.1 Variation and Change

Variability is all around us. Everywhere we look, organisms exhibit differences. Plants, animals, and even microorganisms show high levels of variability, both among and within species. As human beings, we are good at seeing some instances of this variation and not so good at seeing others. For example, we are highly proficient at detecting even the smallest differences between human faces. We can distinguish between siblings and even between identical twins if we know them well enough. We take for granted our ability to detect age and health characteristics of people at a glance. This ability makes sense, considering how much information we can glean from even the slightest change in a person's expression. Most of us have had the experience of knowing what our parent or friend was thinking by reading the tilt of their eyebrow or the curve of their mouth. Detecting humor or affection in a subtle squint of the eyes can be a central part of finding a mate or maintaining a friendship. Despite these keen observational skills, most of us can't look at a group of beetles and detect anything about their sex, health, or age, or even tell one from another. Yet the individual beetles are as different from each other as we humans are: we just haven't been trained to observe those differences. So, variability is all around us, and natural variation is critical to the sorts of questions asked by ecologists.

From the discussion of the Anthropocene in Chapter 1 and the role of climate dynamics and biomes in Chapter 3, it may already be clear to you that ecologists devote much of their attention to understanding variability and change. In this chapter, we concentrate on variation among individuals and the way such variation can lead to change. We specifically focus on changes that are passed down over generations and that can lead to alterations in population traits or to the emergence of new species.

LEARNING OBJECTIVES

- Explain why the transmission of genetic material from parents to offspring is important in the definition of a species.

- Explain the biological species concept.

- Summarize how genes are connected to populations.

- List the four mechanisms of evolution, and explain how each one adds or removes genetic variation from populations.

- Compare and contrast directional, stabilizing, and disruptive selection.

- Explain the principles of sexual selection.

- Explain the process of speciation.

- Distinguish between allopatric speciation and sympatric speciation.

- Identify four ways humans are altering the evolutionary landscape in the Anthropocene.

As you learn more about ecology, it will become apparent that ecologists spend a lot of time thinking about species. They ask how many species breed in an area, how many species are likely to become extinct in the next 100 years, why some species make a big impact on a community, and how individuals of one species depend on the presence of another species to survive and reproduce. In short, the concept of species is central to all the major ecological topics, including predation, competition, mutualism, and ecosystem function. Yet it may not be clear what distinguishes one species from another, if and how species change through time, or what role variation plays in both of these matters.

This discussion takes us into evolution and the process of speciation, fields unto themselves. We will not try to provide a full course in evolution in one chapter. Instead, our goal is to provide the tools to help you understand how and why species evolve and how speciation occurs. Along the way, we illustrate why evolution is important to ecological interactions. We start with some definitions and outline the primary mechanisms of evolution before exploring how evolution can alter populations and species and lead to speciation. We hope that, by the end of the chapter, you will have clear insight into why evolutionary processes are essential to understanding ecology and the natural world.

4.2 Definitions of Species

Defining Species Based on Morphology

Many people have a somewhat intuitive understanding of what a species is, so it may come as a surprise that defining a species can be very complicated. If we asked someone from the general public to explain how to tell if two plants growing near each other belong to different species, that person would probably say something about the two plants looking different. In **Figure 4.1**, one plant (bunchberry) has big white flowers and woody stems, whereas the other (bluets) has small blue flowers and soft stems. Bunchberry has broad leaves with distinct veins, and bluets have small leaves with one central rib. These features and others seem like good ways to distinguish between these two species.

If this is how you would respond, you are in good company. Carl Linnaeus, a Swedish naturalist who lived in the 1700s, was the first to formally recognize physical differences and similarities between individuals and use these differences to define distinct groups that he called *species*. Linnaeus is, in fact, responsible for giving us the scientific tradition of naming every species with a Latin binomial of "genus" and "specific epithet," called *binomial nomenclature*. The **genus** defines a broader group, and the specific epithet identifies a particular species that is part of that broader group. Together genus and specific epithet provide the name for a species, which is often referred to as its **scientific name**.

For example, humans are classified as *Homo sapiens*, indicating that we are of the genus *Homo* (genus names are always capitalized) and in the specific category *sapiens* (specific epithets are never capitalized). There were other species in the genus *Homo*, including *Homo neanderthalensis*, more popularly known as Neanderthals. Most of the other organisms that we see regularly also have scientific names. Our domestic dogs are *Canis lupus*, house mice are *Mus musculus*, and the cockroach hiding in your kitchen is likely to be *Blattella germanica*. The two flowers in Figure 4.1 have Latin names as well; bunchberry is *Cornus canadensis*, and the bluets are *Houstonia caerulea*.

Linnaeus did not name every species, because he had no way of knowing about them all, but in the 10th edition of his seminal work, *Systema Naturae*, published in 1758, he described and gave Latin binomials to an impressive 4,400 species of animals and 7,700 species of plants, including *Canis lupus*, *Mus musculus*, *Blattella germanica*, *Cornus canadensis*, and *Houstonia caerulea*. Today, every scientifically described species is given a binomial name, from the smallest bacterium to the largest animal, whether they are currently living among us or went extinct millions of years ago. Of course, scientists have not been able to identify and name every species on the planet, as we will see in Chapter 11, but this naming system makes it much easier to talk to one another about observations of the natural world, as it forces everyone to use the same name to refer to the same entity, even if we speak different languages.

Linnaeus's basic tool for identifying and naming species was **morphology**, or the organism's appearance (*morpho* means "shape"). Individuals with similar morphology were placed in the same species. Species that had lots of similarities in their morphology were put in the same genus. Groups of genera (plural for *genus*) that had lots of morphological similarities were binned into the same "family," and so on, all the way to the top level of classification—domains. (Domains were added above kingdoms in the 1990s, when scientists recognized that many of the organisms they had thought were bacteria were actually archaea; both bacteria and archaea are single-celled microorganisms lacking nuclei but with significant differences.) Some of you may have learned the entire Linnaean classification categories from domains to species in other science classes, but we all get a bit rusty, so you can check out the full scheme in **Figure 4.2**.

This method of classification innovated the way that scientists looked at and discussed the natural world and earned Linnaeus a prominent seat among the founders of ecology and evolutionary biology. To modern biologists, his classification scheme, and particularly grouping organisms into categories, suggests that the organisms are evolutionarily related to each other. Because Linnaeus developed his classification scheme during the mid-1700s (pre-Darwin), the morphological similarities and differences he laid out did not actually imply evolutionary relationships. Not until 100 years later did Darwin make the connection between past relatives, morphological similarity, and evolution (Darwin 1859).

Figure 4.1 Plants from the same forest may look different in ways that allow observers to classify them into different species. More than 200 years ago, Carl Linnaeus gave these two species their Latin names: **A** *Cornus canadensis* (bunchberry); **B** *Houstonia caerulea* (bluets).

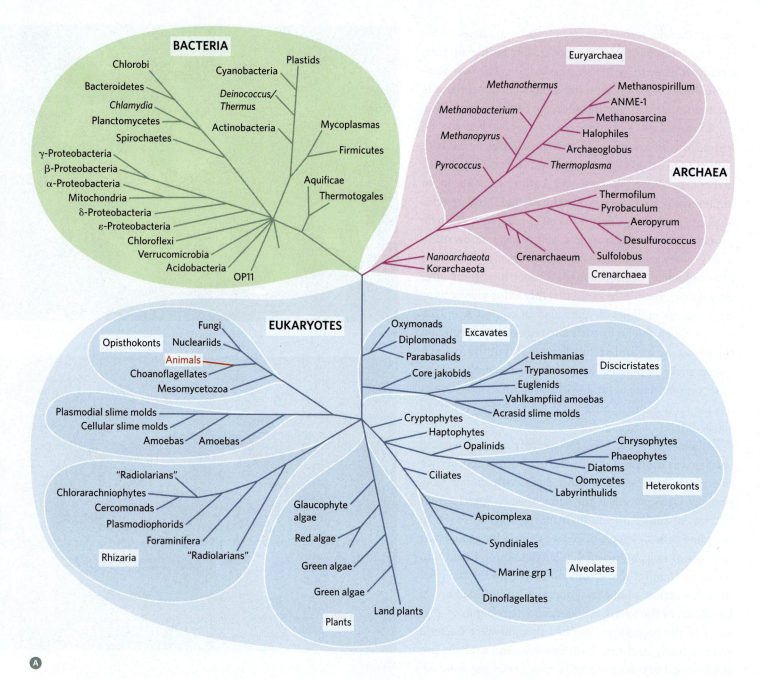

Figure 4.2 Modern classification system for living organisms. **A** The three domains—Bacteria, Archaea, and Eukarya—make up the largest hierarchical level.

The Biological Species Concept

Although Linnaeus's effort to describe and classify species from Sweden and nearby European countries was monumentally difficult and time consuming, he only scratched the surface in describing the Earth's species (see Chapter 11). As people began to follow Linnaeus's lead and formally name species around the world, the incredible diversity and intricacy of the natural world became apparent.

At the same time, the astounding wealth of biodiversity challenged the approach of using only physical characteristics to define species. The more people looked, the more they found groups of individuals with an "intermediate" morphology between two described species. The question became: Should these

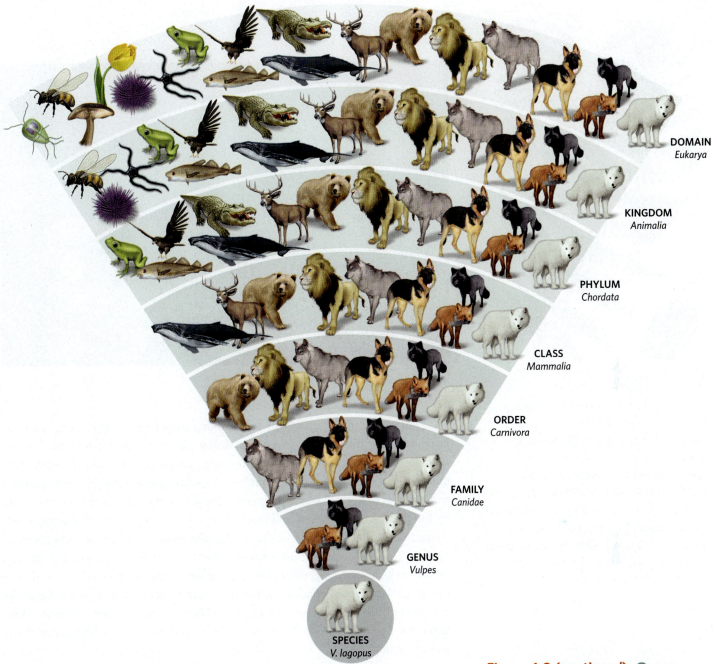

DOMAIN
Eukarya

KINGDOM
Animalia

PHYLUM
Chordata

CLASS
Mammalia

ORDER
Carnivora

FAMILY
Canidae

GENUS
Vulpes

SPECIES
V. lagopus

B

intermediate individuals be grouped with one of the two existing species, or should they be classified as their own separate species?

Answering this question is critical for ecologists because Linnaeus's classification system implies that an organism's physical appearance is tightly linked to what it does in nature. For example, the characteristics of the two plant species in Figure 4.1 reflect each plant's longevity and the way it captures sunlight and grows. The bunchberries have thicker, woody stems because they live for multiple years, and their leaves are darker and bigger because these attributes help them photosynthesize in the shade. The bluets, which grow in small clearings or slight openings, can capture enough sunlight with much smaller leaves and do not need stiff stems because they grow, reproduce, and die in one season.

Figure 4.2 (continued) **B** Within a domain, we can start with a species and move up the hierarchical levels until we include the whole domain. As an example, the Arctic fox (*Vulpes lagopus*) is a species in the same genus (*Vulpes*) as kit foxes and grey foxes, the same family (Canidae) as dogs and wolves, the same order (Carnivora) as lions and bears, and so on up the hierarchical levels. By convention, we often give only the first initial of the genus when naming the species if the genus appears elsewhere, as with *V. lagopus*.

Figure 4.3 The morphology of bracken has remained remarkably unchanged since Linnaeus first described it. **A** Thomas Moore's fern print of the critical features of *Pteris aquilina* (bracken, now renamed *Pteridium aquilinum*) from the mid-1800s shows a pinna (leaflet), a cross section of a stalk, and the sori (or collections of reproductive structures) on the back of the leaflet. **B** In this photograph of modern bracken, we can see the similarity to the bracken of Moore's time.

Looking around us, we can spot other characteristics that seem particularly useful for the species close to us. A dog's large, pointy teeth in the front of its mouth (its canines) reflect ancestors that caught and killed prey with their mouths. A mouse's collapsible shoulders allow it to move through small openings and avoid predators. Cockroach species that live away from humans can usually fly, but the common kitchen cockroaches are wingless: wings get in the way when trying to crawl between the fork drawer and the bread drawer. Every species' morphology reveals something fundamental about how it finds and acquires energy, reproduces, raises young, disperses across a landscape, or avoids death by predators or environmental conditions. For this reason, recognizing that species differ morphologically implies recognizing that different species "do" different things in nature. Ecologists are quite interested in what organisms *do*, both individually and as a group.

Natural scientists in the generations after Linnaeus also noticed that Linnaeus's categories stayed fairly constant. Despite the presence of individuals that sat between categories, the morphological characteristics used to identify a species in one year would also work the next year, and often for many, many years. In fact, descriptions of species made in the 1750s are still accurate for the same species in the 2000s. We can see this consistency if we look at bracken, the most widespread fern on the planet. Linnaeus himself described bracken (and named it *Pteris aquilina*) in 1753. In the mid-1800s, the British botanist Thomas Moore created a print of bracken based on the same description, and that print still matches a modern photograph of bracken (now renamed *Pteridium aquilinum*; **Figure 4.3**).

This constancy, and how organisms fit to their environment, is part of what drove Darwin to think that living things must have some way of passing characteristics from one generation to the next. In fact, Darwin became convinced that the production of offspring was critical to "transmitting" these characteristics (Darwin 1859). He suggested that breeding individuals must choose each other based on similarities that put them in the same species. He also thought such reproduction might help to explain species that lay between categories, because he noticed that offspring tended to look like a mixture of their parents.

Once scientists and naturalists started thinking about how individuals of species mate, they noticed another issue with the morphological classification system, which is that many related individuals do *not* look morphologically similar. For example, young organisms and older organisms often look very different from each other (**Figure 4.4**), and many animals show *sexual dimorphism*, or physical differences between the males and females of the species (**Figure 4.5**). We definitely want to be able to classify juveniles and adults, and males and females, together in one species.

Although Darwin pointed the way toward a modern definition of species, he was more interested in the process of change and differentiation into new groups than he was in defining species. It was not until 1942, when Ernst Mayr proposed the **biological species concept**, that we had a clean, simple definition of species that incorporates the obvious importance of reproduction and solves the dilemma of intermediate morphology.

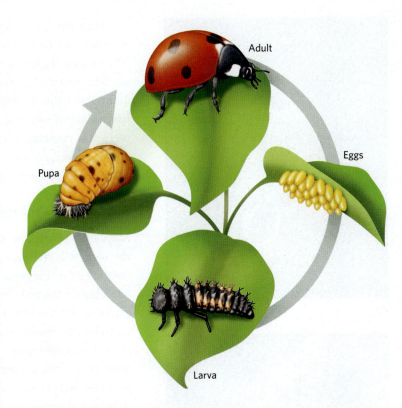

Figure 4.4 Organisms can look very different at different stages of life, but they definitely do not change species as they develop. We can see this in the morphologically different life stages of a ladybug, the common name for many species of lady beetles, including the harlequin lady beetle (*Harmonia axyridis*) shown in the photograph at the beginning of the chapter.

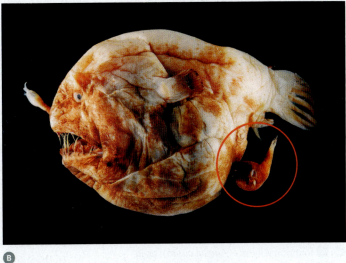

Figure 4.5 Examples of sexual dimorphism can be seen everywhere in nature. **A** In many bird species, males display bright plumage that is used to identify the species, while females often have fairly muted coloration. The male Mandarin duck (*Aix galericulata*), for example, is much more colorful and showy than the female. **B** Sometimes these differences are even more pronounced, and males and females are completely different shapes and sizes. In anglerfish (order Lophiiformes), which live in the ocean depths below light penetration, the female is often an order of magnitude larger than the male (circled). In this anglerfish species, the males are parasitic and must find a female to clamp onto; parasitic males receive sustenance from the females and contribute sperm.

Figure 4.6 Many organisms can create new offspring alone, without a sexual partner. Ⓐ A small, new hydra buds off a larger one; this process does not require a sexual partner. Ⓑ, Ⓒ In moon jellyfish (*Aurelia aurita*), as in most jellies, sexual adults (B) produce planulae, or small larvae, that settle onto a firm surface and become polyps (C). These moon jelly polyps can bud to produce new individuals asexually.

Mayr suggested that individuals could be sorted into species based solely on whether they could mate with one another and produce viable, fertile offspring. The biological species concept limits species membership to only those individuals that can keep the species "going" by contributing to the next generation. This definition helps to explain the morphological consistency across generations, particularly if physical similarities (or differences) play a role in how organisms recognize potential mates. It also addresses the issue of intermediate morphologies because it does not require morphology to identify a species. And it helpfully places all related individuals of any age or stage in the same species.

Beyond the Biological Species Concept

Mayr's biological species concept is not without its problems, however. One of the biggest is that scientists cannot always identify "parents" or even whether creating offspring requires two parent organisms. Being vertebrates, humans are often biased toward thinking that reproduction always requires male and female gametes. But the reproductive process can be quite different for organisms that reproduce by breaking off pieces of themselves (yeasts, bacteria, flatworms, corals, jellyfish, and some plants, to name a few; **Figure 4.6**) or that have no defined sexes (some fish, some amphibians, some plants, and most of the organisms in the previous list) or that can mate with themselves (many plants, via self-pollination). In human societies, we have ample evidence that concepts such as gender and sex are not strictly binary (meaning with only two choices). Given all this, how can we logically apply a strict interpretation of the biological species concept to all situations?

Although most people learned the biological species concept as the definition of species in high school biology, ecologists and evolutionary biologists have continued to debate the best definition since Mayr's time, eventually proposing as many as 12 somewhat unique definitions for the term *species*. Almost all of these definitions acknowledge the features that interested Darwin: namely, that organisms which together form a species have a process for passing genes and similarities to the next generation and for keeping out genes from other species.

Part of the explosion in definitions came from the different tools used in various subfields of biology and even geology. For example, tools for identifying fossil species rarely involve looking for reproductive success, as this information is generally lost in geological time. Modern carbon-dating approaches cannot reveal parentage, but they offer insights into how long species remained static in their physical appearance. Other new technologies and genetic tools, such as using genetic similarity to define evolutionarily unique groups, highlight the degree of reproductive exchange between groups. Each approach focuses on some element of differentiation between species that is important in a particular field, but is there enough overlap between them all to find common ground?

In the early 2000s, this debate quietly began to melt away as scientists focused on the commonalities among species definitions instead of their differences. In general, organisms in the same species must share a lineage in the past and must be able to contribute to changes in their group in the future. This common thread echoes the essential elements of the biological species concept with a focus on how individuals within a species are able to contribute genetic material in a lineage leading to future generations.

This definition is a nod to the commonalities at the core of any understanding of species and **speciation**, the process by which two species arise from one common ancestral species. In general, organisms within the same species must be able to contribute to a common pool of genetic information in the next generation. There is a period of time, however, a sort of gray area, when reproduction and genetic exchange may progressively decrease between two (or more) diverging groups within a species. It is during this time that experts may disagree (based on different species definitions) as to whether or not speciation has actually occurred. In fact, scientific names are sometimes changed as experts recognize different separated groups, even if speciation is not certain or yet agreed upon. Nonetheless, when genetic exchange can no longer occur between two or more groups, all experts (all definitions) would agree that speciation has occurred (**Figure 4.7**).

Regardless of morphological differences or unexpected mating strategies, the key piece of information in all species definitions involves the split into separate species. This split, or speciation, helps scientists to focus on which individuals are in a species and which are not. Individuals that can no longer mate, or create offspring that can mate, have lost the ability to contribute genetic information to a shared future genetic pool. But how does this split happen? To answer that, we need language about the way organisms share genetic material and the way genetic changes are passed along.

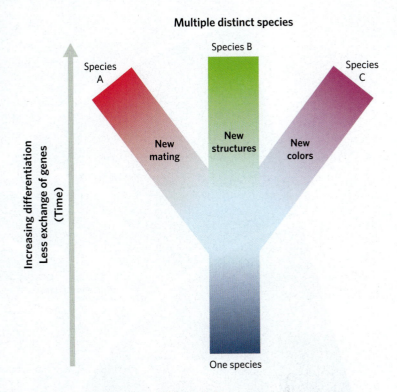

Figure 4.7 Many factors can lead to a decrease in the exchange of genes between groups within a species over time. The critical commonality among all definitions of the concept of species is that individuals of the same species must all be able to contribute individuals to the next generation. When gene exchange decreases to zero between groups of individuals, speciation has occurred.

4.3 Relating Genes to Populations

To start, we need to define the relationship between genes, individuals, populations, and species. These categories describe biological scales of observation, with genes being the smallest unit and species the largest. In addition, they nest within each other like a continuation of Figure 4.2B, with smaller categories below species. All species are made up of one or more populations; all populations are made up of individuals; and all individuals are produced according to the blueprint encoded in their genes. This nesting hierarchy is important to the essential elements of speciation discussed earlier because it points to the way genes lead to changes in individuals, which lead to changes in populations, which lead to changes in species, which are necessary for speciation (**Figure 4.8**). We explore how these changes happen in the paragraphs that follow. As you read, try to think about how the specifics at each level feed into changes that might affect the process of creating new species.

A **gene** is a sequence of deoxyribonucleic acid (DNA) that contains the code for a biological molecule with a particular physiological or behavioral function. DNA is a double-stranded molecule of bonded nucleic acids, packed into tight coils called

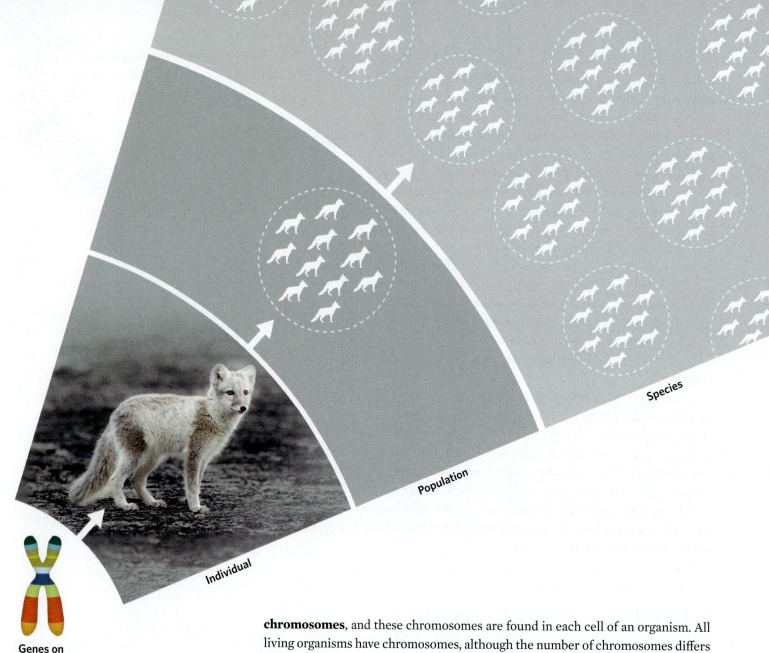

Genes on chromosomes

Individual

Population

Species

Figure 4.8 The hierarchical levels depicted in Figure 4.2B are the essential elements of most classification systems, but hierarchical levels in biology also continue below the species level. Combinations of genes produce unique individuals; collections of interbreeding individuals constitute a population; and all the populations of individuals contributing to a closed breeding set make up a species.

chromosomes, and these chromosomes are found in each cell of an organism. All living organisms have chromosomes, although the number of chromosomes differs among species. Prokaryotes generally have just one circular chromosome, but most eukaryotes have multiple linear chromosomes in each cell. Finally, different cells may have different *ploidy*, meaning different numbers of copies of each chromosome. For example, humans are *diploid*, meaning that we have two copies of each of our 23 chromosomes in most cells. Our reproductive cells, eggs and sperm, are *haploid*, meaning they have just one copy each and combine to create an offspring with two copies of each chromosome, one from each parent.

But let's back up. If a gene is a particular section or sequence of DNA, how does it code for molecules with a particular function, and why does such coding matter? Basically, each gene produces a different protein, and those proteins direct our cells, our development, our appearance, and our functions. In other words, genes lead to *traits*. The collection of genes in one individual, and how they are expressed, dictate how that individual looks, functions, and behaves.

Every species, or rather every individual member of a species, has a characteristic number of genes. For example, each human has about 24,000 genes, each

A picture of a human chromosome post-replication taken with a scanning electron microscope

A gene is a portion of a chromosome that codes for a biological molecule.

Note that, while both chromosomes have the same genes, they can have different forms of the genes called alleles.

An allele, therefore, is an alternate form of a gene.

Pair of homologous chromosomes

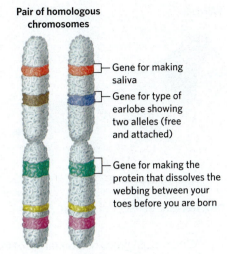

Gene for making saliva

Gene for type of earlobe showing two alleles (free and attached)

Gene for making the protein that dissolves the webbing between your toes before you are born

Figure 4.9 The image on the left is an actual picture of a human chromosome. The schematic drawing on the right demonstrates the concepts of homologous chromosomes, genes, and alleles.

mouse about 22,000, every rice plant about 51,000, and every *E. coli* bacterium about 4,000. The complete set of genes is called the *genome*. All the information needed to create a single individual of a particular species is held within the genome. However, many genes in the genome come in several different versions, with slight differences in the exact arrangement and type of nucleic acids. We call these alternate versions of genes **alleles**, and different individuals can have different combinations of alleles for a particular gene.

In general, genes code for a trait (for example, type of earlobe), and alleles produce variants in the trait (attached or free; **Figure 4.9**). Each variant in the combination of alleles is a **genotype** and codes for a specific observable physical, developmental, or behavioral trait, which is called the **phenotype**. In humans, we generally have two alleles (one on each chromosome) for each gene. There may be far more than two alleles for a particular gene across a population, though. The allele differences lead to genotype differences and create variation in the observable traits (phenotypes) of individuals. This is why no two humans (or other advanced life forms) look exactly alike or behave exactly the same way. You probably learned about these differences in introductory biology. Although the connection between alleles and phenotypes can be extremely complicated, a simple example can help us understand the basics.

Consider how alleles and genes lead to phenotypic trait differences in pigeons. Pigeons are, actually, one of the **taxa** (a general term referring to biological groups at a variety of different organizational levels from subspecies up to orders; the singular is *taxon*) that Charles Darwin focused on in *On the Origin of Species*, where he laid out his theory of evolution (1859). Humans have bred pigeons for centuries, and the practice was very common in Darwin's day. In writing his book, he could discuss pigeon traits, and most of his readers at the time could visualize his descriptions. One useful aspect of extensive pigeon breeding is that breeders are aware of how parent pigeons with different traits can mate to produce a variety of traits in their offspring. What the nineteenth-century pigeon enthusiasts could not do was identify pigeon DNA. Now we have the ability to identify specific pigeon alleles.

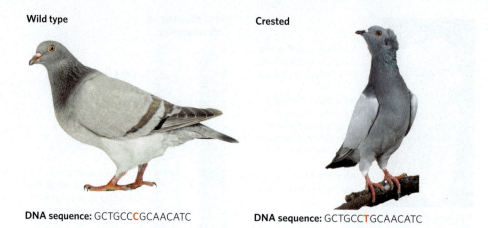

Figure 4.10 Rock pigeons (*Columba livia*) usually have a smooth head (wild type), but some domesticated rock pigeons have a tuft or crest at the back of their head (crest). A single gene controls the presence of the crest in these pigeons, and the crest allele is recessive (Shapiro et al. 2013).

Wild type

Crested

DNA sequence: GCTGCC**C**GCAACATC

DNA sequence: GCTGCC**T**GCAACATC

Wild rock pigeons (*Columba livia*) tend to have smooth heads with no feathery tufts, but domesticated rock pigeons sometimes have a feathery crest (**Figure 4.10**). Michael Shapiro and a team of researchers at the University of Utah looked at the DNA for 42 pigeons, some with crests and some without. They found that only pigeons with two copies of a particular allele had a crest (Shapiro et al. 2013). This observation suggests that the presence of a crest is controlled by one gene and that the crest allele is **recessive**, meaning the trait shows up only if the pigeon inherits the crest allele from both parents (**Figure 4.11**). Individuals who are **homozygous** for a particular gene have two alleles that are the same; in individuals who are **heterozygous**, the two alleles differ. **Dominant** alleles code for their trait even if there is only one copy of the allele (Figure 4.11). Shapiro and his colleagues tested their hypothesis by looking at 130 additional pigeons and found that only those with two copies of the crest allele had a crest. The pigeons with the crest allele can pass that allele on to their offspring and continue the trait in future generations.

Our pigeon example provides a good foundation for understanding how genotypes determine phenotypes because it is simple and clear. However, many genes have more than just two possible alleles, and many traits are controlled by more than just one gene. Those additional alleles and genes lead to more variation, which we can also think of as more genetic diversity and more trait diversity across individuals. This leads to thinking about collections of individuals.

A collection of individuals in one place who can mate together and, thus, pass on their genes (even if they never actually do so) is called a **population**. The mating part (or passing-on-the-genes part) of this definition is critical. By focusing on mating individuals, we avoid lumping together morphologically similar organisms, such as dolphins and sharks, and ensure that only individuals with a chance of sharing genes and passing them down to the next generation are included in the population. As we saw in the discussion of species, the ability to pass genes to future reproducing individuals is a critical aspect of all modern definitions of species. So, a population is a collection of individuals of the same species located in the same place.

Populations are made up of individuals, and all the allele combinations within the individuals lead to an array of unique traits across the population. Because most species have multiple populations spread across a landscape of habitats and biomes, different traits may become common in some populations but not in others. These population-level trait differences reflect differences in allele frequencies in the populations.

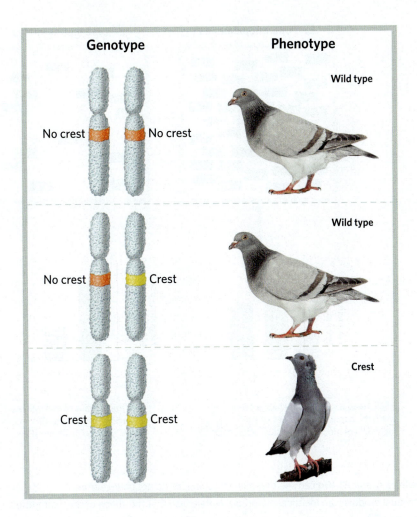

Genotype	Phenotype
No crest / No crest	Wild type
No crest / Crest	Wild type
Crest / Crest	Crest

Figure 4.11 Each pigeon has two copies of every gene by receiving one copy from each parent. A recessive allele only leads to a phenotypic trait if both copies of the gene are that allele. The birds with two copies of the same allele are homozygous (*homo*- means "same" in Greek). Pigeons with one copy of each allele have a heterozygous genotype (*hetero*- in Greek means "the other of two" or "different"), but their phenotype is indistinguishable from the phenotype of the pigeons who are homozygous for the allele for no crest.

But how do these population-level differences occur, and why? The quick answer is **evolution**, which we define as a change in allele frequencies (see next section) within a population over time. This concept is one of the key elements of this chapter. Next, we will explain the mechanisms that produce evolutionary changes, and in the rest of the chapter we will explore the way evolution can alter populations through time and lead to speciation.

4.4 Mechanisms of Evolution

In the discussion of each mechanism of evolution described in this section, we track changes in **allele frequencies** through time. Allele frequency is the number of occurrences of each allele that one would expect to find in individuals in a population. For example, for a gene—let's call it gene A—we might expect to find the allele A_1 three times in 1,000 individuals in the population. Often, evolutionary biologists consider allele frequencies for all the alleles of a gene simultaneously and represent them in frequency distributions, or graphs of the frequencies of each allele (**Figure 4.12**). Frequency distributions for alleles will always have the same shape as a graph of proportions of a population with the same alleles. Because of this similarity, biologists will often refer to allele frequencies without

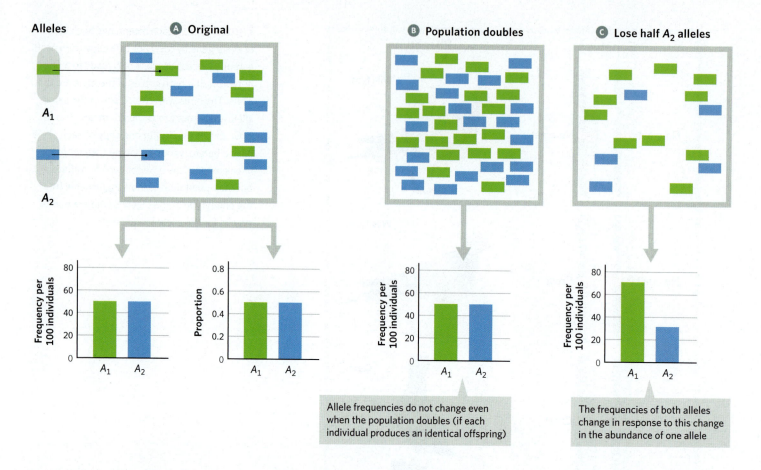

Figure 4.12 Each square represents a population with alleles A_1 and A_2. We can see that allele frequencies do not change from part Ⓐ to part Ⓑ when the population doubles (if each individual produces an identical offspring). Notice also that the shape of the distribution looks the same whether we look at frequencies (left graph for part Ⓐ) or proportions (right graph for part Ⓐ). In part Ⓒ, we have the same number of A_1 alleles as in part Ⓐ, but only half as many A_2 alleles as in part Ⓐ. The frequencies of both alleles change in response to this change in the abundance of one allele.

specifying the number of individuals in the population, with the understanding that the number of occurrences within the population is the same as frequency in the population. For that reason, many of the y-axes on our graphs will be labeled simply "Frequency" without specifying frequency per 100 or per 1,000.

There are only a few ways that allele frequencies can change over time. We can add new alleles or genes, remove alleles, or change the proportion of alleles that are currently in the population (**Figure 4.13**). Any of these changes qualify as evolution, and Figure 4.13 offers a visual representation of the process. As conceptual models go, this one is quite simple, but it is a good starting point. The most important insights come from thinking about the consequences of these changes for populations and species. In the sections that follow, we examine four mechanisms that produce evolutionary change: mutation, gene flow, genetic drift, and natural selection. All of these mechanisms alter allele frequencies. The first two can also add genes or alleles to the population, and the second two can remove alleles in potentially important ways (**Table 4.1**).

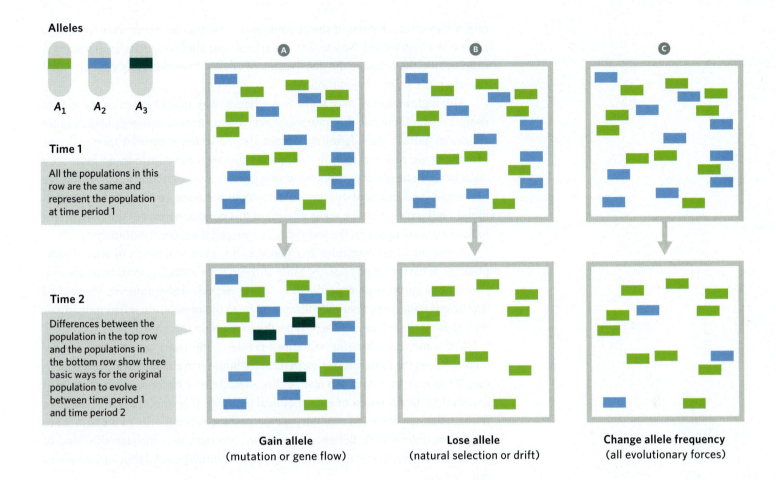

Alleles

A_1 A_2 A_3

Time 1

All the populations in this row are the same and represent the population at time period 1

Time 2

Differences between the population in the top row and the populations in the bottom row show three basic ways for the original population to evolve between time period 1 and time period 2

Gain allele
(mutation or gene flow)

Lose allele
(natural selection or drift)

Change allele frequency
(all evolutionary forces)

Adding Genetic Variation to Populations: Mutation and Gene Flow

The two best-known and best-studied ways to add new genes to populations are mutation and gene flow. **Genetic mutations** occur when there are errors in the replication of DNA within a cell. Every once in a while, the cellular machinery that creates new DNA strands makes a mistake, and one or more nucleic acids are either deleted from the strand or inserted. Mutations generally happen at a slow

Figure 4.13 Evolution occurs when allele frequencies change over time in a population. The squares represent populations of individuals with different alleles (represented by different colors). All the populations in the first row are the same and represent the population at time period 1. Differences between the populations in the top row and the populations in the bottom row show three basic ways for the original population to evolve between time period 1 and time period 2: **A** A new allele and resultant genotype can enter the population, which can happen through mutation or gene flow. **B** The population can lose an allele and genotype through genetic drift or natural selection. **C** All evolutionary forces can change the allele frequency; this specific example shows a loss of alleles, so only natural selection or genetic drift could produce this example of change.

Table 4.1   Mechanisms of Evolution

Mechanism	Effect on genetic variation	Effect on divergence of isolated populations
Genetic mutation	Increases	Increases
Gene flow	Increases	Decreases
Natural selection	Can reduce but cannot add	Increases
Genetic drift	Generally reduces	Increases

rate, although some parts of the genome and some species experience mutations more often than others. Notice that mutations alter allele frequencies in a population because they change the distribution of alleles. Therefore, any genetic mutation is a mechanism of evolution.

Although mutation may seem like a handy way to add genes that code for useful traits, remember that mutations are genetic errors, and organisms cannot choose to mutate; they definitely cannot choose which mutations they undergo. Mutations are rarely useful; in fact, most mutations have little to no impact on traits at all. In nearly every species whose genome has been fully documented, we find that a good chunk of the species' DNA does not code for traits or functions. A mutation (involving one or many alleles) in this noncoding part of the DNA seems to have no real impact on the individual, so we call it a *neutral mutation*.

A fraction of all mutations in a genome, however, will result in altered morphology, behavior, or physiology. Most of these trait-changing mutations are detrimental to survival or reproduction. These *detrimental mutations*, also called *deleterious mutations*, generally disappear from the population when individuals with these traits fail to survive or reproduce (more on this shortly).

Yet, some trait-changing mutations create useful traits, and these *beneficial mutations* are the most interesting in terms of long-term changes in allele frequencies. Beneficial mutations can lead to tiny alterations or big ones, from changes in morphology to alteration of a biochemical pathway. It is important to remember, though, that mutation is a random process, and the vast majority of mutations are neutral or detrimental. Beneficial mutations are rare. It is through this kind of process, however, that genetic mutations add potentially useful trait variation into populations.

The other avenue for adding novel genes to a population is **gene flow**. The term *gene flow* evokes the image of genes floating down a stream or flying through the air, but *flow* is just a poetic word for the movement of alleles from one population to another. Gene flow can happen with the ordinary movement of individuals or even just gametes (reproductive cells from males or females) that carry alleles that may be new to the population. These incoming alleles usually arrive inside individuals who walk, crawl, swim, fly, or otherwise disperse into the population and bring their genes with them (**Figure 4.14**).

Despite this mundane aspect of gene flow, there is some truth to the poetry, and sometimes the flying-gene image is not far off the mark. Readers with hay fever are probably acutely aware of pollen flying through the air in the spring. This pollen carries male gametes or sperm, and those gametes carry genes, with the wind sometimes transporting them long distances. Many aquatic organisms also release gametes into the water, where they can traverse great distances on currents. As long as the moving gamete is united with another gamete and eventually produces viable offspring in the new population, the DNA it transports introduces alleles that may code for novel traits in the population.

Removing Genes from a Population: Genetic Drift and Natural Selection

The process by which alleles are passed from generation to generation is influenced by chance events, such as the random combination of parental alleles into an offspring. In many species, mating partners are random rather than chosen, as

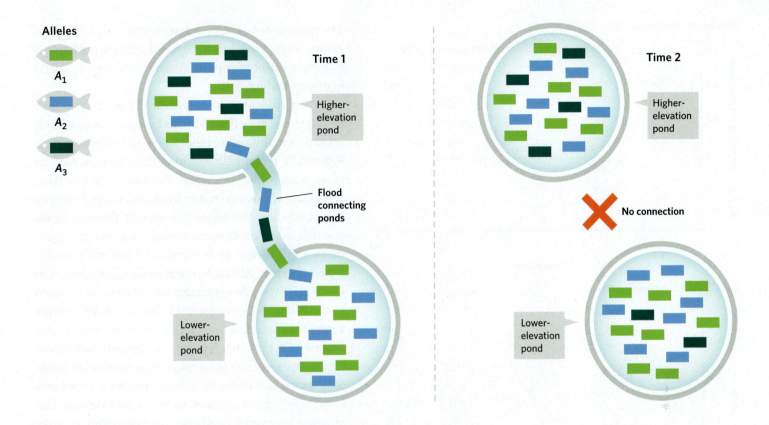

Alleles

A_1

A_2

A_3

Time 1

Higher-elevation pond

Flood connecting ponds

Lower-elevation pond

Time 2

Higher-elevation pond

No connection

Lower-elevation pond

is often the case for species that broadcast their gametes into the wind or water. There is also a random element associated with living and dying that has nothing to do with which alleles or traits an individual possesses. Suppose a tree branch falls on the head of one identical twin, and that twin dies, while the other survives. Because they are identical twins, their genes and traits differ very little, but one was rather unlucky. Their genetic makeup had no influence on which one survived.

In the case of both random mating and random death, haphazard events can remove genes from a population. This chance loss of alleles and genes is called **genetic drift**. Genetic drift clearly influences the allelic and trait distribution of a population, making it the third mechanism of evolution.

Genetic drift can explain the loss of neutral alleles (see previous section), but it can also explain the random loss of beneficial alleles. If we change the example of identical twins to an example of a family in which each individual has a unique and beneficial mutation, then if a tree falls on any member of the family, it would remove one beneficial allele from the population. In most species, it would be far-fetched to imagine that each member of a family would have their own unique allele found in no other member of the population, but the mental image illustrates that a random event with no connection to individual performance can remove beneficial alleles. Beneficial alleles are likely to stick around in a population (for reasons to be discussed), so we generally see the effects of genetic drift on neutral variation.

An important aspect of genetic drift is that chance events have more influence in small populations than in large. For example, it would be highly unlikely that a falling tree would kill all 50 individuals holding a particular allele or trait.

Figure 4.14 Gene flow, one of the mechanisms of evolution, involves the movement of genes or alleles from one population to another. The incoming genetic material alters gene or allele frequencies and may introduce novel alleles; either of these actions leads to evolutionary change. In this example, flooding allows individuals to move between two previously distinct, unconnected ponds. Different colors represent different alleles that confer different genotypes. Any newly transported alleles may stay even after the connection between the ponds disappears.

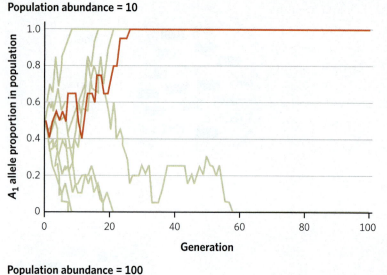

Population abundance = 10

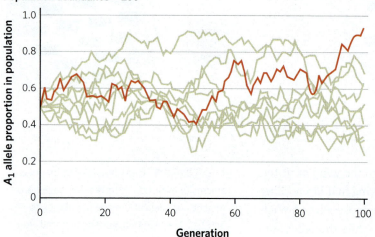

Population abundance = 100

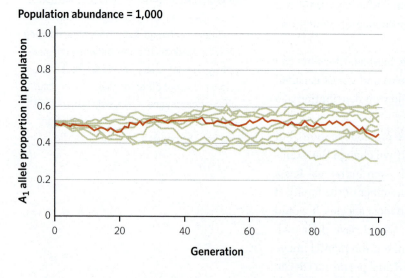

Population abundance = 1,000

The influence of drift, therefore, is amplified when population abundance is low (**Figure 4.15**). Unpacking this important effect of population abundance can be useful in understanding drift.

Many students find genetic drift less intuitive than mutation and gene flow, perhaps because it is less familiar and more difficult to represent with a picture such as Figure 4.14. It may help to think about a familiar random event, such as flipping a coin. We know that any (fair) coin is just as likely to land on heads as on tails. This even probability does not mean that we will alternate heads and tails as we flip a coin repeatedly. Nor does it suggest that we will always see five heads and five tails if we flip a coin 10 times. Although the *probability*, or chance, of heads is 50%, we do not expect 50% of our coin tosses to be heads if we only toss the coin a few times. We may get a run of heads or tails and end up with a few extra in one category. On the other hand, if the coin is tossed 10,000 times, we expect the percentage of tosses that are heads to be very close to 50%. As the total number of tosses gets larger, any random runs of heads or tails balance out, bringing the overall numbers closer and closer to 50%. Even if there are still a few extra in one category, those extras are a tiny fraction of 10,000 tosses.

Now let's relate the flipping of a coin to reproduction in a population. Suppose we have a population with two neutral alleles—that is, alleles with no discernible benefits or negative effects. Because the alleles are neutral, offspring of this population should all be equally likely to survive, regardless of their allele combination (heterozygous or homozygous for one or the other). We can start with equal frequencies of each allele and randomly allow individuals to encounter each other and recombine their alleles. We can also allow the survival of the offspring to be random so that just half survive, and the population never grows or shrinks. The process repeats with each generation, yielding the new proportions of each allele. Over time, small populations will quite quickly eliminate one allele and "fix" the other by chance (Figure 4.15). An allele is considered *fixed* when it is found in every individual in the entire population. In this example, *fixation*

Figure 4.15 The effect of genetic drift is stronger in smaller populations. The three graphs show computer simulations of genetic drift in diploid populations of size 10, 100, and 1,000 over 100 generations. Each simulation starts with the two alleles (A_1, A_2) at equal frequencies with random mating, no mutation, and no migration. Each graph shows 10 different runs of the computer simulation, with one run shown as a dark line for visibility. In each case, drift causes allele frequencies to fluctuate over time, but the fluctuations are far more dramatic in small populations. In every instance when there are 10 individuals in the population, one or the other allele becomes fixed (i.e., reaches 100% frequency or a proportional abundance of 1.0) within the 100 generations (Bergstrom and Dugatkin 2016).

occurs for one allele but not the other, not because one allele is better than the other but because small populations are more affected by chance. Larger populations, on the other hand, will maintain close to even proportions of the alleles for much longer but will eventually fix one allele if no other evolutionary mechanism interferes (Figure 4.15).

What happens when the alleles are not neutral? As you know, mutations can be beneficial or detrimental, so we would expect different outcomes if different alleles lead to phenotypes with different probabilities of reproducing or surviving. Once we start to consider these effects on survival and reproduction, we enter the realm of **natural selection**, the last and most important mechanism of evolution for understanding ecological processes.

Natural selection is the differential survival and reproduction of individuals in a population in response to biotic and abiotic factors in their environment. That means the environment (made up of biotic and abiotic factors) is "selecting" (without any intent or choice) which individuals survive and reproduce and which ones do not. Selection occurs because resources are limited and cannot support all individuals. In a world of limited resources, some organisms have traits that allow them to acquire these resources more effectively than other organisms. The revolutionary element of Darwin's theory was that natural selection not only leads to organisms that, over time, are able to perform better in their environment, but it can also create entirely new species. This mechanism is still one that fascinates and confuses many students of biology, and it plays a much larger role in the change of species than the other mechanisms. We, therefore, give it more attention than the other three.

Any genetically controlled trait that increases the chance that an individual with the trait will survive and reproduce is likely to be passed down to that individual's offspring. The term **fitness** describes the genetic contribution individuals make to the next generation, and any trait that increases survival and reproduction increases an organism's fitness. Any trait that inhibits an individual's survival or reproduction is less likely to be passed to offspring and, therefore, lowers its fitness. Deleterious traits will, therefore, appear less frequently in the next generation simply because there are few, or no, offspring ever created by individuals with the trait. This process can produce a change in allele frequencies across generations, with genes that code for less beneficial traits dropping out of the population and genes that code for beneficial traits becoming more common (Figure 4.16). Deleterious alleles will eventually be lost from the population when the last individual with the allele dies without reproducing.

The idea of evolutionary fitness reflects the fact that unique genomes lead to individuals with different probabilities of survival and reproduction. Ecologists often focus on **relative fitness**, which describes an individual's contribution to the gene pool of future generations relative to the contribution from other individuals in the same population. In a simple sense, then, individuals with higher fitness leave more viable, fertile offspring than individuals with lower fitness. In practice, studying the actual fitness of real organisms is a surprisingly complicated matter.

Example of Natural Selection: St. John's Wort Let's explore a real-life example of natural selection to help visualize this better. But first, we need a little background. Chapter 3 discussed the role climate plays in the physical environment

Figure 4.16 These graphs illustrate how allele frequencies change with natural selection. The alleles are arranged on the x-axis from lower performance or fitness on the left to higher individual performance and, thus, higher fitness on the right. Ⓐ The first graph shows allele frequencies (which will always match the proportion of a hypothetical population with each allele) before selection; Ⓑ the second graph shows allele frequencies after a selective force was applied.

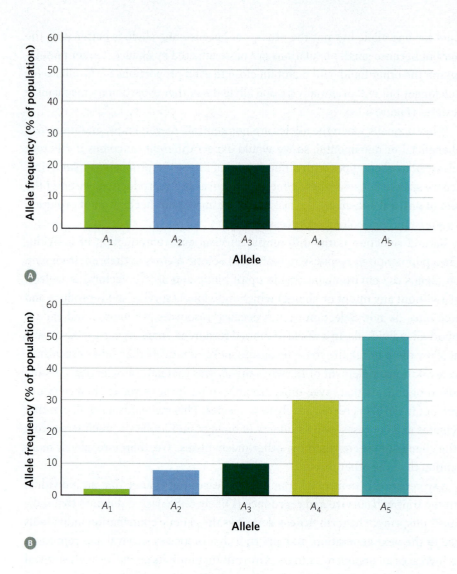

and in the creation of broadly recognizable patterns of terrestrial vegetation (i.e., biomes) around the globe. We started with large-scale effects and then showed ways that these effects might combine with smaller-scale effects to produce localized patterns. If we narrow the focus even more and look at a single geographic location, we can see that the local climate affects the seasonal timing of activities for most organisms in that location. In our own lives, of course, this makes sense. Most of us would not throw on a bathing suit and go swimming in the middle of a winter snowstorm, nor would we put on a winter jacket in the face of a summer heat wave. Instead, we tend to change our clothing and activities as the seasons change, and many terrestrial organisms also change their activities and behaviors seasonally. We call the study of the timing of these seasonal shifts in activity **phenology**.

For plants in temperate environments, one important aspect of phenology is the seasonal appearance of leaves or flowers. Flowers lead to the creation of fruits and seeds, which are obviously critical for reproduction as they pass on alleles (and the associated traits) to the next generation. A plant that can start flowering early may be able to reproduce for a longer portion of the growing season, thus producing more offspring (and more copies of its own genes) in the next generation. Flowering too early, though, could be a costly mistake if the early months of the growing season happen to include freezing temperatures or a killing frost, leading to flower death.

This means that if we gathered a plant from the Mediterranean scrubland biome in Europe and planted it in the temperate forest biome of New England in the United States, flowering time could be a problem. New England stretches from about 41° north latitude in Connecticut to 45° north latitude in Maine. This latitudinal range is approximately equivalent to the area of Mediterranean scrubland found from northern Spain to the southern parts of France. In terms of annual light availability, we would expect very similar patterns in the New England and European locations because of their latitudes. But in terms of annual temperature and water availability, we should expect big differences. These areas of Spain and France have a Mediterranean climate with wet, rarely freezing winters and warm, dry summers. This contrasts starkly with the bitterly frozen winters and warm, wet summers found throughout the temperate forest biome in New England. The arrival of spring in these two locations can also be very different climatically. It is not unusual to have snow on the ground in parts of New England into April and to have nighttime freezes into late May or early June. Few of New England's human residents would turn down a chance to be in Spain in April, as it is generally warmer and milder in the Mediterranean region.

For a plant, different climatic patterns like these would pose significant challenges. There is a lot of experimental evidence to suggest that changes in day length and light availability are the critical cues for plants to break dormancy and prepare for spring growth. A perennial plant's buds start to swell in the spring before leaves unfurl, at the same time the seeds of annual plants take on water and begin to germinate. The similar latitudes of New England and Mediterranean Spain and France produce similar light cues that would likely lead to emergence in freezing temperatures for a Mediterranean plant newly arrived in New England. In addition, the lack of summer rainfall in Mediterranean climates means that plants there often break dormancy *very* early (often in winter) to take advantage of winter rains. For all these reasons, a Mediterranean-adapted plant should not do well in New England. If a plant from the Mediterranean were to survive in New England, it would need to evolve phenological traits that fit its new environment.

An example of a Mediterranean plant that has survived an introduction to New England is St. John's wort (*Hypericum perforatum*), a medicinal plant brought by European settlers to the United States in the nineteenth century. St. John's wort is an herbaceous (meaning nonwoody) perennial plant, native to much of Europe and northern Africa (**Figure 4.17**). To document if and how St. John's wort experienced evolution in flowering phenology in New England, scientists would ideally use long-term observations of its flowering period in one location. Unfortunately, these data do not exist. However, since its introduction in New England, researchers have collected specimens of St. John's wort and stored them in herbaria. An *herbarium* is like a museum that holds specimens of plants collected across many locations. As part of a thesis project, Erin Coates-Connor and Martha Hoopes visited university herbaria in Massachusetts and Connecticut (parts of southern New England) and recorded the date of collection for every specimen of *H. perforatum* collected from 1871 to 1999 (unpublished data, 2009). Although not all of the herbarium specimens were flowering when collected, the collection shows a dis-

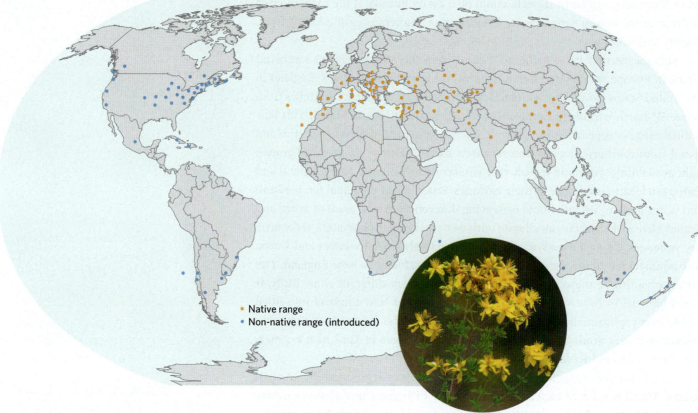

Figure 4.17 The map shows the native and non-native ranges of common St. John's wort (*Hypericum perforatum*, inset). The orange dots indicate its native range; blue dots indicate where it has been introduced (Robson 2002). Dots that appear to be in the ocean indicate islands where the species is present. St. John's wort is invasive in all Mediterranean scrubland biomes outside the Mediterranean regions of Europe and Africa, but it is not invasive in New England, where the climate conditions are quite different. Invasive plants spread rapidly and often have negative effects on native species.

- ● Native range
- ● Non-native range (introduced)

tinct change in flowering period from the end of the nineteenth century to the end of the twentieth century (**Figure 4.18**).

It is important to ask how Figure 4.18 relates to natural selection. It looks quite a bit like a graph for allele frequency, but a flowering date is not the same thing as an allele. On the other hand, plant evolutionary biologists would say that flowering-time phenology is very **heritable**, meaning that the phenotype for flowering time corresponds quite tightly with a genotype. Therefore, we can think of each collected plant in the herbarium collection as an individual with a genotype that allows it to flower at a particular time in the spring. Because the bars represent all the flowering individuals in the herbaria in New England, the distribution of bars represents a frequency of flowering dates, like the frequency distributions that we showed earlier. Remember that the shape of frequency distributions will match the shape of proportional distributions (Figure 4.12), and taller bars show more common flowering dates, while shorter bars show less common flowering dates. A single plant's flowering period might stretch over several weeks, so there is not a one-to-one correspondence between the amount of flowering and a frequency of alleles. Nonetheless, the graph allows us to visualize the range of phenotypes for St. John's wort, and it is easy to see that this range has shifted through time.

When St. John's wort first arrived in New England, it flowered in late May (Julian day 140) and ended flowering by late August (Julian day 240), as would be expected given that it is native to a European Mediterranean climate. However, by the latter half of the twentieth century, the flowering time in New England had extended past the summer months into late October (Julian day 300). Flowering late in the summer might be hard in a Mediterranean climate with no summer rainfall, but extended flowering in New England allows this plant to take advantage of the late summer's warmth and water and, therefore, produce more seeds. More seeds translate into more copies of genes in the next generation. If those seeds lead to viable, fertile adult plants, then the ability to flower over an extended period should increase the copies of the later flowering genes in the next generation. With all the successful copies growing in the next generation, the genes for late-flowering ability would become more common in New England populations over time.

There are clear limitations to what we can infer from these data. First, we expected to see that the earliest flowering might occur later in the year, but there are so few data points in the earliest parts of the flowering season that we cannot make that distinction. Also, at present there is no identified genetic link between a change in flowering time and a change in allele frequencies. If that link exists, though, we can see a clear effect of flowering and seed production on individual plant fitness. There are also issues with data gleaned from 150-year-old herbarium specimens. We have no way of knowing why a collector in 1875 picked and preserved a particular plant, but it is unlikely that the selection was completely random. The collector may have preserved only the biggest and the prettiest plants or the ones with the best flowers (although that particular bias might be okay in this case). Lastly, people do not collect specimens as much now as they did decades ago, and St. John's wort is no longer as rare as it was right after its first introduction. Nonetheless, this example is remarkable in its simplicity. It shows fairly cleanly how climatic forces might shape the distribution of traits in a population and thereby illustrates the straightforward logic behind natural selection.

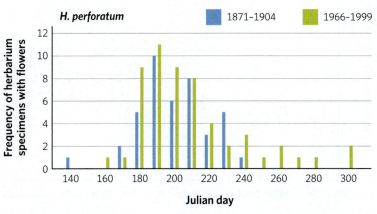

Figure 4.18 This graph shows the frequency distribution of flowering dates among herbarium specimens for common St. John's wort (*Hypericum perforatum*) in southern New England, based on herbarium records. The Julian day refers to the days of the year in order from 1 to 365. In non-leap years, Julian day 140 is May 20; day 240 is August 28; and day 300 is October 27. Note that the range from 160 to 300 (June through October) represents most of the predictably snow-free days. Even though some of the highest bars are still in the same place, notice that the flowering range at the end of the nineteenth century (blue bars) started and ended earlier than the flowering range at the end of the twentieth century (green bars). This indicates that the average flowering period has shifted toward a later date in the plants collected more recently (Coates-Connor and Hoopes, unpublished data, 2009).

4.5 Adaptive Evolution

The St. John's wort example of natural selection is handy because it helps demonstrate the way that a strong selective force can lead to observable phenotypic changes resulting in a population that is more fit, or *adapted*, to its environment. In fact, **adaptation** is the hallmark of natural selection. By *adaptation*, we mean two things. First, an adaptation is a trait that has changed and now represents a phenotype that helps an organism to survive and reproduce in a particular environment. The variety of adaptations in the world is astonishing, including bird calls that help individuals locate mates, fake eye spots on butterfly wings that help them evade predators, fur on the insides of ears to protect from the cold, and so many other examples that we could never name them all.

Second, adaptation is the process of evolution that leads to an adaptive trait's becoming more common or even fixed in a population. The extended flowering time of St. John's wort in New England, for example, is the adapted trait, and the process of changing allele frequencies that led to that shift in flowering time is the adaptive process. The fact that biologists use the same term for both of these meanings can be confusing, but what is meant by the term is generally clear from context. Natural selection is the only one of the four mechanisms of evolution that can adapt

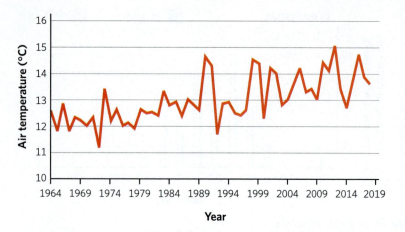

Figure 4.19 This graph represents the annual average high temperature at Harvard Forest (Boose and Gould 2019), a university field station in Petersham, Massachusetts, near the field sites where Coates-Connor and Hoopes conducted their research. Note that the average high temperature increased over a 50-year period stretching from the middle of the twentieth century into the beginning of the twenty-first century. This change in temperature led Coates-Connor and Hoopes to expect to see earlier flowering of native species as snows disappeared sooner.

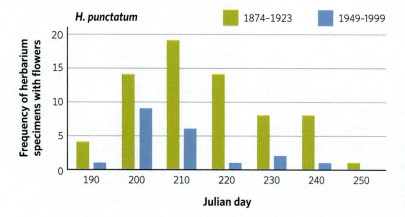

Figure 4.20 This graph shows the frequency distribution of flowering dates among herbarium specimens for native spotted St. John's wort (*Hypericum punctatum*) during a 50-year period starting in 1874 and a 50-year period starting in 1949. There are fewer data points for the second time period (n = 68 for the first time period, and n = 20 for the second), but there is a statistically significant shift in the date at which collected plants were flowering. That shift is somewhat hard to see, but notice that the blue bars generally represent a larger portion of the green bars as you move to the left on the graph. This suggests that the flowering time has moved to the left (i.e., earlier) in the more recent time period (Coates-Connor and Hoopes, unpublished data, 2009).

an organism to its environment, and biologists, therefore, sometimes refer to evolution by natural selection as **adaptive evolution**.

Given how important adaptive evolution can be for a species, it is interesting to note that it can be hard to observe. We described a detectable shift in flowering time for a plant when confronted with a very different climate, but not all selection events are so consistent or distinct through time or across space. Other selective forces, such as droughts, competition, or sublethal diseases, may take a while to leave a mark on the frequency of genes and traits in a population. Or some events may work in one direction, while other selective forces work in different or even opposite directions. It is also fairly common for selective forces to occur so slowly that change is hard to detect during the time span of data collection.

Directional, Stabilizing, and Disruptive Selection

The idea that a selective force can work in two different directions can be seen in the case of St. John's wort. As discussed in Chapter 3, the Earth's average temperature is increasing as a result of climate change. It is only because St. John's wort came from such a different environment that its flowering time shifted to later in the season in New England. In fact, Coates-Connor and Hoopes noted that the temperature in the areas of their study (see Chapter 10) had warmed substantially in the latter half of the twentieth century (**Figure 4.19**). They chose their time periods specifically to compare this warmer period to the cooler era when St. John's wort was first introduced to the region. The researchers expected the newly introduced St. John's wort to flower later, as was observed. But they also expected a closely related native New England species, spotted St. John's wort (*Hypericum punctatum*), to show a shift in flowering time to earlier dates. This shift was expected because the warmer spring temperatures associated with climate change should have made it possible for the native species (*H. punctatum*) to break buds and flower earlier in the year, as the plants were already adapted to the climate. Although they found a statistically significant shift in the direction they expected, this change was harder to see (**Figure 4.20**). There are two reasons why the native species' shift was difficult to observe. First, fewer herbarium data were available for the native species because early New England botanists were less interested in collecting local species than in collecting the newly introduced non-native species. Second, the actual shift in flowering time was more subtle than the shift in the flowering time of the non-native species.

The shifts in average flowering time for the two species are both examples of **directional selection** because in each case the trait was "directed toward" one end of the trait distribution (**Figure 4.21A**), albeit in a different direction for each species. The shift in the non-native species is easier to see and the effect is larger because the selective force was stronger but also because the population must have arrived in New England with enough genetic variation to allow for a shift in the first place.

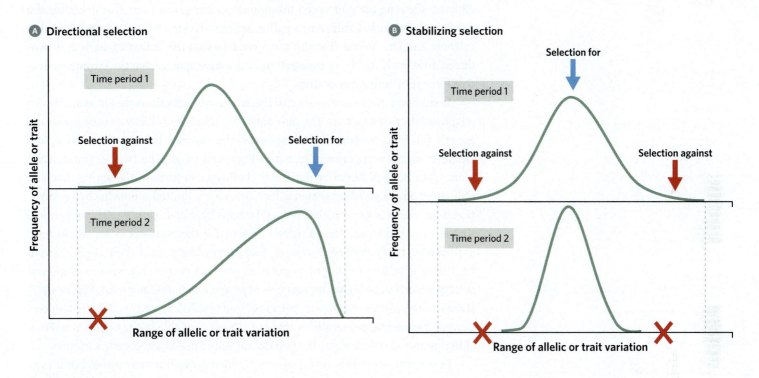

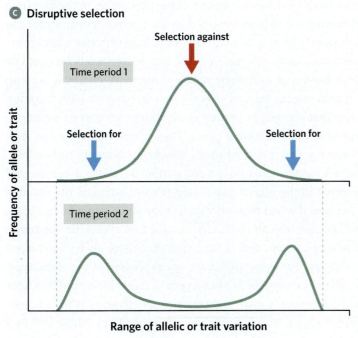

Figure 4.21 Generalized, conceptual graphs of **A** directional, **B** stabilizing, and **C** disruptive selection. For each type of selection, the top graph shows an earlier generation, and the bottom graph shows a later generation after selection. In the top graphs of each panel, blue arrows point to positive selection, and red arrows point to negative selection. Dashed lines indicate the range of variation available in the first time period to make comparisons easier. Notice that both directional and stabilizing selection can lead to loss of variation, marked by red Xs. Notice also that the range of variation never increases between the two time periods.

Let's explore the stronger selective force issue first. The non-native species has two selective pressures. Flowering too early may lead to reproductive destruction, but extended flowering into the wet New England summer and fall can *greatly* increase reproductive success. These combined selective forces shift the trait frequency toward the right side of the graph. This shift is mostly driven by the advantage of flowering later. In fact, the data are so sparse in the early season that it is hard to be sure that anything at the early end has really changed.

At the same time, the native species is moving toward earlier flowering. Selection for the earlier flowering period for the native species comes from the relatively small advantage of a slightly extended flowering period in a warming climate, allowing for more seed production in any given year. Flowering for the native species cannot shift much earlier without risking reproductive failure due to snow and ice. Notice that the same effect, extended flowering period, is produced differently for the non-native and the native species, but the stronger selective force results in a larger shift.

In addition, there is the issue of the amount of genetic variation available for adaptive selection to act on. The non-native St. John's wort likely already possessed genetic variation for later flowering due to the variable, but not often lethal, late summer conditions present in its native European range. The New England native plant confronted different evolutionary challenges in its past. Flowering time can only occur after a plant has grown sufficiently, so it is limited somewhat by germination time, whereas germination is cued by both light and warmth. A plant genome that has evolved in the New England climate for thousands of years would have experienced many very cold winters. Early germinators (and their alleles) would have been eliminated from the population generations ago. No amount of natural selection can lead to higher frequencies of an allele that does not exist. Figure 4.21A shows that a positive selective force cannot lead to a shift in trait or allele frequencies beyond the phenotypes or alleles that are already available in the population. For a shift that includes new traits, the population would need new genetic variation.

In general, with directional selection, the distribution may move, but it cannot move outside the range of variation present in the population in the first time period. The late-flowering individuals observed in the twentieth century for the non-native St. John's wort (Figure 4.18) may have come from very rare alleles present in the European populations. Alternatively, the late-flowering alleles could be the result of random beneficial mutations that became more common because they offered large fitness effects. Natural selection in combination with mutation can lead to organisms that are well fit to their environment. Of course, mutations cannot just be requested like ordering a pizza. They arrive randomly and thus appear in populations very slowly and are rarely beneficial (as mentioned earlier).

Sometimes, there is hidden variation in a population. Recall the dominant and recessive traits observed in the pigeon population (Figure 4.11) and the possible ways they can be expressed when two individuals reproduce sexually. In Figure 4.11, we defined individuals who are homozygous (the two alleles are the same) and heterozygous (the two copies differ) for a trait. Recessive alleles can "hide" in heterozygotes, meaning that these alleles are not expressed in the phenotype. However, recessive alleles can appear in phenotypes of individuals born into later generations if two heterozygotes mate and produce an offspring who is homozygous for the recessive trait. This mechanism allows traits to "show up" in a population over time, and natural selection can act on those traits when they are present. However, the genes for such traits were already in the population even if the traits they code for were previously uncommon or even absent. The reshuffling of the deck of alleles by sexual reproduction just exposed them.

Directional selection is a form of natural selection that generally makes logical sense because it moves a population toward traits that are beneficial. What happens once a population has responded to directional selection and possesses a mean, or average, trait value that fits very well with the range of conditions in an environment? At that point, it is possible to see **stabilizing selection** (**Figure 4.21B**).

Stabilizing selection acts the way its name suggests: it stabilizes the population toward the mean trait value that it already has. We might think that stabilizing selection would lead to no change in allele frequencies and thus no adaptive evolution. Stabilizing selection does often involve positive selection for a modal, common trait, but it may also involve negative selection against a range of deleterious traits. If so, it is possible to lose alleles from the tails of the trait frequency distribution (Figure 4.21B). Even without selection against the tails of the distribution, there is a change in the frequency of alleles that code for a trait as more and more individuals acquire the most common phenotype. We can see the effect of such evolution in the consistency with which humans have opposable thumbs, for example.

What if selection leads to lower fitness for the most common trait? For example, imagine a situation where having the most common trait might become detrimental to survival or reproduction if that trait is a cue to predators or if that trait is appealing to a predator that is new to the area. In either of these situations, the most common trait starts to lead to negative fitness, and we see traits at the extremes of the frequency distribution become more common (**Figure 4.21C**). This type of selection is called **disruptive selection**. Disruptive selection can lead to populations with more than one mode (or peak) in the frequency distribution of traits. Most commonly, there are only two peaks, but we discuss an example with bats later in the chapter involving three peaks.

We have posited many different changes in the environment that lead to changes in fitness and allele frequencies over time. With all of these patterns of natural selection, populations on average become more genetically and phenotypically fitted to their environment. But what if the environment does not offer consistent selective pressure? Natural selection cannot lead to a perfect fit with the environment if selective pressures change frequently because, with frequent change, no single phenotype has a consistent fitness advantage. In fact, if the environment oscillates between two or more very different states, then natural selection can lead to a significant amount of variation being maintained.

Alternatively, if the environment changes consistently and very quickly in one direction, natural selection may not be able to keep up with it. Remember that natural selection can only act on heritable variation present in the population. If the environment moves beyond what the available genetic and phenotypic variation can address, natural selection alone cannot phenotypically fit the population to the new, extreme environment. An influx of new genetic variation from mutation or gene flow would be necessary to provide more extreme phenotypes that may lead to successful survival and reproduction under the new conditions. Situations like this are increasingly likely to occur as the world's biota experiences the dramatic effects of anthropogenic climate change.

Figure 4.22 Male West African fiddler crabs (*Afruca tangeri*) have one very large claw. They do occasionally use the claw for fighting, but the most prominent use is to attract mates. Males wave their claws to attract females, and females tend to mate with males that have the largest claws.

Sexual Selection

If traits can confer important advantages in terms of fitness, why wouldn't all individuals look approximately the same regardless of sex? Can natural selection produce or amplify the differences between the sexes of a species?

These questions posed a serious difficulty for Darwin during the formation of his theory of natural selection. The vast array of sexual ornamentations, structures, vocalizations, and elaborate mating rituals seemed incredibly dangerous for the individuals that possessed them, as it made them more likely to be eaten as prey. As a result, Darwin gave extra consideration to understanding how natural selection allowed such "costly" traits to remain in the population. Even more puzzling was the idea that natural selection may have produced such traits. This led him to suggest that a special form of natural selection, called **sexual selection**, might lead to these traits (Darwin 1871). Sexual selection can be defined as selection for traits that increase mating success rather than survival. Darwin suggested that sexual selection might lead to fairly large differences in phenotypes between males and females (Figure 4.5).

Often the beautiful plumages of birds (see Figure 4.5A), the ridiculously large claws on some fiddler crabs (**Figure 4.22**), and the elaborate mating behavior of some mammals can be attributed to sexual selection. In some species, females choose mates with extreme versions of these traits, leading to more reproduction by males with the traits and, therefore, more copies of genes for extreme traits in the next generation.

Females will often select for exaggerated traits that have no obvious direct survival benefit. In such cases, it may be that the costly, visible traits act as signals for useful, hidden traits. If the useful but hidden traits show up in the offspring and help them to successfully reproduce, the frequency of both the visible and less visible traits should increase over generations. For example, there is evidence that the size of a peacock's tail (a peacock is a male peafowl) and the number of eye patterns on the tail are correlated with larger juvenile offspring and higher offspring survival (Petrie 1994; **Figure 4.23**). In general, sexual selection can lead to some fairly odd physical and behavioral traits.

Figure 4.23 **A** Peacock (left) and female peafowl (right). Males with more abundant eye spots in their tail feathers had **B** healthier, heavier chicks, and **C** offspring with higher survival rates two years after birth (Petrie 1994).

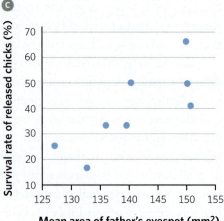

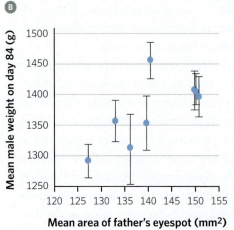

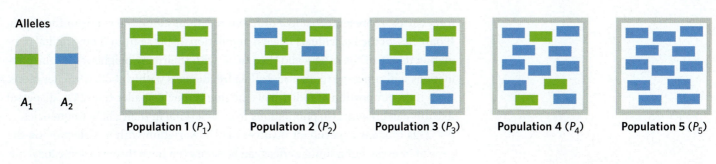

Alleles

A_1 A_2

Population 1 (P_1) Population 2 (P_2) Population 3 (P_3) Population 4 (P_4) Population 5 (P_5)

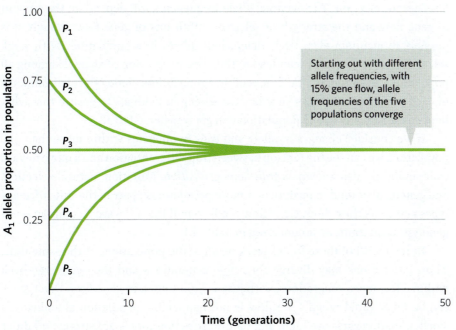

Starting out with different allele frequencies, with 15% gene flow, allele frequencies of the five populations converge

Figure 4.24 Change in allele frequencies through time in response to gene flow. Each line on the graph shows the change in the proportion of allele A_1 over time. We can see the starting proportion for each population along the y-axis (American Phytopathological Society 2021). With 15% gene flow (15% of each population leaves and goes to another population) per year, allele frequencies quickly converge to equal proportions of each allele. Even if almost all reproduction occurs within a population, small amounts of reproduction between individuals from different populations can lead to a loss of the distinct allele frequencies of each population.

Trait Divergence across Populations

The different modes of evolution allow populations to change through time, but in some circumstances, certain populations change differently from other populations of the same species. Remember that populations were defined as groups of individuals that have the chance to interbreed and are physically located in one place. Thus, if there are two or more populations of the same species, this effectively means that there are two or more groups of potentially interbreeding individuals that are separated by space. Each population will have its own unique distribution of allele types and genetic traits and its own potentially unique set of environmental conditions and selective forces. If no individuals move between the populations over long periods of time, then the trait distributions may diverge from one another. This type of *genetic divergence* can lead to *trait divergence*. Such divergence may reflect different environments and selective forces for each population, but it can also occur because random events lead to genetic drift between isolated populations. The essential element for divergence in either case is a lack of gene flow between populations.

Gene flow is a strong force that can override the differentiating effects of both natural selection and genetic drift. For example, in **Figure 4.24** we imagine five local populations of the same species. These five populations begin with different allele frequencies, represented as five different positions along the y-axis of the graph. If only 15% of the individuals in each population disperse from their original population to one of the others and then mate with their new neighbors, all the populations will converge to identical allele frequencies within about 25 generations. Even if the proportion of individuals moving is much smaller than 15%, dispersal like this can prevent two populations from diverging in their allele frequencies.

This thought experiment involves neutral variation with no selection on the gene of interest, but a similar effect can be seen even if we throw in some local differences in selection. The eventual allele frequencies will depend on the amount of gene flow and the strength of selection. With lots of gene flow, though, it is possible to maintain even fairly deleterious alleles in a population. With weak selection, lots of gene flow can lead to the same proportion of an allele among all populations, even though the fitness of that gene differs among the environments of each population. It takes very few dispersing individuals to ensure that what happens genetically in one population will get transferred to all.

If we consider changes in allele and trait frequencies within more than one population, we can see the tension between trait differentiation and a move toward trait similarity, also known as *trait homogenization*. Mutation, natural selection, and genetic drift tend to push two or more populations apart in the suite of genotypes and traits they show. Gene flow tends to pull them toward a common set of genotypes and traits, as summarized in Table 4.1.

In reality, all of these forces work on all of the populations at the same time. Thus, populations may diverge for a few generations and then converge back within a few more generations, with just a slight uptick in gene flow. Divergence can be fairly rapid when gene flow is not happening, population abundance is small, mutation rates are high, and natural selection forces are strong and differ between populations. Divergence can be slow, but still unmistakable, if everything else is the same and only gene flow is removed. This divergence in traits across isolated populations is the starting point for understanding the process of speciation.

4.6 The Speciation Process

At this point, we have all the puzzle pieces necessary to explore the workings of speciation. Recall from the beginning of the chapter that speciation is the process of one species becoming two species. We start by considering a population as a unique evolutionary lineage, hereafter called a *population lineage*, in the sense that every generation is connected to the next through the alleles passed down through generations.

To start the process of speciation, one breeding population lineage needs to divide into two or more isolated lineages with gene flow between them reduced to very low levels. Given enough time and continued limited gene flow, each population lineage will diverge from the other in allele frequencies (see Figure 4.7). A population with locally adapted traits for the species is often referred to as an **ecotype**, reflecting the selective forces of its ecosystem. As discussed at the start of the chapter, eventually the dividing lineages will reach a point when individuals in each lineage are distinct and can no longer interbreed. At this time, each lineage is easily identified as an independent evolutionary unit with unique allele frequencies and ecological distinctiveness, and by any definition of species, we would identify the two lineages as separate species.

As previously explained, there is a "fuzzy" or "gray" phase in this speciation process when differentiation between diverging lineages is happening, but one or more of the various species definitions would fail to identify the diverging lineages as distinct, separate species (the branching area in Figure 4.7). For example, obvious morphological differences between two population lineages may appear early in the process, while failure to interbreed may occur much later. In fact, it is common to see variation in physical traits before observing a complete failure to interbreed.

We see examples of this divergence in nature, such as when changes in size or color occur within a localized population but do not prevent or affect reproduction. Physical changes in such external morphological traits, although dramatic and quite noticeable, may disappear fairly quickly with even small amounts of renewed gene flow. In fact, even though the biological species concept fails to recognize asexual reproduction, its focus on reproduction is still useful because it helps to highlight gene flow within a lineage and the loss of gene flow as lineages diverge. When two lineages can no longer interbreed, gene flow can no longer bring populations that are on diverging pathways back together again. We call this mating barrier to gene flow a **reproductive isolating mechanism**, and it can take quite a long time to evolve, even when starting with ecotypes, or two or more populations that look quite different.

The first step in the evolution of a new species is for one population lineage to split into at least two groups and then for gene flow between the groups to be greatly restricted or eliminated. The speed at which speciation progresses will increase if the two groups experience different and independent natural selection events. But how does an initial population become divided?

Speciation with Physical Barriers

The simplest way for one population to become multiple populations is for a physical barrier to split the initial population. Imagine a species of frog occupying a discrete geographic location, say, a mountainous island in the ocean with two peaks on it (**Figure 4.25A**). Because frogs can move fairly large distances, and there are no real barriers to their dispersal, the one species of frog is found all over the island. Therefore, this island is home to a single population lineage through which alleles are passed from one generation to the next.

Now imagine that anthropogenic climate change continues to melt the polar ice and raise sea levels. The rising ocean water floods the valley between the two mountain peaks, creating two islands where originally there was one (**Figure 4.25B**). If we assume that this frog species (like most amphibians) cannot swim in salt water, then the frogs are not equipped to make the journey between the two newly formed islands. The rising ocean, therefore, prevents gene flow between frogs on the two islands. Given enough time and selection pressure, the two populations may diverge in allele frequencies and trait distributions because of the forces of genetic drift and differential selection on the two islands.

At this point, the two separated lineages of frogs can be thought of as incipient species, meaning two lineages on the verge of becoming two distinct species. The ocean acts as a **geographic isolating mechanism**, which is a physical barrier preventing dispersal and thus gene flow. After many generations, the two lineages

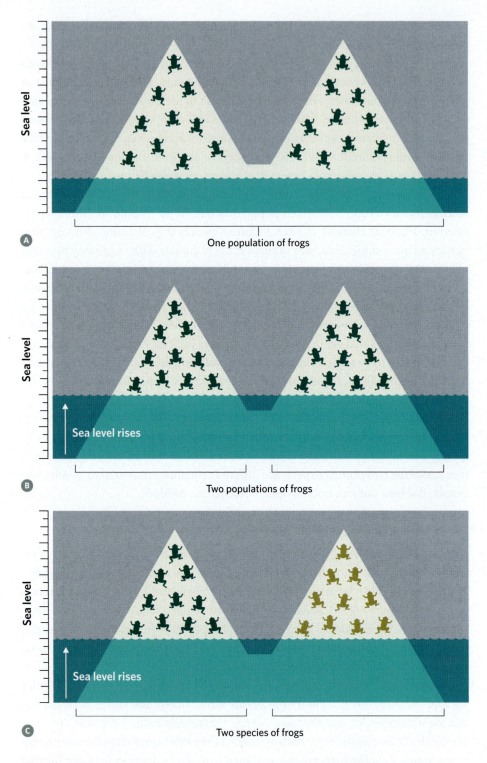

Figure 4.25 If a population exists in an area with significant landscape or habitat variability, an environmental shift that cuts off gene flow can lead to divergence, potentially leading to speciation. Here we have Ⓐ a population of frogs in a mountainous island habitat where Ⓑ a rise in sea level creates a *geographic isolating mechanism* that blocks movement and restricts gene flow between two parts of the population, so that one population becomes two. Ⓒ With the loss of gene flow, natural selection and genetic drift can lead to genetic and trait divergence between the new populations until we have a *reproductive isolating mechanism*, and individuals from one population can no longer interbreed with individuals from the other.

may have diverged so much in morphology, behavior, or physiology that, even if one frog happened to make it to the other island, it would not be able to breed with its neighbors (**Figure 4.25C**). At this point, there is a reproductive isolating mechanism in place, and the trait differences between the two lineages become self-reinforcing.

The island frog example may seem a bit contrived, but see **Figure 4.26** for a real-life example of frog populations isolated by seawater and diverging from each other. Something like this process is likely to have produced many of the

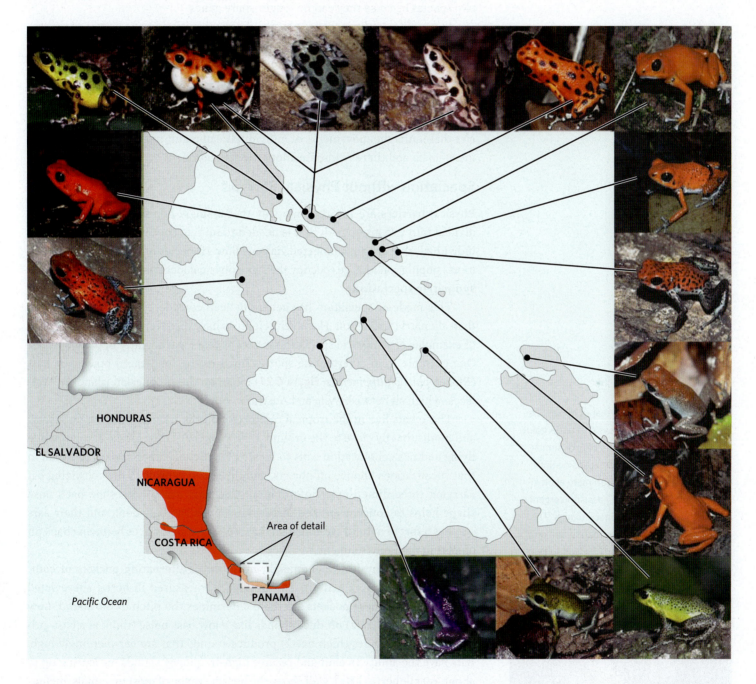

Figure 4.26 A real-life example of divergence among frogs in island populations can be found in the strawberry poison dart frog species (*Oophaga pumilio*) in Panama. The larger-scaled inset map shows the species' geographic range in red, and the more detailed map indicates sample locations in Bocas del Toro, Panama, with lines pointing to examples of color morphs of the species found in those locations (Rudh et al. 2007). Note that all the frogs pictured here are of the same species, regardless of their coloration and patterning. This example has not (yet?) led to speciation, but it has led to very distinctive and recognizable trait divergence.

distinct species we encounter today. When a physical barrier is responsible for initiating the speciation process, the process is called **allopatric speciation**. Remember, though, that a species is all actually or potentially interbreeding individuals, whether or not they are separated by a physical barrier. A geographic isolating mechanism, therefore, reduces gene flow and may aid in the divergence between two populations. Only when a genetically based reproductive isolating mechanism arises, though, do we have two distinct lineages that have become two species because they can no longer share genes.

Before the Anthropocene, physical barriers initiating speciation were created by forces such as continental drift, rising mountains, volcanic eruptions, river meander, lake formation, and desert expansion. Currently, due to the massive forces exerted by humans on the planet, new sets of physical barriers to gene flow are being produced by such human constructs as highways, canals, vast urban landscapes, and the like. Functionally, the emergence of any habitat or barrier that limits the movements of individuals can result in a geographic isolating mechanism and thereby initiate allopatric speciation.

Speciation without Physical Barriers

Physical barriers are not the only way that a single population lineage can be divided into two groups. All that is needed to initiate speciation is two groups of individuals that are not connected via gene flow. If speciation occurs when two (or more) population lineages occupy the same physical location, the process is called **sympatric speciation**.

This mode of speciation has generally been considered much less common than allopatric speciation. However, scientists have discovered a growing number of examples of sympatric speciation in recent years, particularly in plant species. One example from a vertebrate species involves the long-eared horseshoe bats (*Rhinolophus philippinensis*; **Figure 4.27**) found on the Indonesian islands of Wallacea, which sit between Asia and Australia.

These bats live in the tropical rainforests on the islands, foraging on insects and moths that fly close to the ground. Like many bats, they forage at night while flying and use echolocation calls to locate prey. Echolocation is akin to sonar, in that vocalizations bounce off objects, and the sound returns to the vocalizing bat carrying three-dimensional information. The long-eared horseshoe bat's nose shape helps to concentrate and direct the sounds the bats emit, and their ears serve as a sort of natural satellite dish that receives sound waves bouncing back off objects in their environment.

Sound is a physical phenomenon made up of alternating pockets of compressed and decompressed air. The frequency (measured in hertz, abbreviated Hz) with which these pockets alternate determines the pitch of the sound. Low frequencies (low hertz) produce sounds like a low bass noise (think of a bass guitar). High frequencies (high hertz) produce sounds that are ear-piercingly high. Bats have the ability to emit and receive high-frequency sounds measuring up to about 160 kilohertz (kHz), well above the hearing range of most mammals, including humans. For bats, the frequency at which they emit sound is a key determinant of the size of their prey. If a bat specializes on relatively small prey, it will emit high-frequency sounds. If a bat specializes on larger insects, it will emit somewhat lower-frequency sounds.

Figure 4.27 The long-eared horseshoe bat (*Rhinolophus philippinensis*) is undergoing assortative mating between bats of similar sizes. Each bat mating group uses distinct resources based on size, leading to ecological specialization that may assist divergence among the emerging distinct lineages. The combination of assortative mating and diet specialization could potentially lead to sympatric speciation.

Returning to the Wallacean islands, a pair of ecologists documented that long-eared horseshoe bat species came in three body-size morphs: large, medium, and small (Kingston and Rossiter 2004). They also documented that each of these three size morphs emitted distinct sounds while echolocating (**Figure 4.28**). The large individuals emitted a sound at about 28 kHz; the medium individuals emitted a sound at about 40 kHz; and the smallest consistently emitted sounds that were between 50 and 60 kHz. Tigga Kingston and Stephen Rossiter were able to show that each size morph uses its unique frequency range to ecologically specialize on insect prey of a particular size. The small morph–high frequency individuals eat mostly small insects, whereas the large morph–low frequency echolocators eat mostly big insects, with the medium-sized individuals falling in between. These three size morphs represent three distinct ways in which these bats have high fitness, making long-eared horseshoe bats a good example of a species shaped by disruptive selection (Figure 4.21C).

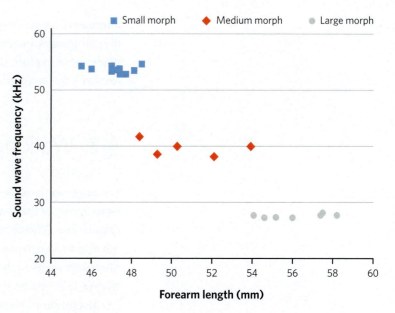

Figure 4.28 The three size morphs of *Rhinolophus philippinensis* bats have distinctly different echolocation frequencies (Kingston and Rossiter 2004).

How do these different behaviors and uses of resources connect to speciation, though? The answer lies in the other reason these bats employ echolocation. When bats mate, they use vocalizations to communicate with each other, and these vocalizations are quite similar to the sounds they use for catching prey. The different echolocation frequencies used to find prey by this species of bat, therefore, lead to *assortative mating*, in which the individuals sort themselves into three groups and mostly reproduce with other individuals in their own "sound" group.

This single species of bat, therefore, has three distinct groups reinforced by variation in morphology, ecology, and potential mating. The individuals use these differences to divide their prey resources within the shared rainforest habitat. At the same time the evidence also suggests that we may have a case of incipient sympatric speciation underway in these bats because they are developing different mating lineages based on their echolocation abilities and reproductive choices.

For the researchers to demonstrate sympatric speciation, however, they would need to show that the groups created by different vocalization frequencies truly lead to independent genetic lineages, with no gene flow between groups. The evidence they have gathered so far indicates a fairly strong differentiation in allele frequencies between groups. They also found that, if the morphs are grouped together, as a whole they are genetically distinct from other species of bat in their genus. Therefore, at this point, the morphs have not differentiated into distinct species, but there is evidence that they are starting to diverge and shift away from functioning as one single genetic lineage. They appear to be on the path shown in Figure 4.7 but are not yet in the section that indicates divergence into separate species.

This example demonstrates how ecological specialization can result in conditions that favor sympatric speciation. The degree to which we see the same mechanism of speciation in other taxa is likely dependent on how much specialization in diet requires a specialization in other traits and the degree to which these other specializations influence mate choice or reproduction. There are alternative forms of sympatric speciation that may be important in different taxonomic groups, but many of them move beyond the scope of this book. Interested readers can explore the topics of polyploidy and assortative mating in a genetics textbook for more information.

4.7 Walking in an Evolutionary Wonderland

We hope we have demonstrated that species are not static entities. Change in traits is part of the evolutionary process and is a constant factor for all species and populations. The dynamic nature of speciation is one of the main reasons we are not always able to place groups of individuals into distinct species. Once your view of ecology shifts to this perspective, it alters how you approach ecological science or even how you experience a walk in the woods.

Evolution is a process that takes place over generations, and speciation can take centuries or millennia. The average life span of a human is currently around 70–80 years, whereas living organisms have inhabited the Earth for at least 3.7 billion years (which is almost 50 million times longer than the average human lives). During these 3.7 billion years, innumerable species have come and gone, often leaving no fossil traces or other concrete evidence of their existence. It seems impossible that we might observe evolution in our tiny blink of time on Earth, but other species are just coming into existence as they start the process of divergence and speciation. Surely many more species will arise in the future.

For each of us, our time as a living creature on this Earth is confined to a minuscule period of this dynamic, long-term process of species formation and extinction. The many nonhuman organisms that we encounter on any given day include species that diverged long ago from their sister species, those that diverged in the last few thousand years, and some that are showing early signs of diverging into two separate species right now. Our pattern-loving human brains want to place every organism we see into clean categories called "species," but this is a quixotic endeavor because evolution is constantly blowing up our categorization schemes.

Human actions today can dictate evolutionary dynamics and trajectories, thereby influencing the species of tomorrow. Given the oversized influence people are having in the Anthropocene, it is hard to ignore the fact that humans are imposing new sets of selective pressures, increasing genetic drift, and restricting gene flow, all of which affect the creation of new population lineages.

Some of the pervasive human-driven factors influencing today's natural world include harvesting practices, the introduction of non-native species, species extinctions, land use changes, and climate change. There is clear evidence that each of these factors is serving as a strong selective force, altering genetic trait distributions across a variety of species. We discuss the ecology of these factors more in Chapter 15, but here we provide some insight into their evolutionary impacts.

When it comes to harvesting practices, humans tend to have preferences for particular traits in the organisms they kill for food or trophies. We can find several examples in fisheries, where selective harvesting of large-sized individuals has decreased the time to sexual maturity and the maximum size these fish attain. There is also evidence that humans alter phenotypes and mating success of fishes, even when we try to use nets that allow certain size classes to escape (Baker et al. 2011). Similarly, hunters often preferentially kill mammals with large horns or antlers because these structures have higher value as trophies. When this preferential harvesting is considerable and continues across several generations, the distribution of horn or antler sizes tends to shift toward the smaller horns (Figure 4.29).

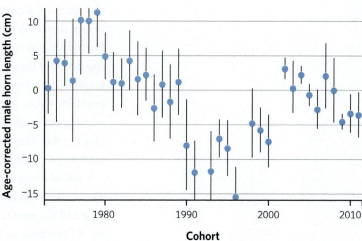

Anthropogenic effects are also leading to large-scale modifications in habitats and landscapes with consequences for gene flow. For example, wood harvesting leads to the fragmentation of a previously continuous forest, creating forest patches that each hold only a part of a previously large, single population. These patch populations can experience evolutionary divergence through time, just as described in the frog example. In addition, roadways, residential development, canals, fences, light pollution, and many other products of urbanization have surprisingly strong negative effects on the dispersal of individuals and, therefore, reduce gene flow. Although ecologists often find evidence that these features divide population lineages into units with little gene flow between them, it can take an extraordinarily long time for these situations to lead to speciation, and evidence that fragmentation is causing speciation has been hard to collect. In Chapter 12, however, we discuss population-level consequences of these broken spatial connections.

Humans can also cause the split of a single population lineage into two or more noninteracting lineages by moving organisms from their native range into new areas or even new continents. Ecologists call such moved organisms **non-native species** and define them as novel populations founded by individuals transported and released well outside their native range by human actions. We used such an example earlier when considering the adaptive evolution of the introduced St. John's wort. In general, a newly established non-native population is nearly always completely cut off from its native source population in terms of gene flow. Also, the biotic and abiotic conditions that the non-native population experiences in its new location can be dramatically different from those experienced in its native range, providing a very different set of selection pressures. The list of non-native plants and animals that have shown marked divergence in their allelic and trait distributions when compared to their native source populations is now quite long and continues to grow (Lockwood et al. 2013).

Perhaps most importantly, as discussed in Chapter 3, the Earth's climate is drastically changing in response to human-derived inputs of greenhouse gases. As a consequence, temperatures tend to be warmer and more variable than in the past, especially nearer the poles, and precipitation levels are changing. These temperature and precipitation shifts are leading to an array of changes beyond

Figure 4.29 Researchers exploring the change in age-corrected mean horn size for male bighorn sheep cohorts at Ram Mountain, Canada, found evidence of a significant decrease in horn size due to hunting pressure between 1973 and 1996. Trophy hunting was substantially limited after 1996, and average horn size subsequently showed recovery. This change after 1996 helped support the scientists' conclusion that hunting played a role in the horn size reduction. Blue dots represent the average horn length (±1 SD) of that year's cohort after correcting for age (Pigeon et al. 2016).

the shift in flowering time discussed previously. Melting glaciers, rising sea levels, more acidic oceans, and altered onset and duration of seasons are all leading to changes in individual species, their predators, mutualistic partners, and the species that they use as resources. The abiotic effects of human-induced climate change, for example, alter the selective regime that organisms experience. For organisms that migrate and conduct their lives in two different latitudes, the climate is generally not changing in both places equally. As higher latitudes warm faster, migrating birds often arrive too late in the spring for their hatchlings to feed on peak insect emergences.

All of these changes impose new selection pressures, and there is a growing list of species that show genetic and trait changes in response. Here we have only touched the surface of these human-induced effects, but we explore elements of the Anthropocene in greater detail in Chapter 15.

SUMMARY

4.1 Variation and Change

- Variation among individuals exists in all taxonomic groups.
- Such variation can lead to changes within groups.

4.2 Definitions of Species

Defining Species Based on Morphology

- Physical differences can be used to distinguish between species.
- This process was formalized in the 1700s by Carl Linnaeus, who used two-part Latin names consisting of a genus and a specific epithet to identify unique species.
- Organisms with similar traits are placed in groups, which are then organized into larger and larger categories, collectively called the Linnaean classification system.

The Biological Species Concept

- Early scientists noticed some species had intermediate characteristics and that physical appearance is linked to what the organism does in nature.
- Morphology captures something fundamental about how species operate in nature.
- Charles Darwin wrote about how traits were passed on to the next generation and that mating individuals likely chose mates based on these traits.
- The biological species concept suggests that if two individuals can mate and produce viable, fertile offspring, they can be considered to be in the same species.

Beyond the Biological Species Concept

- There are problems with the biological species concept.
  - Asexual organisms do not need to mate to produce offspring.
  - Fossil species are exempt from the definition because it is impossible to tell if two species can mate from looking at fossils.
- Current thought focuses on the fact that organisms in the same species must share a common lineage and be able to contribute to changes in the future.
- Speciation, or the forming of new species, depends on genetic divergence over time.

4.3 Relating Genes to Populations

- A gene is a sequence of DNA that contains the code for a biological molecule with a specific function.
- DNA is tightly coiled into chromosomes. The number of chromosomes differs between organisms and sometimes between types of cells in an organism (i.e., reproductive cells versus other cells).
- Genes provide the codes for proteins and thereby lead to organismal traits.
- Alternate forms of genes, called alleles, can produce slightly different versions of the proteins.
- Genetic variation codes for physical variation and combines with environmental factors to produce the immense variability we see in living organisms.

- Individuals who are homozygous for a gene have two alleles that are the same, whereas individuals who are heterozygous for a gene have two alleles that differ.
- With recessive alleles, only individuals who are homozygous for that allele will show the trait. For dominant alleles, individuals who are heterozygous for that allele will show the trait.
- A collection of individuals in one place who can mate together is called a population.
- Allele combinations can be different for populations of the same species.
- Evolution is defined as a change in allele frequency within a population over time.

4.4 Mechanisms of Evolution

- Evolution depends on changes in allele frequencies through time.
- Only a few general mechanisms can alter allele frequencies: genetic mutation, gene flow, natural selection, and genetic drift.

Adding Genetic Variation to Populations: Mutation and Gene Flow

- Mutations are errors in the replication of DNA within a cell.
- Mutations can be neutral (having no effect), detrimental (having a negative effect), or beneficial (having a positive effect).
- Most mutations are neutral; beneficial mutations are extremely rare.
- Gene flow is the movement of genetic material into or out of a population.

Removing Genes from a Population: Genetic Drift and Natural Selection

- Haphazard events can cause the loss of alleles from a population; this is called genetic drift.
- The probability of losing an allele from a population increases when the population has few members.
- Due to the random process of genetic drift, populations can become genetically fixed on a single allele, particularly if the population is small.
- Natural selection is the differential survival and reproduction of individuals in a population in response to interactions between their traits and their environment.

- The process of natural selection can "choose" traits that best fit the environmental conditions.
- If a heritable trait helps increase the likelihood that an individual will survive and reproduce, the alleles that code for that trait will be passed on to the next generation.
- Unique genomes (i.e., collections of traits and alleles) lead to individuals with different probabilities of survival and reproduction, or different levels of fitness.
- **Example of Natural Selection: St. John's Wort**
  - Studying the timing of seasonal behavior is called phenology.
  - A plant native to a Mediterranean climate (St. John's wort) was introduced to New England in the nineteenth century.
  - Over the past 150 years, the flowering season of this plant has extended into the later part of the summer, perhaps because the different climate regime in New England led to different natural selection in New England than in the Mediterranean region.

4.5 Adaptive Evolution

- The St. John's wort example demonstrates evolution because the plant is becoming more adapted to its new environment.
- The term *adaptation* is used for both the trait that has changed and the process of evolution that leads to the trait becoming more common.

Directional, Stabilizing, and Disruptive Selection

- A closely related but native (to New England) plant species called spotted St. John's wort has shifted its flowering time to earlier in the season due to anthropogenic climate warming.
- Both plant shifts in phenology are examples of directional selection because for each plant there is selection toward one end of the trait distribution.
- Stabilizing selection is selection toward the mean trait value, thus narrowing the range of trait variation.
- Disruptive selection is selection against the most common trait value, causing the extreme values to become more common.

Sexual Selection

- Sexual selection occurs when sexually reproducing organisms select for traits that increase mating success rather than survival.

- This can produce sexual dimorphism in a species, meaning the two sexes look very different from each other.

Trait Divergence across Populations

- Different populations of the same species can genetically and/or evolutionarily change in different manners as a result of local environmental differences.
- Gene flow can override these differences and bring about trait homogenization.
- Mutation, natural selection, and genetic drift tend to make two populations of the same species more different in genotypes and traits, while gene flow tends to pull them toward a common set of genotypes and traits.

4.6 The Speciation Process

- Speciation is the process of one species becoming two.
- Isolation and loss of gene flow between two populations is critical to the process of speciation.
- Eventually two populations become different enough that they can no longer breed together, at which point they are considered two separate species.
- An essential element to all speciation is a genetically based barrier to mating that prevents gene flow; we call such a barrier to mating a reproductive isolating mechanism.

Speciation with Physical Barriers

- Physical barriers can prevent gene flow, increasing the potential for divergence between local populations that can lead to a reproductive isolating mechanism and speciation.
- Physical barriers, also known as geographic isolating mechanisms, can include oceans, mountains, rivers, and the like.
- Speciation assisted by a geographic isolating mechanism is allopatric speciation.

- Geographic isolating mechanisms may assist divergence, but speciation does not occur until a genetically based reproductive isolating mechanism arises.

Speciation without Physical Barriers

- Genetic divergence and eventual speciation can occur without physical barriers; this is called sympatric speciation.
- An example can be found in a species of Indonesian horseshoe bat that comes in three body-size morphs—large, medium, and small—each with different echolocation frequencies.
- Echolocation is also used for mating, and the bats tend to mate only with individuals of like size and frequency.
- If this process continues without gene flow between the morphs, it may eventually lead to three different species.

4.7 Walking in an Evolutionary Wonderland

- Human life spans are short, but the process of speciation is often very long; therefore, it is hard for us to observe speciation in nature.
- Human-caused changes in the Anthropocene are producing novel selection pressures that living organisms have to face.
  - Prolonged harvesting of some species is inducing morphological changes in the species.
- Extensive habitat modifications (urban landscapes, canals, roads, and so on) are creating anthropogenic barriers to movement and gene flow where there previously were none.
- The widespread introduction of species to areas outside of their native range is creating novel ecological interactions.
- Climate change is altering the entire abiotic landscape and, therefore, the selection pressures for many species.

KEY TERMS

adaptation	allele frequency	chromosome
adaptive evolution	allopatric speciation	directional selection
allele	biological species concept	disruptive selection

dominant (allele)	geographic isolating mechanism	recessive (allele)
ecotype	heritable	relative fitness
evolution	heterozygous	reproductive isolating mechanism
fitness	homozygous	scientific name
gene	morphology	sexual selection
gene flow	natural selection	speciation
genetic drift	non-native species	stabilizing selection
genetic mutation	phenology	sympatric speciation
genotype	phenotype	taxon
genus	population	

CONCEPTUAL QUESTIONS

1. Give examples of organisms that seem like different species morphologically but that would not be considered different species according to the biological species concept. Explain your reasoning.

2. What ecological roles does natural variation play in both ecology and evolution? Explain what you mean.

3. Define natural selection, and describe how it is a mechanism of evolution. Use an example if this helps.

4. Explain how recessive alleles might make it hard for natural selection to act on all genetic variation.

5. Define and explain the other three mechanisms of evolution outlined in Table 4.1.

6. Define the term *species*, and discuss how one population lineage diverges into two. In your explanation, use the terms *reproductive isolating mechanism* and *gene flow*.

7. Figure 4.23 shows an increase in early juvenile weight and an increase in survival based on elements of the male (father) peacock's tail feathers. This is strong research-based evidence for the conclusion that the tails of peacocks hold information that indicates the fitness of their offspring. But there is still one piece of information missing, based on our definition of fitness. What third panel would you include in the figure to demonstrate higher fitness based on tail feather traits, and what experiment or research method might allow you to get it?

8. A fictional island, with a population of birds and a second population of nonflying beetles, is split by an earthquake that creates a wide channel between the northern and southern portions of the island. There are now two small islands where previously there was one larger island. The channel between them is at least one kilometer across.

    a. Why might this event be more likely to affect speciation for the beetles than the birds?

    b. If divergence and speciation occur in the beetles after this event, would it be sympatric or allopatric speciation? Explain your answer.

9. Colonial bentgrass (*Agrostis capillaris*) is a perennial grass native to China that has been used to revegetate many industrial waste sites in the United States. The plant is widespread throughout much of the country. When *A. capillaris* grows on mine tailings (the soils left from spread mine wastes), it has to deal with high levels of copper. On mine tailings, only plants with a high tolerance to copper survive.

    a. Give two explanations for why some offspring from mine tailing populations might not have high copper tolerance.

    b. In the United States, is colonial bentgrass a non-native species? Explain your answer.

10. Explain why gene flow between populations arranged along a latitudinal (north-south) gradient might help a species adapt to climate change.

QUANTITATIVE QUESTIONS

1. There is a fabulous dataset of measurements collected by the biologist Hermon Bumpus in 1898 for sparrows that died in a cold snap. Explore Bumpus's data for keel length (in.) of female sparrows, available through a Field Museum blog post ("Hermon Bumpus and House Sparrows," by Peter Lowther, September 24, 2014, fieldmuseum.org/blog/hermon-bumpus-and-house-sparrows).

   a. Create graphs of frequency (y-axis) against binned sizes. Do you see a pattern that suggests selection? Use your graphs to explain your answer.

   b. Is this selection stabilizing, directional, or disruptive?

2. In your own words, explain how each mechanism of evolution leads to the increases or decreases in genetic variation and divergence in Table 4.1.

3. Toss a coin four times, and record the number of heads and the number of tails. Do this four more times. Then toss the coin for five sets of 30 tosses, recording heads and tails.

   a. Calculate the proportion of each set that is heads.

   b. Calculate the mean proportion of heads tosses for sets of four tosses and sets of 30 tosses.

   c. Relate the difference between your proportions and means in the four-toss and 30-toss sets to the different effects of genetic drift in small and large populations.

   d. Both of these sets are more akin to results in small populations, so relate your results to what you might expect with 1,000 tosses per set or 1,000 individuals in a population.

4. Look at Figures 4.18 and 4.20, and explain how we can tell that the phenological shift was bigger for the introduced species.

5. Under what conditions will stabilizing selection not lead to a loss of variation? Explain, and draw a graph to illustrate.

6. A geographic isolating mechanism is considered the primary mechanism that initially divides one population lineage into two. From your reading of the news or your life experiences, imagine a scenario in which human actions stop or dramatically slow gene flow across space and thus create two population lineages. Choose a scenario that you think might affect multiple types of species.

   a. Create a theoretical diagram for this scenario starting from a single distribution of one trait (your choice) and proceeding through the steps of speciation to produce two distributions of this trait, one for each newly created species. Be sure to carefully consider all the mechanisms needed to create these new species in your scenario, even if you are only imagining what they are or could be.

   b. Create graphs that qualitatively demonstrate how the speed of speciation would change if your species had a very long life span (decades) or a very short life span (days).

   c. Does your scenario lead to very different selection pressures on each side of the geographic isolating mechanism? Try to explore graphically how the difference in selective pressure or the strength of selection might affect the speed of divergence and speciation.

SUGGESTED READING

Baker, M. R., N. W. Kendall, T. A. Branch, D. E. Schindler, and T. P. Quinn. 2011. Selection due to nonretention mortality in gillnet fisheries for salmon. *Evolutionary Applications* 4(3): 429–443. DOI: 10.1111/j.1752-4571.2010.00154.x.

Darwin, C. 1859. *On the Origin of Species*. London: John Murray.

Kingston, T. and S. J. Rossiter. 2004. Harmonic-hopping in Wallacea's bats. *Nature* 429: 654–657.

Petrie, M. 1994. Improved growth and survival of offspring of peacocks with more elaborate trains. *Nature* 371: 598–599.

Shapiro, M. D., Z. Kronenberg, C. Li, E. T. Domyan, H. Pan, M. Campbell, H. Tan, et al. 2013. Genomic diversity and evolution of the head crest in the rock pigeon. *Science* 339(6123): 1063–1067. DOI: 10.1126/science.1230422.

The common clownfish (*Amphiprion ocellaris*) with its associated sea anemone. An individual clownfish must make choices between foraging for food in the open water or hiding from predators within its anemone.

chapter 5

INDIVIDUALS: PHYSIOLOGY AND BEHAVIOR

5.1 What Is a Trade-Off?

Early ecologists and evolutionary biologists established the fundamental tenet that some resources are always limited. As we saw in Chapter 4, this limitation on resources helps explain differential survival and reproduction, which lie at the heart of natural selection. In this chapter, we show how limitation also creates dilemmas for individuals as they make decisions about their personal use of resources. If an individual spends time or energy satisfying one resource need, that time or energy is unavailable for satisfying other needs.

To illustrate this dilemma in your own life, consider how much time you might need to study today if you want to receive an A on an ecology exam that you have to take tomorrow. If you need a full six hours of study time, then you will not have time to do other things that are important to you, such as work at a job, play video games, or chat online with your best friend in Mexico City. But you have a choice: You can decide to spend only two hours studying and likely receive a lower grade on your exam, leaving you with three hours for your job and one hour for a chat with your friend. If you are not worried about your performance on this particular exam, but you desperately need cash and want to catch up with your friend, these time choices may represent a good trade-off for you. Without ever talking with you, a researcher studying your behavior could deduce your priorities for the day by counting the hours you spend at work and online versus those you spend studying.

This logic can also be applied to the activities of other animals besides humans, such as a female common clownfish (*Amphiprion ocellaris*, shown in the photo). An individual clownfish cannot simultaneously hide from predators in the safety of a stinging sea anemone and venture away from the protection of the anemone to eat the delicious algae on nearby corals. By recording how much time the clownfish spends around the anemone or on the corals, a researcher can deduce whether eating a meal or avoiding predators is more important for her on any given day. We can imagine that her needs may shift over time as her hunger grows or as the density of predators changes.

All living organisms, from the simplest plants to the largest whales, face such dilemmas, and the conscious and unconscious choices they make reflect their ecological and evolutionary context. In fact, the word *choice* is perhaps misleading because organisms are not generally making conscious choices. Instead, certain traits lead to certain outcomes, and more successful outcomes lead to the associated traits' becoming more common, as we discussed in the last chapter. When ecologists and evolutionary biologists look at the array of traits and outcomes in nature, different pathways to success appear, and we will sometimes refer to these pathways and their associated traits as "choices" to highlight the fact that alternative outcomes were possible. It is important to remember, though, that organisms cannot choose their adaptations and, therefore, often cannot make conscious choices around some of these outcomes. In this chapter, we carefully explore the alternative pathways or choices and the trade-offs that individuals make to provide themselves with the highest possible chance of surviving or reproducing. These alternative paths affect individual fitness and play into evolutionary change and natural selection (Chapter 4).

Beyond the connection to evolution, this exploration allows us to touch on two other subject areas that interface with ecology: physiology and behavior. **Physiology** is the study of bodily functions and cellular mechanisms at work within living organisms. **Behavior** encapsulates the way in which individuals of a species physically interact with each other, with individuals of other species, and with their environment (**Figure 5.1**).

In the ecological realm, research on physiology and behavior covers a wide array of topics. We will focus on how individuals get food or energy, how they divide energy among different bodily functions, how they balance water gain with water loss, and how they maintain internal body temperature in the face of sometimes changing external climatic conditions. As we touch on each of these subjects, we introduce ecological principles that manifest at the level of single organisms. However, both physiological and behavioral ecology are rich scientific disciplines that you may want to explore further after completing a basic ecology course.

The term **trade-off** in this context suggests that organisms cannot have it all, that having one trait may preclude having another and that limited resources lead to limited options in behavior and growth. The **optimization** of energy acquisition or energetic performance involves maximizing benefits relative to costs, and

Figure 5.1 A spotted hyena (*Crocuta crocuta*) fitted with a GPS (global positioning system) collar feeds on a wildebeest carcass in Masai Mara National Park, Kenya. Tracking an animal like this allows scientists to follow the animal's overall behavior closely.

we can assess an organism's actions and traits through that lens. In fact, we can think of natural selection as the ultimate judge of the way that actions and traits maximize benefits and minimize costs. In other words, the overall goal of optimization is increased individual fitness.

We explore optimization several times in this chapter. It is worth noting that, in every case, we consider the optimal outcome for only one factor (temperature, food, water, or energy) and that it is not possible for individuals to maintain optimal solutions all the time. In reality, prioritizing one factor leads to making sacrifices in others. For example, putting energy toward reproduction or growth may sacrifice staying warm, and shedding excess heat often sacrifices water conservation. This means that, in real life, an individual is not able to perform at optimal levels for all factors simultaneously. Therefore, we end the chapter by considering how the combined effect of trade-offs helps us understand a foundational ecological concept called the niche.

5.2 The Principle of Allocation

A Simple Energy Model for Animals

A useful place to start when considering trade-offs is with the **principle of allocation**. This principle builds on our previous observation that all individuals have limited access to energy. As a result of this limitation, energy allocated to one of life's necessary physiological functions will reduce the amount that can be allocated to other such functions, so allocation indicates priority. The **first law of thermodynamics** states that energy cannot be created or destroyed. Following this law, the amount of energy taken in by an individual must be equal to the amount that it uses plus the amount that it loses. Let's look at energy intake and allocation more closely and explore ways that individual organisms might optimize their allocation decisions across the variety of energy uses.

A basic model for individual energy allocation might look like this:

$$E_{\text{intake}} = E_{\text{respiration}} + E_{\text{assimilation}} + E_{\text{reproduction}} + E_{\text{waste}} \qquad (5.1)$$

In this equation,

- E_{intake} = the total energy an individual ingests
- $E_{\text{respiration}}$ = the total energy devoted to getting oxygen into the body and cells
- $E_{\text{assimilation}}$ = the total energy converted to living tissue in the organism
- $E_{\text{reproduction}}$ = the total energy used for reproduction
- E_{waste} = the total energy lost in waste or heat or inefficiency

We can calculate each component of this equation through direct observation. For example, to calculate E_{intake}, we add up all the energy that an individual consumes from all available sources in a given time period (a minute, an hour, or a day). In any given day, you may ingest energy in the form of eggs and melon at breakfast, a turkey sandwich and salad at lunch, and salmon with rice and broccoli for dinner. We can calculate the total energy ingested that day by adding up the energy value, or calorie count, of each food item you ate.

As long as we can observe or estimate energy intake over a set time frame for an individual, no matter the form of energy ingested (remember that plants take in energy through photosynthesis—more on this later), we can calculate E_{intake}. We can also try to calculate the energy devoted to cellular respiration, production of living tissue, and reproduction. Lastly, we can calculate the energy lost to waste. These allocation calculations are not as straightforward as tracking energy intake, so we will explain these processes.

Respiration is a key component of all living organisms. Cellular respiration is a metabolic process that breaks large molecules into smaller ones and releases energy and water in the process. Some of this metabolic energy is used for work activities, like walking, digestion, and brain activity. The released energy in metabolism is eventually dissipated to the environment as "waste" heat. Except for waste heat, we have lumped all of these metabolic activities, from thinking to chasing prey, into the respiration term ($E_{respiration}$) in the equation.

Some of the energy not used in cellular respiration is converted into **biomass**. We think of biomass primarily as living tissue or growth, but storage of energy as fat, carbohydrates, or proteins is also very important. These stored energy units can be tapped later when energy intake does not equal energy needed. The assimilation term ($E_{assimilation}$) captures these types of energy.

A portion of ingested energy goes to producing babies or other offspring or caring for them. Many readers of this book may not have kids, but it is astonishing how much energy they take. Energy expenditures for both the production and care of offspring are captured in the reproduction term ($E_{reproduction}$).

A good chunk of an individual's energy intake passes out the back end of the organism in the form of waste, which appears as the last term (E_{waste}) on the right-hand side of Equation 5.1. We may think that there is little energy in this waste material, but one individual's waste, or feces, can become another individual's prized energy source. For example, dung beetles (most of which are in the subfamily Scarabaeinae, the scarab beetles) see animal waste as extremely valuable and will roll it into defensible balls and fight viciously over ownership (**Figure 5.2A**). Humans have also traditionally found uses for animal waste; in many cultures, people rely on dung as fuel for cooking and heating fires (**Figure 5.2B**), as well as for fertilizer. The other major component of the waste term is the heat that organisms lose to the environment, as mentioned in the respiration discussion.

Other than subdividing these major allocation categories, there really is no other place for an individual animal to put the energy that it ingests. The total energy ingested *must* equal the total energy allocated to cellular respiration, reproduction, assimilation, and waste. It is the inescapable balancing of this equation that forces individuals to make trade-offs in the allocation of energy across physiological needs.

You may ask: What if the total energy ingested is really high? Will the individual "escape" these trade-offs? No, probably not. If an individual takes in boatloads of energy, it is true that the organism will have more energy to allocate to these categories; however, it still must make choices (although maybe not conscious choices) about growing larger, reproducing more, respiring more, or creating additional waste. And the individual's evolutionary fitness will reflect those choices. It may be easier to stay alive with more access to energy, but the best alternative will lead to peak fitness and will affect evolution of the species, as discussed in Chapter 4.

Figure 5.2 One individual's waste can be a valuable energy resource for another. Animal dung can be used as Ⓐ food or habitat for dung beetles and other insects or Ⓑ fuel for fire.

Ⓐ

Ⓑ

One may also suggest that to achieve peak evolutionary fitness, the best allocation of energy would be to put all energy toward reproduction. When we observe energy allocation decisions in nature, however, we see that most individuals do not take this approach. Some viruses come very close, but most species need to grow before they reproduce, and growing requires finding food, or energy. In addition, simply existing in the world brings the risk of being eaten. Individuals may need to allocate energy to defensive structures or behaviors in order to live long enough to reproduce. Once an individual grows enough to reproduce, it may need to invest in finding mates, and those mates may have some pretty specific ideas about what they want in a potential partner (see Chapter 4). Those criteria may require investing energy in a large body size or energetically "expensive" structures, colors, chemicals, sounds, or behaviors.

Once these growth, protection, and mate-attracting needs have been met, it might seem as though it would make sense to put all subsequently ingested energy into reproduction. Interestingly, some **semelparous** organisms (meaning organisms that reproduce just once in their lifetimes) do exactly that. In Chapters 12 and 14, we see that many salmon species use such a reproductive strategy. But **iteroparous** organisms (organisms that reproduce multiple times) spread reproduction over multiple years or seasons. This strategy increases the chances of having some offspring arrive in a year or season with environmental conditions that match the organism's specific adaptations. As we will discuss in Chapter 6, differences between juvenile and adult survival generally determine which of these strategies leads to the highest fitness.

A Simple Energy Model for Plants

If we shift our thinking to plants, the energy calculation does not substantially change, even though most plants take in energy through **photosynthesis**, or the capturing of radiant energy from the sun. Although energy from sunlight would seem to have no limit, remember from Chapter 3 that solar radiation is more diffuse at higher latitudes than at lower latitudes. In addition, sunlight is only available during daylight hours and is not available in all seasons in many biomes. Finally, many plants grow in the shade of taller plants, limiting their access to sunlight even more. Overall, the energy equation for plants might be more precisely represented by changing E_{intake} to E_{PSN}, or the energy captured in photosynthesis:

$$E_{PSN} = E_{respiration} + E_{assimilation} + E_{reproduction} + E_{waste} \qquad (5.2)$$

All plant cells respire and use energy, even as the leaves are capturing and storing energy in photosynthesis. Plants store energy in many forms, including ones useful to us humans, such as complex carbohydrates, sugars, and fats, which we enjoy eating when they come in the form of corn, maple syrup, avocados, and other plant foods.

Like animals, plants have an array of potential ways of reproducing. Single-celled plankton or green algae can use energy for simple, single-celled reproductive processes, such as *mitosis* (cell division that creates a copy of a single cell). They can also reproduce sexually through *meiosis* (cell division that creates gametes for reproduction, with half the DNA of the original cell in each newly divided cell) and the subsequent combination of two gametes. Finally, they can

Figure 5.3 The titan arum (*Amorphophallus titanum*), or corpse flower, is the plant with the largest flower in the world. This flower attracts flies and other pollinators by emitting odors that mimic the smell of a rotten animal corpse.

reproduce by fragmenting, as mentioned in Chapter 4. But plants can also construct ornate and exuberant reproductive structures, such as the giant flower of the endangered titan arum (*Amorphophallus titanum*) that lives in the rainforests of Sumatra, Indonesia (**Figure 5.3**). Plants also expend energy in growth and in the production of defenses against herbivory, such as spines or chemical toxins. Because plants do not generally get rid of waste (except for oxygen) as animals do, the E_{waste} term typically takes the form of dead plant tissue, which can be a striking and beautiful sight, as in the autumnal leaf fall from the deciduous trees in many Northern Hemisphere temperate forests (**Figure 5.4**).

As with animals, optimizing the right-hand side of the allocation equation for plants may not be as simple as putting all energy into reproduction. For example, the titan arum tricks flies into pollinating it by mimicking a giant rotting animal (it smells horrible), so it needs to put enough energy into growth and production of airborne chemicals to pull off that deception. In fact, the titan arum usually does not reproduce until it has been growing for at least seven years, making it an excellent example of a long-lived, semelparous plant species.

It is much easier, though, with plants than it is with animals to see how dynamic the process of energy allocation can be. Think about plants growing in a garden. Many of them will grow and grow, as long as conditions are good. That way they can maximize their ability to survive long winters and attacks from herbivores and live to reproduce in another year. If the conditions worsen, however, they may suddenly shift energy allocation away from growth and into reproduction, putting out lots of flowers. You can try an experiment in your own garden to see allocation in action. Spread a little salt in the soil of a tall, nonflowering tomato plant. The salt creates water stress for the plant (because it uses water to flush the salt from its tissues) and may shift its allocation away from stem growth and toward the production of lots of flowers and fruit.

5.3 Optimal Foraging

Animals can change allocation priorities rapidly by changing their behavior. For example, a male coast horned lizard (*Phrynosoma coronatum*; **Figure 5.5**) may normally prioritize reproduction over other energy allocation options and spend precious energy fighting other males for access to females (an example of sexual selection, as discussed in Chapter 4). These expenditures of energy are all allocations within the $E_{reproduction}$ term. But when the lizard gets too hungry to fight, it will shift its energies to looking for

Figure 5.4 Autumnal leaf fall in the deciduous portion of a temperate forest biome.

food, mostly ants. The energy from eating ants will go directly to respiration and next to $E_{\text{assimilation}}$ in the form of stored energy, so the lizard can get back to fighting with other males for reproductive access. We can imagine a variety of similar scenarios in which the lizard must pick and choose its behavior so that its limited energy goes to the process that maximizes its fitness at that moment. Although this example focuses on behavior and choices, we need to remember that many of these "choices" (such as an ability to capture and digest ants) may not be consciously made by the lizard but are more likely hardwired into the organism's evolutionary adaptations.

This example raises the possibility for trade-offs on the left-hand side of the equation, if we break down E_{intake} into all the different potential energy sources. Let's face it, ants are probably not the food of choice for anyone reading this book. So why are they the food of choice for the coast horned lizard? The answer to this question depends on the types of food or energy available, the quantities in which they are available, and the energy that organisms use to get that food. Basically, all organisms need to maximize energy intake relative to the energy spent acquiring that energy. Ecologists and evolutionary biologists sometimes use **optimal foraging theory** to explore the way various strategies affect fitness, adaptations, and behavior.

Figure 5.5 A coast horned lizard (*Phrynosoma coronatum*).

The Theory

Optimal foraging theory posits that an individual animal acts to gain the most energy for the least cost when making its foraging decisions, with the overall goal of maximizing its evolutionary fitness. The theory is based on the idea that foraging animals have choices about what food to eat, with some available food items packing more energetic punch than others. At the same time, to find and physically manipulate some of these food items may require expending a lot of energy.

As an example of the sometimes high energetic cost of food acquisition, picture in your mind a lion hunting a gazelle, all the while noticing the energy it takes the lion to stalk, chase, catch, kill, and then eat the gazelle. Now change the video playing in your head and allow the gazelle to get away, making the lion search and try again. And let the gazelle get away again. And again. And again. Single lions are only successful in catching prey once in every five or six tries, and success depends on the type of prey.

The task of individual foragers is to carefully choose the food they ingest so that they maximize their overall energy intake and minimize the cost of getting it. Equation 5.1 and the principle of allocation allow us to focus on how a fixed amount of energy is split between competing energetic demands. Optimal foraging theory allows the energy intake term (E_{intake}) to vary and assumes that everything but the cellular respiration term ($E_{respiration}$) on the right-hand side of the equation is constant. Take a moment to recall that cellular respiration includes expending energy on movements and thinking, both of which are involved in foraging. The question then becomes one of maximizing energy intake by choosing what food to eat, given that this choice comes at an energetic cost in the $E_{respiration}$ term. With optimal foraging theory, we are, therefore, considering how to "optimize" one side of this equation (E_{intake}) against the other ($E_{respiration}$). Basically, if we pulled the $E_{respiration}$ term over to the left side of the equation and subtracted it from E_{intake}, we would want as much energy left over as possible for reproduction and assimilation (and some unavoidable waste).

A human-based example may help clarify the idea. Imagine that an evil genius trapped you inside a large office building where only two kinds of food were available: candy bars and dried rice cakes (**Figure 5.6**). If you had to survive in the building, and it cost you no energy to find food, you would probably choose to eat the candy bars, as they are more calorie-rich (and tastier). But what if searching for food cost you energy? That is, let's recognize that you have to expend energy in respiration associated with walking and thinking and hiding from the evil genius

Figure 5.6 A candy bar and a rice cake have different appeal in terms of taste and texture, but most importantly (for optimizing energy), they differ in the calories they offer.

in order to find either the rice cakes or candy bars. If that is the case, the relative abundance of the candy bars versus the rice cakes matters. If only 10 candy bars are scattered around the whole building, but 10,000 rice cakes are lying about everywhere, what should your foraging choice be? Is it better to spend your energy searching for the rare but high-calorie candy bars, or should you just eat the more abundant, easier-to-find, low-calorie rice cakes? What would your decision be if you also wanted to have enough energy left over after eating to invest in defeating the evil genius? These are the types of questions posed by ecologists who study organisms through the lens of optimal foraging (although they rarely have to deal with evil geniuses).

A Simple Foraging Model

To answer such questions, ecologists build hypotheses about how foraging trade-offs are managed by individuals living in a natural world with limited food and plenty of dangers. To help visualize these trade-offs, we can build a simple foraging model for predators consuming prey, as expressed in Equation 5.3:

$$P = \frac{E}{C} \tag{5.3}$$

In this equation, the term P represents the **energetic profitability** of the prey consumed. This energetic profitability is equal to the ratio of energy gained from eating that prey item (E) to the energy costs associated with acquiring and eating the prey (C). These costs may be from bursts of respiration while chasing or grabbing prey, but they can also be from the slow burn of energy over time.

It may seem weird to think about energy in terms of time, but remember that respiration happens constantly, so organisms are always burning energy. If you lie on your bed watching puppy videos (or reading ecology textbooks), you are not expending much energy, but you are still using up energy in respiration; you will eventually get hungry and need food. It makes sense, therefore, that profitability (P) increases if lots of energy is acquired from a prey item (if the numerator, E, is big), or if it takes less searching time or capturing energy to get that prey item (if the denominator, C, is small). To make P as large as possible, E needs to be big and C small. When E and C are the same value, the energy expended to get prey is equal to the energy gained by eating it; this is $P = 1$, or the break-even outcome. Foraging can result in a net loss of energy if C is bigger than E. If $P < 1$ for a given food item, we expect predators never to choose this prey because it is not worth their time and energy.

One simple modification we can make to this model is to be more specific about the costs, or the C variable. Before predators can eat prey, they first have to find them, and we can call these costs **search costs**. Search costs are primarily a function of prey density. Some prey may be abundant and, therefore, have low search costs associated with them; others may be rare, requiring long search times and resulting in high search costs.

Once the prey is found, the predator has to capture, manipulate, and consume it; we call these **handling costs**. Think about the amount of energy it takes for you to eat a handful of blueberries versus the amount it takes to break open and consume a coconut, even if you do not need to find either "prey" item. Or consider a fox eating berries from a bush versus catching and eating a rabbit. The fox expends considerably more energy catching and eating the rabbit than it does eating the berries because it needs to catch, kill, and tear the rabbit apart, then separate its flesh from bone and swallow all the bits. The berries do not run away, do not need to be killed, and can be gulped down in one bite. Even though we think of foxes as carnivores, they eat a surprising number of berries, perhaps because of the low handling costs.

We can easily replace the C variable in Equation 5.3 with these more specific cost variables to create Equation 5.4, where S = search cost and H = handling cost.

$$P = \frac{E}{S + H} \tag{5.4}$$

As in Equation 5.3, profitability (P) is maximized when energy intake (E) is large relative to the search and handling costs ($S + H$). With Equation 5.4, we have a quantifiable model that can be used to make testable predictions about how a real-life predator will choose prey to consume.

Examples and Extensions

Suppose a coyote (*Canis latrans*) in the Mojave Desert faces the choice of pursuing and eating two possible prey items: a high-calorie prey (desert tortoise, or *Gopherus agassizii*) and a lower-calorie prey (cactus mouse, or *Peromyscus eremicus*; **Figure 5.7**). Let's call the tortoise *prey 1* and the mouse *prey 2*. We can designate E_1, H_1, and S_1 as the energy value, handling costs, and search costs associated with eating a tortoise, respectively. We can then set E_2, H_2, and S_2 as the same variables for eating a cactus mouse. This creates two versions of Equation 5.4, one for the profitability of the tortoise as prey (P_1; Equation 5.5), and one for the profitability of the cactus mouse as prey (P_2; Equation 5.6).

$$P_1 = \frac{E_1}{S_1 + H_1} \tag{5.5}$$

$$P_2 = \frac{E_2}{S_2 + H_2} \tag{5.6}$$

We can assume that the coyote can compare the profitability of the two prey types and choose the more profitable one, which is essentially saying the coyote is capable of internally calculating the two values of P from Equations 5.5 and 5.6 in real time.

It is fairly reasonable to assume that a tortoise contains more energy than a cactus mouse because the tortoise is much bigger in body size, which translates into more calories for the coyote to ingest ($E_1 > E_2$). If we want to simplify things, we can also assume, this time more naively, that tortoises and cactus mice are equally abundant in the coyote's hunting range. Under this assumption it means we can set $S_1 = S_2 = S$. Lastly, for the sake of simplification, we can make the even more naive assumption that the coyote's handling time is the same when eating a tortoise as when eating a cactus mouse, in which case $H_1 = H_2 = H$.

With these assumptions, the only difference between Equation 5.5 and Equation 5.6 is the energetic reward of each prey item ($E_1 > E_2$). Under these

How does the coyote decide which prey to consume?

$$P_1 = \frac{E_1}{S_1 + H_1}$$

$$P_2 = \frac{E_2}{S_2 + H_2}$$

Figure 5.7 In this scenario, a coyote (*Canis latrans*) in the Mojave Desert is faced with the choice of two prey items: a desert tortoise (*Gopherus agassizii*) and a cactus mouse (*Peromyscus eremicus*). The choice the coyote makes depends on the relative values of P_1 (the energetic profitability of the tortoise) and P_2 (the energetic profitability of the mouse). In these equations, E is the energy acquired from consuming a prey item; S is the search cost, or energy expended to find a prey item, which is partly a function of time spent; and H is the handling cost, or energy expended to capture, manipulate, and consume the prey item (also, in part, a function of time).

very simplified conditions, the coyote's prey choices produce the following inequality:

$$\frac{E_1}{S + H} > \frac{E_2}{S + H} \tag{5.7}$$

This inequality implies that the coyote should always choose to eat the tortoise because of its higher energetic profitability. If we further assume that energy gained from consuming a mouse (E_2) is so low that Equation 5.6 produces $P_2 < 1$, then the coyote should not even bother to search for cactus mice because the searching and handling effort required is more than the small mouse-sized packet of energy that could be caught and eaten.

But suppose cactus mice actually do have a decent amount of meat on their bones, making them worth eating ($P_2 > 1$). What happens if we make the coyote's search time more realistic? In the Mojave Desert, cactus mice are significantly more abundant than the declining populations of desert tortoises, which are threatened with extinction due to habitat loss and climate change. Because there are far fewer tortoises than cactus mice across the landscape, we need to increase the value of S_1 to reflect a much longer search time for tortoises than mice ($S_1 \gg S_2$). If this value is high enough relative to E_1 (the energetic value of the tortoise), then the inequality in Equation 5.7 will be reversed, creating Equation 5.8.

$$\frac{E_1}{S_1 + H} < \frac{E_2}{S_2 + H} \tag{5.8}$$

Given these parameters, the coyote should choose the lower-calorie cactus mouse because it takes significantly less time and energy for a coyote to encounter a mouse than a tortoise. The cactus mouse, therefore, leads to a higher energy reward, or higher profitability.

Taking this one step further, we can make handling times for these prey items a bit more realistic too. Tortoises have large, hard, protective shells, making them much more difficult to consume (or handle) than soft, fuzzy mice. A coyote has to find a way to break through that shell before it gains any energetic reward. This shell drives up the handling cost for a tortoise (i.e., value of H_1). Now that S_1 and H_1

are both higher than S_2 and H_2, it is much more profitable for the coyote to search for and eat cactus mice than it is to hunt tortoises. We can even imagine that tortoise shells may be such a pain to deal with that coyotes might spend hours or even days batting around a tortoise without ever getting inside the shell. In that case, the tortoise becomes a very unattractive prey, with $P_1 < 1$, and the coyote should not bother to search for tortoises at all.

In general, prey choice is dictated by comparing the energetic profitability of potential prey items. We presented scenarios in which the coyote preferred the tortoises over mice, and then in which it preferred the mice over tortoises. In both scenarios, however, the coyote would eat either prey item when encountered if each represented an energetic profit. Of course, it was not hard to think of scenarios in which cactus mice should never be consumed, or the reverse, when tortoises should never be consumed.

The bigger point is that the values of E, S, and H for each prey item determine the order of preference of prey or even whether a predator should completely ignore a prey type. See if you can imagine other outcomes for the coyote example, or try doing this same logical exercise for your favorite predator and its prey (sharks, salamanders, spiders, sloths, and so forth). It is possible to extend these optimal foraging equations to represent foraging choices for herbivores, parasites, or even plants, although the biological meaning of E, S, and H would need to be modified, particularly in the case of plants.

Equations 5.5 through 5.8, and those like them (there are many other formulations and representations of the ideas behind optimal foraging theory), allow ecologists to represent hypotheses about how a real-life predator chooses its prey. If you had the time, money, and fortitude, you could go to the Mojave Desert, observe coyotes, and record the types of prey, the number of each type, the mass or weight of each prey item, the time it takes the coyote to find the prey, and the time it takes to catch, subdue, and eat each prey type. If you make a few assumptions about how much energy is in a gram of prey and how much energy a coyote burns in a minute of searching or handling, you can compare your data to outcomes from the optimal foraging model and see how well optimal foraging predicts coyote prey choices. But you don't have to go to the Mojave Desert to study optimal foraging, as you could do the same thing with birds or squirrels or even spiders found around your living space. Ecologists have employed foraging models in almost any circumstance imaginable.

Theory Meets Reality

Tests of optimal foraging theory often examine one of a few qualitative predictions made from these models, including the following: (1) foragers should prefer prey that yield the most profitability, or energy per unit handling-and-searching time; (2) as the abundance of higher-value prey increases, consumption of lower-value prey should decrease, and lowest-value prey should be dropped from the diet.

One of the first and more elegant tests of the theory came from the laboratory of John Krebs at Oxford University in the 1970s (Krebs et al. 1977). To test the predictions just described, Krebs and his colleagues built a clever experimental setup (**Figure 5.8A**). Their study subjects were birds, specifically great tits (*Parus major*). In a lab enclosure, an individual bird was presented with two sizes of prey in the form of large and small pieces of freeze-dried mealworm. The prey items rolled

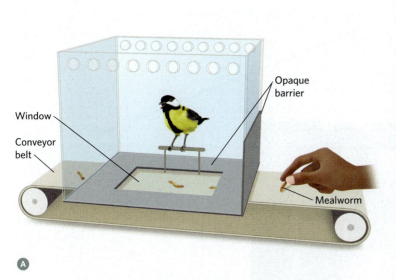

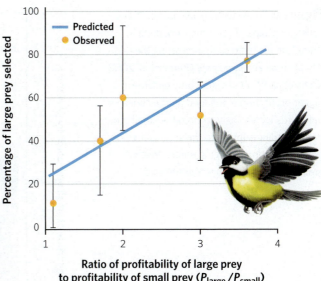

A

B

Figure 5.8 Ⓐ John Krebs and colleagues (1977) conducted a laboratory experiment to test aspects of optimal foraging theory in great tits (*Parus major*). The birds were placed, individually, in an enclosure through which a conveyor belt moved carrying pieces of mealworms of various sizes and at various densities. The birds chose which food to eat partially based on item frequency and caloric value, but the small pieces were also taped to the conveyor belt to increase the handling cost associated with their consumption. Ⓑ Results from the experiment show that as the researchers manipulated the profitability (*P*) of the two prey types (large and small), the birds' selections changed (Krebs et al. 1977). By increasing the handling time for the small prey, the researchers could make the ratio of P_{large} to P_{small} increase. As this ratio increased, the birds became pickier and selected more of the large prey. This outcome supports the prediction from optimal foraging theory that foragers should prefer prey with higher profitability.

past the bird at a constant rate on a conveyor belt under the bird's perch, creating a continuous worm "buffet." The bird could not see what was coming and could only see one prey item at a time.

The researchers varied the density of high- and low-quality prey items presented to the birds by simply increasing (or decreasing) the number of the large and small mealworm pieces on the rolling conveyor belt. To ensure that the smaller pieces truly had lower energetic profitability, the researchers added a piece of tape to each of the small pieces of mealworm to increase the handling time for those pieces. The birds had to tear off the tape to get at the tasty mealworm flesh. Krebs and his colleagues measured each bird's handling time for each prey type.

Based on what you have learned so far, what do you think the pattern of prey selection should be? Krebs and colleagues predicted that the birds would optimize their energy profitability by preferring bigger prey with a higher energy reward and a lower handling time. In order to focus on how the birds' behavior shifted with differences in prey profitability, they looked at a ratio of the profitability of the large prey to the profitability of the small prey and observed how prey choice changed as this ratio changed. In **Figure 5.8B**, we can see that, as the relative profitability of large prey increased, they became a larger proportion of the birds' diet.

Another study (Berec et al. 2003) followed up on this experiment, specifically looking at how selectivity changed as the density of large prey increased. In **Figure 5.9**, we can see that, as the frequency of large prey increased, the birds ate fewer of the small mealworms that moved past them on the conveyor belt. Both of these results support the primary predictions of optimal foraging theory, and subsequent investigations have reinforced these findings.

One interesting observation made in many optimal foraging studies is that predators tend to eat a little more of the low-quality prey than we would expect if they were really, truly eating optimally. This deviation suggests that the model is not perfect. If we think about factors that were not included in the model, we can come up with some good reasons for these deviations.

Figure 5.9 A study by Michal Berec and colleagues expanded on the work done by Krebs and his team by manipulating the density of small and large prey (Berec et al. 2003; Krebs et al. 1977). As large prey became a bigger proportion of the food on the conveyor belt, birds increasingly ignored the small prey items, lending support to the second prediction from optimal foraging theory. Inset: A great tit forages for insects in the wild.

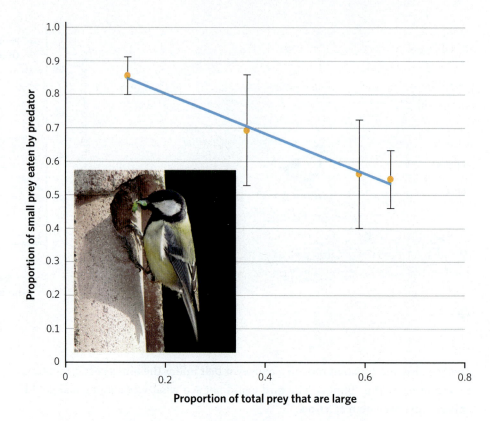

First, as mentioned before, organisms have many different functions that they are trying to optimize. It is possible that predators are *not always* foraging optimally because they may be prioritizing some other need. To build the optimal foraging model, we held constant all the elements on the right-hand side of Equation 5.1 except $E_{respiration}$. In reality, predators probably do not hold these terms constant; they may diverge from their foraging every now and then for a little reproduction or to escape from being eaten themselves.

Second, our optimal foraging model assumes that predators have perfect knowledge of their food choices or the prey items available to them, but real predators do not have such information. Even when researchers make the prey availability very clear to predators, though, they still eat *slightly* suboptimally. Why might this be?

Remember that the process of natural selection hardwires into the organism many of these choices. If there is one truism in ecology, it is that the world changes. Over time, natural selection tends to favor organisms that can survive as environments go through fluctuations. If food availability fluctuates, it becomes less likely that predators will always fulfill the second prediction and overlook lower-quality items in favor of higher-quality ones. In a changing environment, it sometimes makes sense to gobble the unappealing rice cake that comes your way, even if you think there may be a delicious candy bar hiding in another room.

Many tests of optimal foraging theory and its offshoots have been done over the years. A 2001 literature review by Andy Sih and Bent Christensen found that approximately 70% of the 134 studies testing the predictions of optimal foraging found support for the theory (Sih and Christensen 2001). These authors reported that the theory enjoyed its best support in situations when the prey items were completely immobile (e.g., mussels, barnacles, seeds, and plants), but that studies in which prey were highly mobile (e.g., ungulates, small mammals, and fish) generally did not support the theory. They suggested that this latter result was due to the difficulty of testing only a single factor (namely, nonrandom diet choice relative to prey abundance) in the more convoluted and complicated process of predation by mobile animals. As you can imagine, prey-chasing scenarios involve many other interacting factors, such as encounter rates, attack rates, and capture success.

Based on the information on evolution and natural selection from Chapter 4, this difference makes sense. Remember that many of these foraging "choices" are probably not just behavioral choices but may reflect adaptations that maximize foraging efficiency in particular environments. With mobile prey, there may be a variety of behaviors that could optimize energy capture in each environment or for each part of the foraging process (e.g., search, attack, capture, subdue), so it is far less likely that natural selection would produce just one optimal strategy. Sih and Christensen suggest, though, that optimal foraging theory can help to explain adaptations that we see in both predators and their prey (who must evolve anti-predator defenses). We explore more of these effects in Chapter 8, when discussing predator-prey interactions and exploitation.

5.4 Water Balance

Why Water Is Different

Water is a critical component of all life. Although its availability varies from one environment to another (discussed in Chapter 3), it is always important enough that we can take our principle of allocation equation (5.1) and replace energy with water. Changing the energy term (E) in Equation 5.1 to a term for water (W), but keeping the relationships intact, we get Equation 5.9.

$$W_{\text{intake}} = W_{\text{respiration}} + W_{\text{assimilation}} + W_{\text{reproduction}} + W_{\text{waste}} \qquad (5.9)$$

All individuals must balance the two sides of this equation. Just as with energy, we can learn a lot about an organism's priorities by examining how individuals allocate water among the terms on the right-hand side of the equation.

When we considered energy allocation, we did not spend much time on the waste term, or loss. With water allocation and balance, the conversation changes for two reasons. First, water is not only lost to the environment through metabolic reactions (as is energy in the form of heat), but it is also actively used to obtain oxygen for metabolism, to maintain body temperature, and/or to get rid of waste energy. Thus, organisms generally lose water faster than they lose energy, making the loss term a big deal.

Second, unlike with energy, it is possible for an organism to ingest or gain too much water; if that happens, loss or waste becomes useful. Taking in too much water can result in death for terrestrial species, and a change in water salinity can result in death for aquatic species. With energy, we mostly held the left-hand side of the equation constant or thought about how to maximize it. With water, the left-hand side of the equation is constrained to a narrow range, and balancing the left with the right becomes a matter of life and death.

Together these two factors bring into sharp focus the need for individuals to carefully balance their water intake versus loss. The intake part of Equation 5.9 is fairly straightforward (organisms ingest water), although there are a surprising number of ways that organisms can take in water besides drinking it; they can get it through eating food, absorb it through their skin, or draw it up through their roots, to name a few. Generally, people are less familiar with the ways that organisms can change or alter their water loss (waste), so let's explore that element of Equation 5.9.

Water Balance in an Ecological Context

All organisms have some level of control over water loss, and their need to employ these controls varies greatly across ecological conditions. In aquatic and semi-aquatic environments, organisms may actively expel water to maintain an optimal water and mineral balance. In most terrestrial environments, though, organisms need to minimize water loss.

This need is acute in groups that use their skin for breathing. For example, lungless salamanders are an entire family of amphibians that have no lungs and breathe exclusively through their skin (**Figure 5.10**). In contrast to most other amphibians, the majority of species in this family have no aquatic larval stage, and individuals spend their entire lives out of water. In order to breathe, they exchange oxygen and carbon dioxide with the atmosphere through the moist, porous membrane of their skin. When animals with lungs perform this same exchange in their lungs, keeping the tissues moist to encourage osmosis is easy because their lungs are internal and are bathed in water inside their bodies. Lungless salamanders do not have this luxury. The moisture they need in their skin to encourage osmosis is moisture that can evaporate to the air. To avoid drying out, these salamanders choose habitats that provide continually moist microclimates that are well sheltered from winds.

Figure 5.10 This *Ensatina eschscholtzii xanthoptica* is in the family Plethodontidae, the lungless salamanders that breathe exclusively through their skin and tissues lining their mouths. This trait requires them to live in moist, sheltered habitats.

Controlling water loss is also essential for the organisms that reside in deserts. Plants and animals that live exclusively in desert biomes are exquisitely adapted to this **xeric** (low-water) environment. One of the best examples of a desert animal that is a master of water conservation is Merriam's kangaroo rat (*Dipodomys merriami*), which lives in the southwestern deserts of Arizona, New Mexico, Texas, California, and northern Mexico (**Figure 5.11**). As you might guess from the name, Merriam's kangaroo rats move by jumping like kangaroos on extremely elongated back feet. In order to balance their own version of the water equation, kangaroo rats have a suite of adaptive behaviors at their disposal.

One of the most unusual aspects of these rodents' behavior is that, although they live in some of the driest and hottest places in North America, they rarely drink water. This lack of reliance on rainfall helps them survive in the desert but requires a number of other adaptations in order for them to maintain their water balance. Their ingested water comes primarily in the form of metabolically derived water from the seeds that are their main food. They gather dry seeds on the desert floor and store them in large cheek pouches in their mouths. When they return to their burrows, the kangaroo rats cache the seeds in their underground burrows where the ambient humidity is typically higher, due both to moisture in the soil and to recycled respiratory water from their breathing. The dry seeds absorb some of the ambient water and hold onto it. When the kangaroo rats eat these cached seeds, they ingest water absorbed by the seeds, some of which is the rats' own "waste" water from breathing. Research suggests that Merriam's kangaroo rats will sometimes supplement their seed-based diet with insects and succulent plant material, both of which have significant water content (Tracy and Walsberg 2002). This series of behaviors greatly reduces the kangaroo rats' reliance on the scarce and episodic rainfalls of the desert.

Kangaroo rats also have a range of physical adaptations that alter their water balance requirements. We know a lot, both intuitively and from medical science, about the way humans lose water; and humans and kangaroo rats are both mammals, so it is interesting to make comparisons. The average human lying in a bed watching puppy videos (or in a hospital, which is where the numbers originate)

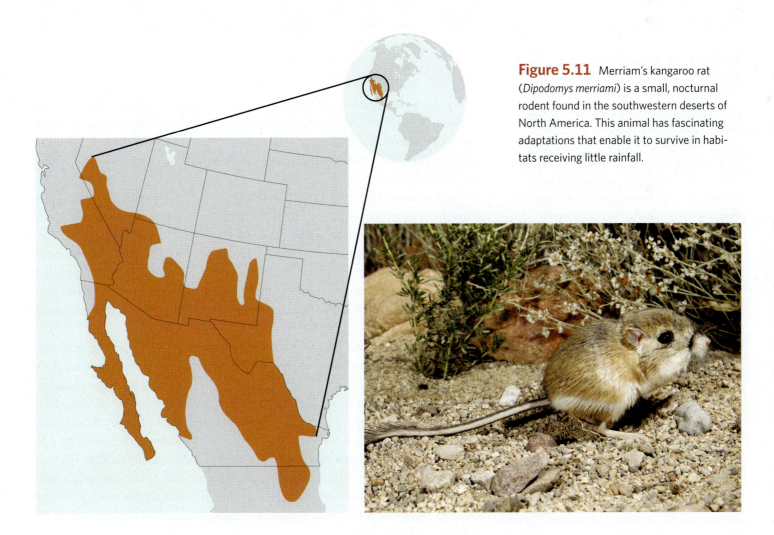

Figure 5.11 Merriam's kangaroo rat (*Dipodomys merriami*) is a small, nocturnal rodent found in the southwestern deserts of North America. This animal has fascinating adaptations that enable it to survive in habitats receiving little rainfall.

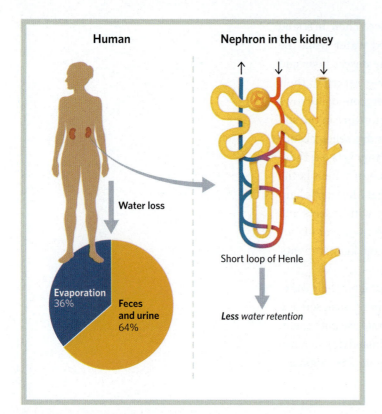

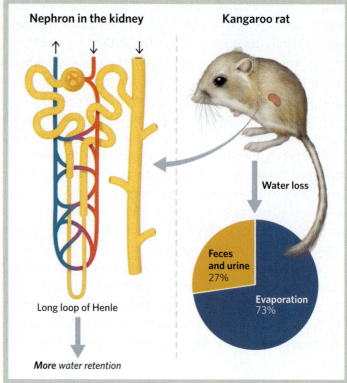

Figure 5.12 Because humans and kangaroo rats are both mammals, it can be interesting to make comparisons, as in this diagram comparing water loss. Humans lose water mainly through excreting feces and urine. Water retention is somewhat regulated by an organism's ability to remove water from wastes in the nephrons (functional units) in the kidneys. This removal occurs through reabsorption of water in the loop of Henle in each nephron. In a nephron in the human kidney, the loop of Henle is relatively short. In contrast, kangaroo rats lose water primarily through evaporation because of an adaptive, extra-long loop of Henle in their kidneys. This physiological change allows the rats to reabsorb more water from waste moving through their kidneys and retain much of the water that would normally be lost to urine (Tracy and Walsberg 2000).

loses about 64% of their water through urine and feces and another 36% through evaporation—as sweat (8%) or during breathing (28%). Kangaroo rats, on the other hand, lose almost 75% of the water they ingest by evaporation during breathing and from their skin, with the rest leaving their bodies via urine and feces (Tracy and Walsberg 2000).

Notice anything interesting? Kangaroo rats lose proportionally much less water from excreting waste as either feces or urine. To accomplish this, their kidneys efficiently extract water through an extra-long *loop of Henle* (**Figure 5.12**). The result is that these kangaroo rats produce the most concentrated mammalian urine known. Additionally, their respiratory system has an unusual countercurrent structure that allows them to reabsorb water in their nasal passages, so that they do not lose as much water to the atmosphere via evaporation when they breathe.

Even with these amazing adaptations, water loss to the air plays a huge role in water balance for kangaroo rats. As a result, these rats behaviorally reduce their water loss by spending most of their daylight hours in an elaborate series of underground burrows. They forage for food and look for mates outside of these burrows only at night, maintaining a nocturnal lifestyle. One of the benefits of their *fossorial* (underground) behavior is that the burrows become humid compared to the desert surface, and this biological microclimate substantially reduces the rat's loss of water to evaporation. To further reduce water loss, while sleeping, a kangaroo rat will sometimes bury its nose in its fur to accumulate a small pocket of moist air from which it breathes, thereby reducing water loss via evaporation off its lungs (Lidicker 1960).

Plants: Trade-Offs between Energy and Water

Plants also adapt to the xeric conditions of deserts by using alterations to their growth and development. For example, for **carbon fixation** (photosynthesis) to occur, a plant needs to acquire carbon dioxide (CO_2) from the air. Plant leaves get CO_2 through stomatal openings in the leaf surface, which we mentioned briefly in Chapter 2. **Stomata** (the singular is *stoma*) are formed by pairs of guard cells, which are like two water-filled balloons joined at both ends (**Figure 5.13**). When water moves into the guard cells, the cells swell and contort, thereby opening a small pore (the stoma) that allows gases to diffuse in and out of the leaf. As water evaporates, the guard cells get smaller, slowly closing the stoma. Regulating the pressure in the guard cells is an intricate process influenced by chemical signals, the plant's internal water pressure, and the availability of soil moisture.

When guard cells swell and open a stoma, CO_2 enters the leaf. With hardly any energy expenditure, this process allows the plant to acquire carbon, which is used to construct sugars that store the Sun's energy captured in photosynthesis. Unfortunately for the plant, although this process has low energy costs, it leads to "expensive" water loss. When leaf stomata are open, water evaporates from inside the leaves and diffuses out into the surrounding atmosphere. In fact, the *concentration gradient* (i.e., difference in water concentration between two locations) driving the diffusion of water out of the plant is steeper than the gradient for CO_2 driving carbon into the plant. Consequently, plants lose somewhere between 100 and 500 molecules of water for each molecule of CO_2 they absorb. This exchange of water for CO_2 is a fundamental but expensive trade-off governing plant growth and photosynthesis.

This aspect of plant biology provides an opportunity to consider trade-offs between water and energy. Because photosynthesis is so crucial to plant life but can lead to such significant water losses, there is strong selective pressure on the photosynthetic process in dry environments. In order to explore this evolutionary trade-off, we need to explain a little more about photosynthesis.

All photosynthetic plants use CO_2 to bind energy into sugars, but there are a few different ways to do this. All the pathways involve the **Calvin-Benson cycle**, in which activated electrons from sunlight provide the energy to build sugars, such as glucose, that contain energy-rich bonds. These sugars then head off to the rest of the plant to power the plant's respiration, reproduction, and assimilation. The pricey part of the process, in terms of water, comes at the beginning of

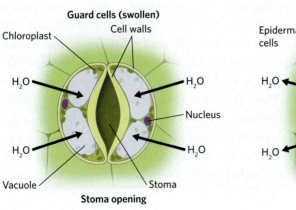

Guard cells (swollen)

Chloroplast

Cell walls

H_2O

H_2O

Nucleus

H_2O

H_2O

Vacuole

Stoma

Stoma opening

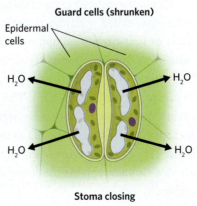

Guard cells (shrunken)

Epidermal cells

H_2O

H_2O

H_2O

H_2O

Stoma closing

Figure 5.13 A stoma opens and allows CO_2 to flow into the plant (and water to flow out) when the guard cells are taut with water. With water loss, the guard cells become less taut and collapse against each other, closing the stoma.

the Calvin-Benson cycle, when the plant combines carbon dioxide with a five-carbon compound (ribulose bisphosphate, or RuBP) to create two molecules of a three-carbon acid (3-phosphoglyceric acid, or PGA). PGA is the basic building block for the next phase of the Calvin-Benson cycle and the glucose and other sugars that the plant will eventually transport to other cells.

You do not need to remember all the names of these compounds, but it is important to know that the enzyme essential for converting CO_2 into PGA is not very good at its job. When CO_2 concentrations drop too low or when temperatures get high, the enzyme (ribulose bisphosphate carboxylase-oxygenase, or RuBisCo) cannot distinguish CO_2 from oxygen (O_2) and it tends to grab O_2 instead of CO_2. Swapping O_2 for CO_2 takes the plant away from gaining energy and into energy consumption. Combining O_2 with sugars is a form of respiration, called **photorespiration** in this context because it occurs during photosynthesis. Photorespiration burns sugars, slows down the rate of energy capture, and reduces plant photosynthetic efficiency. In order to avoid photorespiration, the plant needs to leave its stomata open and capture lots of CO_2, but doing so can lead to the loss of lots of water. As a result, the plant finds itself in a difficult situation with conflicting needs.

Under cool, moist conditions, this dilemma may arise infrequently enough that the plant efficiently captures energy most of the time. During periods of water scarcity or high temperatures, the plant may need to close its stomata, but otherwise photosynthesis proceeds just fine. Approximately 90% of the world's plants are in this position and use the photosynthesis process just described, despite the water issues. We call this form of photosynthesis **C_3 photosynthesis** for the three carbons in the PGA molecule (**Figure 5.14**). But there are two other forms of photosynthesis that have evolved in drier environments for some species of plants. These two photosynthetic pathways are less efficient at fixing energy than C_3 photosynthesis, but they reduce the rate of water loss and the rate of photorespiration when the environment is hot or dry or both.

Let's start with **C_4 photosynthesis**, which spatially separates the Calvin-Benson cycle from the acquisition of CO_2. In C_3 photosynthesis, CO_2 capture and sugar creation both happen in the mesophyll cells, just inside the stomata (Figure 5.14). Opening the stomata allows CO_2 from the environment to flow almost directly into the mesophyll cells. CO_2 moves down a gradient, like water flowing downhill. A low CO_2 concentration in the mesophyll cells increases the difference between the environment and the cell and speeds up the flow of CO_2 into the plant. The stomata can then close faster and protect against water loss.

C_4 photosynthesis creates this gradient by combining CO_2 with a three-carbon compound called phosphoenolpyruvate (PEP). The combination of CO_2 and PEP leads to four-carbon compounds (hence C_4) that head off to the bundle sheath cells, away from the stomata (**Figure 5.15**). In the bundle sheath cells, the four-carbon compounds break down again to feed CO_2 through the Calvin-Benson cycle to create sugars (Figure 5.15). The CO_2-rich environment in the interior bundle sheath cells reduces photorespiration, and the low concentration of CO_2 in the mesophyll cells speeds up CO_2 acquisition and leads to less time with open stomata. There is a little loss of energy in the extra step, but water efficiency is much greater. Common plants that use this form of photosynthesis are corn, sugar cane, and many grasses.

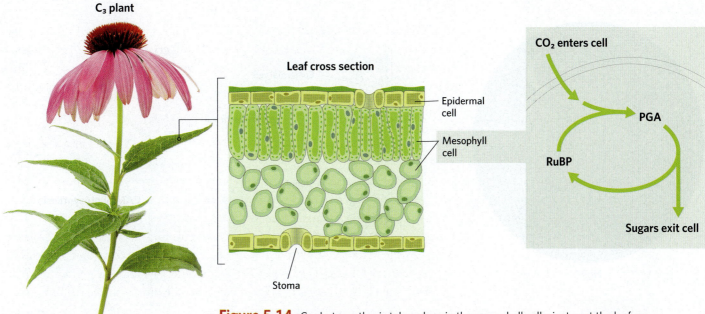

C₃ plant

Leaf cross section

Epidermal cell

Mesophyll cell

Stoma

CO₂ enters cell

PGA

RuBP

Sugars exit cell

Figure 5.14 C₃ photosynthesis takes place in the mesophyll cells, just past the leaf surface, with direct access to the stomata. During the Calvin-Benson cycle, RuBP combines with CO_2 to make PGA, which is then converted into sugar. This pathway is used by approximately 90% of Earth's plant species.

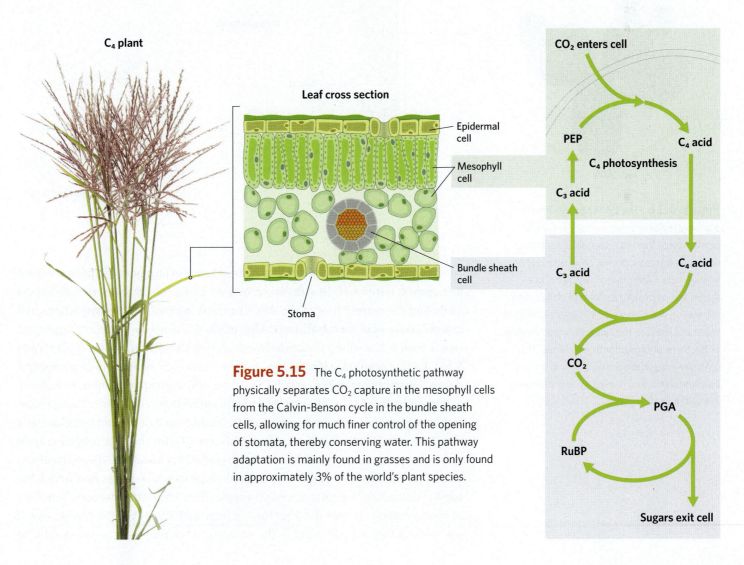

C₄ plant

Leaf cross section

Epidermal cell

Mesophyll cell

Bundle sheath cell

Stoma

CO₂ enters cell

PEP

C₄ acid

C₄ photosynthesis

C₃ acid

C₃ acid

C₄ acid

CO₂

PGA

RuBP

Sugars exit cell

Figure 5.15 The C₄ photosynthetic pathway physically separates CO_2 capture in the mesophyll cells from the Calvin-Benson cycle in the bundle sheath cells, allowing for much finer control of the opening of stomata, thereby conserving water. This pathway adaptation is mainly found in grasses and is only found in approximately 3% of the world's plant species.

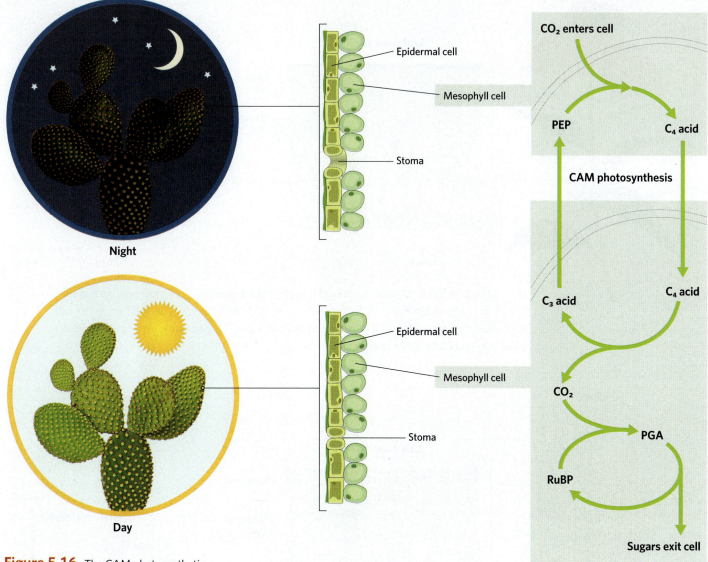

Figure 5.16 The CAM photosynthetic pathway temporally separates the Calvin-Benson cycle from CO_2 capture. CO_2 capture and stomata-opening happen at night, whereas the Calvin-Benson cycle occurs during the day with the stomata closed. This separation is a highly efficient way to conserve water but leads to the slowest carbon fixation of the three photosynthetic pathways. This pathway adaptation is mainly found in cacti and succulents and is used by approximately 7% of the world's plants.

In extremely dry conditions, even this adaptation may not be sufficient to protect against water loss. In a desert, it may be dangerous to open the stomata at all during the searing heat of the day. The third approach to photosynthesis, the **crassulacean acid metabolism (CAM)** method, involves a temporal separation rather than a spatial separation between getting CO_2 and making sugars (**Figure 5.16**). In this method, plants open their stomata only at night, when evaporation is lowest, and take in as much CO_2 as possible through the night. In order to maintain the low CO_2 gradient, CAM photosynthesizers combine CO_2 and three-carbon compounds into four-carbon acids. During the day, the stomata close, and the plants break down the four-carbon acids into CO_2 for the Calvin-Benson cycle and three-carbon compounds for the next night's CO_2 collection. This method was named for the family of plants (the Crassulaceae) in which it was first found. We see it in many cacti, succulents, and pineapples. One interesting aspect of this form of photosynthesis is that the four-carbon intermediary makes the leaves sour. If you were to taste a CAM plant in the morning, it would have a pronounced sour

Table 5.1 Types of Photosynthesis

	C$_3$	C$_4$	CAM
Type of plants	Most plants (about 90%)	Many grasses, corn, sugarcane (about 3%)	Cacti, succulents (about 7%)
Method	Calvin-Benson cycle in mesophyll cells	Calvin-Benson cycle in bundle sheath cells	Calvin-Benson cycle during the day; CO$_2$ acquisition at night
First carbon molecule produced	A three-carbon acid (3-phosphoglyceric acid [PGA])	A four-carbon acid (malic, aspartic, or oxaloacetic acid)	A four-carbon acid (usually malic or aspartic acid)
Efficiency	Most energy-efficient method, but high potential for water loss	Intermediate energy efficiency and water conservation	Lowest energy efficiency but best water conservation
Habitats/biomes	Mostly temperate	Tropical or semitropical, mostly grasslands	Arid/desert

Plants must balance the acquisition of carbon via photosynthesis with the loss of water associated with gaining that carbon in the form of CO$_2$. The climate in which a plant has evolved dictates which of the carbon acquisition strategies it uses: C$_3$, C$_4$, or crassulacean acid metabolism (CAM).

taste that dissipates over the day as the four-carbon acid (usually malic or aspartic acid) is used up in the Calvin-Benson cycle.

One side effect of the spatial separation of the C$_4$ process and the temporal separation of the CAM process is that they both slow down the overall rate of photosynthesis. If conditions are cool and moist, and stomata can be open most of the time, neither of these approaches is as efficient at energy fixation as C$_3$ photosynthesis. Nonetheless, the trade-off between energy efficiency and water efficiency may be essential in hot environments (C$_4$) and dry desert-like environments (CAM). Every plant species has evolved to use just one of these forms of photosynthesis (Table 5.1). The trade-off of lower energy efficiency for higher water efficiency is important enough that C$_4$ and CAM photosynthesis have evolved independently in several groups of plants.

In addition to these photosynthetic trade-offs between water loss and energy acquisition, plants may also make trade-offs in energy allocation that affect their ability to acquire water or light. Here we think back to the principle of allocation and consider how plants allocate their limited energy toward growth and what that might tell us about their needs for peak performance as measured by survival and reproduction. If we think about the drier biomes from Chapter 3, it is clear that water is often in much shorter supply than light. Regardless of how the type of photosynthesis may conserve water, plants still need to acquire water. Roots reach down toward water, and leaves reach up to allow more photosynthesis. Plants can alter their investment in each of these depending on the availability of water or sunlight.

We can examine these allocations to gain insight into water balance. In individual plants, we can compare the biomass of the plant's roots to the biomass of its leaves and stems (i.e., its shoots) and examine the **root-to-shoot ratio**. A high root-to-shoot ratio (>1) suggests a large investment in belowground root growth and architecture (investment in acquiring water and other belowground resources), whereas a low ratio (<1) implies the opposite investment and a choice for leaves and stems over roots (investment in accessing sunlight and acquiring energy). When plants are water stressed, they will invest more in root biomass than in stems and leaves because roots in this context are the limiting factor of plant growth (Reich 2002).

Figure 5.17 Patricia Tomlinson and Paul Anderson (1998) measured root and shoot growth in red oak seedlings (*Quercus rubra*) grown with and without water stress under different CO_2 concentrations. Regardless of CO_2 concentration, the water-stressed seedlings invested relatively more energy in root growth (i.e., they had a higher root-to-shoot ratio).

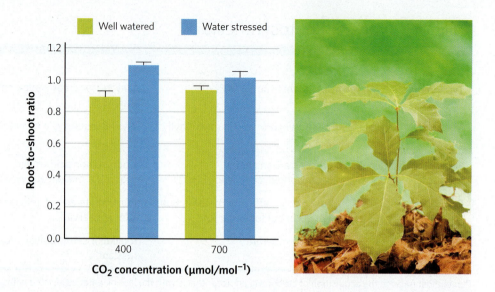

This growth trade-off has been demonstrated in northern red oak seedlings (*Quercus rubra*) grown under conditions of water stress and varying levels of CO_2 (Tomlinson and Anderson 1998; **Figure 5.17**). Regardless of CO_2 levels, when water is harder to acquire, the trees invest more in the roots than in the leaves, thereby physiologically addressing the limiting factor for the plant's growth. Of course, like much in ecology, the situation is often more complicated than the red oak example because root growth is also affected by nutrients in the soil, which may also be a limiting resource.

If we think about water relationships and roots versus leaves on the scale of biomes (Chapter 3), we can hypothesize about the patterns of root-to-shoot ratios across a pronounced water gradient (**Figure 5.18**). For example, we might hypothesize that, in the desert, individual plants should grow deep and extensive root networks and have little aboveground biomass. We might also hypothesize that in many temperate forest environments, water is relatively easy to acquire, and therefore plants' investment in roots and shoots will be nearly equal. Finally, we could

Figure 5.18 Root-to-shoot ratios of plants growing in desert, temperate, and tropical biomes may reflect trade-offs between root and shoot growth driven by water and light availability. If this trade-off outweighs the myriad other selective pressures on plants, relative investment in roots (the root-to-shoot ratio) should be highest in xeric (dry) conditions and lowest in mesic (wet) conditions. Testing these sorts of patterns requires large amounts of data from around the world.

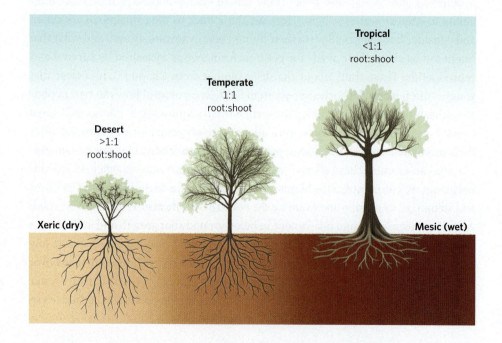

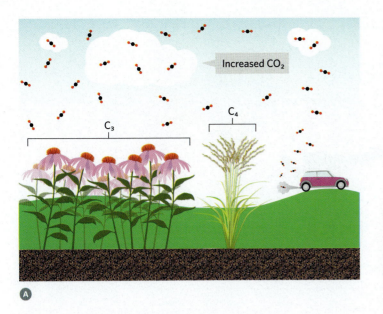

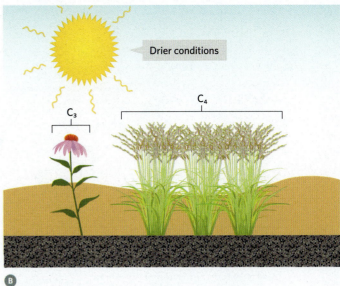

hypothesize that in tropical rainforests, where water is in excess and the important limiting factor may be light availability, the typical plant silhouette might have much more biomass aboveground than belowground. These hypothesized patterns could be tested against existing data on root-to-shoot ratios across biomes or by going into the field and taking these measurements oneself within various biomes. (We think traveling to different biomes and measuring plants sounds more fun.)

Once we start thinking about global patterns, another obvious question is how climate change may alter the distribution of plant species with different photosynthetic pathways. Can we combine an understanding of the trade-offs between energy and water access to predict future changes in the frequency or distribution of plant species? Unfortunately, the short answer is "not really" because so many factors are changing at once that generalizations are difficult. Let's break it down.

If an increase in atmospheric CO_2 makes it easier for C_3 plants to access CO_2 quickly and keep their stomata open less often, then maybe the number of C_3 plants around the world will increase (**Figure 5.19**). The problem with that simple conclusion, though, is that increases in atmospheric CO_2 concentrations are also leading to increases in temperature and decreases in precipitation. Those changes combine to reduce access to water, which favors an increase in C_4 vegetation. Recall that, even with more CO_2, the red oak seedlings grew longer roots when water was scarce. Water stress, therefore, remains very important, even at higher CO_2 levels.

If we look at a map of the world that shows the proportion of local vegetation that is C_4 vegetation (**Figure 5.20A**), we can see that C_4 plants tend to be found most frequently in grassland biomes, particularly in tropical grassland systems (tropical savannas). Global climate change includes not only rising CO_2 concentrations and temperatures but also changes in precipitation. A careful examination of **Figure 5.20B** suggests that parts of the world that currently have large concentrations of C_4 grasses are expected to get drier, which may mean that they will shift toward desertlike conditions.

All in all, we expect to see biomes altered by climate change, and the distribution of plants with different photosynthetic pathways will also change with this shift. All of this makes it difficult to say with any certainty whether C_3 or C_4 species are expected to become more globally common.

Figure 5.19 This simple graphic communicates the effects of global climate change on the abundance of C_3 and C_4 plants. As the global climate changes in response to increasing atmospheric CO_2 concentrations, scientists may find it difficult to predict how the frequency and distribution of C_3 and C_4 plants will change. **A** Elevated concentrations of atmospheric CO_2 favor C_3 plants, but precipitation is also expected to change in many areas. **B** If water availability decreases because of lower precipitation or higher evaporation with higher temperatures, C_4 plant densities should increase.

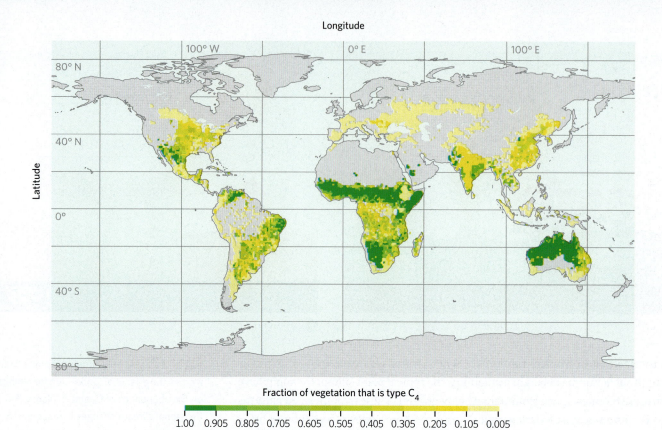

Fraction of vegetation that is type C$_4$

1.00	0.905	0.805	0.705	0.605	0.505	0.405	0.305	0.205	0.105	0.005

A

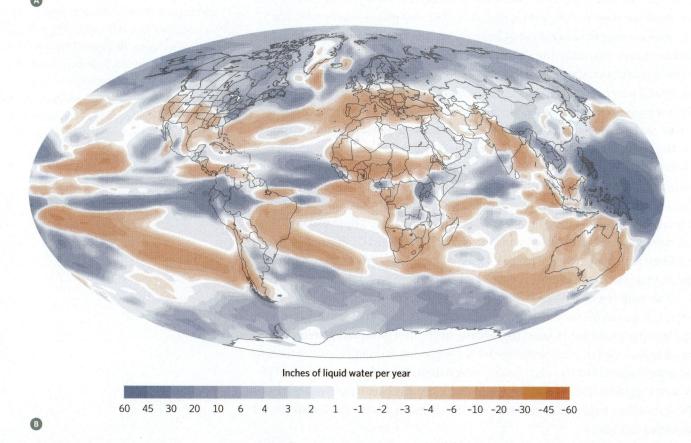

Inches of liquid water per year

60	45	30	20	10	6	4	3	2	1	-1	-2	-3	-4	-6	-10	-20	-30	-45	-60

B

Figure 5.20 C$_4$ vegetation distribution and the effects of global climate change on precipitation. **A** This map shows the global distribution of C$_4$ vegetation (Still et al. 2003). Darker colors indicate that C$_4$ plants constitute a higher fraction of local vegetation. Comparing this map to the biome maps in Chapter 3, we can see that tropical savannas (Figure 3.22A) are almost 100% C$_4$ vegetation, whereas temperate grasslands (Figure 3.25A) are about 50% C$_4$ vegetation. **B** On this map of the projected change in global precipitation by the end of the twenty-first century from NOAA (Delworth et al. 2006), note that many areas at lower latitudes and areas with temperate or tropical grasslands are expected to receive less rainfall. These changes may shift grassland biomes toward deserts and shift vegetation away from C$_4$ grasses toward succulents that use CAM photosynthesis.

5.5 Thermal Trade-Offs

Living within Thermal Limits

Heat enables cellular respiration. When an organism becomes too cold, cellular respiration slows and may cease, leading to death. Exposure to very cold temperatures can also freeze cells. When this happens, the water in cells expands (Chapter 3), bursting the cell walls and destroying the integrity of the organs these cells create. On the other hand, if an individual becomes too hot, cellular structures and DNA break down, which can lead to permanent damage or death. The temperatures at which an individual dies either from heat or cold stress are called the upper and lower lethal temperatures, respectively. These limits impose a basic thermal boundary on the environments and biomes in which an organism can live. Between the lethal limits, there is also a range of performance (survival, growth, and reproduction) that individuals can experience. Natural selection acts strongly on behavioral and structural adaptations that allow individuals to regulate their own internal temperatures, and these features necessarily involve trade-offs.

In the following sections, we explore how animals and plants achieve this regulation. Let's begin with some vocabulary. Although thermal limits can be critical, environments such as aquatic and tropical terrestrial environments offer a fairly narrow temperature range. Organisms in these locations may not need to do much to stay within their thermal limits. When organisms do not regulate their internal temperature but largely allow their internal body temperature to follow external temperatures, they are called **poikilotherms**. This approach may seem impractical or unwise, but many plants, insects, and fish are poikilotherms.

Organisms that regulate their internal temperature practice thermoregulation, and some do it by using only external mechanisms (**ectotherms**), whereas others employ both external and internal mechanisms (**endotherms**). Ectotherms move into sunlight when cold or into the shade when hot; they may also employ a wide range of strategies that allow them to collect or reflect external heat or redirect winds, as we will explore. Many insects, amphibians, and reptiles are ectotherms. Ectotherms, therefore, change with their external environment, but—unlike poikilotherms—they use that environment or external methods to regulate their internal temperature.

Endotherms use many of the same approaches as ectotherms to assist with thermoregulation, but they also use metabolic heat, a range of muscle motions (e.g., shivering, panting), and changes in blood circulation to regulate their internal temperatures. All mammals and birds are endotherms, as well as a small subset of plants (skunk cabbage), insects (some bees), and fish (tuna and swordfish).

Some endotherms use metabolic heat (heat generated by metabolizing food) to maintain their internal temperature at such a consistent level that we call them **homeotherms**. Birds and mammals comprise the vast majority of homeotherms, and they maintain their internal temperature in a narrow, warm band, generally between 36°C (97°F) and 44°C (111°F), though the range is narrower than that for each species. This temperature range is often above the temperature of their external environment, hence the common term *warm-blooded* to describe these species. Homeotherms generally have feathers, hair, or fur; these adaptations help insulate their bodies and hold onto heat or protect against excessive heat. In con-

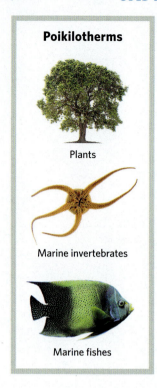

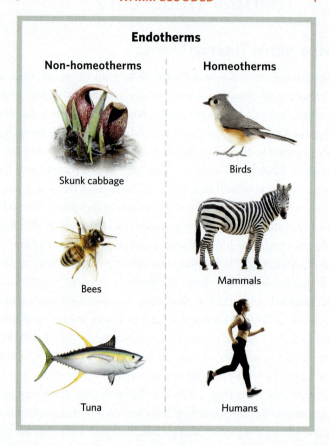

Amount of control over internal body temperature

Figure 5.21 Organisms can be grouped by the degree to which they control their internal temperature (increasing from left to right). Many plants are poikilotherms—organisms that do not regulate their internal temperature—as are a range of marine organisms, which makes sense because the temperature varies less in marine environments than in terrestrial environments. Organisms that use external factors (like sun or shade) to alter their internal temperature are ectotherms; these include reptiles, amphibians, insects, and many freshwater fish. Endotherms use all the regulation processes of ectotherms plus internal processes to regulate their temperature, but only a subset of them (mostly mammals and birds) maintain their temperature in a very narrow range and can be considered homeotherms. Non-homeothermic endotherms include skunk cabbage, bees, and tuna, which can all regulate their temperatures to some extent but do not maintain a constant temperature.

trast, non-homeotherms do not maintain their temperature within such a narrow range but can regulate their internal temperature to some extent.

These terms can get a little confusing, but **Figure 5.21** visually sums them up, which may help. Think first of whether organisms do or do not regulate their internal temperatures—those that *do not* are the poikilotherms; all other organisms do. Within the group that does thermoregulate, many do so solely by moving into warmer areas or altering heat capture by reflecting heat, wind, or water (ectotherms). Those that use internal mechanisms (endotherms) also use all the external tools of ectotherms but add heat from metabolism or changes in the movement of blood and fluids that help hold onto internal heat. Finally, a subgroup of endotherms maintains their internal temperature within a very narrow range (homeotherms) that allows peak performance.

We started this discussion by pointing out that all organisms have lethal thermal limits, both hot and cold, and it makes sense that such limits might impose strong selection for some form of temperature regulation. But why would organisms ever want to maintain a narrow band of internal temperatures, and why would a narrow band allow peak performance? Readers who have some background in chemistry, or who have done much cooking, will recognize that many chemical reactions happen more rapidly with heat, up to a point. Once things get too hot, permanent and often lethal consequences may ensue. Nonetheless, a little heat makes reactions happen faster, which means muscles work better, speed increases, and the ability to capture prey (or, conversely, the ability to escape predators) improves. So, organisms often attempt to maintain their temperatures in a warm, narrow band.

Animal Thermoregulation   In considering the thermal biology of animals, we turn to the classic work of Raymond Huey and Montgomery Slatkin (1976) to graphically explore the benefits of regulating internal temperature. Huey and Slatkin suggested that, as the internal body temperature increased, there were benefits to energy acquisition and overall physiological performance up to a point (**Figure 5.22**). Beyond that ideal temperature (X_{ideal} in Figure 5.22), performance dropped off quite sharply. Because of the humped nature of the curve in Figure 5.22, it would make sense for an organism to regulate its internal body temperature near X_{ideal}, where the energy-gain benefit is maximized (i.e., B_{peak}). If this graph is a realistic representation of what happens in nature, then we should expect to see animals evolve in ways that maintain internal temperatures near X_{ideal}. We do not often know what the exact value of X_{ideal} is for most species (or even what it should be), but we can get a sense of whether such a value exists by checking whether individual animals tend to keep their internal body temperature steady as the external temperature changes. In other words, we can assume that if an individual "does things" to maintain its body temperature near some value, that temperature value must be evolutionarily important.

For an example, let's look more closely at Huey and Slatkin's research. The two studied a lizard called the lace monitor (*Varanus varius*; **Figure 5.23**), which lives in the relatively mild and cool eastern woodlands of Australia. Lizards do not

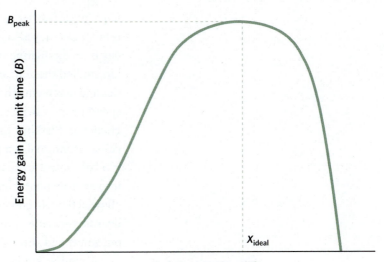

Figure 5.22   In their seminal paper, Huey and Slatkin (1976) explored the physiological benefits of regulating internal temperature. As body temperature increases, physiological benefits increase in terms of energy gain per unit of time, up to a maximum (B_{peak}). At the temperature associated with peak benefits (X_{ideal}), the organism receives its maximum energy return. Beyond X_{ideal}, the physiological benefits start to diminish. Theoretically, organisms should try to maintain their body temperatures at or near X_{ideal}.

Figure 5.23   In a study of the ectothermic Australian lace monitor (*Varanus varius*, inset), the lizard maintained its internal body temperature within a tight range throughout a day by means of behavioral choices, such as basking in the sun and resting in shade (Huey and Slatkin 1976).

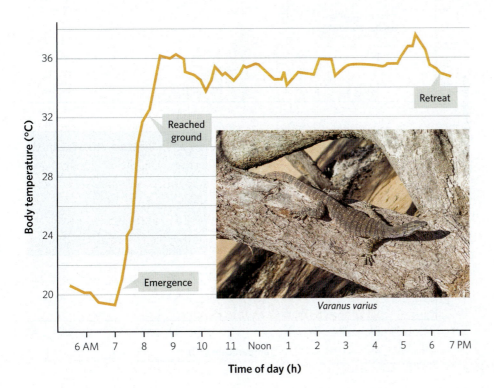

Varanus varius

have fur or feathers to help regulate their metabolic heat, and this lack of insulation means they also have no structural mechanism to prevent heat gained in the day from dissipating into the environment at night when it is cold. Huey and Slatkin studied these lizards by implanting temperature probes into their body cavities and measuring their internal temperature throughout the day. Monitor lizards spend most of their day in trees, hunting prey and escaping their own predators. Huey and Slatkin's data indicate that the lizards were able to maintain a fairly constant internal temperature throughout the daylight hours (Figure 5.23). This level of control arose from the lizard's behavior, which involved physically moving into microhabitats that either added to their internal heat (basking in the sun) or allowed them to cool off (resting in the shade). Through these actions, the monitor lizards were able to regulate their internal temperature quite precisely by seeking out sunny and shady areas in their environment.

A similar study was done with critically endangered Oaxacan spiny-tailed iguanas, or *Ctenosaura oaxacana* (Valenzuela-Ceballos et al. 2015). According to the International Union for the Conservation of Nature, these iguanas are threatened with extinction by the severe reduction of their tropical dry forest habitat due to human development. The researchers studying them compared the iguanas' body temperatures and environmental temperatures across a wide variety of habitats and vegetation types. They found that the iguanas were able to maintain internal body temperatures in a narrower range around their ideal temperature than the wider range of temperatures presented by their environment (**Figure 5.24**).

Figure 5.24 Sara Valenzuela-Ceballos and her colleagues (2015) collected environmental temperatures and body temperatures from Oaxacan spiny-tailed iguanas (inset) in a range of habitat types in the Mexican state of Oaxaca. The top part of the graph shows the frequency distribution of environmental temperatures, and the bottom part shows the frequency distribution of body temperatures. Notice the narrower range of body temperatures, suggesting that the lizards are regulating their temperature toward a preferred or ideal temperature. The graph represents just the observations from pasture during the rainy season, but the researchers collected observations in rainy and dry seasons and in four different habitat types.

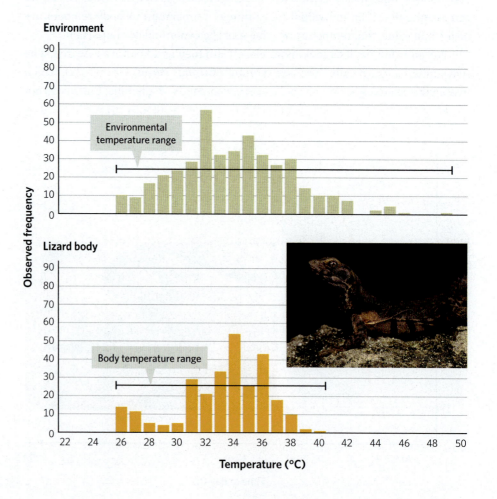

Because reptiles and other ectotherms cannot internally regulate their body temperatures, their survival depends on temperature ranges in their environment that allow them to thermoregulate with behaviors. This dependence makes them uniquely vulnerable to large-scale thermal disruptions. With the onset of anthropogenic climate change, increases in global temperature could compromise the thermoregulatory efficiency of ectothermic animals. A warming planet pushes species into less optimal habitats with respect to temperature and makes their thermal environment more hostile. This exact scenario was investigated by Sara Valenzuela-Ceballos and colleagues. According to their model of the climatic conditions in the Oaxacan spiny-tailed iguanas' geographic range, this species may actually thrive in a warmer world due to their thermoregulatory prowess. Despite this good news, if habitat loss in the dry tropical forest continues at the rate it is going, the species' road to extinction may still be short.

Plant Thermoregulation   Most plants have indeterminate growth, meaning that they do not stop growing as adults. They are also, quite obviously, immobile after they grow out of the seed phase. Given these two facets of their life history, a plant's options for staying within thermal limits must involve features of its growth and development. In this way, plant growth is somewhat analogous to behavior in animals.

As an example, let's look at brittlebush (*Encelia farinosa*), a common shrub that grows in the Sonoran and Mojave Deserts in the southwestern United States. Individuals of this species flourish in this hot, dry, desert biome where summer temperatures routinely exceed 40°C (104°F), and total yearly rainfall can be as low as 75 mm (3 in.). One of the more striking features of this plant, beyond its pretty yellow flowers, is the white cast on its leaves (**Figure 5.25A**), which comes from extensive *pubescence* (hairs) that grow on both the top and bottom leaf surfaces. Pubescence of this sort is an adaptive feature for desert plants and can reduce the leaf's internal temperature by reflecting incoming light off the leaf surface (Ehleringer and Mooney 1978).

In addition to the presence of leaf pubescence, many perennial desert plants avoid the stresses imposed by high temperatures and drought by not having leaves at all (as in the case of cacti), by shedding their leaves during the hot dry summers (as with drought deciduous plants such as the Chilean palo verde tree, *Geoffroea decorticans*, or the California buckeye, *Aesculus californica*), or by producing new leaf types as the seasons progress from mesic (wet) to xeric (dry). This last approach is the one used by brittlebush (Ehleringer and Björkman 1978). Over the course of the growing season, as temperatures move from cooler spring days to ferociously hot summer ones, *E. farinosa* produces new leaves that have more pubescence than early-season leaves. The shift in pubescence allows the plant to change from prioritizing photosynthesis (few hairs, allowing lots of access to light) in the early season to prioritizing temperature control and water retention (lots of hairs to deflect heat and wind and reduce evaporation) in the late season.

Demonstrating that the presence of pubescence significantly affects a leaf's internal thermal regime involved a series of simple, elegant experiments over many years. In one experiment, James Ehleringer and his lab (Ehleringer and Björkman 1978) showed that leaf absorptance of sunlight over the photosynthetically active

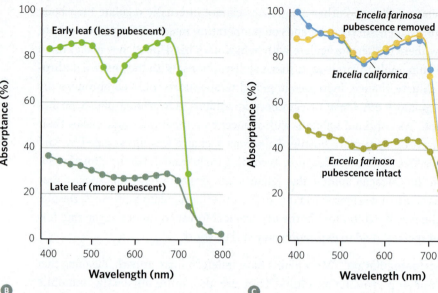

Figure 5.25 **A** Brittlebush (*Encelia farinosa*) is a common shrub in the Mojave and Sonoran Deserts, with leaves that are covered in white hairs called pubescence, an adaptation to reflect incoming sunlight and thereby avoid excessive heating. **B** The less pubescent leaves of the early season showed significantly more light absorptance (i.e., less reflectance) than the heavily pubescent leaves of the late season, over a range of light wavelengths (Ehleringer and Björkman 1978). **C** Researchers removed leaf hairs from a heavily pubescent *E. farinosa* leaf with a razor blade and compared its light absorptance to that of an unmodified leaf and a leaf from a closely related but nonpubescent species (*E. californica*) from a mesic environment. The modified *E. farinosa* and the nonpubescent *E. californica* leaves showed significantly more absorptance (i.e., less reflectance) than the unmodified pubescent *E. farinosa* leaves (Ehleringer and Björkman 1978).

range of solar radiation (400–700 nm) was approximately 50% higher in early leaves (less pubescent, greener) than in later leaves (more pubescent, whiter) (**Figure 5.25B**). In other words, leaves that are less fuzzy allow for more light absorptance and produce higher leaf temperatures.

In another experiment, they removed the late-season pubescence (with a razor) and produced light absorptance levels similar to those of a nonpubescent close relative (*E. californica*) that lives in more mesic coastal areas (**Figure 5.25C**; Ehleringer and Björkman 1978). Although light absorptance is great for photosynthesis, in this case, the absorptance levels in Figure 5.25C suggest that removing the pubescence created a leaf that could no longer limit the amount of radiation it took in, which led to internal temperatures almost identical to those of the leaves from the nonpubescent species. While these higher internal temperatures are tolerable for the nonpubescent species that lives in a wetter, cooler area, they could be deadly for the desert species during the height of summer.

Whether we consider plants or animals, the preceding evidence shows that individuals have ways of keeping their internal temperatures fairly steady despite sometimes broad fluctuations in external temperature. However, maintaining internal body temperature has an energetic cost in addition to a benefit. For the brittlebush, these costs are the energy needed to grow and maintain specialized leaf structures. For the ectothermic lizards, these costs come in the form of deferred foraging time while basking in or hiding from the sun.

For homeotherms, the costs may be even higher because their metabolism produces much of the heat that maintains their consistent and relatively high internal temperatures. This consistency allows high performance but only if these animals can find enough food or shelter. One advantage of a poikilotherm lifestyle is that the consumption costs are much lower. In an environment that does not change too much, this trade-off of lower energy needs for lower performance leads to many organisms having successful lives.

Put differently, the investment in maintaining internal temperatures leads to performance advantages but always at a cost of time or food or energy. Those advantages may lead to success and may explain all kinds of interesting adaptations. But, as the world changes, particularly through human actions—and ecologists should always remember that the world never stops changing—such adaptations may become less advantageous. Let's turn, then, to thinking about why organisms survive and thrive in particular environments.

5.6 The Niche: The Grand Trade-Off

Definitions and Measurement

We have focused our discussion specifically on trade-offs around energy, water, and heat because they are central to an organism's evolutionary fitness. Individuals cannot maximize their performance across all of these factors simultaneously. Instead, they need to optimize their overall performance by making trade-offs, which may compromise performance in one area but offer the best set of adaptations for multiple competing factors simultaneously.

The end result of this grand trade-off is that, over time, each species evolves its own unique suite of physiological and behavioral traits that complement and enhance each other. These traits collectively allow individuals of the species to live and reproduce within a limited set of environmental conditions (Schoener 2009). At any given location on Earth, the conditions for survival of an individual are either present or absent, and how well an individual does in that location depends on how close these conditions come to being optimal.

The specific set of environmental conditions in which an organism can live and reproduce is called its ecological **niche**. This fundamental concept involves determining the critical factors for survival and reproduction for a species rather than identifying a physical location. Nonetheless, a location that provides these conditions can be described as providing appropriate niche conditions. You will see the term *niche* used in many of the chapters to come, so it is worth exploring the concept in more detail.

The word *niche* was first used by Joseph Grinnell, an American zoologist who conducted groundbreaking ecological research in the early twentieth century. He believed that each species had a "role" or a "place" in the natural world determined by its habitat and behavior. Grinnell asserted that ecological niches existed on their own, regardless of whether species "filled" them or not. By the 1930s, Charles Elton, a British ecologist, refined this concept of niche to suggest that the niche includes simultaneously both a species' response to, and its effect on, the environment. This understanding of the ecological niche implied that species have independent agency and can potentially affect and alter their environment in significant ways. An example of this agency can be found with earthworms, which alter the nutrient availability and texture in the forest soils where they live.

The first quantitative description of a niche came from G. Evelyn Hutchinson in the 1950s, when he defined the niche as the range of traits or behaviors that allows a species to persist in a given environment. The Hutchinsonian idea of

Figure 5.26 Ⓐ Devil's Hole pupfish (*Cyprinodon diabolis*). Ⓑ Devil's Hole is a freshwater pool in a narrow rock cleft that represents the entire niche and range of *C. diabolis*.

Ⓐ

Ⓑ

the niche closely allies the concept with our understanding of physiological and behavioral ecology and suggests that aspects of the niche can be precisely measured and quantified.

Let's examine a Hutchinsonian version of the niche in some detail, using the example of the critically endangered Devil's Hole pupfish (*Cyprinodon diabolis*), the vertebrate species with the smallest known geographic range in the world (**Figure 5.26A**). This diminutive fish exists only in a single 20 m² (215 sq. ft.) freshwater pool located in a recessed rock fracture, just outside of Death Valley National Park, in the middle of the Mojave Desert of Nevada (**Figure 5.26B**). As part of the efforts to protect this unique species, ecologists have extensively studied many aspects of the fish's biology and ecology.

Physiologically, we know that the Devil's Hole pupfish lives in water temperatures ranging from 32°C (90°F) to 34°C (93°F). These are the pupfish's thermal limits, above or below which individual pupfish will likely die. But these fish also need a certain amount of dissolved oxygen (measured in parts per million, or ppm) in the water for respiration. For the Devil's Hole pupfish, the range is 1.8–3.3 ppm.

We can use these two physiological tolerances (temperature and dissolved oxygen concentration of water) to construct a graph of the fish's **niche space**, represented by a rectangle (**Figure 5.27**). By *niche space*, we simply mean a region in a multidimensional space constrained by environmental factors that affect the fitness of individuals in that species. In the pupfish's two-dimensional example, the rectangle represents the *fundamental* or *potential niche space* the species can occupy given these two criteria. If, however, the pupfish cannot use all of the conceptual space in the rectangle due to interactions with other species, then the fish may not be able to use the entire environment represented by the rectangle, and its niche space is better represented as some subset within the rectangle (the dotted oval in Figure 5.27). This reduced area is sometimes referred to as the *realized niche space*, discussed further in Chapter 7.

Looking quantitatively at two different physiological requirements simultaneously (as in Figure 5.27) shows us the basic way factors interact in Hutchinson's definition of the ecological niche, but we can extend it to include additional environmental constraints. Including more constraints effectively adds more axes onto Figure 5.27. Formally, Hutchinson defined a species' ecological niche as an **n-dimensional hypervolume**, where *n* indicates the number of dimensions, or different environmental conditions, that determine where an organism can survive. A *hypervolume* is any chunk of space that is defined by more than three axes (a "volume" on hyperdrive). An *n*-dimensional hypervolume is a confusing idea for many of us because we are not good at thinking beyond the typical three dimensions we can touch or draw on paper.

To help envision more dimensions in a niche, let's explore how the two-dimensional niche space figure for the pupfish could be expanded. We could do this by noting that the Devil's Hole pupfish also requires a certain depth of water in which to breed and deposit its eggs. If we add this factor to our graph, we get a three-dimensional niche space, which could be a cube, sphere, or egg shape (**Figure 5.28**). This exercise can be theoretically extended to consider four or even more niche axes, despite the fact that we cannot draw them well in a figure.

Resource Utilization Curves

Because an *n*-dimensional hypervolume is somewhat cumbersome, and because not all elements of the niche are equally interesting or limiting, ecologists often focus on specific aspects of an organism's niche space in which to explore optimal performance. Graphs such as Figures 5.27 and 5.28 show us the outer dimensions of the niche space, but they cannot show us which part of the niche may be optimal or preferred.

In order to explore preferences within one axis of the niche (e.g., the water temperature axis in Figures 5.27 and 5.28), let's again consider the pupfish but this time examine the size of food items that the pupfish might eat. By watching individuals eat, we could construct a frequency histogram with food size on the x-axis and number of individual fish foraging on this food size on the y-axis (**Figure 5.29**). If we assume a smooth curve by watching a very large number of fish, we would likely see a normal distribution (Chapter 2) and could calculate the mean and variance of the size of the pupfish's food. Figure 5.29 indicates that

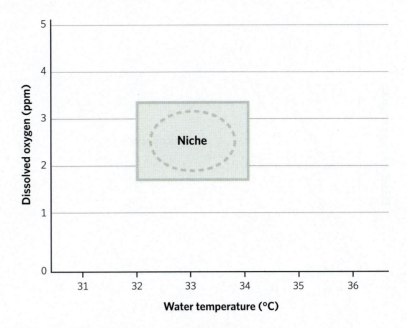

Figure 5.27 This theoretical two-dimensional niche space graph for the Devil's Hole pupfish shows the species' water-temperature and dissolved-oxygen requirements. The rectangle indicates the fundamental or potential niche space, and the oval represents the realized niche, meaning a more limited niche space that may occur due to interactions with other species.

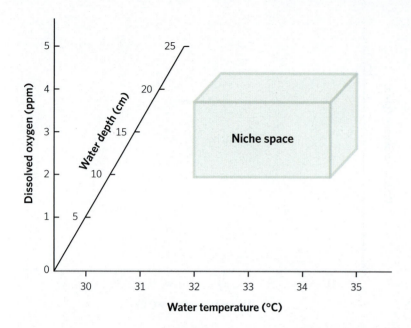

Figure 5.28 Adding another niche-requirement axis (water depth) to the niche space graph in Figure 5.26 produces a three-dimensional space. If we were somehow able to add additional niche axes to this figure, we could approach the idea of Hutchinson's *n*-dimensional hypervolume niche, but we would have to add a lot of axes and figure out how to display them visually.

the pupfish concentrate their consumption on food of a particular size, represented by the mean, but that they also consume food over a range of sizes, represented by the variance. This sort of graph, which shows how an organism uses a particular resource in its environment, is called a **resource utilization curve**. In this context a **resource** is anything that can be used up (and for which organisms might compete; Chapter 7).

In the late 1960s, Robert MacArthur and Richard Levins became the first to quantify the axes of a Hutchinsonian niche by measuring the frequency at which organisms are found at each value along the axis (like Figure 5.29). Their version of the niche breaks the concept into a series of resource utilization curves for each species because, in their view, the niche can be quantified by measuring what living, breathing individual organisms actually do and where they do it (Schoener 2009). So, a species' niche should be based on observing what individuals of a species do, rather than solely on the endpoints of its physiological requirements.

If we were to take the pupfish example one step further and observe the water depth at which individual fish feed, we could combine feeding-depth data with food-size data and construct a histogram to describe the pupfish's foraging preferences across two axes, or within two dimensions of a Hutchinsonian niche (**Figure 5.30**). This graph is a depiction of what a resource utilization curve could look like for the Devil's Hole pupfish. Resource utilization curves can combine data about individual physiological tolerance, food choice, and many other factors to quantify a species' niche. Curves like these allow a more nuanced understanding of the pupfish's niche requirements and improve the ecologists' ability to protect this extremely rare and critically endangered species.

Figure 5.29 In this theoretical frequency histogram of food-size choice for Devil's Hole pupfish, we can see that the mean is also at the peak, or mode, and represents the most commonly eaten and perhaps most preferred food size. The variance is a measure of the range of spread of food sizes around the mean and represents the squared average difference between the mean and the other food sizes. Note that the shape of the histogram approximates a normal distribution, or bell-shaped curve, shown by the orange line. The smoothed version of this is often referred to as a resource utilization curve, with food size as the resource in this instance.

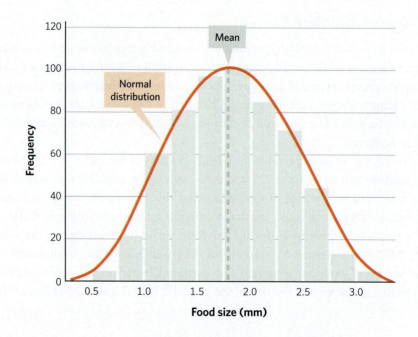

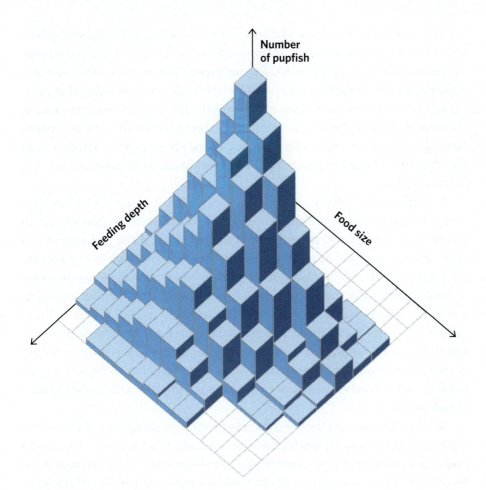

Figure 5.30 Theoretical histogram showing two niche dimensions: feeding depth and food size. The number of pupfish acts as the frequency counter for these dimensions.

The Niche in Context

It is important to recognize that an organism's niche is not completely defined by the physiological and behavioral traits of the organism itself. Every species exists as a member of a larger extended ecological community (Chapters 7–10) and does not live in a vacuum. Species influence each other in both positive and negative ways, as we will explore, and they should not be viewed as static entities, like paintings in a museum. Rather, they should be understood as dynamic and alive with influence, affecting the environment and the other species around them. As discussed in later chapters, the concept of the ecological niche plays a foundational role in our understanding of many other ecological concepts, including competition, predation, mutualism, food webs, biodiversity, and ecosystems. Therefore, we will revisit the concept of the niche and the topics of physiology and behavior many times in this book.

5.7 Physiology and Behavior in a Changing World

Studying ecology from the perspective of single individuals brings to the fore details of energy allocation, behavioral decisions, and the core concept of the niche. These are but a few of the many factors that emerge when considering how individuals respond to their environment. If you found yourself intrigued by these topics, we encourage you to explore the extensive literature on behavioral ecology and physiological ecology. In particular, we suggest you examine how anthropogenic changes affect these factors. How can scientists use knowledge of physiology and behavior to make predictions in a world with a rapidly changing climate, chemical pollutants, species invasions, and extinctions? In all of these circumstances, the degree to which individual organisms can respond to and cope with the changing ecological and environmental conditions is fundamental to their long-term fate.

In Chapter 1, we discussed the origins of the term *ecology*, which derives from a German word (*Oekologie*) meaning the "doctrine of the household." In this chapter, we explored the concept of the ecological niche and arrived at a fairly sophisticated and nuanced definition. Yet at its core, a niche is an organism's unique "position" or "location" within an ecosystem, including the range of conditions for the species to persist (physiology) along with its ecological role in the system (behavior). This sounds a lot like a species' "home address" and aspects of its "household," which hearkens back to the definition of *ecology*. In Chapter 1, we pointed out that in the modern era our "ecological house" is undergoing significant change. Ecological understanding of physiology and behavior can be used to see how we have damaged or altered the niches of individual species.

Perhaps it is easiest to see these human effects through the role of altered climatic conditions on individual behavior, survival, and reproduction. As discussed earlier, allocation of energy across physiological functions can be strongly influenced by local environmental conditions, such as water availability and temperature. If individuals live near their physiological limits, as with the kangaroo rat in the desert, even a slight shift in temperature or water availability can mean the difference between life and death. We know from climate change research that desert ecosystems are likely to experience warmer than average temperatures and relatively large shifts in the timing and amount of precipitation. When such shifts affect individuals of a species, their populations will decline over time, sometimes dramatically (Chapter 6). Given these relationships, it should not be surprising that some of the early "climate change casualty species" are those that live in deserts or other arid ecosystems. Interestingly, ecologists have also found that species living in tropical forests are not very resilient to the effects of climate change. As mentioned in Chapter 3, the tropical rainforest biome experiences little annual variation in climatic conditions. Organisms that evolved in this biome are, therefore, not well adapted to alterations in climate. However, there is also growing evidence that individuals can cope with some degree of climate change through shifts in energy allocation or by altering their behavior. All of these factors need to inform any potential conservation efforts moving forward. Clearly, we still have much more to learn about the physiological basis of species' ecological niches and how readily species can behaviorally respond to shifts in local climate.

SUMMARY

5.1 What Is a Trade-Off?

- Some of the resources an organism needs are only found in limited supply.

- This means that the time or energy an organism spends acquiring one resource is not available to spend acquiring another resource.

- These types of trade-offs are important to studying physiology and behavior.

- Optimization involves maximizing benefits relative to costs in the trade-off process.

5.2 The Principle of Allocation

A Simple Energy Model for Animals

- All individual animals have limited access to energy.

- Acquired energy can only be allocated to certain physiological components, namely, respiration, assimilation, reproduction, and waste.

- A mathematical model of this allocation process can be represented by the following equation:

$$E_{intake} = E_{respiration} + E_{assimilation} + E_{reproduction} + E_{waste}$$

A Simple Energy Model for Plants

- Given that plants acquire energy through photosynthesis, the allocation model for plants replaces E_{intake} with E_{PSN} (the energy captured in photosynthesis):

$$E_{PSN} = E_{respiration} + E_{assimilation} + E_{reproduction} + E_{waste}$$

5.3 Optimal Foraging

- Animals can change allocation priorities through their behavior.

- Organisms try to maximize energy intake relative to energy expended.

- Optimal foraging theory attempts to explore how various strategies affect an organism's fitness and behavior.

The Theory

- An individual acts to gain the most energy for the least cost in order to maximize its evolutionary fitness.

- Optimal foraging theory focuses on how an organism changes its energy intake.

A Simple Foraging Model

- A simple model for predators consuming prey posits that energetic profitability (P) is equal to the ratio of energy gained (E) from consuming prey to the cost (C) associated with acquiring and eating the prey.

- This equation can be modified by breaking down the cost variable into two terms: search costs (S) and handling costs (H).

Examples and Extensions

- When faced with the option of eating either a tortoise or a cactus mouse, a coyote will choose based on the values of the energy gained (E), the search cost (S), and the handling cost (H) for the two prey items.

- In some situations, it is more energetically profitable to eat tortoises; and in others, it is more profitable to eat mice.

- This model allows ecologists to formulate and test hypotheses about animals in the real world.

Theory Meets Reality

- Researchers have conducted many tests of optimal foraging theory.

- One of the first involved a laboratory study of the foraging behavior of the great tit (*Parus major*). The results of this elaborate experiment indicated that the birds foraged optimally.

- To date, about 70% of the experiments that have been conducted provide support for the theory.

5.4 Water Balance

Why Water Is Different

- Water, like energy, is critical for life. We can use our principle of allocation equation for energy and substitute water (W) for energy (E) to get:

$$W_{intake} = W_{respiration} + W_{assimilation} + W_{reproduction} + W_{waste}$$

- Water is different from energy because the loss (waste) term is a bigger deal.

- This highlights the need for individuals to balance water intake versus loss.

Water Balance in an Ecological Context

- All organisms have some control over water loss.

- Terrestrial organisms generally try to minimize water loss, particularly in desert environments.

- The kangaroo rat has a suite of physiological and behavioral adaptations that help it to minimize water loss.

Plants: Trade-Offs between Energy and Water

- Because of the biochemical requirements involved in photosynthesis, plants have to manage water losses that inherently occur with carbon acquisition.

- Three different photosynthetic pathways have evolved, each with its own abilities to conserve water.

  —The C_3 pathway is very energy efficient but results in significant water loss; it is the pathway used by the majority of plants on Earth.

  —The C_4 pathway is better at water conservation; it is used by tropical grasses.

  —The CAM pathway is excellent at water conservation but involves slow growth; it is used by desert succulents.

- Plants can also allocate growth above- or belowground and do so based to some extent on water availability.

5.5 Thermal Trade-Offs

Living within Thermal Limits

- All organisms live within a particular range of environmental temperatures and have various means for dealing with temperature changes.

- Some organisms regulate their internal temperatures, and others (poikilotherms) do not.

- Organisms that regulate their temperature through internal mechanisms are called endotherms, whereas ones that use external mechanisms are called ectotherms.

- Homeotherms maintain a fairly constant internal temperature.

- **Animal Thermoregulation**

  —Physiological performance often peaks at a certain temperature and decreases at either higher or lower temperatures.

  —Ectothermic animals have a variety of ways to maintain their temperature at a fairly constant level.

- **Plant Thermoregulation**

  —Plants also manage their temperature but are more limited in their range of options.

  —Some plants use physical structures to reflect incoming light (and heat).

5.6 The Niche: The Grand Trade-Off

Definitions and Measurement

- Individuals cannot maximize their performance across all resource needs (e.g., energy, water, heat) simultaneously.

- Instead, each species must evolve a unique set of trade-offs involving a whole suite of physiological and behavioral traits.

- Collectively, this suite of traits helps define a species' ecological niche.

- Scientific understanding of the concept of the niche has changed through time.

- We can identify a species' two-dimensional niche space by graphing two physiological requirements together.

- This idea was extended by G. Evelyn Hutchinson to include an *n*-dimensional hypervolume, which includes more than three factors (in fact, *n* factors).

Resource Utilization Curves

- If we construct a histogram of a species' aggregate behavior, we can construct a resource utilization curve that describes behavior or use of a resource along a single axis of a species' niche.

- We can extend such histograms to represent two niche dimensions (plus frequency), but it is very difficult to graphically represent more than two niche dimensions.

The Niche in Context

- A species' niche also involves interactions with other species in the broader ecological community.

5.7 Physiology and Behavior in a Changing World

- The topics in this chapter are important for understanding changes in the world wrought by human actions.

- If ecology is the "study of the household," understanding the physiology, behavior, and niche space occupied by a species is a fundamental first step for conservation actions.

KEY TERMS

behavior

biomass

C_3 photosynthesis

C_4 photosynthesis

Calvin-Benson cycle

carbon fixation

crassulacean acid metabolism (CAM)

ectotherm

endotherm

energetic profitability

first law of thermodynamics

handling costs

homeotherm

iteroparous

n-dimensional hypervolume

niche

niche space

optimal foraging theory

optimization

photorespiration

photosynthesis

physiology

poikilotherm

principle of allocation

resource

resource utilization curve

respiration

root-to-shoot ratio

search costs

semelparous

stoma (*plural*, stomata)

trade-off

xeric

CONCEPTUAL QUESTIONS

1. Explain two major environmental constraints of desert environments, and provide information to show how kangaroo rats have evolved to survive within these constraints.

2. Describe the three different photosynthetic energy-capture pathways. Explain the conditions under which each is adaptive, and give an example of an organism that uses each.

3. Explain optimal foraging theory, provide the formula for the model, and give an example of where it has been shown to work.

4. Describe two physiological methods plants use to deal with high temperatures.

5. Explain how lace monitor lizards have adapted to living in the desert; use figures when appropriate.

6. Butterflies are ectothermic, active during the day, and found from the tropics to the high-temperate latitudes (even in some polar zones).

   a. Would you expect butterflies to bask more in tropical or temperate zones?

   b. In an area with a wide temperature range for a single day, how would you expect basking to vary within a day?

7. Define the term *niche*, and provide a real-life example of a three-dimensional niche from something you see or have seen around you.

8. Explain why resource limitation is important in natural selection. If necessary, review material from Chapter 4 on the details of natural selection.

QUANTITATIVE QUESTIONS

One interesting extension of optimal foraging theory is the marginal value theorem, developed by Eric Charnov in 1976. The theorem deals with an optimally foraging individual in a patchy environment. If we assume that the foraging habitat is always patchy (heterogeneous), then an animal should prioritize foraging in the most profitable patches in order to maximize its energy gain. While in these patches, the forager decreases resource availability, which means that the energy available per unit time is a decreasing function. At some level of resources, the forager should give up feeding in a patch and seek a new location. Staying too long means the forager is eventually wasting time in a patch with low resources. Leaving too soon, though, would lead to the animal's spending too much time traveling and not enough time foraging where food and energy are plentiful. Researchers use this model to predict how long an organism should stay in a patch. The marginal value theorem predicts that the foraging animal should leave the profitable patches as energy becomes so depleted or as food becomes so scarce or so hard to capture that energy per unit time declines to the level of the average across the whole habitat (Charnov 1976).

An avian version of this theory was tested in a laboratory setting by Richard Cowie in 1977. Like John Krebs, Cowie used great tits (*Parus major*) as his study subject. In an aviary, he created a "forest" of wooden dowels with little cups at the end of them. In the cups, Cowie and his colleagues placed mealworms covered in sawdust. They couldn't really manipulate the time between the patches in the aviary, so instead they manipulated the time required to remove a cover from the cup of mealworms and sawdust. An environment with lids that were hard to remove was like an environment with long travel times between patches. The birds were let loose to forage on their own, and the marginal value theorem was used to predict how much time they should spend in patches based on "travel time" (Cowie 1977).

Figure 5.31 shows the results of the experiment along with the prediction from the theorem.

1. Explain why longer "travel time between patches" (which is really "time to remove lids from cups") should lead to longer time in patches.

2. Would you expect to see more mealworms left in cups in the environment with short travel time or long travel time?

3. What can you deduce from the graph regarding *P. major* and their ability to forage optimally?

4. Does this study support or refute the marginal value theorem (or does it do neither)?

5. What factors could account for the variability in the actual results versus the prediction?

6. Would the test of the marginal value theorem have been better or worse if the researchers had varied the density of mealworms in the cups rather than difficulty of removing lids? Explain your answer, including expected results and any pitfalls of this new manipulation.

7. What could be the next steps with this research system?

8. Can you think of any other organism or situations that could be used to extend or continue this research?

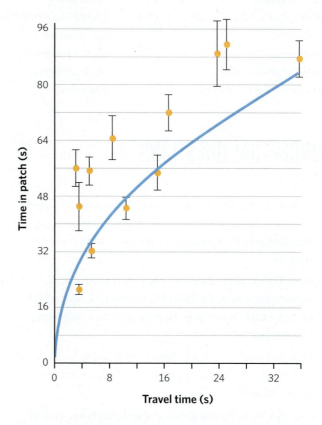

Figure 5.31 Cowie's foraging experiment with great tits (*Parus major*) tested Charnov's marginal value theorem. These results from the experiment plot the time the birds spent in food "patches" (i.e., cups of mealworms in sawdust) against "travel time," or the time required to remove the cover from the mealworm cups (Cowie 1977).

SUGGESTED READING

Berec, M., V. Krivan, and L. Berec. 2003. Are great tits (*Parus major*) really optimal foragers? *Canadian Journal of Zoology* 81(5): 780–788.

Charnov, E. L. 1976. Optimal foraging, the marginal value theorem. *Theoretical Population Biology* 9(2): 129–136.

Sandquist, D. R. and J. R. Ehleringer. 1998. Intraspecific variation of drought adaptation in brittlebush: Leaf pubescence and timing of leaf loss vary with rainfall. *Oecologia* 113(2): 162–169.

Schoener, T. W. 2009. Ecological niche. Pp. 26–51 in S. A. Levin, ed. *Princeton Guide to Ecology*. Princeton, NJ: Princeton University Press.

Tracy, R. L. and G. E. Walsberg. 2002. Kangaroo rats revisited: Re-evaluating a classic case of desert survival. *Oecologia* 133(4): 449–457.

Miami Beach becomes crowded with human beings over spring break. The global human population continues to grow exponentially, and scenes like this are common around the world.

chapter 6

POPULATIONS

6.1 Why Study Populations?

The total population of human beings on the planet has increased from about one billion in 1800 to more than 7.7 billion today. Although none of us is old enough to have experienced this massive increase in human population over a single lifetime, we can still get a sense of how many other humans share our world when we sit in traffic, walk through dense urban development, or have a hard time finding places without human influences. This growth in the human population is one of the hallmarks of the Anthropocene, and although the annual human population growth rate has slowed since 1950, the absolute number of people on the planet today profoundly alters ecological and evolutionary processes. In this chapter, we take our definition of *population* from Chapter 4—namely, a group of individuals of the same species who can mate—and expand on it: a **population** is also a group of individuals of the same species that is spatially distinct from other groups of individuals of the same species. This definition allows us to explore the contribution of one or more individuals to a population and scale up that contribution to explore changes in population abundance across time.

We start with several simplifying assumptions that make it easier to count and to represent population growth with equations. We consider how reproduction and survival affect the numbers of individuals in a population. As the chapter progresses, we explore continuous-time and discrete-time models of population growth, the factors that can stabilize populations at a particular abundance, and how the distribution of individuals across ages or life stages leads to changes in population abundances. These concepts are vital for the remaining chapters of this book because the potential for population growth or decline not only affects species' distributions and interactions but also helps environmental managers avoid extinctions of populations or whole species. At the same time, an understanding of populations allows us to appreciate the challenges humans face as we approach 8 billion people on the planet.

LEARNING OBJECTIVES

- Restate the basic concepts behind mark-recapture sampling.

- Formulate a conceptual model of population growth for a species of your choice.

- Explain the discrete-time population growth model, identify all its variables and parameters, and create a graph that shows growth across time.

- Identify four assumptions behind the geometric, or discrete-time, population growth model.

- Compare and contrast the continuous-time population growth model and the discrete-time population growth model.

- Explain the concept of density dependence, and provide evidence for density dependence in real-world populations.

- Explain the continuous-time logistic growth model, and identify all its variables.

- Distinguish between the continuous-time and discrete-time logistic growth models.

- Identify three different types of survivorship curves, and create graphs for each.

- Explain how population models may be useful in managing rare species with declining population abundances or in managing invasive species with growing populations that are spreading across a landscape.

6.2 Population Definition Revisited

Chapter 4 introduced some of the practical difficulties associated with counting the individuals in a population. We pointed out that sometimes males and females in a species look very different from each other (Figure 4.5) and that juveniles are often quite different in appearance from adults (Figure 4.4). If we focus on life-stage transformations in monarch butterflies (**Figure 6.1**) or oak trees (**Figure 6.2**), we can see that such big differences can cause problems when trying to count all the members of a population. Would you recognize a caterpillar and a butterfly as members of the same population if you did not already know that they were the same species? What would happen if you encountered a species that you did not know well? Even if you recognized individuals as being of the same species, wouldn't it be difficult to find and identify all members of a population when some are flying through the air, others are munching on vegetation, and still others are hidden as pupating larvae? If you could find all of these individuals, would it really make sense to say that individuals at different life stages contribute equally to the population? Would you be comfortable, for example, saying that an acorn should "count" the way an adult oak tree "counts" in a population of oaks?

And what about movement? If ecologists define a population as a collection of individuals all in one place and with the ability to reproduce together, how do they define the place? Do they have to extend the boundaries to include new areas if these individuals move to find a mate? Also, when counting individuals in a population, do they include those that move into, out of, or temporarily through a population?

Figure 6.1 The life cycle of a monarch butterfly (*Danaus plexippus*) starts with an adult's search for a host milkweed plant (*Asclepias* spp.) on which to lay an egg. The stages of development progress from egg to larva (caterpillar) to pupa to adult. Very different estimates of population abundance would occur if we did not recognize all stages as the same species.

There is no one solution to these issues, but there are some useful simplifications. For example, if the goal is to document whether population abundance changes over time, we can focus only on individuals with the potential to create offspring. As an illustration, with oak trees this simplification would mean counting only individual trees that have reached reproductive age. Any acorn that avoids squirrels, germinates, survives past the seedling stage, and grows large enough to create acorns of its own would be counted, while all other oak individuals would be ignored. Later in this chapter, we will move past this simplification, but it is helpful for developing a conceptual model of population change for now.

In terms of the movement of individuals, we need to make additional assumptions. If the number of individuals moving into or out of a population is small enough relative to the whole population, we may be able to overlook movement. Alternatively, there may be as many *emigrants* (ones who leave) as *immigrants* (ones who arrive), resulting in the two types of movement canceling each other, again allowing us to ignore movement. When populations effectively have no immigration or emigration, we refer to them as *closed populations*, but when they experience lots of immigration or emigration, they are called *open populations*. For the rest of this chapter, we will only consider closed populations to keep things simple, but we will return to the movement of individuals in Chapter 12.

Even if we only consider closed populations, we still have to identify the boundaries of the population. In the real world, we usually make an educated guess about what geographic boundaries serve as barriers to individual movement and then limit our counting to within the boundaries. Sometimes, though, we focus on an area with no natural barriers because our interest defines boundaries in useful and simplifying ways.

One interesting aspect of defining geographic boundaries of populations is that the count of individuals delineates both **population abundance** and **population density**, where the former is the number of individuals (sometimes also called *population size*), and the latter is the number of individuals within a defined area. This may seem unnecessarily confusing, but you should look at this dual set of information (abundance and density) as a lucky consequence of our simplifying assumptions, because both pieces of information are needed to create some of the models here and in later chapters. Density also helps to tie observations about populations to our insights about fitness and performance from Chapters 4 and 5. Higher densities often suggest that individuals in a population have high fitness or superior performance in that habitat. This information is important for topics discussed in later chapters, including competition, susceptibility to predators and diseases, and how species move across landscapes.

6.3 Obtaining Population Data

After deciding on a way to define a population and its boundaries, we face the problem of how to go about counting individuals. Counting seems so basic. We have all been doing it since we were very small, but this simple skill becomes challenging when you need to find the things you are counting, and when these things do not stay put or let you place them in stacks of "counted" and "uncounted."

Figure 6.2 The three stages of oak tree growth are acorn, seedling, and an adult tree. Imagine how large the population of oak trees would be if we included every acorn in the population count.

Figure 6.3 The Great Elephant Census of 2014–2015 was the largest wildlife survey in history. This was an Africa-wide survey designed to gather accurate data about the population abundance and distribution of African elephants (*Loxodonta africana*) using standardized aerial surveys. It covered hundreds of thousands of square kilometers.

It helps first to assign a population count to a particular time frame. If the time units match reproductive cycles, then counting should work reasonably well until the next batch of offspring enters the group. A short counting period makes it less likely that new individuals will be born (or mature into) the counted group within a single counting event, and a consistent counting season allows comparisons of abundance or density over time. Such comparisons force us to acknowledge that population abundance changes through time and that counts only represent the population abundance at a given point in time.

After choosing a time frame, we still need to deal with the fact that organisms are not lining up to be counted. Ecologists acknowledge that, for most species, a complete population census is not feasible. Instead, they count a subsample of all the individuals and extrapolate from that subsample to get an estimate of the number of individuals in the whole population using proportions.

For example, how does one actually go about counting the number of polar bears living on the ice shelf off the Aleutian Islands or the number of elephants roaming the grasslands of an African national park? With large organisms, we often—not so surprisingly—do it from a distance, taking aerial counts from a UAV (unmanned aerial vehicle), small plane, or helicopter (**Figure 6.3**) and usually moving along a predetermined and standardized route called a *sampling transect*.

Remember that we defined a population as all individuals located within a preset geographic area. If the abundance across the whole area is relatively consistent, or if the subsampled areas represent the various abundances across the whole area, then we can use proportions to estimate total abundance from our samples:

$$\frac{N_{total}}{A_{total}} = \frac{N_{samples}}{A_{samples}} \qquad (6.1A)$$

which we can rearrange as,

$$N_{total} = \frac{A_{total}}{A_{samples}} \times N_{samples} \qquad (6.1B)$$

where N is the number of individuals and A is area. The subscript *total* denotes all the individuals we want to count or the entire area of interest, and *samples* denotes

all the individuals actually counted in the subsection of area in the transects. Despite the cost of fuel for sampling from UAVs, helicopters, or airplanes, the relationships in Equation 6.1 make *aerial transects* a relatively cheap and fast way to estimate the population abundance of animals that roam across large geographic expanses.

We can extend the transect approach to estimating total abundance of many different types of organisms, including amphibians, reptiles, butterflies, fish, corals, and plants. The transect methodology has many variants, including counting by sound (*sound transects*), which is often used to census arboreal or nocturnal species with distinctive sounds, such as birds, monkeys, frogs, or cicadas. Using sophisticated sound recorders, ecologists also use sound transects to count underwater vocalizing organisms, such as whales and dolphins (**Figure 6.4A**).

It may seem possible that for *sessile* organisms (i.e., those attached to a fixed place and not moving), researchers could actually count all the individuals of a particular species within a bounded area. If the area is too large, though, counting every individual can be close to impossible, even when the organisms do not move. For that reason, researchers still subsample, using *line transects* or *sampling plots* distributed across the target population area (**Figure 6.4B** and **Figure 6.4C**), and use a version of Equation 6.1 to estimate total population abundance.

Anyone who has spent time in the great outdoors has had the experience of seeing signs of mammals, insects, or fish without actually seeing the organisms

Figure 6.4 Ⓐ A researcher uses a directional hydrophone to count sperm whales off the coast of New Zealand. Ⓑ A diver surveys coral organisms at the Ailuk Atoll, Marshall Islands. Ⓒ Native plants are identified and counted in the Santa Monica Mountains National Recreation Area in California.

Figure 6.5 Many government agencies require hunters to purchase hunting permits and clear all captured animals at game stations. Here a mountain goat (*Oreamnos americanus*) is being assessed for weight, sex, and age by a Wyoming Game and Fish Department employee. These surveys allow agencies to collect count data for hunted species.

Figure 6.6 The threatened Florida scrub jay (*Aphelocoma coerulescens*) is the only bird species endemic to the state of Florida. The leg bands identify this specific individual and distinguish it from the rest of the birds in the population; they also allow wildlife managers to track changes in population abundance.

themselves. Because many organisms are *cryptic* (well-hidden) or shy, we may need to trick them into showing themselves or count something else that helps us identify individuals. We are more likely to see scat (ecologists have lots of words for feces, and *scat* is one of them), hair, chewed plant stems, scratched tree trunks, nests, entrances to burrows, or paw prints than the organisms themselves. Sometimes researchers use transects to count such signs of animals, but it can be hard to convert counts of paw prints, for example, into counts of animals because each animal may leave many such signs. Genetic techniques, including the analysis of environmental DNA (eDNA), have greatly increased the ability to distinguish between individuals by using samples of fur or feces, and they can be a useful tool for improving estimates of population abundance of cryptic and shy organisms.

Trapping or hunting can lead to catching and counting a variety of organisms. In addition to counts, these methods may offer detailed information on the age, sex, mating status, body size, and even diet of the killed or captured individuals. Many government agencies keep track of hunting or fishing records as a way to assess population abundance (**Figure 6.5**), but these approaches work for smaller organisms as well. For many insects, the most common way to sample, and thus estimate population abundance, is to set out pheromone or light traps that lure the insects to the trap and hold them or kill them until they can be counted by an ecologist.

One issue with counts derived from hunting and fishing is that reporting is never complete. These methods may indicate, over time, whether population abundance or density is increasing or decreasing, but they do not generally offer reliable counts. For a more accurate assessment of population abundance, less destructive methods generally work better. Captured (but not killed) organisms offer much of the same information as killed organisms, plus we can follow them after release or recapture them later (**Figure 6.6**).

Mark-recapture sampling requires capturing, tagging, and releasing individuals and then resampling. If the tags are unique and allow researchers to distinguish between individuals, recapturing tagged individuals provides detailed information on individual growth, movement, and behavior. Over many years, ecologists can gain information on the survival and reproductive capacity of individuals of different ages. We will use such information near the end of this chapter.

Even if the tags are not unique, recapturing previously marked individuals provides a powerful way to estimate

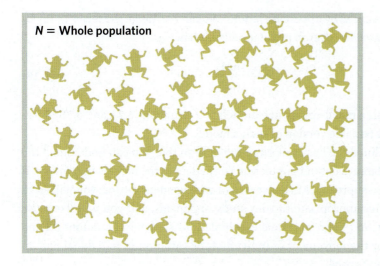

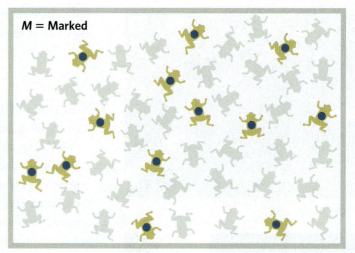

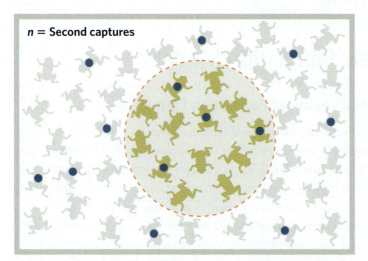

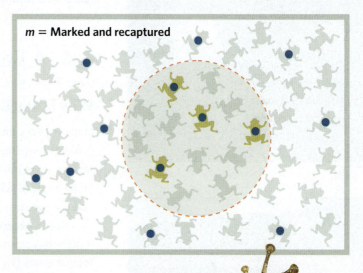

Figure 6.7 Mark-recapture is a two-sample survey approach in which marking and releasing all the individuals in a first sample allow researchers to use the proportion of marked individuals in a second sample to estimate overall population abundance. When the individuals in the original sample are marked (here with a blue tattoo) and released back into the population, they mix with the rest of the population. If the second sample is random, the proportion of tagged individuals in the second sample should be the same as the fraction of the whole population that is tagged. Researchers can then estimate population abundance (N) using Equation 6.2.

overall population abundance. The process again relies on simple proportions. The idea is that the tagged fraction of a second sample should be the same as the fraction of the whole population that is tagged, if tagged and untagged individuals are equally likely to be captured (i.e., if the sample is random). We can represent this relationship with a simple proportional equality, similar to the equality shown in Equation 6.1,

$$\frac{M}{N} = \frac{m}{n} \qquad (6.2A)$$

where, having dropped the subscript terms, we now denote the whole population as N, the whole tagged group of individuals as M, the second set of captured individuals as n, and the previously tagged individuals caught in the second sampling event as m (**Figure 6.7**). Lowercase letters represent a sample, or subset, of the

(A)

(B)

(C)

Figure 6.8 **A** North American beavers are semiaquatic mammals that build dams along waterways. These dams can lead to extensive destruction of forested habitats around them, but the scale of the destruction varies with the biome and evolutionary history of the site. **B** Beaver dams in Algonquin Provincial Park in Ontario, Canada, have actually increased regional species diversity. **C** In South America, however, they have had more destructive regional effects, as seen in this image showing the extensive loss of native *lenga* forest near Ushuaia, Argentina.

individuals represented by uppercase letters. We can rearrange the terms in Equation 6.2A to estimate the overall population abundance as

$$N = \frac{nM}{m} \tag{6.2B}$$

Ecologists call this a **mark-recapture model**, and Equations 6.2A and 6.2B represent the simplest possible version of such a model.

We have outlined the basics of sampling here so that you can understand the assumptions and issues behind population count data in the models that follow. If you ever use these approaches in a research or management setting, you will learn specific techniques appropriate to your study organisms, and you will likely use some refinements of the proportional extrapolations shown in Equations 6.1 and 6.2 to account for the movement of individuals or variation in individual habitat use across time and space.

6.4 A Basic Population Model

Why Ecologists Care about Population Estimates

Once we have an estimate of population abundance, what do we do with it? Often, ecologists are interested in knowing whether the population is shrinking or growing by looking at how the number of individuals changes through time. Let's start with an example of why identifying changing population abundance may be interesting and then venture into how to use this information in a basic population model.

North American beavers (*Castor canadensis*) are native to the continent for which they are named, and their pelts have long been a valuable source of material for fur coats (**Figure 6.8A**). In the 1940s, an entrepreneur transported several of these animals to Tierra del Fuego in South America to start a commercial fur production venture. The company failed shortly after its establishment, and 25 mating pairs of beavers were released into the wild around the fur farm. Although not native to the area, the released beavers found the habitat quite suitable and established a self-sustaining population. With no real predators in the region, their population abundance increased over the years with profound consequences for the local ecosystems (Parkes et al. 2008).

Beavers can be considered *ecosystem engineers*, meaning their traits or behaviors alter habitats in ways that affect the other species that share their ecosystem. In the case of beavers, the changes come from the way they create their dens (usually called dams or lodges; **Figure 6.8B**), which they build in freshwater streams using nearby trees, limbs, and twigs. Beavers eat the soft, growing parts of trees, and they get to these tender pieces by chewing through tree bark and often whole trunks. Their feeding habits and dam creation kill trees, and their dams often convert streams and streamside vegetation into ponds and wetland meadows (Figure 6.8B).

The anthropogenic introduction of beavers in Tierra del Fuego led to the decimation of thousands of kilometers of *lenga* forest (*Nothofagus* spp.) and altered the dynamics of waterways with consequences for dozens of native species (**Figure 6.8C**). Beaver activity has also cost the Chilean and Argentinean governments enormous amounts of money because of the negative effects the dams have on roads and infrastructure (Parkes et al. 2008). The continued growth of beaver populations across Tierra del Fuego has real, practical consequences.

Imagine yourself as a wildlife manager trying to protect a single 1,000-hectare (ha) park (about 2,471 acres) in Tierra del Fuego that contains some wetlands. One of the first things you might do is determine whether the number of beavers in this park is growing or declining. If you conducted annual beaver counts over five years, your (entirely hypothetical) data might look like that shown in **Table 6.1**. From these data, what would you be able to tell your local officials about the beaver population?

From a quick glance at the data, the most obvious conclusion would be that the estimated beaver population is growing. Looking a little more closely, we can see that the estimated population abundance more than doubled in the period of monitoring. For most of us, this is about as far as we can go in describing this dataset when relying on just our wits and internal calculators. But what if we wanted to answer questions such as, "How big will this population be in five more years?" or "How many years will it take before the population reaches 100 individuals?" To answer such questions, we would need to formalize our thought process and write out a simple model of population growth.

Table 6.1 Hypothetical Counts of North American Beaver Abundance in a Fictitious Local Park in Tierra del Fuego

Year	Beaver abundance
2017	10
2018	12
2019	16
2020	20
2021	25

A Conceptual Model of Population Growth

We start this process by building a conceptual model that focuses on how population abundance changes over time. For the beaver example, first we will set one year as our time unit, given that the data were "collected" that way. Next, we need to ask what would cause the number of beavers in the park to increase over the course of a year. The only ways to gain beavers are for adults in the population to give birth and for beavers from another population to move into the park. These two mechanisms of population increase can be called **birth** and **immigration**, respectively. Although the dataset in Table 6.1 shows a growing population, the park may also lose individual beavers over the course of a year. One or more of the beavers may die, or beavers may leave the area. These two mechanisms of population decrease can be called **death** and **emigration**, respectively.

We can put all of these factors together in a full conceptual model of change in population abundance over time (**Figure 6.9**). If we assume that our beaver population is *closed* (no immigration or emigration), we can simplify this conceptual model by ignoring the immigration and emigration arrows in Figure 6.9. This leaves us with only births and deaths affecting how the beaver population abundance in the park changes through time.

A Mathematical Model for Discrete-Time Population Growth

The next stage of model building is to take this conceptual model and quantify the relationships within it using math. With the beaver scenario (summarized in Figure 6.9), coupled with the assumption that the population is closed, we can

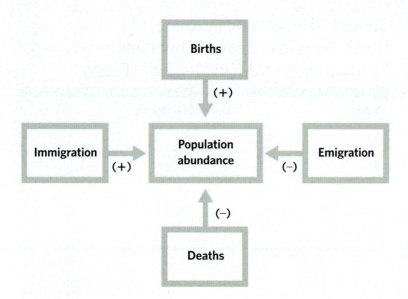

Figure 6.9 This conceptual model of how population abundance changes through time shows the factors that increase (+) and decrease (−) population abundance.

identify the values we would need to track from year to year—namely, population abundance, births, deaths, and time.

As in the mark-recapture model (Equation 6.2), let's indicate population abundance as N. We can label the total number of births in the population B and the total number of deaths in the population D. Lastly, we need to distinguish these variables for one year from the same variables for the next year. Using t to represent time, we can call our starting year t and the next year $t + 1$. The population abundance in a given year is represented by N_t, and the population abundance in the next year by N_{t+1}; we can do the same thing with births and deaths. Putting it all together, we get

$$N_{t+1} = N_t + B_t - D_t \tag{6.3}$$

This equation can be translated into words as follows: "The population abundance of beavers next year (N_{t+1}) is equal to the population abundance of beavers this year (N_t) plus the number of individuals born this year (B_t) minus the number that die this year (D_t)." N, B, and D are all variables that are a function of time, meaning they change with time.

If N, B, and D are known for a given year, then Equation 6.3 is a reasonable way to calculate population abundance from one year to the next. Unfortunately, if N, B, and D change every year, then this equation necessitates counting not only the number of beavers in the wetland park but also the number of births and deaths every year, which of course is more work. Plus, if we were to acquire values for all the variables every year, would this model provide us with any information that we did not already have?

Let's look back at the data in Table 6.1 to see if we can find some simplifying assumptions. The easiest assumption may be that the total number of births and deaths does not change between years, but that is too simple. It is not reasonable to think that 25 adult beavers would produce the same number of offspring as 10 beavers. Maybe, though, it would be reasonable to assume that there is an average number of young beavers produced by each pair of adult beavers every year.

In fact, if we made a guess that, on average, each pair of beavers (they are monogamous) produced one viable baby beaver each year, then we would get five births in 2017 (10 beavers means five pairs that each produce one offspring), six in 2018, eight in 2019, 10 in 2020, and 12 (we definitely will not have a "half beaver" that is viable) in 2021. So, B_t changes each year, even if the number of offspring per pair stays constant. Logically, we can expect that deaths are going to work the same way. Even if each beaver's individual probability of dying stays exactly the same from one year to the next, the total number of deaths will change as the total number of beavers changes.

This logic suggests that we expect B_t and D_t to be functions of N_t. In Figure 6.9, births and deaths affected population abundance, but now we can see there is some interrelationship, or feedback, among the variables. In this situation, population abundance is also going to affect the total number of births and deaths in a year, which then affects population abundance, which then affects births and deaths, and so on.

We can represent this feedback in a new conceptual model (**Figure 6.10**). This model suggests that, when we estimated the reproductive output of each pair of beavers, we were perhaps onto something. We can make Equation 6.3 more generalizable if we know approximately how many births to expect per individual in the population and what the probability of death is for each individual in the population (or what fraction of individuals we expect to die) each year. Such birth rates and death rates are called **per capita rates** because they tell us what we can expect for each individual, or "head" (*capita* is the Latin word for "head"), in the population. The word *rate* can be confusing because it implies time, but here, time is clearly involved because we are talking about births and deaths per individual per year.

If these rates stay constant, we only need to measure or calculate them once. If they never change, we can insert them as **parameters**, meaning they are numbers that vary for different species or locations but do not vary through time (see Chapter 2 for more on the difference between variables and parameters). Conventionally, ecologists use b for expected per capita births in a time unit and d for death probabilities per capita per time unit. We can get total births (B) and total deaths (D) by multiplying per capita rates by the number of individuals in the population.

Returning to the verbal equivalent of Equation 6.3, we can now say, "The population abundance of beavers next year (N_{t+1}) is equal to the population abundance of beavers this year (N_t) plus the number of individuals born this year ($N_t b$) minus the number that died this year ($N_t d$)," giving us the following equation:

$$N_{t+1} = N_t + N_t b - N_t d \qquad (6.4)$$

We can rearrange Equation 6.4 so that N_t appears on the right-hand side only once:

$$N_{t+1} = N_t(1 + b - d) \qquad (6.5)$$

You may have noticed that this rearrangement emphasizes not just birth and death but also birth and survival. All individuals who are alive must either stay alive (survive) or die, so one minus the per capita death rate equals the **per capita survival rate**, which is usually represented as s. If we translate Equation 6.5 into words, it says that the population abundance next year will be equal to all the individuals from this year who survive ($N_t [1 - d]$), plus all new births ($N_t b$). As a final common simplification, ecologists often refer to the combined birth and survival rate term with a single parameter, generally labeled as λ (lambda) and defined as the **finite per capita rate of population growth**, or

$$\lambda = 1 + b - d \qquad (6.6)$$

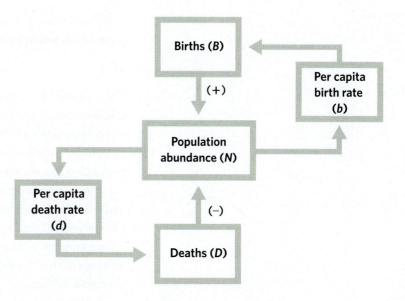

Figure 6.10 This revised conceptual model of population growth assumes a closed population and, therefore, does not include immigration and emigration (Figure 6.9). Each arrow indicates an effect of one model component on another, showing how population abundance changes the birth rate and death rate, and that these rates alter the total number of births and total number of deaths. Those totals then alter the population abundance. A plus or minus sign indicates that the effect increases or decreases the abundance, respectively. Arrows without a plus or minus sign signify a more complex relationship, to be explored later in the chapter.

If we substitute λ into Equation 6.5, we get the commonly used **discrete-time population growth model**:

$$N_{t+1} = N_t \lambda \tag{6.7}$$

This is called a *discrete*-time model because we calculate the population abundance or density once after each full year and do not consider what happens in each moment throughout the year; in other words, we consider time as discrete (coming in "chunks"), not continuous. In our example, the chunk is exactly one year, and new population abundance estimates are evaluated once a year at the same time.

The λ parameter in Equation 6.7 is very useful, particularly for conservation and wildlife management in a changing world. It encapsulates the per capita population growth rate succinctly, offering crucial information with one parameter. Take a moment before reading further to think about how population abundance will vary with different values of λ. You should be able to come up with three distinct possibilities.

It may seem obvious that the three general ways that population abundance can change from one year to the next are to stay the same, grow, or shrink. These outcomes are useful because they provide a few more insights about our model. If population abundance stays the same (i.e., $N_{t+1} = N_t$), this means that $\lambda = 1$. It also means that each death is balanced by a birth, or that $b = d$. If the population grows, then $\lambda > 1$. Again, we can also look at this outcome in terms of the relative sizes of the per capita birth and death rates and recognize that there are more births than deaths (i.e., $b > d$) when the population grows. If the population shrinks, then $\lambda < 1$, which means that the death rate is greater than the birth rate (i.e., $b < d$).

Visually Representing Population Change through Time

Take a moment to think about what these changes in population abundance would look like through time. On a graph with time on the x-axis and population abundance on the y-axis, what would the line or curve representing population abundance over time look like when $\lambda < 1$, $\lambda = 1$, or $\lambda > 1$? From the numbers in Table 6.1, we can see that the growing beaver population generally increased a little more each year than it had the year before, so that the population growth appears to accelerate as the number of individuals increases. That makes sense given the feedback between population abundance and birth and death rates, and we can translate that observation into the upper curve in **Figure 6.11**.

This population growth curve ($\lambda > 1$) would be bad news for Tierra del Fuego wildlife managers, because an increasing beaver population would mean more negative effects on the native ecosystem it inhabits. Our basic population model suggests two options for reducing the beaver population in the future: either lower the expected per capita birth rates (b) or increase the per capita death rates (d). In a practical sense, this would mean that a wildlife manager could administer a birth control drug to the beavers or institute an annual hunting program. But how much hunting or birth control would stabilize or decrease the population? The discussion of λ already told us that birth and survival rates must be reduced individually or in combination so that $\lambda = 1$ (for a stable population abundance) or $\lambda < 1$ (for a decreasing population abundance).

Let's assume that annual beaver hunts alone can drive λ values below 1, meaning that the population would decline year after year through hunting. You may be

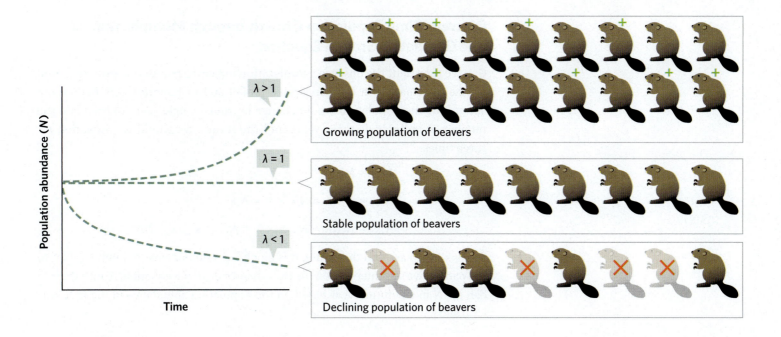

Growing population of beavers

Stable population of beavers

Declining population of beavers

surprised by the curve that describes such a shrinking population (the lower curve in Figure 6.11) because you might expect the curve for a decreasing population to be a mirror image of the curve for a growing population. Remember, though, when $\lambda < 1$, the population loses a fraction of itself with each time step. That lost fraction is bigger in absolute terms when the population is bigger, so the drop in population abundance is steepest in the beginning and then slows.

If λ stays below one, then the population keeps decreasing, eventually reaching zero, or *population extinction*. Once a closed population reaches zero individuals, that is the end; it will always be zero. If the closed population contains the entire species, then population extinction is the same as *species extinction* (Chapter 11).

If birth control is added to the beaver population management methods, then perhaps λ can be made far less than one, which would make the population shrink even faster (**Figure 6.12**). Notice that more extreme values of λ just lead to steeper curves for population growth or decline.

Figures 6.11 and 6.12 are useful for scientists because they underscore a fundamental insight about population growth. Populations do not grow by the same numbers every year, even when birth and death rates stay the same. We had already figured this out when we realized that we could not use constant values for B and D, but take a moment to think about what this pattern of population growth means. Larger numbers of individuals should have more offspring, and fewer individuals will lead to fewer offspring. But most of us forget that this means that the actual change in population abundance will get bigger or smaller every year *even if the per capita birth rate stays exactly the same*. From Figures 6.11 and 6.12, we can see that growing populations that change very little in the early years can explode in later years, and populations that decline sharply at first may then decline quite slowly toward extinction.

Figure 6.11 This graph represents three population dynamic behaviors from the discrete-time population growth model (Equation 6.7) with different values of λ. In terms of the beaver example, $\lambda > 1$ produces a beaver population that is growing in abundance over time, whereas $\lambda < 1$ produces a beaver population that is declining through time.

Figure 6.12 When the values of λ are more extreme than in Figure 6.11, the population growth dynamics change by producing steeper (faster) gains or losses.

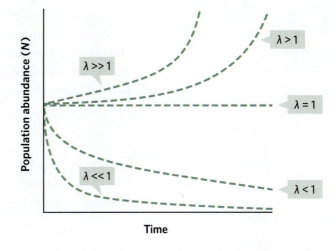

Discrete-Time Population Growth through Multiple Years: The Geometric Growth Equation

Can we represent the observed acceleration in beaver population growth, based on the information in the conceptual model and in Equation 6.7? Let's think about how the population would change beyond a single year. To calculate the number of individuals three years into the future, we would use Equation 6.7 three times:

$$N_{t+1} = N_t \lambda$$

$$N_{t+2} = N_{t+1}\lambda = N_t\lambda \times \lambda = N_t\lambda^2$$

$$N_{t+3} = N_{t+2}\lambda = N_{t+1}\lambda \times \lambda = N_t\lambda \times \lambda \times \lambda = N_t\lambda^3$$

To look 100 years into the future, it is much less cumbersome to jump straight to the pattern we see emerging on the right-hand side of the equation. Notice that we can get N in any future year based on the population abundance in the first year (N_0) and λ^t.

$$N_t = N_0\lambda^t \tag{6.8}$$

We call Equation 6.8 the **geometric growth equation**. This equation produces curves like those in Figures 6.11 and 6.12 in that we expect equations with exponents to grow faster and faster, or to shrink more and more slowly. You can explore this relationship for yourself quite easily (a spreadsheet makes such explorations go much faster), and we offer a few suggestions for doing so in the Quantitative Questions at the end of the chapter.

Using Data to Inform Population Models: A Test Drive

We now have both the data from our estimates of beaver abundance and a general equation that describes how population abundance changes through time. The next step combines the data and the equation to answer questions that wildlife managers might pose for situations like that of the beavers in the Tierra del Fuego park.

First, we have to decide what information from Table 6.1 is represented in Equations 6.7 and 6.8. Although Table 6.1 provides N_t, it does not provide birth rates, death rates, or λ. How do we get λ without having to go into the field and physically count all the beaver births and deaths?

One advantage Equation 6.7 has over Equation 6.5 is that, using 6.7, we can calculate λ by knowing the population abundance in one year and the population abundance in the next year, even if we do not know the actual per capita birth and death rates. All we need are at least two years of census data to explore the beaver population dynamics:

$$\lambda = \frac{N_{t+1}}{N_t} \tag{6.9}$$

Because we have several consecutive years of abundance estimates, we can calculate λ for the beaver population from each year to the next. **Table 6.2** shows these λ values. Although the year-to-year λ values in Table 6.2 are close, they are not the same. So, perhaps the assumption of constant b, d, and λ is incorrect. Before we toss out the model, though, let's think about how constant b, d, and λ values would play out in the real world.

Real-world populations come in packets of whole individuals, which means per capita average rates of births and deaths must be rounded numbers (e.g., no one encounters 0.36 of a living beaver). Plus, per capita rates truly are averages, and they will never lead to perfect predictions. If the beaver population really had one value for the finite per capita rate of population growth (λ), could we figure out that value from our data? If we calculate λ from Equation 6.8 instead of Equations 6.7 and 6.9, we can determine the average λ that allows a population to grow from a starting population abundance to a population abundance at a later time, t. Rearranging Equation 6.8 (and remembering that a fractional exponent represents a root—e.g., $A^{1/2} = \sqrt{A}$, and $A^{1/t} = \sqrt[t]{A}$), we get

$$\lambda = \left(\frac{N_t}{N_0} \right)^{1/t} \qquad (6.10)$$

Using Equation 6.10, we can plug in 10 beavers for N_0, 25 beavers for N_t, and 4 for t, and calculate that the λ value that would have taken our beaver population from 10 beavers in 2017 to 25 beavers in 2021 is approximately 1.257. **Table 6.3** shows the predicted number of beavers for each year, using this value of λ with the geometric growth model. Notice that the model's predictions of abundance and the field-estimated abundances are repeatedly very close to each other (i.e., within one whole number). This match means the simple model works pretty well in representing reality. In general, model predictions work better for bigger populations, in which small deviations from the averages can balance each other out.

We now finally have all we need in order to answer some of the questions posed by the wildlife managers in Tierra del Fuego. As you recall, we were interested in knowing how large this population would become in five years. We now have the tools to answer this question. Plugging 1.257 into Equation 6.8 for λ, we can say that there would be 78 beavers in our population in 2026 if the per capita growth rate stays constant. The equation actually tells us there will be 78.45 beavers, but (as we mentioned before) partial beavers do not survive very well in the real world.

This simple model of population growth is very useful and applicable to real-life situations. It is not difficult to collect the data needed to "drive" this model, and it

Table 6.2 Hypothetical Counts of North American Beaver Abundance in a Fictitious Local Park in Tierra del Fuego and Finite Per Capita Rates of Population Growth (λ) Calculated from These Data

Year	Beaver abundance	λ for year t to year $t + 1$
2017	10	1.2
2018	12	1.33
2019	16	1.25
2020	20	1.25
2021	25	λ

Table 6.3 Comparison of Hypothetical Counts of North American Beaver Abundance in a Fictitious Local Park in Tierra del Fuego and Abundances Predicted by a Simple Geometric Model of Population Growth

Year	Time since first count (years)	Counted beaver abundance	Predicted beaver abundance
2017	0	10	10
2018	1	12	12.57
2019	2	16	15.81
2020	3	20	19.88
2021	4	25	25

The two abundance estimates (counted and predicted) match closely, indicating that the model is doing a good job of representing reality.

is surprisingly accurate in many situations. Nevertheless, we should always use models like this with a little caution.

Assumptions in Simple Models Limit Applications

Despite its usefulness, the geometric growth model is a cartoon or simplification of real life, as are all models (Chapter 2), and it is important to be aware of its limitations. A few particularly important assumptions associated with the geometric growth model are worth noting for the ways they limit its usefulness.

First, we made the assumption that the individuals in the population breed during a discrete time period, which in the hypothetical beaver example was every spring. The one-year time unit was a good choice because it fit the sampling schedule and because beavers actually do give birth once a year in the spring. In fact, each female is only fertile for about 12 to 24 hours in any given year. Many species have distinct periods of time during which they reproduce, and even deaths may occur more often during particular times of the year. But not all species work this way. For many species, a more continuous pattern of births and deaths through time is the norm. Consider, for example, humans. We could snap our fingers to a fast beat and have each snap represent global human population growth, with new humans being born (and dying) at every snap. How can we represent such a species and its population growth? Later in this chapter, we work through the steps we just followed for beavers but instead apply them to species with more continuous patterns in population growth, like humans.

A second big assumption of the current model is that nothing changes from year to year except population abundance (N). The fact that we are not allowing birth and survival rates to vary through time should strike you as a huge and potentially problematic simplification. From Chapter 4 and Chapter 5, we know that survival and reproduction change with fitness and with the environment, and it seems reasonable that such changes might alter average per capita population growth rates. For example, although we know that beavers alter their environment, we have included nothing in the model about the way a growing beaver population might deplete environmental resources through overcrowding. Can birth and survival rates remain constant if beavers remove so many trees or food sources that some individuals cannot get enough to survive or reproduce? A population can grow so large that the quality of life for each individual begins to diminish. We have not captured that effect in the equations, but we will do so later in this chapter.

Third, our model assumes a single birth rate and a single death rate for all individuals in the population, regardless of age, sex, health, or other differences. You may again feel that this is unreasonable. As pointed out at the beginning of the chapter, individuals of different ages are likely to have different expectations of giving birth and of dying. Even if we count only mature individuals, we know they are not all equally fertile or robust. This pattern should be obvious from your own experience. In humans, it is impossible for anyone below puberty to reproduce, and the chance of dying goes up pretty dramatically past the age of about 75 years. Later in the chapter, we explore the effects of age or stage of life on birth and death rates and the subsequent consequences for population growth.

Fourth, the model treats males and females as the same. Does that seem acceptable? It may be a reasonable assumption to make in the case of asexual species that reproduce by budding or self-fertilization, or if the sex ratio is equal and pairs of individuals are monogamous, as is the case with beavers. But what about species with harem mating systems in which one male mates with several females, or pack animals in which one male and one female mate and the other pack members mostly do not? There are ways around these problems, too, but the mathematics for dealing with all the different reproductive and mating strategies lead to very specific models for particular species and populations.

Our simple model assumes that we can average the different birth and death rates across all the members of the population. That assumption may be good enough as long as the population maintains a distribution of individuals that still fits this average. If the age distribution or sex ratio becomes skewed, or if the population has a very particular mating system, then these simple models will perform less well. In this next section, we work through some of these issues, one at a time.

6.5 A Model for Continuous-Time Population Growth

What if we want to explore population growth for a species that produces young at any time of the year? Humans are a great example of this, but there are many others, such as cats, mice, cockroaches, and many species of algae. In practice, species in temperate biomes are often more seasonal than those in tropical biomes, but we can find organisms with continuous breeding throughout the world.

A model that incorporates births (and deaths) occurring throughout the year allows us to think about populations that change continuously through time. If you are not sure why such change would be important, think about a bank account offering interest. Would you rather get 5% interest on your savings calculated once at the end of each year or have the bank calculate a daily fraction of 5% interest (5% divided by 365 days in the year) and deposit that amount in your account every day? The daily-fraction method means you would get interest on your interest as the days went on, making your money grow faster than if interest were calculated only once per year. **Continuous-time models** work the same way, as they enable populations to grow the way daily interest rates allow bank accounts to grow. Continuous-time models can keep track of all the little changes in population abundance that accumulate through time.

Our continuous-time model will still have inputs from per capita birth rates and losses from per capita death rates, as in Figure 6.10, but now we will zoom in on much smaller changes over much shorter time frames. Revisiting Equation 6.4, we can see that the discrete-time model added a chunk of new individuals each year (or each time step). Rearranging Equation 6.4 to focus on that added chunk per year, we get

$$\text{Change per year} = N_{t+1} - N_t = N_t(b - d) \qquad \text{(6.11A)}$$

Because we now want to add individuals more frequently than once per year, we need to make the actual time step more explicit and focus on the fact that we have been looking at per capita *rates*:

$$\frac{\Delta N}{\Delta t} = \frac{N_{t+1} - N_t}{1 \text{ year}} = N_t(b - d) \tag{6.11B}$$

The added delta (Δ) in front of N and t on the left-hand side of this equation is mathematical shorthand for "change," meaning this equation calculates the change in population abundance over a change in time. Notice that there is no explicit mention of time on the right-hand side of the equation because b and d are annual per capita birth and death rates, so time is already included.

Now let's try to shrink the period of time in which we are looking at the population change. As we choose shorter "slices" of time, it is important to recognize that no single, tiny unit of time will always fit perfectly. If we insert a specifically sized time step, it will not work for all species. Instead, what we need is a way to look at the change in the population abundance over the smallest possible slice of time, so that we know it is small enough to work for any species (even bacteria or yeast cells). Borrowing from calculus, we know that derivatives are the smallest possible slices of a variable, and by convention we represent derivatives with a d in front of the variable. As our ΔN and Δt become super tiny, we can represent the change in the population that happens over the smallest possible slice of time as the following:

$$\lim_{\Delta t \to 0} \left(\frac{\Delta N}{\Delta t} \right) = \frac{dN}{dt} = rN \tag{6.11C}$$

The important thing to note here is that r represents the **instantaneous per capita rate of population growth** and is equivalent to the instantaneous birth rate minus the instantaneous death rate (i.e., $b_{inst} - d_{inst}$). The limit expression (i.e., $\lim_{\Delta t \to 0}$) is shorthand for "the limit of allowing the change in time (Δt) to approach (but never quite reach) zero." In other words, dt is the smallest possible slice of time, and dN is the tiny change in abundance that occurs in that slice of time. The new per capita rate of population growth, r, is "instantaneous" because it takes place in that tiny instant. Because the time step has changed, the values of b and d that go into the instantaneous per capita rate of population growth (r) will be very different from what they were in the discrete-time equations, which is why we use the *inst* subscript. We can read Equation 6.11C as, "The change in the population abundance during the smallest slice of time is equal to the population abundance (N) times the instantaneous per capita rate of population growth (r)."

We now have a way to represent population changes during tiny slices of time, and we want to think of a way to use such minuscule changes in population to answer the same sorts of questions that we asked about the beavers, such as how big the population will be at some time in the future or when the population will cross a particular threshold size. Using integration (a form of calculus that allows mathematicians to calculate the area under a curve), we can add up all the small changes from a starting population abundance (N_0) to an ending population abundance (N_t) and arrive at the following general equation:

$$N_t = N_0 e^{rt} \tag{6.12}$$

The only new part of this equation is e, which is Euler's number and the base of the natural logarithm ($\log_e = \ln$), which you may have seen in previous science and math courses. Like π, e is not represented as a number because it is irrational: it cannot be represented as a simple fraction and has endless decimal places. You do not need to know the exact value of e, though it is approximately 2.7183. As a mathematical placeholder, e is useful because

$$\frac{d}{dx}e^x = e^x$$

For the purposes of this book, it is completely fine to just accept it as a number like π.

We used calculus to get Equation 6.12, but you do not have to understand calculus to use this equation. Let's start by thinking about how the population is going to change with different instantaneous per capita growth rates (different values of r). As we did with λ in Equation 6.7, we can think about the value of the per capita birth rate that will lead to no change in the population abundance. We would expect a growing population on one side of that value and a shrinking population on the other side.

Looking at Equation 6.11C, we can see there is no change in population abundance over time ($dN/dt = 0$) when $r = 0$. That works well in Equation 6.12, too, because $e^0 = 1$. So, we can expect the population to shrink when $r < 0$, stay the same when $r = 0$, and grow when $r > 0$. If we had the same per capita birth and death rates in two populations, we would expect a population with continuous breeding to grow faster than a population with discrete breeding and discrete population growth because the continuously breeding population accrues "interest" on new births throughout the year. Using the same logic, we expect a continuously changing population to shrink more slowly than a discretely changing population because early deaths reduce the population from which the next fractional deaths will be deducted.

Although both the discrete and continuous models have time as an exponent, ecologists generally call the *continuous-time* version a model of **exponential population growth** and the *discrete-time* version a model of **geometric population growth**. If you get confused about which is which, remember that the continuous, or exponential, growth model actually has the growth rate, r, as an exponent.

6.6 Models of Density Dependence

In discussing the assumptions made by simple models, we pointed out that constant per capita birth and death rates are pretty unrealistic. To explore why these fixed rates are a problem, let's examine the models of exponential and geometric population growth in more detail. The first thing to notice is that these models inevitably predict growth to enormous population abundances.

Returning to the beaver example, let's calculate the population abundance of beavers in our hypothetical park over 30 years. We will use the number of individuals in 2021 (25 beavers) as the starting point and the value of λ derived from the

Figure 6.13 This graph shows the predicted growth of our hypothetical beaver population over 30 years using the discrete-time equation (Equation 6.7). The scale is so skewed by the final large numbers that the early years all appear to have almost no beavers, when in fact, the model predicts a population of more than 200 beavers by 2035 and more than 700 by 2040. On the ground, such numbers would be very noticeable.

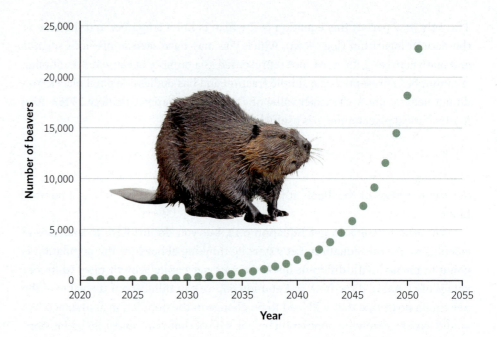

hypothetical census data (1.257; **Figure 6.13**). In Figure 6.13, we use Equation 6.7 to estimate population abundance for every year through 2050. Take a close look at the y-axis. The number of beavers starts at 25 and increases to more than 20,000 individuals over the course of 30 years. Although there may be room for more than 20,000 beavers in all of Chile or Argentina, it stretches the imagination to envision 20,000 beavers in one 1,000 ha park.

This type of unbelievable population explosion would result from both the geometric and exponential models when $\lambda > 1$ or $r > 0$. In fact, such per capita growth rates would eventually lead to infinite population abundances in both models. If these models were always correct, the world should be covered in mosquitoes, frogs, dandelions, beavers, and many other species. Since we are not drowning in mosquitoes, frogs, dandelions, or beavers, these per capita growth rates must change as population abundance changes, which means that birth and death rates (b and d) have to change.

Let's think about why birth and death rates might change while still focusing on beavers. North American beavers rely on trees for food and woody material to maintain their dams. The dams serve as a place to raise their young and provide critical shelter during the cold winter months. If a beaver population were to grow without limit within the space of a single wetland, eventually the wetland would run out of trees for food and dams. Without enough food or shelter, the per capita birth and survival rates for beavers would be expected to drop. When survival rates go down, death rates must go up.

In general, when a critical resource becomes scarce, individuals that use the resource will reproduce less and have lower probabilities of survival. These reductions will slow population growth through time. Ecologists have named this phenomenon **density dependence** because the per capita rates of birth and death vary with population abundance, meaning they show a "dependence" on the population density (remember that population abundance can also be expressed as a density because all populations are in a defined area).

If resource levels affect per capita birth and death rates, we can think about every patch of habitat as having an upper limit to how many individuals it can hold, based on the supply of resources available in that patch. We could even measure these resources in units of the species of interest and use them to define a **carrying capacity (K)**, meaning the number of individuals that the resources in a habitat patch can sustain over the long run. The carrying capacity provides both a target population abundance and a measure of resources.

The resources that eventually limit population growth are specific to the needs of the species, and individuals within a population compete for them. Because these individuals are all in the same species, we call the competition that leads to density dependence **intraspecific competition** (*intra-* meaning "within," and *specific* here meaning "relating to a species," so "within a species"). As you can imagine from what you learned in Chapter 5, these limiting resources will vary tremendously between species, habitats, and circumstances. For many animal species, food and shelter are common limiting resources. Although plants are autotrophs, they still need energy; therefore, light availability is a common limiting factor for plants. In addition, for plants and many sessile (immobile) invertebrates, space itself can be limiting because it offers access to light and nutrients.

Density-Dependent Population Growth in Continuous Time: The Logistic Model

The density-dependent model that we are going to build in this section is called the **logistic growth model**. There are several ways to insert the idea of a carrying capacity into a basic population growth equation, but it is easiest with the equation for instantaneous change in continuous time (Equation 6.11C). Let's continue to assume that all individuals in the population have the same birth and death rates. In that case, as population density increases, the resources provided by the habitat are split among more and more individuals, meaning the per capita access to food and shelter should decrease. The simplest way for the per capita birth and survival rates to change with increasing population density (meaning a decreasing per capita resource share) would be if they decreased linearly.

When a population is tiny and population density is so low that individuals do not even notice each other's use of resources, the population will have its maximum per capita growth rate, which is called the **intrinsic rate of increase**, designated by r_{max}. For each species, r_{max} reflects something intrinsic about the species and the way the individuals of this species react to their habitat. To use a more complete term, we could call r_{max} the *intrinsic per capita population growth rate*. Like the carrying capacity, r_{max} can vary across different habitat patches because resource quantity and quality differ from location to location.

As individuals use up the available resources in a single patch, their birth rates or survival rates decline in proportion to the available resources until the rate of population growth reaches zero. Remember that when the population is not growing (i.e., $r = 0$ in Equation 6.11C), the population abundance remains constant, neither growing nor shrinking. So, at this population growth rate of zero, the resources are not depleted. They are, instead, at a level that allows births

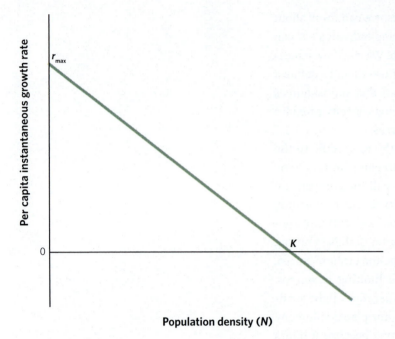

Figure 6.14 With density dependence, the per capita instantaneous rate of population growth changes with population density (N), starting at a maximum per capita growth rate (r_{max}) set by the species and the habitat, and decreasing linearly to zero when the population density reaches the carrying capacity (K).

Figure 6.15 In this conceptual sketch of how the population density might change over time with density-dependent per capita population growth, the inflection point is the point in the curve at which the population growth starts to slow down. You can recognize it because the curve switches from being a curve that is concave up to being a curve that is concave down.

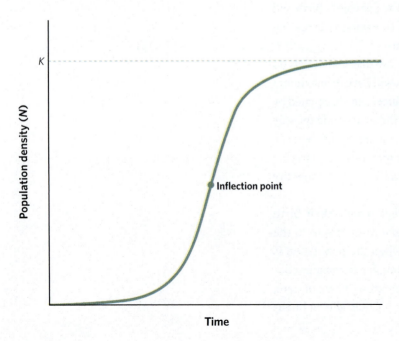

and deaths to balance each other exactly. This level of resources keeps the population stable at exactly the size that can be supported by the local environment—in other words, at the carrying capacity, which we represent as K. So far, this is just a conceptual model, and we can produce a graph like **Figure 6.14** to represent the simplest way for the per capita population growth rate to change with population density.

We can also use our conceptual model to imagine how population abundance would change over time with density dependence. Look at Figure 6.13 and try mentally to combine the constant per capita growth in that figure with the changing per capita growth from Figure 6.14. It seems unlikely that a density-dependent population would grow straight up to the carrying capacity and then stop, as though it had a constant per capita growth rate and then hit a ceiling. Instead, we might expect a graph that initially looks similar to Figure 6.13, but it seems reasonable that the change in the size of the population would slow down before it levels off at a constant population abundance. That slowing down would fit with the changing per capita rate of increase that we see in Figure 6.14. One reasonable possibility for showing such slowing down is a curve that looks a bit like an S (**Figure 6.15**).

You may notice in Figures 6.14 and 6.15 (and in other figures later in the chapter) that we have switched to the term *population density* for N. N can represent both abundance and density because it is a number of individuals (abundance) that are in a specific area (density). For most of this chapter we use the term *abundance*, but now we want to emphasize that the population is not spreading out over a larger area (we will discuss movement out of a habitat patch as N increases later in the chapter). Instead, as abundance increases, the density in the location of the population also increases, so it is appropriate to use N to represent density here.

To model change in population density with a mathematical function, the easiest starting point is Equation 6.11C. We can replace r with r_{max} and add a factor that modifies r_{max} to reduce the per capita population growth rate as density increases. Because we define the carrying capacity in units of the species of interest, we can compare the sizes of N and K and get a sense of the fraction of resources that are left before the per capita population growth rate hits zero. If K represents the total number of individuals that the resources in a habitat patch can support, then $(K - N)/K$ is the fraction of the carrying capacity remaining when the population density is N. This fraction will be close to one when N is small and will approach zero as N increases. We can use this fraction to modify r_{max} so that the population grows at the maximum per capita rate when N is tiny and slows to zero when N hits the carrying capacity (i.e., when $N = K$). Putting all this together gives us

$$\frac{dN}{dt} = Nr_{max}\left(\frac{K - N}{K}\right) \tag{6.13}$$

Equation 6.13 is called the **logistic equation**. It was derived and named in 1838 by Pierre-François Verhulst and rederived by several other mathematicians and ecologists studying population growth in later decades. Equation 6.13 fits our conceptual model remarkably well. Comparing Equation 6.13 to Equation 6.11C, we can see that the constant population growth rate (r) from exponential growth has been replaced in the logistic equation by a growth rate that changes because $(K - N)/K$ changes.

Equation 6.13 predicts population growth like that shown in Figure 6.15. Initially, the growth appears to be exponential, but then it slows at the inflection point, halfway to the carrying capacity. Notice that the population grows very slowly at first, even though the per capita population growth rate is highest with small population abundances. This pattern makes sense if you think about it carefully. Returning to our banking analogy, you can see that even a ridiculously high interest rate of 100% would only produce a single dollar in interest if the account only holds one dollar. Once the account contains $1,000, though, a 2% rate of interest will produce $20, or 20 times as much money, even though the interest rate has dropped significantly.

The logistic equation does the same thing, with small total additions to the population (dN/dt) at small population densities and increasing additions as the population gets bigger, even though the per capita rate of population growth decreases. These total additions increase as population density increases with maximum total instantaneous population change occurring when the population is halfway to the carrying capacity, after which dN/dt begins to decline again and becomes zero when $N = K$ (**Figure 6.16**).

The carrying capacity is also a **stable equilibrium** for the population. The term *equilibrium* in this context means a value is unchanging. An equilibrium point is considered stable if the system returns to that point after being knocked away from the equilibrium. For Equation 6.13, the population abundance or density (N) reaches equilibrium at the carrying capacity (K), where it does not change. If we perturb the population by removing individuals, the population quickly grows back to K. If lots of individuals suddenly enter the habitat patch and raise the population density so that $N > K$, then the $(K - N)/K$ term becomes negative, leading to a declining population until the population abundance again stabilizes at K.

The logistic growth model is a foundational concept for many areas in ecology, and we will revisit it in Chapters 7, 8, and 9. Also in later chapters, we will come back

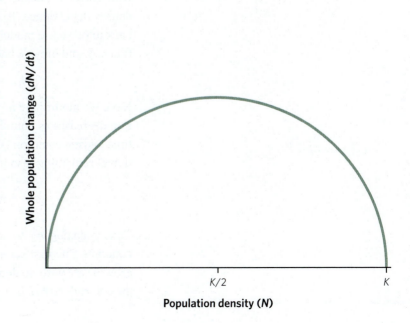

Figure 6.16 Here, we see the contribution of population density to whole population change (dN/dt) with the logistic model. When N is small, the actual whole population change is also small even though the per capita population growth rate is large and close to r_{max}. The peak change in whole population abundance with the logistic model comes when $N = K/2$. At the origin, when $N = 0$, the change in population density is zero, and at the carrying capacity (when $N = K$), the change in the whole population density is again zero.

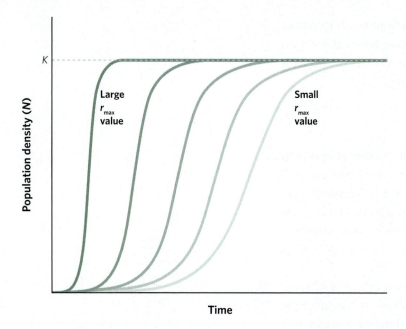

Population density (N)

K

Large r_{max} value

Small r_{max} value

Time

Figure 6.17 Logistic population growth leads to steeper S-shaped curves if r_{max} is bigger. Darker lines represent larger r_{max} values. As r_{max} increases, the population reaches its carrying capacity (K) faster.

to ideas about mathematical equilibria and whether they are stable or unstable. If you are struggling to understand the material presented in this chapter thus far, either seek assistance from your instructor or try reading the first part of the chapter again.

Another noteworthy feature of the logistic equation is that it will never allow the population to grow beyond the carrying capacity. In continuous time, the feedback from resource scarcity affects population growth instantly, so that the population always adjusts and responds to the current resource level. If the peak per capita growth rate is very big, meaning r_{max} is very big, the population just reaches K faster, leading to a steeper S curve (**Figure 6.17**).

Yet, in the real world, it is unlikely that the effect of density dependence on population growth is instantaneous. Effects on survival, for example, may take some time to appear if individuals slowly starve to death or if resource scarcity is only an issue in certain seasons. In fact, if organisms give birth in one season and offspring do not use many resources independently in the early stages of life, the effects of density may not be apparent until the next year or until the juveniles mature. The logistic model does not include any of these real-world **time lag** features, but we can explore the effects of lag times in density dependence by exploring a discrete-time density-dependent equation.

Density-Dependent Population Growth in Discrete Time

The **discrete-time density-dependent population growth** equation can take many forms, but here we focus on the one that is most similar to our continuous-time formulation. When we built the continuous-time equation (see Equations 6.11A–C), we defined

$$r = b_{inst} - d_{inst}$$

which means r encapsulates the two per capita rates that we expect to change with density dependence. To build a discrete-time density-dependent equation, let's go back to our basic model of geometric population growth in one time step (Equation 6.7), and break λ back into "1 + b − d" (Equation 6.5).

$$N_{t+1} = N_t\lambda = N_t(1 + b - d) = N_t + N_t(b - d) \quad \text{(6.7 restated)}$$

Next, we modify the b − d portion of this equation so that the per capita growth rate decreases as abundance increases. Using similar logic as in the continuous-time logistic equation (Equation 6.13), we multiply the constant b − d by the fraction of remaining resources available to share across individuals, $(K - N_t)/K$.

$$N_{t+1} = N_t + N_t(b - d)\left(\frac{K - N_t}{K}\right) \quad \text{(6.14A)}$$

This is analogous to what we did in the continuous-time density-dependent equation. The first set of parentheses, (b − d), contains the maximum per capita growth rate with no density effects, a rate similar to r_{max}. The changing per capita growth rate comes from multiplying (b − d) by the second set of parentheses as

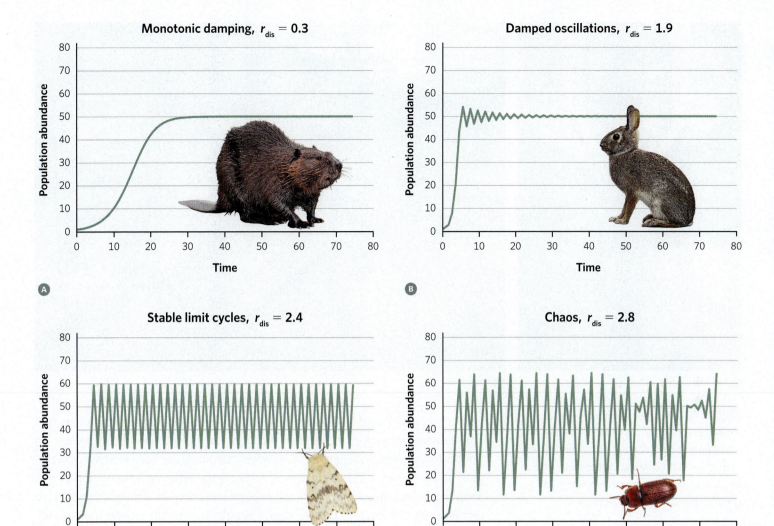

Figure 6.19 The discrete-time logistic model can produce complicated population dynamics. For all the models, $K = 50$ and the starting population abundance is 1. **(A)** $r_{dis} = 0.3$ (monotonic damping); **(B)** $r_{dis} = 1.9$ (damped oscillations); **(C)** $r_{dis} = 2.4$ (stable limit cycles); and **(D)** $r_{dis} = 2.8$ (chaos). The inset animals represent examples of species that exhibit these dynamics. Remember that these behaviors are only possible with discrete-time population growth. The continuous-time logistic equation always leads to monotonic damping, no matter how large r_{max} is.

fluctuations because of their random appearance. We discuss the factors that can lead to cycles much more in Chapter 8, when examining predator-prey dynamics. However, it is important to know that such cycles can come from density-dependent population growth with a time lag between resource availability and reproduction or death, and that discrete-time models have a built-in time lag that captures this real-life possibility.

6.7 Real-World Evidence for Density Dependence

Conceptually, it seems fairly obvious that populations do not grow to infinite abundances and that something in the environment eventually puts the brakes on growth rates. However, evidence for specific environmental brakes has been difficult to produce because in the wild many factors can affect survival rates, reproductive rates, and population growth. Using an experimental approach can be very helpful when exploring this aspect of population ecology.

Figure 6.20 Ⓐ An adult male North American elk. Ⓑ Fences like this one help maintain an experimental North American elk population in the Starkey Experimental Forest and Range in Oregon and Washington, which is run by the US Forest Service.

As you may recall from Chapter 2, ecological experiments manipulate specific explanatory factors to see their effect on a response variable of interest. In the context of density dependence, we want to control (or keep relatively consistent) all the factors that influence survival or reproduction in a population except abundance (and thereby density) and explore the effects of changing abundance (explanatory variable) on the per capita population growth rate (response variable). Several ecologists have done this in the laboratory with tiny organisms, such as paramecia, and we explore some of these results in the context of competition in Chapter 7. Here we focus on an experiment involving larger organisms in a fairly natural setting.

There are several reasons why we should expect to see density-dependent effects in populations of large-bodied mammals. These species tend to live a long time; they vary in physical size and reproductive output with food availability; and they fiercely defend their access to food. Once a population of these large mammals reaches an abundance where food resources become limited, the individuals show resource stress in terms of their nutritional status, mortality rate, and reproductive rate for several years.

Looking for evidence of density dependence, John Kie and his colleagues experimentally manipulated population densities of North American elk (*Cervus elaphus*) on the high-elevation rangelands of the Blue Mountains in Oregon and Washington (Stewart et al. 2005; **Figure 6.20A**). The US Forest Service maintains an experimental site called the Starkey Experimental Forest and Range where researchers can set up large-scale experiments to explore how elk and other large mammals influence the nearby forests and rangelands and how human management of these habitats, in turn, affects the biology of the large mammals.

The entirety of this experimental area is fenced, so that the elk cannot (easily) escape (**Figure 6.20B**). The fenced area is large enough, however, for the resident animals to move around freely, finding food and mates as they would in the wild. The only exception to this rule is that, under natural circumstances, elk migrate away from these rangelands in the winter and move to lower-elevation sites that provide grasses and other plants to sustain them through the winter months. Because the enclosed elk cannot migrate, the managers provide a winter feeding pen with food and shelter to keep the elk alive until the spring.

Kie and his colleagues used this experimental rangeland to create two populations of elk, one that was kept at low density (approximately four elk per km²) and another that was kept at high density (approximately 20 elk per km²), although the specific number of elk in each enclosure varied a bit across years. The two groups were kept in separate, large, fenced enclosures during all nonwinter months from 1998 to 2001. The high-density population was near the maximum abundance ever recorded for elk living in the wild, so it is fairly certain that elk in the high-density treatment experienced some amount of resource stress.

The researchers hypothesized that the high-density treatment would reduce both the availability and intake of food for the elk. As a result, they expected female elk in this enclosure to have reduced fat storage and lower pregnancy rates than elk in the low-density setting. Because the elk were herded into the wintering area every year, the researchers had a chance to tag and measure each animal and determine its physiological condition and reproductive status. They used a sonogram to scan the rumps of adult and yearling females for fat because the rump is the primary location for fat storage among ungulates. In addition, for each adult female, they recorded estrogen levels to determine if she was pregnant.

After three seasons of following the experimental elk populations, the researchers found that female elk in the high-density population had lower fat stores, suggesting that there was intraspecific competition among elk for resources (Stewart et al. 2005; **Figure 6.21A**). The rump fat in yearling and adult females in the low-density setting was almost twice the amount of fat in individuals in the high-density treatment. In addition, the fraction of pregnant females declined as population density increased (Stewart et al. 2005; **Figure 6.21B**). In the high-density population, only about half the females became pregnant in each year of the study. However, in the low-density setting, 60%–80% of females became pregnant.

Nonetheless, there was a fair amount of variability from year to year in both measures within the low-density situation. This variability almost certainly comes from factors that the researchers could not control, such as weather events. In many populations these more unpredictable events that influence resource levels or influence mortality and reproduction can overwhelm the influence of density alone. These types of factors collectively are called **density-independent factors**

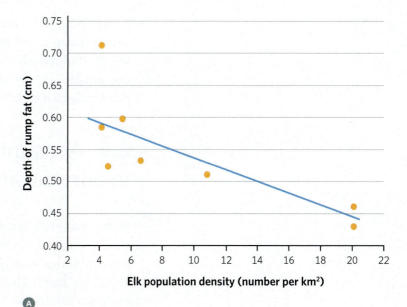

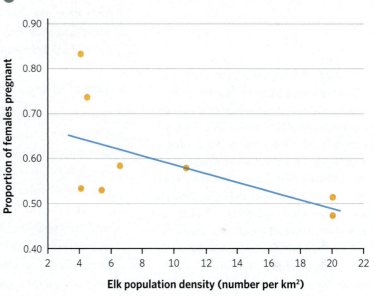

Figure 6.21 These graphs, based on data from John Kie and colleagues' experiment (Stewart et al. 2005), illustrate the differences in health and reproductive status for elk in low- and high-density experimental treatments. **A** Fat reserves significantly decrease with increasing elk population density. **B** The proportion of pregnant females also decreases with increasing population density. Note that the lines in both graphs were drawn by us to highlight data trends and are not lines of linear regression.

because they act on a population regardless of its density. The prevalence of density-independent factors can make it hard to detect density dependence, but the elk study clearly shows that such experiments are possible and that population density can have a fairly rapid impact on per capita rates of reproduction and potentially on population growth rates.

6.8 Accounting for Individual Differences in Survival and Reproduction

So far in this chapter, we have assumed that all individuals in a population have equal chances of surviving and reproducing. Although this assumption may hold true for viruses, bacteria, and a few single-celled organisms, it is untrue for most species on Earth. There are many different ways in which age or life stage may influence survival and reproduction. When such individual differences matter, ecologists say that the population is *structured*. Here we explore age and stage structure, how they relate to survival and reproduction, and how to include these individual differences in basic population models.

Life History

We can describe how reproduction and survival vary throughout an organism's life for many of the world's species. Doing so allows scientists to compare species and consider how such changes over a lifetime determine the patterns of population growth. A species' **life history** is the temporal sequence of events that determines survival and reproduction from an individual's birth until its death.

In terms of reproduction, some species, like mosquitoes, mature quickly and are capable of breeding very soon after birth. Such species tend to produce several young per breeding period and are thus described as having high annual **fecundity**. For other species, it can take decades before individuals become sexually mature and reproduce, and even when they do, they only produce one or a few young per year. For example, humans do not become sexually mature until after about 15 years of age, and females can only reproduce until approximately age 45. For species that are highly social, sexual maturity may arrive long before individuals are capable of defending a territory or attracting a mate. Social structure may, therefore, dictate that most individuals delay reproduction for several years beyond when they are first physiologically capable of reproducing. Whether because of physiological or social limitations, the number of young produced by these species in any given breeding event will be small. Such species often have low annual fecundity.

Age or stage of life can also have a dramatic effect on the chances of dying in any given time period. We can describe lifetime patterns in survival probability using **Figure 6.22**. In this figure the surviving proportion of a **cohort** (group of individuals all born at the same time)

Figure 6.22 Organisms can be divided into three groups based on "survivorship curves" that reflect changes in their mortality rates throughout their lives. Lots of parental care can lead to a downward-facing curve, with high juvenile survival and a sudden increase in mortality rates at old age (Type I); a constant mortality rate throughout most of life leads to a fairly straight line (Type II); and high juvenile mortality leads to a steep drop followed by a fairly consistent mortality rate until old age (Type III). Note that the y-axis can be either the number or the proportion of a cohort that is surviving, but it must be on a log scale to produce these patterns.

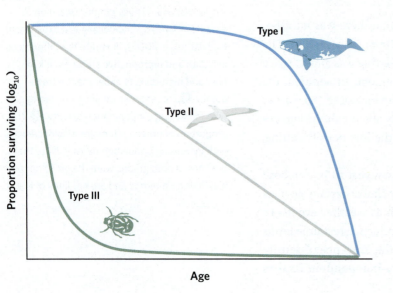

is plotted on a log scale against age. Anywhere that we see a straight line on this graph, individuals are experiencing a mortality rate that does not change with age. Steeper slopes indicate a higher annual mortality rate.

In Figure 6.22, the blue line represents a *Type I survivorship curve*; this type of curve means that individuals of the species have a fairly high rate of annual survival until they reach old age, when mortality increases dramatically. Humans, whales, and elephants tend to have Type I survivorship curves. The green line, in contrast, represents a *Type III survivorship curve*, which means that individuals have a very high chance of dying in the earliest part of life but a much higher rate of survivorship after this initial juvenile period. Many tree species, amphibians, insects, and marine invertebrates exhibit Type III survivorship curves. Finally, the gray line represents a *Type II survivorship curve*, the intermediate condition, in which individuals experience a fairly constant annual survival rate throughout their lives. Many seabirds, tide pool anemones, and some reptiles have Type II survivorship curves. These three types of survivorship curves are idealized, and most species fall somewhere in between.

Age and Stage Structure

Now that we know survival and reproduction change throughout an organism's life, we can focus on how the individuals in a population experience these changes and how populations are structured. We understand humans well, so let's start with ourselves. As a species, humans have a distinct **age structure**, meaning survival and annual fecundity rates vary with age group (**Figure 6.23**). *Age pyramids*, like the hypothetical one in Figure 6.23, can help us to see that different numbers of individuals are in each age group. These pyramids can also help translate the differences into expectations for population growth. In any given year, we would expect only the individuals represented by the light green bars in Figure 6.23 to

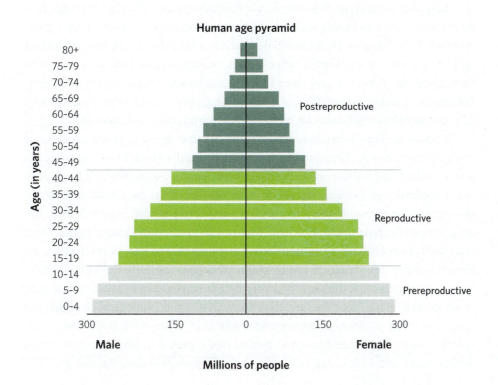

Figure 6.23 This hypothetical age pyramid for humans shows reproductive and nonreproductive segments for males (left side) and females (right side). This pyramid shape (wide at the bottom and narrow at the top) indicates a population that is growing fairly quickly through time.

Figure 6.24 Age structure in the fir tree *Abies forrestii* var. *georgei* (also known as the Smith fir tree, *Abies georgei*) from a single plot on the Tibetan plateau (Wang and Liang 2016).

reproduce, not those represented by the gray or dark green bars. We can see age structure in other species as well, including trees (**Figure 6.24**).

Age, of course, is not the only factor that determines whether reproduction happens in humans. People make the decision to reproduce based on their partner's wants and desires, their financial situation, where they are in their education and careers, as well as a host of other factors. In addition, we know that mortality in humans may reflect factors other than age, such as access to medical care, marital status, economic advantages, or even housing opportunities. For humans, a lot of what we call social factors are tied up with reproduction and mortality.

Most organisms reproduce based more on size or some other measure of maturity than on age. For example, body size is usually much more important to survival than age because having a large body can increase chances for escape from predators or increase access to vital resources. In fact, body size better describes how reproduction and survival vary between individuals in both the snapper and fir tree populations from Figure 6.24. Sometimes maturity can affect survival (or reproduction) without being strongly tied to age. In such cases, we do not refer to age structure but to **stage structure**, because individual differences in reproduction and survival vary by life stages. These stages often span unequal amounts of time but are clear from size or morphological changes. For organisms that undergo morphogenesis (distinct changes in body form with development), like amphibians and insects (see Figure 6.1), stage structure often makes more sense than age structure. As with age structure, we can quantify

Bottlebrush
stage

Sapling
stage

Grass
stage

Longleaf pine
life stages

Mature
tree stage

Seed stage

Seedling stage

stage structure by counting the total number of individuals in each life stage in the population.

To understand the difference between age and stage relative to reproduction and survival, consider the fascinating life cycle of the longleaf pine (*Pinus palustris*; Figure 6.25). Longleaf pines grow in fire-prone ecosystems in the southeastern United States, and their life cycle reflects evolutionary adaptations to the threat of death from fire. For this reason, longleaf pine survival is better represented using life stages rather than an individual's age.

Longleaf pine seeds drop from the parent tree in October or November each year, and, if they fall on nutrient-rich soil, the seeds germinate and grow into seedlings within a few weeks. Yet, seedlings are very susceptible to death by fire. If they survive, they develop into what is known as a "grass" life stage. At the grass stage, an individual tree is very short in stature with a clump of needles at the growing tip, making what is actually a tree look a lot like a clump of grass (Figure 6.25).

These needles help protect the young tree from fire; at this life stage, death by fire is very rare. For a while, the little "grass" tree does not grow much in an upward direction but instead grows downward, investing in an extensive root system. Longleaf pines will stay in the grass stage for one to seven years, with some individuals remaining in this stage for as long as 20 years.

Once a longleaf pine tree has established a strong root system, it will "bolt" into and through the bottlebrush stage (Figure 6.25). At this point, the tree invests heavily in upward growth, often extending up in height by a few feet in just a few months. At this stage, the tree is again very susceptible to fire because its growing central stem is unprotected by needles. If the tree can grow tall enough that the growing tips escape flames, then it will survive most fires. Finally, as the tree grows beyond this stage, it will eventually become tall enough to be considered a sapling, at which point it will extend its branches outward and upward to capture sunlight in a crowded canopy. Beyond the sapling stage, the tree is largely immune to death by fire. It is only at this mature life stage that a longleaf pine becomes sexually mature and produces cones and seeds. A mature longleaf pine can live up to 200 years.

Survival rates for longleaf pine populations are low in the seed and seedling stages, become high in the grass stage, low in the bottlebrush stage, and then high again when mature. Reproduction is zero during every stage except at maturity. Because a single tree will spend various numbers of years within each stage, describing this life stage pattern using calendar age is cumbersome and misleading. For example, two life stages (seed and seedling) occur within the first calendar year of life for longleaf pine. Collapsing them into one calendar year will misrepresent how important these stages are to population growth. At the other extreme, individuals will have the same survivorship for several years when the tree is in the grass stage or is mature. It is far more efficient to record the survival and reproduction using stages rather than noting that these rates stay the same for one to 20 calendar years (ages).

Stage Structure and Population Growth

Age and stage structure have the capacity to affect the way populations grow. Once we jettison the assumption that all individuals have the same per capita survival and reproduction rates, we need to keep track of individuals in different ages or stages and apply age- or stage-specific annual fecundity and survival rates. Although this may seem daunting, we can use an expanded version of our discrete-time population growth model for structured population growth. In the next section, we work through the creation of a stage-structured model for loggerhead sea turtles, a globally occurring marine species that is showing steady population declines across most of its geographic range.

Loggerhead Sea Turtle Life History Loggerhead sea turtles (*Caretta caretta*) are found in tropical or subtropical oceans in the Atlantic, Pacific, and Indian Oceans. They spend almost their entire lives at sea, foraging on fish and invertebrates. Only the females venture onto land, and they do so only long enough to crawl from the surf up onto a sandy beach to dig a large hole, deposit a few dozen eggs into that hole, and then cover up the eggs (**Figure 6.26A**). Loggerhead sea turtles have no parental care. After laying their eggs, the females crawl back into the surf and swim to deeper waters, their annual act of motherhood complete.

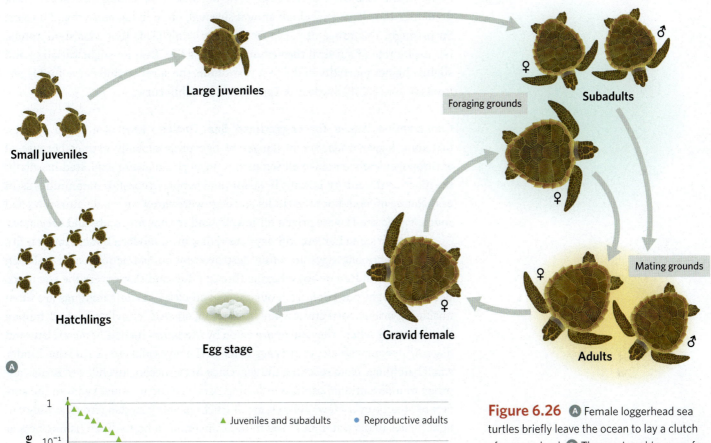

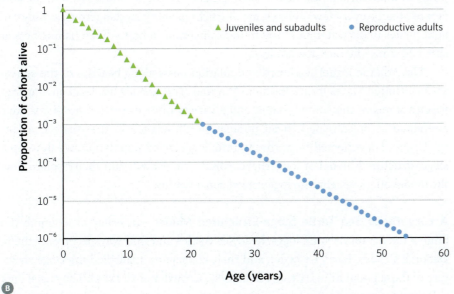

Figure 6.26 **Ⓐ** Female loggerhead sea turtles briefly leave the ocean to lay a clutch of eggs on land. **Ⓑ** The survivorship curve for loggerheads shows a linear decline, suggesting an almost constant mortality rate and a Type II survivorship curve. Only about one of every 1,000 baby turtles survives long enough to reproduce (Frazer 1983).

Loggerhead eggs remain buried in the sand for approximately 80 days until they hatch, at which point the hatchlings emerge as a group, orient toward the ocean, and crawl toward the surf. The eggs and hatchlings are vulnerable to predation by gulls, raccoons, foxes, and even ghost crabs. Once the hatchlings reach the water, they will stay close to floating wrack (i.e., floating seaweed and kelp) and forage on invertebrates in the wrack for most of their first months of life. During this early stage, they continue to suffer relatively high mortality from sharks and large, predatory fish. If they survive this predator gauntlet, eventually they get big enough and mobile enough to swim near the shore and forage on invertebrates living in the shallow waters. Once they become large enough to escape most predators, they have fairly high annual survival, often living more than 50 years. Surprisingly, this range of perils from egg to adult leads to a consistent annual per capita rate of survival throughout a turtle's life. Despite slight variation and slightly higher mortality in the first year of life, the survivorship curve for loggerheads (Figure 6.26B) suggests a Type II survivorship curve.

Conservation Issues for Loggerhead Sea Turtles   Most sea turtle species, including loggerheads, are in danger of becoming globally extinct because of anthropogenic effects. Like all sea turtles, loggerheads were exploited heavily in the nineteenth century, initially by sailors and people in coastal communities as an easy source of food. Both the adults and eggs were eaten and made into stews and soups. Turtle shells were prized for jewelry and as rims for combs and eyeglasses.

Although some harvest still happens today, most modern human impacts are more unintentional. Eggs are often destroyed by human actions such as beach raking or by children or dogs digging through the sand that covers the egg nests. Because of light pollution from our towns and cities, turtle hatchlings are often unable to orient correctly toward the ocean, instead crawling toward human developments where they die or are eaten by predators such as domestic cats and dogs. All-terrain vehicles crush eggs while in the nests and create ruts that hinder small hatchlings from reaching the sea. Once in the ocean, juvenile sea turtles are prone to unintentional capture in fishing nets. Sea turtles must swim to the surface to breathe every few minutes, making fishing nets very dangerous, as they can hold turtles underwater and drown them. Juveniles tend to die in fishing nets at about 20 times the rate that adults do.

The Atlantic loggerhead turtle population nests along beaches of the southeastern United States from Florida to Virginia. The species has legal endangered species status in the United States, and a variety of conservation programs have attempted to bolster loggerhead population abundance in the western Atlantic. Using data collected from the Atlantic loggerhead population, we develop a stage-structured model of loggerhead population growth and then explore how the model offers insight into loggerhead conservation.

A Loggerhead Sea Turtle Stage-Structured Model   In order to examine the human impact on Atlantic loggerhead sea turtles and find ways to protect them, Deborah Crouse, Larry Crowder, and their colleagues quantified the stage structure of these populations (Crouse et al. 1987; Crowder et al. 1994). They used data collected by James Richardson and Nat Frazer, who tagged loggerhead sea turtles near Little Cumberland Island, Georgia, for more than 20 years. Frazer used these observations to estimate age-specific survival and fecundity rates for loggerheads (Frazer 1983; Figure 6.26B). Note that this type of accounting is also referred to as *demography*, which is the scientific study of the growth and regulation of populations. Although here we focus on turtles, demographic studies like this are extremely important for humans in terms of socioeconomic issues, political planning, and even life and health insurance.

Crouse, Crowder, and colleagues noticed that Frazer's mortality rates and observations of sea turtle behavior were very consistent, suggesting they could group turtles into discrete stages according to survival rates and exposure to

Table 6.4 Stage Classes of Loggerhead Sea Turtles

Stage class (notation)	Stage duration (years)	Annual survival rate (s)	Annual fecundity rate (f)
Egg/hatchling (h)	1	0.6747	0
Small juvenile (sj)	7	0.75	0
Large juvenile (lj)	8	0.6758	0
Subadult (sa)	6	0.7425	0
Adult (a)	>32	0.8091	76.5

Based on data from Crowder et al. (1994).
The notation letters h, sj, lj, sa, and a are the subscripts we will use to indicate each stage class in our model.

particular perils (Table 6.4). Note that the duration of each stage varies, from one year for the egg/hatchling stage to more than 32 years for the adult stage. These life history stages also match well with what we know of loggerhead turtle fecundity, namely that females lay eggs only as adults, when they are more than 20 years old.

Figure 6.27 depicts the stage structure of a loggerhead sea turtle population. We could track the entire life of one individual loggerhead sea turtle with this figure. This hypothetical individual would start as an egg and emerge from the egg as a hatchling. If this individual survived its first year, it would "graduate" into the next stage as a small juvenile, where it would stay for approximately seven years. If it survived through the small-juvenile life stage, it would graduate to become a large juvenile. If we followed this individual long enough, we might find it reaching the adult stage and staying there until it died.

Of course, progressing that far would require surviving each stage, and not just for one year but sometimes for many years before graduating to the next stage and dealing with a new suite of dangers. How many turtles manage to "thread the needle" to adulthood? To answer that question, Crouse, Crowder, and colleagues examined Richardson's and Frazer's 20-year dataset and looked for the proportion of turtles in each stage that survived to stay in that stage as well as the proportion of turtles that survived and graduated to the next stage.

Figure 6.27 Life history model for loggerhead sea turtles (based on Crowder et al. 1994). Based on Table 6.4, we would not expect to see any fecundity for subadults (f_{sa}), but the inclusion of that arrow is explained on the next page.

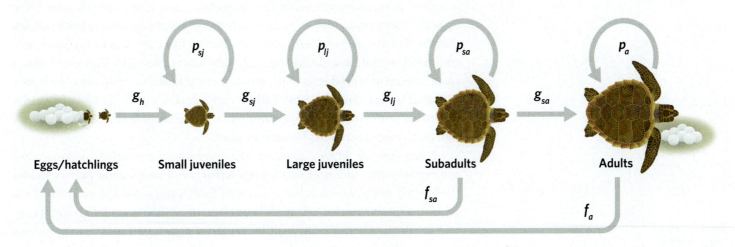

To do this, they divided the annual per capita survival (Table 6.4) into two pieces: surviving to remain *in place* (*p*) and surviving to *graduate* (*g*). In Figure 6.27, we use arrows that start in a stage and loop back to that stage to show in-place survival, and we use straight arrows between stages to show the proportion of survival that involves graduating to the next stage. Note that loops for staying in place only occur for stages that last more than one year, so there is no loop for the egg/hatchling stage. If hatchlings survive the first year, they become small juveniles. Also, the survival for adults all goes into the loop because adults have no other stage into which they can graduate.

To clarify how these loops and arrows align with actual observations of loggerhead turtles (Table 6.4), consider a three-year-old and a seven-year-old turtle in the small juvenile stage. The three-year-old will grow in body size over the course of a year, but Richardson's and Frazer's observations suggest that its chance of survival does not change noticeably with this growth and that it will follow the loop if it survives the year. The arrow to the next stage (i.e., large juveniles) shows the path the seven-year-old turtle is likely to take if it survives the year. Because Crouse, Crowder, and team were lumping the three-year-old and the seven-year-old into one stage (and because some individuals grow in body size faster than others), the chances of following the loop or following the arrow to the next stage do not really work well with ages but instead represent the proportion of the whole stage that follows one path or the other. Remember that the loop and the arrow together represent annual survival of turtles that started the year in that stage. Dying is represented by not following either the loop or the arrow to the next stage. The process works the same way at every stage except for the adult stage, where the only option is to survive in that stage or to die.

Figure 6.27 provides a good conceptual model for the life history of a loggerhead sea turtle, but we need to incorporate reproduction. We know that only adults reproduce, so we show reproduction as an arrow for fecundity (*f*) from adults to the egg/hatchling stage. You will notice, though, that the diagram also includes an arrow from the subadults to the eggs. Why would we include that arrow if no reproduction happens until turtles hit the adult stage?

When Crouse, Crowder, and colleagues divided up their stages, they realized they had a slight accounting issue. We discussed logistical issues associated with counting individuals at the beginning of the chapter and pointed out that the time of year when counting occurs is crucial. For turtles, the best time of year to count is right after reproduction, when researchers can assess how many females came up onto a beach to lay eggs. Crouse, Crowder, and team decided to acknowledge that the "brand-new" adults (the ones graduating from the subadult category into adulthood) would have just reproduced, right before the count. This means that a small amount of reproduction happens in these graduating subadults right before they start being counted as adults. Given that reproduction is the sign of maturity for female turtles, this feature works well for researchers in the field as well, as a sign that a tagged turtle has become an adult.

Using this life history diagram, the researchers converted turtle survival and reproduction numbers into a **stage-structured population matrix model** (Lefkovitch 1965), which is like a series of linked, stage-specific versions of the geometric growth model. Using this stage-structured matrix model, ecologists can forecast the number of individuals in any given stage in the next year by looking

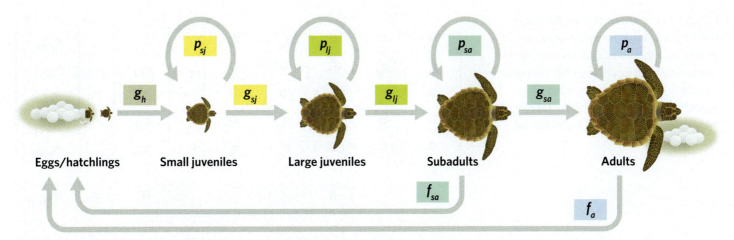

Figure 6.28 This figure combines the life stage diagram from Figure 6.27 with a life stage matrix for loggerhead sea turtles (based on Crowder et al. 1994). The colors indicate contributions from a particular life stage. Parameters with the same color come from the same life stage in both the diagram and the matrix.

$$\begin{pmatrix} 0 & 0 & 0 & f_{sa} & f_a \\ g_h & P_{sj} & 0 & 0 & 0 \\ 0 & g_{sj} & P_{lj} & 0 & 0 \\ 0 & 0 & g_{lj} & P_{sa} & 0 \\ 0 & 0 & 0 & g_{sa} & P_a \end{pmatrix}$$

at how many individuals stay in that stage from this year and how many new ones enter from another stage.

For example, the number of adult loggerhead turtles in the population next year (time $t + 1$) can be predicted by multiplying the number of adults this year (time t) by the adult survival rate, and then adding any subadults that survived and grew enough to enter the adult life stage. This approach is similar to the way we kept track of survival and births in our previous equations representing whole population growth, only now we are following the fate of groups of similar individuals rather than assuming the growth of the whole population can be described with one parameter.

Keeping track of every individual in the population clearly means we have a lot of numbers to worry about. To do this efficiently, we use notation to refer to each stage, as in Figure 6.27 and Table 6.4, so that survival within a stage (x) is p_x, surviving and graduating into a new stage is g_x, and fecundity for a particular stage is f_x. In all these parameters, x refers to one stage, so it takes the value of h, sj, lj, sa, or a. Remember, annual survival in a stage is split between the probability of staying in that stage and the probability of graduating to a new stage. If we add p_x, and g_x for any stage, we should get s_x, the total survival for that stage.

One way to keep track of all the calculations necessary to do this math is to use a mathematical notation called a matrix. The matrix keeps all the parameters ordered to facilitate the multiplication and addition needed to estimate how population abundance changes from one year to the next. We can link the elements of Figure 6.27 to the position of these values in the stage-structure matrix shown in **Figure 6.28**.

We also need new notation for population abundance because we want to keep track of how many individuals are in each stage. We can represent the abundance of a stage at time t as $n_{x,t}$. The abundance of the whole population, N_t, is just the sum of all the individual $n_{x,t}$ values. In order to do the math, we multiply our

Figure 6.29 Shown here are the matrix and vectors needed to project population abundance, and their equivalents from the geometric growth model. Note that the matrix represents λ, the current population vector represents N_t, and when multiplied together, they produce the population vector for next year (N_{t+1}).

$$
\begin{pmatrix}
0 & 0 & 0 & f_{sa} & f_a \\
g_h & p_{sj} & 0 & 0 & 0 \\
0 & g_{sj} & p_{lj} & 0 & 0 \\
0 & 0 & g_{lj} & p_{sa} & 0 \\
0 & 0 & 0 & g_{sa} & p_a
\end{pmatrix}
\cdot
\begin{pmatrix}
n_{h,t} \\
n_{sj,t} \\
n_{lj,t} \\
n_{sa,t} \\
n_{a,t}
\end{pmatrix}
=
\begin{pmatrix}
n_{h,t+1} \\
n_{sj,t+1} \\
n_{lj,t+1} \\
n_{sa,t+1} \\
n_{a,t+1}
\end{pmatrix}
$$

$$
\boldsymbol{\lambda} \qquad\qquad \cdot \qquad \boldsymbol{N}_t \quad = \quad \boldsymbol{N}_{t+1}
$$

matrix by a *population vector*, which is a single column that lists all the $n_{x,t}$ values. Each $n_{x,t}$ is an *element* of the population vector. That matrix multiplication gives us our population vector for next year, meaning all our $n_{x,t+1}$ values (**Figure 6.29**).

Doing math using a matrix is called linear algebra; but don't worry, the skills needed for this model are just addition, multiplication, and some careful bookkeeping. Keeping track of what parameter from the matrix is multiplied by what vector element is confusing until you try it on your own. Take a look at **Figure 6.30** if you want to learn the general approach. But you do not need to be able to do matrix algebra to understand the influence of stage (or age) structure on populations and how these models are used by today's practicing ecologists. We illustrate the relevant insights with visualizations like those in Figures 6.27–6.30.

A key thing to notice about the stage-structured matrix model in Figures 6.29 and 6.30 is that it mimics the discrete-time population growth model (Equation 6.7). In this stage-structured model, the matrix plays the same role as λ, which makes sense because both λ and the matrix keep track of per capita birth and survival rates. By keeping track of life stage differences for survival and fecundity, we essentially expand the λ term. In fact, a stage- or age-structured matrix model can produce a characteristic population growth rate that corresponds to a single λ value for the geometric growth equation.

Let's work through an extreme example to understand how stage structure affects population growth. Consider a population that is 90% eggs/hatchlings, 5% small juveniles, and 5% large juvenile turtles. What will the population abundance and stage densities be the following year, at $t + 1$? Without even knowing the specifics of the matrix, we can tell from Figure 6.30 that the population will be smaller next year, meaning the year-to-year λ is less than one. How do we know that? First, none of these three age classes are able to produce eggs, so we know that no new eggs or hatchlings will enter the population. We also know that survival is not 100% for any of these age classes, meaning some turtles will die. Taken together, these observations mean that the population will definitely decrease between the first year and the next year. A decreasing population abundance means that λ is less than one for that particular year.

But these tiny turtles will eventually grow, and some will survive. Over time, hatchlings will become juveniles, and some of the small and large juveniles will grow into subadults and then mature into reproductive adults. At that point, the population will produce new eggs and hatchlings and perhaps begin to increase in size again. Given enough rounds of reproduction, growth, and survival, the

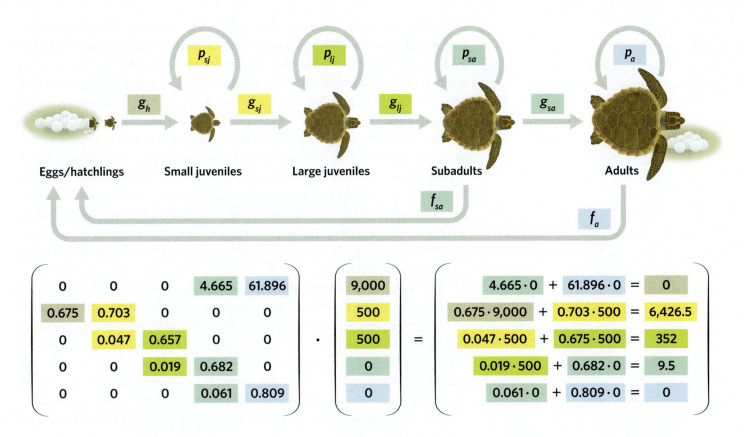

$$
\begin{pmatrix}
0 & 0 & 0 & 4.665 & 61.896 \\
0.675 & 0.703 & 0 & 0 & 0 \\
0 & 0.047 & 0.657 & 0 & 0 \\
0 & 0 & 0.019 & 0.682 & 0 \\
0 & 0 & 0 & 0.061 & 0.809
\end{pmatrix}
\cdot
\begin{pmatrix}
9{,}000 \\
500 \\
500 \\
0 \\
0
\end{pmatrix}
=
\begin{pmatrix}
4.665 \cdot 0 + 61.896 \cdot 0 = 0 \\
0.675 \cdot 9{,}000 + 0.703 \cdot 500 = 6{,}426.5 \\
0.047 \cdot 500 + 0.675 \cdot 500 = 352 \\
0.019 \cdot 500 + 0.682 \cdot 0 = 9.5 \\
0.061 \cdot 0 + 0.809 \cdot 0 = 0
\end{pmatrix}
$$

proportions of individuals in each stage will stabilize, bringing the population to a **stable stage distribution**, in which the proportions of individuals in each age class remain constant over time. The stable stage distribution is dictated by the survival and fecundity parameters in the matrix.

Once a population settles into a stable stage distribution, population growth continues based on the characteristic value of λ associated with the stage-structured matrix model. If a population does not have a stable stage (or age) distribution, then the value of λ will vary tremendously as one forecasts population abundance across years. The time it takes to achieve a stable stage distribution depends on how unbalanced the initial stage distribution was and how long the species lives. The turtle population in our example (90% eggs/hatchlings) would take about 30 years to hit a stable stage distribution. Longer-lived species take more time to recruit individuals through the stage classes than short-lived species.

Beyond this insight into how the distribution of individuals across life stages influences population growth, matrix models are of enormous use for evaluating activities that ecologists and conservation groups hope will ensure a population's persistence. Crouse, Crowder, and colleagues found that their original stage-structured model, based on survival and reproduction on Little Cumberland Island, predicted a declining population, equivalent to a population growth rate (λ) of 0.95. For this population to persist into the future, the per capita growth rate needs to be greater than one. One enormous advantage the researchers had in using this model is that they could explore the effect of stage-specific management and conservation strategies on sea turtle population growth.

Figure 6.30 In this sample stage-structured matrix model, we start with 9,000 loggerhead eggs and hatchlings, 500 small juveniles, and 500 large juveniles (90%, 5%, and 5% of the population, respectively). To determine the number of individuals in each stage in the next year, we multiply the first row of the vector (9,000) by each parameter in the first column of the matrix (0, 0.675, 0, 0, 0), the second row of the vector by each parameter in the second column of the matrix, and so on, multiplying only elements and parameters of the same color. We find the new population vector by adding all the multiplication products within a row of the matrix. For simplicity, we do not show the multiplication by every zero in the matrix. There are no subadults or adults to produce eggs, so the population decreases from one year to the next because there are deaths but no births.

In fact, they used a version of this model to explore conservation approaches that would allow the Atlantic loggerhead sea turtle population to stabilize or even grow. Today, the two most common sea turtle conservation approaches are to protect eggs and nestlings on beaches, which is called headstarting (**Figure 6.31A**), or to reduce bycatch deaths by equipping fishing nets with "turtle excluder devices," or TEDs (**Figure 6.31B**). A TED is like a trapdoor in a large fishing net; it pops open if something as large and heavy as a sea turtle bumps against it, thus allowing turtles to escape. The door stays closed, though, when smaller organisms hit it, thus allowing nets to catch and hold smaller fish and shrimp. When used in real-life fisheries, TEDs substantially reduce the number of subadult turtles that drown, while only minimally reducing the fisheries' catch. The question Crouse, Crowder, and colleagues posed was, Which approach would provide the most benefit to the turtles over the coming decades, headstarting or TEDs?

The team explored these questions by changing the matrix values to reflect expected increases in survival from these two conservation actions. Headstarting primarily affects the survival of eggs and hatchlings, so they evaluated increased survival in this stage. TEDs have the largest effect on the younger oceangoing life stages, such as juveniles, but also affect the survival of subadults and adults. For conservation efforts using TEDs, they evaluated increases in the survival rate of all these stages and included a slight bump in reproduction to match real-world observations of increased fecundity when TEDs were required in fisheries.

Based on what you now know about the life history and population structure of loggerhead sea turtles, which conservation action do you think produced the greatest increase in annual per capita population growth? If this is unclear, go back and think about what happens to population growth for loggerhead turtles when there are many eggs and hatchlings in the population but not enough adults.

The matrix models that Crouse, Crowder, and colleagues explored demonstrated that headstarting did very little to bolster the loggerhead population abundance, measured as the number of females in the population over time. Although headstarting clearly increased egg and hatchling survival, the model showed exactly what you may have guessed based on Figures 6.28 and 6.30. Even though there were many more young turtles after headstarting, most of them did not survive in the open ocean to reach maturity and reproduce. In fact, the matrix models showed that, even if every egg produced in a single year survived to become a small juvenile, the population still would not grow. All the great volunteer efforts on the beaches protecting nests and hatchlings simply did not translate into enough breeding individuals to make the population stable.

On the other hand, the matrix model clearly indicated that protecting juveniles and subadults by using TEDs had a positive impact on population growth. In particular, if TEDs could increase the survival rate of large juveniles by even a small percentage, the loggerhead population would begin to grow ($\lambda > 1$). The juvenile turtles that managed to escape fishing nets through TEDs were the lucky ones that had already survived the gauntlet of early life in the sea. If their lives were also spared from fishing nets, they had a much better chance of reaching sexual maturity and leaving behind more offspring.

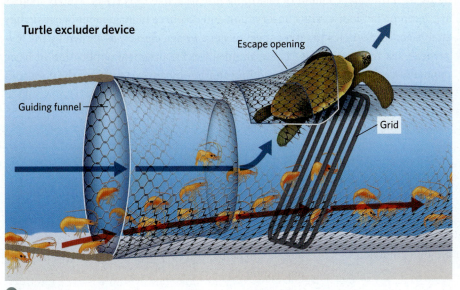

Figure 6.31 Ⓐ Headstarting programs monitor the nests of sea turtle eggs and protect them from human disturbances. As hatchlings emerge, volunteers sometimes assist them toward the ocean, as urban light pollution can attract hatchlings away from the water. Ⓑ Sea turtles sometimes get caught in fishing nets, but the use of turtle excluder devices greatly reduces mortality.

Turtle excluder device
Escape opening
Guiding funnel
Grid

6.9 The Importance of Population Ecology

An understanding of populations is foundational to all of ecology. You will encounter the definitions, concepts, and models described here in nearly every other chapter in this book. It is particularly important that you have command of the basic population model, including the various discrete- and continuous-time versions of this model. The density-dependent versions of these models are used heavily in the next three chapters. Therefore, if you are uncertain of how these models work, take some extra time with this chapter and the conceptual and quantitative questions at the end to be sure you are comfortable before moving forward.

A knowledge of population ecology (and the associated models that describe changes in population abundance through time) has far more value than just as a building block for other ecological concepts. As the examples presented in this chapter demonstrate, a deep understanding of population growth and decline is critical for developing responses to the myriad ecological changes associated with the Anthropocene. The populations that we want to manage for increased growth include those representing native species threatened with extinction (like loggerhead sea turtles) or those that provide food, fiber, or fuel sources for human societies. The populations we want to manage for decreased growth include those representing the thousands of invasive species (like North American beaver in Tierra del Fuego) and pathogens that reduce agricultural yields, harm native forests, and harm wildlife or humans.

The use of mathematical models to aid our efforts to ensure population growth or decline include versions of every equation provided in this chapter. For example, a version of Equation 6.13 is the foundation for managing nearly all commercially harvested fish populations worldwide. Another version of the same equation is used to set harvest and logging limits for birds, mammals, and trees. In addition, population modeling is foundational for the field of epidemiology and helps with the management of emerging diseases, like COVID-19. Matrix models are common in biodiversity conservation and resource management because they enable action agencies to target specific stage or age classes, as shown by the loggerhead sea turtle example. The examples for the practical application of basic population ecology go on and on, adding another incentive to master the core concepts and equations we have detailed in this chapter.

SUMMARY

6.1 Why Study Populations?

- Populations are groups of individuals of the same species that are spatially distinct from other groups of individuals of the same species.

- This chapter focuses on how population abundance changes through time and explores the factors that can affect that change.

6.2 Population Definition Revisited

- Counting individuals in a population can be challenging because of issues related to life stages and movement.

- At times it is convenient to make assumptions that discount or ignore movement into and out of a population and assume a closed population.

- Because all populations are defined as being in a specific location with geographic boundaries and a defined area, ecologists often treat population abundance and population density as synonymous.

6.3 Obtaining Population Data

- Assigning a time frame within which to survey a population helps with counting the individuals in that population.

- It is often necessary to count a subsample of the individuals in a population by using a sampling regime or transect and then using ratios to estimate total population abundance.

- Many different methodologies are available for population sampling, depending on the organism under study.

- The use of mark-recapture sampling allows for fairly accurate assessment of population abundance using proportions.

6.4 A Basic Population Model

Why Ecologists Care about Population Estimates

- Ecologists generally care about whether a population is growing or declining.

- Beavers introduced to South America provide a useful example for a growing population.

A Conceptual Model of Population Growth

- A population can increase because of either births or immigration, and it can decline because of deaths or emigration.

- From these general observations, we can build a simple conceptual model of population growth.

- In order to simplify the model, we assume the population is closed and, therefore, choose to ignore both immigration and emigration.

A Mathematical Model for Discrete-Time Population Growth

- Using the conceptual model, it is possible to build a mathematical model of population growth.

- The simple mathematical model is $N_{t+1} = N_t + B_t - D_t$, where N_{t+1} is the population abundance next year, N_t is the population abundance this year, B_t is the number of births this year, and D_t is the number of deaths this year.

- It is possible to modify this model to include the concepts of per capita rates of birth, death, and survival.

- If these concepts are included, the birth and survival rates can be combined into a single term λ (lambda), which is the finite per capita rate of population growth.

- Substituting λ into the population model produces a simplified model: $N_{t+1} = N_t\lambda$.

- The λ term encapsulates population growth in a single parameter and is useful for conservation efforts because if $\lambda > 1$, the population grows; if $\lambda < 1$, the population shrinks.

Visually Representing Population Change through Time

- Making a graph that shows how population changes over time is informative if you put time on the x-axis and population abundance on the y-axis.

- By examining graphs, it is possible to see how a population grows or declines, sometimes to extinction, using the discrete-time population growth model.

Discrete-Time Population Growth through Multiple Years: The Geometric Growth Equation

- Sequential use of the discrete-time population growth equation points to a better model to predict population abundance in the future.

- The geometric growth equation is $N_t = N_0\lambda^t$, where N_t is the population abundance in the future, N_0 is the population abundance in the first year, λ is the finite per capita rate of population growth, and t is time.

Using Data to Inform Population Models: A Test Drive

- An example from beaver population growth offers insight into the use of the discrete-time and geometric growth models.

Assumptions in Simple Models Limit Applications

- Four assumptions made by the geometric growth model are worth noting for the limits they place on its usefulness.

- First, the model assumes that the population breeds or reproduces once per time period (e.g., once a year), but not all species follow this pattern.

- Second, it assumes that birth and survival rates are constant through time, but again, not all species and situations work this way.

- Third, it assumes that every individual in the population has the same birth and survival rate, which is unlikely to be true for most organisms.

- And fourth, the model treats males and females identically when, in reality, for some species the sexes count either more or less in terms of adding to births or deaths.

6.5 A Model for Continuous-Time Population Growth

- Some species produce offspring at any time and do not fit one of the four assumptions of simple models.

- Calculus offers a way to model continuous-time population growth over a fixed time period.

- The continuous-time model is $N_t = N_0e^{rt}$, where N_t is the population abundance in the future, N_0 is the population abundance in the first year, e is the natural logarithm, r is the instantaneous per capita rate of population growth, and t is time.

- The continuous-time model produces exponential population growth, whereas the discrete-time model produces geometric population growth.

6.6 Models of Density Dependence

- If taken to their logical conclusions, continuous- and discrete-time models predict explosive population growth, with population numbers approaching infinity.

- This situation is not realistic in nature, as populations always face something that limits their growth.

- Often, the limit is due to scarcity of one or more resources (e.g., space, food, habitat).

- In response to scarcity, the rate of population growth will slow; this phenomenon is called density dependence.

- The amount of resources provided by a system determines the number of individuals that can be supported by that system; this number is the carrying capacity.

Density-Dependent Population Growth in Continuous Time: The Logistic Model

- The logistic growth model incorporates density dependence by including a carrying capacity term into the mathematical model.

- The logistic equation is

$$\frac{dN}{dt} = Nr_{\max}\left(\frac{K - N}{K}\right)$$

where dN/dt is the change in population density over time, N is the population density, $r_{\max}$ is the intrinsic rate of increase, and K is the carrying capacity.

- The $(K - N)/K$ term acts as a discounting term and slows population growth as the population density approaches the carrying capacity.

- The population density under this model levels off at the carrying capacity and produces a stable equilibrium.

- A graph of this type of population growth produces an S-shaped curve.

- The population can never overshoot the carrying capacity in this model.

Density-Dependent Population Growth in Discrete Time

- A discrete-time density-dependent population growth equation can assume many forms, but we provide one form:

$$N_{t+1} = N_t + N_t r_{\text{dis}}\left(\frac{K - N_t}{K}\right)$$

where N_{t+1} is the population density at the next time step, N_t is the current population density, r_{dis} is the maximum per capita rate of population change, and K is the carrying capacity.

- Under this model, the population can overshoot the carrying capacity and then quickly return to that value.

- The behavior of this model changes depending on the value of r_{dis}. If $r_{\text{dis}} < 1$, the equation leads to monotonic damping in the population growth; if $1 < r_{\text{dis}} < 2$, it leads to damped oscillations; if $2 < r_{\text{dis}} < 2.57$, it leads to stable limit cycles; and if $r_{\text{dis}} > 2.57$, it leads to chaotic dynamics.

6.7 Real-World Evidence for Density Dependence

- Two North American elk populations were established in an experimental setting, one that was kept at a low density and another at a high density.

- The elk were examined for changes in body condition (i.e., fat storage) and pregnancy rates.

- The researchers found that elk living in the high-density population had lower fat storage and were less likely to become pregnant than those in the low-density population, suggesting density-dependent population regulation.

6.8 Accounting for Individual Differences in Survival and Reproduction

Life History

- A species' life history is the temporal sequence of events that determines survival and reproduction from an individual's birth until its death.

- Three different types of survivorship are recognized and are best understood in a graphical form:

  — Type I: high early survival until a certain age, and then survival declines rapidly (e.g., humans, elephants, whales)

  — Type II: constant survival throughout its lifetime (e.g., some seabirds and reptiles)

  — Type III: very low initial survival during early stages until a certain age is reached, and then high survival till death (e.g., trees, amphibians, insects, marine invertebrates)

Age and Stage Structure

- Many species have a distinct age structure to their populations, meaning that survival and reproduction vary with age group.

- This can be shown with age pyramid graphs.

- Populations of other organisms are stage structured, meaning the stage of the organism's life history is more important to reproduction and survival than is numerical age.

- For example, longleaf pines have six distinct stages in their life cycle but can remain in a particular stage for variable amounts of time. Each stage has distinct reproductive and survival probabilities.

Stage Structure and Population Growth

- Age and stage structuring can affect how populations grow or decline, and this variability can be captured in a population growth model.

- **Loggerhead Sea Turtle Life History**

  — Loggerhead turtles have five distinct stages to their life history.

  — Each stage has unique survival and fecundity values, which best fit a Type II survivorship curve.

- **Conservation Issues for Loggerhead Sea Turtles**
  - Loggerheads were overharvested in the nineteenth century as food and for their shells.
  - Today, they are in danger of extinction because of egg destruction, light pollution, and unintentional capture in fishing nets.

- **A Loggerhead Sea Turtle Stage-Structured Model**
  - A stage-structured population matrix model allows scientists to examine which stage of the turtle's life is the most sensitive and which is the most amenable to conservation measures.

  - By manipulating the matrix model based on real data, scientists discovered that juvenile and subadult survival were more important for turtle population growth than survival of the egg/hatchling stage.

6.9 The Importance of Population Ecology

- The concept of a population is fundamental to ecology.
- Understanding populations is crucial to responding to human-caused ecological changes.
- The population models in this chapter are similar to the ones used to manage populations of interest in the real world.

KEY TERMS

age structure
birth
carrying capacity (K)
cohort
continuous-time model
death
density dependence
density-independent factor
discrete-time density-dependent population growth
discrete-time population growth model
emigration
exponential population growth

fecundity
finite per capita rate of population growth (λ)
geometric growth equation
geometric population growth
immigration
instantaneous per capita rate of population growth (r)
intraspecific competition
intrinsic rate of increase (r_{max})
life history
logistic equation
logistic growth model

mark-recapture model
parameter
per capita rate
per capita survival rate
population
population abundance
population density
stable equilibrium
stable stage distribution
stage structure
stage-structured population matrix model
time lag

CONCEPTUAL QUESTIONS

1. Think about a population of a common species in your area.
   a. Describe the difficulties you might encounter if you tried to count all the individuals in the population.
   b. How might you best go about estimating the population density?
   c. How would you define the boundaries of the population?

2. Explain why it is important when conducting mark-recapture estimates that the marked individuals mix well with the rest of the population and are not more (or less) likely to be captured than other individuals in a second sampling effort.

3. Choose a species that interests you.
   a. Based on your knowledge (or additional knowledge from a quick internet search), determine whether this species has discrete-time or continuous-time population growth.
   b. Based on your knowledge of the factors that affect population growth, are there places where you would expect to see a growing population and other places where you would expect to see a decreasing population?

4. Based on your understanding of λ, explain why this parameter might be important to conservation managers.

5. Identify each of the following sources of mortality as either density dependent or density independent. If you are not sure into which category to place a factor, describe what additional information you would need to make a decision.

   a. Annual storms that wipe out half the population

   b. A fixed amount of food that leads to food shortages and weak babies who are less likely to survive

   c. A fixed number of shelters that leads to increased mortality at high densities

   d. A lethal disease, the transmission of which increases with contact

   e. A cold spell that kills individuals in several populations across a region, killing more individuals in bigger populations but always the same fraction of the population regardless of the population abundance

6. Look at Figure 6.26B.

   a. If this were instead a Type III survivorship curve, how would it change?

   b. If each female turtle laid 1,000 eggs (on average), but the number that made it through the hatchling stage (i.e., survived past the first year) were the same as in the real-life model investigations discussed in this chapter (Crouse et al. 1987; Crowder et al. 1994), would this have any qualitative effect on the insights from their models?

   c. Can you tie your answers to questions (a) and (b) to our discussion of counting organisms?

7. Human population demographers have noted that human population abundance would not stabilize right away even if each woman of reproductive age had only two children, one to replace herself and one to replace a male partner. This phenomenon is called *demographic inertia*. Can you explain it, based on Figure 6.23? (Hint: Note the sizes of each age class, and remember that humans generally have a Type I survivorship curve, with high rates of annual survival in most age classes until old age.)

8. Population projection matrices add stage structure to the simple population growth model, $N_t = N_0\lambda^t$. Describe any issues you might encounter when using these models with a population that experiences density dependence. Given these issues, why might matrix models still be useful in conservation biology?

QUANTITATIVE QUESTIONS

1. Suppose you decide to estimate the number of cockroaches in the basement of an apartment building. You put out bait, catch 21 cockroaches, mark them with a dab of blue enamel paint, and release them. Three days later, you return, put out more bait, and capture 28 cockroaches, six of which have a dab of blue paint.

   a. Assuming this is a good mark-recapture sampling scheme, what is your estimate for the abundance of the cockroach population in the basement?

   b. Do you see any issues with the sampling scheme? If so, what are they?

2. Use a spreadsheet or a calculator to explore Equation 6.7 through 10 time steps for $\lambda > 1$ and then for $\lambda < 1$. Start with 100 individuals each time, and use at least two values of $\lambda > 1$ and two values of $\lambda < 1$. Relate your findings to Figure 6.12.

3. Start with 25 beavers in 2021 and use Equation 6.8 and $\lambda = 1.257$ to show that the expected size of the beaver population would be 78 in 2026.

4. Equations 6.8 and 6.10 are also useful for calculating the amount of time required for a population to reach a particular size.

   a. Using either equation, calculate the number of years it will take for the beaver population to reach 100 individuals with $\lambda = 1.257$ if you start with a population of 10 individuals (N in 2017). Notice that you need to rearrange terms and that you may find yourself taking a logarithm to solve for t in an exponent position. It should not matter whether you use a base 10 logarithm ($\log_{10}$) or the natural logarithm (ln), but convince yourself that this statement is true by trying it each way.

   b. What if you start with a population of 25 individuals (N in 2021)? Does a different starting year yield a different expected year that the beaver population should exceed 100 individuals?

5. Using Equation 6.14B, recreate Figure 6.18, starting with a population abundance of 25 in 2021, $K = 100$, and $r_{dis} = 0.257$. Do you see the classic S-shaped growth curve? Why or why not? Given the population abundance in 2017–2021, does it make sense to use the average per capita population growth rate in this period as an estimate of r_{dis}? Explain your answer.

6. Using the following data, determine whether this population shows density-dependent birth or death rates or both. Illustrate your answer using graphs. (Hint: If your y-axis is per capita birth and death rates [b and d], what would your x-axis be?)

Year	Population abundance	Deaths	Births
2016	100	10	70
2017	160	15	50
2018	195	20	40
2019	215	20	30
2020	225	25	25

7. Explore Equation 6.14B. For this exercise, we recommend that you use a spreadsheet with a column for year and a column for population abundance. Above both columns, put a cell with the starting population abundance (N_0), another with the maximum per capita growth rate (r_{dis}), and a third cell with the carrying capacity (K). If you refer to the cell with N_0 and put in a formula for the population abundance at N_1, you will find that you need to include both the r_{dis} and K and that you can copy the formula down to create N_t for each of your years. Using this approach will allow you to vary r_{dis} just by changing one cell.

a. Can you produce monotonic growth, damped oscillations, limit cycles, and chaos, as in Figure 6.19?

b. What happens if you change K or N_0? Does either change alter the overall dynamic behavior and bump it into a different category of monotonic growth, damped oscillations, limit cycles, or chaos?

SUGGESTED READING

Crouse, D. T., L. B. Crowder, and H. Caswell. 1987. A stage-based population model for loggerhead sea turtles and implications for conservation. *Ecology* 68(5): 1412–1423.

Parkes, J. P., J. Paulson, C. J. Donlan, and K. Campbell. 2008. *Control of North American Beavers in Tierra del Fuego: Feasibility of Eradication and Alternative Management Options.* Landcare Research Contract Report: LC0708/084. Prepared for Comité Binacional para la Estrategia de Erradicación de Castores de Patagonia Austral. Lincoln, New Zealand: Landcare Research New Zealand.

Stewart, K. M., R. T. Bowyer, B. L. Dick, B. K. Johnson, and J. G. Kie. 2005. Density-dependent effects on physical condition and reproduction in North American elk: An experimental test. *Oecologia* 143: 85–93.

A mother grizzly bear (*Ursus arctos*) and her cubs consume sockeye salmon in Katmai National Park and Preserve, Alaska. Grizzly bears and black bears (*Ursus americanus*) have largely overlapping resource needs and are therefore likely to compete strongly when both species are present.

chapter 7
COMPETITION

7.1 Competitive Interactions Have a Negative Effect on All Competitors

As you look around the natural world, it should be fairly clear that no species exists in a vacuum. Instead, every individual comes in contact with dozens or even hundreds of other species throughout its day. Each of these exchanges has the potential to be significant in the organism's life. Take, for example, the efforts to restore populations of grizzly bears (*Ursus arctos*) to the North Cascades, the section of the Cascade mountain range that straddles the US-Canadian border. Grizzly bears are often an inspiring sight for intrepid tourists along the west coast of North America, as they are an iconic large predator of the northern temperate biomes. Unfortunately, their populations declined dramatically in the twentieth century due to human activities, such as persecution and hunting. Several populations bounced back, thanks to solid wildlife management efforts and the restoration of some coastal river salmon populations. However, grizzly bear population growth seems to have fizzled in the interior North Cascades, despite hunting restrictions and the release of hundreds of grizzlies back into these habitats over the years. Why aren't grizzly bears making their return to these wild places?

One possible answer is that, in these inland habitats, the grizzly bear's niche overlaps with that of the local black bears (*Ursus americanus*). Black and grizzly bears are both generalist omnivores that will eat roots, fruit, fish, and other mammals. Interestingly, black bears are generally smaller than grizzlies and may be scared away when individuals of the two species spar over high-protein foods such as salmon. In the inland forested habitats of the North Cascades, however, salmon are rare, and other food resources are more dispersed and have lower nutritional value. In these locations, grizzly and black bears rarely directly interact, but their food resources overlap much more than in coastal habitats. Because black bears are smaller than their grizzly counterparts, they can use foods with lower nutritional value more efficiently to support themselves. This physiological difference, combined with the large numbers of black bears in the interior North Cascade forests, means black bears consume far more of the available fruits and high-energy plants than grizzlies. The greater overlap in shared resources leaves the grizzly bears unable to consume enough food to reproduce. The inland grizzly populations, therefore, never increase, even with the helping hand of wildlife managers.

The example of grizzly and black bears, beyond being one that excites the imagination, brings us to a discussion of **community ecology**, or the interactions among multiple species. These interactions are fascinating but complex, and for new students they can be somewhat overwhelming at first. Therefore, we start by focusing on two-species competitive interactions and their outcomes. In Chapter 6, we explored intraspecific competition, which is competition among individuals of the same species in a single population, particularly with respect to limited resources and a population's carrying capacity (K). In this chapter, we expand the scope and investigate **interspecific competition**, which is competition for resources among individuals of different species.

When two species use a common resource that is in limited supply, both species can be negatively affected in terms of their population density or evolutionary fitness. Ecologists often use the notation "$(-, -)$" as a shorthand to represent two-species competition. The negative signs on each side of the comma represent negative effects for each species. Note that competitive effects do not have to be equally negative for both species for interspecific competition to occur. One species may be an inferior competitor and experience more significant negative effects than the other. Also, individuals of two species can use the same resource but not compete for it. If the resource is superabundant in the environment, then the amount used by individuals of one species will have little to no fitness impact on individuals of the other species, even if the resource is critical to both species' fitness. For example, terrestrial organisms all require oxygen but rarely compete for it, despite its critical importance.

Ecologists sometimes split interspecific competition into two general types, resource and interference. **Resource competition**, also called exploitative competition, occurs when individuals of one species more efficiently consume or use up a shared resource. This action lowers the availability of the resource for individuals of the other species, thereby affecting their fitness. This type of competition is sometimes referred to as an *indirect interaction* because competing individuals do not directly interact but instead compete through a shared resource (we will discuss this more in Chapter 10). Although all competition involves competing for resources, sometimes it also involves physical interactions between individuals. With **interference competition**, individuals directly interact with each other through aggressive behavior or behavioral displays in order to increase access to a limiting resource. These direct interactions result in fitness consequences for all individuals involved. Notice that resource competition is at the heart of both types of competition, but interference competition involves physical interactions on top of the competition for resources.

In this chapter, we build on the ideas of the niche and resource scarcity from Chapters 5 and 6 to understand interspecific competition. In Chapter 5, we used resource utilization curves to quantify and graphically represent a species' niche space (see Figures 5.26–5.29). We saw that access to resources was essential for survival and reproduction, and that the quality and type of available resources could determine an individual's traits. In Chapter 6, we discussed how population growth rates can vary with resource availability. In this chapter, we explore overlap in niche use with examples and graphs and present two different ways to model interspecific competition. After we establish the basic concepts, we examine the long-term consequences of interspecific competition.

7.2 Competition and Niche Space

One of the first studies to examine interspecific competition was conducted in the 1930s by Georgii F. Gause, who used two species of paramecia (*Paramecium aurelia* and *Paramecium caudatum*) as his subjects. Paramecia are tiny, unicellular, ciliated protozoans commonly found in freshwater ponds and lakes. Their ubiquity in natural waters and ease of care in the laboratory have made them unwitting participants in scientific research for hundreds of years (**Figure 7.1**). Gause wanted to study how interspecific competition played out in these two species of paramecia with very similar feeding habits (Gause [1934] 2019).

Figure 7.1 Two species of paramecia used in the interspecific competition experiments by Gause: Ⓐ *Paramecium aurelia* and Ⓑ *Paramecium caudatum.*

To tackle this question, he initially grew individuals of each species alone in separate test tubes, each of which contained the same physical and biological conditions. That is, both test tubes were exposed to the same temperature and amount of light and had healthy populations of bacteria to serve as food. When reared in separate test tubes, each species exhibited classic logistic population growth, initially showing slow growth, followed by a rapid rise in population abundance, with an eventual stabilization at a carrying capacity (**Figure 7.2**). From reading Chapter 6, one would expect this type of population dynamic for growth in a closed system, where the consistent and limited supply of food in the form of bacteria leads to intraspecific competition, thereby limiting the population's growth.

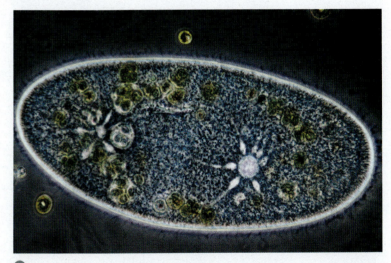

Ⓐ

The interesting new information came when Gause placed populations of the two species together in the same test tube. He found that the population of one species, *P. aurelia*, grew more slowly in the two-species experiment but eventually reached the same carrying capacity as it did when grown alone. But the population of the second species, *P. caudatum*, went extinct (Figure 7.2). Gause repeated these experiments with the same conditions several times, and in every iteration *P. caudatum* was always on the losing end.

So, does this fit the idea that interspecific competition has a negative effect on both species? Although the effect on *P. caudatum* is more obvious (it went extinct), notice that *P. aurelia*'s population growth was slower in the presence of its competitor and only approached its

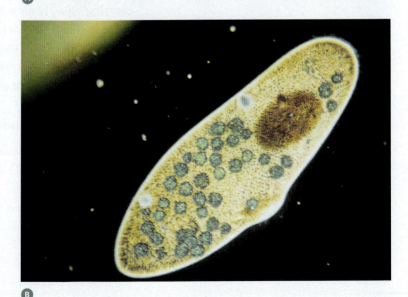

Ⓑ

Figure 7.2 Gause grew two species of paramecia, separately and together, and looked at the population growth in each situation. When the two species were grown apart, they exhibited logistic growth; when grown together, there was interspecific competition, in which *P. aurelia* "won" and *P. caudatum* "lost" (Gause [1934] 2019).

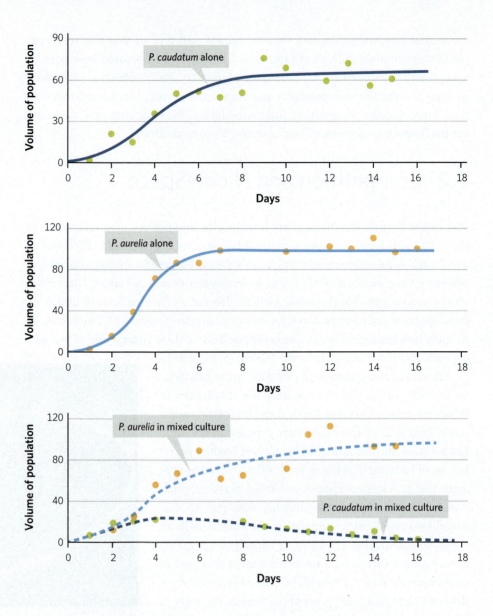

carrying capacity once *P. caudatum* died out (i.e., when interspecific competition ceased). What, then, determines which species survives in an interspecific competition situation? And does one species always lose? To answer these questions, we need to explore what leads to the negative effects Gause observed, which takes us back to the discussion of resources and niches from Chapter 5.

A Conceptual Model of Competition

Resource utilization curves provide a way to explore overlap in resource use by two different species. **Figure 7.3** shows three ways individuals of different species can use a single resource. Figure 7.3A indicates a range or gradient for a single resource. The two species use the same resource but they do not overlap in the exact types of this resource that they need. An example of this kind of resource division might come from two bird species that both eat seeds but do not eat the same-sized seeds or seeds from the same plant species. Alternatively, we can imagine cases in which the two species overlap in resource use, either a little (Figure 7.3B) or substantially (Figure 7.3C).

Current range

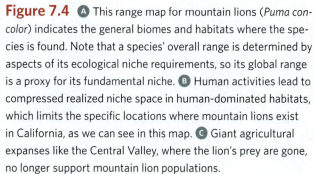

Figure 7.4 ⒜ This range map for mountain lions (*Puma concolor*) indicates the general biomes and habitats where the species is found. Note that a species' overall range is determined by aspects of its ecological niche requirements, so its global range is a proxy for its fundamental niche. ⒝ Human activities lead to compressed realized niche space in human-dominated habitats, which limits the specific locations where mountain lions exist in California, as we can see in this map. ⒞ Giant agricultural expanses like the Central Valley, where the lion's prey are gone, no longer support mountain lion populations.

Current California range

Central Valley

are not found everywhere in this broad range of habitats and biomes, because some food items are more essential than others. Although on occasion lions will consume many different types of small mammals, they need large ungulate prey such as deer, elk, and sheep for long-term survival. In places like the Central Valley of California, more than 100 years of intense human agriculture inadvertently eliminated almost all the large prey as the landscape was altered (**Figure 7.4B** and **Figure 7.4C**). This anthropogenic transformation constricted the available prey and, therefore, the potential habitats available as niche space for the mountain lions. For this species, the realized niche in the presence of humans leads to a reduced range of geographic locations compared to the fundamental niche, which includes all the areas where mountain lions could exist if humans did not reduce prey availability.

We can apply the ideas of fundamental and realized niches to Gause's paramecia experiments. Paramecia eat bacteria, algae, and yeasts, so the fundamental niche for both paramecia species includes a wide range of food sources. In his experiments, though, Gause inoculated his test tubes using only a single taxon of bacterial food (hay bacillus, *Bacillus subtilis*). Both species of paramecia, therefore, experienced a strongly reduced realized niche even when in isolation. When Gause raised the two species together with only a single food resource, their resource utilization curves overlapped completely. Because *P. aurelia* consumed the hay bacillus more efficiently, the realized niche for *P. caudatum* was reduced even further, leaving no resources on which it could maintain a population.

Gause's experimental design is clearly a simplified version of the real world. Recall from Chapter 5 that Hutchinson described the ecological niche as an *n*-dimensional hypervolume encompassing essentially an unlimited number of niche axes. Resource utilization curves take into account just one of these axes for the limiting resource over which the two species compete. By reducing the pool of resources to just one, Gause explored what happens when the resource utilization curves completely overlap. Although his experiments explored a greatly simplified niche space, they offer some essential insights.

First, his experiments—and later those by Garret Hardin (1960)—suggest that, if two species are too similar in their resource use, they cannot coexist at the same

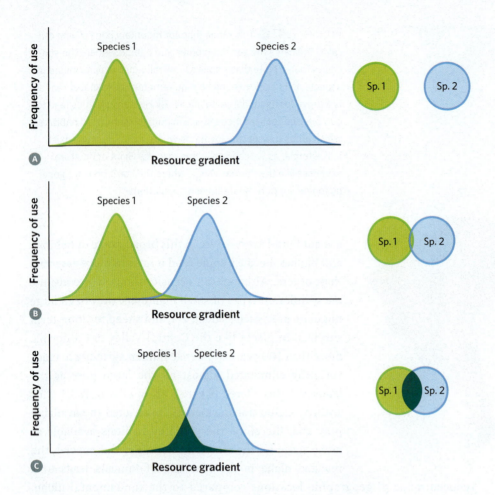

Figure 7.3 These graphs illu[...]
overlap between two species in [...]
situations: **A** no overlap in reso[...]
B slight overlap; and **C** subst[...]
The y-axes show resource use [...]
each species; the x-axes show [...]
the resource along a gradient—[...]
if the resource is seeds, this co[...]
variation in seed size, lipid cor[...]
ness, and so forth. When there[...]
resource use (A), the species [...]
When there is slight overlap ([...]
have some competitive effect[...]
As the degree of overlap incr[...]
effects of interspecific comp[...]
The Venn diagrams are anoth[...]
trate this overlap in resource[...]

Based on the discussion of resource use in Chapter 5, which situation is likely to lead to significant interspecific competition? When there is no overlap in resource use, we should see no competitive effects, and minimal overlap should lead to small negative effects on each species. But when there is substantial *niche overlap* in resource needs between the two species, the presence of individuals from one species will reduce the availability of resources for individuals of the other species. This will affect survival, reproduction, and population growth in ways similar to what we saw for intraspecific competition in Chapter 6.

This comparison brings us to the idea of a species' **fundamental niche**, which is the full range of conditions (biotic and abiotic) and resources in which individuals of a species can survive and reproduce. The fundamental niche is often contrasted with the **realized niche**, which, in the context of interspecific competition, is the niche space that individuals of a species can access in the presence of their competitors.

We discussed fundamental and realized niches in Chapter 5, but an example here may help to demonstrate how a reduced niche (or a reduction in the breadth of resources) can alter where a species is found. The mountain lion (*Puma concolor*) has a fundamental niche that includes an array of prey species, allowing the lions to occupy a geographic range covering much of western North America and most of South America (**Figure 7.4A**). This includes habitats as diverse as the coniferous forests of the Pacific Northwest, the dry mountainous regions in Mexico, and the tropical rainforests of Central and South America. Yet the lions

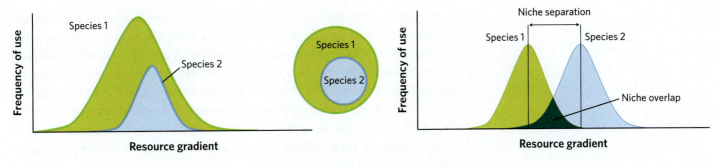

Figure 7.5 The competitive exclusion principle states that when there is complete overlap in resource use (or complete niche overlap), two species cannot coexist due to competitive interactions. In this illustration, species 2 has no resource it can access outside the influence of species 1.

Figure 7.6 Ecologists have done extensive research on pairs of competing species to measure and specifically quantify aspects of shared and unshared niche space (e.g., niche separation and niche overlap).

time in the same place. This idea is called the **competitive exclusion principle**. When two species substantially overlap in their resource use, even a slight advantage in acquiring the resource by individuals of one species will impose fitness costs on individuals of the other species and drive the second population extinct.

Second, Gause's experiments shed light on which species will "win" in a competitive situation. Let's return to our earlier questions. Does one species always become locally extinct, and what determines which species persists? For the first question, ecologists have focused on the extent of overlap in resource use. If the resource utilization curve for one species sits inside that of another species, as in **Figure 7.5**, then it seems reasonable that the species that can use the larger resource pool (i.e., the one represented by the wider curve) should persist.

Can we use this insight and the competitive exclusion principle to predict when species should coexist? Is there a threshold limit to the amount of overlap in resource use that allows coexistence? In other words, if we compare two resource utilization curves with progressively less and less overlap, is there a point at which survival for both species suddenly becomes possible?

These sorts of questions led ecologists to do detailed work on the extent of niche overlap between species (**Figure 7.6**) and the exact dimensions of resource utilization curves along axes of shared resources. Over the years, ecological scientists tried to quantify *limiting similarity*, or how similar two species can be in their resource use and still coexist. The problem with such investigations is that niches based on an *n*-dimensional hypervolume have a ridiculously high, and potentially ever-changing, number of dimensions for potential overlap. For example, **Figure 7.7** shows resource use in two-dimensional niche space, but at what point should ecologists assume they have measured all relevant niche dimensions?

In order to approach that question, let's focus again on which species "wins." Looking carefully at Figure 7.2, we can see that, when the two species were grown separately in Gause's experiment, the initial exponential

Figure 7.7 In this configuration showing two species' resource use along two different resource axes, there is no overlap between the two species along the resource B gradient but some overlap along the resource A gradient.

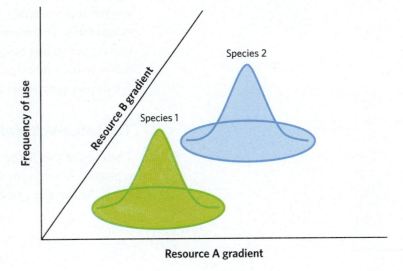

growth phase of the logistic curve was steeper for *P. aurelia* than for *P. caudatum*. This observation helps explain why *P. aurelia* always persisted in the competitive interaction with *P. caudatum*. Individuals of *P. aurelia* were able to acquire resources (food) and reproduce faster than *P. caudatum* individuals. They always won by getting a head start of sorts. This type of information does not show up on the x-axis of our resource utilization curves, but we can glean it by looking at the height of the curve along the y-axis. If frequency points to a species' ability to access a resource, then perhaps the species that can remove the resource faster will have a bigger effect.

Is it possible to come up with something more specific than this? In the discussion of intraspecific competition (Chapter 6), we noted that each new competing individual reduced the per capita rate of population growth by reducing access to the resources in the habitat. Based on what Gause observed in his experiments, it seems as though interspecific competition works in a similar manner, with individuals from one species "using up the carrying capacity" of the other. If so, and based on Equation 6.13, it makes sense that Gause's paramecia populations would grow more slowly when grown together than when grown separately. It also makes sense that the two competing species would settle at lower equilibrium densities when grown together than when grown alone, because part of their carrying capacity has been hijacked by the competitor species. Let's see if we can quantify these ideas and use the resulting model to help answer questions about when species might coexist.

7.3 Conditions for Coexistence

One of the useful things we noted about mathematical models in Chapter 2 is that they help clarify the basic ideas from conceptual models, thereby offering insights into ecological outcomes. When we explore models of interspecific competition, we want to discover the outcomes of this competition and under what conditions competing species can coexist.

One avenue for approaching these questions is a model based on density-dependent population growth that was first developed by two nineteenth-century mathematicians, Alfred J. Lotka and Vito Volterra. Lotka and Volterra worked independently yet arrived at the same model for exploring two-species interactions. This model is conceptually straightforward and has been a handy tool for generating ecological insights for generations. We revisit versions of this model in the context of predation in Chapter 8, so it is doubly useful here to take the time to understand how it works and what information it can provide.

The Lotka-Volterra Model

The core of this model is the logistic equation from Chapter 6 (Equation 6.13, repeated here and renumbered as Equation 7.1). As before, N is population density, r_{max} is the intrinsic rate of growth, K is carrying capacity, and t is time.

$$\frac{dN}{dt} = Nr_{max}\left(\frac{K - N}{K}\right) \tag{7.1}$$

Because interspecific competition occurs between two species, we start out by making two logistic growth equations (Equations 7.2 and 7.3). In order to differentiate between species, we label species 1's variables with a subscript of 1 and species 2's with a subscript of 2. Remember that the $(K - N)/K$ part of the equation means "the population will grow up to, but not exceed, the carrying capacity." So, for each species, we are explicitly including aspects of *intra*specific competition for the resources at play.

We cannot assume that two competing species will have the same carrying capacity, even when they use the same resources. This is, in part, because the two species have different fundamental niches and also because the carrying capacity for each species is in units specific to that individual species. Also notice that time (t) is the same for both species, so the variable t does not get a subscript. These equations have lots of subscripts, so we are going to drop the *max* from the r_{max} term for each species; but remember that r_1 and r_2 in these equations represent the maximum per capita growth rate for each population without inter- or intraspecific competition.

$$\frac{dN_1}{dt} = r_1 N_1 \left(\frac{K_1 - N_1}{K_1} \right) \tag{7.2}$$

$$\frac{dN_2}{dt} = r_2 N_2 \left(\frac{K_2 - N_2}{K_2} \right) \tag{7.3}$$

We now have two equations that allow us to track both species' logistic population growth up to their respective carrying capacities. But these equations are for two species growing alone. In order to examine competition between the species, we need to incorporate their interaction into each equation. This is where our conceptual model comes in. We already know that we want individuals of the two species to have a negative effect on each other; that is, we need a $(-, -)$ interaction. In other words, in each equation, we want to represent the way the competing species "hijacks" part of the focal species' carrying capacity. (In each equation, the focal species is the one indicated by the subscript in the dN_x/dt expression on the left-hand side of the equation.) To represent this negative effect mathematically, we can add individuals of the competitor species into the numerator of the term in parentheses on the right-hand side of the equation. In this way, we essentially subtract the population density of the focal species and the population density of the competitor species from the carrying capacity in order to modify how much of the maximum growth rate each focal species can achieve. (Remember from Chapter 6 that a population is always defined in a specific place and that N_1 and N_2 represent both abundance and density in that place. To remind ourselves that the population density is increasing and reducing access to limited resources, we will continue to use the term *density* in situations where the goal is to highlight this competition for resources.)

Now we need to think about units. It is easy to forget that these two equations differ in their units because they represent different species. Although time is the same in all of them, the species on the left-hand side of Equation 7.2 is different from the species on the left-hand side of Equation 7.3. If these two equations were for Gause's paramecia, one would be for *P. caudatum* and one for *P. aurelia*, and we would need to ensure that the units on the right-hand side of each equation matched the units of the left-hand side.

So how can we "convert" individuals of one species into individuals of the other species? Using Gause's paramecia, we can start by looking at how each species used up resources. *P. aurelia* did it faster than *P. caudatum*, so perhaps each individual of *P. aurelia* is "worth more" than each individual of *P. caudatum*. Using a conversion term—also called a **competition coefficient**—we can convert individuals of *P. aurelia* into the "currency" of *P. caudatum*, much like the way we use 1 m/100 cm to convert centimeters into meters. For now, we do not need exact numbers. We can just use a parameter and then figure out the value of that parameter for individual species-specific competitive interactions.

Keeping the subscript notation the same as in Equations 7.2 and 7.3, we can add individuals of species 2 into the equation for the population growth of species 1 (Equation 7.2) like this:

$$\frac{dN_1}{dt} = r_1 N_1 \left(\frac{K_1 - N_1 - \alpha N_2}{K_1} \right)$$

(7.4)

The variable N_2 represents the number of individuals of species 2, and the parameter α converts those individuals into their equivalent number of individuals of species 1, in terms of how much of species 1's carrying capacity they use up. We make the αN_2 term negative to show that the competitors use up, or remove, part of the resources (carrying capacity, K_1) of species 1.

The αN_2 term is key to linking the two species in interspecific competition. It reduces species 1's population growth rate (hence the minus sign) when the number of individuals of species 2 is high *or* when the per capita effect of species 2 on species 1 (value of α) is large. For example, if species 1 is *P. caudatum* and species 2 is *P. aurelia*, and if the population density of *P. aurelia* is small (i.e., N_2 is a small number) but the effect of these individuals on *P. caudatum* is large (i.e., α is a large number), then the value of αN_2 can still be substantial, and *P. aurelia* can have a noticeable negative impact on the population growth rate for *P. caudatum*. Of course, the effect will be much more noticeable if the population density of *P. aurelia* is big. Take a moment to multiply large and small numbers together, and think about how population density and conversion factor (α) numbers fit into Equation 7.4.

Let's try visually representing α, because understanding the competition coefficients both theoretically and mathematically is important for the next steps. Picture a small patch of habitat where individuals of two plant species are competing for space. The carrying capacity for this habitat is represented by the dark gray area in **Figure 7.8**. This means there is a limited resource (space to grow and intercept light) available in this location.

Imagine that individuals of species 1 consume some amount of this space, represented in the figure by the light gray circles with a green leaf. Note that all the individuals of species 1 use approximately the same amount of space; that is, the circles are all roughly the same size. But what happens when species 2 (represented by light gray circles with a tan leaf) is added to this niche space? You can see that individuals of species 2 occupy approximately four times as much space as individuals of species 1; that is, the circles for species 2 are four times the size of the circles for species 1. Each individual of species 2 is, therefore, equivalent to four individuals of species 1. In this situation α would have a value of 4. This simple example is useful for visualization, but remember that the resource does not have to be space. It could be a food source, nesting habitat, or any other lim-

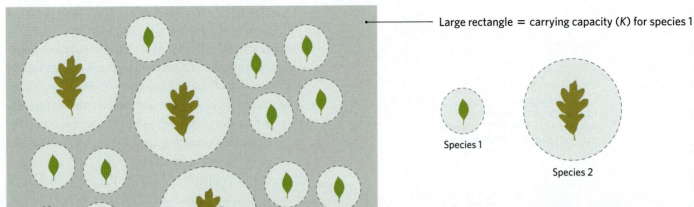

Large rectangle = carrying capacity (*K*) for species 1

Species 1

Species 2

Individuals of species 2 each use four times the resources of individuals of species 1, so α = 4.0.

Figure 7.8 Visual representations like this help explain the competition coefficients (α or β) for two plant species occupying the same habitat. If the resource is space, then each individual of species 2 monopolizes or uses four times the resources that an individual of species 1 uses. Thus, the competition coefficient (α) to convert resource use by individuals of species 2 into units of resource use by species 1 is equal to 4.

iting resource, and size is not always going to determine which species is a better competitor.

Finally, we need to reexamine the linkage in the other direction and make an equation for species 2 that includes the competitive effects from species 1. We do so by adding similar terms to the basic logistic equation for species 2 (N_1 and β), where β is the competition coefficient that translates individuals of species 1 into individuals of species 2.

$$\frac{dN_2}{dt} = r_2 N_2 \left(\frac{K_2 - N_2 - \beta N_1}{K_2} \right)$$

(7.5)

Given that α converted each individual of species 2 into four individuals of species 1 (based on Figure 7.8), it might seem logical that β would convert each individual of species 1 into 0.25 individuals of species 2, but this system does not work so symmetrically. Remember that the gray rectangle in Figure 7.8 represents the carrying capacity of species 1. From our previous niche figures (i.e., Figures 7.3, 7.5, 7.6, and 7.7), we know that the individuals of any two species use niche space differently and have different carrying capacities in the same habitat. In order to determine the value of β, we need a new figure that represents the niche space of species 2, and we need to compare the relative use of species 2's carrying capacity between the two species in order to calculate the value of β. For the moment, though, we do not need numbers because we are trying to explore the model generally.

We now have two equations representing the population growth of two different species, and the equations are linked (i.e., they share parameters and variables), which means they affect each other. How do we proceed from here? Recall that, in Chapter 6, we visualized population growth using a standard graph setup in which the x-axis represented time and the y-axis represented the number of individuals, or the rate of change in the number of individuals (dN/dt). Now, with two populations, we can simply depict both populations in the same type of graph, much the way Gause did for his paramecium species (Figure 7.2). The problem with this sort of figure is that the shapes of the curves depend on the density of each population when the populations encounter each other. What we need is a way to look at how

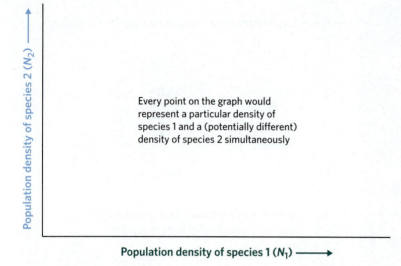

Population density of species 2 (N_2)

Every point on the graph would represent a particular density of species 1 and a (potentially different) density of species 2 simultaneously

Population density of species 1 (N_1)

Figure 7.9 This is the basic layout of a phase-plane graph for the Lotka-Volterra two-species competition equations. Each axis represents the population density of one competitor species (i.e., either N_1 or N_2), and any point on the graph represents a population value for species 1 and a population value for species 2 simultaneously.

each species affects the other along the whole range of population densities for the two competitors.

In order to visually represent both population densities simultaneously, ecologists make **phase-plane graphs**, with the density of one species on each axis (**Figure 7.9**). In phase-plane graphs (often just called "phase planes"), every point on the graph represents a particular density of species 1 and a (potentially different) density of species 2 simultaneously. For example, the point (102, 37) means that species 1's population contains 102 individuals, and species 2's population contains 37 individuals.

In a phase-plane graph, time is an assumed but hidden additional axis, like an invisible z-axis that comes out of the page. Because the two species affect each other, any point on this graph represents the densities at one time point, but those densities can change. At each new time point, we could represent the new densities with a new point. We could connect all the new points to see the two species' population growth trajectories over time, like combining each species' graph from Figure 7.2. What we want, though, is not just any trajectory. We want to see what happens at the end of the story. Where do the population densities of the two species end up when they stop changing, when they hit equilibrium?

As explored in Chapter 6, population abundance stopped changing when a population hit its carrying capacity ($N = K$) or went extinct ($N = 0$). Each of those endpoints is a potential point of equilibrium in single species' population growth. For our two competing species, we want to know what the equilibrium possibilities are, and we can get them from our equations by looking for places where the change in the population growth rate for each species becomes zero.

Although eventually we will want to know what happens when the population densities of both species simultaneously stop changing, we have equations for each species' individual rate of population growth, so let's start by looking at each species' equation alone. We will start off simply with each species alone in order to show you how the phase plane graphs and isoclines work. We will build through all the outcomes of competition from Figure 7.9 through Figures 7.16 and 7.18. To find the conditions for zero growth, we solve the equation for the condition $dN/dt = 0$. In Equations 7.4 and 7.5, three terms are multiplied together: a maximum per capita population growth rate (r), current population density (N), and the term inside parentheses. If any one of these factors becomes zero, then dN/dt becomes zero.

But not all of these possibilities are equally interesting. If the population density is zero, then we have no population and no possibility for competition, which means our questions are irrelevant. Additionally, if the maximum rate of population growth in the habitat is zero, then the population can never grow, which suggests it would never increase toward a carrying capacity and probably would not be in the habitat for any significant amount of time; again, not very interesting. So, the only interesting way to achieve zero population growth is for the numerator in the term in parentheses to be equal to zero. For species 1, this would mean

$$0 = K_1 - N_1 - \alpha N_2 \tag{7.6}$$

You may notice that Equation 7.6 is an equation for a straight line, which is fairly handy. That means we can use a line to represent all the interesting points on our phase-plane graph where species 1 has zero growth. We call such a line an **isocline**, because it represents a place where the change (*-cline*) is even, or unchanging (*iso-*). We can graph a straight line by finding any two points on that line. To graph the zero-growth isocline, the easiest points to find are the ones where $N_1 = 0$ (and N_2 does not) and then where $N_2 = 0$ (and N_1 does not).

$$N_1 = K_1, \text{ when } N_2 = 0 \qquad (7.7)$$

$$N_2 = \frac{K_1}{\alpha}, \text{ when } N_1 = 0 \qquad (7.8)$$

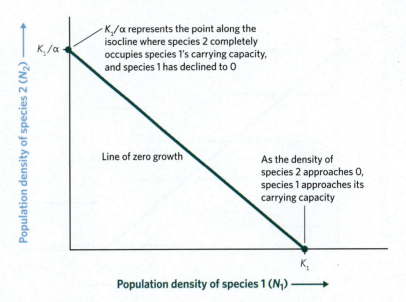

Equations 7.7 and 7.8 define the endpoints of our line of zero growth, or zero-growth isocline, for species 1 (**Figure 7.10**). At the endpoint where $N_1 = K_1$, there are no individuals of species 2 ($N_2 = 0$). Without individuals of species 2, we expect species 1 to reach equilibrium at its carrying capacity, K_1, based on intraspecific competition.

The other endpoint of the isocline is where $N_2 = K_1/\alpha$. Notice that this point essentially represents the carrying capacity of species 1 in units of species 2, meaning all of species 1's carrying capacity has been used up by individuals of species 2. If species 1's population density goes to zero, it cannot start growing again and never returns (extinction is forever even in our modeled world).

The species 1 isocline represents all the combinations of species 1 and species 2 population densities for which the population growth for species 1 will be zero. Note that this line does not mean that the population density of species 1 is zero. It is just the line along which population density of species 1 does not change. Note also that this line does not say anything about the population growth rate for species 2. We will get to that soon by looking at the isocline for species 2.

Now let's use the isocline to examine how the population density of species 1 changes in different portions of the phase-plane graph. Remember from Chapter 6 that population density grew when it was below the carrying capacity and decreased when it was above the carrying capacity. The isocline in Figure 7.10 represents the carrying capacity of species 1, filled with species 1 individuals at the point on the x-axis and with species 2 individuals at the point on the y-axis. All the other points on the isocline have a mixture of species 1 and species 2 occupying (or using up) the carrying capacity of species 1.

With this in mind, we can use the isocline to identify the portions of the phase plane where the population of species 1 grows or shrinks. When the population density of species 1 is greater than the line of zero growth (i.e., above and to the right of the species 1 isocline), then the population density of species 1 will decrease through time until it hits the line again. Because we record population density for species 1 only on the x-axis, all changes over time in the population density for species 1 will show up as arrows to the left or right. Essentially, the isocline divides the graph into two halves, with the population density of species 1 (N_1) increasing on the left half and decreasing on the right.

Figure 7.10 At any point along this zero-growth isocline for species 1, the population density of species 1 is neither increasing nor decreasing through time (time is implicit in these graphs). The endpoints of the line are derived from the carrying capacity for species 1 (K_1) and the number of individuals of species 2 that would need to be present to use up all the resources in species 1's carrying capacity (i.e., K_2/α).

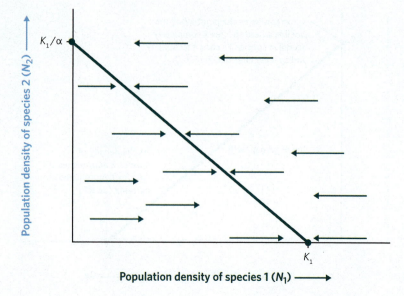

Figure 7.11 Population trajectories (represented by arrows) above and below the zero-growth isocline for species 1 show the expected direction of change in population density of species 1 in different portions of the phase plane. Arrows pointing to the right indicate that the population density of species 1 increases through time; arrows pointing left indicate that the population density decreases through time. Each arrow starts at a random spot that represents a combined density of species 1 and species 2.

Figure 7.12 The zero-growth isocline, endpoints, and trajectory arrows for species 2 are shown on this phase-plane graph. Because the population density of species 2 is represented on the y-axis, the trajectory arrows point up to show an increase in population density through time and down to show a decrease.

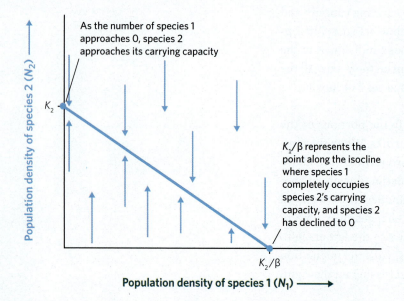

We can show this behavior using vector arrows parallel to the x-axis, pointing in the direction that species 1's population will change in the next instant of time (**Figure 7.11**).

We can repeat this whole exercise for species 2, getting analogous solutions.

$$0 = K_2 - N_2 - \beta N_1 \qquad (7.9)$$

$$N_2 = K_2, \text{ when } N_1 = 0 \qquad (7.10)$$

$$N_1 = \frac{K_2}{\beta}, \text{ when } N_2 = 0 \qquad (7.11)$$

If we create another phase-plane graph and look at just the isocline for species 2 and the conditions for population change in species 2, we get **Figure 7.12**. Notice that all change in the population density for species 2 involves arrows that point up or down because we record the density of species 2 with the y-axis.

So far, we have not really advanced beyond what Gause could see in Figure 7.2. By putting these isoclines together on a single phase-plane graph, though, it is possible to explore all the potential outcomes of competition and perhaps answer our previous questions. When does species 1 win? When does species 2 win? What are the conditions for coexistence?

When we draw two lines within a space defined by two axes, there are only a few possible configurations. The two lines can cross or not cross. If they do not cross, we can ask which one is "on top of" the other. If they cross, we can ask where the intersection of the two lines sits in the graph. For the purposes of our two-species discussion, we can then ask what type of equilibrium exists at the intersection point and whether the eventual stable equilibrium will be at the intersection point (where the two species coexist) or along the axes (where one species will have outcompeted the other). The answers to these questions point to four outcomes of interspecific competition from this model.

Outcomes of Competition in the Lotka-Volterra Model

Let's begin with competitive exclusion, the situation in which one species outcompetes the other, as in Gause's paramecia example. We will start with species 1 as the *superior competitor* and species 2 as the *inferior competitor*. In this scenario, the isocline for species 1 will always be above the isocline for species 2 on the phase-plane graph

(**Figure 7.13**), meaning species 1 can always access more resources than species 2. Even when we translate the carrying capacity of species 1 into individuals of species 2, that number is higher than the carrying capacity for species 2. Because species 2 will not grow above its carrying capacity, it can never "use up" species 1's carrying capacity (i.e., it can never reach K_1/α). This situation suggests that individuals of species 1 have a stronger competitive effect on species 2 than individuals of species 2 have on species 1.

Using vector arrows to explore the growth of each population, as we did in Figures 7.11 and 7.12, we can examine the growth of both populations simultaneously by placing the tail of a species 1 arrow (see the dotted arrow in **Figure 7.14A**) at the head of a species 2 arrow and drawing a line to complete the triangle (treating them like vectors). The new hypotenuse (tan arrow)

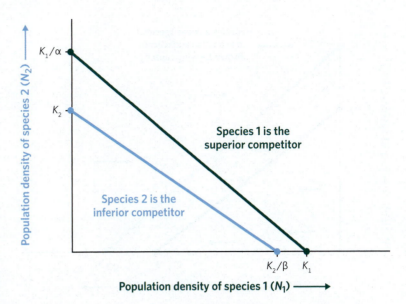

is called the **resultant vector**, and it indicates the direction in which the combined population densities will move over time. Generally, vectors show not only direction but also size, which here would mean how much a population grows or decreases through time. But because we want to represent general outcomes, we are not being careful or specific about the sizes of population density changes. We, therefore, will continue to refer to arrows as trajectories, and it is more important to notice the direction than the size of the arrows. When we add resultant trajectories all over the graph, we can see how the populations of the two species will interact in different portions of the phase plane in which we have different combinations of both species (**Figure 7.14B**).

Figure 7.13 This phase-plane graph represents competitive exclusion, with species 1 outcompeting species 2. Note the relative values of K_1 versus K_2/β and K_1/α versus K_2. The population density of species 1 can increase through time in every section of the phase plane where the density of species 2 increases, plus species 1 can increase in density in sections of the phase plane where species 2 cannot.

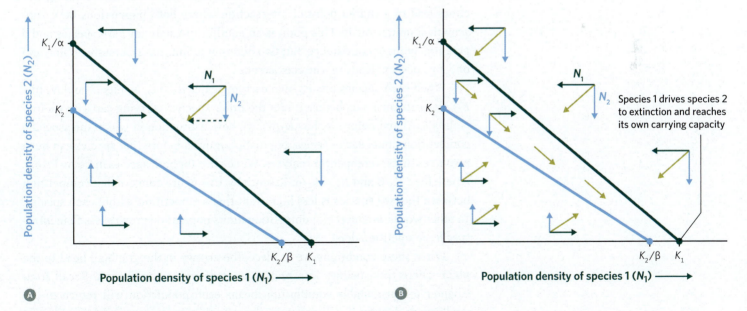

Figure 7.14 **A** Building on Figure 7.13 (competitive exclusion with species 1 outcompeting species 2), we can include population density trajectory arrows for each species at any point. Dark green arrows indicate the direction of population change through time for species 1; blue arrows indicate the same for species 2. The combined trajectory (tan arrow) shows the direction of movement of the two population densities together. **B** With a number of tan arrows in place, we can see that all the trajectories lead to point K_1, meaning that, given enough time, species 1 reaches its population carrying capacity and drives species 2 to extinction.

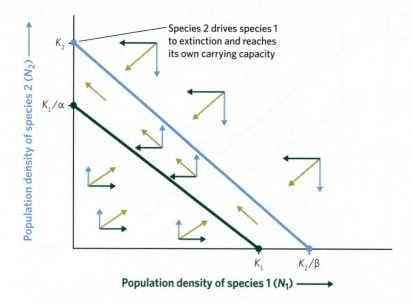

Species 2 drives species 1 to extinction and reaches its own carrying capacity

Figure 7.15 In this phase-plane graph, the isocline for species 2 lies above the isocline for species 1, meaning species 2 always outcompetes species 1—the opposite of what is portrayed in Figures 7.13 and 7.14.

The arrows above the species 1 isocline and below the species 2 isocline direct us into a central area between the two isoclines. In this central region, the resultant arrows all point down to an equilibrium at species 1's carrying capacity (K_1) on the x-axis (Figure 7.14B). As long as species 1 exists in abundances greater than zero, the only possible outcome of competition with these conditions is that species 1 drives species 2 to extinction and expands in population density until it reaches its own carrying capacity. You can imagine that, at every point on the graph, species 1 is exerting its competitive dominance and outcompeting species 2 until all individuals of species 2 are gone. The exact opposite would occur if species 2 were the superior competitor and species 1 the inferior competitor (**Figure 7.15**).

The outcome of Gause's competition experiments with *P. aurelia* and *P. caudatum* fits these scenarios, with *P. aurelia* outcompeting *P. caudatum*. What we could not see in Figure 7.2 was that this outcome would happen regardless of what the densities of the two paramecia species were at the start of the experiment. The outcome of the paramecia experiments fits very well with Figure 7.5, where the resource utilization curve for one species lies completely within the curve for the other species, and where one species has an advantage of either accessing more resources (wider resource utilization curve) or accessing the same resources faster and more efficiently (taller resource utilization curve). Using the Lotka-Volterra phase-plane graphs, we can now see the inevitability of the competitive exclusion outcome that Gause found.

Now consider the situation when the two isoclines cross. Two crossing isoclines lead to a shared point of intersection where both populations have zero growth simultaneously. This point is an equilibrium state for both species and a potential point of coexistence. But two different graphs have a crossing point, and only one of them leads to true coexistence.

When each species hits its own carrying capacity (i.e., N_1 hits K_1 and N_2 hits K_2) along its own axis before it hits the other species' carrying capacity (before N_1 hits K_2/β and before N_2 hits K_2/α), we have a situation in which intraspecific competition limits each species' population growth before that species can burn resources that its competitor requires. We see this for both species in **Figure 7.16A**, where $K_1 < K_2/\beta$ and $K_2 < K_1/\alpha$. Essentially, this figure conveys that competition between the two species is less important than competition within each species. In other words, *interspecific* competition limits population growth less than *intraspecific* competition does.

When these conditions are in place, the arrows in the graph all head to the point where the isoclines intersect, a point of stable equilibrium. Recall from Chapter 6 that a stable equilibrium means each population will return to the equilibrium density if a disturbance moves either species away from this point (**Figure 7.16B**). The intersection point defines the point of **stable coexistence** between the two competitors.

This outcome answers one of our initial questions: Is it possible for two competing species to coexist? From Figure 7.16, we can see that coexistence is possible

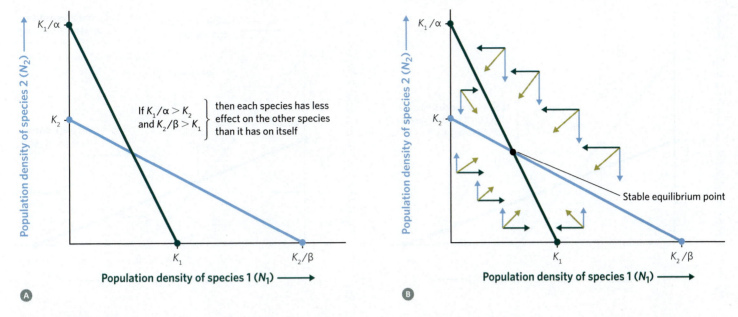

If $K_1/\alpha > K_2$
and $K_2/\beta > K_1$ } then each species has less effect on the other species than it has on itself

A

B

Stable equilibrium point

Figure 7.16 **A** This isocline arrangement will produce stable coexistence of two competing species in the Lotka-Volterra model. Each population will stop growing at its own carrying capacity before it completely occupies the other species' carrying capacity (i.e., intraspecific competition will limit population growth before either species can competitively exclude the other). **B** When we add the trajectory arrows, we see that all arrows lead toward the intersection point of the two isoclines, meaning the two population densities will return to this stable equilibrium point after a disturbance. Each species exists at a lower density in this stable equilibrium than without interspecific competition (i.e., each species is at a density lower than its own carrying capacity).

and that the two coexisting, competing species still experience the negative effects that we said were a hallmark of all interspecific competition. Notice that each species in Figure 7.16B is at a lower population density at the stable equilibrium point than if it existed alone (and could reach its carrying capacity). We have drawn these two lines in a very general way and with no numbers, but that part of the result is consistent, no matter how one draws isoclines that cross this way.

We see such stable coexistence in an array of ecological situations, although not when the niches overlap as completely as they do for Gause's paramecia. Coexisting competitors tend to have much narrower niche overlap, as suggested in Figure 7.3B. One good example of this comes from the understory of temperate forest biomes. One might think that the tall overstory trees would remove all the light and outcompete the smaller understory plants. Instead, these smaller "spring ephemeral" plants produce flowers early in the season before the trees' leaves have fully emerged (**Figure 7.17**). They take advantage of a time when trees use less light. As we discuss in Chapters 11 and 13, the factors that go into community coexistence are a bit more complicated than our two-species model, but limited overlap in use of resources definitely contributes to stable coexistence among species.

It may seem as though we have addressed all the potential outcomes of this model, with one or the other species always winning and with stable coexistence. But we still have one more way to draw two isoclines, and this last outcome is the most interesting and unexpected. When the two lines cross the other way, so that each species hits its competitor's carrying capacity before it hits its own ($K_1 > K_2/\beta$

Figure 7.17 Understory plants, such as this wild blue phlox (*Phlox divaricata*) blooming under the forest overstory trees in the Great Smoky Mountains National Park of Tennessee, take advantage of the available sunlight early in the spring season.

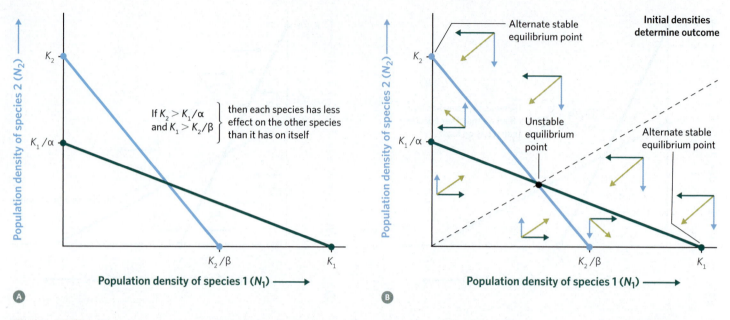

A If $K_2 > K_1/\alpha$
and $K_1 > K_2/\beta$ } then each species has less effect on the other species than it has on itself

B Initial densities determine outcome

Alternate stable equilibrium point

Unstable equilibrium point

Alternate stable equilibrium point

Figure 7.18 **A** Unlike in Figure 7.16, this isocline arrangement produces unstable coexistence. In this situation, each species will use up the other species' carrying capacity before it reaches its own carrying capacity. **B** When we add trajectory arrows, the arrows between the isoclines lead away from the intersection point of the isoclines, meaning the population densities will not return to that point if they leave it, making this an unstable equilibrium point. We have, instead, two stable equilibrium points, and which one the system reaches depends on the initial population densities of the competitors. If the combined densities are below the dashed line, species 1 will exclude species 2. If they are above the dashed line, species 2 will exclude species 1.

Figure 7.19 *Tribolium confusum* and *Tribolium castaneum* are two different species of flour beetles with very similar niches. The most easily observable difference between the species is in their antennae. Flour beetles like these were used as an example of a species that can produce chaotic population dynamics in Figure 6.19.

Tribolium confusum

Tribolium castaneum

and $K_2 > K_1/\alpha$), we have **unstable coexistence** (**Figure 7.18**). This term can be confusing because the lack of stability leads to a lack of coexistence. We would not choose to use the term here, except that it is entrenched in the ecological literature.

In Figure 7.18B, the arrows between the isoclines around the intersection point lead away from the intersection. Although the exact point of intersection is an equilibrium, it is an unstable one. If a disturbance moves either species' population density away from that specific point, the two species' populations will not return to that point. Instead, one species will eventually outcompete the other. Because change is a constant aspect of ecology, species with such intersecting isoclines generally do not reach the equilibrium point and do not stay there if they do reach it. They, therefore, do not coexist for extended periods in the real world.

In order to understand this outcome a little better, look at each axis. Notice that each species at low densities can consume its way through the shared limited resources and blow right past the other species' carrying capacity before stabilizing at its own carrying capacity. This means that each species can exclude the other species. But which one will do so? The answer depends on the initial densities of each competitor. If we start with nonzero densities for each species, the outcome will be determined by where the two population densities sit in relation to a line between the origin and the point of intersection. Below the dashed line in Figure 7.18B, species 1 wins; above the dashed line, species 2 wins.

Why would this ever happen? A good example is when two species are so similar that they share the same niche and, therefore, cannot coexist. There is no reason in such a situation for one species to always outcompete the other. Instead, the advantage of numbers goes to the species that arrives first at a resource-filled site. Ecologists call this a *priority effect* (Chapter 13), meaning the species that arrives before the other has the upper hand in competitive situations. Priority effects help ecologists understand why, in nature, sometimes one competitor wins but at other times the other competitor wins.

Ecologists who have tried to repeat Gause's investigations of competition with other species in very simple systems have found priority effects. For example, Thomas Park (1948) examined two closely related species of flour beetles, *Tribolium confusum* and *Tribolium castaneum* (**Figure 7.19**), under conditions where he controlled the amount of flour (food), the temperature, and the humidity. The two species had almost identical niches and did not coexist in any of his experimental competition trials. He found temperature and humidity combinations that could determine the winner, as in competitive exclusion, but he also found combinations in which the eventual winner depended on which species got the upper hand early in population growth, as with unstable coexistence and priority effects (**Figure 7.20**).

Figure 7.20 Thomas Park maintained populations of *Tribolium confusum* and *Tribolium castaneum* in a number of different combinations of temperature and humidity. The two species never coexisted in the long term, although the time to extinction of one species or the other varied. At certain temperature and humidity combinations, the eventual "winner" of competitive interactions depended on population densities early in population trajectories (Park 1948).

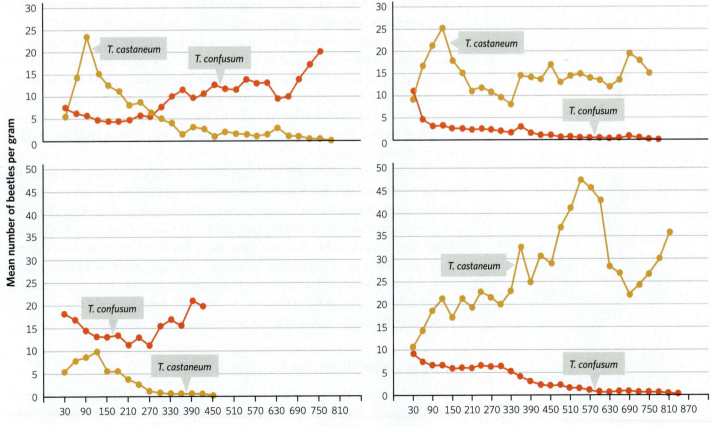

A Few Issues with the Lotka-Volterra Model

Despite the intuitive logic of the Lotka-Volterra model, it diverges from reality in a few ways worth mentioning. For one, it is based on the continuous time growth models from Chapter 6 and, therefore, assumes that time comes in very small slices. This assumption means that a small change in species 1's population density leads to an instantaneous response in species 2's population density, and vice versa. Again, the natural world is often much more nuanced and complicated than this. In general, individuals of a species need time to react to changes in their environment, which means that significant time lags are built into many (if not most) real-life species interactions.

Another issue is that the model suggests that populations steadily grow (or decrease) until they reach an equilibrium. Such population trajectories assume that population growth and species interactions follow a single set of parameters without change. As discussed in Chapter 6, many real-world factors, such as the carrying capacity (K), change over time, so population densities and interactions also probably change in the real world.

A corollary of this problem lies with one of the central tenets of competition theory, namely that the disputed resource is always in limited supply. If the resource is not in limited supply, interspecific competition will be so weak that it will not affect the survival and reproduction of individuals of either species. In natural environments, resource availability is constantly in flux and may go through periods of limitation and abundance. What happens then? In reality, the balance of competition may change over time (see Chapter 13). The Lotka-Volterra model is called an equilibrium model, and it is useful for looking at potential outcomes, but real-world disturbances and changes can keep a system from reaching equilibrium outcomes. We explore some nonequilibrium dynamics in Chapter 13.

Yet, the most common criticism of the Lotka-Volterra model concerns the competition coefficients (α and β). We described these as having a fairly intuitive meaning. We said each coefficient represents the area of shared overlap in the niche spaces of two species. Put another way, competition coefficients are a way of translating the relative effect of an individual of one species into individuals of the other species. But how exactly can one measure this overlap and its ecological effects in real situations?

The visualization example in Figure 7.8 seems straightforward, but resources in the real world are not always readily visible and do not always get used in similar-sized chunks. As we mentioned, ecologists cannot measure every resource in an n-dimensional niche, and the Lotka-Volterra model offers a way to step back from such endless niche measurements. But what if ecologists do have a pretty good idea about what the shared resources are? There must be a better way to look at how competitors actually use those known, limiting resources. There is, and later in this chapter we explore an alternate competition model that addresses how competitors affect each other by removing specific resources.

Insights from the Lotka-Volterra Model

Clearly, the Lotka-Volterra model is not perfect, but we can still gain some important and useful insights from its exploration. First, the model helps answer some of the questions about coexistence posed at the beginning of this chapter and helps us to see the realm of potential competitive outcomes. In looking at the natural world, it is clear there should be competitive exclusion and coexistence, but the existence of a distinct set of conditions that lead to unstable coexistence or priority effects is something of a revelation.

The model gives us conditions we can test experimentally (e.g., for a given situation, is $K_1 > K_2/\beta$, and is $K_2 > K_1/\alpha$?). If we go outside and measure values in nature and find they fit the conditions for competitive exclusion or for coexistence, then we have a more comprehensive understanding of the interactions. Even if the natural situation fails to conform to the model's predictions, we still have gained understanding, and negative cases can lead to new and interesting lines of inquiry. In hindsight, one of the main accomplishments of the Lotka-Volterra model is that it stimulated decades of solid ecological research and fostered a much greater understanding of the role that competition plays in the natural world (McIntosh 1985).

A second important feature of the model is its prediction that stronger competitors should consistently outcompete weak competitors, regardless of their respective population abundances. According to the model, if a strong competitor enters a new area, it should eventually take over and outcompete a competitively weaker species. This is a useful and testable prediction when studying situations such as range expansions or invasive species.

Third, a critical insight from the model is that stable coexistence is predicted to occur when intraspecific competition is stronger or more important than interspecific competition. One of the aspects of the natural world we explore in great detail in Chapters 10 and 11 (food webs and biodiversity, respectively) is that there are a lot of species coexisting together on the planet. If our theoretical and mathematical understanding failed to predict conditions in which coexistence is likely, we would have to seriously question the usefulness of the science. Coexistence is everywhere, and showing that this competitive outcome is possible with simple additions to a basic population growth model is important.

7.4 A More Mechanistic Model with a Focus on Accessing Resources

The Lotka-Volterra model assumes competition for unspecified but shared resources, which is a useful approach for structuring our thoughts about the potential effects of competition. In the real world, though, what if we know exactly what resources are shared and limiting for two species? What if competition is for more than one resource? Or what if two shared resources are not equally limiting for each species? If each species has unique aspects to its niche, then it makes sense that two species might use the same resources but in slightly different ways (Figure 7.7).

The Tilman Model, or the Resource-Ratio Hypothesis

In the real world, organisms use many resources and likely share multiple resources with other species. These resources rarely exist in isolation, completely unaffected by other resources. David Tilman, a community ecologist who did some of his

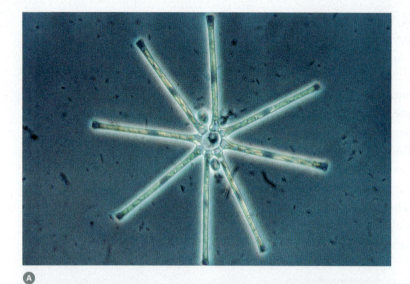

Figure 7.21 David Tilman used two freshwater diatom species (Ⓐ *Asterionella formosa* and Ⓑ *Cyclotella meneghiniana*) to investigate interspecific competition mediated through resource use.

earliest research on single-celled algae in Lake Michigan, formalized a very useful way of exploring competition for shared resources (Tilman 1977). This second model of interspecific competition provides a unique set of ecological insights in relation to the Lotka-Volterra model, and it forms the basis of a large and growing ecological literature on competition. If you continue your ecological education past this course, this model of competition is the one you will encounter most in the more current ecological literature.

Tilman and colleagues ran 76 long-term experiments on resource competition between two species of freshwater diatoms, *Asterionella formosa* and *Cyclotella meneghiniana* (**Figure 7.21**). Diatoms are extremely common forms of microalgae and represent the predominant members of many marine and freshwater phytoplankton communities (i.e., the photosynthetic primary producers in water). Beyond their obvious need for sunlight for photosynthesis, diatom survival is also critically linked to the availability of silica. Silica is ingested by the algae in the form of silicic acid, $Si(OH)_4$, and is used to create a silicon dioxide shell called a frustule. Diatoms also require phosphorus (in the form of phosphate, PO_4^{3-}), which is essential for cell membranes and biomolecules like DNA. Silica and phosphorus are typically limited in the diatom's natural environment, and thus individuals of two (or more) diatom species are likely to compete for access to these essential nutrients.

As expected, Tilman and colleagues found clear evidence that *Asterionella* and *Cyclotella* strongly compete for these two resources. However, they found that individuals of *Asterionella* and *Cyclotella* had slightly differing physiological requirements for the two resources. This observation led Tilman (1980) to develop a mechanistic model of competition in which two species use limited resources in slightly different ways. Tilman's model has both a mathematical and a graphical solution, but we explore only the graphical aspects, as they tend to be more intuitive. The model has three essential components: the zero net growth isoclines (ZNGIs, Tilman's term for an isocline) and their combination (R^*), an environmental resource supply point, and species-specific consumption vectors. While this may sound like a scientific tossed salad of words, it will begin to make sense as we work through the components of the model one by one.

Zero Net Growth Isoclines and *R**

In our discussion of phase planes in the section on the Lotka-Volterra model, we said that each species had an isocline for zero population growth. Tilman suggested that we can use a similar concept but think about zero net growth isoclines that are associated with various resource levels. All species on the planet have very specific physiological resource requirements (Chapter 5), but this approach works for any resource that is essential for survival or reproduction for a focal species. For any individual in that focal species, there is typically a threshold resource level below which death may be likely and growth and reproduction do not occur. Exactly at the resource threshold, individual growth and survival are possible, but reproduction is not. Above this threshold, survival and reproduction increase as resource availability increases. Basically, this threshold represents a point of zero net population growth. Tilman called this resource threshold $R*$ (pronounced "R-star").

Based on the *Tribolium* work by Park, we also know that $R*$ for a particular resource may not remain constant as other aspects of the environment change. As we noted, Park found that food consumption and survival in flour beetles changed with temperature and humidity. We could explore this change by plotting each $R*$ value for one resource along a gradient of some other environmental variable (e.g., temperature, pH, the availability of another resource). For example, we could look at the minimum amount of food necessary to stay alive ($R_{calories}*$) in environments of different temperatures. For humans and most other heterotrophs, the $R*$ for calories should get bigger as temperature goes down (i.e., more calories are needed at colder temperatures). The line or curve formed by all those $R*$ values along a gradient would be a line of zero net growth. Such lines are best thought of as isoclines, like the ones we saw in the phase-plane graphs. When the resource level is exactly on the line, we expect to see zero net population growth through time. To distinguish these isoclines from phase-plane isoclines, and to make it clear that they are associated with access to specific resources, Tilman called these isoclines **zero net growth isoclines (ZNGIs)**. The most common way of looking at ZNGIs is to explore two or more resources of interest rather than one resource and a changing abiotic variable (**Figure 7.22**).

Returning to the diatoms, let's put phosphate availability on the x-axis. The simplest possible ZNGI would be a vertical line that represents an unchanging amount of phosphate (R_X* in Figure 7.21) needed to balance reproduction and death. Although we do not show population abundance on this graph, we know that the population will grow on the right side of the ZNGI (when phosphate levels are above the threshold, R_X*) and decrease to the left of the line. The simplest possible second ZNGI would come off the y-axis and represent the $R*$ levels for silicate as phosphate availability increases. If we again assume that the uptake of phosphate does not affect the use of silicate, then the silicate ZNGI will be a horizontal line that represents a constant silicate $R*$ (labeled R_Y*). The same rules apply, so above this threshold line the population grows, and below the line it declines (**Figure 7.23**).

These are the simplest possible ZNGIs, and in the real world there is no reason for them to be completely vertical (the phosphate ZNGI) or horizontal (the silicate ZNGI). If the use of one resource changed with the availability of the other resource, these lines could have slopes or be curves. These simple ZNGIs are useful to show how the $R*$ approach works.

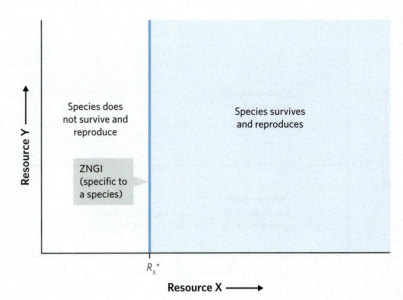

Figure 7.22 The amounts of two resources of interest to a species in an environment are shown along the axes. R_X* is the threshold level of resource X at which births and deaths are balanced and the population of the species does not grow. The blue line represents the species' ZNGI for resource X as resource Y changes. If the species uses the resources independently, R_X* does not change as Y changes, and the ZNGI is a vertical line. If the environment has more of resource X than R_X* (i.e., the light blue area), the species can survive and reproduce. If it has less, the species cannot live in the area.

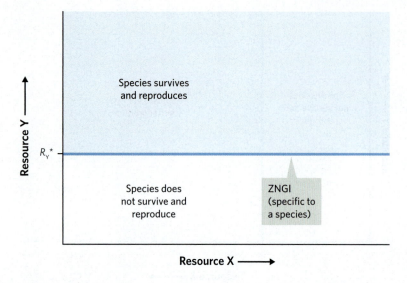

Figure 7.23 The blue line represents the species' ZNGI for resource Y. If the environment has more of resource Y than R_Y^* (i.e., the light blue area), the species can survive and reproduce. If it has less, the species cannot live in the area. In the diatom example, the ZNGI for resource Y and R_Y^* represents threshold levels of silicate, but this figure works for any two independently used resources.

Figure 7.24 The combined ZNGIs for resources X and Y for a single species identify resource combinations that allow the species to survive and reproduce (light blue box) and other combinations that do not (area outside the box). The lowest survivable level for each resource individually is shown with the ZNGI. The point (R_X^*, R_Y^*) shows the critical resource levels for that species; below this point, the species does not have enough of either resource to survive.

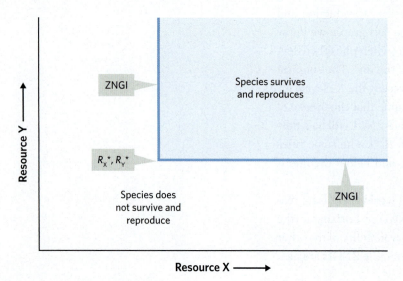

When we put both of these resource requirements together for a single diatom species (**Figure 7.24**), we get a defined space, sometimes called an envelope (shaded area in Figure 7.24), where individuals of this diatom species happily survive and reproduce and where their population grows through time. Resource levels outside the envelope are too low to allow population growth (the white area of Figure 7.24). The failure of the population to grow can come from a lack of either resource (phosphate or silicate) or a combination of too little of both. The point where the two resources are both at the ZNGI for that species represents the lowest survivable level for the two combined resources for that species. Technically, the lowest point at which a population of the species can survive with a resource, which is represented as (R_X^*, R_Y^*) in Figure 7.24, is also the point at which the population of that species is neither growing nor shrinking, meaning it is at equilibrium. That means the point (R_X^*, R_Y^*) should represent the amount of resources available at the carrying capacity. Notice also that the envelope defined by the two resource ZNGIs is very similar to the two-axis niche graph we saw in Chapter 5 (Figure 5.27).

The Resource Supply Point

The second component of the Tilman model is the resource supply point. So far, we have considered the individual resource needs of a single species. Without saying it explicitly, we have implied that the availability of each resource is constant through time. That is, we assumed that the amount of phosphate or silicate, for example, lies either above or below the two resource ZNGIs in an unchanging way. Such unchanging resources are unrealistic, but let's explore why in a little more detail.

First, individuals of our focal diatom species use phosphate and silicate to stay alive; therefore, their resource use decreases the amount of each resource that is available. Second, resource levels can decrease simply because ecosystems are inherently "leaky," meaning they gradually lose some of their resources from the movement of water, wind, and other forces into other ecosystems. Conceptually, this is simply recognizing that natural systems are not test tubes with impermeable glass boundaries that keep everything in (or out). For example, in a lake ecosystem, phosphorus molecules can simply drain away with water that flushes into an outlet stream or river. Third, and related, resource levels can also increase through time due to inputs from outside sources. In terms of lakes, the amount of phosphorus can increase because of the actions of microbes within the water column or sediment, or through human release of phosphorus into streams that flow into the lake. In other words, resources can enter and leave an ecosystem in a variety of ways.

Changes in resource supplies through time are called fluxes. For diatoms, phosphate and silicate flow into and out of a lake, making these resources variably available to the species that live in that lake. If inputs are higher than losses at any point in time, the amount of a resource will go up, and vice versa. However, if the flow of resources out of the system *equals* the flow of resources into the system, we have an equilibrium in resource availability. We define that equilibrium in our model as the environmental **resource supply point (S)** (**Figure 7.25**). Basically, *S* identifies a steady availability of the resource and acknowledges that resources can be used and resupplied. Exactly where the resource supply point sits in a graph plays an important role in the outcome of competitive interactions, as you will see.

Consumption and Renewal Vectors

The third and final component of the Tilman model is the rate at which individuals of a species consume available resources (i.e., species-specific consumption vectors). Consumption of resources by the species of interest will draw the resource level below the resource supply point (*S*) because *S* represents the balance of inflows and outflows in the environment without consumption by the focal species. We can show consumption as a vector, which represents a change in the quantity of a variable through time. Vectors are very much like the trajectory arrows used in the phase-plane graphs. Consumption of resource X will lead to a vector pointing left, and consumption of resource Y will lead to a vector pointing down. No change through time is represented by a single point.

The **consumption vector**, therefore, represents the change in the resources through time, where the change stems from individuals of the species consuming the resources. As an example, consider an office where individuals eat doughnuts and drink coffee. If the workers started with five pots of coffee and five boxes of doughnuts, the consumption vector would point toward lower amounts of coffee and doughnuts. The exact location of the vector depends on the consumption rate, which could be quite high if the workers stayed up late the night before and needed to stay awake today. Or it could be low if they are trying to cut back on caffeine and carbohydrates. Consumption vectors generally need to be empirically measured, and they vary from species to species.

In Figure 7.25, we drew hypothetical consumption vectors for diatom use of phosphates and silicates. The consumption vectors represent the rate of consumption of resource Y and the rate of consumption of resource X. If we put these two together the way we put our trajectories together in the phase-plane graphs,

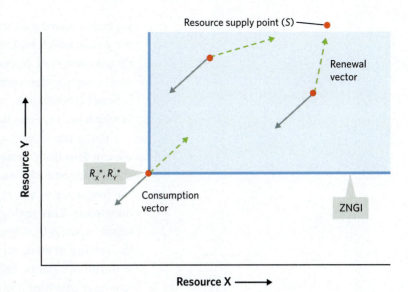

Figure 7.25 The resource supply point (*S*) represents the natural availability of the resources in the environment. The consumption vectors represent one species' characteristic consumption rate, so they always have the same slope and magnitude, regardless of where they are on the graph. The resource renewal vectors always point toward *S* because environmental forces, without consumption by the species, move availability back toward *S*. The equilibrium point (R_X^*, R_Y^*) lies along the ZNGI, where consumption and renewal vectors point in opposite directions and the population cannot consume more without decreasing in abundance.

we end up moving a certain distance along the x-axis and a certain distance along the y-axis, which basically gives us a slope (*change in Y/change in X*). Notice that the consumption vectors all have the same slope no matter where they originate. Consumption vectors are essential to the $R*$ approach because the balance between consumption and renewal determines the $R*$ for the resource, or the level at which each species has zero net growth.

Using the same logic, and recalling how resources may enter our system, we can also draw **renewal vectors**. We show a few hypothetical examples of these in Figure 7.25. In our office example, the renewal vector is controlled by the frequency at which someone brews new pots of coffee and brings in more doughnuts. This analogy breaks down a little here because, in nature, a resource supply point is in a dynamic equilibrium, always losing a little and gaining a little because of environmental forces. These environmental forces keep the loss and gain in balance and maintain a constantly renewed resource supply. If our hypothetical office had elves to keep replenishing the coffee and doughnuts and goblins to keep eating and drinking, even when the staff was not in the office, so that there were *always* five pots of coffee and five boxes of doughnuts, then the analogy would work. If only all offices worked that way! With the background environmental inflow and outflow, all renewal vectors point directly from the current resource level toward S because that is the point to which the system returns automatically.

The Full Model and Outcomes of Competition

Given the basics of the model, we can apply it to a situation with populations of two different species competing over two resources. It is important to recognize that each species is going to have its own specific ZNGIs and consumption vectors for the two resources. The ZNGIs are likely to be different among species due to differences in physiology and niche requirements. If we think about superimposing two species' ZNGIs, there are at least four ways the ZNGI envelopes can interact, which should not surprise us, given the four outcomes of Lotka-Volterra competition.

Let's look at the simplest situation, the one in which one species' ZNGI lies completely below the other species' ZNGI along one (or both) resource axes (Figure 7.26). This graphical representation of niche overlap shows that species A can survive at lower levels of both resources than species B. More formally, this representation says that the ZNGI envelope for species A covers all of species B's ZNGI envelope, and more. Based on the discussion of the Lotka-Volterra model, you should suspect that such a situation will lead to competitive exclusion, with species A winning because it can reduce the resources to a point below the survival threshold of species B.

This model provides a way to assess if this is the most likely outcome of competition between the two species, given these conditions. Note that in Figure 7.26 there are three areas defined on the graph due to the locations of the ZNGIs, and these represent three possible regions for the resource supply point (S) to be located. If S is located anywhere in region 1, then the ecosystem does not contain enough of one (or both) of the resources to sustain either species, and neither will survive. If S falls into region 2, then species A will out-compete species B because species A can draw resource levels down below the ZNGIs for both resources for species B (below the two-resource $R*$ for species B).

Now for the big question: What happens if S falls in region 3? Again, species A will win by competitively excluding species B. Remember that each species can (and will) draw resources down to its $R*$, where births and deaths will balance. That drawdown is like having a population grow to carrying capacity and draw resources down to an equilibrium carrying capacity level, meaning species A will draw the resource levels out of region 3 and into region 2, right down to its own ZNGIs. In this case, species A competes better for both resources and outcompetes species B.

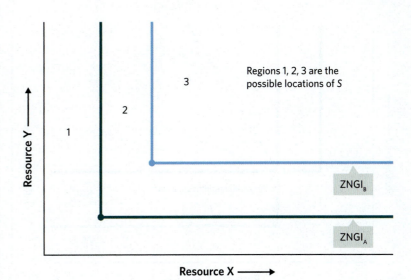

Figure 7.26 When two species each have a ZNGI for two resources, and one species' ZNGIs lie completely below those of the other species, the ZNGIs create three different areas on the graph (regions 1, 2, and 3), and the equilibrium and potential for coexistence of the two species depend on which region contains the resource supply point (S). If S lies in region 1, both species go extinct. If S lies in region 2, only species A can survive. And if S lies in region 3, species A will again outcompete species B. The ZNGIs and associated $R*$s for species A sit completely below the ZNGIs and associated $R*$s for species B, which means that species A competes better for both resources and will always win.

We could create the exact opposite scenario by simply switching the roles of the two species, so that species A's ZNGIs would lie within species B's. In that new scenario, species B would "win" the competitive interaction and exclude species A. These two ways of arranging the ZNGIs for the two species give us competitive exclusion like the exclusion that we saw in the Lotka-Volterra model when the isoclines did not intersect.

Now let's see if we can find conditions that lead to coexistence or unstable coexistence as in the Lotka-Volterra model. When the two species' ZNGIs intersect (Figure 7.27), it means each species can experience positive population growth at a lower resource level than the other species, for one of the two resources. In other words, each species is a better competitor for one of the resources. For example, in Figure 7.27A, species A can maintain positive population growth at a lower level of resource X than species B can, but species B can grow at lower levels of resource Y. Because neither species can commandeer both resources simultaneously, the two species can coexist. Note, though, that coexistence depends critically on the location of S.

In Figure 7.27A, we have drawn the consumption vectors for the two species (C_A and C_B; more on these shortly), creating six different regions where S can be. We already know what happens if S is located in region 1: the environment doesn't support either species' physiological needs, and both drop out of the ecosystem. If S is located in region 2, species A can draw down resources along the x-axis to a level that leads to the exclusion of species B. Species A maintains positive population growth at lower levels of resource X than species B can, and thus species B drops out of the ecosystem. The reverse outcome, where species A is lost, occurs when S is located in region 6.

The other three regions (3, 4, and 5) need additional consideration, as they all lie within the physiological envelopes that allow survival (bordered by the ZNGIs) for both species. Before we continue, we need to say a bit more about C_A and C_B in Figure 7.27A. These consumption vectors are drawn through the intersection point of the two ZNGIs. That intersection point represents the resource levels that lead to zero net growth for both species simultaneously, meaning equilibrium with both species coexisting. We can, therefore, call the intersection point a *potential*

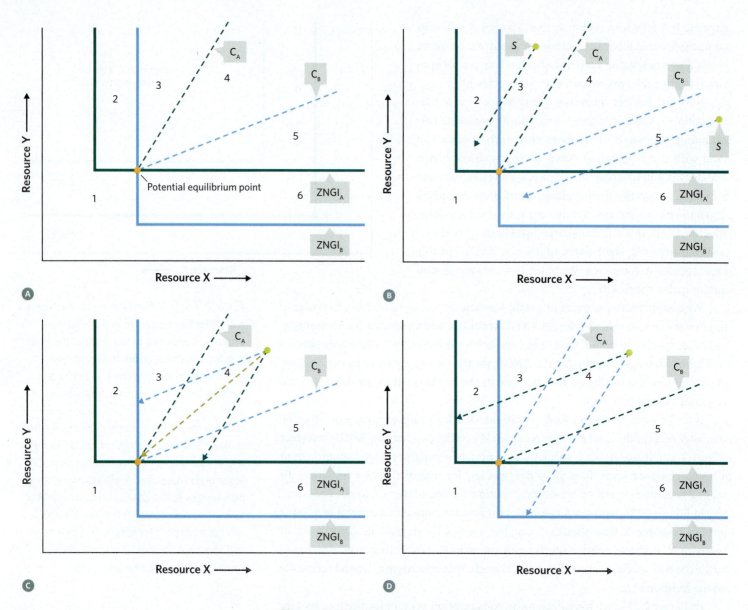

Figure 7.27 With two intersecting ZNGIs, each species is a better competitor for one of the two resources. The point of intersection (orange circle) represents a potential equilibrium point at which the species experience zero net growth simultaneously. When the ZNGIs overlap like this, coexistence will depend on where on the graph the resource supply point (*S*) lies. **A** Consumption vectors for species A (C_A) and species B (C_B) show consumption from within the overlapping survivable space. Region 1 has resource levels too low for both species, so neither species survives. Regions 2 and 6 each have one resource below the level that allows for survival of one species. Species A wins in region 2, and species B wins in region 6. **B** In regions 3, 4, and 5, each light green circle represents a sample *S* in that region, and the dashed arrows show how each species can consume resources, pulling resource levels outside the "survivable envelope" for the other species. Species A wins if *S* is in region 3, and species B wins if *S* is in region 5. **C** If *S* is in region 4, neither species is able to draw down resources to a level where the other species will have negative population growth. Instead, the species in combination pull the resource levels along the tan dashed arrow to the equilibrium point, resulting in coexistence. **D** What if each species consumes the resource for which it is a better competitor more quickly? That case alters the slopes of C_A and C_B and leads to unstable coexistence. Before hitting its own ZNGI, each species can draw resource levels down to a point where the other species has negative population growth.

equilibrium point, representing a resource level where both species coexist. But we need to explore the conditions for coexistence a little further.

Remember that C_A and C_B represent consumption of both resources and that the slope is the ratio of the rates of consumption. C_A illustrates that consuming a chunk of resource X allows it to acquire proportionally more of resource Y. This relationship holds because individuals of species A grow, survive, and reproduce with consumption of resource X. The more individuals of this species there are, the more they will consume resource X *and* resource Y.

Look at the graphs and think about the way each species can draw down resources. With what S do you expect to see coexistence? Remember that the consumption vectors have the same slope from any S, so you can mentally draw a line parallel to each consumption vector in each region to see the level and direction toward which consumption will drive resource levels (Figure 7.27B). If S falls in region 3, species A draws resource X below the level where species B can maintain positive population growth, and thus species B is competitively excluded from the ecosystem (i.e., species A wins). The same would happen for species B if S fell within region 5, where species B would outcompete species A.

If S lies in region 4, though, each species pulls resource levels down to a point where it meets its own limit of survival before it is able to lower the resources too much for the other species. In other words, neither species is able to draw down resources to a level where the other species will have negative population growth (Figure 7.27C). In this case, the model predicts that the two species will coexist. As in the Lotka-Volterra model, notice that we can also interpret this situation as one in which *intraspecific* competition for the resources is stronger than *interspecific* interactions.

At this point we have seen all of the same outcomes from the Tilman model that we saw from the Lotka-Volterra model except unstable coexistence (i.e., priority effects). But what if each species is actually a faster consumer of the resource for which they can survive at lower levels (i.e., the resource for which they are a better competitor)? This is the situation in Figure 7.27D, where the consumption vectors are shown with different slopes than in Figure 7.27C. These consumption vectors lead to unstable coexistence, or priority effects, in region 4. Whichever species arrives first will win because it can reduce resources to a level where the other species cannot survive. If species A arrives first, it will draw resource levels down into region 3, then into region 2, and exclude species B. If species B arrives first, it will draw resource levels down into region 5 and then region 6 and exclude species A.

Now that we have explained the whole model, it should be interesting to note that this model is also called the **resource-ratio hypothesis**. There are two reasons for this name. First, the consumption vector is basically a ratio of the consumption rates of each resource for each species. Second, our analysis to determine whether or not we would have coexistence relied explicitly on the location of S, but that location is generally dependent on a ratio of the availability of resource X and resource Y. If Y is low and X is high, that can put us in region 5 of Figure 7.27. If they are similar, then the resource ratio puts us in region 4, and so forth. The consumption vectors and resource supply points basically determine

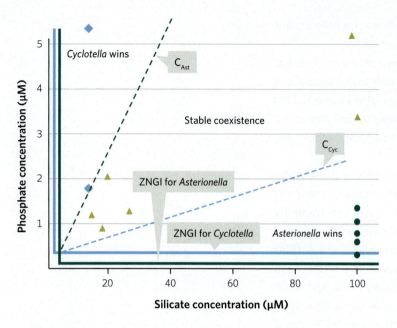

Phosphate concentration (μM) / Silicate concentration (μM)

Figure 7.28 The results of Tilman's experiments with diatoms support the resource-ratio hypothesis, or *R** model, which suggests that species can coexist when resource levels fall in the region between the species' consumption vectors (i.e., region 4 in Figure 7.27), but that other regions should lead to only one species (or no species) surviving. When Tilman set resources at the levels indicated by blue diamonds, *Cyclotella* dominated; at the levels indicated by dark green circles, *Asterionella* dominated; at the levels indicated by tan triangles, the species coexisted (Tilman 1980).

the outcome, and Tilman emphasized the importance of resource ratios in determining competitive outcomes.

A test of the *R** model came from Tilman's own work with the two diatoms, *Asterionella* and *Cyclotella*, which differ in their physiological requirements for silicate and phosphate (Tilman 1977, 1980). Their respective ZNGIs overlap and cross (**Figure 7.28**), suggesting the potential for coexistence. In order to test the model, Tilman grew the two diatoms together in the laboratory while experimentally manipulating resource levels of both silicate and phosphate. He found that each species won some of the time in the competitive interactions, and which species won depended critically on the resource levels in the experiment. He also found conditions for coexistence (Figure 7.28). Indeed, the experimental results fit very well with the predictions of his model.

7.5 Long-Term Competitive Outcomes

Both the Lotka-Volterra and Tilman models produce similar outcomes for two species' coexistence with interspecific competition, but what do these models tell us about competition over longer time scales? Let's examine the longer-term outcomes, which we call the three *E*s: extinction, exclusion, and evolution.

Extinction

Extinction is perhaps the easiest outcome to understand, but it is also one of the more challenging to demonstrate. The Lotka-Volterra and Tilman models suggest that one competitor may sometimes drive another to extinction. If this is a common outcome of interspecific competition in nature, we should be able to point to many examples. However, such examples are few and far between. Does this mean that extinction is an uncommon outcome of interspecific competition? Maybe, but consider that many millions of species have existed and gone extinct throughout Earth's history, and most have left no evidence behind of their passage. How often, then, should we expect to find evidence that one species went extinct through a competitive interaction with another?

Let's consider an example. Imagine that squirrels and chipmunks coexisted for centuries within a particular forest, where they both survived by eating acorns (**Figure 7.29**). In this hypothetical, head-to-head matchup, the squirrels always behaviorally dominated the chipmunks and "won" any contested acorns. But during these several hundred years, acorns were plentiful, and each species was instead limited by some other resource, such as nesting sites. Because squirrels nest in trees and chipmunks nest underground, the two species did not compete for their limiting resource (i.e., nesting sites) and coexisted.

Now suppose that during the most recent decade, many oak trees have failed to reproduce because of an unfavorable change in climate and have died more often due to the emergence of an invasive pathogen (e.g., sudden oak death, a disease from an invasive pathogen, is reducing oaks across the western United States in a fashion that fits this scenario). Together, these two stresses have caused the oak trees to produce far fewer acorns every year than in the decades before. Under these new conditions of acorn scarcity, access to acorns has become the critical issue for chipmunk and squirrel survival, reproduction, and fitness. Because squirrels will win any hypothetical contest for acorns, the competitive disadvantage for the chipmunks has caused many individual chipmunks to die or fail to reproduce. Over time, we could witness declining chipmunk abundance, with chipmunk populations being driven extinct in locations where they previously had thrived. (For readers who are worrying about global chipmunk extinction, rest assured that chipmunks are doing fine. This hypothetical example mimics some actual observations, as the effects of declining oak populations are real, but as far as we know, squirrels are not driving chipmunks extinct.)

Figure 7.29 Ⓐ A western gray squirrel (*Sciurus griseus*). Ⓑ A least chipmunk (*Neotamias minimus*).

Now imagine hiking in our hypothetical forest after the chipmunks have lost the competitive battle for acorns to squirrels. You see lots of squirrels as you walk, but never a chipmunk. Being an inquisitive ecologist, you wonder whether interspecific competition has played a role in determining which species of small mammals you are encountering on your hike. If all you know is what you observe, you would have no idea that chipmunks had ever been there or that competition with squirrels led to their extinction. There is no smoking-gun ecological evidence that chipmunks once scurried along the forest floor (e.g., no huge piles of decaying chipmunk poop or great graveyards of fossilized chipmunk bones), and no evidence that squirrels beat the chipmunks to the acorns (e.g., no videos of squirrels wrestling chipmunks for acorns). The chipmunks disappeared without leaving a shred of evidence of their existence or the cause of their demise.

If this example is at all representative of the natural world, we should expect to observe very little evidence of extinction due to interspecific competition, because observing such evidence would require detailed, long-term knowledge of every ecosystem on the planet. Ecologists call this disappearance phenomenon the **ghost of competition past**. In other words, when competition has a strong structuring effect on an ecological community in ways that we cannot see because the competitors are no longer interacting (or interacting in ways that have changed), we call that the ghost of competition past.

Even when we have some evidence of past interspecific competition, we are still seriously limited in our ability to infer that the interaction directly caused the extinction of one of the species. Continuing with the squirrel and chipmunk scenario, let's assume that we *do* have past observations or fossil evidence to suggest that chipmunks were once quite common in this forest. Maybe we even know that they shared the acorn resource with squirrels, suggesting that competition may have been a factor in their population dynamics. What we most likely do not know, however, is how intense this competition was, whether it truly had an impact on chipmunk fitness, or whether the acorns were a scarce resource (and for how long).

Without a detailed understanding of the ecosystem, we cannot firmly conclude that interspecific competition with squirrels directly drove chipmunks extinct. Because chipmunks are no longer there, we are also unable to conduct experiments to test whether squirrels may have been the agents of the chipmunk demise. It could be that the current distribution of squirrels and chipmunks arose because a predator consumed all the chipmunks but had no taste for squirrels; or because a chipmunk-specific disease wiped out the little critters but did not seriously affect the squirrels; or even that flooding from a very wet year killed the ground-nesting chipmunks but not the tree-nesting squirrels. Just observing that a species is absent is not sufficient to prove extinction due to competition.

Finally, extinction generally implies that a species disappears not just from one ecosystem or habitat but from a whole region (Chapter 1 and Chapter 11). That would be like chipmunks disappearing across all of North America. We sometimes use the terms *local extinction* or *extirpation* to refer to extinction from a single ecosystem or locale (such as one forest), but this is essentially synonymous with the term *exclusion*, which is discussed next. Species tend to exist in many places, and the outcome of competition depends strongly on local conditions, as described by the Tilman model. For one species to lead to global or regional extinction of another species through competition, the first species would have to be the superior competitor across a wide range of situations or conditions—in fact, across *all* conditions in which the other species could exist. That is rare. More often, we see extinction from a type of habitat or from a local area.

Nonetheless, some studies do point strongly to interspecific competition driving a species to extinction in a region (larger than one ecosystem). For example, Michael Marchetti documented that the introduction of non-native bluegill sunfish (*Lepomis macrochirus*) at least partially contributed to the complete extinction of the native Sacramento perch (*Archoplites interruptus*) across the Central Valley of California (Marchetti 1999). Other examples also tend to come from instances of invasive species or well-studied ecosystems. Because humans have such large effects on the world, the Anthropocene is also full of examples in which humans cause extinction of other organisms due to overharvesting or competition (Chapter 15).

Exclusion

The loss of all individuals of a species from an entire ecosystem is an extreme outcome of interspecific competition. As we know, a species' niche is multidimensional in nature (Chapter 5). Having many resources means that reduced access to one resource may not drive the species to extinction. A reduction in one resource may, however, restrict where we find individuals of the species, as we saw earlier in the mountain lion example. This outcome is called **exclusion**, and it can occur through many different mechanisms.

Exclusion is fairly well documented in the natural world (Sax et al. 2007), perhaps because both of the competing species are still in the region, making it possible to document their interactions and the fitness consequences of these interactions. A particularly interesting example comes from the work of Kurt Fausch and colleagues (1994). In the 1990s, they studied competition between two species of closely related salmonid fish species in Japan, the white-spotted char (*Salvelinus leucomaenis*) and the Dolly Varden trout (*Salvelinus malma*; **Figure 7.30A**).

The two salmonids exist across a variety of aquatic habitats, including freshwater streams and rivers and in near shore marine habitats. They also each show

White-spotted char
(*Salvelinus leucomaenis*)

Dolly Varden trout
(*Salvelinus malma*)

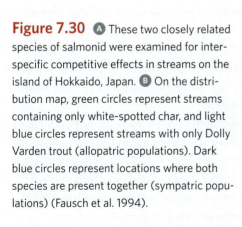

Figure 7.30 Ⓐ These two closely related species of salmonid were examined for inter-specific competitive effects in streams on the island of Hokkaido, Japan. Ⓑ On the distribution map, green circles represent streams containing only white-spotted char, and light blue circles represent streams with only Dolly Varden trout (allopatric populations). Dark blue circles represent locations where both species are present together (sympatric populations) (Fausch et al. 1994).

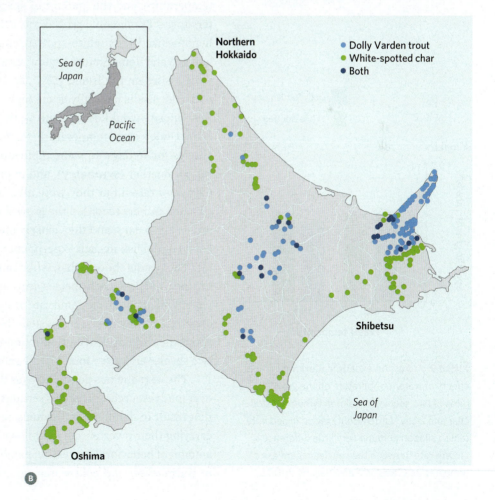

flexibility in their diet and foraging behaviors, feeding at times on smaller fish but mostly on various invertebrates in the water column. These feeding observations suggest that there is no single food resource that either species needs to persist, making it unlikely that competition between them (when it occurs) will drive one species totally out of an ecosystem. However, the species overlap enough in their physiological requirements and food sources that Fausch and colleagues believed that interspecific competition was likely to some degree.

The researchers conducted an extensive survey for the two salmonids across all the streams on the mountainous island of Hokkaido, the northernmost island in the Japanese archipelago. They found that the two species occurred in the same river drainage systems but only occasionally in the same locations (**Figure 7.30B**). The white-spotted char was found almost exclusively in low-elevation sites, where water temperatures were warm, whereas the Dolly Varden trout was found in high-elevation streams, where the water was cool. Although not perfect (as there are some areas where the two overlap), on the surface this distribution looks like a straightforward case of competitive exclusion. What was not clear from these observations was the specific resource the two species were competing for, and what effect this competition had on the fitness of individuals of each species.

Further investigation was needed, so Shigeru Nakano and his graduate student at the time, Yoshinori Taniguchi, built a series of replicate streams in a laboratory facility that allowed them to strictly control food availability, water temperature, and the amount of space available for each fish to set up feeding territories (Taniguchi and Nakano 2000). Based on their field observations, they hypothesized that white-spotted char would outcompete Dolly Varden trout when water temperatures were as warm as they are at natural low-elevation stream sites on the island (12°C, or 54°F). They predicted that when the water was cold, as it is naturally at the high-elevation sites on the island (6°C, or 43°F), the opposite would occur, with Dolly Varden trout outcompeting white-spotted char. Under each temperature regime, they raised young individuals of each species for 76 days. In some replicate streams, individuals of both species were raised together (**sympatry**), and in other streams, the same number of individuals were raised but they were all of only one species or the other (**allopatry**). The researchers recorded the growth rates of the individuals in each experiment across all 76 days, and they closely observed the territorial behavior and feeding habits of the fish in each experimental stream.

The results showed that, when raised only with their own species (allopatry), the individuals of each species grew at about the same rate in the warm and cold water and at about the same rate as each other (**Figure 7.31**). In other words, the experimental evidence suggested that water temperature explains very little about the disjoint distribution found in nature. So, what did determine why the two species are rarely found in the same streams on Hokkaido?

The next obvious possibility was that individuals of the two species competed over food resources. In natural settings, salmonids establish behavioral hierarchies that result in dominant individuals occupying the prime real estate where water carrying their invertebrate food flows past. Taniguchi and Nakano compared the amount of behavioral aggression exhibited by individuals of the two species when in warm-water and cold-water environments. When raised only with others of

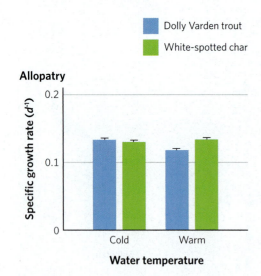

Figure 7.31 An experiment examining the effects of water temperature on the growth rates of two species of salmon (white-spotted char and Dolly Varden trout) raised in separate tanks (allopatry) found very little difference in growth rate between the species regardless of temperature (Taniguchi and Nakano 2000).

their own species, individuals aggressively fought for available food more often if they were raised in warm water than if they were raised in cold water (**Figure 7.32**). Keep in mind that these are *intra*specific aggressive interactions; in the allopatric streams, only individuals of the same species are competing for food. In the cold-water streams, individuals of both species were about equally aggressive, and in the warm-water streams individuals of both species became more aggressive. The white-spotted char responded to a change in water temperature more strongly than the Dolly Varden trout, though.

Now compare this pattern to what happened when individuals of each species were raised together (sympatry) and foraging aggression was observed. In cold-water streams, nothing really changed. Individuals of each species were somewhat aggressive in their interactions over food, but this level of aggression was not that different from when they were living exclusively with individuals of their own species. But when the water temperature was raised, the white-spotted char became very aggressive when establishing foraging positions in the water column (Figure 7.32). This suggests that the disjoint distribution observed in streams across Hokkaido is the result of substantial interspecific competition mediated through aggressive interactions, particularly at the low-elevation (or warm-water) sites.

The experiment was telling, but it still did not identify the ecological mechanism at work that prevented the white-spotted char from using the upstream habitat dominated by the Dolly Varden trout. Taniguchi and Nakano suggested that their experimental low temperature of 6°C (43°F) might not have been cold enough to provide the Dolly Varden trout a competitive edge. Alternatively, they suggested that young Dolly Varden trout may be better able to resist starvation than similarly sized white-spotted char and could, therefore, dominate sympatric encounters during the long periods of low food availability that occur at the

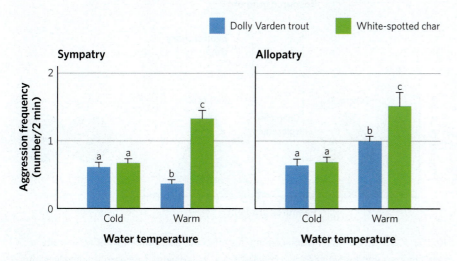

Figure 7.32 An experiment examining the effects of water temperature on aggressive behaviors in white-spotted char and Dolly Varden trout when raised together in tanks (sympatry) and in tanks by themselves (allopatry) found increases in aggressive behaviors at warm temperatures for both species when raised alone. Moreover, the Dolly Varden trout were far outpaced in levels of aggression at warm temperatures when the two species were raised together. Significant differences are indicated by the letters above the bars (i.e., *a* indicates statistically different responses from *b* and *c*) (Taniguchi and Nakano 2000).

high-elevation locations. Put another way, the Dolly Varden trout do not acquire the resources as quickly, but they can survive at a lower resource level (or R^*) than the white-spotted char. Based on the resource-ratio model, this suggests that the Dolly Varden trout should competitively exclude the white-spotted char in habitats where food levels occasionally sink to very low levels. For our purposes, it is important to realize that competitive exclusion, when it does occur, is not always easy to observe or demonstrate experimentally.

Evolution

The third and final possible outcome of interspecific competition is evolution. Recall that evolution by natural selection is defined by changes in allele frequencies in a population over generations due to differences in fitness (reproduction and survival) based on heritable variation in traits among individuals in a population (Chapter 4). By this definition, any interspecific interaction that is strong enough to cause a change in fitness is a potential agent of natural selection.

All competitive interactions must have some negative effect, which suggests that interspecific competition always has the potential to act as a selective force. If we return to measuring the amount of niche overlap between two species, we can imagine that two strongly competing species each impose hefty fitness costs

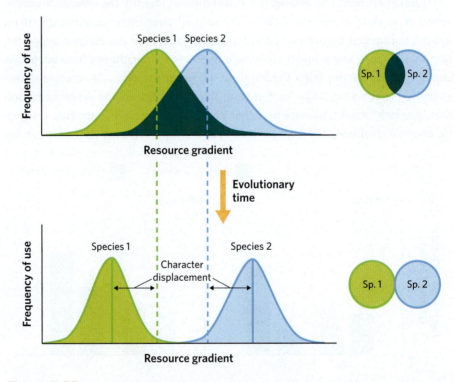

Figure 7.33 In character displacement, a shift in the distribution of heritable traits is caused by competition between two species, because individuals who experience less interspecific competition will have greater access to limiting resources and thus higher fitness. Over time, this evolutionary selective pressure leads to a change in the average trait value in each population (the average trait, or "character," in the population is changed, or "displaced") and a decrease in niche overlap. We can see this shift in the average trait by noticing the way the peak of each frequency distribution shifts. Dashed lines show previous peak values. Note that no new trait variation is created through this process, but instead the trait frequencies within the existing variation change substantially.

on individuals of the other species; therefore, any individual that uses resources in such a way that it experiences a reduction in this niche overlap would have a fitness advantage (**Figure 7.33**). Individuals with traits that reduce overlap will have better access to resources and will go on to leave more descendants, who will also have the traits that allow them to exploit resources that are not shared. Over generations, this selection will lead to a general shift in the average traits found in populations of the two competing species. This trait shift will act to reduce the strength of the competitive interactions between the two species in a phenomenon called **character displacement** (Figure 7.33).

Clearly, an evolutionary response requires enough time for individuals to live, die, and have offspring (at least one generation, usually many more). Not surprisingly, evidence that two species evolved in response to competition for shared (and scarce) resources has been hard to gather. A breakthrough realization came when ecologists demonstrated that evolution can readily be observed even after one generation if one tracks key traits within most, if not all, individuals of a population.

One of the most famous examples of character displacement comes from investigations by Rosemary and Peter Grant (and their long roll call of students), who have spent more than 40 years studying the finches in the Galápagos Islands. One of their many findings was that some species of finches diverged in their beak size when they coexisted on the same island (Schluter et al. 1985). Two of the more common finch species in the Galápagos, *Geospiza fortis* and *Geospiza fuliginosa*, each occur on some of the islands by themselves (e.g., on Daphne Major and Los Hermanos, respectively), and the individuals living on these islands have a distinctive range of beak sizes. For finches, beak size is tightly coupled with the size and thickness of the seeds they can crack and eat. Birds have no teeth or other means of opening seeds (or other foods). Big beaks are useful for cracking larger seeds, and small beaks are useful for smaller seeds. Thus, for birds like Galápagos finches, really small differences in beak size and shape can mean the difference between eating enough to live another day and have offspring, or dying and failing to reproduce. Having the right beak size for the seeds that are available on an island is key to realizing maximum fitness.

To understand the situation better, we need to look at the distribution of beak sizes when the two species of finch live alone on their own islands (allopatry on Daphne Major and Los Hermanos) versus when they coexist on a single island (sympatry on Santa Cruz). Although the largest individuals of *G. fortis* are at such low densities that they are hard to detect on the graph for Daphne Major, the x-axis has the same potential range of beak sizes for a given species across all three graphs in **Figure 7.34**. Any difference in the distribution

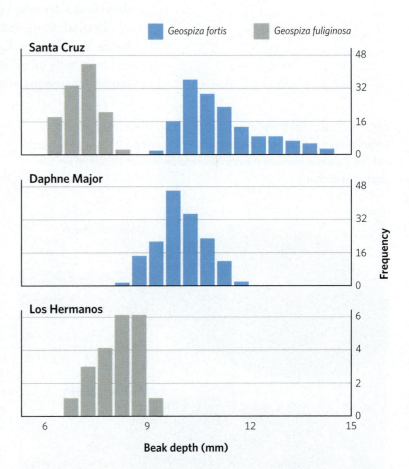

Figure 7.34 This series of graphs shows the distribution of beak sizes between two finch species (*Geospiza fortis* and *Geospiza fuliginosa*) across three of the Galápagos Islands. When the two species are found by themselves on Daphne Major and Los Hermanos islands, significant overlap in their beak size is observed. Yet when the species coexist on the same island (Santa Cruz), their beak sizes shift, producing little to no overlap (Schluter et al. 1985).

of beak sizes is attributable to the individuals on that island specializing on particular seed sizes. When the two species live alone, each on their own island, individuals of *G. fuliginosa* and *G. fortis* concentrate on intermediate-sized seeds and, thus, have a large amount of overlap in the distribution of beak sizes across their populations. With this degree of overlap in their fundamental niche, and based on the experimental and modeling work described thus far, you would expect the two finch species not to coexist if they found themselves on the same island. Yet, they do coexist on Santa Cruz Island. How is this possible?

Looking at the top graph in Figure 7.34, you can see that the distribution of individual beak sizes for each species on Santa Cruz Island has shifted, with the beaks of *G. fuliginosa* becoming smaller and those of *G. fortis* becoming larger. This shift in beak size in sympatry is a product of natural selection. On Santa Cruz, the individuals of *G. fuliginosa* with relatively large beaks competed for available seeds with individuals of *G. fortis* with relatively small beaks. These competing individuals, from both species, suffered lower fitness as a consequence and were more likely to either die or fail to reproduce, removing the alleles that coded for these beak sizes from each population.

In contrast, the individuals who foraged for seeds that were too small or too big for the other species to eat were able to capture more calories, survive better, and have more offspring. The alleles for these beak sizes thus got passed on disproportionately in the next generations. It does not take too many generations for this type of natural selection to effectively minimize the foraging niche overlap between two species in a situation such as this one on Santa Cruz Island. In this case, the Grants and their colleagues detailed this process of evolution in real time. We urge you to delve more deeply into the Grants' amazing body of work (particularly through the book *The Beak of the Finch: A Story of Evolution in Our Time*, by Jonathan Weiner [2014]), as it provides both a fascinating ecological and evolutionary story and insight into the long and varied careers of these two stalwart ecological luminaries.

Another excellent example of evolution in response to interspecific competition comes from a series of studies by David Pfennig and colleagues (2000, 2006) on two related species of pond-breeding toads, the plains spadefoot toad (*Spea bombifrons*) and the New Mexico spadefoot toad (*Spea multiplicata*). The habitats of these two toads overlap in the deserts of the southwestern United States, and the two species compete for resources in the aquatic tadpole stage. Ponds in these deserts are a rare and ephemeral commodity. For either species to thrive, individuals must survive through the tadpole stage, which means they must accumulate resources fast enough to metamorphose before the ponds dry up.

In both species, there are two different physical types, or morphs, of tadpoles. All tadpoles of both species start out with a small, detritus-feeding mouth mor-

Figure 7.35 The physical divergence in body shape in spadefoot toad tadpoles (genus *Spea*) is based on feeding morphology. The carnivore morph eats fairy shrimp, whereas the omnivore morph focuses its diet on benthic organic sediment.

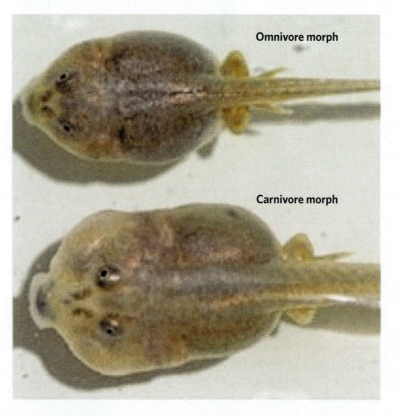

Omnivore morph

Carnivore morph

phology, which is characterized by small-sized jaw muscles. Individuals with this morphology eat a variety of dead material in the benthic sediments (i.e., bottom mud) and thus are referred to as the omnivore morph (**Figure 7.35**). In both species, though, some tadpoles transform into a carnivore morph after eating fairy shrimp. The carnivore morph has larger jaw musculature and a more toothed mouth. Not all tadpoles that encounter fairy shrimp will make this transformation, but once a tadpole has morphed into the carnivore form, it specializes on eating fairy shrimp. Although the omnivore form can eat fairy shrimp from time to time, they mostly eat benthic detritus.

Pfennig and colleagues posit two evolutionary pressures that maintain these morphs in each species when that species is alone. First, by having two morphs that specialize on different food sources, each species can reduce intraspecific competition between tadpole offspring. Second, two morphologies offer two different ways to succeed in an environment with variable and unpredictable water availability. These species need as many of their young as possible to grow quickly before the ponds dry up, and two different morphs lead to two different means to achieving successful maturation. The fairy shrimp morph eats higher-calorie food and develops faster, which can be good in years with little water and shallow ponds. The omnivore morph develops more slowly but tends to metamorphose into an adult at a more opportune date in years with plenty of water. Not surprisingly, the fraction of the tadpoles to transform in each species is not the same in deep ponds as it is for shallow, ephemeral ponds. This divergence appears to be genetically encoded because the difference persisted when tadpoles were reared away from their ponds in controlled conditions.

All of these observations describe the tadpoles of the two spadefoot toad species when they are alone (allopatry) in a pond. But a walk through the desert in the southwestern United States in the rainy season would reveal some ponds with tadpoles of both of these spadefoot species (sympatry). If the two morphs reduce *intra*specific competition, and there is a genetic basis for the morphologies they exhibit, then perhaps *inter*specific competition creates selective pressure that alters the tendency for each of the feeding morphologies.

Tadpoles of both species eat the same foods (detritus for the omnivore morph and fairy shrimp for the carnivore morph), so we know there is overlap in resource use. From the tight time period for growth and metamorphosis, it is clear that these shared resources are limiting, so we have the elements in place for interspecific competition. When each species is alone, it tends to have about 20% of its tadpoles transform into the carnivore morph (**Figure 7.36**). But when the species coexist in the same pond, *S. bombifrons* produces more than 90% carnivore tadpoles, whereas the carnivore percentage for *S. multiplicata* drops to near 5% (Figure 7.36). This result suggests a shift in a heritable trait (feeding morphology) in the presence of the competitor, which implies character displacement.

Figure 7.36 Ecologists examined how the percentage of the population that is carnivorous changed in two species of spadefoot tadpoles in the wild in sympatry (when the species grew together in the same pond) or allopatry (when the species grew in separate ponds). Error bars represent the standard error of the mean (Pfennig et al. 2006).

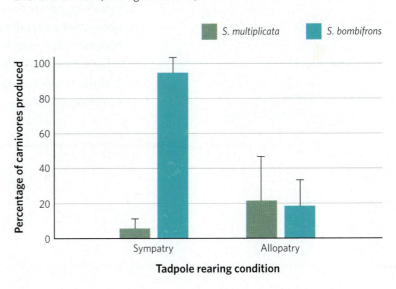

One reason it can be so difficult to observe competition in nature is that individual organisms do everything they can to avoid interactions that produce negative effects. Individuals that do not avoid negative interactions tend to have lower fitness and pass on fewer genes, so avoidance itself is often a trait that confers higher fitness. Character displacement is the evolutionary end result of individuals concentrating on the resources that they acquire best and avoiding the portions of resource space where they must compete.

We call this splitting of resources between competitor species **resource partitioning** because competing species sometimes divide up general resources based on the ones they can access most easily. We see this sort of partitioning in the Galápagos finches (based on seed size) and in the spadefoot toad tadpoles (based on eating detritus or fairy shrimp), so how do we know whether we have an evolutionary response in the form of character displacement or just a temporary shift?

When organisms alter their resource usage in the short term, they reduce their occupied niche space from their fundamental niche to a limited or realized niche, as we mentioned at the start of the chapter. We call this behavioral reduction in niche usage a **niche shift**, or a temporary alteration in the way individuals in a population use resources in the presence of a competitor species.

Individuals with the particular traits that are well suited to local environments are going to excel at acquiring resources. Access to unshared resources should confer an evolutionary fitness advantage because well-provisioned individuals should contribute more genes to the future gene pool. As long as the advantageous traits are genetically heritable, over time they should become more common in the population, leading to a shift in the average trait (or character) of the population. And *that* shift is what we call character displacement. Notice that character displacement is a natural product of resource partitioning and niche shifts, but at its core, the resource shift must have a heritable genetic component.

Behavior-mediated resource shifts can be hard to distinguish from character displacement. To distinguish between them, we can look for a shift in a genetically based trait (such as beak shape or tadpole morphology), or we can try removing the competitor to see if individuals in the population go back to using the shared resources. If the change is genetic, it is much harder for individuals to revert to using previously shared resources because genetic changes occur over generational time. Again, we see why ecologists speak of the *ghost of competition past*, because past competition can shape current interactions. It is hard to identify what traits come from character displacement unless we can find the competing species both alone and in sympatry and are able to observe evolution in action.

7.6 Next Steps

The role that interspecific competition plays in structuring natural communities has been discussed and debated by ecologists for decades. At one point, the general consensus was that competition was the most important process across all of nature (McIntosh 1985), and as a result, we see a lot of theoretical and empirical research on the subject discussed in ecology textbooks. Today, after decades of even more research, most ecologists see competition as one of a suite of forces acting to structure natural communities. In this chapter, we examined the basic tenets of interspecific competition that ecology students should have on their tool belts. Yet the role of competition in communities is often nuanced and subtle, because it sometimes works in concert with other factors, as we explore in Chapters 10, 11, and 12.

A critical feature of interspecific competitive interactions that we have only touched on so far is the shifting nature of the environment and environmental conditions. Because interspecific competition requires a limited resource, any force or action that alters resource levels will, in turn, alter the interaction between competitors. This factor was introduced in the Tilman model, and it is one reason the model is common in the literature today and included in this textbook.

When the environment is altered by factors such as floods, droughts, earthquakes, fires, algal blooms, 17-year cicada emergences, mayfly hatches, oil spills, or forest clear-cutting, competitive interactions that appeared stable and unchanging can suddenly be altered or even curtailed. In some scenarios, and with certain competitors, the presence and effects of interspecific competition are cyclical (i.e., every spring or during peak breeding seasons) or even random (see Chapter 13). This observation does not negate or undermine the importance of competition, but rather it adds a critical level of sophistication and complexity to our understanding of the natural world. This insight also means that, as scientists, we often need a more nuanced approach to our research methodologies and experimental designs in order to tease out the role competition plays in the environment.

SUMMARY

7.1 Competitive Interactions Have a Negative Effect on All Competitors

- Community ecology is the study of multispecies interactions.

- Intraspecific competition is competition between individuals of the same species for resources.

- Interspecific competition is competition between individuals of two or more different species for resources.

- Interspecific competition always has negative effects on both species.

- Resource, or exploitative, competition is interspecific competition over a shared resource.

- Interference competition is interspecific competition involving direct interaction (aggression or behavioral displays) between species, resulting in fitness consequences.

7.2 Competition and Niche Space

- Gause studied interspecific competition using two species of paramecia.

- He found that when grown together, both species experienced negative fitness consequences.

A Conceptual Model of Competition

- Resource utilization curves help us to conceptualize and understand the process of interspecific competition.

- The amount of overlap between two species' resource utilization curves indicates the relative strength of the competitive interaction. More overlap means greater competition.

- The fundamental niche is the full range of conditions and resources that a species needs to survive and reproduce; a realized niche is a reduced niche space that is available in the presence of a competitor.

- The competitive exclusion principle says no two species can occupy the same niche space. If they do, one of the species is driven to extinction.

- The concept of limiting similarity explores how similar two species can be in their resource use and still coexist.

7.3 Conditions for Coexistence

The Lotka-Volterra Model

- The core of the Lotka-Volterra model is the logistic growth equation, and it uses the same nomenclature.

- Two competing species require two equations linked by resource use.

- The key terms in these equations are the competition coefficients (α and β) that "translate" individuals of one species into individuals of the second species.

- Phase-plane graphs represent the densities of both species simultaneously, with one on each axis, allowing the examination of dynamic population density of two linked populations.

- Isoclines (where population density does not change) are used on phase-plane graphs to explore the potential outcomes of competitive interactions and the conditions under which they occur.

Outcomes of Competition in the Lotka-Volterra Model

- Competitive exclusion occurs when the two isoclines do not intersect. The superior competitor's isocline sits above that of the inferior competitor.

- Competitive exclusion is similar to what Gause observed in his paramecia experiment.

- Stable coexistence between species occurs when each species has less effect on the other species than it has on itself. This leads to a stable equilibrium point, represented on a phase-plane graph by the point of intersection of the two species' isoclines.

- Unstable coexistence (also called priority effects) between two species occurs when each species competes more strongly with the other species than it does with itself. In this situation, the initial population densities determine the competitive outcome of the interaction.

A Few Issues with the Lotka-Volterra Model

- The continuous-time nature of the model means the responses of the species are instantaneous, which is not realistic in the natural world.

- The parameters in the model, such as the carrying capacity (K), are assumed to be constant, allowing for a smooth movement toward equilibrium; but in the real world, such parameters are variable, and this assumption is often violated.

- Measuring and quantifying competition coefficients in the real world is challenging.

Insights from the Lotka-Volterra Model

- The model provides a basis for both competitive exclusion and coexistence.

- It suggests that stronger competitors should always outcompete weak competitors, regardless of their population densities, which is testable.

- The model also suggests situations in which stable coexistence is the norm, and these conditions match what we see in nature.

7.4 A More Mechanistic Model with a Focus on Accessing Resources

- The Lotka-Volterra model does not suggest a mechanism for how competitive interactions occur or what happens when there are more than two resources.

The Tilman Model, or the Resource-Ratio Hypothesis

- The model is based on Tilman's experiments with freshwater diatoms that require two resources for positive population growth, silica and phosphate.

Zero Net Growth Isoclines and R*

- The graphical solution of the Tilman model is based on the idea of a zero net growth isocline (ZNGI) and the concept of R^*.

- For every resource, there is a threshold resource level (R^*) below which positive population growth is prevented. These thresholds define the ZNGIs for individual species.

- The Tilman model plots ZNGIs for two resources, X and Y (on x and y axes, respectively).

- The R^*s are the threshold levels along the two-resource axes. The point where the two ZNGIs intersect creates an envelope that defines positive population growth.

- With two species, there are two different sets of ZNGIs because each species has its own resource requirements.

The Resource Supply Point

- The environment naturally renews resources and removes them, producing an equilibrium resource availability called the resource supply point (S).

Consumption and Renewal Vectors

- The rate at which individuals use up or consume their resources draws resource levels below the natural resource supply point.

- The way that each species consumes resources is represented as a consumption vector on the two-resource graph.

- Even with consumption by competitors, environmental factors will tend to move the system back toward the resource supply point; this movement is represented by a renewal vector.

The Full Model and Outcomes of Competition

- When one species' ZNGIs completely envelop (i.e., are lower than) the other species' ZNGIs, the model predicts competitive exclusion because the first species draws resource levels below the physiological tolerances of the second species.

- Competitive exclusion (either species can competitively dominate) is predicted when the two ZNGIs cross, and S is at levels where one of the species can draw one of the resources down below the physiological tolerance of the second species.

- Stable coexistence is predicted when the two species' ZNGIs cross and S is at levels where neither species can draw resources down to exclude the other.

- Unstable coexistence occurs when the two species' ZNGIs cross, and each species shows inefficient use of the resource for which it is a superior competitor. Even if S lies in a region that appears to have adequate resources for both species, the first to arrive draws resources down to a level where the other species cannot survive.

- The Tilman model provides a mechanistic explanation based on resource densities for the same competitive outcomes seen in the Lotka-Volterra model.

7.5 Long-Term Competitive Outcomes

Extinction

- Extinction is easy to understand but, logically, is not the most common end result of interspecific competition because, as we know, the natural world is filled with coexisting species. If extinction were the norm, we would see very few coexisting species.

- Extinction from interspecific competition does occur but is hard to observe. The ghost of competition past is the idea that we can see the effects of competition in extinction, but we cannot see the interaction because the competitors are no longer interacting.

Exclusion

- When one species is restricted from a location because of interspecific competition, we call this exclusion.

- This phenomenon is documented in the natural world.

Evolution

- Interspecific interactions that are strong enough to affect fitness in a heritable trait can cause evolutionary changes.

- When two species' niches overlap, interspecific competition can cause a shift in one or both species' resource use and thereby decrease niche overlap. This is called character displacement and is observed in natural examples.

- When species split or divide resources to avoid or lessen competitive overlap, the process is called resource partitioning.

7.6 Next Steps

- Ecologists initially thought that interspecific competition was the most important force structuring communities but now see it as one of many forces.

- Environmental fluctuations and forces of change can alter and reshape competitive outcomes and are important factors in our understanding of ecological communities.

KEY TERMS

allopatry
character displacement
community ecology
competition coefficient
competitive exclusion principle
consumption vector
exclusion
extinction
fundamental niche

ghost of competition past
interference competition
interspecific competition
isocline
niche shift
phase-plane graph
R^*
realized niche
renewal vector

resource competition
resource partitioning
resource-ratio hypothesis
resource supply point (S)
resultant vector
stable coexistence
sympatry
unstable coexistence
zero net growth isocline (ZNGI)

CONCEPTUAL QUESTIONS

1. Identify which of the following situations involve competition. If there is competition, indicate if it involves interference competition or resource competition:

   a. Garlic mustard (*Alliaria petiolata*) is a non-native species spreading through New England. It takes over new areas by killing native plant species with chemicals that it exudes from its roots.

   b. Lions attack a pair of hyenas and steal a gazelle that the hyenas have killed.

   c. Lions enter a new area, and their predation lowers gazelle population densities. Subsequently, hyena densities in the same area decrease. The lions and hyenas never encounter each other.

   d. Male frogs call from breeding ponds in the spring, and predators home in on these sounds, often capturing and consuming both male and female frogs.

   e. In the western United States, ground squirrels and tiger salamanders will use the same burrows at different times of the day or year, with tiger salamanders out in breeding ponds at night and in the wet spring when ground squirrels are in burrows.

2. If fruit-eating bird communities on South Pacific islands are structured by (i.e., are strongly affected by) interspecific competition,

   a. How much rotten fruit will there be on the ground?

   b. What will happen to fruit densities (increase, decrease, stay the same) and resource utilization curves (wider, narrower, stay the same) if a new competitor invades?

   c. What will happen to fruit densities and resource utilization curves if a native frugivorous (fruit-eating) bird goes extinct?

   Be sure to explain your answers to all parts of this question.

3. The idea of the ghost of competition past is tantalizing as an explanation for current patterns in species distribution and resource use. Yet, at the same time, it is also rather toothless, as it can often represent a sort of "just-so story" of the past. Can you envision and describe a situation in which a ghost-of-competition-past explanation is fairly robust and perhaps testable?

4. One by-product of humans introducing non-native species to an area is interspecific competition with native species. Explain how this might occur using an example of a non-native species in your local area or one you know something about. Try to describe the type of native species the non-native might compete with and the possible resource or resources that might be involved.

5. Discuss three pros and three cons of the Lotka-Volterra competition model.

6. Can you envision and describe a scenario involving interspecific competition in which the Tilman model would not apply? Can you explain the factors that make it hard or impossible to apply the model to your example?

7. Common mynah birds have been introduced to Australia, and they use nesting sites similar to those preferred by rosellas (a type of Australian parrot). Mynahs will perch on rosella nests and harass breeding rosellas, occasionally pecking at incubating eggs. Rosellas' preferred nesting sites are tree hollows, but they will avoid hollows in eucalyptus trees when mynahs are in the area. Explain how the following terms fit or do not fit this situation:

   a. character displacement

   b. niche shift

   c. fundamental niche

   d. realized niche

   e. resource partitioning

QUANTITATIVE QUESTIONS

1. Species 1 and 2 compete, with $K_1 = 100$ and $K_2 = 50$. Using the Lotka-Volterra competition equations, express the outcome of competition between these species, and draw the phase-plane graphs for the following sets of parameters.

   a. $\alpha = \beta = 0.25$

   b. $\alpha = 0.75$, $\beta = 0.50$

2. Explain what happens to the strength of competition with a niche shift. Explain what happens to the strength of competition with character displacement. How do the two outcomes differ? How would you distinguish between them? Use either the Lotka-Volterra model or the Tilman model to explain your answer.

3. Demonstrate how stable coexistence can occur under the Tilman model. Can you explain in words how this model demonstrates a mechanism for interspecific competitive interactions?

4. For each phase-plane graph shown in **Figure 7.37**, state whether the species coexist. If they do not, who survives?

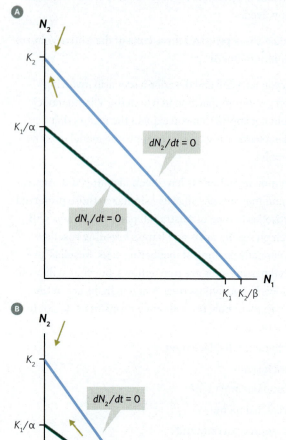

Figure 7.37

5. Suppose that mosquito wing lengths determine the height at which adult mosquitoes fly and encounter blood meals. Longer-winged mosquitoes can fly high and feed on birds, and shorter-winged mosquitoes fly low to the ground and feed on mammals. **Figure 7.38** shows how mosquito species A (blue bars) is found alone in location 1, and mosquito species B (green bars) is found alone in location 2. In location 3, both species are found. Based on these data, would you expect the mosquitoes in location 2 to be feeding more on birds or on mammals? Why? Explain why the two species have similar wing length morphology when their distributions do not overlap but different morphology when they co-occur. If a third mosquito species arrives in location 3, how might it feed in order to coexist with species A and B?

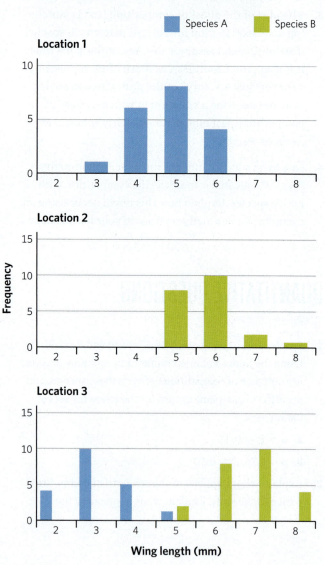

Figure 7.38

SUGGESTED READING

Gause, G. F. (1934) 2019. *The Struggle for Existence: A Classic of Mathematical Biology and Ecology*. Mineola, NY: Dover. First published by Williams and Wilkins, Baltimore.

Hardin, G. 1960. The competitive exclusion principle. *Science* 131(3409): 1292–1297.

Park, T. 1948. Interspecies competition in populations of *Tribolium confusum* Duval and *Tribolium castaneum* Herbst. *Ecological Monographs* 18(2): 265–307.

Schluter, D., T. D. Price, and P. R. Grant. 1985. Ecological character displacement in Darwin's finches. *Science* 227(4690): 1056–1059.

Tilman, D. 1977. Resource competition between planktonic algae: An experimental and theoretical approach. *Ecology* 58(2): 338–348.

A raft spider (*Dolomedes fimbriatus*), common to wetlands across northern Europe, has caught and begun to consume a three-spined stickleback (*Gasterosteus aculeatus*). Raft spiders are interesting arachnid predators in that they hunt by gliding on the surface of the water and find prey by sensing their aquatic vibrations.

chapter 8

EXPLOITATION

8.1 Everybody Has to Eat

We hope you have eaten something today. Whether your meal was big or small, vegetarian or not, you—like all living organisms—must take in energy in order to survive. For species that photosynthesize (autotrophs), energy intake involves the capture of sunlight. For the rest (heterotrophs), the need for energy requires the capture, consumption, and assimilation of energy contained in another organism's tissues. This need for energy creates ongoing attack-and-defense interactions between the eaters and the eaten. The natural world itself has been characterized as "red in tooth and claw," meaning that deadly struggles for existence are all around us.

From sea anemones that feed on tiny plankton in the deepest ocean trenches to polar bears that hunt seals at the North Pole to caterpillars that chew the leaves of the green-bean plants in your garden, individual species around the world eat or parasitize one another. The most general term for these interactions is **exploitation**, which we define as an interaction in which the individuals of one species increase in fitness by consuming individuals (or parts of individuals) of another species, who experience a decrease in fitness. In Chapter 7, we indicated the negative-negative interaction of competition with the shorthand (−, −), and we can use (+, −) as an analogous representation of the lopsided nature of exploitation.

From the above examples of exploitative interactions, it should be clear that "red in tooth and claw" may be an overly dramatic depiction of exploitation, except in the case of animals eating animals. Certainly, there is not much of a "struggle," and hopefully little "blood," involved in our "catching" and eating a green salad for lunch. Whether the struggle results in the loss of blood or chlorophyll, though, the ecological and evolutionary dynamics of exploitation are central to our understanding of nature.

Exploitation comes in many forms, but the common factor among them is that individuals of one species benefit by consuming individuals of another species (or parts of them). The most familiar form is *predation*, in which a **predator** kills and

LEARNING OBJECTIVES

- Define exploitation, identify the simple model of the interaction, and distinguish among the three main forms of exploitation.

- Differentiate and compare the four possible outcomes of exploitation.

- Use phase-plane graphs to explore outcomes of exploitative interactions, and evaluate the conditions under which they occur.

- Explain how adding density dependence for the prey affects the stability of the Lotka-Volterra model.

- Differentiate among the three types of exploiter functional responses.

- Explain how Allee effects influence the Lotka-Volterra model.

- Differentiate among primary, secondary, and tertiary defense strategies.

- Explain how the evolutionary responses of exploiters can lead to coevolution of predators and prey.

(A)

(B)

(C)

Figure 8.1 Not all of the feeding interactions in a stream system are predator-prey or exploiter interactions. (A) A Chinook salmon (*Oncorhynchus tshawytscha*) acts as a predator when it feeds on aquatic invertebrate prey such as (B) caddisflies (order Trichoptera). (C) Caddisflies consume fallen leaves that have no fitness relevance for their source tree. Consumption of these fallen leaves qualifies as detritivory, not exploitation or predation.

consumes **prey**. This is the type of exploitation often seen on nature programs or when your cat brings you a half-eaten mouse. Generally, a predator will consume many different prey individuals in its lifetime. This form of exploitation is so common and familiar that *predation* is often used as a synonym for *exploitation*, even among ecologists, and *prey* is the common term for the loser in an exploitative interaction.

Another common form of exploitation is *parasitism*, in which **parasites** obtain energy from their **hosts** by attacking the host's body or organs and consuming tissues or fluids, often without killing the host. This category is quite broad and includes disease organisms (or **pathogens**) and parasites that can live either inside (*endoparasites*) or outside (*ectoparasites*) their hosts. Parasites are generally smaller than their hosts (although a tapeworm stretched to its full length would usually be much "taller" than its host), and each parasite tends to use only one host for its whole life, or at least for one whole stage of its life. Examples of parasitic exploiters include mosquitoes, tapeworms, bacteria, and many others.

Finally, we have *herbivory*, in which the exploiters consume plants, either whole plants or select parts of them. Like parasites, **herbivores** do not typically kill the plants they eat. Like predators, though, they can (usually) eat many individual plants (or the parts of many individual plants) in their lifetime. Rabbits, snails, goats, and grasshoppers are great examples of herbivores. We explore all three types of exploitation in this chapter. In whatever form it takes, exploitation is an extremely strong ecological force that produces a suite of fascinating community dynamics and evolutionary adaptations.

Terminological Issues

Think for a moment about a lion competing with a hyena for a zebra carcass, and notice that the "shared resource" in this interspecific competition is also "prey" for both the lion and the hyena. In Chapter 5, we defined a resource as a niche element that can be used up, so this overlap between resources and prey makes some sense. But we want to caution that not all resources are prey. Some resources are not biotic and, therefore, are not ingested; examples include light and space in the case of plants, and shelters, nests, or burrows in the case of many animals. Even if we focus on ingested, biotic resources, there are many types of exploitation. The common thread for all of them is that the consumer individual receives a fitness benefit, and the consumed individual experiences a decrease in fitness. For this to hold true, the prey must be alive when the consumer encounters them.

For example, when juvenile Chinook salmon (*Oncorhynchus tshawytscha*) feed on caddisfly larvae (order Trichoptera), the salmon are the exploiters and the caddisflies are the prey, and the interaction clearly fits the definition of exploitation (**Figure 8.1A** and **Figure 8.1B**). When a caddisfly larva consumes leaves that have fallen into a stream, though, it is eating bits of vegetative material that no longer have any bearing on the fitness of the plant that produced the leaves (**Figure 8.1C**). The leaves are already separated from the plant and cannot contribute to its growth, reproduction, or survival. The way caddisflies consume leaves is better defined as *detritivory* (or scavenging). In the same manner, the *saprophytic* fungus (*saprophytic* means obtaining food by absorbing dead organic matter) that produces the jack-o'-lantern mushroom (*Omphalotus olearius*; **Figure 8.2**) does not fall under the rubric of exploitation because, although it is consuming organic matter, that material is already dead. We would call the fungus a **detritivore**, just as we would the caddisfly.

At times, this issue with terminology can become more challenging. For example, consider a scenario in which a gray wolf (*Canis lupus*) attacks and kills an old and enfeebled elk (*Cervus canadensis*) that is past its reproductive age. One could argue that this is not technically exploitation because there is no evolutionary fitness effect on the elk, if we define fitness strictly as the ability to contribute to the gene pool of future generations. The problem in this situation is that we cannot always know that an elk is past reproductive age. We also cannot know whether the elk may still have some fitness advantage to offer by protecting young relatives who could go on to reproduce. Partially because of these complications, when defining exploiters and prey, ecologists usually focus on whether the consumed resource is alive (or part of a living organism) and not on its reproductive status. This assumption works pretty well in practice when defining what counts as prey because many organisms in the wild do not live much past reproductive age.

Figure 8.2 Jack-o'-lantern mushrooms (*Omphalotus olearius*) are the fruiting bodies of a saprophytic fungus that breaks down dead organic matter in the soil.

8.2 Possible Outcomes of Exploitation

The (+, −) schematic works as a simple conceptual model and suggests patterns that we might expect to see as a result of exploitation. First, we expect exploitation to lead to increases in the population density of the exploiter species and decreases in the population density of the exploited or consumed species. These predicted changes seem fairly obvious, but let's explore some examples to see whether these patterns emerge.

The mountain yellow-legged frogs of California are a species complex (recently identified as two species rather than one: *Rana muscosa* in the south and *R. sierrae* in the north) that was once very common but which experienced a precipitous range-wide population decline in the twentieth century. In fact, mountain yellow-legged frogs have decreased in abundance so much that both species were listed in 2014 as federally endangered under the US Endangered Species Act.

The high mountain lakes where the tadpoles of these frogs develop are naturally fishless. But in the early 1900s, significant efforts were made to introduce trout to these lakes in order to entice tourists and anglers to the area. Roland Knapp and Kathleen Matthews wondered if the observed decline in frog densities might be a response to the introduced, non-native rainbow trout (*Oncorhynchus mykiss*). In earlier work, Knapp and Matthews observed that rainbow trout will eat mountain yellow-legged frog tadpoles and occasionally even consume adult frogs.

To test their hypothesis, they examined the abundance of frogs and trout in a series of small lakes in the Sierra Nevada (Knapp et al. 2001). **Figure 8.3** shows the data they collected for adult frogs and trout from 22 lakes located within a mountain basin. Notice that lakes with many trout have few frogs, and lakes with few trout have many frogs. Does this pattern make ecological sense given the prior

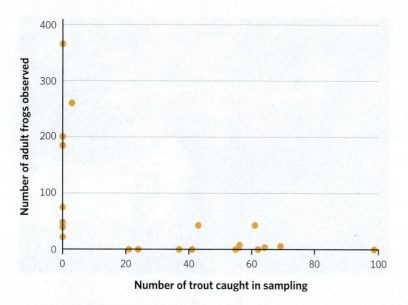

Figure 8.3 The abundance of adult mountain yellow-legged frogs (*Rana muscosa* and *Rana sierrae*) versus the number of rainbow trout (*Oncorhynchus mykiss*) caught across 22 lakes in the Sierra Nevada, California (Knapp et al. 2001).

Figure 8.4 The abundance of mountain yellow-legged frog larvae in never-stocked, stocked-now-fishless, and stocked-fish-present lakes (533 lakes in total), measured as the number of larvae per meter of shoreline and expressed as $\log_{10}$(abundance + 1); the bars indicate a +1 standard error (Knapp et al. 2001). NS indicates no significant difference, and $P < 0.001$ indicates a difference that is very significant.

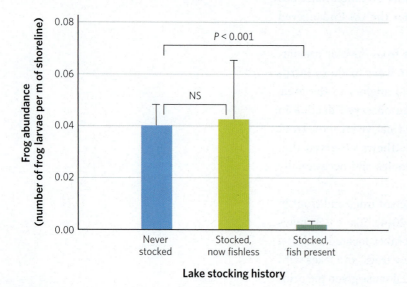

observations of predation? It definitely seems to support the prediction that exploiter populations would increase in abundance, and exploited (prey) populations would decrease. But the data are only a snapshot in time and do not clarify how the two populations might change in response to one another.

To break this down, if the trout eat frogs, and the frogs are a necessary trout resource, would we expect to see high densities of trout where frog densities are low? Shouldn't low frog densities lead to low trout densities? Or would high trout densities lead to low frog densities? We have a chicken and egg problem. How are these two species going to affect *each other*?

We need a way to think about effects that go back and forth. With interspecific competition, this was easier to conceptualize because the negative effects worked in both directions and led to declines in population abundance for both species. But with exploitation, it is more complicated. If we apply the Tilman model, we can imagine that the trout drive frog population abundances down to the lowest level that can support trout (R^* for trout using frogs as a resource). In other words, the exploiter species should drive the prey abundance down to a consistently low level, but only if there is a consistent resupply (S, or resource supply point in the Tilman model) of prey (i.e., frogs). If the trout eat frogs faster than frogs can reproduce, then perhaps trout will just drive frogs extinct.

We can see a more general pattern if we look at trout and frog population abundances collected across 533 lakes in the John Muir Wilderness and the Sequoia and Kings Canyon National Parks, also in the Sierra Nevada (Knapp et al. 2001; **Figure 8.4**). Knapp and Matthews divided the lakes into three categories: (1) never stocked with trout ("never stocked"), (2) stocked with trout historically but without a current trout population ("stocked, now fishless"), and (3) presently stocked with trout ("stocked, fish present"). From these data, it is clear that frog (prey) population abundance declines to extremely low levels in the presence of trout. This evidence strongly suggests that trout lead to declines in frog population abundance and possibly even to frog extinction (notice in Figure 8.3 how many lakes with trout have no frogs).

It is not unusual for predators to cause local extinction of their prey. Georgii Gause (introduced in Chapter 7) carried out predation experiments in which the prey were *Paramecium caudatum* and the predators were a ciliate (*Didinium nasutum*) that eats paramecia almost exclusively (Gause [1934] 2019). In his first experiment,

Gause let populations of *P. caudatum* grow for a few days (feasting on bacteria) in closed, experimental flasks and then introduced the *D. nasutum* predators into the flasks. In all experimental replicates, the predators quickly reduced prey numbers, eventually driving the prey population to extinction. This outcome fits what we saw with the frogs and the trout. But ciliates and paramecia population dynamics progress much faster than population dynamics of trout and frogs, so Gause saw a next chapter to the story. The extinction of the paramecium left the *Didinium* predators without food, and the *D. nasutum* population eventually also went extinct in every one of his experimental replicates (**Figure 8.5**).

Is this outcome of prey extinction followed by predator extinction typical of all exploitative interactions? Should we expect eventually to see both frogs and trout go extinct in the Sierran lakes? It makes sense that exploiters can overconsume prey and drive them extinct and then go extinct themselves, but is this the only possible outcome? If it were, we would see almost no exploitative interactions in the natural world, as all the participating species would have gone extinct by now. Because the world is demonstratively full of exploiters and prey, there must be some unrecognized factors that allow coexistence but that do not appear in Gause's experiment.

Gause eventually recognized his omission and added a level of spatial complexity to his experimental design and reran the experiment. This time, he added tiny bits of oats to the system to serve as "refuges" for the paramecia (they hide from the *Didinium* on and around the oats). With this addition, the predator species went extinct because prey density got low enough for predators to have trouble finding them. The prey did not go extinct, however, because they had a refuge (**Figure 8.6A**).

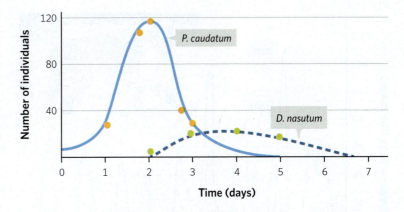

Figure 8.5 In Gause's first set of predation experiments, the population of the prey (*Paramecium caudatum*) was driven to extinction by the predator (*Didinium nasutum*) in a homogeneous microcosm without immigration. Once there were no more prey to eat, the predator also went extinct (Gause [1934] 2019).

Figure 8.6 Gause's subsequent experiments demonstrated **A** how a predator population (*D. nasutum*) goes extinct when its prey (*P. caudatum*) have a refuge, and **B** how prey and predator populations cycle together when the environment is homogeneous with immigration (or steady introduction) of prey (Gause [1934] 2019).

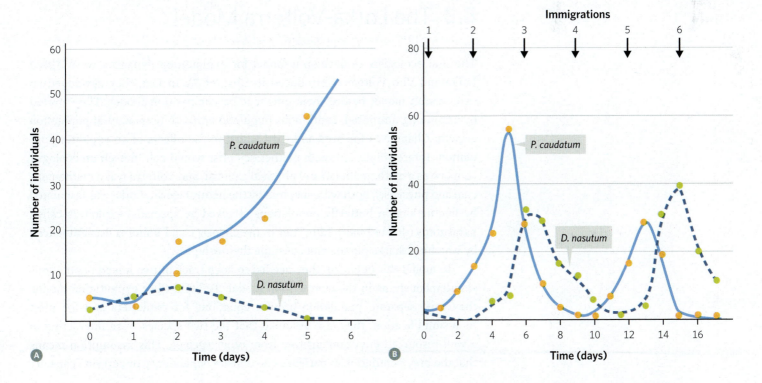

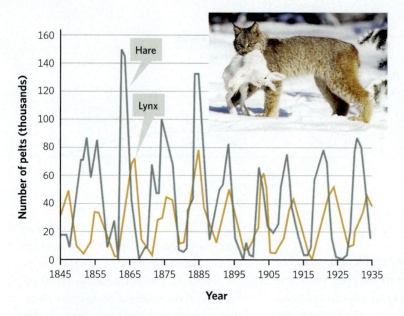

Figure 8.7 Note the tight coupling and cyclic nature of Canadian lynx (*Lynx canadensis*, a medium-sized wild cat) and snowshoe hare (*Lepus americanus*) populations. Data are from North American trapping records, and the number of animal pelts was used as a surrogate for population abundance (MacLulich 1937). Inset: a lynx with a captured hare.

In a third set of experiments, Gause added new prey individuals (i.e., immigration) every few days throughout the experiment. Under these conditions, the predators and prey began a complex cycle in which, as prey numbers decreased, predator numbers also dropped (**Figure 8.6B**). When predator population abundance dropped, the prey populations rebounded, causing a corresponding increase in predator numbers. The resulting **exploiter-prey cycles** were fairly regular, evenly spaced, and steady through time. But do we see these cycles outside of test tubes or in systems where there is no artificial immigration of prey? In fact, these sorts of cycles seem *more* common outside of test tubes and show up in several exploiter-prey systems (such as that of the lynx and snowshoe hare; **Figure 8.7**) without external additions of prey.

Lastly, ecologists and land managers in a number of ecosystems around the world have observed exploitative interactions with predator and prey populations that do not cycle. Fluctuations in weather or other external variables may lead to large changes in the abundance of either population but without cycles or imminent risk of extinction. Common examples include raptors (predatory birds) and their prey species since the banning of the pesticide DDT, coyotes and rabbits in much of North America, and spiders and flying insect prey.

From these examples, we can see four possible outcomes of exploitation: prey extinction that leads to exploiter extinction, exploiter extinction with surviving prey, exploiter-prey cycles, and stable coexistence. In the next section, we use a model to explore the ecological conditions leading to these four outcomes.

8.3 The Lotka-Volterra Model

The first ecologists to develop a model for exploitation dynamics were Alfred Lotka and Vito Volterra (introduced in Chapter 7). In fact, they developed an exploitation model before their interspecific competition model. They started by examining unlimited, continuous population growth (exponential population growth; Chapter 6) for prey and then added in the effects of an exploiter population. Interestingly, although neither scientist would call himself an ecologist (Lotka was a mathematician and physical chemist, and Volterra was a mathematician and physicist), both were inspired by the natural world. Lotka was fascinated by fluctuations in butterfly populations attacked by specialist exploiters called parasitoids (defined more fully later in the chapter), and Volterra was influenced by watching sharks pursue and consume their fish prey.

To model this behavior, they needed to keep track of two linked populations simultaneously, as in the competition model. To build their exploitation model, they used separate population growth equations for exploiters and for prey. As a simplification, they also assumed that the two species were interacting in a well-resourced environment free from other species. This assumption means that the prey population experiences no density dependence or carrying capacity and that the only factor that keeps the prey population from growing to infinity is exploitation. For the exploiter, their only resource is the prey, and population growth depends on capturing and consuming prey individuals. Lotka and Volterra also used the simplest possible assumption about consumption rates—namely, that exploiters capture a constant fraction of the prey they find.

The Prey Equation

Let's explore their model, starting with the prey. The model for the prey needs an intrinsic rate of population growth that reflects natural birth and death rates without the influence of an exploiter species. We can, therefore, begin with an exponentially growing prey population (Equation 6.11C) and then add the exploiters as an additional factor causing death. Let's name the quantities we want in the model and express the equation with words.

$$\begin{pmatrix} Change\ in\ Prey \\ Population\ Density \end{pmatrix} = \begin{pmatrix} Exponential \\ Population\ Growth\ of\ Prey \end{pmatrix} - \begin{pmatrix} Number\ of\ Prey \\ Lost\ to\ Exploiters \end{pmatrix} \quad (8.1)$$

Before we convert this to a mathematical expression, it is worth noting some of the important aspects of each element in the equation. First, population growth equations always explore population growth over some chunk of time, which is not explicitly stated in Equation 8.1. Because we start with the exponential, continuous population growth equation, the left-hand side of the equation represents the change in the population in the smallest possible slice of time.

As we saw in Chapter 6, a population grows at a faster rate when there are more individuals in the population that can reproduce. Thus, the growth of the prey population depends on its intrinsic rate of increase (r) and the prey population density (N_p). Notice that, as in Chapters 6 and 7, we are using the term *density* instead of *abundance* for N values to remind ourselves that these organisms are in a specific place that has a set area, and the interactions will alter as the density of organisms in the area changes. Because we are working with two different species with different population densities in this model, we identify the prey density with a subscript letter p. The population density and intrinsic rate of increase were the only components of the exponential growth equation, and they show up again here in the first element on the right-hand side of the equation. Although N_p has no time associated with it, r is the instantaneous per capita growth rate (with a unit of number of individuals born per individual per unit time), so the units are still consistent (number of individuals per unit time).

For the exploitation element in Equation 8.1, we need to consider two things: how often the exploiters and prey encounter each other and the average number of prey individuals consumed in each encounter. In Chapter 5, we discussed some of the dynamics of how exploiters (or foragers) find and choose prey when they can choose from several prey species. We mentioned that search costs reflect the density of the prey on the landscape and that predators choose prey based partially on how easy it is to find them. In that discussion, the forager population was at the same density whether the forager (the coyote) was searching for mice or tortoises. Here, we want the population densities of the exploiter and the prey to be able to vary, but we know from optimal foraging theory that density across the landscape will affect how likely exploiter and prey are to bump into each other.

Figure 8.8 Prey capture rates vary tremendously among organisms. Ⓐ A whale shark (*Rhincodon typus*) can consume millions of planktonic organisms as it swims around with its mouth open. Ⓑ A Burmese python (*Python bivittatus*), on the other hand, will not have to eat again for a long time after consuming a young alligator (*Alligator mississippiensis*).

Lotka and Volterra also noted the importance of density in determining encounter rates for exploiters and prey. Both men turned to their understanding of physics and chemistry to observe the way that molecules in an enclosed space bump into each other. The speed of a chemical reaction is dependent on the concentrations of the chemicals, because a higher concentration means the chemical elements are more likely to randomly bump into each other. We call this *mass action*, or the law of mass action. Mass action offers a good first estimate of exploiter-prey encounter rates. It seems reasonable that organisms "bump" into each other based on their density in a habitat. We can, therefore, multiply the densities of the exploiters (N_e; note the subscript *e*, for "exploiters") and the prey (N_p) to get an estimate of the number of encounters.

But how many of those encounters result in consumed prey? Optimal foraging theory tells us that consumers may successfully capture prey in only a small portion of their encounters with prey individuals, and that the time needed to handle or process the prey depends somewhat on how hard that prey item is to find, catch, swallow, or remove from its protective shell. This suggests we want to incorporate some information about the exploiter's mode of capturing and eating prey.

Some predators can eat many prey individuals at once, as in the case of whale sharks (*Rhincodon typus*), which feed on millions of plankton just by opening their mouths and swimming in a straight line (**Figure 8.8A**). Other predators eat only a single prey individual at a time, with each predation event occurring once every few weeks, as in the case of non-native Burmese pythons (*Python bivittatus*) and their alligator prey (*Alligator mississippiensis*) in the Florida Everglades (**Figure 8.8B**). In our model, we can represent this combination of capture success and eating mode as the **capture rate (*f*)**. As in the competition equation, this term functions as a conversion factor and helps to make the units consistent because we are interested in the proportion of encounters that are successful (i.e., proportion per predator per chunk of time). The actual number of prey captured will equal the capture rate multiplied by the number of encounters between exploiters and prey.

Putting all this together, we get the following mathematical expression of our conceptual model:

$$\frac{dN_p}{dt} = rN_p - fN_pN_e \tag{8.2}$$

The term dN_p/dt is the instantaneous change in prey population density over continuous time; r is the intrinsic rate of increase for the prey population; N_p is the prey population density; N_e is the exploiter population density; and f represents the capture rate. Although this equation describes prey dynamics (which we know because the N_p is on the left-hand side of the equation), notice that both N_p and N_e are present, just as we saw both species represented in each of the Lotka-Volterra competition equations. Also note that the term on the far right ($-fN_pN_e$) places limits on the prey's population density as prey density increases.

The Exploiter Equation

The situation is a little different for the exploiter equation because the exploiter population will not grow at all without prey. (Remember, we are starting by assuming that our prey species is the only source of food for the exploiter.) So, let's build the exploiter equation the way we built the population growth equations in Chapter 6, by thinking about births and deaths.

$$\begin{pmatrix} \text{Change in Exploiter} \\ \text{Population Density} \end{pmatrix} = \begin{pmatrix} \text{Births Possible from} \\ \text{Prey Consumption} \end{pmatrix} - \begin{pmatrix} \text{Natural Deaths} \\ \text{of Exploiters} \end{pmatrix} \quad (8.3)$$

As with the prey equation, we are considering the instantaneous rate of population growth for the exploiter, so all the changes in exploiter numbers occur in the smallest possible slice of time.

The right-hand side of Equation 8.3 looks like the conceptual models for population growth in Chapter 6, so we can handle deaths the same way we did there. Deaths are caused by a range of natural factors, some of which may have nothing to do with the exploiter-prey system. Therefore, deaths should simply be equal to a per capita death rate (d) multiplied by the population density of exploiters (N_e).

It would be great if exploiter births worked the same way, but we have a bit more information for the exploiter than we had for the populations in Chapter 6. We know that each exploiter individual can only reproduce if it consumes enough food, but exploiters cannot convert every prey item they consume into new exploiter offspring. For example, when a red fox (*Vulpes vulpes*) eats a rabbit, it does not immediately give birth to a new kit (a baby fox; **Figure 8.9**). In fact, it may take a whole season of eating rabbits for a red fox to accumulate enough resources to produce a litter of kits. Nonetheless, we can imagine that a cumulative amount of food (or rabbit prey) is necessary for reproduction and that more food leads to more reproduction.

Figure 8.9 Ⓐ A red fox (*Vulpes vulpes*) eating a single rabbit does not translate directly into reproductive output. Ⓑ It takes the consumption of many prey and the passage of several months for a red fox mother to produce a litter of kits.

Based on this reasoning, we can even come up with a conversion factor, which we call the **exploiter conversion factor (c)**, that allows us to translate consumed food into an expected number of offspring. We can determine the births in the exploiter equation by multiplying the number of captured prey individuals from the prey equation (fN_pN_e) by c to convert consumed prey into expected births of exploiters. In other words, the birth rate for exploiters is dependent on prey density (N_p) and the exploiters' ability to capture (f) and convert prey into offspring (c). Note that we dropped the negative sign because the captured prey have a negative effect on the prey population growth but a positive effect on the exploiter population growth.

Putting all this together, we get the following equation for population growth in the exploiter species.

$$\frac{dN_e}{dt} = cfN_p N_e - dN_e \tag{8.4}$$

If we compare this equation to exponential population growth in Chapter 6 (Equation 6.11B), the cfN_p term is like the instantaneous per capita birth rate, b, but it is not a constant the way b was. Here, the birth rate is dependent on the density of prey.

In Chapter 6, we could predict whether a population would increase or decrease based on whether b minus d was greater than or less than zero. We can see that the same is true in this equation and that the population of exploiters will only grow if births are greater than deaths—that is, if $cfN_p > d$. That insight helps us to think about the equilibrium outcomes of exploiter-prey interactions.

Equilibrium Conditions

We can now look at Equations 8.2 and 8.4 to determine the conditions that produce the four possible equilibrium outcomes of exploitation described earlier. To do this, we use phase-plane graphs again. As in Chapter 7, we place the population density of each focal species on a different axis, so that we can look at both densities simultaneously (**Figure 8.10A**). For all of the Lotka-Volterra exploiter-prey phase-plane graphs, prey density is on the x-axis, and exploiter density is on the y-axis.

A first step when using these phase-plane graphs is to determine the conditions under which either the prey population or the predator population is neither increasing nor decreasing. The mathematics involved in finding the equilibrium conditions are the same as in Chapter 7. For each species, we set its growth equation equal to zero and solve to find the conditions under which that population does not grow. For each equation, the solution gives us a line or curve that represents the joint population densities of each species that allow the density of the focal species to remain unchanging.

As in Chapter 7, we call such lines *isoclines*, or *lines of zero growth*, because the population density of the *focal species* neither grows nor declines along these lines. Remember that each equation only predicts the growth of the focal species (the one on the left-hand side of the equation), so the prey equation will lead to a prey isocline, and the exploiter equation will lead to an exploiter isocline, resulting in a pair of isoclines on the phase-plane graph.

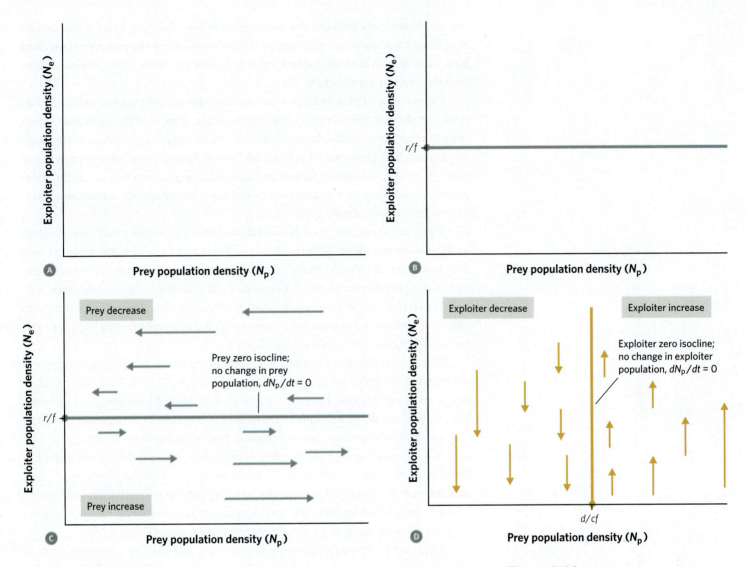

A Exploiter population density (N_e) / Prey population density (N_p)

B Exploiter population density (N_e) / Prey population density (N_p)

r/f

C Prey decrease

Prey zero isocline;
no change in prey
population, $dN_p/dt = 0$

r/f

Prey increase

Exploiter population density (N_e) / Prey population density (N_p)

D Exploiter decrease   Exploiter increase

Exploiter zero isocline;
no change in exploiter
population, $dN_p/dt = 0$

d/cf

Exploiter population density (N_e) / Prey population density (N_p)

Figure 8.10 Building Lotka-Volterra exploiter-prey phase-plane graphs. **A** We begin with the two axes that will allow us to simultaneously examine the densities of both exploiter (i.e., predator) and prey populations. **B** The horizontal line is a prey zero-growth isocline. Notice that it is at a constant density of exploiters and does not specify a specific prey density. **C** The prey zero-growth isocline and arrows show population growth behavior for the prey on either side of the isocline. **D** The exploiter zero-growth isocline and arrows show population growth behavior for the exploiter on either side of the exploiter isocline.

In Chapter 7, we found each isocline by finding two points of the line, but the solutions are a little different in this model. In Equations 8.2 and 8.4, the density of the focal species appears in both terms on the right-hand side of the equation. This means that, once we set the population change to zero, we can simplify each equation by dividing by the density of the focal species and then finding the solution for the remaining variable. Let's work through this solution for the prey.

$$\frac{dN_p}{dt} = rN_p - fN_pN_e$$

$$0 = rN_p - fN_pN_e$$

$$0 = r - fN_e$$

$$fN_e = r$$

$$N_e = r/f \qquad (8.5)$$

Starting with Equation 8.2, we set the change equal to zero to represent no change (i.e., equilibrium). We then divide both sides by N_p to simplify the equation. We can divide this way because the assumption is that the prey density will not be zero if we have any prey in our system. We then subtract the negative term from both sides and divide by f (capture rate) to isolate our value of N_e (exploiter density) that leads to equilibrium.

The solved equation indicates that equilibrium for the prey has nothing to do with prey density but is *entirely* dependent on the density of the exploiter! When the exploiter's population density is exactly r/f, the prey's population is unchanging through time (constant). Looking carefully at the second to last equation in our solution, we see this result makes mathematical sense, even if it is a little unexpected. At this density, consumption by the exploiter offsets the exponential population growth of the prey ($fN_e = r$).

If you are scratching your head and wondering how this could work biologically, you are not alone. This result comes from assuming that the exploiter eats prey purely based on prey density, with each exploiter consuming a constant fraction of the prey it encounters (f), regardless of whether it encounters three prey individuals or three million. We will address the lack of reality in this assumption later in the chapter. For now, let's continue with these traits for the exploiter and see what insights this simple model offers.

We can visualize this isocline on a phase-plane graph as a line of zero growth anchored at the point r/f on the exploiter axis (**Figure 8.10B**). Remember that, at any point along this line, the prey's population is neither increasing nor decreasing. You may recall from Chapter 7 that lines such as this divide a phase plane into regions of population increase or decrease. Below the isocline, the prey's population will increase because the number of predators is low and does not offset the intrinsic growth rate of the prey (r), while above the line the prey's population will decrease because consumption by so many predators overwhelms the prey's intrinsic population growth. We can represent this behavior on the graph using arrows that point along the x-axis (prey population axis) (**Figure 8.10C**).

When we set the exploiter equation to zero and use the same approach to solve for the conditions that lead to no change in the exploiter population density, we get the following:

$$\frac{dN_e}{dt} = cfN_pN_e - dN_e$$

$$0 = cfN_pN_e - dN_e$$

$$0 = cfN_p - d$$

$$cfN_p = d$$

$$N_p = d/cf \tag{8.6}$$

The zero-growth isocline for the exploiter has nothing to do with exploiter density and is solely dependent on prey density. This time the result is less surprising because we already know that, for zero growth, the birth rate has to offset a constant death rate for exploiters. This suggests that we should have a constant per capita birth rate, which would require a constant density of prey.

To explore the dynamics on a phase-plane graph, we can put the isocline on the axes of exploiter and prey density (**Figure 8.10D**) with a line of zero growth anchored at the point *d/cf* on the x-axis. The isocline is a line of zero growth between a section of the phase plane where the exploiter population density increases and a section where it decreases. We can show these changes with arrows that point along the y-axis. The relationship is somewhat different this time because the exploiter needs prey to survive. When there are too few prey to eat, the exploiter population declines, and when there are plenty of prey to eat, the exploiter population grows.

This Lotka-Volterra model offers the most insight into exploiter-prey dynamics when the two equations are examined together (i.e., if we combine Figures 8.10C and 8.10D to create **Figure 8.11**). The first thing to notice about Figure 8.11 is that the isoclines will always intersect, meaning there is always an *equilibrium point* (at the intersection) where neither prey nor exploiter densities are changing.

As discussed in Chapter 7, we cannot assume that all equilibrium points are stable. This one is *neutrally stable*, meaning that, if a disturbance knocks the prey and exploiter away from their equilibrium densities, they will neither return to the equilibrium nor move further from it. The system will just stay at that "distance" from the equilibrium point. This can be confusing, so let's explore why densities do not return to equilibrium by looking at what happens in each quadrant of the phase plane and then focusing on what neutral stability looks like for each species through time.

As in Chapter 7, we can look at how the combined densities of exploiters and prey change in each section of the phase plane by combining (placing tip to tail) the arrows from Figures 8.10C and 8.10D. The resultant vector indicates the expected dynamics by pointing toward the densities of each species that we expect in future time steps (Figure 8.11). These arrows focus on the direction of change. Because we are not being careful about the magnitude of change, we should think of the arrows as trajectories rather than as true vectors. Notice that the two intersecting isoclines create four quadrants (A, B, C, and D) in the phase plane, with different dynamics in each quadrant. Also, the resultant arrows all point toward an adjacent quadrant, forming an overall counterclockwise "loop." If we follow each resultant arrow to the next quadrant, it looks as though exploiters and prey would cycle around the phase plane repeatedly, going through each quadrant in order, but never returning to the equilibrium point.

Let's think about the dynamics biologically and see if that makes sense, starting with the top right quadrant (quadrant A). Here, prey numbers are high enough to support exploiters, so the population density of exploiters increases as they eat prey. Not surprisingly, this causes the density of prey to decrease, as lots of exploiters eat the prey. That decrease in prey density moves us left into quadrant B, where there are too few prey to support the exploiters, so the exploiter density also decreases. Decreased exploiter densities carry the dynamics into quadrant C, where low exploiter densities allow prey numbers to increase; the prey, though, are still at low densities, making them hard for the exploiters to find, so the exploiter densities continue to decrease. As the prey densities increase, though, they cross a threshold where they can now support the exploiter population

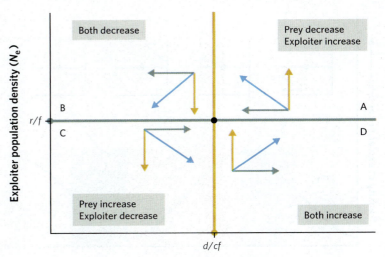

Figure 8.11 When we combine exploiter and prey isoclines on a phase-plane graph, the dot at the intersection of the isoclines represents a neutrally stable equilibrium point. Blue arrows indicate the resultant trajectories from changes in the exploiter (gold arrows) and prey (green arrows) population densities. The letters in each quadrant (A, B, C, and D) refer to the explanation of population dynamics in the text.

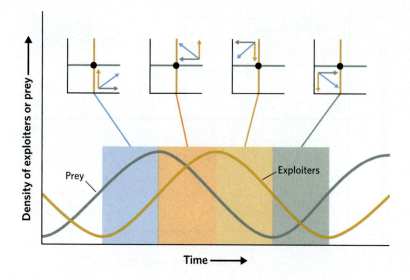

Figure 8.12 The cycling of exploiter and prey populations over time is predicted by the phase-plane graphs. The four insets show the phase-plane graphs with exploiters represented by the gold arrows, prey by the green arrows, and the resultant trajectory by the blue arrows.

Figure 8.13 Neutral stability results in exploiter-prey cycles of constant amplitude. The trajectories in this figure are slightly more symmetrical than the model actually produces, but they emphasize that the size of the disturbance away from the equilibrium point does not increase or decrease with each cycle. Larger disturbances (darker colors) move the densities further from the equilibrium point and lead to larger circles.

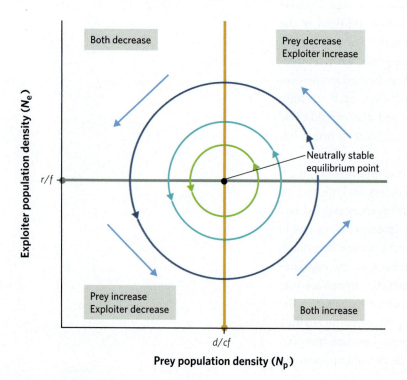

growth (quadrant D), and both populations increase until the exploiter population becomes big enough to start to limit the prey population again (quadrant A).

This pattern of each species responding to the density of the other species leads to repeating counterclockwise circuits around the quadrants of the phase plane. When the densities of both species are plotted over time, we see cycles in the population density of both exploiter and prey (**Figure 8.12**), as described earlier. These are the same types of cycles that Gause observed in his study with paramecia and their ciliate predators, so the model exhibits one of the behaviors we see in real populations.

Of note is that the exploiter densities follow the prey densities, about one-quarter cycle behind them (Figures 8.7 and 8.12). That makes sense because exploiters need time to respond to changes in prey populations. It also makes sense if we look at the four quadrants in Figure 8.11 and notice that exploiter density changes the way that the prey density changes, but one quadrant behind the prey in the circuit around the phase plane.

Let's return to the idea of *neutral stability* to help us understand the amplitude (magnitude) of our cycles. We know that the intersection point of the isoclines represents densities of exploiters and prey that allow zero population growth in both species. In addition, that equilibrium point represents exact densities of exploiters and prey. Any small shift in the system, therefore, carries the exploiters and prey away from this minuscule realm of stable densities. One of the few consistent generalizations across all ecological systems is that change happens, so we do not expect to see exploiters and prey sitting right at the equilibrium point in the real world. The model suggests that the size of the shift or disturbance away from equilibrium determines the amplitude of the cycles (represented in **Figure 8.13** by different-sized circles) because the system stays on a single track through its circuit around the phase plane. Once moved off of the equilibrium point, the population densities never return to the equilibrium and never spiral further away regardless of the number of cycles around the graph.

It is useful to remember that, in the natural world, the exploiter and prey populations sometimes go extinct. Can the Lotka-Volterra model produce such dynamics? Returning to the phase planes, imagine a fairly large shift away from the equilibrium point, and think about how the amplitude of cycles associated with that shift might affect population densities of each species

(**Figure 8.14**). In this graph, extinction comes when the trajectory (i.e., circle) intersects an axis, indicating that one species has dropped to zero density (extinction). We can see this happen if the population cycles are too large. Notice that we can lose just the exploiter from the system (incomplete green circle ending in a green star), and the prey density will increase exponentially, as predicted from Equation 8.2 (which becomes a model of exponential population growth if we take away the effect of the exploiter). If we lose the prey population, though, then the exploiters also go extinct (incomplete gray circle ending in a gray star). These two scenarios mimic the two types of extinction that Gause found with the *Paramecium* and *Didinium*.

The simple Lotka-Volterra model, therefore, replicates three of the outcomes we saw in the real-world exploiter-prey examples. Yet, not surprisingly, some of the simplistic assumptions we built into the model produce problems. Cycles like the ones depicted in Figure 8.12 have been found repeatedly in nature, but even the most uniform of cycles observed in real life fall woefully short of the regularity predicted by this model. In addition, not all coexisting exploiter-prey pairs demonstrate cyclical behavior. Is there some element of the real world that we could add to this model that might stabilize the dynamics and allow exploiters and prey to coexist without cycles?

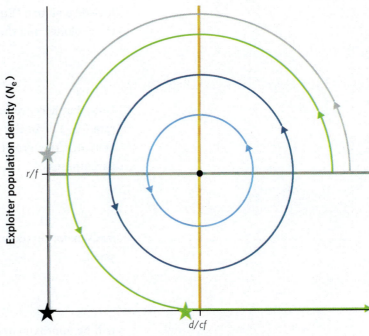

Figure 8.14 This exploiter-prey phase-plane graph shows the potential for extinction with large amplitude cycles. Stars indicate extinctions. The second to largest circle (incomplete green circle) intersects the x-axis, meaning that the exploiter density goes to zero and exploiters go extinct, in which case prey increase indefinitely, as shown by the arrow along the x-axis. With very large amplitude cycles, the prey may go extinct first (incomplete gray circle), in which case the exploiter density decreases until exploiters also go extinct.

8.4 Added Realism Stabilizes Exploiter-Prey Dynamics

As in previous chapters, we will now introduce more realism into our simple model. We address a few of the unrealistic aspects of the Lotka-Volterra model and explore whether altering some of the simplifications allows the model to capture more real-world exploiter-prey dynamics, and specifically whether adding these factors stabilizes the dynamics and increases the tendency for the exploiter and prey to coexist without cycles.

Exploiter-Prey Model with Density Dependence in the Prey

One of the bigger assumptions of the Lotka-Volterra exploiter-prey model is that the prey population experiences no density dependence from resource scarcity. This might work in the real world if the prey exist at very, very low densities (and, thus, experience undetectable levels of resource pressure), but it seems more reasonable to assume that prey densities are generally higher in nature. If so, then without the exploiter, the prey should experience density dependence in the model just as we would expect in real-world populations (Chapter 6).

To incorporate density dependence into our model, we can modify the prey equation by adding a carrying capacity. The easiest way to do this is to use the logistic equation (Equation 6.13) to describe the prey population growth without the exploiter, and then subtract consumed prey.

$$\frac{dN_p}{dt} = r_{max}N_p\left(\frac{K - N_p}{K}\right) - fN_pN_e \tag{8.7}$$

Note that here we have not included a subscript letter p in the intrinsic growth rate (r_{max}) or in the carrying capacity (K), but know that both of these parameters describe growth of the prey population. As in Equation 6.13, the factor in parentheses,

$$\frac{K - N_p}{K}$$

leads to density-dependent population growth. When the prey population is very small,

$$r_{max}\left(\frac{K - N_p}{K}\right)$$

will be approximately equal to r_{max}. But as the prey population approaches K, the per capita growth rate will approach zero. If the prey population goes above K, the density-dependence term will lead to negative growth, making the population decrease. In the basic model (Equations 8.2 and 8.3), exploiters were the only factor that controlled the prey population or kept it from exploding. Equation 8.7 includes two sources of control, both the exploiter and density dependence.

When we set Equation 8.7 equal to zero and divide by N_p, we can see that the isocline for this prey equation is no longer a constant.

$$0 = r_{max}N_p\left(\frac{K - N_p}{K}\right) - fN_pN_e$$

$$0 = r_{max}\left(\frac{K - N_p}{K}\right) - fN_e$$

$$fN_e = r_{max}\left(\frac{K - N_p}{K}\right)$$

$$fN_e = \frac{r_{max}K}{K} - \frac{r_{max}N_p}{K}$$

$$fN_e = r_{max} - \frac{r_{max}N_p}{K}$$

$$N_e = -\frac{r_{max}N_p}{fK} + \frac{r_{max}}{f} \tag{8.8}$$

Instead, we have an equation for a line (notice that the equation has the form $y = mx + b$) with a negative slope ($-r_{max}/fK$). As in the competition model, solving for the two axis intercepts gives us the two points needed to plot a straight line. To do this, we set N_e to zero and solve for N_p, then set N_p to zero and solve for N_e. The result is that the prey density increases to the prey carrying capacity ($N_p = K$) when no exploiters are in the system, and that the other end of the line is anchored

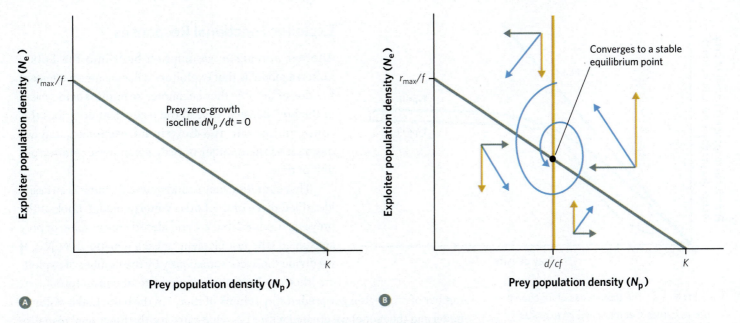

A

B

Figure 8.15 Adding density dependence to Lotka-Volterra exploiter-prey models. **A** Adding a carrying capacity (K) for the prey to the model leads to density dependence or self-regulation in the prey. Notice that the prey density now plays a role in the prey isocline. Also, without the exploiters, the prey are limited by their carrying capacity (indicated by the intercept at K on the x-axis). **B** If we combine the population-growth arrows for the exploiter and the prey, we see that adding density dependence for the prey creates a phase-plane graph that spirals in toward a stable equilibrium. This spiral suggests that these population densities over time will cycle with smaller and smaller amplitude and eventually stabilize at constant densities of prey and exploiters.

at r_{max}/f (**Figure 8.15A**). The exploiter equation is the same as in the basic model, so the exploiter zero-growth isocline remains unchanged from that model.

We can explore the dynamics in this modified model by looking at the phase-plane graph with both exploiter and prey isoclines together. Not surprisingly, adding density dependence changes the model outcomes. We still have four areas in the graph, but they are now of unequal sizes (**Figure 8.15B**). When we combine all of the zero-growth lines and arrows, the exploiter and prey populations still cycle, but this time the densities slowly spiral in toward the single stable equilibrium point in the middle of Figure 8.15B.

Translating these spirals onto a graph with time on the x-axis and population densities on the y-axis, we see cycles of decreasing amplitude as the distance away from the equilibrium point gets smaller with each spiral (**Figure 8.16**). These are **damped oscillations**, like those for discrete time, density-dependent population growth (when $1 < r_{dis} < 2$ in Figure 6.19). Eventually, both exploiter and prey populations cease to cycle as each population arrives at its equilibrium density. Any factor that can reduce cycles this way is called a *stabilizing factor*, and a prey carrying capacity is the most common factor that acts to stabilize exploiter-prey dynamics. Given that we expect many organisms in the real world to experience density dependence, it makes sense that many of the exploitative interactions we observe in nature do not cycle.

Figure 8.16 This graph depicts exploiter-prey population dynamics through time when the prey has a carrying capacity. Both populations experience *damped oscillations*. The amplitude of the cycles gets smaller and smaller over time until the two species stabilize at equilibrium densities. While cycling continues, the prey is always one-quarter cycle ahead of the predator.

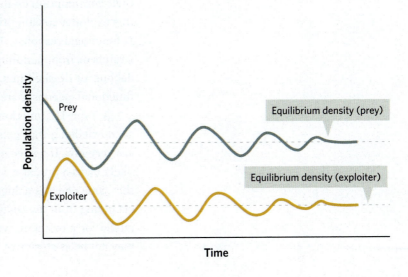

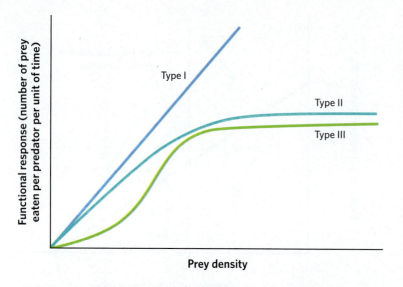

Figure 8.17 The three types of functional response curves that represent an exploiter's rate of prey consumption.

Exploiter Functional Responses

Another unrealistic assumption built into the Lotka-Volterra model is that exploiters will consume a constant fraction of the prey they encounter, regardless of increases in the prey density. We call the factor that describes this consumption rate the **functional response**, and we define it as the number of prey eaten per exploiter per unit of time.

That definition may sound unwieldy, but it is an easily identified piece of the Lotka-Volterra model. Look at the prey equations 8.2 and 8.7, and identify the number of prey consumed (the last element in each equation, fN_pN_e). If we divide those consumed prey by the number of exploiters, then we have a per capita rate of consumption (i.e., number of prey eaten per predator in a chunk of time). In the basic Lotka-Volterra model and the model we created with a carrying capacity, the functional response is a constant function of prey density (fN_p). That means that, as prey density goes up, the number of prey eaten per predator goes up linearly *and never slows down.*

This assumption seems pretty unrealistic, and it is true that we rarely see such constant, unrelenting consumption of prey by exploiters in nature. One of the few exceptions might be filter-feeding organisms (e.g., some zooplankton or whale sharks), which spend no time or effort handling individual prey. These predators consume prey *en masse*, and consumption could conceivably increase proportionately as the prey population increases. Even so, eventually gut size limits the amount of prey that an exploiter can consume and digest. Even if exploiters love eating prey, at some point no more prey fit in their guts, and they are satiated and eat no more.

What, then, are some more realistic options for the functional response? Ecologists focus mainly on three common functional responses, creatively named Type I, Type II, and Type III (**Figure 8.17**). (Note that these curves should not be confused with the unfortunately same-named survivorship curves in Chapter 6.) The functional response in the simple Lotka-Volterra model is Type I. This type of response is mathematically easy, but organisms in the real world cannot increase their consumption endlessly, as we just explained. This natural constraint means that exploiter consumption levels off at high densities of prey, which we call a Type II functional response. The plateau in a Type II response can come from gut constraints or from handling time that limits how many prey an exploiter can chase, debone, or de-shell in a chunk of time (Holling 1959). In the real world, Type II functional responses are much more common than Type I functional responses.

A Type III functional response introduces one more element of reality by acknowledging that consumption may be slow when prey population densities are very low. If a species is rare, it is harder to find, and at very low densities it likely takes a predator longer to search for the uncommon prey. (For some prey, this same effect is achieved by hiding from predators—think of Gause's paramecia hiding from the *Didinium* among the oats.) If we step outside our two-species model for a moment, we can imagine that a hard-to-find, uncommon prey type may not be as attractive, in optimal foraging terms, as an easy-to-find, more com-

mon prey type (Chapter 5); so when the density of a particular prey is low, an exploiter may shift to a more available prey type, even if it is energetically less desirable (Holling 1959; Chapter 5). Either way, prey are consumed at a lower rate when present at very low densities. As prey densities start to rebound, however, consumption increases rapidly and eventually levels off to a constant rate once high prey densities are reached.

How does each of these new functional responses affect the stability of the exploiter-prey system? We could mathematically add either functional response to the Lotka-Volterra prey equation, and go through the same steps to create a new equation. Alternatively, we could look at the isoclines on the phase-plane graph and skip the math (**Figure 8.18**; if you would like to see the math, we recommend reading Nicholas Gotelli's *A Primer of Ecology* [1995]).

Figure 8.18 shows the isoclines for all three functional responses modeled without a carrying capacity. A Type I functional response leads to neutral stability and cycles. A Type II functional response divides the graph into four uneven sections. We saw uneven quadrants in Figure 8.15B, but there the largest sections had one species growing and the other shrinking, which allowed the whole system to spiral toward the equilibrium. With a Type II functional response, the larger sections of the graph are the portions where both species are growing or both are shrinking, leading to spirals away from the equilibrium point and cycles that increase in amplitude. With such cycles, we are likely to see extinction of one or both species, since the spiraling outward increases the chance that one species' trajectory will reach zero (i.e., hit one of the two axes). This result may be surprising. Although a Type II functional response is common in the real world, adding this feature to the model makes it less stable and makes the two species *less* likely to coexist.

This unexpected result makes sense in light of the ideas we have discussed. If the exploiters cannot keep up with increases in prey density, then they cannot control prey growth. Without a carrying capacity, nothing else is controlling prey growth, so the system becomes unstable. Whether this happens in the real world is a different question. In natural settings, prey typically have a carrying capacity, and exploiters are more likely to exhibit a Type II than a Type I functional response. This combination can lead to a range of potential dynamics, some of which are stable and some unstable, depending on prey densities and exploiter efficiency.

The consequences of adding a Type III functional response to the model are fairly straightforward. If the exploiter isocline crosses the prey isocline anywhere except at low prey densities (Figure 8.18), then we see a pattern similar to a Type II functional response—that is, cycles of increasing amplitude and extinction of one or both species. If the exploiter isocline intersects the prey isocline at low densities, though, we see that the angles of the isoclines are similar to the angles with a prey carrying capacity, and thus we see damped oscillations.

A pattern is emerging here. If the exploiter isocline intersects the prey isocline at a place where the prey isocline has a negative slope, we see damped oscillations and stability. If the prey isocline has a positive slope at the intersection point, we

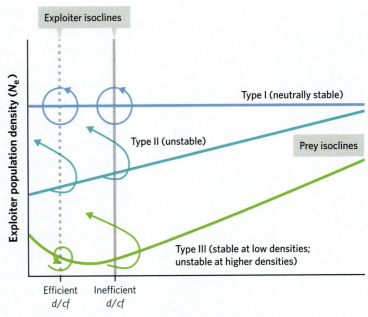

Figure 8.18 Functional responses change prey isoclines in ways that can affect exploiter-prey stability. A Type I functional response leads to neutral stability and cycles. A Type II functional response always *decreases* stability because it leads to cycles of increasing amplitude that lead to the extinction of one or both populations. For an efficient exploiter, a Type III functional response increases stability and leads to damped oscillations. When the exploiter is inefficient, a Type III functional response is like a Type II response and increases instability.

see expanding spirals and increasing amplitude cycles, producing instability and extinction of one or both species. If the isoclines meet where they are perpendicular, as in the basic Lotka-Volterra model, we see neutral stability and cycles.

This pattern replicates all the outcomes expected from exploiter-prey dynamics: damped oscillations and stable coexistence (prey isocline with negative slope), cycles (perpendicular isoclines with the prey isocline parallel to the x-axis and each species' stability dependent solely on the density of the other species), and extinction of one or more species from expanding oscillations (prey isocline with positive slope).

In general, in the Lotka-Volterra exploiter-prey model, the more independent the exploiter and prey population are from one another (i.e., the less their carrying capacities are tied together), the more stable the interaction is between them. This result becomes clear if we consider that, when either exploiter or prey populations are limited by more than just a single factor, the dynamics become less volatile. This result also makes sense if we consider that these models assume that the exploiter and prey are in a tight, two-species-only interaction. Yet many (perhaps most) exploiters consume multiple prey species, which decreases their dependence on one prey species and makes Type III functional responses more likely.

Allee Effects

So far, we have explored different aspects of how interactions change between exploiters and prey as their densities change, but we have not covered one important and somewhat counterintuitive effect. Sometimes individuals will group together, or aggregate, meaning that their average density across a large space may be low, but their local density is high. In other words, they are clumped in one place. In the Type III functional response, we saw that, from the perspective of the prey, low densities can lead to lower per capita risk of exploitation and death, but what about high densities?

W. C. Allee, an ecologist from the early 1900s, was intensely interested in why individuals aggregate in some species but not others. While working at the Woods Hole Oceanographic Institution in Massachusetts, he studied marine mammals that formed groups. His observations led him to ask why these species aggregate, and thereby increase their local density, when we know from other research and models (Chapter 6) that individual fitness declines when population density is high.

If the density dependence we saw in Chapter 6 holds across all situations, it would never be advantageous for individuals to purposefully aggregate. Yet often they do. Not only mammals aggregate, but many species of fish, birds, and insects also aggregate (**Figure 8.19**). In fact, the more ecologists explore this question, the more examples of aggregation they find. The list now includes plants, bacteria, and other organisms that we would never imagine might have a reason or capacity to purposefully aggregate. This begs the question: Why do they do this?

The answer lies in understanding how aggregation offers fitness benefits. If density-dependent growth ruled

Figure 8.19 The picture shows a large winter aggregation of beetles in the family Coccinellidae, which are commonly referred to as ladybugs, lady beetles, or ladybird beetles.

Figure 8.20 A single Cape buffalo (*Syncerus caffer*) may be vulnerable to predation by a pride of lions (*Panthera leo*), but a whole herd of buffalo is intimidating to lions and reduces the chance that any single buffalo will be caught and killed.

the world, individuals in low-density situations (nonaggregated) would enjoy virtually unfettered access to all needed resources, resulting in high survival and reproduction rates. Yet Allee documented that sometimes individuals living in very-low-density conditions suffered high mortality and low reproductive rates. In fact, when density was very low in these species, per capita reproduction and survival rates dropped below the rates seen in super-high-density situations. In other words, being alone resulted in worse fitness than being overcrowded.

This pattern, in which fitness declines as density declines, has come to be known as an **Allee effect**, and it occurs almost solely at low densities. Sometimes the pattern is also called inverse density dependence, or undercrowding. One mechanism that produces Allee effects is the avoidance of predation. Being in a group means that there are more potential scouts or protectors. Instead of having one pair of eyes, ears, or chemoreceptors on the lookout for predators, there are hundreds, which allows for faster evasive action and higher per-individual survival rates.

We see these warning, protective, and evasive behaviors in schools of fish, swarms of birds and insects, alarm-calling ground squirrels (and other ground-nesting organisms), and herds of herbivores. A pride of lions (*Panthera leo*) can take down even a healthy adult Cape buffalo (*Syncerus caffer*) if its lions can separate it from the herd, but together, a herd of Cape buffalo can generally protect all members from predator attacks (**Figure 8.20**). Synchronized movement by a school of fish or a swarm of birds or insects can confuse and overwhelm predators, sometimes to the point where the predator just gives up.

Even if the predators break through defenses and manage to eat one or more members of an aggregation, a large population reduces any single individual's chance of being consumed. For example, a red-winged blackbird (*Agelaius phoeniceus*) in a flock (**Figure 8.21**) has a lower per capita chance of being eaten by a predator than a bird that is alone or in a small group. If a predator is going to take only one bird, the probability of an individual bird's being eaten when flying with four other birds is approximately one in five (a 20% chance of being eaten) but

Figure 8.21 Flocks of birds like these red-winged blackbirds (*Agelaius phoeniceus*) protect members from predators by diluting the capture risk per individual.

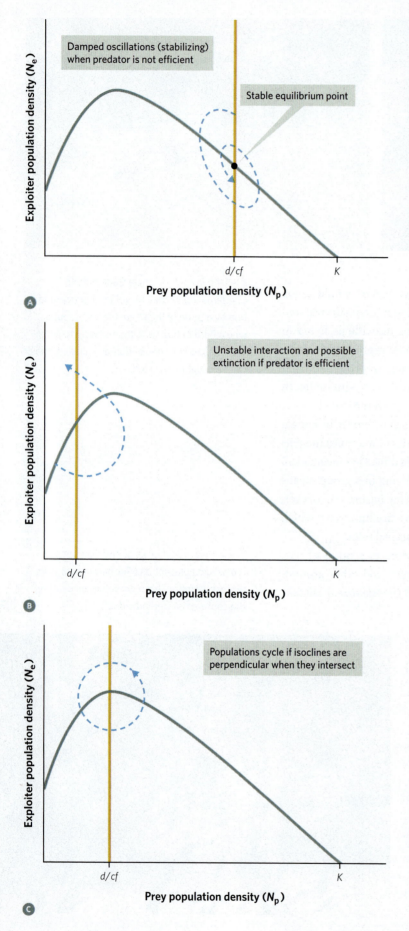

A

Exploiter population density (N_e)

Damped oscillations (stabilizing) when predator is not efficient

Stable equilibrium point

d/cf K

Prey population density (N_p)

B

Exploiter population density (N_e)

Unstable interaction and possible extinction if predator is efficient

d/cf K

Prey population density (N_p)

C

Exploiter population density (N_e)

Populations cycle if isoclines are perpendicular when they intersect

d/cf K

Prey population density (N_p)

drops to one in 1000 (a 0.1% chance) when flocking with 999 others. In addition, such a swarm can confuse a predator, which may lower capture rates even more. Even if a predator eats more than one blackbird, given the huge numbers in the flock, the predator is eventually going to be satiated, thereby further lowering each individual bird's chance of being eaten.

Allee effects may also result from increased reproduction as density increases from very low levels. This idea seems counter to all of our resource and density discussions since Chapter 5, but it is possible. Consider an organism that exists at very low density on a landscape where resources are not limiting but where finding a mate is very difficult. We see such situations for some rare large mammals (e.g., mountain lions, *Puma concolor*), some birds (e.g., kakapo, *Strigops habroptilus*), and many plants that require pollinators to reproduce. Martha Groom (1998) explored these plant effects in red ribbons (*Clarkia concinna*), a type of wildflower, and found that plants in very small groups were unlikely to attract pollinators and thus had very low seed production rates.

Because Allee effects lead to lower predation rates, adding them to the Lotka-Volterra model should influence the outcomes. A phase-plane analysis of Allee effects suggests three outcomes: stability, extinction, or cycles, depending on how the isoclines intersect (**Figure 8.22**). Allee effects can be strong or weak, so the specific shape of the prey isocline can change, but it should always have a positive slope to show inverse density dependence (survival or reproduction increases with prey density) at low prey density, and then a negative slope as the Allee effect fades and density dependence takes over. The section with the negative slope is the same as what we saw with a prey carrying capacity (Figure 8.15). The positive slope at low prey densities makes sense because the isocline represents a series of exploiter and prey densities where death imposed by the exploiter balances population

Figure 8.22 An Allee effect in the prey isocline can influence the outcome of exploiter-prey dynamics in three ways. **A** When the exploiter and prey isoclines cross to the right of the hump in the prey isocline, we see damped oscillations and stability. **B** When they cross to the left of the hump, we get instability and possible extinction. **C** When they cross at the apex of the hump, we see neutral stability and cycles. (Each example shown includes an Allee effect at low population densities and a prey carrying capacity at high population densities in the prey isocline.)

growth by the prey. A positive slope suggests that more predators are required to either kill the prey as density increases (e.g., Cape buffalo) or to balance increased reproduction as prey density increases (e.g., red ribbons).

Notice that the outcomes from adding an Allee effect again depend on where the predator isocline intersects with the prey isocline. If the predator isocline crosses the prey isocline to the right of the hump in the prey isocline (i.e., the prey isocline has a negative slope at the intersection point), population cycles decrease in amplitude and spiral toward stability (Figure 8.22A). These decreasing cycles give us damped oscillations and stable coexistence, as we saw at low densities with a Type III functional response or with a Type I functional response and a prey carrying capacity. If the isoclines cross to the left of the hump in the prey isocline (i.e., the prey isocline has a positive slope), cycles increase in amplitude and one or both species is likely to go extinct (Figures 8.22B). Only if the isoclines intersect at right angles, crossing at the apex of the hump, do we see neutral stability and population cycles (Figure 8.22C).

8.5 How Do the Models Fit the Real World for Different Types of Exploitation?

Although the models we have explored offer insights into the three types of exploitation described at the beginning of the chapter (predation, herbivory, and parasitism), each type of interaction has elements that distinguish it from the others.

Predation

The models in this chapter have been useful in exploring the way that true predators—**carnivores** (organisms that eat animal tissue)—affect their prey species. The trout that eat mountain yellow-legged frogs and the *Didinium* that ate Gause's paramecia are both examples of true predators. Unfortunately, it is often challenging to get data for the predator-prey models once we move from the lab to the field. Observing population dynamics for large-bodied species can take decades or even centuries, which is too long for most ecological researchers. In addition, predators and prey may be of very different sizes and may move around their environment in unique ways, making them difficult to study. Yet there are certainly examples of experimental work that support many of the insights from the Lotka-Volterra models.

For instance, cyclical patterns between exploiters and prey arise fairly often in nature. They are seen frequently in predator-prey interactions that occur at higher latitudes, where less biodiversity (Chapter 11) produces simplified systems in which it is not uncommon for one predator to eat mostly one type of prey. The most commonly discussed example of this is the lynx–snowshoe hare interaction (see Figure 8.7), and there are also many studies that explore cycling of voles (small rodents) and their owl or mammalian predators. In addition, a good body of research in the past two decades has explored whether single-species dynamics (like the population cycles discussed in Chapter 6), plant-herbivore dynamics, climatic effects, or predator-prey effects play the predominant role in producing predator-prey cycles. This work has produced a fair amount of evidence that predators and prey do contribute to cycles in each other's population, even if they are not always the sole cause of those cycles (Graham and Lambin 2002; Korpimäki et al. 2005; Brommer et al. 2010). Field research has also demonstrated many cases of extinction that arise from predator-prey interactions. Because many of these interactions and outcomes are caused by human influences in the Anthropocene, we discuss them further in the final chapter of the book.

Herbivory

Herbivory is the feeding of individuals of one species on individuals, or parts of individuals, of a primary producer species (see the discussion of primary productivity in Chapter 3). The primary producers are usually terrestrial plants, but ecologists often also refer to organisms that eat seaweeds and algae as herbivores. To ease our discussion, we refer to all of these interactions as herbivory. Like all forms of ecological exploitation, herbivory leads to a $(+, -)$ interaction for the participating species, yet there are two aspects of herbivory that distinguish it from other forms of exploitation.

First, herbivores typically only consume portions of individuals. When they graze, browse, or nibble, herbivores enjoy a positive fitness benefit, and the attacked plant suffers negative fitness consequences, but those consequences often do not cause the plant to die. We still see the same general range of population dynamics for herbivores and their plant "prey," partially because chomping on a plant may mean that the plant's offspring (i.e., seeds) or its reproductive structures (i.e., flowers or new vegetative stems) have been removed, directly reducing the plant's reproductive fitness. But even if an herbivore feeds solely on nonreproductive parts, like leaves, this will reduce the energy a plant can gain through photosynthesis, thereby reducing its fitness.

The second distinguishing aspect of herbivores is the degree of specialization that we see among them. Although predators of animals may have preferences as to which prey species they eat, they rarely refuse available prey organisms, and their diet tends to include a diverse array of prey (Chapter 10). We call such species **generalists** with respect to diet. Although plenty of herbivores are also generalists, many are not and instead consume a very narrow range of species, often sticking to just one family or genus. We call organisms with narrow diets **specialists** with respect to diet.

Why do we see more diet specialization in herbivores? Any explanation must include the evolutionary relationship between herbivores and their plants, and we explore that relationship more later in the chapter. A simple and proximate explanation for herbivore diet specialization may result from the interaction between plant defenses (also discussed later) and the many ways an herbivore can consume plant material. Because herbivores do not eat whole individuals, they can avoid competition with other herbivores and circumvent some plant defenses by specializing on a particular part of their "prey." The term *specialist* can also be used to refer to an herbivore that focuses on a very specific portion of a plant. Some herbivores may feed exclusively on the leaves, fruit, pollen, or flowers of a plant, whereas others may eat only the seeds, bark, roots, sap, or woody parts (**Figure 8.23**). This specialization is possible because some herbivores—particularly insect herbivores—are very specific in the method by which they feed, only consuming

Acorn eating
Aphelocoma californica
Scrub jay

Leaf eating
Phryganidia californica
California oakworm caterpillar

Sap feeding
Platycotis vittata
Oak treehopper

Gall forming
Andricus kingi
Red cone gall wasp

Quercus lobata
Valley oak

Wood boring
Tremex columba
Pigeon tremex

Leaf and branch browsing
Odocoileus hemionus
Mule deer

Root eating
Thomomys bottae
Botta's pocket gopher

Wood eating
Chrysobothris mali
Pacific flatheaded borer

Root eating
Phyllophaga sp.
June bug

plant tissues through chewing, sucking, or boring. Insect herbivore specialization is so common that many insect species show lifetime fidelity not only to a single plant species, but even to a single, individual plant (**Figure 8.24**). Because of their reliance on one species for their existence, specialist insects have lifelong relationships with their plant hosts.

Being a specialist or a generalist brings with it a range of other traits. For example, insect (invertebrate) specialist herbivores tend to be small, have high metabolic rates, and are often able to reach large population abundances. These characteristics allow them to inflict damage that ranges from very small to massive. Vertebrate generalist herbivores, on the other hand, tend to have a more direct negative impact on the plants they consume, owing to their larger body size relative to the plants they eat and their greater mobility. They also tend to be at lower population densities than insect herbivores. On a population level, both

Figure 8.23 Herbivores often concentrate their exploitation of a plant on particular structures such as bark, roots, new leaves, or sap. Thus, one tree can have many different specialized herbivores feeding on it, as depicted here.

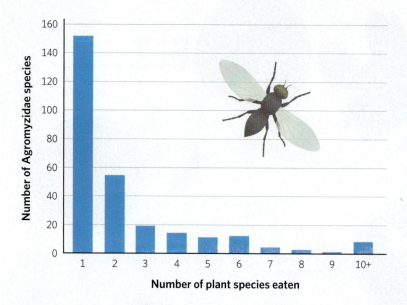

Figure 8.24 As is apparent from this bar graph, most species of leaf-miner flies (family Agromyzidae) are highly specialized and feed on just a single plant species (Spencer 1972).

vertebrate and invertebrate herbivores can inflict a range of damage on prey populations, but specialists are less likely to drive prey populations extinct. Evolutionarily, such relationships are often less successful if the herbivore kills its host or plant prey.

Nonetheless, it is possible for both types of herbivores to sometimes eat their host plants to extinction. Goats introduced to certain islands have eaten several species of plant to extinction, and high densities of insects can lead to the decimation of crops and forests across whole regions. Anyone who is a gardener may have had experience with insect and mammalian herbivores that arrive and decimate certain plants or crops.

We also see population cycling with herbivores. For example, some ecologists have speculated that the lynx–snowshoe hare cycles could be due to cycles between the hares and their plant food resources. The woody shrubs that hares feed on may be inducing the cyclical behavior, in which case the snowshoe hares are the predators and the plants are the prey. In this context, the lynx population cycle is occurring only because the lynx carrying capacity (in the form of snowshoe hare density) is cycling. Moths provide another example, with both gypsy moths (*Lymantria dispar*) and eastern tent caterpillars (*Malacosoma americanum*) cycling quite predictably in mid-Atlantic and northeastern US forests and potentially contributing to cycles in tree **recruitment**. *Recruitment* here refers to a combination of factors that leads to new trees entering the adult age class. The cycles come from combined effects on reproduction (with decreased flower and seed production in years with high densities of insect larvae), adult survival (larvae munching on leaves can reduce photosynthesis so much that it occasionally leads to periods of synchronized tree mortality), and sapling growth (from increased light and nutrient availability after adult mortality).

In addition, there is growing evidence to suggest that herbivores may be one of the main reasons for the incredible number of tree species in tropical rainforests (Forrister et al. 2019). Specialized insect herbivores keep any one tree species from becoming too common. How can 650 different tree species coexist in a single hectare in the tropical rainforests of Ecuador? It is unlikely that Ecuador is home to 650 distinct ecological niches for trees with respect to light, water, and nutrients. Instead, ecologists are beginning to recognize that each plant has a different niche space with respect to its specialist herbivores. Herbivory may, therefore, be a key driver of plant species diversity (Chapter 11).

Parasitism

Parasitism clearly falls within the definition of exploitation as an interaction in which one species gains in fitness and another loses fitness. A parasite is an organism that obtains its nutrition from one, or a very few, host individuals. It causes harm to the host but often does not kill the host. In this way, parasitism is a bit like herbivory in that the fitness costs to hosts are often lower than they are in predator-prey interactions. At the mild end of the effects on hosts, we can think about a human reaction to most mosquito bites; at the severe end, we can think about the effects of virulent human pathogens such as the COVID-19 virus or Ebola virus.

A first useful distinction to make when considering parasitism is between macro- and microparasites. **Macroparasites** are large enough to be seen with the naked eye; mosquitoes are an example. They may live on or in a host, but they often reproduce with infective stages outside the host. Animal macroparasites live on the body or in the gut cavities of their hosts. In plants, macroparasites often penetrate the host's tissues in order to tap into the xylem and phloem. Common examples of animal macroparasites include tapeworms, ticks, lice, and fleas. Common plant macroparasites include mistletoe, dodder, and many types of gall-inducing insects.

In contrast, **microparasites** are tiny and often occur intracellularly. They reproduce directly within their host, where they can achieve very high densities. Examples include common disease agents such as viruses, bacteria, and protozoans. Microparasites can be further subdivided into ones that are transmitted directly from host to host (such as the many viruses that cause the common cold, which can be transmitted directly from one human to another), and ones that require an intermediate host, or vector, for transmission (such as Zika virus and the malaria parasite, which are generally transmitted to humans by mosquitoes). If a microparasite produces symptoms that are clearly harmful, the host is said to have a disease, and the term *pathogen* is sometimes applied to parasites that give rise to a disease (e.g., the lyssaviruses are the pathogens that cause rabies).

Parasitoids are insect exploiters that fall somewhere between specialist predators and parasites. They attack other insects and lay eggs or larvae inside the host. These eggs eventually develop, consume, and kill the host insect in the process. Like true predators, they tend to kill their prey or hosts. Like parasites, they tend to have just one prey or host in their lifetime. To make parasitoids even more interesting, ecologists have found instances of *hyperparasitoids*—parasitoids that attack other parasitoids! In general, parasitoids are the subset of exploiters with which people are least familiar, and our favorite pop culture analogy is found in the series of *Alien* movies directed by Ridley Scott. The fictional aliens in these movies use humans as incubation chambers and food for their young. The young aliens develop inside humans and eat them from the inside out, which is exactly the way parasitoids use their hosts. In general (but not always), one parasitoid emerges from an attacked host, and most insects have specialist parasitoid predators.

Some parasites have fairly simple life cycles, requiring only food to consume and a place to grow and reproduce; mosquitoes, viruses, and bacteria are examples. Other parasites, such as the lancet liver fluke (*Dicrocoelium dendriticum*), have extremely complex life cycles (**Figure 8.25**). An individual fluke lives its adult life and reproduces in the liver and bile ducts of a **definitive host**, which is the host that carries the reproductive stages of the parasite. In this case, it is typically a sheep or cow but can be a human. The fluke makes resistant eggs that are excreted in the sheep's feces. The first **intermediate host** (i.e., an organism that supports the non-adult form of a parasite), a terrestrial snail in this case, ingests the fluke's eggs while consuming the sheep's fecal matter. The larval forms of this parasite settle in the gut of the snail, where they grow to become juvenile flukes. The snail's immune system tries to protect itself and walls off the juveniles in cysts. These cysts are subsequently shed via the respiratory system of the snail, coming out as little mucus balls left behind in the snail's slime trail.

These cysts are then consumed by a second intermediate host, an ant. The ant uses the snail's slime trail as a source of moisture and inadvertently consumes the encysted juvenile flukes. If a parasite enters the ant's subesophageal ganglion, which is a cluster of nerve cells under the esophagus, the parasite can hijack the ant's nervous system and drastically alter its behavior. An infected ant will climb up to the top of a blade of grass, clamp down with its jaws, and hold on, remaining immobile. Ostensibly, this behavior puts the ant (and its parasite) in the best location to be accidentally ingested by a grazing sheep, thereby starting the cycle all over again.

Some of the most active research in exploiter-prey dynamics has been done with parasites, partially because some parasites have significant effects on human lives and also because their small body sizes and short generation times allow

Figure 8.25 The complex life cyle of the lancet liver fluke (*Dicrocoelium dendriticum*), a parasite of multiple sequential hosts.

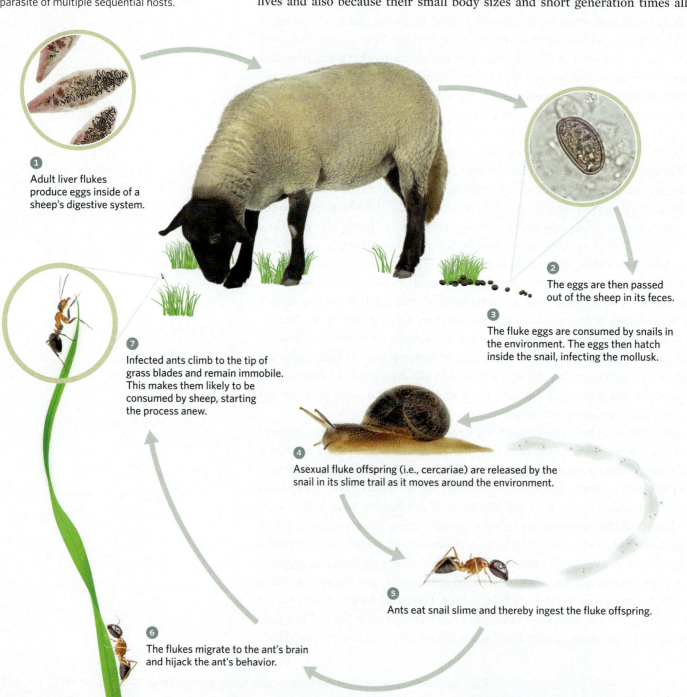

1 Adult liver flukes produce eggs inside of a sheep's digestive system.

2 The eggs are then passed out of the sheep in its feces.

3 The fluke eggs are consumed by snails in the environment. The eggs then hatch inside the snail, infecting the mollusk.

7 Infected ants climb to the tip of grass blades and remain immobile. This makes them likely to be consumed by sheep, starting the process anew.

4 Asexual fluke offspring (i.e., cercariae) are released by the snail in its slime trail as it moves around the environment.

5 Ants eat snail slime and thereby ingest the fluke offspring.

6 The flukes migrate to the ant's brain and hijack the ant's behavior.

faster exploration of population dynamics. Given this intensity of focus, the models for host-parasite and disease dynamics are much more developed and often more complicated than the ones presented in this chapter.

One common thread between these simple and complex models is the importance of the population density of each species. For example, we know that when humans are densely packed together, we need to worry more about disease transmission. We are more likely to notice a coughing stranger who is standing next to us in a crowd than one who passes us on the other side of the street. Clearly this issue came to the fore during the COVID-19 pandemic, hence the restrictions on gathering in groups, emphasis on social distancing, and the wearing of masks. Density effects such as these lead to some simple take-home messages about diseases: for example, a threshold density of hosts is necessary to maintain a disease in a population. This echoes one of the fundamental insights we gleaned from the simple Lotka-Volterra model—namely, that the predator isocline results from a specific density of prey.

8.6 Evolutionary Outcomes of Exploitation

Exploitative interactions are some of the strongest evolutionary forces in nature because of their direct influence on the fitness of both species involved. As a result, species evolve defensive traits that lessen the fitness impacts of their exploiters, and exploiters evolve traits that allow them to overcome these defenses. Scientists call this process **coevolution**, which is evolution by natural selection through reciprocal selective effects between two or more closely interacting species. We explore this concept here with regard to exploitation and revisit it in the context of mutualism in Chapter 9. Coevolution in response to exploitative species interactions has led to the emergence of new species and some spectacular traits within species.

Exploiter-Prey Coevolution

An individual exploiter that successfully eats prey is more likely to survive and reproduce. A prey individual that successfully avoids being eaten by an exploiter is also more likely to survive and reproduce. Given these consequences, it should not be surprising that exploiters can exert significant evolutionary selective pressure on their prey and that the prey can exert selective pressure back onto the exploiters.

Yet the evolutionary consequences for prey are decidedly more dire than for exploiters. This basic asymmetry is sometimes referred to as the *life-lunch concept* because each encounter for the prey is a matter of life and death (or at least the loss of a "limb"), but for the exploiter it is usually only a matter of eating a meal or not. The strong evolutionary pressure on the prey to avoid being consumed has produced a wonderfully dazzling array of adaptations. We shoehorn this huge diversity of adaptations into three broad categories of defenses: *primary, secondary*, and *tertiary* (based loosely on categories from Denno and Lewis 2009).

Primary defensive strategies serve to prevent the exploiter from ever finding the prey. Secondary strategies often draw attention to the prey but also send a message that deters the exploiter from attacking. Tertiary strategies help the prey

Figure 8.26 A painted frogfish (*Antennarius pictus*) camouflaged by an orange sponge in the waters around Sulawesi, Indonesia.

Figure 8.27 Burchell's zebra (*Equus quagga burchellii*) Ⓐ on the open plains and Ⓑ behind trees in Etosha National Park, Namibia. Notice how well the zebra stripes provide camouflage when the animals stand by woody stems rather than out in the open.

to escape or fight back if the exploiter does attack. Species generally evolve toward only one of these strategies, not all three, although we provide some examples of species that use different strategies at different ages or stages of their lives, or even use a combination of two strategies at the same time.

Primary Prey Defense Strategies   An effective way to avoid being eaten is not to be seen in the first place. Many organisms use some form of **camouflage** to avoid being seen by their predators. This defense is often referred to as **crypsis**, or *cryptic coloration*. For example, marine fish species in the family Antennariidae, commonly known as frogfish, are renowned for their impressive use of camouflage (**Figure 8.26**). Some frogfish resemble coral or rocks, while others successfully imitate sponges, seaweed, sea urchins, or sea squirts. Many frogfish are able to slowly change their color to match almost any background, and some even adorn their bodies with living algae or invertebrates to look more like background rocks.

These elaborate physical traits and behaviors serve two distinct yet vital purposes. First, they allow the frogfish to avoid being seen by predators. Frogfish are readily consumed by predatory wrasses, eels, and even other frogfish when flushed from their hiding places. Second, extreme camouflage acts to hide frogfish from their own prey by allowing them to blend in with the background until the unsuspecting prey approach close enough to be eaten easily. We cannot forget that many species are both exploiters and prey, and that some adaptations may serve them in both roles.

Camouflage does not always have to be as dramatic as in frogfish. More common are relatively simplistic traits, such as the black and white stripes on a zebra (**Figure 8.27A**). The stripes at first seem to confer a disadvantage, as they surely make individuals stand out against the muted browns and greens typical of African savannas. However, a zebra hiding among shrubs all but disappears (**Figure 8.27B**). Although the savanna has a preponderance of grasses, there are also places with low shrubs and scattered trees, and a zebra's stripes may allow it to hide in almost plain view. Interestingly, the stripe pattern on zebras has also been shown to

Ⓐ

Ⓑ

Figure 8.28 **Ⓐ** When the caterpillar of the elephant hawk moth (*Deilephila elpenor*) is threatened by a predator, it can widen the anterior part of its body, emphasizing the eyespots and making them more conspicuous. Research suggests this ability allows the caterpillar to resemble a snake, which avian predators are more likely to avoid. **Ⓑ** A polyphemus moth (*Antheraea polyphemus*) has distinct eye spots on its hind wings that are often covered by the front wing and then revealed suddenly in an attempt to startle would-be predators.

reduce the number of insect bites the animal receives. This outcome is partially due to the stripes acting as camouflage against the insects, which apparently have trouble seeing a safe place to land on a striped pattern (Caro 2016).

Variations on the general theme of camouflage are plentiful across the plant and animal world. Examples include the use of countershading to hide from predators foraging above or below and hiding one's eyes in an attempt to reduce overall conspicuousness. Notice that many of these defenses work best for prey avoiding true predators, rather than those avoiding herbivores or parasites.

Secondary Prey Defense Strategies   Once an exploiter has spotted a prey individual, different secondary defense strategies kick in to reduce the likelihood of capture or increase the chance of escape once captured. One common strategy is for prey to announce themselves, not as themselves but as something other than what they are. Numerous Lepidopterans (moths and butterflies) in their caterpillar (larval) or adult forms may project the appearance of big predatory eyes (**Figure 8.28**). Sometimes prey will mimic another species. Broad-headed bugs (family Alydidae) are herbivorous or parasitic organisms that suck on plant tissues. Most of them look innocuous to predators, but some have developed adaptations that make them look like ants, which are not at all innocuous (**Figure 8.29**). Ants not only have massive strength for their size, good mandibles (jaws), and the ability to sting, but they are also social insects that can mobilize a whole army to protect themselves or their colonies. A broad-headed bug that looks like an ant is like a grade school student convincingly wearing a US Navy SEAL costume.

Figure 8.29   Herbivorous broad-headed bugs in the family Alydidae mimic ants to appear more dangerous. These two broad-headed bugs, *Hyalymenus tarsatus*, are from Central Texas and look remarkably like ants, but you can tell they are true bugs by their piercing/sucking mouth parts.

Figure 8.30 The Io moth (*Automeris io*) uses more than one type of prey defense strategy. As a caterpillar, its stinging and venomous spines, bright colors, and contrasting red and white stripes are secondary defense strategies. An adult moth has no venom but can project a startling image by revealing the large eye spots on its hind wings, also a secondary strategy. Alternatively, an adult can camouflage itself by covering its hind wings in a primary defense strategy.

These examples of secondary defense strategies all involve sending a message to the exploiter to fool it into thinking the prey are dangerous. Sometimes, though, the prey really are dangerous, and advertising their dangerousness helps deter exploiters from trying to eat or taste them. The Io moth (*Automeris io*; **Figure 8.30**), for example, is a North American moth whose larvae (caterpillars) have the ability to sting, and they advertise this nastiness with bright colors and obvious spines. In its adult stage, this moth can expose its hind wings to reveal big eye spots (secondary defense strategy), or it can camouflage itself (primary defense strategy). Although insects offer excellent examples, advertising a defense is also seen in marine fireworms, sea urchins, porcupine fish (also known as blowfish), terrestrial porcupines, and many other creatures (**Figure 8.31**).

These advertisements of dangerousness work best when predators encounter the signal frequently and consistently enough to understand the message. The stark lines of contrasting colors on the Io caterpillar are an example of **aposematic coloration**, or warning coloration. Such contrasting coloration almost always signals danger or warning to a potential predator. Think of the contrasting black and white stripes on a skunk, the contrasting black and yellow bands on bees and

Figure 8.31 Defensive strategies such as spines can be extremely effective protection against potential predators. Ⓐ Bearded fireworm (*Hermodice carunculata*). Ⓑ North American porcupine (*Erethizon dorsatum*). Ⓒ Purple sea urchin (*Strongylocentrotus purpuratus*). Ⓓ Long-spined porcupine fish (*Diodon holocanthus*).

wasps (**Figure 8.32**), or the contrasting orange and black of distasteful monarch butterflies. All of these species are toxic or noxious but also surprisingly vulnerable. When attacked, they may easily die, but along the way they will inflict enough damage that exploiters are unlikely to attack that species again. Their safety, therefore, depends at least somewhat on having exploiters recognize them as dangerous.

That danger signal is reinforced when multiple species use similar signals or advertisements. If you are unsure of how this mechanism works, think about whether you would be more likely to pick up a caterpillar with contrasting coloration and obvious spines or one that looked as though it was trying to blend into the background. If you had encountered a spiny caterpillar in the past, and lived to scream about the pain it inflicted, you would remember its distinctive appearance and would not pick up another one. However, if you picked up a camouflaged caterpillar, you probably would not experience anything bad and would not remember its appearance.

Such reinforcement works with aposematic coloration in general, but it also works on a much finer scale. Ecologists have long noted that toxic or venomous

Figure 8.32 Many bees and wasps show a similar and easily recognized body coloration (i.e., alternating yellow and black stripes). This pattern reinforces a selective advantage to all the toxic species at once and makes it easy for predators to avoid them. Such color patterns are an example of aposematic coloration and Müllerian mimicry.

species that advertise their danger often occur with other species having a very similar appearance. By mimicking each other, these species have done the equivalent of taking out more advertising time on the airwaves. As a group, they can send their "beware" message to potential predators much more effectively. **Mimicry** allows each individual and species to spread the risk of delivering the message across more individuals and even across unrelated species. There is inherent risk to aposematic coloration, because attackers or exploiters may not know that the color signal indicates danger until they attack one individual. That individual may give up its life to send the message, but that exploiter is not going to mess with any other prey that look that way.

This type of mimicry is called **Müllerian mimicry**, after the eighteenth-century German naturalist Fritz Müller, who studied butterflies in the Amazon. While working in Brazil, Müller noticed that two or more unpalatable butterfly species (with noxious chemicals or stinging cells) would often closely resemble each other and have aposematic coloration. Müller hypothesized that, over evolutionary time, there was a selective advantage for both species to closely resemble each other, thereby gaining protection by reinforcing predator learning. If a predator tried to eat an individual of one of the species and found it noxious, it would learn to avoid any other prey species that closely resembled the first.

We find many examples of Müllerian mimicry in the natural world, including the narrow waist, aposematic coloration, and obvious stingers present in most bees and wasps (Figure 8.32). Many potential predators have learned from painful stings to avoid any and all insects that have yellow and black bands. Overall, Müller's general insight was that common unpalatable forms have lower mortality than novel unpalatable forms as a result of learning on the part of predators. This hypothesis has been well supported by field experiments (Sherratt 2008).

In fact, such mimicry can be so useful that some species employ a second type of mimicry that involves sending a *false* danger signal while piggybacking on the real danger associated with noxious species. This false-signal type of mimicry is called **Batesian mimicry**, after the eighteenth-century naturalist Henry Walter Bates. Bates also conducted research on Amazonian butterflies. Like Müller, he noticed that some of the species he observed looked strikingly like one another in their color patterns, but Bates noticed that some of the species had very different traits. He concluded that some species with no toxins or noxious defenses mimicked the colors of toxic or noxious species to reduce their own threat of exploitation. This strategy worked if the noxious species were at high enough densities to send a strong message to potential predators.

Whereas Müllerian mimics *reinforce* a message of toxicity, Batesian mimics *dilute* the message. Having too many Batesian mimics in one area can remove the protective status that the warning signal offers. Nonetheless, there is strong selection for Batesian mimicry. As long as predators encounter the noxious species often enough to learn to steer clear of any individuals with those colors, all the Batesian mimic has to do is show a similar signal, and its chances of becoming lunch drop dramatically. We define Batesian mimicry as an adaptive trait that allows a palatable species (i.e., the Batesian mimic) to look like an unpalatable, noxious, or toxic species (i.e., the model), thereby conferring on the mimic an evolutionary advantage in the form of a perceived threat to the predator.

We find examples of Batesian mimicry in many species beyond Amazonian butterflies. One of the more interesting examples involves a few species of snakes. New World coral snakes in the genus *Micrurus* possess one of the most potent venoms of any snake in North and Central America. They have strikingly bright colors and banding patterns to warn off potential predators (Greene and McDiarmid 1981). "False" coral snakes in the genus *Pliocercus*, on the other hand, are nonvenomous or only mildly venomous but have evolved to look remarkably like their lethal counterpart (**Figure 8.33**), thereby gaining a significant evolutionary advantage.

Figure 8.33 False coral snakes in tropical southern Mexico and Central America exhibit Batesian mimicry. The venomous (and lethal) species are true coral snakes in the genus *Micrurus*. The nonvenomous or mildly venomous species are false corals of the genus *Pliocercus*. Note the extreme similarity in color and banding patterns. The nontoxic species (the Batesian mimic) gets the benefit of the toxic species without having to produce the venom (Greene and McDiarmid 1981).

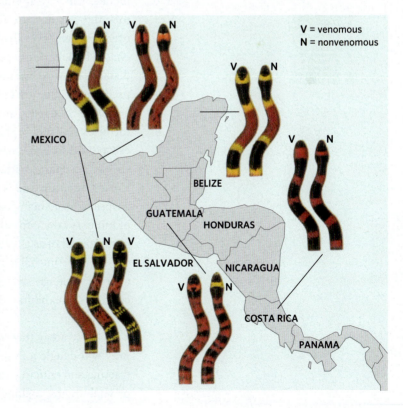

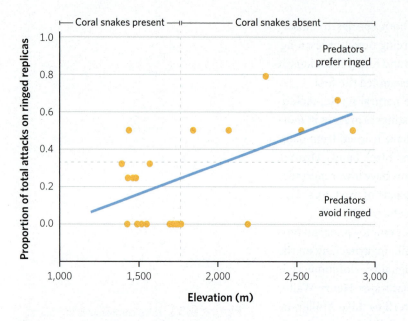

Figure 8.34 In Pfennig and colleagues' manipulative experiment on Batesian mimicry, their coral snake replicas were attacked by predators much less often in areas where coral snakes naturally occur than in areas where they do not occur (Pfennig et al. 2001).

In a fascinating bit of research, David Pfennig and colleagues (2001) used fake coral snakes (replicas) to test how protection from predation varied with the density of real versus fake signals in a population. The researchers placed the fake snakes, painted with rings to look like coral snakes, along a transect that included sites of low elevation (where real coral snakes occur) and high elevation (where coral snakes do not occur). The researchers then looked for marks on the fake snakes to indicate attacks by predators. They found that the replicas were attacked significantly less often in areas where coral snakes naturally occur than in locations where they do not occur (**Figure 8.34**). This result implies that natural selection on the predator to avoid the Batesian mimic is indeed sensitive to the frequency of individuals with the true noxious trait.

Tertiary Prey Defense Strategies   In general, a tertiary defense involves a way to escape or fight off an attack once it has already started. Sometimes the line between a secondary and tertiary defense is blurry, as it is hard to tell whether the strategy prevents attack or interrupts the attack. For example, advertising toxins with aposematic coloration can warn away predators, but having toxins, with or without warning coloration, can also halt an attack in progress by causing physical distress or even death to the predator.

The most common tertiary defense strategies involve physical defenses that allow a species to escape or fight back. Some good examples of escape behavior include the way aphids drop off plant leaves in the presence of ladybug predators; the way fleas exercise their extreme jumping ability when agitated by their predators; and how flying fish leap out of the water and glide long distances when chased by a marine predator. All of these escape mechanisms allow the prey to put distance between themselves and the predator as quickly as possible, even after a predator has initiated an attack.

Other defensive strategies, though, can involve staying in place. Turtles and tortoises stay where they are but use their hard protective covering as a form of escape, as do clams, armadillos, pangolins, pill bugs, and many others. Some prey species respond to attacks not by fleeing but by playing dead. Such behavior is called *thanatosis* and is essentially a form of bluffing in which an animal mimics a corpse to avoid being eaten by predators seeking living prey. Because this version of mimicry (thanatosis) generally only comes into play after an attack, it is really best thought of as a tertiary defense strategy.

Aggregating behavior (forming flocks, shoals, schools, or swarms) can also be a tertiary defense strategy if the whole group takes defensive action to distract or confuse the exploiter. Such action makes it less likely that any single individual will be eaten.

One strategy that many people find surprising is the tactic of sacrifice. Some prey will dangle a false target, such as a long or brightly colored tail that is distracting and enticing, to draw the exploiter's attention away from the prey's head or vital organs, making attacks by predators less likely to be deadly. Even more surprisingly, some prey will sacrifice a body part to a predator to remove the threat

Figure 8.35 This mountain lion warning sign suggests that fighting back is a good strategy to avoid becoming prey.

of death. Sea stars can shed limbs, and some lizards and amphibians can drop their tails when attacked, leaving the predator with a less than optimal meal and the prey individuals with enough body parts to move on and reproduce.

Finally, fighting back can be an excellent tertiary defense strategy, particularly for big prey. Buffalo fight (in groups, as we mentioned before) against lions, and elephants and hippos are often safe even when not in a group. It is important to notice that all these defense strategies make it harder for the predators to catch prey by making the capture rate (f) smaller, which also acts to shift the isocline in Figure 8.22 to the right, producing more stability. Incidentally, fighting back is the suggested advice if you are ever attacked by a shark or a mountain lion (Figure 8.35).

Evolutionary Responses in Exploiters We have focused on the effects that exploiters have on prey evolution because the life-lunch principle imposes asymmetric fitness effects that lead to stronger selection on prey. Yet exploiters also experience adaptive changes in response to their prey. This evolutionary response to each other results in coevolution, or a fitness push and pull between exploiters and prey. Coevolution does not lead to a single, perfect set of adaptations for each species. Instead, it requires constant adaptation in order to keep up with the innovative new traits in the other species. Biologists call this constant cycle of trait evolution an **evolutionary arms race**.

In an evolutionary arms race, some fraction of prey individuals will have a genetic variation that allows them to escape or avoid predation better than their peers in the same population. Because these individuals produce more offspring, over many generations natural selection increases the frequency of this prey defense strategy in the prey population (Chapter 5). If some exploiter individuals have a trait or genetic variation that allows them to outmaneuver this prey strategy, those exploiters will presumably capture more prey, be healthier, and produce more offspring. Again, natural selection will allow this advantageous trait or genetic variation to become more common in the exploiter population. In order to escape predation, the prey now need to evolve a new adaptation, and the cycle begins again.

Figure 8.36 The kapok tree (*Ceiba pentandra*), native to Mexico, Central America, and South America, has much of its trunk covered in defensive thorns that ward off many potential herbivores.

Figure 8.37 Latex oozes from the cut stem of a common dandelion plant (*Taraxacum officinale*). This kind of secondary chemical defense is extremely useful against localized insect attacks.

The evolutionary arms race between exploiters and prey is an example of the **Red Queen hypothesis**, which proposes that all species must constantly evolve in order to maintain their fitness in the face of a changing environment full of other evolving species. The term comes from Lewis Carroll's *Through the Looking-Glass and What Alice Found There*, in which the character of the Red Queen tells the protagonist, Alice, that she must run as fast as she can just to stay in place. As ecologists learn more about exploiter-prey relationships, it seems that evolutionary arms races are possible only when the interaction between the predator and prey is stable and constant over evolutionary time (Brodie and Brodie 1999). In most cases, an arms race is not likely because most predators are generalists and switch to a new prey species if their current one becomes rare.

Plant-Herbivore Coevolution Because of the sedentary nature of plants, and because individual plants are often simultaneously attacked by a suite of herbivores rather than just one, the adaptations developed by plants to avoid consumption are of a somewhat different character than those of animals. A basic exploration of plant defenses begins with recognizing that some defenses are always present, called *constitutive defenses*, while other defenses are present only when needed, called *inducible defenses*. An advantage of the latter type is that the overall energetic cost to the plant can be lower because these defenses are only mounted when needed. A disadvantage of inducible defenses is that, with some herbivores, a delay in activating a defense can be costly.

Toxins are a common plant defense and are often described as **secondary metabolites**, or organic chemicals that are typically not involved in the plant's basic growth and development but can be essential to its survival. Secondary metabolites, such as lignin, cellulose, or silica dioxide, act to reduce the digestibility of the plant tissue. These metabolites can be present in relatively high concentrations throughout the plant. Secondary plant chemicals come in many varieties, including alkaloids, such as caffeine, nicotine, morphine, and quinine; terpenoids, such as camphor, citronella, menthol, and tetrahydrocannabinol (more commonly known as THC); and phenols, such as tannin and lignin. Many of our drugs and pharmaceuticals take advantage of these plant defenses, as do quite a few of our much-loved recreational and dietary products, such as coffee, tea, hot chili peppers, and chocolate.

Plants also have a host of mechanical defenses available to them to deter herbivores. Some of these are structures, such as defensive hairs (i.e., trichomes), spines, or thorns (**Figure 8.36**), that may contain poisons or irritants. Others are substances, such as latex or resins, that are released at the exact location of a plant's injury from herbivory (**Figure 8.37**). These substances cause the mouthparts of insect herbivores to gum up or even glue shut, which is obviously not ideal for the insect. Some plants place a lot of physical protection over their seeds and fruits, as anyone who has tried to open a coconut by hand can attest.

Plant defense strategies can add a level of complexity to herbivorous interactions by attracting enemies of their own enemies. For example, some plants, upon being damaged, will release volatile organic chemicals that attract insect predators that arrive and prey on the plant's herbivores (Allmann and Baldwin 2010). This antiherbivory strategy may be fairly common, and so this three-way interaction

Figure 8.38 An ant of the genus *Pseudomyrmex* gathers Beltian bodies (nutritionally rich detachable tips) from a bullhorn acacia tree (*Vachellia cornigera*).

is likely present in many situations because congregations of prey often attract predators. In a similar way, some trees (e.g., *Acacia* sp. and *Macaranga* sp.) act as both host and hotel for specialized ant species that live in close association with the plant. The ants get places to live (hollowed out stems and thorns) and at times receive nutrient rich food (nectar or Beltian bodies; **Figure 8.38**) provided by the tree. In return, the tree is richly rewarded by the protection from herbivores that only a swarming colony of stinging and potentially biting ants can provide. We will discuss such defenses involving other species in much greater detail in the chapter on mutualism (Chapter 9).

8.7 Exploitation in the Anthropocene

Most students of ecology appreciate learning about exploitation interactions simply because it enriches their daily interactions with the natural world. For example, understanding that the common black-and-yellow-striped coloration of bees is a signal to predators to avoid eating them can make your picnics in the park a bit more entertaining, or at least make for an interesting conversation.

Similarly, knowledge of exploitation dynamics can make you a more informed consumer of science and health media. For example, the role of plant secondary metabolites has recently come to the fore in the form of the debate over the value of organically grown versus non–organically grown produce. Reducing the synthetic chemicals we ingest through fruits and vegetables is one of the classic arguments for organically grown produce. Yet people have also suggested there are nutritional benefits to organically grown food above and beyond what is found in conventionally grown crops. Interestingly, nutritional benefits have been harder to demonstrate scientifically until recently. An analysis of more than 300 scientific studies across multiple crop species found evidence that organically grown produce contains higher concentrations of antioxidants and other compounds that may be beneficial to humans (Barański et al. 2014). Although the link between organic crops, pest damage, and nutrient availability requires more research, these chemicals may be the result of herbivore-induced stress. Organic crops are often exposed to higher levels of insect damage, which causes a stress response in the plant in the form of antioxidants and other secondary metabolites. In other words, allowing natural herbivores to eat some tiny fraction of a crop's biomass may produce healthier food for humans. As a consumer of both food and media reports on organic foods, you should have a better picture of the ecological backstory of this debate having read this chapter.

In the Anthropocene, exploitation dynamics have become extraordinarily important for the conservation of biodiversity. As detailed in Chapter 15, the global increase in species extinctions is fueled in large part by the loss of predators that sit atop food webs (e.g., sharks, lions, wolves). The loss of these predators has consequences for the prey they would normally consume. You can probably predict some of these effects from taking a close look at the Lotka-Volterra models, but we explore this topic in greater detail in Chapter 10.

On the other hand, one of the defining features of the Anthropocene is the exponential rise in the number of non-native species (Chapter 15). Species introduced to a new geographic range by humans are considered **invasive species** when they spread rapidly in their new environment and potentially have large impacts on native systems and species. Although only a small subset of non-native species become invasive, extinction of native prey species via exploitation by a non-native predator is one of the most dramatic impacts of invasive species. Biodiversity losses from species extinctions caused by invasive species often provide some of the clearest evidence that exploitation can have lasting and powerful effects on natural ecosystems (Lockwood et al. 2013).

Finally, the movement of people and animals around the globe has resulted in a massive increase in the number of *emerging infectious diseases* affecting both people and domesticated animals (cows, chickens, goats, horses). An emerging infectious disease is one that recently appeared within a population or is rapidly expanding in its geographic extent. These diseases are caused by a variety of macro- and microparasites that are transported into new host populations through global travel or the introduction of non-native species (e.g., mosquitoes). Emerging infectious diseases have had enormous impacts on human societies, as we have witnessed firsthand with the outbreaks of COVID-19, Zika virus disease, Ebola virus disease, encephalitis, and other deadly illnesses. We are also seeing a large number of emerging infectious diseases in plants (e.g., sudden oak death, rapid ohia death) that pose potentially harsh outcomes on forests, including the urban woodlands that may be located near where you live. An understanding of the models and outcomes of exploitation provides a critical foundation for combating disease and building scientifically informed policies to prevent the occurrence of future disease outbreaks.

SUMMARY

8.1 Everybody Has to Eat

- Exploitation, or the consuming of one species by another, is a common and important force in nature.

- A simple model of exploitation is a (+, −) interaction.

- There are three main forms of exploitation: predator and prey; parasite and host; and herbivore and prey.

Terminological Issues

- Defining the key terms and concepts involved in exploitation can be challenging, but some basics to remember are

  — The exploiter receives a fitness benefit from consuming the prey.

  — The prey experience a loss of fitness because they have been eaten or partially consumed.

  — Predation does not occur if the individual consumed is already dead. Consumption of dead organisms is detritivory.

8.2 Possible Outcomes of Exploitation

- The four potential outcomes are

  — extinction of the prey, followed by extinction of the predator;

  — extinction of the predator, followed by increases in the prey;

  — cyclical population dynamics for the exploiter and prey; and

  — stable coexistence between exploiter and prey.

8.3 The Lotka-Volterra Model

- The model is based on the exponential growth equation (Equation 6.11C).

- Two interacting species are linked using separate equations for each species' population growth.

The Prey Equation

- The prey equation includes an exponential growth term and losses to predation based on prey capture rates.

The Exploiter Equation

- The exploiter equation includes a birth term based on conversion of prey into exploiter offspring and a death term based on natural mortality.

Equilibrium Conditions

- Phase-plane graphs allow us to examine the dynamics of two linked populations and the conditions for equilibrium.

- Using the zero-growth isocline, we can explore outcomes of exploitative interactions.

- Where the two species' isoclines intersect, we have a neutrally stable equilibrium point.

- Phase-plane analysis suggests cyclical dynamics of predator and prey populations is common.

- Extinction is possible if the amplitude of the cycles increases to the point where one of the species' numbers drop to zero.

8.4 Added Realism Stabilizes Exploiter-Prey Dynamics

Exploiter-Prey Model with Density Dependence in the Prey

- The model is based on the logistic equation (Equation 6.13).

- Adding density dependence to the prey produces damped oscillations and is a stabilizing force.

Exploiter Functional Responses

- The consumption rates of prey by predators are called functional responses, which identify the number of prey consumed per predator per unit of time.

- A Type I functional response is a constant or linear rate of prey consumption regardless of prey density.

- A Type II functional response occurs when predators experience satiation at high prey density, resulting in a plateau in prey consumption per predator per unit of time.

- A Type III functional response occurs when predator consumption is low at low prey densities, then increases, and eventually levels off as predators satiate.

- Adding a Type II functional response to the Lotka-Volterra model decreases stability due to larger cycles.

- Adding a Type III functional response to the Lotka-Volterra model decreases stability due to larger cycles at high prey densities; but at low prey densities, it produces damped oscillations and greater stability.

Allee Effects

- At low population densities, some species experience reduced fitness, known as an Allee effect.

- With Allee effects (also called inverse density dependence), per capita growth rates increase with density, but only at low densities.

- Allee effects can be added to the Lotka-Volterra model, producing destabilization, neutral cycles, or stabilizing damped oscillations.

8.5 How Do the Models Fit the Real World for Different Types of Exploitation?

Predation

- Typical predation involves eating animal tissue, and the Lotka-Volterra model applies well to this type of exploitation.

Herbivory

- In typical herbivory, the whole prey organism is not killed or consumed.

- Specialist feeding is common among herbivores (particularly insects).

- Herbivory produces coexistence, cyclical behavior, or extinction of one or both species.

Parasitism

- In parasitism, individual parasites often do not kill or completely consume their host but instead feed or obtain nutrition from them.

- Macro- and microparasites are defined by their size.

- Parasitoids are insects that attack other insects by positioning eggs or larvae to develop inside a host's body, which eventually kills the host.

- Parasites have fairly complex and intricate life histories.

- Parasite dynamics are studied because of their relationship to human health.

8.6 Evolutionary Outcomes of Exploitation

- Exploitative interactions often lead to coevolution (reciprocal adaptive evolution between two or more strongly interacting species that affect each other's fitness).

Exploiter-Prey Coevolution

- The life-lunch idea suggests an asymmetrical evolutionary pressure on prey to avoid being eaten, whereas for exploiters, the pressure is for a single meal.

- **Primary Prey Defense Strategies**

  — The primary defense strategies of prey (i.e., camouflage and crypsis) serve to prevent encounters with predators.

- **Secondary Prey Defense Strategies**

  — Secondary defense strategies send a message to predators to reduce the likelihood of capture or increase the likelihood of escape.

  — These defenses may involve aposematic (warning) coloration or mimicry (Müllerian or Batesian).

- **Tertiary Prey Defense Strategies**

  — Tertiary defense strategies in prey serve to deter an attack once it has started and involve many different structures or strategies.

- **Evolutionary Responses in Exploiters**

  — An evolutionary "arms race" between exploiter and prey can occur.

- **Plant-Herbivore Coevolution**

  — Because plants are sedentary, they have different defensive traits from those found in animals.

  — Constitutive defenses are defenses that are always present; inducible defenses are synthesized when needed.

  — Toxic chemicals (toxins) are a common plant defense; many plant toxins have properties that are valued by humans.

  — Plants also have physical or mechanical defenses.

8.7 Exploitation in the Anthropocene

- Understanding exploitation is increasingly important as we confront the novel challenges of the Anthropocene, including

  — Humans are driving many predators extinct, which is having strong repercussions on their prey.

  — Predatory invasive species are causing extinctions in native species that are prey.

  — A massive increase in emerging infectious diseases is posing a serious threat to people and organisms around the globe.

KEY TERMS

Allee effect
aposematic coloration
Batesian mimicry
camouflage
capture rate (f)
carnivore
coevolution
crypsis
damped oscillation
definitive host
detritivore
evolutionary arms race

exploitation
exploiter conversion factor (c)
exploiter-prey cycle
functional response
generalist
herbivore
host
intermediate host
invasive species
macroparasite
microparasite
mimicry

Müllerian mimicry
parasite
parasitoid
pathogen
predator
prey
recruitment
Red Queen hypothesis
secondary metabolite
specialist

CONCEPTUAL QUESTIONS

1. Define the following terms, and provide examples of each in the natural world around you: *predation, herbivory, parasite, pathogen, parasitoid.*

2. Explain three factors that might lead to oscillations in the abundance of an herbivore. Give an example for each. Recall that we encountered oscillations in single-species population growth (Chapter 6).

3. Evaluate the fit of each type of predator to the Lotka-Volterra exploiter-prey model. Do any of the additions we made to the model to add realism alter your assessments?

4. What is an Allee effect? Offer three examples of situations in which you might see an Allee effect. Can you identify any species in your local area that might have Allee effects?

5. Describe four general prey defense strategies that animals have, and provide real-life examples of each from your local environment.

6. Plants have many adaptations to avoid predation. Describe three types of defensive strategies that plants in your local environment employ, and provide specific examples.

7. Herbivory often involves both specialists and generalists. Describe some of the many ways to be a specialist herbivore using examples from your local environment.

8. Exploiter-prey interactions often lead to what has been described as an "arms race" between the players.

   a. Explain this statement and the process it describes, and provide a specific example from your local environment to illustrate the concept.

   b. How does this evolutionary arms race fit with the Red Queen hypothesis?

QUANTITATIVE QUESTIONS

1. Work through the Lotka-Volterra exploiter-prey equations and identify the units for each element in each equation. Make sure that you can make the units for each term on the right-hand side of the equations match the units on the left-hand side.

2. Using both words and phase-plane graphs, explain how the dynamics of the Lotka-Volterra predator-prey model can produce stable population cycles.

3. Draw one or more phase-plane graphs to illustrate how our models can produce each of the real-world outcomes we see for exploiter-prey interactions (i.e., exploiter extinction, prey and exploiter extinction, cycles, and stable coexistence).

4. Recall the three types of functional responses that predators may have to prey densities.

   a. Draw a graph with prey density on the x-axis and number of prey eaten per predator per unit of time on the y-axis to illustrate these functional responses.

   b. Why would you expect to see any of these?

   c. What types of functional response would be required for a predator to control a prey species that increases to very high densities?

   d. What type of functional response might lead to density-dependent mortality in the prey?

   e. To explain your answer to part (d), draw another graph of the relevant functional response, with prey density on the x-axis and the proportion of prey population eaten per predator per unit of time on the y-axis.

   f. Explain why such a graph illustrates density-dependent prey mortality.

5. Using words and phase-plane graphs, explain how adding a Type II functional response with a carrying capacity could affect the Lotka-Volterra exploiter-prey model dynamics.

6. Consider a hump-shaped isocline for prey in the Lotka-Volterra model. Explain how three different outcomes can occur depending on where the predator isocline intersects the prey isocline.

7. Non-native species often enter communities and maintain small, contained populations for decades or generations before their population growth suddenly explodes and they begin to spread across a landscape.

   a. Explain this pattern based on the basic pattern of population growth from Chapter 6.

   b. Can you see a way that Allee effects might contribute to this pattern?

8. We often use slightly different models than those discussed in this chapter to explore diseases, but we could use the Lotka-Volterra model to explore the dynamics of humans and one of our pathogens.

   a. Can you think of a pathogen that seems to cycle through human populations?

   b. Describe a Lotka-Volterra pathogen isocline in the context of human populations and the density required to maintain a pathogen in a population.

   c. How do you think global human travel might affect the dynamics of this system?

SUGGESTED READING

Brodie, E. D., III and E. D. Brodie Jr. 1999. Predator-prey arms races: Asymmetrical selection on predators and prey may be reduced when prey are dangerous. *BioScience* 49(7): 557–568.

Denno, R. F. and D. Lewis. 2009. Predator-prey interactions. Pp. 202–212 in S. A. Levin, ed. *The Princeton Guide to Ecology*. Princeton, NJ: Princeton University Press.

Gause, G. F. (1934) 2019. *The Struggle for Existence: A Classic of Mathematical Biology and Ecology*. Mineola, NY: Dover. First published by Williams and Wilkins, Baltimore.

Groom, M. J. 1998. Allee effects limit population viability of an annual plant. *American Naturalist* 151(6): 487–496.

Korpimäki, E., L. Oksanen, T. Oksanen, K. Norrdahl, and P. B. Banks. 2005. Vole cycles and predation in temperate and boreal zones of Europe. *Journal of Animal Ecology* 74(6): 1150–1159.

A honeybee visits a flower to acquire food and becomes covered in pollen in the process. As it forages, the bee inadvertently transfers pollen from one flower to another, thereby providing cross-pollination for the plant population. Two-species interactions like this, where both species benefit, are defined as mutualisms.

chapter 9

MUTUALISM AND FACILITATION

LEARNING OBJECTIVES

- Distinguish among the different types of beneficial interactions between species.

- Differentiate between obligate and facultative mutualism using examples.

- Summarize the major benefits of mutualism.

- Explain why variations on the Lotka-Volterra models do not work well for examining mutualism.

- Summarize the game theory approach to understanding mutualism.

- Express in words and figures a basic understanding of an economic trade model of mutualism.

- Explain facilitation and how it relates to foundation species.

- Summarize how success in restoration ecology may be enhanced by an understanding of positive ecological interactions.

- Explain invasional meltdown and how it relates to mutualism.

9.1 Why Can't We All Just Get Along?

Sometimes when we look out at the natural world, instead of fighting, death, and predation, we find examples of collaboration and cooperation. What could be more emblematic of this harmony than watching the amazing interplay between flowers and bees on a sunny summer day? We all know that the bees facilitate plant reproduction by moving pollen among individual flowers. But the interaction is not just good for the plants; the bees get something out of it too. The bees consume the sweet nectar produced by the flowers and feed their larvae a rich diet of pollen collected from the plants' stamens (i.e., male reproductive parts). Both species participating in this interaction benefit. Yet as we will see, this seemingly simple interplay between two species is way more complex and interesting than you might imagine if you casually watch bees buzz between flowers in your garden or at the park.

Until now, the multispecies interactions we have focused on result in negative fitness outcomes for at least one of the two species involved, with (−, −) interactions in interspecific competition and (+, −) interactions in exploitation. As discussed in Chapters 7 and 8, such interactions create strong selective forces and can be crucial in structuring which species are part of an ecological community. Yet, interactions with beneficial effects for one or more of the species, (+, 0) or

(+, +) in our notation, are also common and are an extremely important ecological force. The general term for interactions that benefit at least one species and have either a positive or net zero effect on the other species is **facilitation**. Within the larger category of facilitative interactions, **mutualism** describes interactions in which both species experience a benefit, and **commensalism** describes interactions that benefit one species but neither benefit nor harm the other. The former are represented as (+, +), and the latter as (+, 0).

Facilitative interactions are common in nature and are often pivotal in terms of creating the world we see around us (Stachowicz 2001). Without facilitative interactions, eukaryotic cells, with their beneficial mitochondria "power plants," would not have evolved. (You may remember from general biology classes that mitochondria were originally engulfed prokaryotic cells.) On land, mutualistic interactions such as pollination helped drive the evolutionary diversification of flowering plants throughout the Earth's history. In our oceans, some of the most productive and diverse habitats stem from strong mutualistic interactions that allow the formation of coral reefs. These facilitative interactions, along with many others, are essential to the maintenance of much of the world's biodiversity.

Although the ecological literature contains many examples of mutualism and facilitation, much of this published research focuses on the biology of the interactions rather than on the larger theory governing the dynamics and evolution of positive interactions (Bronstein 1994). As a result, our theoretical understanding of mutualism and facilitation lags behind our understanding of interspecific competition and exploitation. This gap is partially because the classic modeling approaches used in Chapters 7 and 8 suggest that mutualistic interactions are less stable than exploitative or interspecific competitive interactions. Unstable model outcomes imply that the interactions are unlikely to persist and, therefore, unlikely to be important in natural systems. But the natural world holds many examples of facilitation and mutualism, and in this chapter, we address some of the shortfalls of earlier models and explore how recent ones suggest that mutualism makes critical contributions to the success of many interacting populations. We start by categorizing mutualistic interactions, also called mutualisms, based on their biological outcomes, then review several recent efforts to model them, and end by considering the substantial role that facilitation plays in applications of ecological science.

9.2 Obligate and Facultative Mutualism and Partner Specificity

One of the difficulties in studying mutualisms arises from the fact that mutualistic interactions exist along a continuum in the natural world. This range of variation means that when we categorize or label interactions, we artificially impose a discrete framework on something that is more continuous in nature. It should come as no surprise, then, that a vast array of interactions falls under the category of mutualism. These range from strong, necessary-for-survival interactions between two specific species to weaker, more diffuse interactions that sometimes occur and sometimes do not and that may involve a number of potential partners. In the broadest sense, a mutualistic interaction can be defined as either an **obligate mutualism**, meaning it is necessary for the survival of at least one of the participating species, or a **facultative mutualism**, meaning that the interaction improves the fitness of one of the species but is not necessary for its continued existence.

One potential source of confusion is that a mutualism may be obligate for one partner but not for the other, meaning we could have *obligate-obligate*, *obligate-facultative*, or *facultative-facultative* mutualistic interactions. This range of types means that mutualisms produce a multitude of different interactions that vary widely in their importance. Not surprisingly, all this variation has also slowed down ecologists' ability to produce broadly applicable models and generalizations about mutualism.

Some of the most familiar mutualisms fit into these categories of obligate and facultative. For example, warm-water corals, with their beneficial interaction between coral polyps and zooxanthellae, are an excellent example of an obligate-obligate mutualism in that each species needs the other to survive for more than short periods of time. (Coral can exist for short periods in a "bleached" state without zooxanthellae, but long periods without their partners lead to whole reefs dying.) As for obligate-facultative mutualisms, many flowering plants rely heavily on other species for pollination, but their pollinating insects are often (though not always) able to choose from a wide variety of food resources and do not need any particular plant species. Finally, generalized seed dispersal offers an example of a facultative-facultative mutualism. Many animal dispersers (e.g., bears, foxes, jays) can survive without ingesting the fruit or seeds they disperse, and most plants can reproduce without dispersal, but the mutualism increases the fitness of individuals of both partners. Let's explore each of these types of mutualism with some examples.

Leafcutter ants (of which there are more than 40 species in the genera *Atta* and *Acromyrmex*; **Figure 9.1A**) cultivate belowground fungal gardens and rely on these fungi for nourishment. All species of leafcutter ants tend fungi from the Lepiotaceae family. The ants nourish their fungus garden with freshly cut plant leaves and keep it free from insect pests and molds (**Figure 9.1B**). In return, the fungus grows nutritious and enlarged hyphal structures, called gongylidia, produced in bundles that feed the ants. In addition to providing digestible protein, the gongylidia are also rich in enzymes, which the ants do not digest. These undigested enzymes come out in the ants' feces, which they deposit on top of the fresh leaves. The fecal enzymes help break down the leaves, making their nutrients more accessible to the fungus. The fungus cannot survive without cultivation by the ants, and ant larvae require nutrition from the fungus to survive (De Fine Licht et al. 2013). Evolutionary biologists have speculated that this mutualistic partnership between ant and fungus has existed for somewhere in the neighborhood of 30 million years. Other examples of clear obligate-obligate mutualisms include the algal and fungal partners that make up lichen, many figs and their wasp pollinators, and some ant-plant associations.

The leafcutter ants and their fungal mutualists display a common feature of many obligate-obligate mutualisms, namely, species specificity of the relationship. Although there are many species of leafcutter ants and many species of tended fungal mutualists, the relationship between each pair tends to be very specific. This specificity makes sense. If each species needs the other to survive, then they are unlikely to be generalists and have these beneficial relationships with multiple species. For the leafcutter and fungi example, when ant queens establish a

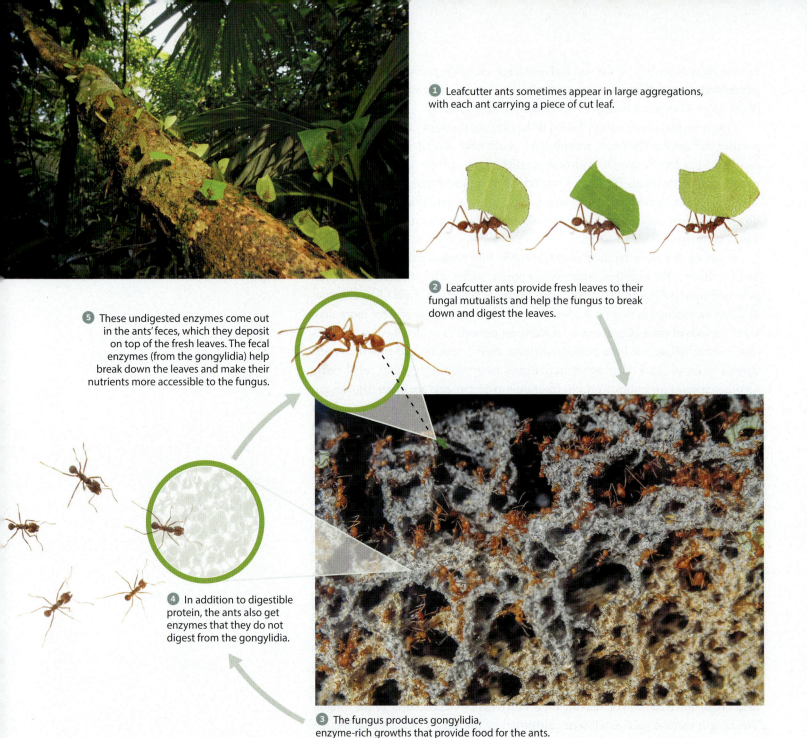

① Leafcutter ants sometimes appear in large aggregations, with each ant carrying a piece of cut leaf.

② Leafcutter ants provide fresh leaves to their fungal mutualists and help the fungus to break down and digest the leaves.

⑤ These undigested enzymes come out in the ants' feces, which they deposit on top of the fresh leaves. The fecal enzymes (from the gongylidia) help break down the leaves and make their nutrients more accessible to the fungus.

④ In addition to digestible protein, the ants also get enzymes that they do not digest from the gongylidia.

③ The fungus produces gongylidia, enzyme-rich growths that provide food for the ants.

Figure 9.1 Leafcutter ants in Ecuador sometimes appear in large aggregations, with each ant carrying a piece of cut leaf. The relationship between leafcutter ants and their fungi gardens is an obligate-obligate mutualism (De Fine Licht et al. 2013).

new colony, they take a chunk of fungal material with them, maintaining their relationship with that one species. Research suggests that their fecal matter (used as fertilizer) does not work well with other fungal species and that the gongylidia of other species do not produce the specific nutritional requirements targeted to that ant species (De Fine Licht et al. 2013).

When relationships between mutualistic partners involve some degree of **generalism**, meaning at least one partner is able to swap and use a different species as its mutualistic partner, the mutualism is more likely to be facultative for at least one partner. Somewhere between 10% and 60% of flowering plants are able to self-pollinate, but visits by pollinating insects increase fitness and genetic

(A)   (B)   (C)

Figure 9.2   (A) Sweet or common violets (*Viola odorata*) are pollination generalists and produce early season flowers that are pollinated by a range of bee species. (B) Late in the season, though, these plants produce smaller, closed flowers that sit close to the ground and are self-pollinating. The closed flower forces the anthers to touch the stigma (top of the pistil, or female portion of the flower) and leads to self-pollination. (C) This plant also has a mutualistic interaction with ants. The seeds have a fatty appendage called an elaiosome that is a reward for ants that carry off the seeds.

variation in their progeny. For example, sweet violets (*Viola odorata*) produce two types of flowers: ones that attract a range of pollinator species and ones that self-pollinate (**Figure 9.2**). For most of the season, they produce insect-pollinated flowers, but late in the season, self-pollinating flowers develop if insect pollination has not succeeded. In the absence of the generalist pollinating insects, the self-pollinating flowers still provide positive fitness benefits for the plant; therefore, the pollination mutualism is facultative and generalistic.

But what about for the pollinators? We mentioned that many pollinators can choose from a range of food resources and may not need specific plant species to survive. Remember that common pollinators range from bees, flies, and other insects to birds and bats. Let's focus on ones that require the mutualism. For many bees, the process of pollination provides something essential to survival, whether it is nectar, pollen, or oils. The *Osmia* bee species in Figure 9.2A is a solitary bee (so all females can produce offspring, unlike in colonial bees, which have one reproducing queen), and males use pollen to support the offspring. The fact that they require the products of the mutualism for survival makes the mutualism obligate for them. Although some bee species are specialists, the ones we see and hear about the most—such as honeybees and bumblebees—are generalists and will visit many plant species.

Facultative-facultative mutualisms can be somewhat context dependent, may exist between surprising partners, and are even more likely to involve generalists. For example, when a black bear (*Ursus americanus*) eats manzanita berries (*Arctostaphylos* spp.) and then walks several miles and excretes the seeds (**Figure 9.3**), both the manzanita and the bear benefit. The bear obtains energy from the manzanita berry, and the manzanita gets its seeds dispersed to a new location. Both species can maintain their populations without the mutualism and both can potentially interact with other mutualist partners, but individuals of both species certainly enjoy fitness benefits from the interaction.

There is one last semantic issue concerning mutualisms that can be confusing. Many of the commonly discussed mutualisms involve **symbiotic relationships**, or *symbioses*, and in everyday speech or articles written for the general populace, the terms *mutualism* and *symbiosis* are often used interchangeably. Symbiotic

Figure 9.3   Black bears (*Ursus americanus*) and manzanita shrubs (*Arctostaphylos* spp.) are facultative partners in a loose mutualism. When bears eat manzanita berries, they receive a caloric benefit, and depositing the seeds far from the parent plant offers a transportation or dispersal benefit to the manzanita plant.

Figure 9.4 Many symbiotic interactions are detrimental or pathogenic, like the interaction between the bacterial symbiont *Vibrio shiloi* and the colony-forming coral *Oculina patagonica*. On the left, the zooxanthellae impart color to the coral; on the right, coral with symbiotic *V. shiloi* is bleached and without color due to loss of zooxanthellae. The bleaching can lead to coral death.

relationships, though, are characterized by close physical contact, often with integrated bodies and physiology, and only some of them involve beneficial interactions. Lichen (symbiosis of algae and fungi) and reef-building coral species (symbiosis of coral polyps and photosynthesizing zooxanthellae) are both examples of symbiotic mutualisms. However, corals also form detrimental or even pathogenic symbioses with some species. For example, the bacterium *Vibrio shiloi* can form an antagonistic symbiotic relationship with the coral *Oculina patagonica*, and the presence of the bacterium kills the zooxanthellae that give corals their characteristic colors, with resultant bleaching and coral death (**Figure 9.4**). Detrimental symbioses fall into the categories of pathogens and microparasites (discussed in Chapter 8). From the examples given here, though, it is clear that a considerable number of mutualisms involve free-living organisms that only come in contact periodically, making them far from symbiotic. When mutualisms are symbiotic, they are often also obligate-obligate and very partner-specific.

9.3 Benefits of Mutualism

An alternative approach to categorizing the diversity of mutualistic interactions is to focus on the ecological benefits of the interaction. First, though, it is important to point out that mutualistic interactions are not instances of **altruism**, meaning the goal is not to offer benefit to other individuals or to the other species. Mutualisms are created by two self-interested individuals from two different species, with each individual trying to gain a fitness advantage over other individuals in its own species. In fact, mutualistic partners likely have no idea that they are providing a benefit to the other species. In addition, mutualistic interactions between two species generally have both costs and benefits, and the balance of costs and benefits can make mutualisms more variable in nature and potentially harder to find and characterize. We address this balance more in later sections of this chapter, but here we characterize the sorts of exchanges we see in mutualistic interactions. Most mutualisms work because the two partners receive different benefits. We can divide the benefits into three general categories: (1) transportation, which primarily affects reproduction and dispersal; (2) energy or nutrients; and (3) protection or defense.

Transportation: Reproduction and Dispersal

One of the most familiar examples of mutualism is pollination. The obvious benefit to the plant involves receiving pollen from, or transferring pollen to, another flower. The transported pollen enables cross-pollination and sexual reproduction. As we discussed earlier, even in plants that can self-pollinate, transported pollen introduces new genetic material and can lead to the production of seeds with higher fitness than those produced by self-pollination. And the bees and other pollinators benefit by obtaining food, usually in the form of pollen, nectar, or oils.

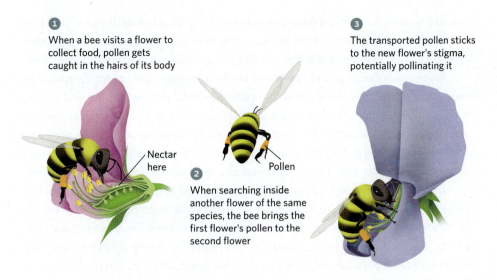

① When a bee visits a flower to collect food, pollen gets caught in the hairs of its body

② When searching inside another flower of the same species, the bee brings the first flower's pollen to the second flower

③ The transported pollen sticks to the new flower's stigma, potentially pollinating it

Nectar here

Pollen

Figure 9.5 Bees consume nectar and feed pollen to their larvae. In the process of drinking nectar and filling pollen sacs on its legs, a bee will bring its hairy abdomen and thorax in contact with sticky pollen grains. When the bee moves to a new flower, it accidentally transfers pollen from its body to the stigma (female structure) of the new flower. Some flowers have evolved physical structures to facilitate this process.

Yet it is important to recognize that a bee pollinator does not transport plant gametes (pollen) with the goal of helping or enabling the plant's sexual reproduction. Instead, it is enticed into close proximity with the plant's gametes by the offer of an energy source within the plant's flower structure. That energy source may be nectar for the bee itself or pollen that it collects for larvae. While searching inside flowers for nectar, oils, or pollen, additional pollen gets stuck to the bee's hairy abdomen and thorax (**Figure 9.5**). When the bee then visits the flower of another plant of the same species, it inadvertently transports the stuck pollen to the flower of that plant, enabling cross-pollination. Many plants have structures that facilitate the transfer of pollen onto pollinators and then onto the female reproductive structures of the next plant, thus increasing the plant's sexual reproduction (Figure 9.5).

More than 100,000 different animal species are known to serve as pollinators for the nearly 250,000 species of flowering plants, including recently discovered marine invertebrates that pollinate seagrasses (Van Tussenbroek et al. 2016). Most of these pollinators are insects (bees, wasps, butterflies, moths, and beetles), but more than 1,500 species of birds and mammals have also been recorded transferring pollen.

Another common form of transportation benefit comes from the dispersal of seeds to new locations, particularly when the seeds are heavy and do not move well by riding on the wind or water currents. Transport may be accidental in these dispersal mutualisms, because unnoticed seeds are often attached to animal fur, hooves, or nest-building material. Such "hitchhiking" seeds may be deposited in new locations far from the parent plant. This movement allows the plant species to disperse (see Chapter 12) and avoids intraspecific competition between parents and offspring. When the animal gets no benefit from this transportation, this dispersal is better described as a commensal interaction (commensalism), rather than a mutualistic interaction.

Dispersal is often critical for plants, to the extent that many plants evolve adaptations that encourage animals to take their seeds to distant locales. In tropical forests, somewhere between 50% and 75% of the tree species produce fruits whose seeds are transported by animals. Often animal dispersers eat and consume the seed itself, because the plant produces enticing fruits to facilitate such consumption. Seeds are tiny, complete organisms, so it may seem strange that plants would encourage other organisms to eat their offspring, but such mutualisms are extremely common. In addition to the example of the bear and manzanita, other examples include many fruits that we also eat, such as raspberries and tomatoes.

Sometimes the seeds are destroyed in the disperser's gut, as we destroy most seeds that we eat, but many seeds will pass through a consumer's gut and remain viable. In fact, some plant seeds are actually more viable after they pass through an animal gut because the interaction with digestive acids breaks the seed coat and brings the seed out of dormancy. This process is called scarification, and it can act as a cue that the seed has been moved to a new habitat and should germinate.

Other plants entice dispersers with coatings or appendages that are rich in carbohydrates or lipids. In such cases, the disperser does not generally eat the seed, just the surrounding fruit or fatty deposit. You have likely been a seed disperser yourself on many occasions as you gathered fruits (cherries, peaches, watermelons) from a grocery store, garden, backyard, or farm, and then discarded the seeds in a garbage can or compost bucket, or along the side of the road.

Many seeds are adapted specifically to ant dispersal and have a lipid-rich outer coating that ants eat after carrying the seed to their colony or to a site away from the parent plant. These lipid bodies are called *elaiosomes*, and we find them on the seeds of *Viola odorata*, the sweet violet mentioned earlier (see Figure 9.2C). The seeds of many violet species are dispersed by ants in exchange for nutritional elaiosome rewards.

Acquisition of Energy and Nutrients

In the examples just described, the mutualistic benefit for plants is the ability to move genes or offspring to new habitats and thus avoid competition and start new populations. For the animal dispersers, though, the benefit is general energy or nutrients. This type of benefit is one of the most common mutualistic benefits, yet in many mutualistic interactions, the transfer of energy is less obvious than it is in pollination or seed dispersal. We only need to look as far as our own intestinal tracts to see examples of such mutualisms.

The human digestive system supports numerous species of microbes, collectively referred to as its **microbiota**. These bacterial species help us digest our food, often breaking down cell walls and allowing us to absorb carbohydrates, fats, and essential vitamins we could not access on our own. In exchange, our gut provides a protected environment for them, and we conveniently feed them multiple times a day by bathing them in the chewed food we consume.

The relationship between humans and the microbial organisms, or flora, in our gut is a complex and multifaceted mutualism involving many different taxa. We can see how important and dynamic this mutualism is by looking at the colonization of the intestinal tract in newborn babies. Work by Jeremy Koenig and colleagues (2011) shows that, from birth through the first two-and-a-half years, the diversity of microorganisms in a human baby's gut increases steadily, approaching that of the mother (**Figure 9.6**). The importance of this mutualism is also

Figure 9.6 The diversity of microorganisms found in a newborn child's intestine from birth through 850 days of age, as measured through stool samples, steadily increases, approaching that of the mother. The measure of diversity used here is called phylogenetic diversity, which is discussed in detail in Chapter 11 (Koenig et al. 2011).

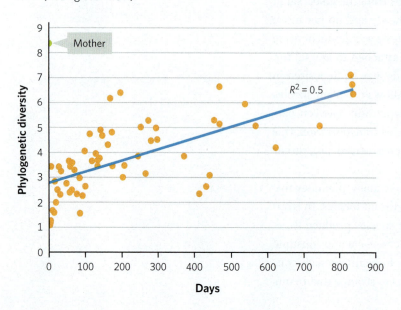

painfully evident when we disrupt it through the consumption of pathogens, such as those from tainted food or water, or even when we combat those pathogens with antibiotics. As anyone can attest after a bout of food poisoning, having unhealthy gut microbiota is a fairly miserable experience.

Interestingly, in some cases our gut microbiota can competitively exclude unhealthy microbiota and pathogens. In recent years, scientists have shown that recolonizing damaged microbiota using a dose of healthy bacteria can alleviate the symptoms of a number of debilitating diseases, such as irritable bowel syndrome, colitis, fatty liver disease, and multiple sclerosis, among others (**Figure 9.7**). In this medical procedure, called fecal microbiota transplant, fecal matter is collected from a healthy donor, mixed with a saline solution, and placed in a patient's lower gastrointestinal tract by colonoscopy or enema. A few decades ago, Western medicine would have considered such a procedure to be insanely dangerous and a source of contamination (and some of you may be feeling a bit squeamish right now too). Scientists now recognize, though, that fecal microbiota transplant can replenish mutualistic bacteria that have been killed or suppressed in a patient's gut, whether by strong antibiotics or pathogens (Smits et al. 2013). It is important to keep in mind that our understanding of the gut microbiome and its host-biota interactions is in its infancy, and the future likely holds many additional interesting discoveries.

Intestinal tracts are not the only places where organisms exchange nutrients for mutual benefit. Plants and fungi offer one of the most commonly observed examples of mutualistic interactions in which both partners receive nutrients or energy (**Figure 9.8**). Many species of soil-living fungi are excellent at acquiring nutrients such as nitrogen and phosphorus from soil. These fungi, known as *mycorrhizal fungi* or *mycorrhizae*, live among and sometimes within plant roots and can provide these nutrients for their mutualistic plant partners. The plants in turn supply the fungi with carbohydrates from photosynthesis. These fungi-plant mutualisms have been in existence for at least 400 million years, as indicated by the fossil record (Remy et al. 1994). Mycorrhizal mutualisms are present in nearly 80% of all plant species and are perhaps the most common mutualistic association found in the plant kingdom.

Protection and Defense

Finally, mutualisms can involve the protection or defense of one or both of the mutualistic partners. Our gut flora receives this mutualistic benefit from us (in addition to energy and nutrients), but many other examples exist in nature as well. One of the most well-known involves ants in the genus *Pseudomyrmex* and Neotropical acacia trees in the genus *Vachellia*. The tree produces swollen hollow thorns for the ants to use as places to nest, while also providing food

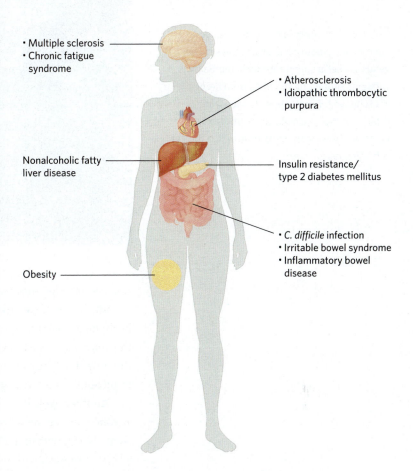

- Multiple sclerosis
- Chronic fatigue syndrome

- Atherosclerosis
- Idiopathic thrombocytic purpura

Nonalcoholic fatty liver disease

Insulin resistance/ type 2 diabetes mellitus

- *C. difficile* infection
- Irritable bowel syndrome
- Inflammatory bowel disease

Obesity

Figure 9.7 Several medical disorders are associated with alterations to the microbiotic community of the human intestine and could be treated by fecal microbiota transplant (Smits et al. 2013).

Figure 9.8 Mycorrhizae are soil fungi that can form symbiotic associations with plant roots to exchange resources. In general, the plants offer carbohydrates to the mycorrhizae, and the mycorrhizal fungi offer critical resources such as phosphorus to the plant.

Figure 9.9 Ants in the genus *Pseudo-myrmex* and Neotropical acacia trees in the genus *Vachellia* sometimes form mutualisms, with ants protecting the trees from herbivory in exchange for protection (homes within thorns) or food such as nectar from extrafloral nectaries (the green bump from which the ant is feeding here).

sources for the ants (the Beltian bodies shown in Figure 8.38, or nectar from extrafloral nectaries like the one in **Figure 9.9**). The ants protect the trees from herbivores, particularly insects, by attacking anything that touches the tree. At the smallest disturbance, the ants swarm out from the acacia thorns and attack the intruder. They also protect the trees from encroaching competitors by trimming back the leaves of neighboring plants with their mouthparts.

Another well-known defense-protection mutualism is probably familiar to anyone who has seen the movies *Finding Nemo* and *Finding Dory*. In these movies, Nemo is a clownfish in the family Pomacentridae. Species of this family form mutualisms with sea anemones (**Figure 9.10**). The fish gain protection from predators by spending most of their time within the center of a vast arsenal of stinging nematocyst cells on the anemone tentacles. The fish's mucus provides protection from the anemone's stinging nematocysts. In fact, research suggests that chemical aspects of the clownfish's mucus cause the anemone to "see" the fish biochemically as an extension of itself. The anemone, in return, is defended by the clownfish from its own predators and parasites. In addition, the clownfish's feces are a source of nutrients, particularly nitrogen, that aid the anemone in its growth and reproduction.

Figure 9.10 Clownfish in the family Pomacentridae form mutualistic interactions of the defense-protection type with sea anemones.

9.4 Models of Mutualism

A Conceptual Model of Mutualism

Despite all these different ways to trade benefits, there are some basic commonalities to all mutualistic interactions that we can use to structure our thoughts about how the interaction works and to understand how mutualisms increase fitness for individuals of both species. Let's start by thinking about the elements that we know are needed in a mutualism model.

First, any model of mutualism needs to show a positive interaction. For a basic model, it seems reasonable that benefits to individuals of each species would increase as the density of the mutualistic partner increases. Given that each species is acting for its own benefit and only tangentially or accidentally assisting the mutualistic partner, this increase may not be linear. In fact, it makes sense that there may be limits to the benefits of mutualists, but we can start with a simple linear model.

Eventually, we may want to consider a model that allows the interaction to shift from positive to negative to highlight the fact that mutualistic interactions are not static and may vary by location or with the density of competitors. In fact, some mutualisms can shift over to exploitation given the right conditions. A model like this would also be useful for exploring the evolution of mutualisms. An additional consideration is that we may need different models for facultative and obligate mutualisms. The ability to survive without a mutualistic partner leads to some fundamental differences.

Because mutualisms are so varied, the potential ways to address them are also more diverse than for competition and predation. We start with the most basic models that address these points and build on models familiar from our explorations of interspecific competition and exploitation. Because these models have some fundamental flaws, we move to other formulations that allow us to explore the conditions for successful or stable mutualistic interactions and offer some insight into evolutionary pressures.

Lotka-Volterra Approaches

In the preceding chapters, we used Lotka-Volterra variations of the logistic growth model to examine interspecific competition and exploitation to good effect. When we turn to modeling mutually beneficial interactions, it is logical to start with the same base model.

A Mutualism Model Based on the Competition Model   The Lotka-Volterra two-species competition equation reflects our schematic model for competition $(-, -)$. Is it possible to alter this same model to explore a $(+, +)$ interaction instead? Recall the mathematical expressions of the Lotka-Volterra competition model (Equations 7.4 and 7.5, repeated here and renumbered as Equations 9.1A and 9.1B, respectively):

$$\frac{dN_1}{dt} = r_1 N_1 \left(\frac{K_1 - N_1 - \alpha N_2}{K_1} \right) \quad (9.1A)$$

$$\frac{dN_2}{dt} = r_2 N_2 \left(\frac{K_2 - N_2 - \beta N_1}{K_2} \right) \quad (9.1B)$$

Notice that the negative effects of the other species appear with the competition coefficients multiplied by the density of the other species (αN_2 and βN_1). Recognize, also, that these terms effectively reduce the amount of the carrying capacity (K) that is available to the focal species. Put simply, in interspecific competition, species 2 has a detrimental effect on the population growth of species 1, and species 1 has a detrimental effect on the population growth of species 2. Remember that, as in Chapters 7 and 8, we are using the term *density* instead of *abundance* for N_1 and N_2. This usage helps to remind us that the populations of both species are in a bounded, specific area, and that any changes in the number of organisms in that area may have density-dependent effects.

For mutualisms, we want to flip the competition interaction because individuals of species 2 will have a positive effect on the individuals of species 1, and vice versa. We can easily make this transformation by altering the sign (from negative to positive) of the coefficients:

$$\frac{dN_1}{dt} = r_1 N_1 \left(\frac{K_1 - N_1 + \alpha N_2}{K_1} \right) \tag{9.2A}$$

$$\frac{dN_2}{dt} = r_2 N_2 \left(\frac{K_2 - N_2 + \beta N_1}{K_2} \right) \tag{9.2B}$$

This altered set of equations says that the addition of the mutualist species will provide a benefit for the focal species, essentially by allowing it to increase its carrying capacity. Increasing the carrying capacity seems most appropriate for mutualisms with nutrient and energy benefits and potentially less useful for those with a transportation or protection benefit. Before we worry about this specificity, let's see how well this model works for this subset of mutualistic interactions.

We can explore the outcome of this mutualism model with a phase-plane graph. As in previous chapters, we can find the zero-growth isocline for each species by setting the growth equation for that species to zero and exploring the situations that lead to no change (i.e., zero population growth). For Equations 9.2A and 9.2B, our isoclines are going to look very much like our competition isoclines (Equations 7.6 and 7.9) but with a different sign in front of the coefficient for the effect of the other species:

Species 1: $\qquad 0 = K_1 - N_1 + \alpha N_2 \qquad$ or $\qquad N_1 = K_1 + \alpha N_2 \qquad$ (9.3)

Species 2: $\qquad 0 = K_2 - N_2 + \beta N_1 \qquad$ or $\qquad N_2 = K_2 + \beta N_1 \qquad$ (9.4)

As with the isoclines for competition, each of these equations is an equation for a line. The difference here is that the equations have positive slopes because the terms representing the effect of the other species are positive. Therefore, each isocline will only have one axis intercept, and that point will represent the density of the focal species when it is alone.

For facultative mutualisms, that intercept should be the carrying capacity, where we expect each population to stabilize without the other species. The two isoclines from these intercepts can behave in two basic ways: either they cross, or they do not cross (**Figure 9.11**). Let's think about when we expect to see each of these outcomes. If we compare Figure 9.11A and Figure 9.11B, we can see that the isoclines that look "steeper" will intersect and actually represent a weaker mutualistic interaction. This interpretation is a little counterintuitive, so we will walk you through it.

Start with the isocline for species 1, and think about the fact that population growth for species 1 is measured along the x-axis. If species 2 has a strong positive effect on species 1, then a small change in the population density of species 2 (a small move up the y-axis) would lead to lots of growth in the population density of species 1 (a big change along the x-axis), or a species 1 isocline that lies close to the x-axis, as in Figure 9.11B. Similarly, for species 2's isocline, a strong mutualistic effect would mean that small increases in species 1 would lead to big changes in the population density of species 2, or an isocline that lies closer to the y-axis (again as in Figure 9.11B). Based on this reasoning, we can see that Figure 9.11A, with the intersecting isoclines, represents a weak facultative mutualistic interaction, and Figure 9.11B represents a strong facultative mutualistic interaction.

Now let's consider population growth for each species in each section of the phase-plane graphs in Figure 9.11. As in Chapter 7, these isoclines show the place where positive population growth has slowed to zero. If we start on the x-axis and move toward K_1, we know from Chapter 6 that the population density of species 1 should grow until we pass K_1 and should start to decrease after we pass K_1. That means that the population density of species 1 *decreases* below (or to the right of) the dark green line and *increases* above (or to the left of) the dark green line. The population density of species 2 will increase below K_2 (below the blue line) and decrease above K_2 (above the blue line). Our eyes interpret "above" and "below" a little differently with these positive slopes, so these trajectories are at first a little surprising. It makes intuitive sense, though, that increases for each species above the carrying capacity would only occur in the presence of enough mutualistic partners to support that increase. These growth patterns lead to the arrows in Figure 9.11 and show the following two outcomes: with weak facultative mutualistic interactions, we should always get a stable equilibrium (Figure 9.11A); and with strong facultative mutualistic interactions, the populations grow without limit and essentially "explode" to infinite population density (Figure 9.11B). Such growth cannot happen in the real world, however, which suggests that this model may not be the best one to use for mutualism.

An additional problem with these Lotka-Volterra mutualism equations arises when we try to model obligate mutualisms. Any species for which the mutualism is

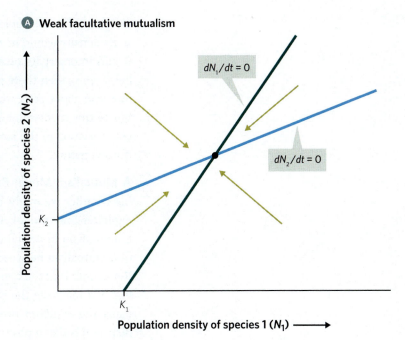

A Weak facultative mutualism

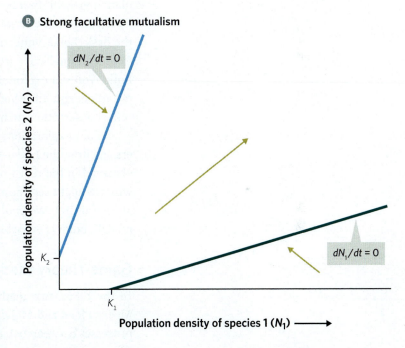

B Strong facultative mutualism

Figure 9.11 A mutualism model based on the Lotka-Volterra competition model leads to two outcomes for facultative mutualisms depending on the strength of the interaction (i.e., the α and β from Equations 9.3 and 9.4). **A** If the interaction is weak, then we observe a small increase in the population density of the focal species with a large or moderate increase in the population density of the mutualistic partner. The arrows lead to a stable equilibrium point with each population above its carrying capacity. **B** When the interaction is stronger, each mutualistic partner increases so much in the presence of the other species that the two isoclines do not intersect, and there is no stable equilibrium; instead the populations of the two species increase exponentially without limit.

obligate will not be able to survive without the mutualistic partner by definition, so its density would have to be zero in the absence of the partner, and its isocline would intersect its own axis at (or below) zero. If the isoclines intersect at or below zero, then there is no stable, nonzero population for either species. If they intersect above zero, we get an unstable equilibrium point (as shown for strongly interacting species in Figure 9.11) with unfettered growth. In other words, an obligate variation of the model also does not produce outcomes that reflect observations in nature.

A Mutualism Model Based on the Exploitation Model You may be wondering whether we might have better success if we started with the Lotka-Volterra exploiter-prey model. It makes sense that such a model might work well for a mutualism in which one species consumes a part of the other, as in the case of a mutualism between a plant and an animal that disperses the plant's seeds. Remember that the benefit of dispersal is often accidental on the part of the animal and may cost the plant some consumed seeds or pollen. The problem with using the exploiter-prey modeling approach for mutualistic consumption and dispersal is that a positive effect of such consumption removes the limits to population growth. In a model without such limits, both populations grow to infinity again. Recall that predators were the only limit to prey population growth in the basic Lotka-Volterra exploitation model. Even if we add a carrying capacity for one mutualistic partner (akin to a prey carrying capacity) to such a modified Lotka-Volterra exploiter-prey model, the two populations explode to infinity or fail to coexist. In other words, this model also fails to produce outcomes that reflect observations in nature.

In our explorations of interactions between competitors or between exploiters and prey, the Lotka-Volterra models were useful because they led to outcomes observed in the real world and offered easily digestible insights into where and why we might see these outcomes. For mutualisms, the same model approaches are not useful. We could try adding many more features and dimensions to the models, but it is probably better to try a different approach.

Game Theory Models

In the 1950s, two mathematicians working at the RAND Corporation, Merrill Meeks Flood and Melvin Dresher, produced a model that examined cooperation processes between two parties. This model was eventually dubbed the **prisoner's dilemma** because it works well as a metaphor for two prisoners weighing their options while being interrogated by police for a crime they committed together. The model falls into the general category of mathematical **game theory**, which is the study of conflict and cooperation between intelligent, rational decision-makers. The approach is used in such diverse fields as economics, political science, psychology, and biology.

The prisoner's dilemma model was first applied to biology in the 1980s by Robert Axelrod, who studied the evolution of cooperation. Ecologists and evolutionary biologists began using game theory to explore why mutualisms evolved and the possible outcomes of mutualistic interactions. We use a slightly altered crime metaphor here to describe the basics of a game theory model and its outcomes and then translate these outcomes to questions of ecological mutualisms.

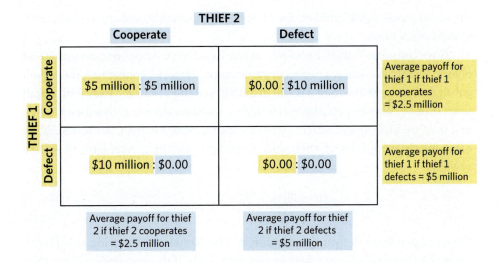

THIEF 2

	Cooperate	Defect
THIEF 1 Cooperate	$5 million : $5 million	$0.00 : $10 million
THIEF 1 Defect	$10 million : $0.00	$0.00 : $0.00

Average payoff for thief 1 if thief 1 cooperates = $2.5 million

Average payoff for thief 1 if thief 1 defects = $5 million

Average payoff for thief 2 if thief 2 cooperates = $2.5 million

Average payoff for thief 2 if thief 2 defects = $5 million

Figure 9.12 Here we see the prisoner's dilemma payoffs for two thieves who stole and then stashed $10 million, which they are going to retrieve at a later date. Yellow represents decisions and payoffs for thief 1; blue represents decisions and payoffs for thief 2. Defecting (turning the other person in to the police) leads to a higher average payoff for each thief.

Suppose you have two criminal partners who steal $10 million from a bank. Because the police will quickly find out about the robbery and begin actively looking for suspects, the thieves jointly decide to bury the money in a hidden location. Working as a team, their idea is that when the police "heat" is off, they will go together and dig up the money, split it evenly between them, and then go their separate ways. While waiting for police activity to subside, however, they have no way to communicate with each other. At this point, you probably have a television or movie plot in your head that illustrates the dilemmas facing the two thieves.

While sitting around and waiting, each thief has two choices. They can cooperate with each other as originally planned, wait out the police investigations, not go to jail, and get half the money ($5 million each). Alternatively, each thief has the option to defect from the cooperative agreement, anonymously turn the other thief over to the police, and quickly run off with the full $10 million. The risk for each thief is jail time. The reward is money. The chance of going to jail for each thief and the amount of money each takes home are dependent not only on each individual's own decision but also on the decision of the other thief (**Figure 9.12**). If one thief defects and the other one does not, the defecting thief will get all the money and freedom, while the second thief gets no money and goes to jail. If both thieves defect, neither gets the money, and both go to jail. But if the two thieves cooperate as originally planned, they will each get some money, and no one ends up in jail. What is a thief to do in this situation?

One way to rationally evaluate the options is to calculate the average reward each thief will receive by choosing a given strategy. To get the average, we also have to consider what the other thief chooses (Figure 9.12). Consider the risks and rewards from the perspective of thief 1. Thief 1 does not know what thief 2 is going to do. With no additional information, the easiest assumption is that there is a 50% chance that thief 2 will stick to the agreement and a 50% chance that thief 2 will defect. With that assumption, if thief 1 honors the original agreement, there is a 50% chance of making off with $5 million and no jail time, and a 50% chance of jail time with no money. If we average these two reward payoffs,

$$\frac{\$5 \text{ million} + \$0}{2} = \$2.5 \text{ million}$$

Figure 9.13 A honeybee and a hummingbird rob nectar from flowers. **Ⓐ** Notice the dark patches in the flowers, indicating slits or holes where nectar has already been removed. Although honeybees are one of the most successful generalist pollinators, nectar robbery rarely results in pollination. **Ⓑ** A hummingbird steals nectar from a flower by piercing the corolla to get nectar and bypassing the pollen-laden anthers.

the mean payoff is $2.5 million for choosing to cooperate, and the risk is a 50% chance of jail time.

Now consider if thief 1 chooses instead to defect, again assuming a 50-50 chance that thief 2 decides to cooperate or defect. If thief 2 cooperates, thief 1 will get all the money while thief 2 is still in hiding, and thief 1 can get away with no jail time. If thief 2 also defects, then both thieves will have double-crossed each other, and they are likely to get caught by the swarming police in the ensuing scuffle. In other words, if both defect, then neither thief gets anything. Averaging these two rewards,

$$\frac{\$10 \text{ million } + \$0}{2} = \$5 \text{ million}$$

the mean payoff for defecting is $5 million. The 50% chance of going to jail is the same risk that thief 1 faced when cooperating. Thus, if the assumption about thief 2's decision is accurate, the logical strategy for thief 1 is to defect because overall the average reward is higher than when choosing to cooperate. Keep in mind that, no matter what decision is made by the thief, the risk of jail time is 50%.

The interesting part comes if we morph this scenario into the context of a mutualistic interaction. The two thieves become two species. The rewards and risks are measured in terms of fitness outcomes with higher or lower survival or reproduction. If the prisoner's dilemma holds as described, the two species should never cooperate. Each species should always try to exploit the mutualistic interaction, and it should devolve toward exploitation or competition.

If we look for examples of mutualisms that devolve this way in nature, we can find quite a few. For example, both bird and bee pollinators sometimes "rob" or "steal" nectar from flowering plants. Birds do this by using their pointy bills to pierce the flower tube (or corolla) at its base, and bees do this by chewing a hole in the nectary (the portion of the flower that holds the nectar) and accessing the nectar while bypassing the pollen and reproductive structures (**Figure 9.13**). The flower loses its energetically expensive nectar while failing to obtain any fitness benefit through pollen exchange and sexual reproduction. The bird or bee, on the other hand, gains the entire energetic nectar reward and potentially even accesses more nectar than if it had to navigate to the floral nectar source the standard way. The fitness difference for the nectar robber due to its "cheating" in the mutualism is zero, or potentially even a fitness gain, but the fitness cost to the plant from this cheating is huge.

We can look at another example, this one involving cleaner fish, or fish that remove and eat parasites off other species. In this relatively common type of marine mutualism, a small cleaner fish will set up a "cleaning station" on a rock outcrop or a reef. Interested fishes will come by, assume a particular position, and allow the cleaner to remove dead tissue and parasites from its body and sometimes even from inside its mouth (**Figure 9.14**). As you can imagine, the fitness differential of "cheating" in this situation is particularly high. A cheating large fish may simply eat its mutualistic partner after cleaning is complete. The large fish gains the mutualist fitness benefit of reduced parasite loads and also gains energy and nutrients from eating the cleaner fish. Meanwhile, the cleaner fish initially gains energetically from ingesting the parasites and dead tissue from the large fish as food, but then loses its life—a clear net loss in fitness.

If the prisoner's dilemma accurately describes species interactions in the natural world, examples of cheating in mutualistic interactions should be very common. In fact, one could argue that the outcomes of the prisoner's dilemma model dictate that mutualisms should never have evolved in the first place, as there should be no net fitness gain for the individuals of the two participating species. Clearly, that result does not fit with the many examples of mutualism that we find around the world. In addition, when scientists have looked, they have generally found fewer examples of cheating than we might expect.

This apparent paradox between real-world observations and model outcomes intrigued biologists for years, spurring several theoretical explorations of when and why cooperation or mutualism evolves and devolves. One answer came from Robert Trivers, who showed that cooperation was more likely to evolve if the two individuals in the prisoner's dilemma "game" had multiple opportunities to interact through time, the way that species do in an environment. This new variation is called the **iterative prisoner's dilemma (IPD)**, and it allows the "players" to adapt their behavior according to how the other player just behaved.

If the initial decision of one of the players is to cooperate, the two players gain a minor initial reward. However, if the players repeatedly cooperate over several decision events through time, these minor rewards add up. Thus, if each player expects cooperation from the other, through time they both receive relatively higher rewards than if they never cooperated. Trivers demonstrated that this form of self-reinforcing reward for cooperation, which he called **reciprocal altruism**, could evolve between individuals of two different species, effectively explaining the conditions necessary for ecological mutualism to evolve (Trivers 1971).

Robert Axelrod and William Hamilton (1981) extended Trivers's work by looking for player strategies that facilitated cooperative outcomes. To do this, they staged an IPD tournament among mathematicians and game theorists. Participants submitted different one-player strategies, which the scientists paired with other participants' one-player strategies. The submissions included strategies such as (1) the player always cooperates no matter what the other player chooses; (2) the player never cooperates; and (3) the player shifts between cooperating and not cooperating (with various rules for this interaction). A single "contest" was a computer simulation that ran for 200 iterations, wherein benefits to each player could accrue across iterations. All strategies were paired with all others in a round-robin IPD tournament format. To determine an overall tournament winner, Axelrod and Hamilton devised a scoring system and then tallied the total points accumulated by each one-player strategy across all tournament pairings.

The winning strategy was remarkably simple and was called **tit-for-tat**. This dictated that player A should always cooperate on the first move but thereafter copy player B's most recent move. For example, player A would start by cooperating but would only cooperate on the second turn if player B cooperated on the first turn. However, if player B started by not cooperating, player A would not cooperate on the second turn. In the tournament, this process simply repeated itself across all 200 IPD model iterations.

Figure 9.14 A bluestreak cleaner wrasse (*Labroides dimidiatus*) cleans the mouth of a tiger grouper (*Mycteroperca tigris*) in Sodwana Bay, South Africa. Cleaning inside the mouth this way exposes the cleaner fish to the risk of predation.

Tit-for-tat produced higher scores than all other strategies and often resulted in the two players cooperating through nearly all IPD iterations. Axelrod and Hamilton attributed tit-for-tat's success to two factors: niceness and forgiveness. The tit-for-tat strategy was "nice" in that it always cooperated on the first turn, and it was "forgiving" because it resumed cooperation as soon as the other player returned to cooperating.

If we imagine the scores in Axelrod and Hamilton's tournament as fitness rewards, we can translate the outcome of their tournament to the evolution of cooperation in natural systems. The tit-for-tat strategy outperforms all other interaction scenarios, as measured by fitness accrued by individuals of each species. Because natural selection always works toward maximizing fitness, cooperative strategies such as tit-for-tat should evolve regularly across the natural world.

Later expansions of IPDs to include multiple species produced more realistic representations of commonly observed mutualisms. Thus, although standard game theory does not do a great job of representing natural ecological systems, its derivatives such as IPD do. Game theory continues to inform several aspects of ecological science, including allowing researchers to anticipate what will happen when mutualistic interactions are altered or broken by agents of global change.

An Economic Model

Game theory models, such as the prisoner's dilemma and its offshoots, provide ecologists with insights into the evolution of mutualistic interactions. However, from our initial review of mutualism types, we know that they can grade between being obligate to being facultative. We also know that the types of fitness rewards vary across species and that the quality of these rewards may also vary across systems or locations. The next step for modeling mutualisms is to identify a way to explicitly account for the value and variation of these rewards.

One useful model framework comes from examining the conditions for beneficial trade agreements in market economics. A common **economic model of trade** assumes that two independent groups differ in their abilities to access resources. If each group (or species) is better at acquiring one resource than the other resource and if their differences complement each other (meaning they have opposite resource-acquisition strengths), then it may be advantageous for them to trade some of the resource they are good at obtaining in exchange for the resource that is more difficult for them to acquire. This approach seems like a particularly good one for examining mutualisms.

Mark Schwartz and Jason Hoeksema (1998) saw this connection between economic theory and ecology and applied such a model to mutualistic interactions in which the partnering species trade energy or nutrients. An excellent example of such a mutualism is the exchange of carbohydrates (such as glucose) and phosphorus between a tree and a mycorrhizal fungus, and we'll examine this model using these two organisms.

Unfortunately, for this model to work, it needs consistent units, or currency, for each resource. In economics, such currencies can be money, but equivalent units are a little harder to explain or model in ecology. Even though phosphorus and glucose are both consumed for growth, their uses are so different that it is hard to apply a common unit like calories. Nonetheless, we can think about the annual or daily amounts of each resource that might be used.

How we set up such units may be a little easier to understand if we use a human example. We know that human bodies do not need the same daily amount of calcium as they do protein. We could measure our recommended calcium intake per day in milligrams and our recommended protein intake per day in grams. Despite these differences in quantity, both are essential, and we could come up with units of calcium and units of protein that represent 5% of our total daily needs. The physical amount of calcium in a unit of calcium would be different from the physical amount of protein in a unit of protein, but we would need 20 units of each resource each day. Similarly, in this mutualism model, although each organism may place a different value on each benefit, the model works as long as the units are consistent across the organisms.

Schwartz and Hoeksema defined resource currencies for carbon and phosphorus so that trees needed an equal number of units of each resource. Because the absolute amount of carbon per unit does not have to be the same as the absolute amount of phosphorus per unit, a useful way to envision this is to consider the amount of carbon or phosphorus that a tree can access with a set amount of energy.

The model also requires a trade-off between the two species for acquisition of resources. This is where mathematics come in, but the model also has a graphical solution that is easier to interpret, so we focus our understanding there. We start by assuming that the tree requires units of both carbon and phosphorus to survive. We can then graphically represent these requirements by plotting carbon on the x-axis and phosphorus on the y-axis (**Figure 9.15**).

As you learned in Chapter 5 when we discussed the principle of allocation, organisms have limited energy or resources. If organisms devote all their energy to one task, they will not have any for another task. Therefore, although the tree in our example has a maximum amount of phosphorus (P_{max}) and a maximum amount of carbon that it can acquire (C_{max}), it cannot simultaneously access both amounts at their maximum. If the tree puts some energy toward acquiring phosphorus, it will need to give up some ability to capture carbon. This is just another way of saying what we said in Chapter 5: plants can either grow more aboveground structures and have greater access to sunlight and higher photosynthesis and carbon capture rates, or they can put more energy toward roots and have better access to water and soil nutrients, including phosphorus. What the plants cannot do is maximize both carbon capture and phosphorus acquisition at the same time.

If we examine our graph in Figure 9.15, which has the extraction of one resource on one axis and the extraction of the other resource on a second axis, the line connecting the maximum extraction points on each axis represents all the ways that the tree alone (in isolation, away from its mutualist partner) can acquire both resources simultaneously. We call this line the *isolation acquisition isocline* (IAI). In order to identify what is the best strategy for a tree, we need to know the relative value of carbon and phosphorus to the tree.

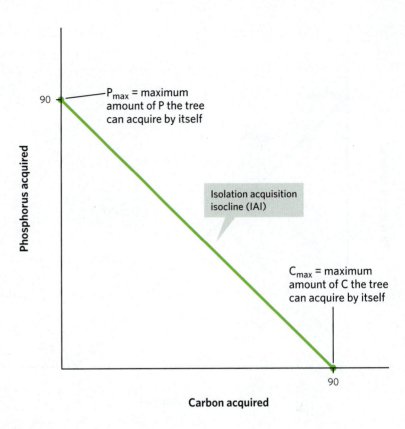

Figure 9.15 Organisms such as trees must continuously trade off allocation of scarce resources. The isolation acquisition isocline (IAI) for a tree represents all the combinations of ways that the tree can use its resources to acquire carbon (C), phosphorus (P), or both if the two resources do not interact (meaning if acquiring one does not block or facilitate acquiring the other).

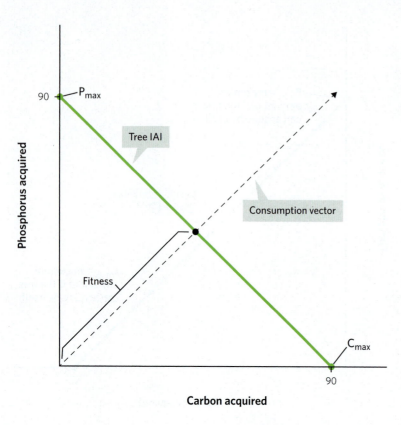

Figure 9.16 A consumption vector (dashed arrow) shows the equilibrium amounts of phosphorus and carbon that the tree acquires. The point at which the consumption vector and IAI intersect is a measure of evolutionary fitness because if one individual tree has an equilibrium point further from the graph's origin point than other trees in the same population, the tree acquiring more resources will grow, survive, and reproduce better than the others.

As we saw in the R^* model in Chapter 7, organisms consume resources at a particular pace that we call the consumption rate. Given the fitness implications of the units for each resource in the basic Schwartz and Hoeksema model, the tree will use the two resources equally. That means the consumption rate will start at the origin and point upward with a slope of 1 (reflecting equal consumption of phosphorus and carbon; **Figure 9.16**). The point where the consumption vector intersects the IAI indicates how much carbon and phosphorus the tree will acquire. Because all individuals need resources to survive and reproduce, the intersection point of the consumption vector and the IAI is a measure of fitness. A point further from the origin would indicate higher resource access and higher fitness.

We can sketch out a similar set of conditions for a mycorrhizal fungus (**Figure 9.17**). Fungi have no chlorophyll and cannot photosynthesize, so they must acquire all their carbon from external sources, such as detritus and decomposing organic matter in the soil. At the same time, fungi are fairly efficient at acquiring phosphorus because they have long branching filaments called *hyphae* that stretch out belowground like tree roots. These hyphae are super tiny with lots of projections that increase their surface area, allowing for relatively efficient uptake of phosphate from the soil.

Based on these features, and for the purposes of this model, we can represent the fungus's differential access to carbon and phosphate as threefold, meaning the fungus

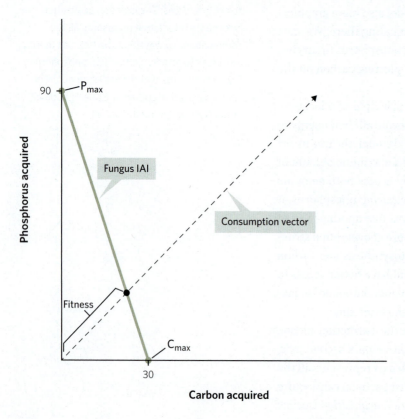

Figure 9.17 Resource acquisition and the IAI for mycorrhizal fungus. Notice that the fungus acquires phosphorus more easily than carbon. As with the tree, an intersection point that lies further from the graph's origin indicates higher resource access and thus higher fitness.

can use its energy to acquire 90 units of phosphorus (P_{max}) or 30 units of carbon (C_{max}) or some combination along the IAI (Figure 9.17).

By setting up the model this way, we have dictated that the mycorrhizal fungus has an easier time acquiring phosphorus than it does carbon, whereas the tree acquires both resources equally but has an easier time securing carbon than the fungus. The fungus, like the tree, maximizes its fitness when it acquires both resources in an equal ratio. When growing in isolation, each species will acquire the two resources in equal proportions to maximize their fitness (**Figure 9.18**). That means the tree will get 45 units of carbon and 45 of phosphorus, while the fungus will access 22.5 units of each.

This basic asymmetry in resource acquisition makes it advantageous for each species to enter into a resource trade and thereby acquire resources more efficiently. The trade isoclines in **Figure 9.19** represent all possible combinations of the two resources if the two species become specialists on one resource and trade for the other. Note that the slope of the trade isocline can vary but will always be greater than those of the IAIs for each species on its own. For example, if the tree specializes in acquiring carbon through photosynthesis and forgoes collecting any phosphorus at all, it will take in 90 units of carbon at C_{max}. At this point, let's suppose the tree trades 30 units of carbon to the fungus (Figure 9.19). This leaves the tree with 60 units of carbon, which is more than the 45 units it had at its peak fitness in isolation (the intersection of the consumption vector and the IAI in Figures 9.18 and 9.19). But can it get enough phosphorus from the fungus to make this trade worthwhile?

Figure 9.18 Resource acquisition and IAIs for a tree and a mycorrhizal fungus in isolation from one another. Note the differential abilities of the tree and fungus to acquire both carbon and phosphorus, as well as the relative amounts of each nutrient the organisms acquire in isolation.

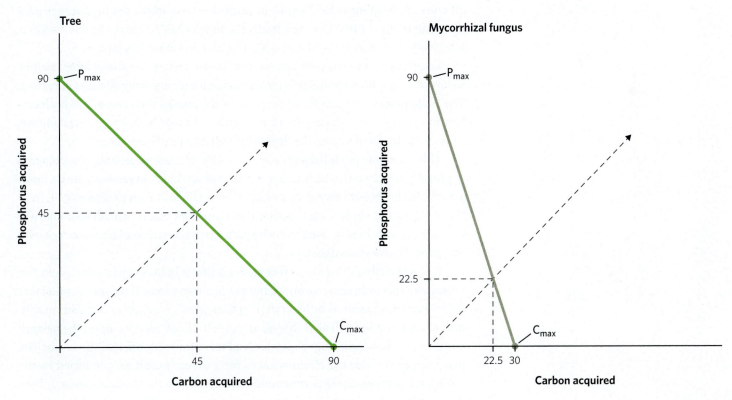

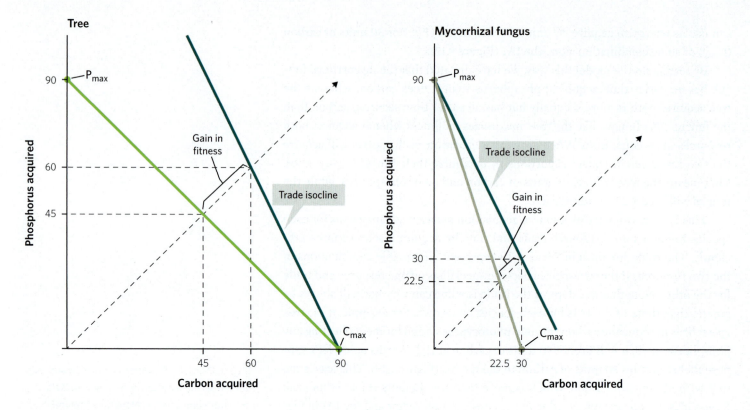

Figure 9.19 Conditions for trade between a tree and a mycorrhizal fungus and the resultant benefits for both organisms.

If the mycorrhizal fungus also specializes and spends all its energy acquiring phosphorus, it can acquire 90 units of phosphorus at P_{max}. If the fungus then trades 60 units of phosphorus for 30 units of plant-derived carbon, the fungus sees a fitness gain (Figure 9.19). After the trade, the fungus has 30 units of each resource, whereas in isolation, it could only acquire 22.5 units of each resource.

Returning to the tree, we can see that it also comes out ahead after trading, as it receives 60 phosphorus units in exchange for giving away 30 units of carbon. The trade produces a significant increase in the overall total amount of both carbon and phosphorus available to the tree with no change in its energy expenditure. This trade clearly increases the fitness for individuals of both species.

This economic model of mutualism is pretty interesting, as it allows ecologists to quantify the circumstances under which they are likely to see mutualistic interactions. In the tree-mycorrhizae example, both species receive a fitness gain from the trade, so the model suggests that the mutualism can be established and will probably be sustained, as long as each species continues to acquire resources with the same relative specializations.

One interesting thing about this model is that it helps explain why facultative mutualisms occur in some locations but not in all locations. If resource availability differs between locations or habitats, it is likely that the resource acquisition abilities of each species will also change as a result. If that change alters the balance of the trade, then trade may no longer lead to a fitness payoff, and the mutualism may not occur. Once the circumstances for a fitness payoff are identified by the model, researchers can experimentally test or validate the model to identify both the conditions for mutualistic interactions and aspects of the model that are oversimplifications.

From experiments and other explorations, we know that this economic model of mutualism has some limitations. Such explorations have suggested that the model, as explained here, presents a simplified version in which resources are acquired independently (Hoeksema and Schwartz 2003). If the resources instead somehow interact, so that acquiring carbon either facilitates or impedes the acquisition of phosphorus, for example, then the IAIs will curve: they bulge out if the two resources facilitate access to each other, or they droop toward the origin if the two resources impede each other. This more complex version of the model is the one applied most commonly, and it has held up quite well for looking at interactions between plants and a range of belowground microbiota, from fungi to nitrogen-fixing rhizobia.

In addition, the model has some practical shortcomings. In particular, it is really only appropriate for resource mutualisms, and even for these mutualisms, the units can be very challenging to define. And for a mutualism that trades food resources for protection from predation, it is much harder or even impossible to identify common currencies. Similarly, the model does not work for cases of obligate mutualism because the two species involved, by their very nature, cannot exist independently of each other. Therefore, the assumption that resources can be acquired in isolation does not hold.

What Have We Learned about Mutualism?

There is no doubt that our ability to generalize about mutualism lags behind our theoretical understanding of competition and exploitation. Notice that the last two models we discussed offer insight into the conditions under which we might find mutualism in nature and when mutualism might be more or less beneficial. These models may even provide relative or concrete measures of fitness benefits from such mutualistic interactions. But what these models do not offer is a formula for exactly how the population densities of mutualist partners change when the species interact. The population insight from the Lotka-Volterra competition and exploitation models seemed pretty useful. How do we feel about not having it for mutualism?

In many instances, the specificity of the population-dynamics models may be misleading. As pointed out in Chapters 7 and 8, insights from the Lotka-Volterra models are less about specific values of population density and more about understanding population dynamics or the conditions for coexistence. If we are looking for insights into coexistence and improvements in population performance, the game theory and economic models of mutualism, though more theoretical, actually offer useful information about when (and perhaps where) to expect stable mutualistic interactions.

A final issue, which is not unique to models of mutualism, is that the economic model assumes only two species are interacting. In the real world, of course, many species can interact simultaneously in a community or ecosystem. That fact does not negate the value of two-species models, but with mutualisms (particularly the resource-exchange mutualisms in our economic trade model) there is evidence that organisms often interact with many mutualistic partners. For example, the soil is full of fungi, bacteria, and other microbes, and a plant interacting with these soil organisms may be sharing resources with many species at once while at the same time competing with another suite of species for resources or trading partners.

Figure 9.20 In this association between two remoras and a green sea turtle (*Chelonia mydas*) in Akumal Bay, Yucatán, Mexico, there is little benefit or cost to the sea turtle, making this interaction a commensalism. Yet the sea turtle facilitates the remora by swimming, accessing food, and avoiding predators.

Figure 9.21 Most humans engage in a commensal relationship with tiny mites (*Demodex folliculorum*) that live on the hair follicles on our faces. This colored scanning electron micrograph shows human hair follicles with the mites (colored red) associated with individual hairs (colored green) in a human eyelash.

It is possible to extend this model (and the competition and exploitation models) to more than two species. We discuss models that include multispecies interactions in the next chapter, but in the next section we explore some of the ways that one species can have positive effects on another species without receiving benefits. Many of these positive effects involve more than two species.

Despite the challenges inherent in the mutualism models, we have managed to produce some generalizations about mutualistic interactions. First, we know that many mutualisms involve both benefits and costs. Mutualistic partners are not acting altruistically but are instead acting to improve their own fitness. In order to entice participation, they often have to give away something of value.

Second, mutualisms may be context dependent, but the costs and benefits associated with the interactions can help us evaluate when and where we find mutualisms. Because resources, predators, and physical forces vary with location, the value of resource-exchange and protection mutualisms, in particular, may vary with location. Mutualistic partners may end the interaction if the mutualism is not locally beneficial. Many plant and fungal species create mutualistic interactions in some settings but not in others, and a large library of ecological studies explores the global patterns and predictors for such interactions.

Finally, the combination of both context dependence and costs can lead to cheating and instability. As seen with the nectar-robbing bees, mutualist partners may cheat and find ways to accrue benefits without trading, and it is reasonable to assume that both species in a mutualism are constantly evolving toward a more efficient or beneficial use of their mutualist partner. Because mutualists often interact closely, it is possible that each attempt at an advantage by one species leads to *coevolution* in the other species (see Chapter 8 for the main discussion of coevolution). All of these effects make mutualism extremely interesting and potentially very important for the overall structure and dynamics of ecological communities.

9.5 Facilitation

As mentioned at the start of the chapter, *facilitation* is the general term for interactions that benefit at least one species and have a positive effect or no effect on one or more other species. Mutualism is a special type of facilitative interaction in which both species benefit.

Examples of facilitation beyond mutualisms abound, particularly in ecosystems where many species coexist. For example, green sea turtles (*Chelonia mydas*) and remoras, which are ray-finned fish in the Echeneidae family, engage in a facilitative relationship in which the remora hitches a ride with and feeds near the turtle but does not have much of an effect on the turtle (**Figure 9.20**). These sorts of two-species interactions in which one species benefits at little or no cost to the other are often called commensalism (+, 0), another subset of facilitative interactions. Another common example is the relationship between humans and *Demodex folliculorum*, a microscopic mite (**Figure 9.21**) found only in the hair follicles of humans and typically only on the face. This mite, which is present on almost everyone, consumes the sebum (oils and waxes) secreted by our skin but generally has little to no effect on our health.

Even more broadly, we can consider an example of a tree growing tall and providing shade to its understory. The environment the tree creates is beneficial to many animal species and some plant species that require cooler temperatures, higher moisture, or less sunlight. The tree, on the other hand, likely receives diffuse and difficult-to-measure benefits or competitive effects from the understory plants and animals. Similarly, a line of trees along an agricultural field (called a hedgerow) can benefit the crops not only by shielding them from wind but also by hosting an array of predatory insects that eat insect herbivores and thus reduce crop damage (**Figure 9.22**).

Some species can facilitate huge collections of other species. For example, within coral reef biological zones, it is the corals themselves that produce the hard and complex three-dimensional structures that facilitate the presence of the sharks, anemones, shrimp, and seahorses that are typical of these aquatic communities (**Figure 9.23**). If the corals die because of bacterial symbionts or anthropogenic factors, the entire foundation of the biological zone is removed, and the whole ecosystem collapses.

Ecologists call such species **foundation species** because they provide the foundation or infrastructure for a whole community or ecosystem. Foundation species often make up a dominant portion of the biomass in their habitat, and they often are primary producers or involve a symbiosis with primary producers (as is the case with coral polyps and their symbiotic, photosynthesizing zooxanthellae). Interestingly, these same foundational coral species often depend on the fishes and invertebrates they shelter to keep them from becoming overgrown with algae and to protect them from predators and competitors. Coral reefs are a prime example of the importance of indirect facilitative interactions in maintaining entire biological zones or biomes. In other ecosystems, foundation species might include hemlock trees in a hemlock forest, one or more species of grasses in a grassland or prairie, or even kelp in a nearshore marine kelp forest.

Figure 9.22 Hedgerows provide habitat for a wide range of species and may protect adjacent agricultural crops by increasing the densities of predaceous insects that eat herbivore agricultural pest insects, thus facilitating higher crop yields.

Figure 9.23 Intact coral reefs, like this one in Fiji, create a three-dimensional structure that offers hiding places, settling habitat, and protection from waves, thus facilitating the survival and reproduction of a wide range of other species. The reef-building coral in this community is the foundation species.

Restoration Ecology and Facilitation

Because foundation species are crucial for the survival and reproduction of other species in a habitat or community, they can determine the types of species we find in any location. This fact, and a focus on facilitation, can play a critical role in the efforts of land managers to mitigate the effects of human land use, resource extraction, and habitat destruction in the Anthropocene. **Restoration ecology** addresses the ecological recovery of degraded or damaged ecosystems through active human intervention, generally termed *management*. Although the goals of restoration projects vary tremendously, a central theme is enabling the return of native species to a community from which they were lost as a consequence of human actions, such as mining, acute or chronic chemical deposition, intensive agriculture, or altered water flows or wave action.

To restore an entire community of species in the wake of such damage often requires beginning from the ground up. A central challenge for restoration ecologists is figuring out which species to reintroduce first in order to facilitate the return of several others. An ecological community, like a good soccer team, does not spring into existence by throwing random "players" together (see Chapter 13). Instead, the community has to be built around a few centrally important species, with others included only after a sustainable foundation has been established. It may seem strange, but the foundation species may not be the first species introduced. In fact, sometimes restoration ecologists need to think about slowly shifting a degraded ecosystem through several transitional communities before arriving at their target community species composition (see the section on succession in Chapter 13).

From the soccer team analogy, it should be clear that restoration ecologists must identify which species play a strong facilitative role in the "team" or community they wish to create. Identifying these species is not always straightforward, but the study of facilitative interactions, and mutualisms in particular, provides significant clues for ensuring success.

For example, damaged ecological communities often lose pollinator species, which can lead to low or no seed production in many plant species (Menz et al. 2011). The failure of a large fraction of plants to reproduce creates a multitude of other problems for the damaged community. To restore these degraded systems, one of the primary challenges is attracting and retaining native pollinators. The suite of pollinators in any community typically includes a small number of rare, specialized taxa; many moderately specialized taxa; and a few common, generalist taxa that provide the majority of pollination services. Several restoration studies have shown that planting vegetation that attracts and sustains a broad group of pollinators has a number of positive restoration benefits. With the future of ecological restoration estimated to be a trillion-dollar global business, establishing animal-mediated pollination services will be hugely important for ensuring success in the restored plant communities.

But pollinators and foundation species are not the only species with facilitative effects. Sometimes, restoration ecologists think about introducing species that can physically alter or maintain a system. At times, it is critical to think about the physical factors in an ecosystem. In Chapter 6, we discussed the destructive role that beavers play in Tierra del Fuego in South America, but in much of North

America, where they are native, beavers can be important in creating wetland habitat and allowing gaps in forests that lead to healthy tree regeneration. Similarly, in arid environments, plants that can survive in dry conditions can be important for stabilizing soil, reducing erosion, and enhancing soil water-holding capacity. The restoration of facilitative interactions demonstrates that management choices at local scales can have profound effects on the diversity of ecological interactions across larger landscapes.

Invasional Meltdown

Although it may seem counterintuitive, the role of facilitation can also be important in understanding the degradation of some natural systems in the Anthropocene. As an example, sometimes the presence of one invasive species can facilitate the establishment of many other invasive species. This situation has been called **invasional meltdown** and is more formally defined as the process by which two or more non-native, invasive species facilitate each other's establishment or exacerbate each other's impact on native species (Von Holle 2011).

Daniel Simberloff and Betsy Von Holle first proposed the existence of invasional meltdowns in 1999 in order to highlight how some invasive species affect native communities beyond a direct interaction with native species (Simberloff and Von Holle 1999). The concept has been vigorously debated in the years since, partially because large-scale invasional meltdowns do not seem to be very common. Despite the low incidence of documented examples, some situations clearly show that meltdowns can fundamentally alter entire communities.

One such example comes from Christmas Island, a small Australian territory in the Indian Ocean just south of Java, Indonesia. The yellow crazy ant (*Anoplolepis gracilipes*; **Figure 9.24**) was unintentionally introduced to the island decades ago in ship cargo. The species initially established small local populations but eventually formed gigantic supercolonies. These massive colonies contained more than 1,000 queens and could extend for kilometers (Green et al. 2011). The establishment of these ant supercolonies was promoted by a mutualism with a group of non-native, honeydew-secreting scale insects (*Tachardina aurantiaca* and *Coccus* spp.).

Scale insects are widespread tropical pests that feed on the sap of trees and excrete honeydew, a sugary liquid. Yellow crazy ants consume the honeydew and protect the scale insects from their enemies. This mutualism does not exist elsewhere in the world because these species did not coevolve. Instead, these two non-native

Figure 9.24 Ⓐ Yellow crazy ants (*Anoplolepis gracilipes*) are invasive to Christmas Island, Indonesia. The novel mutualism between yellow crazy ants and invasive scale insects (*Tachardina aurantiaca* and *Coccus* spp.) led to the decline of Ⓑ the native red land crab (*Gecarcoidea natalis*) and the native forests.

Ⓐ

Ⓑ

species encountered each other and established a new mutualism on Christmas Island. The ants also transported the scale insects between trees on the island, establishing new scale-insect populations and, in due course, new ant supercolonies. This mutualistic interaction resulted in a persistent and widespread scale-insect problem on the island and large numbers of yellow crazy ant supercolonies.

The mutualism between ants and scale insects also facilitated the invasion of a third species, the giant African snail (*Achatina fulica*). The snail had been introduced by island residents as a form of biocontrol more than a century before the yellow crazy ant or the scale insects arrived on the island. For many years, the snail had not spread into the inland native forests on the island because of the presence of a predaceous, native red land crab (*Gecarcoidea natalis*) that ate the snails.

The explosion of yellow crazy ant supercolonies, facilitated by the scale insects, decreased the native red land crab populations. Before the invasional meltdown, red land crabs migrated from the inland forests to the coast each year as a part of their mating routine. Although crazy ants do not bite or sting crabs, as a defense against being eaten they spray formic acid at the crabs' eyes. The sheer number of crazy ants on the island thus led to high levels of formic acid at ground level, which eventually overwhelmed and blinded large numbers of native crabs. The population decline of the red land crabs from the inland native forests allowed the invasive giant snail to move unimpeded into these forest habitats. The meltdown continued because blinded crabs eventually died from dehydration and exhaustion, and their carcasses fed the crazy ant supercolonies, which led to more ants to tend more scale insects.

To make matters still worse, the red land crabs had played the role of a keystone species (Chapter 10) in Christmas Island native forests. The crabs' burrows aerated the soil, and their feces served as fertilizer. The loss of crabs produced substantial and widespread structural changes to the island's forests, including allowing a suite of non-native plants to invade the native forest interior (**Figure 9.25**).

To combat this spectacular invasional meltdown, the island's national park personnel executed a widespread yellow crazy ant bait-and-kill program in 2009. The park rangers used a helicopter to treat the ant supercolonies with a very low concentration of ant-specific pesticide. Subsequent monitoring showed a 99% reduction in crazy ant densities. Today, it is not yet clear whether the reduction of supercolonies has led to the recovery of the crab population or reversed the island's ecological meltdown.

Facilitation: A Larger View

Clearly, facilitation can play an important role in ecological communities, and it is a concept that we touch on again in future chapters as we explore multispecies interactions. Although we have highlighted several situations in which facilitation involves a direct interaction between one or more species, the more common situation may be for facilitation to act indirectly by changing physical factors (as when tall trees provide shade or wind protection to the understory) or by altering the balance of some other direct interaction. Many examples of the former type appear in Chapter 13, where we examine ecological succession and community change. We will explore some examples of the latter type in Chapter 10, where we

Figure 9.25 The unintentional introduction of the invasive yellow crazy ants to the island triggered an invasional meltdown. The meltdown is due to the invasive ant forming key ecological linkages (including a mutualism) between several other invasive species on the island, which together create facilitative interactions that severely alter native forests on the island (Green et al. 2011).

AFTER introduction of yellow crazy ants

Invasive yellow crazy ants

Mutualism

Invasive scale insects

1 Ants protect scale insects and eat the honeydew the scale insects produce

2 CREATION OF ANT SUPERCOLONIES

3 Ants spray formic acid, blinding and eventually killing crabs

4 Invasive snail populations increase

5 Non-native plants invade understory

BEFORE introduction of yellow crazy ants

Native red land crab

1 Crabs keep plants and invasive snails under control

Intact forest before invasional meltdown

Forest after invasional meltdown, with increased non-native understory

examine food webs and multispecies interactions. One reason to highlight facilitative interactions in this chapter is that the large majority of ecological studies have focused on negative interactions and the role they play in structuring communities. Yet moving forward in the Anthropocene, it is vitally important to recognize the critical role that beneficial interactions can play.

SUMMARY

9.1 Why Can't We All Just Get Along?

- When an interaction between two species benefits at least one of the species, it is called facilitation.

- When facilitation benefits both species, it is called mutualism.

- When facilitation benefits one species but has no effect on the other, it is called commensalism.

- Our understanding of facilitative interactions lags behind what we know about competition and exploitation.

9.2 Obligate and Facultative Mutualism and Partner Specificity

- Obligate mutualisms are mutualisms that are necessary for the survival of at least one of the participating species.

- Facultative mutualisms improve the fitness of both species but are not necessary for the continued survival of at least one participating species.

- There can be a range of obligate and facultative mutualisms for one or both partners.

- Some well-known mutualisms are obligate-obligate, like that between coral and zooxanthellae.

- Others, like those between plants and their pollinators, can be obligate for one of the partners but not the other (i.e., obligate-facultative).

- Seed dispersal is often a facultative interaction for both partners (i.e., a facultative-facultative mutualism).

- Many obligate mutualisms are highly species specific, meaning each of the two species involved needs the other to survive.

- Facultative mutualisms can sometimes be fairly general, meaning at least one partner is able to swap and use a different species as its mutualistic partner.

- Symbiotic relationships are characterized by close physical contact and can range from parasitic to mutualistic. Only a subset of mutualistic interactions are symbiotic.

9.3 Benefits of Mutualism

Transportation: Reproduction and Dispersal

- Moving pollen from one flower to another, which facilitates plant reproduction, is not an altruistic act on the part of the pollinator, but it generally benefits both species.

- Dispersal of plant seeds by animals is sometimes accidental, but plants often produce fruits (which benefit dispersers) to encourage the consumption and subsequent dispersal of seeds.

Acquisition of Energy and Nutrients

- Often, for a pollinator species, mutualism does not involve transportation benefits but rather energy or nutrient benefits, as they get to consume the nectar, pollen, or fruit of the plant.

- Plants and mycorrhizal fungi at times can trade resources (nutrients or carbohydrates).

Protection and Defense

- Some mutualisms can result in the protection or defense of one or both partners, as in the relationship between some species of ants and acacia trees.

9.4 Models of Mutualism

A Conceptual Model of Mutualism

- Models of mutualism must show a positive interaction.

- There are many ways to model this type of interaction.

Lotka-Volterra Approaches

- **A Mutualism Model Based on the Competition Model**

  - Taking the two-species competition equations from Chapter 7, we can change the signs on the "competition coefficients" and make them positive rather than negative to create a mutualism model.

  - Phase-plane graphs reveal problems with this model; we see cases where the two species' populations grow without limit, which does not accurately represent the natural world.

- **A Mutualism Model Based on the Exploitation Model**
  - Alternatively, we can use the Lotka-Volterra exploitation model from Chapter 8, but it also causes populations to grow toward infinite density.

Game Theory Models

- The prisoner's dilemma is a game theory scenario in which two thieves who have stolen money have to decide whether to cooperate or defect (i.e., rat out the other one).

- Average payouts suggest the rational strategy is to defect and not work mutualistically.

- The prisoner's dilemma model was applied to biology to examine how a mutualism could form between two partner species.

- An iterative prisoner's dilemma (IPD) game allows the two partners (species) to interact on multiple occasions through time.

- Evolutionary biologists staged a tournament in which different IPD strategies were pitted against each other in an attempt to facilitate the formation of a mutualistic interaction.

- The winning strategy was called tit-for-tat and involved initial cooperation followed by doing what your partner did the previous time.

- The IPD model suggests that cooperative strategies between species can evolve over time if the conditions are right.

An Economic Model

- Ecologists can apply an economic model of trade interaction and resource specificity to mutualistic interactions in nature.

- This approach works best when there are two resources to trade and the resources have a common currency that can be used to assess benefits.

- When applied to the mutualism between a plant and a mycorrhizal fungus, the model suggests that each partner should specialize in acquiring one of the resources and then trade that resource to the other partner.

- When this occurs, both species benefit and have higher fitness than if they existed alone.

- The model works well for certain kinds of mutualisms but not for those involving services like pollination or seed dispersal.

What Have We Learned about Mutualism?

- Unlike competition and exploitation models, mutualism models do not explore changes in population density for mutualists.

- The focus on costs and benefits, though, may offer insight into when and where we should expect to see mutualisms and suggests that some mutualistic interactions may vary spatially.

- This may be due to context dependency, the characteristics of mutualisms, and the diversity of mutualistic interactions possible in the natural world.

- These characteristics and the close interactions between mutualists create the potential for coevolution between mutualist partners.

9.5 Facilitation

- *Facilitation* is a more general term for interactions that benefit at least one of the partners; it encompasses mutualism (both species benefit) as well as commensalism (one species benefits but the other is unaffected or minimally affected).

- Some facilitative interactions are quite important in the natural world, such as the way a tree provides shade for the many organisms in the forest understory.

- This understanding has led to the idea of foundation species: species that provide an ecological foundation for many other species, even though they might not benefit from the interactions.

Restoration Ecology and Facilitation

- To restore a degraded landscape, it may be necessary to include both facilitative and mutualistic species in the restoration mix.

- These kinds of positive community interactions are extremely beneficial when building a community from the ground up.

Invasional Meltdown

- Sometimes, pairs of invasive species can cause significantly more damage together than either would on its own.

- This process has been called invasional meltdown and can hasten the degradation of invaded ecosystems and communities.

Facilitation: A Larger View

- Facilitation is important in ecological topics such as food webs and succession, which we discuss in later chapters.

KEY TERMS

altruism	game theory	obligate mutualism
commensalism	generalism	prisoner's dilemma
economic model of trade	invasional meltdown	reciprocal altruism
facilitation	iterative prisoner's dilemma (IPD)	restoration ecology
facultative mutualism	microbiota	symbiotic relationship
foundation species	mutualism	tit-for-tat

CONCEPTUAL QUESTIONS

1. Define mutualism and commensalism. How are they the same, and how do they differ? Give examples of each. Explain the two general types of mutualism and provide examples.

2. Why are there costs associated with mutualism?

3. What problems have ecologists encountered when trying to model mutualism? How did they get around them?

4. Why might mutualisms vary with environments, resource availability, and predator or exploiter density?

5. Explain how game theory has been applied to studying mutualism.

6. Pick an example of a mutualism from the chapter, and try to explain how evolution might have produced this mutualism from an exploitative interaction. This exercise is easier for mutualisms that involve transportation or protection rewards in exchange for energy or nutrient rewards. Can you explain why?

QUANTITATIVE QUESTIONS

1. Re-explore the model expressed by Equations 9.2A and 9.2B, and derive the isoclines for yourself using Equations 9.3 and 9.4.

2. Explain the phase-plane trajectory arrows in Figure 9.11.

3. Figure 9.26 shows a phase-plane graph for an obligate mutualism based on a Lotka-Volterra mutualism derived from the Lotka-Volterra competition model. For obligate-obligate mutualisms, each species cannot exist without the other. Here we show the more common obligate situation, in which each species cannot survive in a habitat until the obligate mutualist reaches a threshold level and in which the effect of the mutualistic partner is strong. This means that the isocline starts where the mutualistic partner's density is above zero. In the area shaded green, neither species can survive because its mutualistic partner's population density is too low. The two species cannot survive in the green zone, but above that zone, the population densities of both species explode. Justify the arrow trajectories to explain why the intersection of the two isoclines represents an unstable equilibrium.

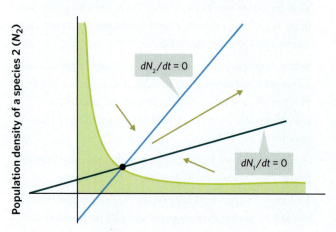

Figure 9.26

4. An evolutionarily stable strategy is one that, if adopted by a population in a given environment, is impenetrable, meaning that it cannot be taken over or "invaded" by any alternative strategy that is initially rare. The tit-for-tat strategy in the iterative prisoner's dilemma has been called evolutionarily stable. Discuss how you might test this idea.

5. Explain the market (economic) model of mutualism using an example that includes figures and words.

6. Redraw Figure 9.18 for a scenario in which the acquisition of the two resources is not independent and, specifically, in which each resource facilitates the acquisition of the other. Your IAI lines should curve away from the origin. Can you explain why? Will such interactions between the resources make the mutualism more or less likely to occur? Think about resource acquisition in isolation and how different it will be from resource acquisition with trading. Figures 9.16 and 9.17 may help if you are having trouble organizing your thoughts.

SUGGESTED READING

Axelrod, R. and W. D. Hamilton. 1981. The evolution of cooperation. *Science* 211(4489): 1390–1396.

Koenig, J. E., A. Spor, N. Scalfone, A. D. Fricker, J. Stombaugh, R. Knight, L. T. Angenent, and R. E. Ley. 2011. Succession of microbial consortia in the developing infant gut microbiome. *Proceedings of the National Academy of Sciences* 108(Supplement 1): 4578–4585.

Schwartz, M. W. and J. D. Hoeksema. 1998. Specialization and resource trade: Biological markets as a model of mutualisms. *Ecology* 79(3): 1029–1038.

Simberloff, D. and B. Von Holle. 1999. Positive interactions of nonindigenous species: Invasional meltdown? *Biological Invasions* 1(1): 21–32.

Stachowicz, J. J. 2001. Mutualism, facilitation, and the structure of ecological communities. *Bioscience* 51(3): 235–246.

Sea stars add a burst of color to a tide pool at Point of Arches, Olympic National Park, Washington. Tide pools support complex food webs that have fascinated scientists and naturalists for hundreds of years.

chapter 10

MULTISPECIES INTERACTIONS AND FOOD WEBS

LEARNING OBJECTIVES

- Explain the differences between food chains and food webs.

- Summarize the ideas involved in trophic energy transfer efficiency.

- Explain the concept of trophic pyramids.

- Compare top-down and bottom-up control of trophic systems.

- Identify some of the evidence used to support top-down and bottom-up control.

- Compare two of the ways food web models are assembled.

- Explain the concept of trophic subsidies.

- Evaluate the different concepts involving strongly interacting species.

- Describe how indirect effects can influence trophic complexity.

10.1 Embracing Complexity

Perhaps you are one of the lucky ones who has spent time poking around in a tide pool at the ocean. Or maybe you have observed a similar environment while visiting an aquarium or aquatic park. Or perhaps you have watched one of the countless nature shows on TV featuring tide pools. Regardless of where or how you encountered tide pools, you undoubtedly noticed that one of their most visually striking features is the sheer abundance of different life forms all crammed together in one small spot. Sea stars, algae, sea anemones, snails, sea urchins, kelp, tunicates, crabs, fishes, and many other organisms form the intricate community that inhabits a tide pool. The vast number of species present in this one location may cause you to pause and think about how they all get along. Are they somehow competing, preying on each other, or beneficially interacting with each other? Is it possible to make any ecological sense of such a riot of different living organisms? How can we, as scientists, even begin to approach such diverse and complex communities?

The previous three chapters on community ecology focused on single interactions between two species and the impacts of those interactions on species distributions, abundance, and dynamics. Although those chapters include some complex ideas and outcomes, the focus on just two species was clearly a simplification of the intricate web of interactions we see in the real world. This chapter shifts the focus directly onto whole communities, which means we will be highlighting a diverse set of interactions among a broader range of species. In doing so, we tackle networks, indirect effects, and complex trophic (feeding) webs. We start by defining indirect effects and then characterize multispecies feeding connections. We then expand these concepts to better reflect reality. Because all organisms require energy, feeding interactions offer an intuitive avenue to explore weblike linkages, and we use some examples to demonstrate that these linkages can have large effects on communities. In the final sections of the chapter, we return to indirect effects, exploring a wider array of situations in which they play a role in ecological communities.

10.2 Indirect Effects

To begin, we must acknowledge that any of the two-species interactions from Chapters 7 through 9 are likely to alter the way species experience their environments. Recall the example of character displacement in spadefoot toads (*Spea bombifrons* and *S. multiplicata*) from Chapter 7. The tadpoles of these two species competed for food resources, including fairy shrimp. Although we discussed how resource scarcity can lead to interspecific competition, we never delved into the population dynamics of the spadefoot toads' food (fairy shrimp). In a similar manner, the example of the rainbow trout (*Oncorhynchus mykiss*) that ate tadpoles of mountain yellow-legged frogs (*Rana muscosa* and *R. sierrae*) in Chapter 8 may have caused some readers to wonder if these trout ever ate spadefoot toad tadpoles.

The Sierra Nevada lakes where the trout consume frog tadpoles are a long way from the New Mexico ponds where David Pfennig and colleagues (2006) studied the spadefoot toads, but in both locations, fish are known to eat tadpoles. In fact, fish of various species will eat tadpoles across many different aquatic ecosystems. Therefore, it is not too hard to imagine a hypothetical pond where tadpoles eat fairy shrimp, and the fish, in turn, eat the tadpoles. In such a pond, there would be two separate but linked feeding interactions, and these two interactions would most certainly affect each other. A large population of fish could decrease tadpole numbers, thereby allowing many fairy shrimp to escape predation. Alternatively, an increase in the fairy shrimp population might lead to an increase in tadpole numbers, providing more food for fish, leading to an increase in fish population density.

The effects of linked interactions among more than two species are called **indirect effects**. Indirect effects occur when an interaction between two species is affected by a third species. The idea is that one species affects a second species through a "change" in an intermediate, or mediating, species. If fish eat tadpoles that eat fairy shrimp, the tadpoles are the mediating species that "passes" the indirect effect between the fairy shrimp and fish. The "change" in the mediating species may occur through its abundance, behavior, habitat use, or even morphology.

This concept may seem a bit confusing, but remember that most of our examples thus far describe situations in which organisms respond to the density, behavior, or traits of another species. As we move through this chapter, we revisit some of these examples and highlight several of the most important types of indirect effects. Quantifying and even identifying indirect effects can be challenging, because of the inherent difficulties with simultaneously measuring and controlling a suite of interspecific reactions. Nonetheless, indirect effects play an important role in community dynamics and have a significant place in modern ecology (Strauss 1991; Wootton 1994). To ease into this topic, we begin by thinking about feeding interactions as a chain.

10.3 Food Chains Offer Simple Conceptual Models

Feeding interactions, or **trophic interactions**, were introduced in Chapter 5, but it is useful to revisit them here. The word *trophic* comes from a Greek word meaning "nourishment." We can link trophic interactions in a **food chain**, like the predatory interactions we described between fish, tadpoles, and fairy shrimp. The most common food chains differ slightly from this example by including a **primary producer** at the bottom (foundation) level. Primary producers are autotrophs (self-feeders) because they produce their energy from inorganic sources rather than from eating other biological organisms. Autotrophs are generally plants and algae (which use photosynthesis to capture energy) or bacteria and archaea (which use chemosynthesis to capture energy). Any organism that mainly eats plants or other autotrophs is a **primary consumer** (herbivore), which in turn is consumed by a **secondary consumer** (predator). Each step in this chain represents a **trophic level**, meaning a feeding or eating level. We can see the chain aspect of this basic three-level model if we link each level with an arrow to indicate the upward direction of the **energy flow** (**Figure 10.1**). We can also see the logic behind some of the terms, because energy flows from the primary producer to the primary consumer and then to the secondary consumer.

One of the nice things about this simple, conceptual, food chain model is the ability to add complexity. For example, using the tadpole example, we may want a model that has three levels of animal consumers: an herbivore, a predator that consumes the herbivore (secondary consumer), and a **top predator** (sometimes known as a **tertiary consumer**, or an apex predator) that consumes the secondary consumer. The conceptual food chain model is adaptable by just adding another arrow and species to the top of the chain.

The Coachella Valley in Southern California offers a real-life example of a four-level food chain (**Figure 10.2**). In this desert environment, plants are consumed by herbivorous insects (e.g., grasshoppers), which are eaten by scorpions, which in turn are eaten by kit foxes (*Vulpes macrotis*). Notice in Figure 10.2 that numerical ranks are assigned to the trophic levels in the food chain consecutively, starting with the primary producers and following the flow of energy up the chain. Although the Coachella Valley example has four trophic levels, many terrestrial food chains are described with just three trophic levels. Part of this trend comes from a narrow focus on verte-

Figure 10.1 In this simple conceptual model of a food chain, the arrows indicate the direction of energy flow, not who is eating whom.

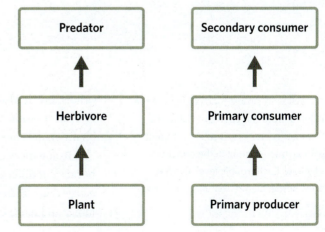

Figure 10.2 A simple food chain from the Coachella Valley in the Southern California desert. Numerical ranks indicate the trophic level, with arrows pointing in the direction of energy flow. Each trophic level also has alternative nomenclature that indicates level or type of organism.

brates, but it also reflects the more dominant trophic interactions in many terrestrial systems. We return to this point shortly, as we explore ways to make food chain models more realistic.

In aquatic ecosystems, four- or even five-level food chains are common, partially because of differences in the primary producers found in aquatic versus terrestrial ecosystems. Terrestrial plants grow with very stiff cell walls and are generally much larger and more fixed in place than the tiny mobile algae called phytoplankton that

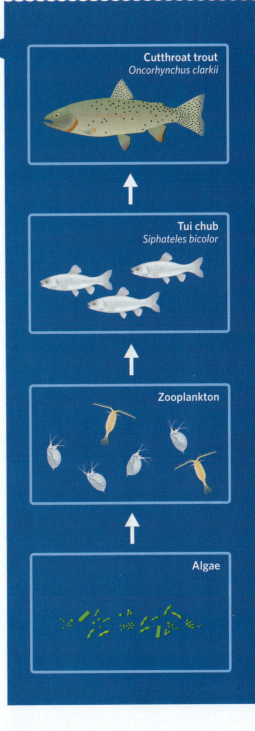

dominate in aquatic systems. A food chain for a lake or an ocean will *always* include primary consumers that are zooplankton, so a common aquatic food chain might be

Phytoplankton → *Daphnia* → Small fish → Predatory fish

For an example of an aquatic food chain, we can consider Lake Tahoe, which was discussed in Chapter 2 in the context of its non-native signal crayfish (*Pacifastacus leniusculus*) populations. The introduction of non-native species started in Lake Tahoe in the 1870s and has included many more species than just the signal crayfish. Jake Vander Zanden, Sudeep Chandra, and others have reconstructed the lake's historical food chain from 1872, before the introduction of non-native species (**Figure 10.3**; Vander Zanden et al. 2003). The dominant food chain at that time consisted of algae that were eaten by zooplankton, which in turn were consumed by tui chub, which were then eaten by cutthroat trout—four taxa, like the food chain in the Coachella Valley.

Although the food chains in Figures 10.2 and 10.3 draw our attention to particular organisms at each trophic level in a specific community, the overall organization of the chains suggests commonalities across communities. Without focusing on specific taxa, it is possible to ask questions about the density of individuals at different trophic levels. It is even possible to ask whether a food chain in the ocean behaves like a chain in a lake, or whether three-level chains behave like four-level chains. Because the idea of a food chain helps formulate reasonable ecological questions or, even better, helps answer such questions, food chains can be considered conceptual models.

10.4 Insights from Food Chain Models

Trophic Energy Transfer Efficiency

Let's start by thinking about the flow of energy implied by the arrows in the food chain model. In Chapter 5, we emphasized that individual organisms must balance energetic acquisitions with energetic losses and needs. These energetic trade-offs affect the way individuals survive, behave, reproduce, and accumulate biomass.

Figure 10.3 Lake Tahoe sits in the Sierra Nevada between Nevada and California. Its native aquatic food chain in 1872 included algae, zooplankton, tui chub, and cutthroat trout. Non-native species were added to the lake in the later 1870s, irrevocably altering this food chain (Vander Zanden et al. 2003).

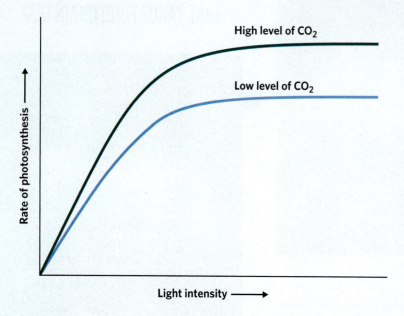

Figure 10.4 Photosynthesis does not afford plants access to unlimited energy. At low light levels, light intensity itself limits photosynthesis and, therefore, energy acquisition. At high-intensity light levels, plants lose water through open stomata, but closing the stomata limits access to CO_2. Eventually all plants hit a plateau in photosynthetic rate and energy acquisition.

Here, we can build on those ideas and suggest that such constraints also affect the way biomass accumulates within any particular trophic level.

In Chapter 5, we discussed how individuals have some control over energy allocation through their physiological and behavioral processes. However, there are inherent energetic constraints that eclipse all of an individual's traits and allocations. First, primary producers have access to a huge store of energy from the Sun, but we know that this source is not unlimited. The availability and intensity of sunlight in different habitats, the availability of carbon dioxide or water, and the speed of chemical reactions all place physical limits on the rate of photosynthesis and, therefore, on primary producers' energy acquisition (**Figure 10.4**). For heterotrophs (organisms that acquire energy by eating other organisms; see Chapter 5), energy acquisition is limited by the abundance of their food. This trophic connection suggests that the biomass of primary producers should affect the biomass of herbivores, which in turn should affect the biomass of predators.

But access to energy is just the first piece of this puzzle. Organisms can only control some components of their energy allocation because there are certain inherent costs to being alive. We introduced the principle of energy allocation in Chapter 5 to highlight the balance that must exist between energy intake and energy use for an individual:

$$E_{intake} = E_{respiration} + E_{assimilation} + E_{reproduction} + E_{waste} \quad \text{(5.1 restated)}$$

where

- E_{intake} = the total energy an individual ingests
- $E_{respiration}$ = the total energy devoted to getting oxygen into the body and cells
- $E_{assimilation}$ = the total energy converted to living tissue
- $E_{reproduction}$ = the total energy used for reproduction
- E_{waste} = the total energy lost as waste or heat or inefficiency

In the context of food chains, it is possible to put values into this equation and track energy as it moves through a food chain one individual at a time. For instance, imagine that a young boar consumes 100 calories of acorns (**Figure 10.5**). What is the fate of that total amount of energy, remembering that calories are a measure of energy content? For simplicity, we can lump $E_{assimilation}$ and $E_{reproduction}$ together because they both represent the use of energy to create new tissue or biomass. With this simplification, the 100 calories must be divided among three categories: respiration, new biomass, and waste. But how many calories go into each category?

Of the 100 calories the boar consumed, approximately 35 calories go to respiration, meaning they are lost to the environment as heat due to the nature of chemically digesting food and the constraints of energy transfer. Approximately 50 calories pass through the boar as waste. This allocation leaves only 15 of the

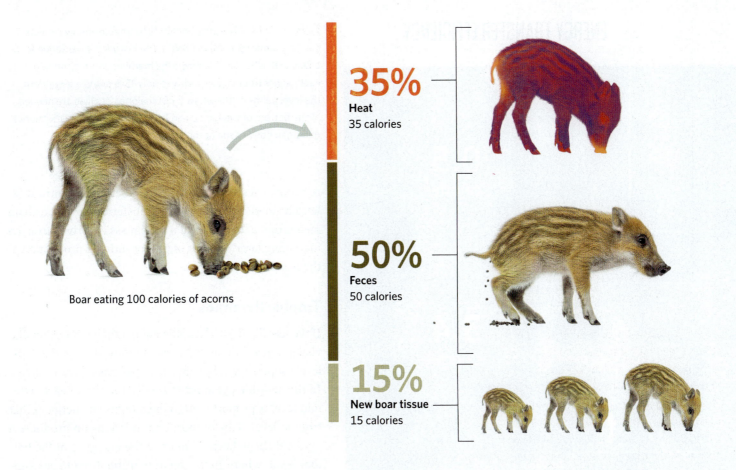

Boar eating 100 calories of acorns

35%
Heat
35 calories

50%
Feces
50 calories

15%
New boar tissue
15 calories

original 100 calories eaten by the boar that can be allocated to making new boar biomass. A boar is not special in this regard, because as a rule, most species only convert about 10%–20% of the calories they consume into biomass.

In the context of a food chain, this 10%–20% of the total calories the boar ingested represents the amount of energy that is available to a predator that eats the boar. This percentage of energy transfer to the next trophic level is not specific to boar and their predators but is generally consistent across all ecological systems (Pianka 1999). This low *energy transfer efficiency* across trophic levels places a physical limitation on biomass accumulation that holds for the vast majority of nonphotosynthetic organisms in the world and thus has huge implications for all life on Earth.

For example, let's look at how energy might move up a food chain such as the historical one in Lake Tahoe (**Figure 10.6**). If phytoplankton create or fix 1,000 calories of algal biomass through photosynthesis, and we assume 15% energy transfer efficiency, then zooplankton feeding on the algae will produce zooplankton biomass from only 150 of those original 1,000 calories. This same transfer efficiency is applied at every trophic exchange. At the level of the tertiary consumer (cutthroat trout), the available biomass will be less than 3.5 calories from the original 1,000 calories fixed via photosynthesis by the algae. If we go one step further up the food chain and include a human from the Washoe tribe (which inhabited the area around Lake Tahoe for about 10,000 years before European immigrants arrived) who catches and eats a trout, almost 60,000 calories of algae would be necessary

Figure 10.5 In this hypothetical example of energy allocation within a young boar after consuming 100 calories of acorns, only 15% of the total energy consumed is turned into new boar tissue and made available for use (consumption) by predators at the next higher trophic level.

ENERGY TRANSFER EFFICIENCY

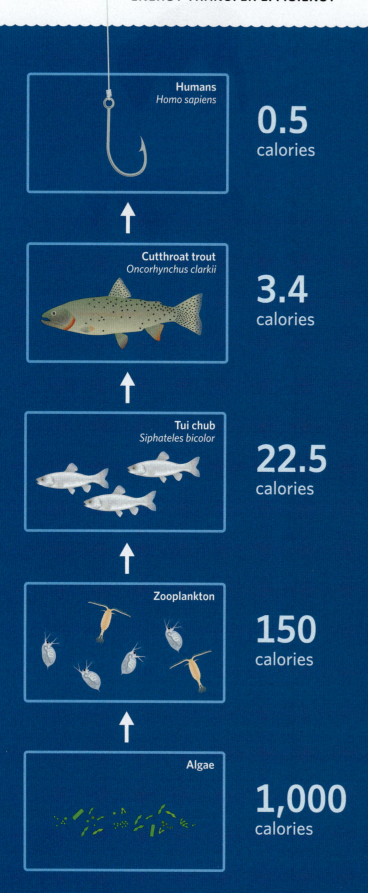

Humans
Homo sapiens

0.5
calories

Cutthroat trout
Oncorhynchus clarkii

3.4
calories

Tui chub
Siphateles bicolor

22.5
calories

Zooplankton

150
calories

Algae

1,000
calories

Figure 10.6 The inherent limits to trophic energy conversion can be examined in the context of the historical Lake Tahoe food chain, with a human Washoe tribe member as a top predator. Each higher trophic level receives only 15% of the energy from the level below it. Therefore, 1,000 calories of algae are needed at the bottom of the food chain to produce a meager 0.5 calories of human body tissue at the top.

to produce a single, modestly sized 200-calorie trout fillet. The inefficiency of energy transferred up a food chain means that a top predator's body mass is derived from an incredibly large amount of photosynthetic production at the base of the food chain.

Trophic Pyramids

If we use the flow of trophic energy up a food chain and visualize it as a pyramid, we can see how the lower trophic levels support and bolster the upper ones (**Figure 10.7A**). In this **trophic pyramid** of energy, the size of each pyramid level is proportional to the amount of energy in that trophic level. This proportional relationship produces a pyramid shape because most of the energy is in the bottom level, where huge amounts of photosynthesis support the upper trophic levels. Because of the inefficiency of energy transfer, each successive level is much smaller than the level below it on the food chain. The end result is a pyramid shape with a tiny amount of energy sitting atop all other trophic levels.

If we use the same visualization but change the currency that each level represents, we can construct a trophic pyramid of biomass. The same basic pyramidal shape is produced when visualizing biomass across trophic levels, even though the units for biomass transitions (g/m^2) are different from the units for energy transitions ($kcal/m^2/y$) up the food chain (**Figure 10.7B**). This concordance suggests that the amount of biomass held in one trophic level is positively correlated to the amount of energy contained in that level.

The connection between biomass and energy in trophic pyramids is the rule in terrestrial food webs, but the relationships can be a little different in some aquatic food webs, where most of the primary producers are phytoplankton. These tiny, often single-celled organisms reproduce so rapidly that the actual biomass of phytoplankton at any moment can be lower than the biomass of herbivorous zooplankton that graze on them (**Figure 10.8**). When this happens, you get a slightly inverted biomass

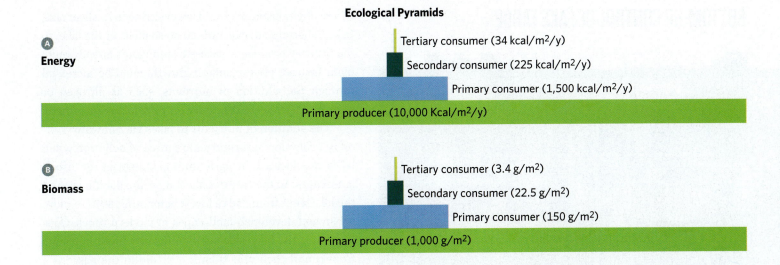

Ecological Pyramids

A **Energy**

Tertiary consumer (34 kcal/m²/y)

Secondary consumer (225 kcal/m²/y)

Primary consumer (1,500 kcal/m²/y)

Primary producer (10,000 Kcal/m²/y)

B **Biomass**

Tertiary consumer (3.4 g/m²)

Secondary consumer (22.5 g/m²)

Primary consumer (150 g/m²)

Primary producer (1,000 g/m²)

pyramid relative to the energy pyramid. This inversion comes from looking at a single snapshot in time and is possible only because the phytoplankton reproduce extremely fast. They compensate for lower biomass at any given moment by constantly and rapidly producing biomass *through* time. If we included all the phytoplankton alive over a primary consumer's generation, the aquatic food pyramid shape would closely resemble the shape of the terrestrial food pyramid.

Bottom-Up and Top-Down Control

Given the inherent pyramidal structure of biomass across food chains, it is easy to assume that the basal level of the food chain can affect the abundance of organisms higher up the chain. But is it also possible to imagine effects in the other direction, down the food chain? In the beginning of the chapter, we discussed why trophic relationships might cause species at the top of the food chain to affect species at the bottom. Can we actually see such indirect effects? Ecologists have argued for decades about how energy flow in food chains is controlled. Does the control come from the bottom up or the top down? Proponents of **bottom-up control** suggest that organisms at the basal trophic levels (autotrophs) act as gatekeepers to the upward flow of energy, thereby constraining the abundance of species at the higher trophic positions. In this scenario, more photosynthesis translates into more energy that can be moved up to higher trophic levels, much as we saw in the trophic pyramids.

In the Lake Tahoe food chain with bottom-up control, humans could have access to more fish if they increased the biomass of phytoplankton (primary producers). One way to do this would be to increase the amount of sunlight reaching the waters

Figure 10.7 **A** Energy and **B** biomass pyramids built using a 15% energy transfer between trophic levels. Note that either unit of energy currency produces the same basic shape.

Figure 10.8 Aquatic ecosystems can produce slightly inverted biomass pyramids. The small size and rapid reproductive capacity of the phytoplankton mean that a small biomass can support longer-living zooplankton if measured at a single point in time. If we examine the biomass of phytoplankton over a zooplankton or tui chub generation, however, we find the same pyramidal shape that we see in terrestrial systems.

Cutthroat trout

Tui chub

Zooplankton

Phytoplankton

Aquatic System

Fox

Scorpion

Grasshopper

Creosote bush

Terrestrial System

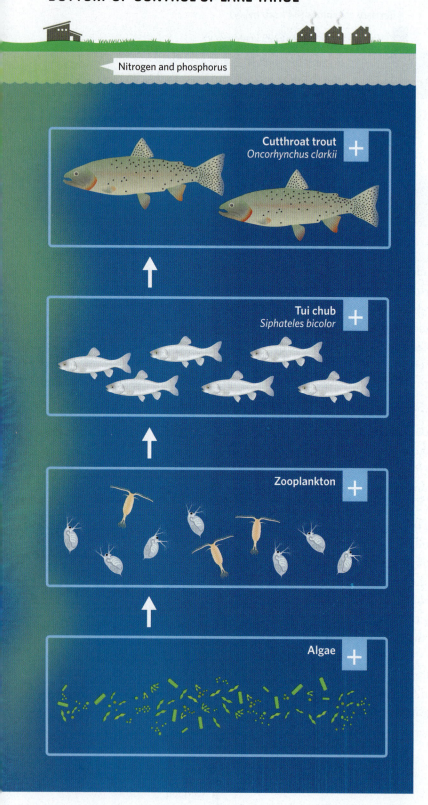

Nitrogen and phosphorus

Cutthroat trout
Oncorhynchus clarkii
+

Tui chub
Siphateles bicolor
+

Zooplankton +

Algae +

where phytoplankton grow. However, due to its sheer size, Lake Tahoe already has little to no shade over the lake, so the primary producers' access to sunlight is already quite high. Instead, phytoplankton growth could be increased through the addition of nutrients, such as nitrogen or phosphorus, that are necessary for phytoplankton growth.

The addition of nutrients to lakes via rainwater run-off is a common anthropogenic source of pollution when lakes are located in agricultural, industrial, or urban landscapes. In the case of Lake Tahoe, the habitat around the lake went from native forest before the 1800s to more urban and developed landscapes in modern times. Consequently, the lake receives significantly more nutrients from runoff each year than it ever did in the past. As a result, the phytoplankton populations have increased. Not surprisingly, given what we know about food chains, the number of zooplankton (herbivores) has also increased, which in turn has produced larger populations of small fish (i.e., tui chub, the secondary predator). More tui chub provide a larger food base, which should have produced more cutthroat trout (top predator; **Figure 10.9**). In reality, though, the Lake Tahoe ecosystem is significantly more complicated because of the anthropogenic introduction of many non-native species at multiple levels in the food chain. These non-native species have created less-direct trophic connections between phytoplankton, tui chub, and cutthroat trout. Nonetheless, the basic principle holds: increases at the bottom trophic levels of the Lake Tahoe food chain increased biomass at the higher trophic levels.

By contrast, proponents of **top-down control** posit that the upper trophic levels control the biomass and abundance of organisms at lower trophic levels by eating them, thus exerting a strong influence on energy flows down the food chain. Control in this direction is easiest to understand if we start with just three trophic levels and think about what happens if a top predator is added, creating a fourth level.

What if a food chain similar to Lake Tahoe's had only three trophic levels, occupied by phytoplankton, zooplankton, and tui chub? If cutthroat trout were then introduced, we could follow the resultant effects down to the lower trophic levels (**Figure 10.10**). The newly

Figure 10.9 Increased urban development around Lake Tahoe has led to the addition of the nutrients nitrogen and phosphorus to the lake. This anthropogenic fertilizer has stimulated the growth of algae, which in turn has increased the density and biomass of organisms at each level up the food chain. The plus signs indicate an increase in biomass or population density at each trophic level.

TOP-DOWN CONTROL OF A LAKE FOOD CHAIN

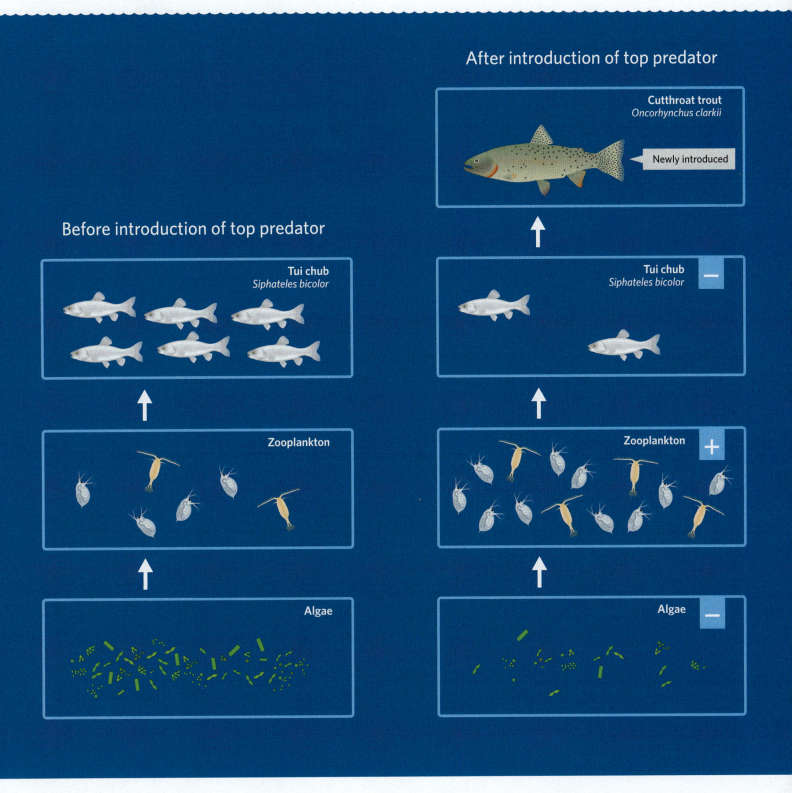

After introduction of top predator

Cutthroat trout
Oncorhynchus clarkii

Newly introduced

Before introduction of top predator

Tui chub
Siphateles bicolor

Tui chub
Siphateles bicolor −

Zooplankton

Zooplankton +

Algae

Algae −

Figure 10.10 In this hypothetical lake, the introduction of a top predator to a three-level food chain leads to a trophic cascade. The new arrivals (increase in population density) at the top trophic level lead to alternating decreases and increases in the densities of organisms at the lower levels, shown by minus and plus signs, respectively.

COMPARISON OF TOP-DOWN AND BOTTOM-UP CONTROL

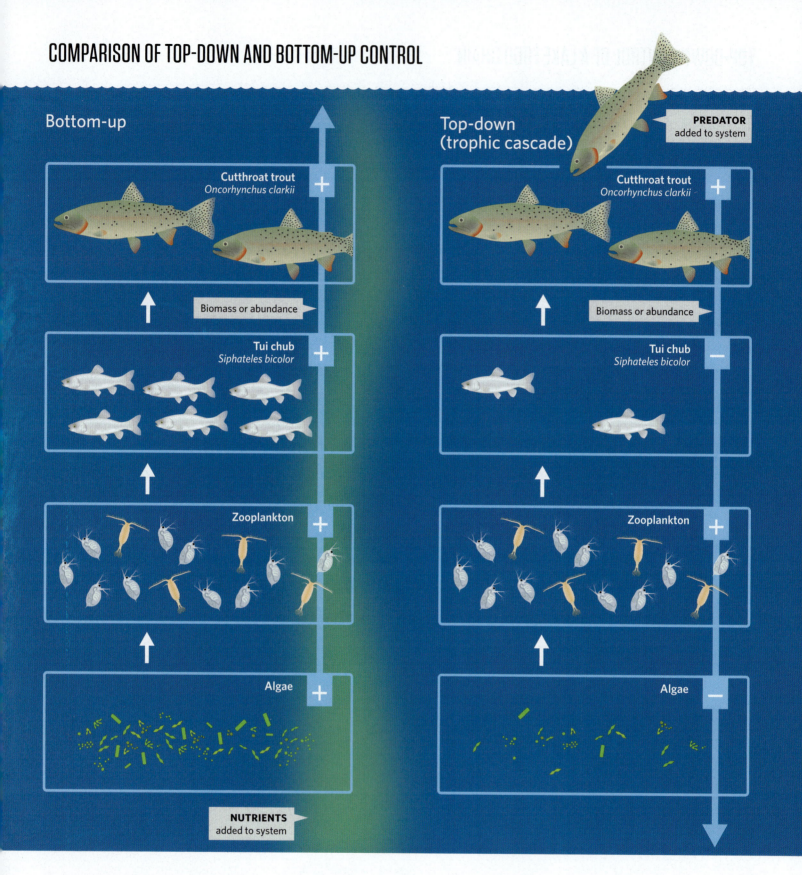

Figure 10.11 This visual provides a succinct comparison of bottom-up and top-down control of trophic systems. The blue arrows indicate the direction of trophic control; white arrows indicate the direction of energy transfer; and plus or minus signs indicate changes in biomass or population abundance at each trophic level.

introduced trout population would grow from eating tui chub, and the chub population would decline because of the new predation pressure from trout. With fewer chub around to consume zooplankton, the zooplankton population would increase. The newly abundant zooplankton would put intense grazing pressure on the phytoplankton and thereby reduce phytoplankton abundance.

Notice in this hypothetical example that the top predator (trout) not only has effects on the lower levels, but it actually controls the energy flow at lower trophic levels. The effects of this top-down regulation appear to "cascade" down the food chain with alternating increases and decreases in biomass and abundance of organisms at each level of the chain. This alternating effect on abundance and biomass is commonly called a **trophic cascade**. You may also notice that this cascade fits nicely with some of the intuitive ideas mentioned at the beginning of the chapter.

When compared side by side, these two views about the control of food chains appear directly at odds with one another (**Figure 10.11**), which suggests that one should be right and the other wrong. It may not be surprising that the truth lies somewhere in the middle. From the descriptions of food chains and trophic pyramids, both scenarios seem possible, and, indeed, the ecological literature provides evidence to support both bottom-up and top-down control.

Evidence for Bottom-Up Control A large body of evidence for bottom-up control comes from observations made by *limnologists*, scientists who study the chemistry, physics, and ecology of freshwater lakes, ponds, rivers, and streams. One of the fundamental ways ecologists categorize lakes is by trophic status, which they can determine by the clarity of the water. Lakes that have very clear, blue water generally support only a few primary producers and thus have a low overall biomass per unit area. Such clear lakes are called **oligotrophic** (the prefix *oligo-* means "few" in Greek; **Figure 10.12**). Lakes that have more opaque, green water generally support an abundance of basal autotrophic species (hence the green color) and thus have high biomass per unit area. These lakes are called **eutrophic** (prefix *eu-* means "good" or "true" in Greek; Figure 10.12). Although we give them two distinct names, most lakes fall somewhere between these two extremes. And, despite the Greek roots of the words, ecologists do not necessarily consider eutrophic lakes to be in better condition than oligotrophic lakes.

Figure 10.12 Crater Lake in Oregon is oligotrophic; it is very blue, indicating little plant life in the water. Copco Reservoir in California, on the other hand, is eutrophic; its green and murky appearance indicates an abundance of phytoplankton.

Oligotrophic

Eutrophic

Bottom-up control of a food chain plays into the categories of eutrophic and oligotrophic. In lakes with lots of biomass at the bottom of the trophic pyramid (i.e., lots of phytoplankton), and thus high rates of photosynthesis, there tend to be large amounts of biomass at each trophic level, making the lakes eutrophic. In lakes with few phytoplankton and low rates of photosynthesis and productivity, we find very little standing biomass across all trophic levels, making the lakes oligotrophic.

Limnologists have observed that lakes can shift from oligotrophic to more eutrophic with the addition of nutrients to the water, usually in the form of nitrogen or phosphorus runoff from terrestrial sources (e.g., agriculture), as previously mentioned. In an effort to test whether lakes in general can change from oligotrophic to eutrophic, Lev Ginzburg and H. Reşit Akçakaya (1992) compiled biological data from a variety of North American lakes with differing levels of phosphorus input. The authors found that, at each trophic level (i.e., phytoplankton, zooplankton grazers, and predatory fishes), biomass increased as the amount of added phosphorus increased (Figure 10.13). This pattern strongly suggests that the addition of limiting nutrients, such as phosphorus, can lead to increases in biomass all the way up a lake food chain. As a result, it appears that lake food chains can be controlled from the bottom up.

Extreme levels of anthropogenic nutrient additions can also complicate our food chain predictions. When nutrient input gets too high, phytoplankton blooms can limit sunlight's ability to penetrate water, leading to lower overall levels of photosynthesis, low oxygen concentrations in the water, and low biomass at higher trophic levels. Part of this effect also occurs because, at such high abundances, the phytoplankton overwhelm the herbivores, who are unable to keep up with eating them. This leads to masses of decomposing phytoplankton, and the decomposition process takes up more oxygen. Lower oxygen levels from reduced photosynthesis and increased decomposition reduce the performance of some of the consumer species, thereby further reducing consumption of phytoplankton and leading to lower abundance and biomass at higher trophic levels.

Clearly, all these effects can get quite complicated, but nutrient additions definitely do not always lead to better lake conditions or more organisms at every trophic level. Anthropogenic nutrient additions to Lake Tahoe over the past 50 years or more have led to the conservation slogan "Keep Tahoe Blue" for the lake. Many of you may have passed a construction site and seen mesh tubes of sand placed to restrict the flow of material downhill, or seen signs in stormwater drains warning people to limit nutrient inputs into waterways. These measures reduce anthropogenic nutrient inputs and avoid intense eutrophication of lakes and streams with their attendant alterations of aquatic food chains.

Evidence for Top-Down Control The early ecologist and philosopher Aldo Leopold first proposed the idea of a trophic cascade in his book *A Sand County Almanac* in 1949 (see Chapter 1), but the idea really took off when Nelson Hairston, Frederick Smith, and Lawrence Slobodkin published a paper titled "Community

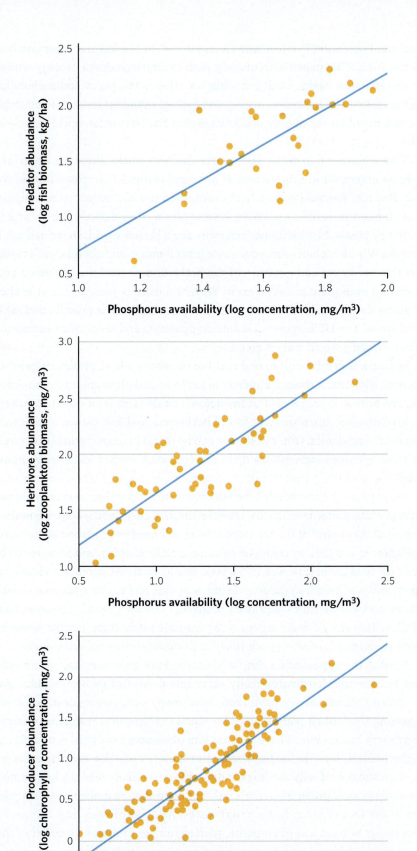

Figure 10.13 Ginzburg and Akçakaya (1992) plotted the relationships between biomass at different trophic levels (producers, herbivores, and predators) and resource availability (phosphorus) in lakes across North America. Increasing phosphorus input into lakes produces more biomass at each trophic level. Each data point on the graphs represents a single lake.

Structure, Population Control, and Competition" in the journal *American Naturalist* in 1960. This paper is frequently read by undergraduate ecology students because Hairston, Smith, and Slobodkin (or HSS, as the paper and authors have come to be known) created a concise, compelling argument for top-down trophic cascades based on the simple observation that the terrestrial world is generally green.

Take a moment to think about why this extremely mundane observation might be important in this context. If plant populations (i.e., green things) were controlled and limited by their herbivores, then the herbivores would consume as much plant material as possible, leaving the world relatively brown and less covered by plants. But we do not generally see a brown world, so we can ask the question, Why don't herbivores decimate plant biomass and abundance? HSS suggests that predators and parasites of the herbivores control the herbivore populations and keep them at lower levels. With herbivore populations kept in check, the plants they eat proliferate until competition checks their growth, making the world green. The HSS argument is amazingly simple and seductive, because most of the terrestrial world truly is green.

In fact, it is not difficult to find real-life situations where predators have been removed, allowing herbivore numbers to explode, and plant abundances to plummet. The brown, mucky world of an enclosed cattle farm is an excellent example of a predator-free situation with lots of herbivores and few plants, all due to the exclusion of predators. Other examples can be found in more-natural ecosystems, as human activities increasingly drive top predators toward local or even global extinction.

Other research has extended the HSS theory to suggest that control by competition or predation alternates trophic levels on the food chain, with competition always working as the control at the top trophic level, followed by predation at the trophic level below it, and then by competition at the next level down, and so on to the bottom of the food chain (Fretwell 1977). Food chains with odd numbers of levels have high densities of primary producers (a green world), and chains with even numbers of levels have low densities of primary producers (a brown world). Hairston, Smith, and Slobodkin started their argument by assuming only three trophic levels and, therefore, limited themselves exclusively to discussing terrestrial ecosystems.

Despite the appeal of its simple logic, the HSS argument has several problems. First, would herbivores really eat plants down to tiny brown nubs, given the chance? Plants, as we have seen, have many defenses against herbivores, suggesting that not all plant material is equal in terms of transferring energy to the primary consumers. There may be lots of competition between herbivores for the juiciest, least-defended leaves and almost none for less desirable bits. But consumption of only the juicy, vulnerable parts likely leaves a large amount of well-defended, green plant material that continues to make the world green, even in the face of many hungry herbivores. Also, herbivores may compete for other things beyond just plant tissue, such as refuges or mates. In either of these situations, herbivores may be controlled by competition and not by predation, and the world would stay green.

Perhaps most importantly, though, it is not clear that top-down and bottom-up forces must act independently. Stephen Fretwell's extension of HSS to ecosystems with different numbers of trophic levels (1977, 1987) suggests that the number of trophic levels is determined by the primary productivity of the ecosystem. His work actually combines both bottom-up and top-down control. Mary Power (1992) has suggested that there is no reason to see these two forces as incompatible. Both productivity and top predators can have simultaneous impacts on the abundance of organisms at different trophic levels, and both forces depend on indirect interactions.

All these arguments ignore one critical issue with HSS and many of the arguments about control in food chains: What about all the other interspecific interactions that are not captured in a food chain? The very fact that HSS and others wrestle with ideas of control by competition or predation suggests that food chains present an incomplete story. The concept of a food chain assumes that all individuals within a given trophic level are the same, but we know they are not. For example, an insect herbivore is clearly very different from a vertebrate herbivore. In addition, some species do not fit cleanly into a single trophic level. For example, how should ecologists categorize an organism, such as a raccoon, that eats from more than one trophic level? In order to address these issues, the simple food chain model needs to be modified.

10.5 Food Webs: Adding Complexity

One of the biggest shortcomings of food chains is that a simple "chain" model leaves out so many possible feeding interactions. For example, the Coachella Valley food chain model assumes that a kit fox only eats scorpions, and the Lake Tahoe food chain model suggests that cutthroat trout only eat tui chub. From detailed observations of kit foxes and cutthroat trout, we know this is not true. Kit fox and cutthroat trout eat individuals of many other species, as do most top predators. Of course, it is possible to generalize the chains to include many species at a single trophic level, but then the model ignores individual species that eat more or compete better, or ones that eat foods from different trophic levels. In fact, species labeled as "predators" in food chains often eat plant parts.

Species that acquire energy by consuming across multiple trophic levels are called **omnivores**. Omnivory is so common that species that feed only from one trophic level are singled out as **trophic specialists**. Food chains omit the concept of omnivory in the service of simplicity, but omnivores are more the rule than the exception. Reams of data show that a large majority of organisms eat many different things, even when we initially believed them to be trophic specialists (Polis 1991). More importantly, some organisms eat such a varied diet that we could never assign them to a specific trophic level. Bears and humans are good examples of such organisms. Yet bears and humans can easily have big effects on entire food chains. Human effects are probably fairly obvious to all readers, but Chapter 14 describes the way bear consumption can move nutrients across ecosystem boundaries (from rivers to forests). Humans and bears are not the only omnivores with big effects on energy flows, which suggests we need a way to include omnivores in our discussion of indirect effects and trophic interactions.

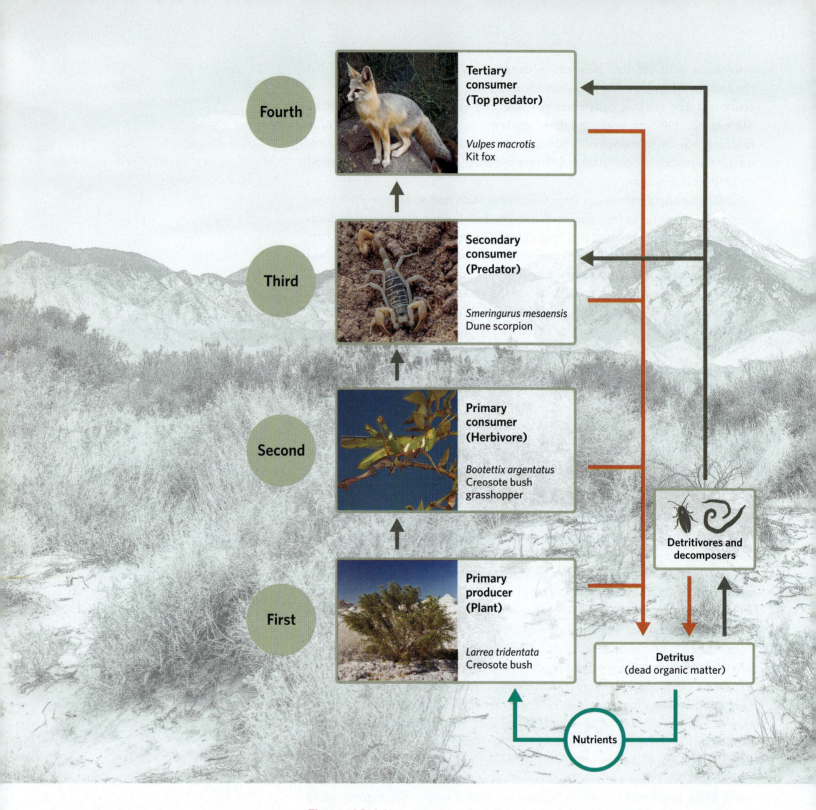

Figure 10.14 A hypothetical Coachella Valley food web in which energy from dead organisms is recycled within the web, a process accelerated by detritivores and decomposers. The darkest arrows show consumption across trophic levels. The orange pathway shows that dead organisms at each level become detritus. The teal pathway shows that detritus eventually becomes available as nutrients for plants.

The food chain model also fails to recognize that many organisms consume alternative foods at different stages of their lives. Think for a minute about what you ate as a baby. For much of your first year of life you consumed breast milk or formula as your sole form of nutrition. Eventually, you were weaned onto solid food, and your diet breadth grew and perhaps has continued to grow as you have aged. These types of developmental changes, referred to by ecologists as *ontogenetic diet shifts*, are very common throughout the natural world and are generally left out of food chain models.

Finally, food chains overlook an essential concept for all of biology and ecology, namely "the circle of life" (to quote *The Lion King*). Everything that lives will eventually die, and the energy contained in once-living creatures is recycled into energy for another creature to use. This death-to-life concept is pretty clear in the arrows that move up the food chain, but what happens to the organisms at the top? The return of their energy to the chain involves a whole suite of additional species called **detritivores** (insects, earthworms, and fungi) and **decomposers** (bacteria, fungi, and protists) that feed on dead plant and animal parts called **detritus**. These species ingest decaying life as their source of energetic nutrition (**Figure 10.14**).

These and other problems with the food chain model led ecologists to a more complex trophic model called a **food web**. A food web is a graphical representation of the energetic feeding relationships among species that coexist in a community. Food webs work as more complex conceptual models of multispecies interactions.

Examples of Food Webs

Figure 10.15 shows a simple food web diagram for the Coachella Valley, indicating far more detail than the simple food chain depicted in Figure 10.2 or the simple web in Figure 10.14. Instead of a one-dimensional connection among the species, we have a web of relationships that includes omnivory, ontogenetic diet shifts, and the detrital portion of the food web. Although Figure 10.15 is a more realistic accounting of the trophic relationships in the valley, it is still a gross oversimplification of the true food web for the Coachella Valley. To borrow the form of a standardized-test analogy, Figure 10.15 is to a real food web as a child's sketch is to a painting by Claude Monet.

Gary Polis spent 16 years studying the food web of the Coachella Valley and decided that he could not capture the feeding interactions in a single web. Instead, he produced a series of complex subwebs with dozens of species present in just the portion of the web where, for example, scorpions served as a primary predator. He found the same complexity when producing portions of the web where vertebrates were the main predator, or the portion that represented the subterranean web in sandy soils (Polis 1991).

The single arrow from the grasshopper to the scorpion in both Figure 10.2 and Figure 10.15 was perhaps one of the biggest simplifications Polis made. When he tried to explore what scorpions actually ate, he found they consumed more than 100 different species of arthropods, with no indication that the number of prey types had all been identified (Polis 1991). In fact, the number of scorpion prey species

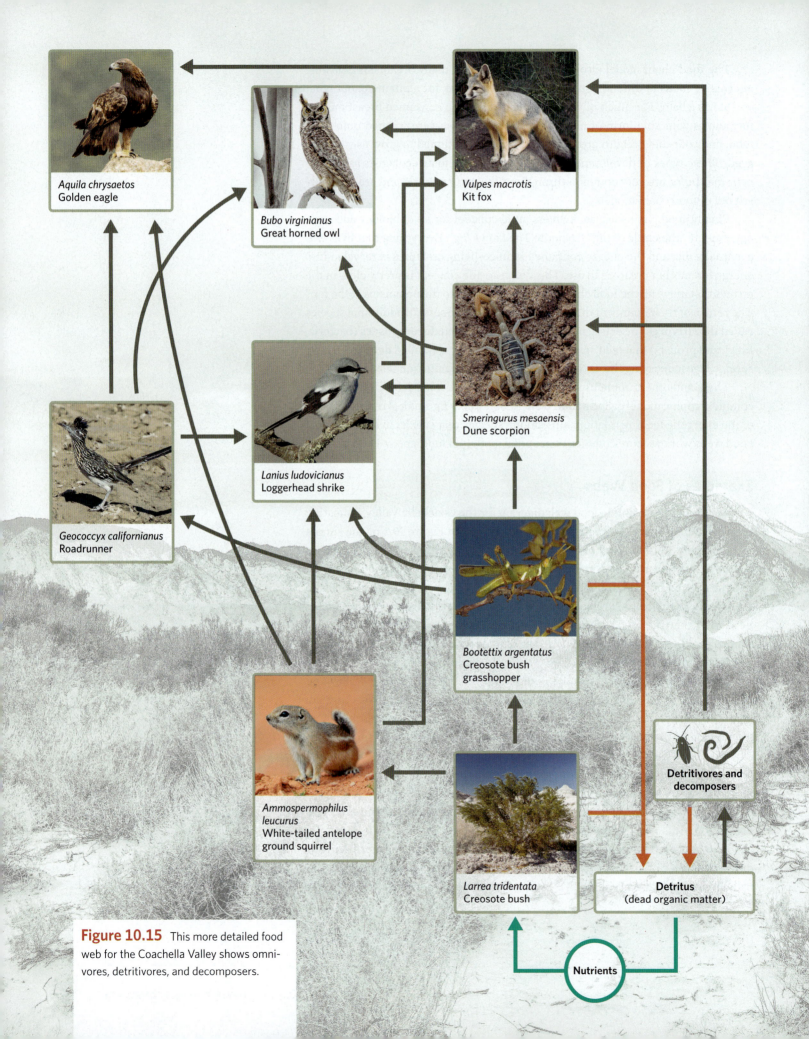

Aquila chrysaetos
Golden eagle

Bubo virginianus
Great horned owl

Vulpes macrotis
Kit fox

Smeringurus mesaensis
Dune scorpion

Lanius ludovicianus
Loggerhead shrike

Geococcyx californianus
Roadrunner

Bootettix argentatus
Creosote bush
grasshopper

Detritivores and
decomposers

*Ammospermophilus
leucurus*
White-tailed antelope
ground squirrel

Larrea tridentata
Creosote bush

Detritus
(dead organic matter)

Nutrients

Figure 10.15 This more detailed food
web for the Coachella Valley shows omnivores, detritivores, and decomposers.

increased with about half the surveys he conducted and showed no signs of leveling off after 200 surveys (**Figure 10.16**), suggesting scorpions have an extremely broad diet.

One of the reasons Polis did not try to represent the complete Coachella Valley food web pictorially was that, in the early 1990s, he could not fit the hundreds of species and trophic interactions that he documented into a single, readable figure. If he had, it would have looked even more complicated than **Figure 10.17**, which depicts the marine food web off the Northwest Atlantic coast. This food web was assembled by a team of researchers to aid in the management and sustainable harvest of fisheries in the area (Lavigne 1996). Each line in the web represents a *trophic link*, meaning an instance of consumption in this complex net of relationships. Note that using a food web instead of a food chain addresses some of the concerns raised earlier about the inclusion of omnivory and ontogenetic diet shifts. A close look at Figure 10.17 reveals dozens of examples of omnivory. Food webs like this even include cannibalism, or *looping* (see examples on the left side of Figure 10.17), in which an animal eats its own kind.

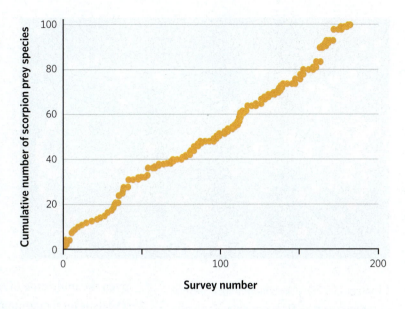

Figure 10.16 Cumulative number of prey species plotted against survey number for the first 100 arthropod species that Polis identified as being consumed by the scorpion *Smeringurus mesaensis* in California's Coachella Valley (Polis 1991). Note that about half the times Polis looked, he found new species that were consumed by the scorpion.

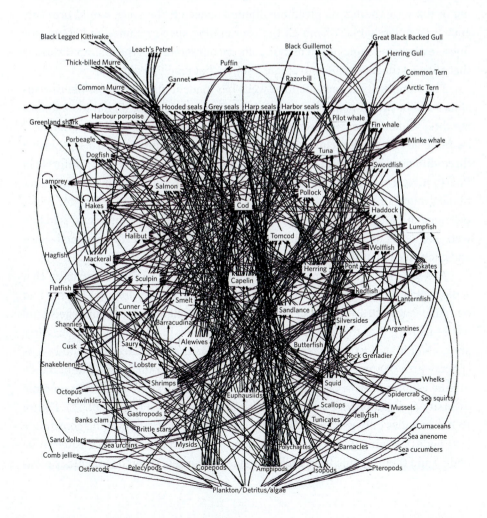

Figure 10.17 This diagram of the Northwest Atlantic marine food web was assembled by a team of researchers in the mid-1990s (Lavigne 1996). Note the extreme complexity of this web, as each arrow represents an identified trophic link.

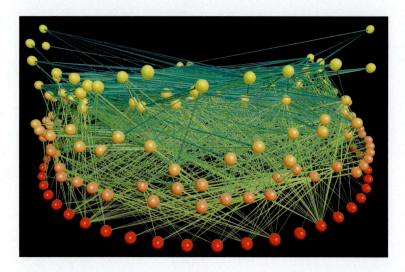

Figure 10.18 This ball-and-stick diagram of the food web for El Verde rainforest, Puerto Rico, is organized vertically, with nodes (balls) representing individual species at a particular trophic level. Red nodes represent basal species, such as plants; orange nodes represent intermediate species; and yellow nodes represent top predatory species. Each line (stick) represents a trophic interaction between two species. In a three-dimensional digital version of this food web, individual nodes can be examined to reveal more details.

Despite its overwhelming visual complexity, Figure 10.17 is still not complete, as it leaves off the detrital part of the web. Notice that the arrows depicting trophic links generally only point upward in this figure. A food web that depicts detritivores and decomposers will have arrows pointing downward, representing energy flowing toward detritivores. Also note that some species in the Northwest Atlantic are strongly linked with a whole host of other species, as is the case with Atlantic cod (*Gadus morhua*) and grey seals (*Halichoerus grypus*). We will revisit this issue later in the chapter, but for now it is enough to notice that not all species in this web are equally connected.

In addition, this food web contains a number of other major simplifications. You may ask, How can that be, given the multitude of trophic connections? But a close inspection of Figure 10.17 reveals several taxonomic groups that act as generic placeholders, such as shrimps or gastropods or mussels. There are at least 30 of these generalized groups, located mainly in the lower portions of the diagram. Also note the potentially huge category of "plankton, detritus, and algae" at the very bottom of the diagram. Every one of these placeholder categories likely encompasses tens or even hundreds of species, with each species having its own unique net of trophic links that are not depicted. If we were to expand the details of these groups and give a full accounting of the web, the figure, given our limited space on the page, would probably appear completely black from all the overlapping species names and connector lines—exactly the same problem that Polis encountered. It is these simplifications that allowed the researchers to create a somewhat readable diagram.

The complexity of real-life food webs has made it almost impossible to visualize webs in two dimensions. As a result, scientists now use digital tools to create three-dimensional representations of food webs in order to differentiate and examine all the trophic connections at once. For example, **Figure 10.18** shows a three-dimensional ball-and-stick model of the food web for El Verde rainforest in Puerto Rico. This depiction color-codes species (represented by balls) into functional groups, or trophic nodes. In this case, there are separate colors for basal, intermediate, and top predatory species. In order to see all the interactions in this web, one needs to be able to rotate the digital model in three-dimensional space and click on each of the balls to identify individual species.

And yet, this *still* does not convey the full extent of trophic complexity that ecologists have uncovered in real ecosystems. For example, Jennifer Dunne and her colleagues (2013) note a significant accounting flaw in most published food webs—namely, that they leave out parasites. We saw in Chapter 9 that parasites are a ubiquitous feature of the natural world, so their absence from food webs is noteworthy. Dunne and her team use a food web diagram for the Punta Banda marine estuary in Baja California, Mexico, to demonstrate the added taxonomic diversity and complexity associated with parasite species (**Figure 10.19**). Figure 10.19A includes only primary producers (green nodes) and free-living consumers (red nodes). Figure 10.19B is the same food web with all known parasites (blue nodes) and their trophic links included. As you can see, when one includes para-

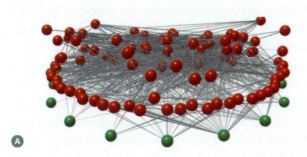

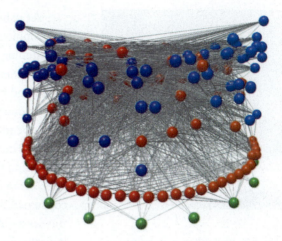

Figure 10.19 Food webs for the Punta Banda marine estuary in Baja California, Mexico. **A** The web with basal and consumer species (green and red, respectively) but no parasites, and **B** the web with parasites added (blue). Note the extreme increase in both complexity and diversity after the addition of parasites (Dunne et al. 2013).

sites, the food web becomes blazingly complex and hard to visualize even in three dimensions. Yet this level of complexity is likely present in almost all food webs. From here forward, whenever you hear the term *food web*, you should think of Figure 10.19B, because that is the reality of the natural world.

These more complex models offer a sophisticated glimpse into the trophic interactions within real ecological communities. The data for such complex models are difficult to collect, though, and generalizations can be hard to draw from the complexity. Given these practical issues, ecologists often fall back on using food chains or simplified food webs as a matter of necessity, but clearly these more detailed web models are preferable if we can obtain them.

Moving beyond Trophic Levels

Given enough observations of trophic interactions in a community, the notion that food webs have discrete trophic levels goes up in smoke. How is it possible, with webs like those shown in Figures 10.17–10.19, to partition taxa into clear-cut trophic levels such as herbivore, predator, and top predator? The autotroph level is probably the only one that we can safely assume is, in fact, a basal level, since most (but certainly not all) autotrophs do not consume other plants or animals. Beyond that assumption, the truth is that most organisms feed on a variety of other organisms, and there is no clear way to differentiate "levels."

Ecologists for decades simply ignored this complexity because obtaining empirical evidence of a species' total diet breadth was a monumental task. The historical and least technical way to examine trophic relationships was to observe individuals of a focal species and accurately record everything they ate. This approach sounds very reasonable, but it produces a bias toward cataloging only the most common dietary items.

For example, it is likely that around 90% of the diet of any given squirrel species is fairly consistent. Squirrels are the classic nut and seed eaters of children's stories, and if you were to spend several days or even months recording the diet of squirrels of a particular species in a particular tree, your data sheets would be filled with descriptions of various nuts and seeds. If these were the only observations used to categorize the species' trophic status, it would fit quite nicely in the category of primary consumer.

Figure 10.20 Squirrels as a group are omnivores and consume a wide variety of food types beyond nuts and berries, including animal protein (e.g., spiders, snakes, small birds, and small mammals) on occasion.

Yet using more sophisticated methods for tracking diets points to more varied trophic relationships, even for organisms like squirrels. For example, instead of watching what individual organisms consume in the limited places where we can see them, ecologists can quantify everything these individuals have already consumed by dissecting or flushing their guts to examine the contents. Such *gut content analyses* are time-consuming (and potentially icky) but provide more complete documentation of what individuals of various species consume.

Alternatively, sometimes it is harder to catch and handle live organisms than it is to find their waste, so ecologists use fecal material to document trophic linkages. From the earlier discussion of energy balance, we know that about 50% of consumed food exits as waste, and *fecal analysis* takes advantage of the fact that this waste often carries identifiable traces of the consumed items. This approach is particularly useful for identifying hard tissues, such as seeds and bones, and is used frequently for many carnivores. Even if prey species are not visually identifiable, the fecal material contains DNA from consumed prey. A technique called DNA metabarcoding allows researchers to use previously identified DNA sequences (DNA barcoding) to identify many species (hence *meta*barcoding) in a single sample. Using this approach, ecologists have learned that squirrels will occasionally include spiders, snakes, small birds, and even other squirrels in their diets (**Figure 10.20**).

Fecal and gut content analyses reveal a far more complex and nuanced understanding of the trophic connections within a community and serve as the basis for much of the information in the detailed food webs we have described. But these approaches also have limitations, especially when ecologists are trying to summarize the trophic position of a species in a complex food web.

One such limitation is that there are often significant gaps in the information gleaned. For example, if we were to examine your stomach contents (let's assume by flushing rather than by dissection), we would likely find a range of different food items across an entire day, a week, or throughout a year. If we flushed your stomach at 1:00 p.m. on a weekday, we would likely find the remains of what you commonly eat for lunch. If we sampled your digestive system at 1:00 a.m. on New Year's Eve, however, a very different "diet" might emerge. Clearly, if we wanted a full accounting of your diet for an entire year, then sampling only during holidays or special feasts would bias the interpretation of your diet. Therefore, gut content and fecal analyses need to be approached with a keen awareness of the potential gaps and variability over time.

We also have the problem of variation among individuals of the same species; in other words, not all members of a species eat the same things because individuals have unique phenotypes (Bolnick et al. 2011). This observation means that, in order to get a representative accounting of diet breadth within a taxonomic group, one has to examine many different individuals at different life stages and at different times of the day and year, which clearly adds considerable time and significant cost to a food web project.

Finally, gut content and fecal analyses pose a host of methodological challenges related to quantifying and counting diet items. Think, for example, about how easy it might be to identify a spoonful of corn in a human's gut contents or fecal matter in light of how resistant corn kernels are to complete digestion. Then consider how impossible it would be to tell whether an individual drank a liter bottle of soda along with the corn. Which of these food items contributes more calories to the person's daily food consumption, and thus which food item is the larger source of energy? Many times, and in many situations, the answer is not clear from just sampling and analyzing gut or fecal contents.

One way to assess the proportion of the diet that comes from different sources is to use chemistry and **stable isotopes**. Isotopes are simply variants of an element. Although all isotopes of an element have the same number of protons, they can have different masses because of differing numbers of neutrons. There are two general types of isotopes, radioactive and stable. Radioactive isotopes decay with a burst of released energy, whereas stable isotopes do not decay. It is the latter that are useful for understanding trophic relationships because they do not change as they "move" through a food web. Using an instrument called a mass spectrometer, ecologists can measure the ratio of different isotopes in a sample of tissue drawn from an individual of interest. The sample, and the stable isotopes it contains, provide information to help determine that individual's source of food. The reasoning behind this necessitates an additional dip into chemistry.

Both carbon and nitrogen have stable isotopes, called ^{13}C and ^{15}N, respectively. The superscript numeral on each symbol indicates the atomic mass of the isotope. The most common isotope of carbon has an atomic mass of 12 (^{12}C), so the ^{13}C isotope has one extra neutron added. Similarly, the common form of nitrogen has an atomic mass of 14 (^{14}N), so ^{15}N has one neutron added. Carbon isotopes are conserved between organisms, meaning that when a predator eats prey, it takes in and assimilates the carbon isotopes of that prey. Because not all primary producers have the same carbon isotopic ratio (i.e., the ratio of ^{13}C to ^{12}C), measuring this ratio in a tissue sample from an individual can indicate the source of an individual consumer's carbon.

Nitrogen isotopes behave differently from carbon isotopes in that they accumulate as food moves "up" a food chain. Approximately 3.4 units of ^{15}N accumulate in the next higher trophic level from the consumption of an item in a lower trophic level. Therefore, the ratio of ^{15}N to ^{14}N in an organism's tissues can indicate that organism's trophic position. Put simply, for carbon "you are what you eat," and for nitrogen "you are what you eat plus 3.4 units" (Fry 2006).

Using these properties of carbon and nitrogen isotopes, ecologists can construct simple food web diagrams based on stable-isotope data taken directly from the tissues of the individuals themselves. For example, in an aquatic system like a pond, algae come in two forms: **periphyton** is algae from the shallow-water zone, and it attaches to some substrate, whereas phytoplankton floats in the open water. The two types of algae have naturally different carbon isotopic signatures. We can, therefore, deduce which type of algae an aquatic herbivore eats by taking a bit of its tissue, looking for different ratios of ^{13}C to ^{12}C, and comparing those ratios to the ratio of an established standard. The comparison is expressed as a δ value (δ element = [(ratio of heavy to light isotopes in sample / ratio of heavy to light isotopes in the standard) − 1] × 1,000). If we do this for every species in the system, we can determine whether organisms are getting their carbon from the shallow-water periphyton or the open-water phytoplankton.

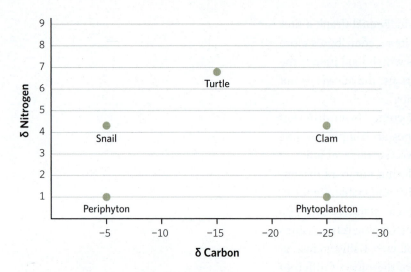

Figure 10.21 In this hypothetical stable isotope graph and presumptive food web for a simple freshwater community, the dots represent measurements from individuals. Carbon isotopes are conserved when an organism eats a food item, whereas nitrogen isotopes accumulate up a food chain. The δ notation denotes measurement from a tissue sample made relative to an isotopic standard. The δ values are calculated as δ element = [(ratio of heavy to light isotopes in sample / ratio of heavy to light isotopes in the standard) − 1] × 1,000.

Nitrogen isotopes generally indicate a species' relative trophic position within a food web. Because stable-isotope measurements vary over a continuum and do not just occur with set values, ecologists are not limited to using discrete trophic levels. For example, an organism can have a trophic position of 3.6 (between the third and fourth trophic levels) rather than having to assume it is in exactly (and only) the third trophic level. The resolution from nitrogen isotope analysis allows quantitative comparisons of the trophic levels for different organisms or for organisms in different ecosystems.

Combining both carbon and nitrogen isotope data can provide a two-dimensional description of an organism's diet. By graphing carbon isotope ratios on the x-axis and the nitrogen isotope ratios on the y-axis for individual organisms (**Figure 10.21**), scientists can create a picture of the trophic linkages among organisms. For example, in Figure 10.21, we can see that the clam is eating phytoplankton because the clam's carbon isotope ratio matches the phytoplankton's carbon isotope ratio. We know which of these species eats the other because the clam is 3.4 nitrogen ratio units above the phytoplankton, indicating trophic enrichment. In the same way, we can see that the snail eats periphyton.

What do the isotope ratio values for the turtle tell us? First, the turtle's carbon isotope ratio value suggests that its carbon originates from a 50-50 mixture of periphyton and phytoplankton because the turtle sits pretty evenly between the two primary producers along the x-axis. But how does the turtle acquire that carbon? Is it eating one or both types of algae or eating clams and snails? To interpret trophic position, we look at the nitrogen isotope ratio values for the turtle. Here we see that the turtle is *almost* a whole trophic level above both the clam and the snail. Because the turtle does not sit exactly 3.4 nitrogen units above the clam and snail, we intuit that the turtle's diet must include items lower than the clam and the snail on the food web. In this case, the turtle is likely ingesting some form of plant material, which also suggests that the turtle is omnivorous.

One of the many ways stable isotope analysis is used in food web studies is for detecting changes in trophic architecture. For example, Megan Layhee and colleagues (2014) used stable isotopes to examine food web differences in two similar streams on the island of Kauai, Hawai'i. One of the streams has few invasive species and minimal human disturbance (Limahuli), while the other has many invasive species and shows extensive anthropogenic modifications ('Ōpaeka'a). Using stable isotopes, Layhee and her colleagues show that invasive species have added an entire trophic position to the altered food web in the 'Ōpaeka'a stream (**Figure 10.22**). In this figure, the y-axis indicates trophic position, rather than nitrogen units. The maximum trophic position in the Limahuli stream is approximately 3.5, whereas in 'Ōpaeka'a it is approximately 4.5. The basal flow of energy (i.e., carbon source) has also shifted in the presence of invasive species. Without invasive species, the basal energy is largely derived from periphyton (so carbon values cluster on the left side of the figure), whereas the basal energy in the presence of invasive species is produced more by phytoplankton (so the carbon values shift to the right; Figure 10.22).

This analysis of and insight into these Hawaiian food webs is possible after just a few days of work collecting and processing tissue samples from the two streams. Before we had stable-isotope methodologies, a detailed analysis like this would have been almost impossible and likely would have led to significant simplifications in the depiction and understanding of these food webs. In a sense, stable-isotope biochemistry has altered the way many scientists approach food web studies, as well as our understanding of ecological interactions in communities.

This is not to say that food web models built using stable-isotope measurements are without their problems and limitations. One of the most serious issues is that the approach works best if the food web under study is relatively simple, meaning it does not include many species. In the freshwater streams on Kauai, there were only seven different taxa in the less-disturbed Limahuli and only 16 in the more-disturbed 'Ōpaeka'a. Such simple freshwater food webs are fairly common on oceanic islands such as the Hawaiian Islands. When food webs get more complex, it becomes challenging to discern ecological patterns using stable isotope analysis, because many of the taxa will occupy a similar "trophic space" within the food web. As a result, stable isotopes are not often used to rectify especially complex webs, which unfortunately may include most food webs. Instead, stable isotopes can be used to address specific questions, such as whether mountain lions and bobcats occupy similar trophic positions, or whether all the herbivores in the boreal forest eat the same plants.

10.6 Food Web Insights

Given the complexity of food webs, one question nagging ecologists has been whether there are, or ever can be, consistent patterns that emerge from comparing food webs across ecosystems and whether these patterns offer insights beyond those gained from food chains. For example, are some trophic connections consistently more important than others? Do trophic connections have the same strength or effect through time, or do they vary? Can we identify important elements of food webs and expect them to be consistently influential across different ecosystems? Are particular species always (or, at least, often) providing crucial links within food webs?

If some trophic connections are stronger, more consistent, or in some way more important than others, then

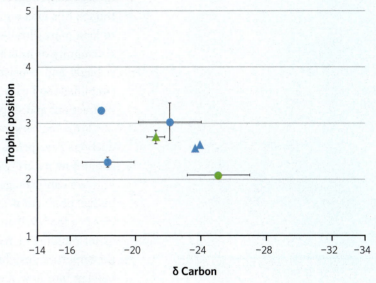

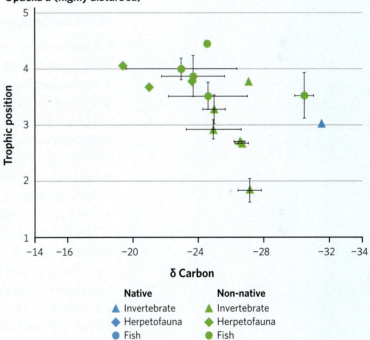

Figure 10.22 Trophic position and δ carbon isotope values for two different streams on the island of Kauai, Hawai'i, are shown. The Limahuli is a relatively pristine stream dominated by native organisms, whereas the 'Ōpaeka'a is loaded with non-native species. Note that the y-axis here differs from the y-axis in Figure 10.21 and shows trophic position, not δ nitrogen. These figures indicate that the primary source of carbon changed with the introduction of invasive species (i.e., more negative δ carbon values), and an entire trophic level was added to the top of the food web in the 'Ōpaeka'a (Layhee et al. 2014).

these strong connections may control energy flow through the web. It may feel as though this idea points us back to food chains, but food webs definitely allow us to look more closely at the relative importance of individual species in a whole community of interacting species. From a practical perspective, the presence of strongly and consistently interacting species means scientists can construct a simplified food web (like Figure 10.15) using only the most important community interactions and skip collecting information on some of the less critical details.

From the perspective of basic ecological insights, knowing which trophic connections have primary control over energy flow tells us something fundamental about how nature works, how an ecosystem processes energy, and (potentially) how we can manage food webs for human purposes (e.g., fisheries production, biofuel production, agricultural systems). We may also be interested in which species in a food web are most critical to the recovery of a system after a disturbance or for building or restoring an ecological community. Next, we explore some of the important insights from food webs, building on insights from food chains and moving into new ideas that we can only begin to understand by looking at the larger suite of interacting species.

More Evidence for Top-Down Control

When discussing evidence for top-down control in food chains, we took an easy approach by offering the logical argument from HSS rather than using empirical data or a specific case study. Yet compelling evidence for top-down control and trophic cascades has been observed across a variety of ecosystems. Much of the initial work on trophic cascades was done in aquatic environments, partly because biomass at the various trophic levels tends to respond rapidly due to the short generation times of aquatic primary producers. Single-celled phytoplankton clearly reproduce much faster than trees and shrubs. As a result, we saw an initial deluge of stream and lake examples demonstrating trophic cascades in the 1970s and 1980s, but since then the phenomenon has been observed in some terrestrial systems as well.

One of the most discussed terrestrial examples comes from the jewel of the US National Park System, Yellowstone National Park, and involves one of its top predators, the gray wolf (*Canis lupus*). Throughout human history, wolves have been feared, mostly because of the risk they pose to livestock, but also because of their perceived threat to human lives. It should not be surprising, then, that wolves were hunted to extinction in Yellowstone and the surrounding ecosystem by 1926, with the entire larger ecosystem remaining wolf-free for the subsequent seven decades. In a big restoration effort starting in the winter of 1995–1996, wolves were reintroduced by wildlife biologists into Yellowstone National Park and have been closely monitored ever since. This reintroduction was done largely to reestablish a wild wolf population, but the ecological effects appear to have rippled across the entire Yellowstone food web (Ripple and Beschta 2012).

As predicted by the population growth models from Chapter 6, the initial release of a few wolves into the park led to slow population growth at first and then a steady increase in the wolf population abundance (**Figure 10.23A**). As the population grew, the wolves spread across a larger area, and eventually the population settled at around 40 wolves between 2009 and 2016 (Beschta and Ripple 2016). If the greater Yellowstone food web experiences top-down control, then the reintroduction of a top predator should influence species in all the lower trophic levels.

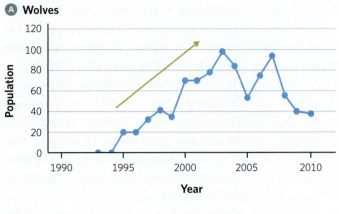

A Wolves

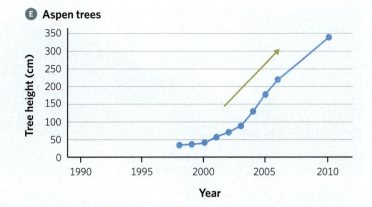

E Aspen trees

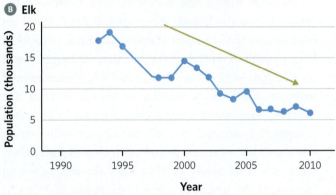

B Elk

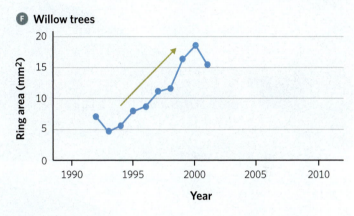

F Willow trees

C Cottonwood trees

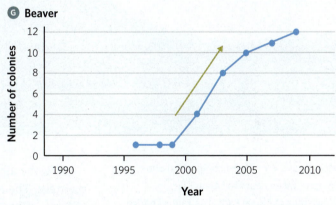

G Beaver

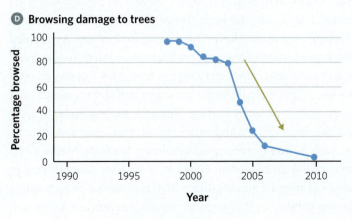

D Browsing damage to trees

Figure 10.23 After the reintroduction of wolves to Yellowstone National Park, the abundance of **A** wolves, **B** elk, **C** reproductive-sized cottonwood trees, and **G** beaver colonies all changed. In addition, over the same time period, **D** browsing damage caused by elk decreased; **E** the height of aspen increased; and **F** the area of annual growth rings of willow trees increased. Arrows indicate the general direction of change (Ripple and Beschta 2012). DBH = diameter at breast height.

Figure 10.24 These comparison photographs, taken near the Lamar River in Yellowstone National Park, illustrate a change in the density of willows (the shrubby vegetation seen only in the 2001 and 2010 photos) after the 1995–1996 wolf reintroduction. As of 2010, both willow height and plant cover had increased compared to the time before wolf reintroduction (Ripple and Beschta 2012).

In the 70 years before wolves were reintroduced, Yellowstone elk (*Cervus canadensis*) populations exploded. Following the return of the wolves, elk populations in the park decreased significantly (**Figure 10.23B**). This decline in elk worked in synergy with a change in elk behavior; namely, the elk altered where and when they browsed on their preferred trees—aspen (*Populus tremuloides*), cottonwood (*Populus* spp.), and willow (*Salix* spp.). This change in elk behavior presumably was to avoid predation by wolf packs, but it effectively lowered herbivory pressure on the tree populations.

William Ripple and Robert Beschta monitored the aspen, cottonwood, and willow trees. They found an increase in the abundance of cottonwood trees that were large enough to reproduce (i.e., diameter at breast height greater than 5 cm; **Figure 10.23C**), and this change seemed to be the result of less damage to trees of all sizes. For example, they found a significant decrease in browsing damage (i.e., damage from elk nibbling or browsing on trees) over the same time period in which wolf populations were increasing (**Figure 10.23D**). They also saw an overall increase in the height of aspen trees (**Figure 10.23E**) and in the annual growth rings of willow trees (**Figure 10.23F**).

These effects on Yellowstone's vegetation can be seen in pictures from before and after the wolf reintroduction. Long-term photographic evidence shows conspicuous differences in the landscape (i.e., aspen and cottonwood seedlings) within a few years of the reintroduction of wolves (**Figure 10.24**). The data and photographs suggest that adding a top predator physically altered the appearance of the Yellowstone landscape via a top-down trophic cascade.

Yet, the apparent cascading effects of wolf reintroduction did not end with the trees. The return of willows and the increased growth of other trees appear to have facilitated the expansion of beaver (*Castor canadensis*) populations from Custer Gallatin National Forest into the park (**Figure 10.23G**). Willows, cottonwoods, poplars, and aspen trees are preferred food items for beavers. As detailed in Chapter 6, beavers also use some of the uneaten parts of downed trees to build their dams. Beavers and their associated engineering feats play important roles in flowing water systems, as they help decrease the erosion of stream banks, raise local water tables, and modify nutrient cycling. Through these effects, beavers ultimately influence plant, invertebrate, and vertebrate diversity and abundance.

When a species, such as the beaver, produces large, landscape-wide alterations, it is sometimes called an *ecosystem engineer*, which we discuss later in the chapter. Chapter 6 examined a situation in which non-native beavers had a negative influence on the habitat. In Yellowstone, though, beavers are native, and their engineering prowess altered the system in ways that increased the density of many other native taxa. This chain of events suggests that the return of wolves to Yellowstone affected not only terrestrial species across different trophic levels but also the region's hydrologic (water) cycle, as well as the distribution and abundance of reptiles, amphibians, and fish (Ripple and Beschta 2012).

As you can see, the pattern suggests that the reintroduction of wolves produced a trophic cascade in the Yellowstone ecosystem food web. Of course, the evidence is observational, and many other factors changed around the time of the reintroduction. The end of an extended drought coincided with the wolf reintroduction, and higher water levels could help explain the increased tree growth; however, a new drought set in about five years after reintroduction, and the effects have, nonetheless, persisted. Beschta and Ripple (2009) have found evidence of trophic cascades in several other terrestrial systems in North America, and they suggest that other regions of the world (Asia, Europe, Oceania, and Australia) are likely to show similar patterns upon the reintroduction or population recovery of their top predators.

It is worth noting that trophic cascades do not always involve the addition of a top predator. If a top predator is removed from a system, the lower trophic levels may respond in a similar cascading fashion. It is likely that a significant trophic cascade occurred when wolves were initially removed from the Yellowstone ecosystem, but at that point no ecologists were recording all the changes in abundances of different species. Yet from looking at tree rings and aging adult trees currently in the park, we do have evidence to suggest that trees did not produce many seedlings during the period without wolves (**Figure 10.25**). The 70 years of high elk density and low tree recruitment (i.e., the establishment and survival of new trees to adulthood) may have been caused by the loss of the primary mechanism of elk population control, the wolf.

Food Web Energy Limits

A food web model also allows for greater exploration of how energy moves through trophic levels than does a food chain. For example, have you ever wondered why there are no large vertebrate predators that prey on only mountain lions, great white sharks, bald eagles, or crocodiles? First, it is scary to imagine how big such a predator would be and what adaptations it would need to be able to feed on these already large and fierce creatures. Beyond the size consideration, understanding the inefficiency with which energy is transferred upward through a food web provides a good explanation for why such predators do not exist.

Consider for a moment a predator that feeds primarily on great white sharks. Great whites have relatively small population abundances compared to those of species at lower trophic levels, and they spread themselves out over a huge expanse of ocean. Given this extremely low density of sharks, a predator that focused on great white sharks would have to cover the entire range of great whites in order to find enough prey to sustain itself. It would also have to swim at least as fast as the great white in order to catch one. The energy needed to swim that far and that fast would be enormous.

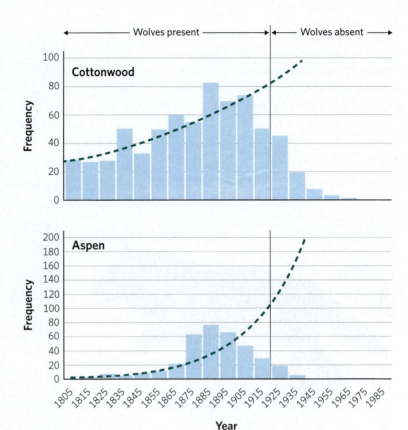

Figure 10.25 Using tree rings, scientists can determine the year a tree first started growing. These graphs show the number of trees (frequency) by the decade they started growing for cottonwood ($n = 674$) and aspen ($n = 330$) in Yellowstone's northern range before wolf reintroduction. The dashed lines represent the expected frequencies of tree establishment if the two populations were growing exponentially. These data strongly suggest that the frequency of recruitment increased early in the nineteenth century and then decreased after the extirpation of wolves (Beschta and Ripple 2016).

Figure 10.26  **A** A Yellowstone wolf (*Canis lupus*) in winter colors. **B** Yellowstone National Park is surrounded mostly by federally managed lands (e.g., USDA Forest Service or Bureau of Land Management land).

Now consider the inefficiency with which energy is transferred upward in the food web to support the great white shark population and its (mythical) huge predator. Even with a highly productive ocean, it is unlikely that there would be enough energy left over at the top of the marine food web to support a gigantic predator that feeds solely on great white sharks and is able to travel very far and very fast. The inefficiency of trophic energy transfer seems to set an upper limit on the number of trophic levels a food web can support. Empirically, we know from documenting many food webs that most peak at five or six trophic levels.

These energetic constraints have important conservation implications in the Anthropocene. For example, the reintroduction of wolves to Yellowstone was not without its share of controversy. For starters, wolves cannot read park boundary signs. As the wolf population grew, some wolf packs foraged outside the park and targeted livestock as prey, bringing them into conflict with ranchers and livestock owners. One explanation for why the wolves did this was that, although Yellowstone is quite large (8,983 km², or 3,468 mi.²; **Figure 10.26**), it may not be large enough to support a self-sustaining population of a top predator like the gray wolf. The land area may be insufficient to support large enough populations of aspen, willow, and cottonwood trees (i.e., the basal food web species) to support an elk population that is, then, large enough to accommodate a stable population of wolves within the boundaries of the park. This huge protected area may be just too small for a healthy and lasting population of wolves.

In a related way, the energetic constraints imposed by the transfer of energy between trophic levels are also an issue for feeding the world's human population. The same constraints that apply to supporting wolves apply to supporting humans. Huge numbers of people require lots of food. The energy required to grow enough grains and vegetables to support human populations is less than one-tenth of the energy required to grow plant material to feed to herbivores (e.g., cows and chickens) in order to produce meat or poultry for us to consume (Pimentel and Pimentel 2003). If the meat-heavy diet of wealthy cultures in the United States, Canada, Europe, Australia, Argentina, and Japan is embraced by the remainder of the world's human population, energetically the planet as a whole will have a very difficult time feeding everyone, particularly as the human population approaches eight billion. This insight into food web energetics has provided a rationale for some people to switch to a more sustainable plant-based diet and is a fundamental tenet of how ecological systems function.

Trophic Subsidies

Ecologists have also used the idea of energy transfer to uncover situations in which food webs receive energetic inputs from external sources. As ecologists have expanded their understanding of the way energy flows through a food web, they have noticed some anomalous situations in which the basal portions of a particular food web could not possibly be the sole source of energy for the upper levels. These observations suggest that sometimes a food web acquires a significant amount of its overall energy from **allochthonous inputs**, meaning inputs from an "outside source." These inputs underpin the entire food web because they act as energetic trophic subsidies.

Gary Polis and colleagues stumbled upon an example of trophic subsidies while doing research on island food webs in the Gulf of California off Baja California, Mexico. On several islands, they encountered extraordinarily high densities of the primary predators, which were spiders, mostly orb-weaving *Metepeira arizonica* (**Figure 10.27A**). These spider communities, in turn, were supported by an extremely simplified, terrestrial island food web (**Figure 10.27B**; Polis and Hurd 1995). Could the limited amount of energy that flows through these simple food webs support so many spiders?

Following an intensive research effort, the team determined that the islands' terrestrial food webs were reinforced from energy sources beyond what was coming from the lower levels of the terrestrial food web. Marine carrion (dead marine animals), algal wrack (dead seaweed), and waste from seabird colonies (fish scraps, dead chick carcasses, and bird parasites) all washed ashore and added massive amounts of energy to the island's terrestrial food web (Polis and Hurd 1996). The researchers were able to quantify this subsidy across 19 islands by examining the ratio of marine-derived energy input to terrestrial input in a food web (measured in grams of biomass per square meter per year). They found that allochthonous inputs exceeded indigenous food sources on 16 of the 19 islands, with some islands having marine inputs that were 10 times greater than their terrestrial inputs (**Figure 10.27C**).

Energetic trophic subsidies are also quite common in **lotic**, or flowing-water, ecosystems. Such ecosystems are odd for a number of reasons, one being that they are long, thin, aquatic environments embedded in a wholly terrestrial landscape. Sometimes these lotic systems are quite narrow, as is the case with headwater streams that run through high-elevation mountain forests. In other cases, the lotic system will be very wide, as is typical of rivers flowing past salt marshes and other coastal estuary habitats.

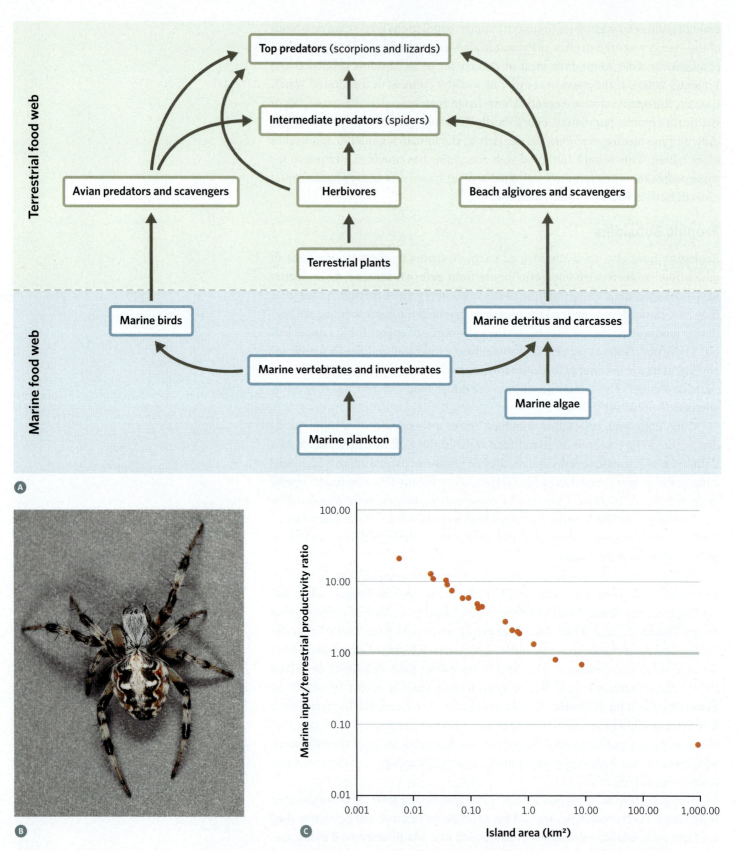

Figure 10.27 **A** Polis and Hurd (1995) constructed this Baja Island food web. Note the linkages between the terrestrial and marine parts of the system. **B** *Metepeira arizonica* is an orb-weaving spider common to the United States and Mexico. **C** Polis and Hurd (1996) examined the ratio of allochthonous marine inputs to local terrestrial inputs on 19 Baja islands and found that marine inputs exceeded terrestrial inputs on 16 of the 19 islands. Note that smaller islands receive a larger portion of their inputs from external marine sources.

It turns out that the major source of energy in forested headwater streams generally comes from the vegetation that lines the stream (also called *riparian* vegetation). Due to the shade provided by the riparian trees, little sunlight reaches the streambed, thereby limiting photosynthesis by periphyton. But these same trees drop their leaves and branches into the stream, which provides an **exogenous** (outside-of-system) energy source necessary to support the aquatic food web. In a sense, the cast-off detritus (dead leaves and sticks) from riparian vegetation takes the place of the primary producers for the stream's food web. Without this energy subsidy from terrestrial plants, the biomass of stream-dwelling organisms would be greatly reduced.

Interestingly, this cross-boundary movement of energy becomes less critical as one moves downstream, where the river channel widens and allows more sunlight to reach the water. Therefore, in the midreaches of rivers, **endogenous** (within-system) primary production takes over as the main energy source supporting the aquatic food web (**Figure 10.28**). The varying role of exogenous versus endogenous energetic input into lotic ecosystems is part of a larger concept referred to as the *river continuum hypothesis*. This hypothesis explains the predictable series of aquatic food web and energetic changes you encounter as you move downstream from the headwaters to the mouth of a river. Energetic trophic subsidies like this are presumably quite common in nature, which adds a fascinating, albeit complex, element to understanding food webs.

Figure 10.28 Notice the difference between Ⓐ shaded headwater streams and Ⓑ open midreach (midsize) streams. The amount of sunlight reaching the water dictates how much primary productivity (e.g., photosynthesis by algae) occurs within the stream. In the shaded stream, most of the energy for the aquatic organisms must come from exogenous sources (e.g., terrestrial trees and shrubs).

10.7 Indirect Effects Add Complexity

At the beginning of this chapter, we raised the topic of indirect effects and pointed out that many of the two-species interactions from Chapters 7–9 are potentially more complicated than they first appear because multiple species may be interacting and having indirect effects on each other. Food webs allow us to see many of those complex interactions and indirect effects, but there are other types of ecological interactions we want to consider, ones that are not based on organisms consuming other organisms (i.e., not food-based) or in which the community-level impacts come more from indirect effects than from direct consumption. All our discussions of top-down and bottom-up control could apply to interactions of this type, but here we want to focus on effects from particular species whose impacts ripple through entire communities.

Strongly Interacting Species

Previously, we mentioned that one of the most interesting advantages of using a food web model is that we can consider the full array of species interactions and perhaps identify critically important interactions. If you step back for a moment, it is pretty amazing to think that something as complicated and networked as one of the food webs in Figures 10.17–10.19 can respond in a predictable manner to a fairly simple manipulation, such as the addition or subtraction of a single species (e.g., a top predator). The existence of strong and consistent linkages among food web members has been recognized since 1969, when Robert T. Paine first coined the term **keystone species** to describe the robust influence one or a few species had in the rocky intertidal food webs he studied along the Washington State coastline (Paine 1969).

The idea of a keystone species is a metaphor rooted in stone masonry. In building a stone arch for a doorway, the most important single stone (and the last one placed) is the wedge-shaped one at the top that holds it all together, and thus that stone is called the keystone (**Figure 10.29**). This stone is under the least amount of physical pressure of any of the stones in the arch, yet without it, the arch would collapse. The metaphor suggests the importance of a single species in supporting an entire community.

By definition, a keystone species (1) must have a large effect on community structure, function, and diversity, and (2) that effect must be disproportionately large relative to the biomass or abundance of the species (Power et al. 1996). From this perspective, the gray wolves of Yellowstone National Park are clearly a keystone species. Other examples include predatory sea stars in the tide pools that Paine studied, sea otters (*Enhydra lutris*) in the kelp forests along the Pacific coast, tiger sharks (*Galeocerdo cuvier*) in the marine food webs of Australia's Shark Bay, and mountain lions (*Puma concolor*) in the Yosemite National Park ecosystem (**Figure 10.30**).

This list focuses on keystone predators, in part because they have received much of the research attention. Yet, as we have seen, control of a food web does not have to come from the very top. Mutualisms can also play a critical role in communities and in food webs. When ecologists use the term **keystone mutualist**, they typically mean a generalist mutualist with a large enough range of partners that removal of the mutualist species can lead to a collapse of the food web or community. Many keystone mutualists are pollinators (e.g., bees, butterflies, bats) or seed dispersers (e.g., monkeys, birds) that maintain gene flow and dispersal throughout an ecosystem.

Figure 10.29 The keystone in a stone arch is the critical center stone at the crown of the arch that balances the pressure from the two sides and holds the structure upright.

Keystone

For example, in the South American grasslands of Patagonia, the green-backed firecrown hummingbird (*Sephanoides sephaniodes*; **Figure 10.31**) is considered a keystone mutualist. The hummingbird pollinates around 20% of the local plant species, many of which have evolved so that only the green-backed firecrown can successfully pollinate them. In turn, the plants supply most of the hummingbird's diet in the form of sugary nectar. Without green-backed firecrowns, large swaths of the existing Patagonian food web would collapse because no other pollinator is equipped to pollinate many of the plants.

A keystone mutualist can also be a specialist species if it supports a foundation species, meaning a species of massive importance to an ecosystem or community but with its importance proportional to its biomass (see Chapter 9). A foundation species thus carries the same importance as a keystone species but not in as surprising a way. Foundation species are often the dominant primary producers in a system, such as hemlocks in hemlock forests or kelp in the underwater marine forests off the Pacific coast of North America. Foundation species are often, but not always, the most abundant species in the ecosystem. It, therefore, makes sense that a specialized mutualist that is critical to the survival or reproduction of a foundation species might also be considered a keystone mutualist.

The effect of keystone species comes from strong indirect interactions. By critically interacting with one or a few important species, the impacts of the keystone species ripple out toward many other taxa in the community. We started the chapter by thinking about indirect effects, and here we can see that such nonlinear effects can sometimes be strong enough to alter the structure and function of an entire food web or community. Let's focus a little more on how indirect effects work for keystone species.

A

B

C

Figure 10.30 Ⓐ Sea otters (*Enhydra lutris*), like this one eating a sea urchin, are a keystone species in nearshore marine environments of the eastern Pacific Ocean. Without sea otters, sea urchin abundances can grow to such high densities that they decimate the kelp forests that support much of the nearshore coastal environment. Ⓑ Tiger sharks (*Galeocerdo cuvier*) are thought to be keystone predators in the marine food web of Shark Bay, Australia. Ⓒ Mountain lions (*Puma concolor*) are thought to be potential keystone species in the greater ecosystem around Yosemite National Park.

Figure 10.31 The green-backed firecrown hummingbird (*Sephanoides sephaniodes*) is a keystone mutualist in the ecosystem of the South American Patagonian grasslands.

Keystone predators often indirectly protect the foundation species or a wide array of other species. Usually that effect comes from one strong interaction. The original keystone predator in Robert Paine's intertidal study was a sea star (*Pisaster ochraceus*) that increased the diversity of the invertebrate intertidal community by consuming lots of the competitively dominant mussels. By reducing the mussel population abundance, the sea star allowed other invertebrates to avoid competitive exclusion by the mussel. This effect is called **predator-mediated coexistence**, and predators that are not keystone predators can also have this effect.

To explore keystone predators that protect foundation species, we can consider the sea otter (*Enhydra lutris*) (see Figure 10.30A). Remember that these species are keystone predators because their effect on their community is much bigger than would be predicted by their biomass. Sea otters are furry, marine members of the weasel family with the densest fur of any mammal. The fur helps to insulate them in the frigid waters of the North Pacific, but it also made them a valuable target for fur traders in earlier centuries. By the mid-nineteenth century, sea otters were virtually extirpated (wiped out) from the coastal waters of North America as far north as Alaska. A ban on hunting has allowed small populations to rebound, and otters have been reintroduced to many areas where their populations had dwindled. Today more than 10,000 otters frolic off the coast of Alaska and many more entertain tourists as far south as Monterey, California.

As with the extirpation and reintroduction of wolves, this history of disappearance and reintroduction offers ecologists insight into the role of otters in coastal food webs and the health of those food webs without otters. James Estes has made a career of studying otters and other large marine organisms. His research is largely responsible for our understanding of the keystone role otters play and a trophic cascade that almost led to the complete loss of kelp forests following the decline of otter populations.

Otters eat a wide variety of marine invertebrates, but their preferred food is sea urchins. Sea otters are often seen floating on their backs and cracking shells to feast on the soft bodies of the invertebrates inside them (Figure 10.30A). Given that they sometimes use kelp to assist with this floating, their lives seem pretty low energy and low stress. But maintaining their internal body temperature in the cold waters of the North Pacific requires massive caloric intake, even though they have the world's best fur. This huge caloric requirement means that a small number of otters can keep the rapidly reproducing sea urchin populations in check. Without otters, sea urchins graze kelp to such an extent that Alaskan islands without otters had no kelp forests, and islands with otters had healthy kelp forests (Estes and Palmisano 1974). Because kelp is the foundation species for the nearshore marine community, the presence of otters is critical to the maintenance of the entire food web and the kelp forest community. In recent years, it has also become clear that otters play an additional important role in combating climate change by protecting kelp, thereby increasing overall marine photosynthesis and carbon sequestration.

Other Important Forms of Indirect Effects

Based on the situations just described, is it possible to identify additional types of indirect effects? The otter, after all, does not protect the kelp by floating in it or by swimming patrol around it. Instead, the act of eating herbivorous sea urchins

RESOURCE COMPETITION

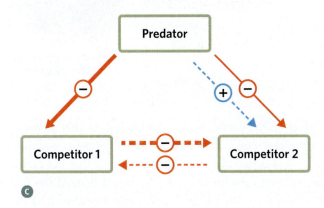

Ⓐ

TROPHIC CASCADE

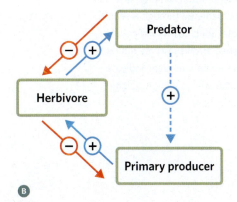

Ⓑ

PREDATOR-MEDIATED COEXISTENCE

Ⓒ

APPARENT COMPETITION

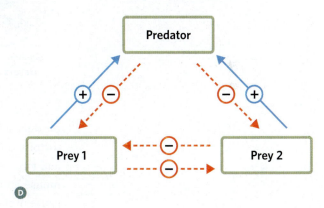

Ⓓ

reduces grazing pressure on the kelp. This keystone predator effect is similar to the situation mentioned at the beginning of the chapter, in which a predator can protect a primary producer (or an organism two trophic levels below it) by reducing the abundance of an herbivore.

Another keystone predator effect comes from reducing the abundance of a dominant competitor and thereby allowing inferior competitors to thrive. There are clearly many other ways for species to produce such indirect effects. **Figure 10.32** explores conceptual models for four of the more common indirect interactions: predator-mediated coexistence; trophic cascades (both discussed earlier); resource competition between two species for a shared, limiting resource (discussed in Chapter 7); and apparent competition. Apparent competition points to how predators can alter the interaction between prey species in a way that mimics the population density effects of competition. In exploring food webs, it is important to be aware that focusing only on trophic interactions can make it hard to see other interactions.

To understand **apparent competition**, let's revisit the coyote, desert tortoise, and cactus mouse: the trio of species examined for optimal foraging in Chapter 5. In that situation, a single predator (coyote) attacks two different prey species, both of which are negatively affected by the predation. Note that coyote abundance will increase as coyotes eat tortoises. This increase in coyotes will then

Figure 10.32 Conceptual models of four common types of indirect effect interactions have solid arrows for direct effects and dashed arrows for indirect effects. Red arrows and minus signs indicate negative effects; blue arrows and plus signs indicate positive effects.

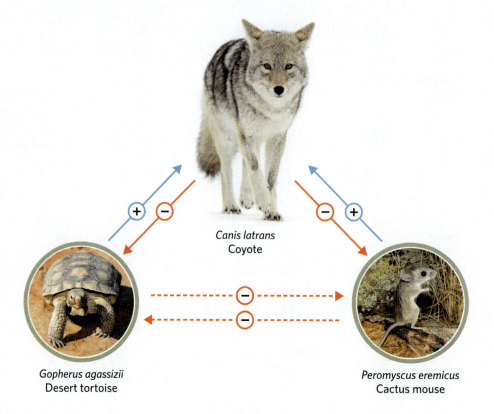

Figure 10.33 In this scenario, an indirect effect (apparent competition) among tortoises and cactus mice is mediated through a shared predator, the coyote. Solid arrows indicate direct effects; dashed arrows indicate indirect effects. Red arrows and minus signs indicate negative effects; blue arrows and plus signs indicate positive effects.

Canis latrans
Coyote

Gopherus agassizii
Desert tortoise

Peromyscus eremicus
Cactus mouse

produce a large negative effect on the cactus mouse population. In this scenario, tortoise abundance has an indirect negative effect on cactus mouse abundance, and vice versa (**Figure 10.33**). To a casual observer, it might appear that the cactus mouse and the tortoise are competing with one another, when instead the real cause of their negative effects on each other is an indirect effect produced by the coyote. This resemblance to competition is the reason for the term *apparent competition*, but keep in mind that, despite its name, apparent competition does not usually involve any competition—just a shared predator.

It does not take great insight to realize that entangled interactions such as these are fairly likely, given the complex, connected nature of food webs and their dense networks of trophic linkages. In fact, any food web that has three or more members has the potential for indirect effects, which includes almost every food web you can imagine. Historically, ecological science focused intensely on pairwise interactions (e.g., competition, predation, and mutualism), just as we did in Chapters 7–9. However, recognizing that interactions are not limited to pairwise situations can both enrich and greatly complicate the interpretation of ecological data and experimentation.

As we move beyond keystone species and think about the indirect effects that other strongly interacting species can have on food webs, it is important to recognize single species that can alter whole ecosystems. **Ecosystem engineers** are species that create, modify, maintain, or destroy a physical habitat. As we discussed already in this chapter and in Chapter 6, North American beavers can be considered ecosystem engineers, as their dams alter the hydrology and geomorphology of streams and wetlands.

Additionally, African bush elephants (*Loxodonta africana*) on the Serengeti plains of Tanzania are often considered ecosystem engineers. Elephants have the

Figure 10.34 African bush elephants (*Loxodonta africana*) are ecosystem engineers of the African savannas.

propensity to destroy and uproot trees, which in turn creates conditions conducive to the growth of many grass species (**Figure 10.34**). Despite all we said about the importance of climatic factors in Chapter 3, elephants actually play a surprisingly large role in the maintenance of some savanna systems. In the absence of elephants, wetter sections of the African savanna would turn into more heavily vegetated woodlands. In that sense, the elephants play an enormous role in maintaining the entire ecosystem despite the fact that their numerical abundance is quite low.

Perhaps one of the greatest benefits of recognizing indirect effects, creating food web models, and identifying strongly interacting species is that these pursuits allow ecologists to convey an image of the interdependent nature of life on Earth. This focus on interdependency, in turn, emphasizes the potential for seemingly unimportant taxa to have critical functional and trophic roles within a community and highlights the importance of indirect effects in the tangled web of interactions that make up an ecological network. Studies of community ecology, interactions within communities, and food webs have increased exponentially in the last several decades. It is not easy to summarize the myriad ecological elements that create a functioning community, but we hope this chapter helps foster a greater understanding of how single species combine to form complex and beautifully intricate ecological systems.

SUMMARY

10.1 Embracing Complexity

- Entire communities, which are full of multispecies interactions, are important to study as a whole.

10.2 Indirect Effects

- Two-species interactions are only part of the story because, when three or more species interact, there is the potential for indirect effects.

10.3 Food Chains Offer Simple Conceptual Models

- A simple chain showing who eats whom and the flow of energy is a good first model to describe feeding relationships.

- Often, food chains are depicted with three levels but may include more.

10.4 Insights from Food Chain Models

Trophic Energy Transfer Efficiency

- Only 10%–20% of the energy an organism takes in is available for movement to the next trophic level because the remainder is lost as heat or waste.

- This rate of transfer is the general rule for all living organisms and has large implications for ecological systems.

Trophic Pyramids

- Because of limits on energy transfer efficiency, ecological systems need significantly more energy at the lower trophic levels in order to support the upper trophic positions.

- You can visualize these relationships as a pyramid of trophic energy.

- If you replace energy with biomass in this model, you can draw a trophic biomass pyramid.

- Aquatic systems with tiny, fast-reproducing primary producers can sometimes show a partially inverted pyramid.

Bottom-Up and Top-Down Control

- Proponents of the idea of bottom-up control of food chains suggest that the basal trophic levels control the amount of energy and biomass at the upper trophic levels.

- Proponents of the idea of top-down control suggest that the upper trophic levels can influence and control the energy and biomass at levels below them.

- Top-down control can also be referred to as a trophic cascade, as the effects of the top levels "fall" down the food chain.

- **Evidence for Bottom-Up Control**

  - One line of evidence comes from the study of lakes, where the overall trophic status of a lake is determined by the amount of nutrients available for the basal levels.

- **Evidence for Top-Down Control**

  - One line of evidence originates in the simple observation that the terrestrial world is mostly green, from which it is argued that the basal organisms (plants) must not be controlled by herbivores, meaning that the herbivores themselves must be controlled by their predators.

  - There are some problems with this line of evidence.

10.5 Food Webs: Adding Complexity

- A food chain model leaves out much of the complexity and interactions we see in nature, including omnivory, developmental (i.e., ontogenetic) diet shifts, and the detrital parts of food chains.

Examples of Food Webs

- Food webs can be fairly simple or blindingly complex.

- Some examples of complex food webs come from the North Atlantic marine system, a rainforest in Puerto Rico, and an estuary in Baja California, Mexico.

- When we include parasites in the food webs, they become even more complex.

Moving beyond Trophic Levels

- Diet analysis is the most direct way to study feeding relationships.

  - The simple approach is to watch what an organism eats.

  - Alternatively, one can analyze gut contents or fecal material to get clues to an organism's diet.

  - Although straightforward, these approaches have their challenges, such as the fact that stomach contents change with the time of day or year and individual preferences.

  - Also, it can be difficult to count individual diet items in stomach or fecal samples.

- Stable isotope analysis can help to track aspects of an organism's diet.

  - Ecologists measure stable isotopes of carbon and nitrogen in organisms to identify where food originates as well as how high up a food chain a particular organism eats.

  - Stable isotope analysis can sometimes be used to construct simple food webs.

  - It can also be used to examine changes in diet over time and space, or changes caused by altered environmental conditions.

10.6 Food Web Insights

- Ecologists have wrestled with the idea that some food web linkages may be stronger or more important than others.
- The presence of more important linkages affects how conservation efforts proceed or how one assembles a new community.

More Evidence for Top-Down Control

- The reintroduction of wolves into Yellowstone National Park appears to have caused a significant trophic cascade.
- Wolf predation caused a decrease in elk populations, which in turn caused an increase in aspen, willow, and cottonwood trees.
- The return of the trees allowed beavers to recolonize the area, which led to changes in both the physical landscape (e.g., how rivers flow) and the associated aquatic communities.

Food Web Energy Limits

- Most food webs do not contain enough energy at the top trophic levels to support populations of even-higher-level predators.
- For example, Yellowstone National Park is likely not big enough in area to house a sustainable population of wolves, so wolves are leaving the park.
- The same kind of energetic limits may prevent the human race from being able to feed everyone if all humans adopt a meat-heavy diet rather than a plant-heavy diet.

Trophic Subsidies

- Some ecological systems appear to support more biomass than can be accounted for by their biomass of basal organisms.

- These food webs acquire significant additional energy inputs from outside the system, called subsidies.
- An example from a Mexican island indicated that marine sources (i.e., dead marine algae and animals) provided 10 times more energy to the ecosystem than terrestrial sources.
- Many stream systems generally depend on terrestrial vegetation (i.e., leaves) as an energy source in the upper parts of watersheds.

10.7 Indirect Effects Add Complexity

Strongly Interacting Species

- Strong or important species in food webs are sometimes called keystone species, suggesting that their presence is vital to the functioning of the entire system.
- At times, these may be top predators or even keystone mutualists.

Other Important Forms of Indirect Effects

- The most common indirect effects include predator-mediated coexistence, trophic cascades, resource competition, and apparent competition.
- Not all indirect effects involve eating interactions. They can take many forms, including apparent competition, in which two species appear to compete but the interaction is actually driven by a predator.
- Ecosystem engineers are important, strongly interacting species that can alter whole ecosystems.
- Indirect effects, food web models, and strongly interacting species all highlight the interrelation and overall complexity of natural communities.

KEY TERMS

allochthonous input
apparent competition
bottom-up control
decomposer
detritivore
detritus
ecosystem engineer
endogenous
energy flow
eutrophic
exogenous

food chain
food web
indirect effect
keystone mutualist
keystone species
lotic
oligotrophic
omnivore
periphyton
predator-mediated coexistence
primary consumer

primary producer
secondary consumer
stable isotope
tertiary consumer
top predator
top-down control
trophic cascade
trophic interaction
trophic level
trophic pyramid
trophic specialist

CONCEPTUAL QUESTIONS

1. Explore the idea of trophic ecology and discuss reasons why a food web model is superior to a food chain model for representing the realities of nature.

2. Discuss which force you believe is stronger in organizing food chains and webs: top-down control or bottom-up control. In your answer, explain what top-down and bottom-up control mean, and try to incorporate concrete examples to make your argument.

3. Very real limits exist on energy transfer between trophic levels. Use this fact to explore (a) the concept of eating from a lower trophic level and (b) the energetics of a predator that eats mainly bald eagles.

4. Discuss some of the pros and cons of the three main methods (gut content analysis, fecal analysis, and stable isotope analysis) that ecologists use to investigate trophic relationships in food webs.

5. Can you explain how stable isotopes of carbon and nitrogen can be used to investigate trophic relationships and create simple food webs? In your answer, try to include both figures and examples.

6. Explain how resource competition is an indirect effect. Draw an interaction model like the ones in Figures 10.32 and 10.33 for one of the examples of exploitative competition in Chapter 7.

7. Look at the food web model in Figure 10.15 and answer the following questions about indirect effects.

    a. If the grasshopper is the most common heterotroph (by biomass), and the eagle is the least common (by biomass), which species in the food web is most likely to be a keystone species? Explain your answer.

    b. Identify at least one trio of species that might represent apparent competition, and explain your answer.

    c. Find at least five examples of resource competition.

    d. Removal of which species might lead to a trophic cascade? Explain your answer.

QUANTITATIVE QUESTIONS

1. Think back to the example of the 200-calorie trout fillet described earlier in the chapter.

    a. How many calories of plants would be necessary to support eight billion humans eating only vegetables versus eight billion humans eating only meat? (Hint: Assume that humans are eating meat from the third trophic level).

    b. When eating from marine systems, we often eat at the fourth or fifth trophic level because we prefer predatory fish. Based on the comparison in part (a), discuss the relative effect of eating at the third trophic level versus the fourth or fifth trophic level.

2. Now that we understand that the two-species interactions from Chapters 7–9 are all part of larger multi-species interactions with indirect effects, we can explore the effects of primary producer productivity on the biomass or population densities of herbivores and predators. **Figure 10.35** is a modification of the phase-plane graphs from Chapter 8. The main difference is that

Figure 10.35 offers three prey isoclines (green and blue lines) based on low, medium, and high primary producer productivity. A second difference is that the predator (or exploiter) isocline (black curve) shows a predator that is limited by factors beyond prey density, such as territories or denning sites, water availability, or human culling. As in Chapter 8, r_{max} is the maximum instantaneous rate of reproduction for the prey; f is the exploiter or predator capture rate; d is the exploiter death rate; and c is the rate at which exploiters convert consumed prey into new predators or exploiters.

    a. Explain how primary producer productivity alters (i) the carrying capacity for the prey (i.e., how changes in primary producers lead to K_{low}, K_{medium}, and K_{high}), (ii) the equilibrium density of prey, and (iii) the equilibrium density of predators.

    b. Explain how this graph relates to the ideas of top-down or bottom-up control.

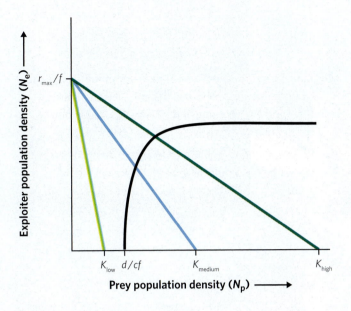

Figure 10.35

3. The text says that Figure 10.16 suggests that scorpions have a very broad diet. Figure 10.16 offers more information, though, and actually suggests that Polis had only seen the tip of the iceberg in terms of the diversity of prey that scorpions consume. Can you explain what about Figure 10.16 leads to this insight? If you feel stumped, look at species-area curves in Chapter 11, and then return to this question.

4. Explain Figure 10.27C. Why would we expect bigger proportions of the total productivity on smaller islands to come from external sources?

5. If two herbivore species can affect each other's densities through a shared predator, how might these population density effects differ from the effects of two herbivores that compete for a shared resource? Examine the phase-plane graphs in Chapters 7 and 8 for exploiter-prey interactions (Figures 8.1, 8.15, and 8.18) and competitive interactions (Figures 7.14–7.17), and identify community-level effects that an ecologist could measure to distinguish between apparent competition and resource competition.

SUGGESTED READING

Hairston, N. G., F. E. Smith, and L. B. Slobodkin. 1960. Community structure, population control, and competition. *American Naturalist* 94(879): 421–425.

Polis, G. A. 1991. Complex trophic interactions in deserts: An empirical critique of food-web theory. *American Naturalist* 138(1): 123–155.

Power, M. E. 1992. Top-down and bottom-up forces in food webs: Do plants have primacy? *Ecology* 73(3): 733–746.

Ripple, W. J. and R. L. Beschta. 2012. Trophic cascades in Yellowstone: The first 15 years after wolf reintroduction. *Biological Conservation* 145(1): 205–213.

Wootton, J. T. 1994. The nature and consequences of indirect effects in ecological communities. *Annual Review of Ecology and Systematics* 25(1): 443–466.

A perfect example of biodiversity, this coral reef in the Red Sea is brimming with a variety of life. Almost all of what you see here is a living organism, from the fishes to the sponges, to the anemones, to the coral itself. All of it is alive, and all of it is working together to make up a community.

chapter 11

BIODIVERSITY

11.1 What Is Biodiversity?

Your first encounter with a coral reef can be a life-expanding experience, whether your interaction is through a thick pane of glass at an aquarium or out snorkeling in the ocean. The sheer abundance of different types of fish and marine life can be bewildering yet enchanting, like stepping out of your reality into another world. The variety of colors, shapes, movements, species, and behaviors all in one place produces a riot of organic life that we rarely encounter in our daily lives. These gorgeous layers of variation make coral reefs an enticing illustration of the beauty and complexity of biological diversity, or **biodiversity**.

For many people, the very term *biodiversity* conjures beautiful images, not only of reef systems but also of dense, verdant tropical rainforests humming with bird and insect calls. These vast collections of species are clear and visible aspects of the planet's biodiversity. In this chapter, however, we will explore how the definition of biodiversity extends beyond simply the number of species, known as **species richness (S)**. The scientific idea of biodiversity includes assessments of how evenly individuals are spread among species, their ecological functions, and even their genetic and phylogenetic diversity (to be defined later). The most widely accepted definition of biodiversity comes from the United Nations (UN) Environment Programme's Convention on Biological Diversity. Its three-part definition, published in the 1990s, includes the total number of all species living on Earth (global species richness), the evolutionary diversity represented by these species, and the diverse communities and ecosystems these species build (United Nations 1992). What is striking about the UN definition is that it recognizes the basic "currency" of species numbers, while also including evolutionary and ecological elements that are vitally important to generating and maintaining species.

We start this chapter with the basic question of how many species are on Earth today and then dive into the details of how this biodiversity manifests at smaller-than-global scales by examining spatial patterns, rarity, and the way biodiversity is distributed across the continents and oceans.

LEARNING OBJECTIVES

- State the three-part definition of biodiversity.

- Compare the methods scientists use to determine how many species there are on the planet.

- Explain why it is so difficult to accurately assess the total number of species on the planet.

- Distinguish between the effects of abundance, area, and heterogeneity on species richness.

- Summarize the differences between alpha, beta, and gamma measures of diversity.

- Explain the Shannon-Wiener diversity index and compute H' with relevant data.

- Differentiate between the seven forms of rarity.

- Explain how functional diversity and phylogenetic diversity each differ from species diversity.

- Identify the major geographic pattern found in biodiversity, and explain one of the hypotheses for the pattern.

Chordates
60,000 sp.
3.4%

Unicellular
organisms
70,000 sp.
3.5%

Fungi
100,000 sp.
5.2%

Plants
310,000 species
16.3%

Other invertebrates
360,000 species
18.9%

Insects
1,000,000 species
52.6%

Figure 11.1 This infographic depicts data from 2009 for the numbers of taxonomically described organisms grouped by major category, with percentages of the total number of described species at the bottom of each box (Chapman 2009).

11.2 Global Species Richness

One of the first questions to arise when discussing biodiversity is *How many species are on our planet?* The related question *How do we know?* follows any answer to the first. On the surface, these seem like fairly simple and straightforward questions, and most people assume scientists have a good handle on the answers, but these questions are actually very challenging.

Formal Species Descriptions

An appropriate first step in determining how many species exist on the planet is to add up the number of species scientists have formally described. A proper species description requires the publication of a peer-reviewed, scientific paper that provides a clear characterization of the new species and explains how and why it differs from closely related taxa already described in the literature. This paper should also list the new species' formal Latin name, or scientific name (genus and species).

In 2009, Arthur Chapman made a detailed list of extant, described species by scouring the published literature, finding just shy of two million (**Figure 11.1**). Chapman's data have some interesting features. For example, the majority of described species are insects, followed by other invertebrates and plants. Chordates (which include fish, amphibians, reptiles, birds, and mammals, including humans) make up a tiny portion of the described species. The chordate portion is even smaller than groups that we tend to notice less, like the prokaryotes and fungi. Biologists have long recognized that insects and other invertebrates appear to dominate the Earth's biodiversity. For example, J. B. S. Haldane, a celebrated evolutionary biologist of the early twentieth century, remarked that "the Creator,

Figure 11.2 The extreme diversity in beetle color and shape patterns is astounding.

if he exists, must have an inordinate fondness for beetles," because the beetles (taxonomic family Coleoptera) make up nearly half of all the described insect species (**Figure 11.2**).

So, if essentially two million species have been described by science, how many species are left undescribed? This is a difficult question, and ecologists and taxonomists have been working to answer it for decades. Estimates of the number of species not yet formally described vary from two million to 100 million, with a recent study even suggesting that the numbers may reach into the billions (Larsen et al. 2017). Why do scientists have such a hard time estimating this number?

Ecologists face five main difficulties in estimating global species richness. The first is the problem of **synonymy**, which is a formal way of saying that often many names are used to refer to the same species. That there are synonymous names for the same species should not be particularly surprising because the descriptions of new species come from different scientists, working in different places, over different time periods, and often published in different languages. For example, within the taxa that comprise the flowering plants (Angiospermae), the largest estimate of the number of described species is essentially double that of the lowest. Depending on the plant group, more than 60% of the species have more than one "official" name in the literature (Scheffers et al. 2012). This degree of synonymy suggests that estimates of angiosperm diversity based on scientifically described species may be too high.

If more than half of all flowering plants have two or more names, how prevalent is this problem among other taxa? We know that the situation is pretty bad for fungi, with an estimated two-thirds of described species having more than one name. We can visualize one of the challenges inherent in this problem by revisiting the *Ensatina* salamander (*Ensatina eschscholtzii*) discussed in Chapter 2 (see Figure 2.9). This highly variable species was first described in 1850, and in many instances, different color patterns and subspecies of this salamander were assumed to be entirely different species because of variation in their skin pigment. Unfortunately, there is no quick answer or easy way to fix the problem of synonymy. Clarifying species synonyms requires considerable time, effort, and funding in order to rectify the situation on a species-by-species basis. This work often involves specialized taxonomic knowledge and the use of sophisticated genetic tools.

The second problem with estimating global species richness stems from the fact that scientists have historically relied only on morphological or physical characteristics to describe species. For large organisms, such as mammals, birds, reptiles, and amphibians, morphological traits work reasonably well to differentiate between species. However, as we can see with the *Ensatina* salamander, even in these groups, morphology can be misleading. But as the science and technology behind mapping genomes has progressed, so too have scientists' ability to discern less obvious differences among species. In addition to identifying issues with synonymy, the growth of genomic tools has led to the discovery of huge numbers of so-called **cryptic species**, which are two or more species "hidden" under a single species name. This commonly occurs when groups of closely related species have very similar morphological traits (**Figure 11.3**).

Often, cryptic species are only apparent when genetic data demonstrate that morphologically similar individuals do not interbreed—a hallmark of a species' boundary (Chapter 4). Good examples of cryptic species are the western grebe (*Aechmophorus occidentalis*) and the Clark's grebe (*A. clarkii*), two bird species that nest around freshwater lakes in western North America. Individuals from each species look remarkably similar and have very similar mating dances. Until the 1980s, they were both identified as western grebes. However, it turns out they do not interbreed and are clearly separate species, according to the biological species concept presented in Chapter 4. The only notable morphological difference from a human's perspective is that the black "cap" on the head extends below the eye on the western grebe but not on the Clark's grebe (**Figure 11.4**). To most casual observers, these differences are minor and often difficult to spot. Thus, individuals of these two species could easily be lumped together and counted as one species in a biodiversity survey.

More than 60% of recently published descriptions of new species come from detailed explorations of morphologically similar species complexes. The problem of cryptic species leads to errors in the opposite direction, so to speak, from those of synonymy, but the two types of errors do not cancel each other out, because they occur in different taxa and at different rates. In other words, identifying many hidden or cryptic species does not make up for having many species names for the same species (synonymy). In fact, these phenomena

Figure 11.3 After 25 years of field observations in the Guanacaste Conservation Area in Costa Rica, Paul Hebert and colleagues (2004) were not convinced that *Astraptes fulgerator* was really just one species of butterfly. Using DNA, they were able to identify at least 10 distinct species—pictured here in their caterpillar stage—lumped into *A. fulgerator*. Individuals from each species look remarkably similar, but the caterpillars feed on different plants, and the different species are often in different ecosystems and do not interbreed.

have nothing to do with each other. For poorly studied taxa, the number of cryptic species may be as much as 10 times higher than the number currently described (Scheffers et al. 2012).

A third problem with estimating undescribed species is a mundane one: most species are hard to see. The vast majority of species on the planet are small in size, which means it is hard to find and then describe them. Even among the better-studied groups, such as vertebrates, scientists continue to find and formally describe new species, with these "newcomers" often being relatively tiny (**Figure 11.5**). In addition, many species are concealed from view, enigmatic, or just plain difficult to identify with their color patterns or behavior. These traits make it very unlikely that scientists will randomly encounter and formally describe them.

To make matters worse, small, retiring organisms, such as insects and nematodes, are often quite localized in their distribution, meaning they are found in only a few small bits of habitat. It is also worth recognizing that the vast majority of the species on the planet appear to be microscopically small organisms in the domains Archaea and Bacteria (Larsen et al. 2017). These types of species do not fit well into the biological definition of species, so scientists have to use advanced genetic and genomic techniques to differentiate among taxa. When biodiversity estimates include the multitudes of these microscopic living organisms, it changes our view of diversity and radically alters how we think about life on this planet (**Figure 11.6**). The distribution of types of species in Figure 11.1 is very different from the distribution in Figure 11.6 because we are much better at identifying the species that we can see well and that we encounter often, such as plants, vertebrates, and insects. Once we recognize the true diversity in the Archaea and Bacteria (prokaryotes), we can see that the proportion of the world's diversity that resides in the remaining groups is much smaller than we depicted in Figure 11.1. We can particularly see the shrinkage in plants' share of the world's species richness, as we are quite good at identifying plants and have not combined them with other groups in Figure 11.6.

A fourth problem with estimating species is that scientists have spent a long time studying and cataloging species in well-populated and easily accessible biomes, such as temperate forests and Mediterranean hillsides. They have not comprehensively investigated locations that are more difficult to reach, such as the Sahara desert, the Congolese rainforest, or the Mariana Trench, which lies 10,994 m (6.83 mi.) below the ocean's surface.

For example, even with all the hikers, strollers, runners, and general human traffic in the woods of the northeastern United States, it still took decades and DNA evidence to confirm the presence of mountain lions (*Puma concolor*) in New England. The eastern populations of the lions were declared extinct in 2011 after no official sighting for almost 80 years. Since then, western mountain lions have slowly moved east, and some breeding populations have established east of the Mississippi River in Tennessee. Sightings were few and far between and hard to verify in New England, though. An animal struck by a car in Connecticut and mountain lion hair found near a killed horse have confirmed that there are some

Figure 11.4 Western grebes (*Aechmophorus occidentalis*) and Clark's grebes (*A. clarkii*) were formerly identified as one species.

Figure 11.5 *Brookesia micra* is a recently discovered species of chameleon from Madagascar. Its diminutive size likely "hid" it from researchers for years.

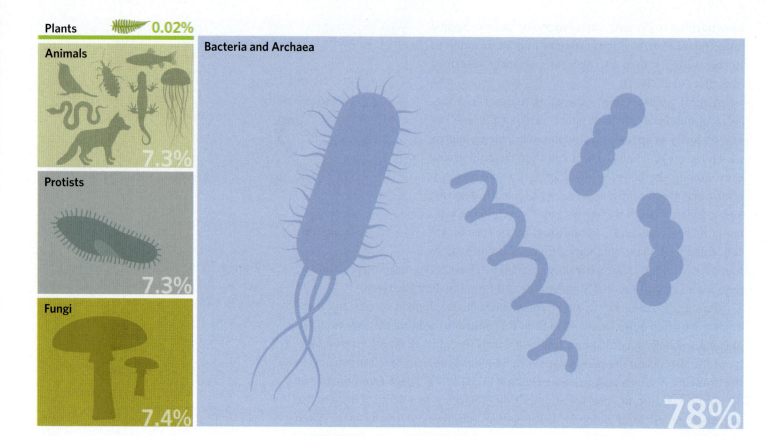

Plants 0.02%

Animals 7.3%

Protists 7.3%

Fungi 7.4%

Bacteria and Archaea 78%

Figure 11.6 This updated infographic showing the distribution of different types of organisms on the planet is based on recent estimates of the numbers of species of Archaea and Bacteria (prokaryotes). It shows projected richness percentages for different taxonomic groups (Larsen et al. 2017). The groups in this graphic are combined into larger categories than in Figure 11.1, so that the category of animals, for example, now encompasses insects, other invertebrates, and chordates. Notice that one of the smallest groups from Figure 11.1 (prokaryotes) is now not only the largest group but also more than three-quarters of the total!

lions in New England. It still is not clear, though, whether they are stray males or breeding populations have been established.

Given the difficulty in confirming the presence of a large top predatory mammal in one of the most populated places in the United States, how much harder do you think it might be to find or document new species of moss or beetle living 45 m (148 ft.) above the ground in the canopy of the Brazilian rainforest? And how many people do you think are up there looking for new species in the canopy? The uneven distribution of search efforts combined with the difficulty of finding or identifying species means that scientists are profoundly ignorant about the number of species living in vast regions of the globe.

Finally, estimating the total number of undescribed species is hampered by a deficit of trained taxonomists to catalog all of the planet's species. Scientific-funding institutions around the world are more focused on understanding the way the world works (particularly as those workings affect humans) than in cataloging the rare tiny species of the world. As a result, we face an acute crisis in taxonomic expertise; this deficit is sometimes called the *Wallacean shortfall*, after Alfred Russel Wallace, a pioneering taxonomist of the nineteenth century who codiscovered the theory of natural selection with Charles Darwin.

As an example, consider species in a family of aquatic flies called Chironomidae. These small flies look remarkably similar to one another in their dominant larval stages. Among the larvae, which are found in streams, there can be color variations, but they all look to the untrained eye like tiny, squirming, wormlike things. Not only would most observers not be able to distinguish different species, but they might not even be able to distinguish subfamilies or even know that they are definitely looking at an organism that belongs in the Chironomidae family. In fact, an easy trait like coloration in Chironomids can change within a species depending on the oxygen availability in the water.

The example of tiny fly larvae may seem too obviously difficult, but we need more qualified specialists examining organisms at this level in order to tackle this problem. Most of us would expect only highly trained experts to have the skills necessary to identify insects to the species level. However, there is a growing trend in biodiversity science in which engaged nonscientists, so-called citizen scientists, are using digital photographs and tools available on smartphones and other mobile devices to help find and identify new species. According to Benoît Fontaine and colleagues (2012), more than 60% of the new species discovered in Europe during the last few decades were described by nonprofessional scientists. Perhaps this problem has a solution that will engage more nonscientists in the business of cataloging global biodiversity as we move forward into the Anthropocene.

Estimating Global Species Richness

All the issues detailed in the preceding section suggest that it will be a very long time before every species on Earth has a formal description (if it ever happens), particularly because so many species are going extinct every year due to human causes. Clearly, then, ecologists cannot know precisely how many species the globe currently holds. But it is possible to estimate global species richness using a variety of extrapolation methods.

The most common approach is to tally the number of species found in particular locations or in distinct taxonomic groups and then extrapolate up to the whole planet. One famous example of this approach was the fieldwork done by Terry Erwin while working for the Smithsonian Institution on tropical beetles in Panama in 1982. From 19 individual trees of the same species, he collected more than 1,200 species of beetles and weevils, using a technique similar to that shown in **Figure 11.7**. If the sheer magnitude of this number does not seem extraordinary, remember these were not individual insects but entirely different species within just two insect orders found in only 19 trees! Take a moment to think about this.

Employing some back-of-the-envelope calculations and extrapolations, Erwin estimated that there were 41,389 unique beetle species per hectare within the Panamanian tropical forest he was studying (Erwin 1982). Of course, not everywhere on Earth is as diverse as the tropical rainforests of Panama. Nonetheless, by determining the portion of the collected beetles and weevils that were specialists and noting that similar specialization was likely on all 50,000 species of tropical trees in the world, Erwin calculated that there might be more than eight million species of beetles around the planet. But he didn't end his calculations there, because beetles only account for approximately 40% of insect species. Extrapolating to

Figure 11.7 A researcher fogs trees with insecticide in a rainforest in Panama to collect insects.

Table 11.1 Planetary Biodiversity Estimates

Source	Estimated number of species
May (1988)	10–50 million
Purvis and Hector (2000)	14 million
Chapman (2009)	11 million
Mora et al. (2011)	8.7 million (± 1.3 million)
Costello et al. (2013)	5 million (± 3 million)
Larsen et al. (2017)	1–6 billion

other groups, Erwin suggested that, globally, there could be up to 20 million species of insects. As we saw in Figure 11.1, insects make up a large portion of the world's macrobiodiversity, so Erwin calculated that global species richness could easily exceed 30 million species!

In the decades since, some of Erwin's insect calculations have come under fire for being overextrapolations, and subsequently a number of other approaches have been used to estimate the extant number of species globally. Interestingly, these alternative estimates do not fully agree with each other (**Table 11.1**). The previously mentioned efforts (Larsen et al. 2017) to estimate the numbers of Archaea and Bacteria suggested a far larger number of species worldwide (one to six billion) than anyone else has claimed before. When looking across all the publications on this topic, we are left with estimated global species richness values that range from as low as five million to as high as six billion species. We urge interested students to delve into the methodologies used by the researchers in Table 11.1, as they differ greatly and involve some really interesting biology and ecology.

Not surprisingly, for taxonomic groups like chordates and plants, we have a fairly good estimate of the global number of species, with nearly 80% of these species already described in the literature (**Figure 11.8**). Although this is impressive, there are still tens, if not hundreds, of thousands of species of chordates and plants to be discovered and naming errors to be corrected. Other groups are not so well known, and thus our estimates of their global species richness are highly uncertain. The standout groups in this regard are nematodes (tiny roundworms), fungi, prokaryotes, and, again, the insects (Figure 11.8). The biological features linking these groups are that they all tend to be small in size and they are visually or behaviorally cryptic. These groups also have few taxonomic experts devoted to their study. For example, in 2018, the Society of Nematologists had approximately 180 members, whereas the American Society of Mammalogists had 2,500 members, and the Botanical Society of America had 3,000 members. Clearly, for anyone interested in discovering new species, taxonomy and systematics are subdisciplines of ecology that offer the potential to make great scientific contributions.

Figure 11.8 This bar graph shows the estimated numbers of species occurring in different taxonomic groups, with the number of scientifically described species superimposed in green (Chapman 2009). Note that scientists have identified and described a large percentage of the estimated total number of species of chordates and plants but have identified a low proportion of the total species in many other groups. Recent estimates by Larsen et al. (2017) put the numbers of prokaryotes (i.e., Archaea and Bacteria) at orders of magnitude higher than this graph suggests.

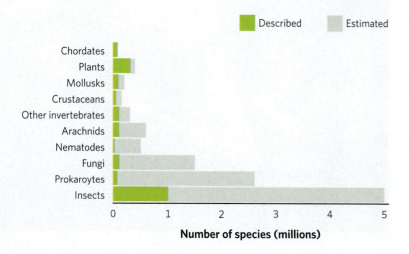

11.3 Effects of Abundance, Area, and Heterogeneity on Species Richness

Oddly, in order to assess species richness, it is often important to focus on abundance. It may seem strange to worry about the number of individuals when we are trying to count whole species, but we can only identify a species if we see or capture at least one individual of that species. Once we start counting, we quickly realize that there are big differences in the numbers of individuals of

Sample A

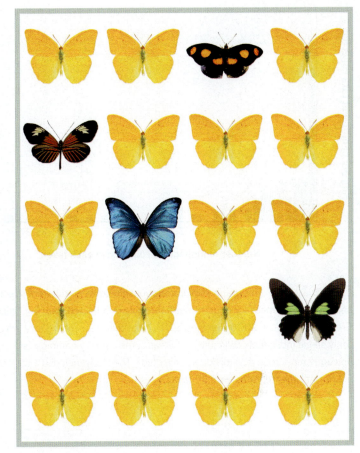

Sample B

Figure 11.9 Two insect communities, sample A and sample B, have the same species richness but very different species evenness. Which community looks more diverse to you?

each species in a habitat. We also find that sampling more habitat leads to higher species richness counts. The exact nature of these relationships is important and not completely intuitive. Next, we explore how these relationships affect our ability to assess species richness and the way richness increases with sampling intensity. These topics also lead us to the concept of species evenness.

Species-Abundance Relationship

As a first step to understanding these concepts, consider two samples, sample A and sample B, that each contain 20 individual insects (**Figure 11.9**). In terms of species richness (number of species), sample A and sample B both contain five different species. However, even a cursory glance at the two samples tells us that they are not equivalent in an important way. Sample B includes 16 individuals of one species of butterfly and single individuals of the four other species. Sample A, on the other hand, includes four individuals of each of the five species. The two samples have the same species richness, but sample A is more "balanced" than sample B. To our eyes, therefore, sample A appears more diverse, an impression that we explore further in the next section.

The difference between the two samples is due to what ecologists call **species evenness**, which is a measure of how equally individuals are distributed among species in a sample. The highest evenness comes when all species have the same number of individuals, as in sample A. The lowest evenness results when one

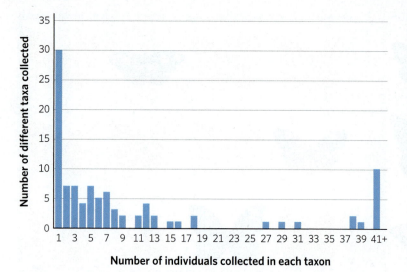

Figure 11.10 This graph shows a relationship between different species and their abundances for aquatic invertebrate taxa collected from Big Chico Creek, California, in 2006 (Benigno 2011). Note that many of the taxa are rare (only single individuals collected), whereas some are very common (significantly more than 40 individuals collected). The study included a total of 5,726 total invertebrates from 99 taxa.

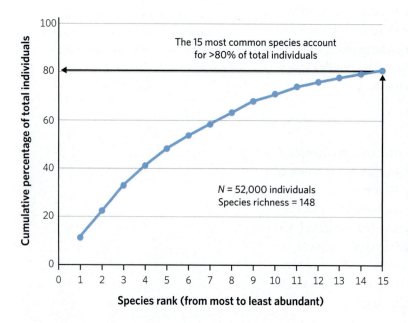

Figure 11.11 Common species often account for the bulk of any taxonomic sample. Of 52,000 individual insects, representing 148 species collected from Breitenbach stream in Germany, just 15 species accounted for more than 80% of all the collected individuals (Illies 1971).

species has many individuals and every other species has just one individual, as in sample B.

Extending this example a bit further may help to conceptualize species evenness. Suppose you cut out each of the 40 insect pictures in Figure 11.9. Now imagine placing all the individuals from sample A into one paper bag, and all from sample B into another. Finally, imagine reaching into one of these bags and randomly choosing an individual (a single piece of paper in this analogy). What species do you expect that individual to represent from the sample A bag? How about from the sample B bag?

In sample A, there is an equal probability of choosing an individual of any of the five species because all five species have the same number of individuals (four each). The probability of pulling an individual of any one species is 20% (4/20 = 0.20). An individual insect pulled from sample B would be a yellow butterfly most of the time because 16 of the 20 individuals in the bag are yellow butterflies. This translates into an 80% (16/20 = 0.80) chance of grabbing a yellow butterfly and only a 5% (1/20 = 0.05) chance of grabbing a blue butterfly, indicating a low level of evenness.

A common way to represent these differences is to create a graph of **species abundance**, which represents the numbers of individuals in a species and the number of species found with that many individuals in the sample or community. Such graphs produce what is commonly called a *species-abundance distribution* (**Figure 11.10**). These distributions generally look more like sample B than sample A in Figure 11.9. In the majority of communities, we find that some common species account for the bulk of individuals and that the majority of the other species are generally represented by only a few individuals (Figure 11.10).

For example, let's look at a long-term dataset from a single stream (the Breitenbach) in Germany, where researchers collected 52,000 individual aquatic insects and identified them as belonging to 148 distinct species (Illies 1971). A graph showing the 15 most abundant taxa reveals that those 15 species accounted for more than 80% (42,115 individuals) of all the individual insects collected (**Figure 11.11**). The other 133 species together produced only 9,885 individuals (19%) of the total collection. In fact, for many species, only a few individuals were ever found in the stream samples. The species-abundance relationship has profound implications for accurately sampling biodiversity. If only a single individual is present out of 52,000 collected insects, what are the odds of finding one of these in a single routine sample of the stream? In case this answer is not obvious, the chances would be pretty slim.

Frank W. Preston, an English-American engineer and ecologist whose day job was running a glassmaking business, was interested in this relationship and wondered if a logarithmic transformation of the abundance data might offer more information on the very rare species, which barely show up with untransformed data. Preston created abundance bins, each of which was twice the size of the next smallest bin. Such binning creates a base-2 log transformation. In other words, each bin starts at 2^b, where b is the log of 2^b. Along the x-axis, b increases by one in each bin, but the actual abundances increase by a factor of two (e.g., 1, 2, 4, 8, 16). He then placed each species into the bin that best represented its abundance. He could have labeled each bin with the log, b, or with the actual number. Either way, by making the distance between 4 and 8 (i.e., between 2^2 and 2^3) the same as the distance between 16 and 32 (i.e., between 2^4 and 2^5), he produced a log-transformed frequency distribution of species abundances because the x-axis was on a log scale (**Figure 11.12**). The resulting plot of species abundance approaches a normal distribution. This representation of the species-abundance distribution is often called a *log-normal plot* (because performing a log transformation on the abundance values achieves a normal distribution). Because the bins in Figure 11.12 go down to fractions or numbers less than one, "abundance" in this case is likely to be biomass or proportion of a sample.

The assumption that the species-abundance relationship follows a log-normal distribution has been challenged through the years because accurately representing this distribution is fundamental to testing ecological and evolutionary hypotheses. However, this distribution holds up remarkably well in a wide range of habitats and situations. An important insight from this log-normal distribution is that we tend to find only a few species that are common and many species that are rare at any single location or time.

More practically, Preston suggested that this distribution implies that we will have a hard time finding species that are on the left side of the distribution. He posited that low abundance acts as a "sampling veil," or a curtain that essentially "hides" species at low abundances from the view of observers (Figure 11.12). Like the non–yellow butterfly species in sample B of Figure 11.9, low-abundance species are rarely encountered in biodiversity inventories. If we do not see them, they are hidden behind the veil. This veil can be "lifted" through more sampling, which requires more effort, time, or money. In fact, ecologists often use the log-normal distribution as an expectation of what a complete and thorough survey effort should attain. If their sampling efforts do not accurately follow this distribution, they assume they have not sampled enough.

Failing to lift the sampling veil can have profound consequences if biodiversity information is used for applied environmental policy decisions. For example, in California (as in most US states) the state's Department of Fish and Wildlife has instituted a biodiversity monitoring program to keep tabs on the water quality across the state's streams and rivers. Aquatic insect species are highly sensitive to changes in water temperature, dissolved oxygen, sedimentation, nutrients, and pollution,

Figure 11.12 Frank Preston developed a log-normal species-abundance distribution and posited that low abundance acts as a sampling veil. The number of species is plotted for different abundance intervals, with each interval being twice the preceding one. The portion of the graph (blue) to the left of the veil line is theoretical, depicting those species that are expected to be present but whose low abundance makes it unlikely for them to be present in an empirical sample (Verberk 2011).

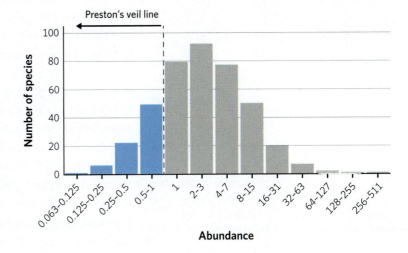

Figure 11.13 Under the California Department of Fish and Wildlife's Surface Water Ambient Monitoring Program, or SWAMP, researchers and managers sample macroinvertebrates across a range of streams and use the resulting abundance data for key indicator species, such as the stonefly pictured here (family Perlidae), to help determine water quality.

which means that macroinvertebrate biodiversity surveys act as a kind of report card for river and stream health (**Figure 11.13**). Assaying the biodiversity of macroinvertebrates over time, therefore, can help detect sudden changes in water quality and enhance the ability of environmental managers to investigate toxic spills or other anthropogenic contamination issues. These kinds of sampling efforts require whole teams of researchers and are often a great way to get involved in ecological fieldwork if you are looking to dip your toe (as it were) into ecology research.

One of the big difficulties with these *bioassessments* is the always-present issue of species rarity. Many of the numerically common aquatic macroinvertebrates tend to be the most tolerant of poor water quality (e.g., crayfish, Asian clams, leeches, blackfly larvae), whereas some of the rarer taxa (e.g., mayflies, stoneflies, caddisflies) are the least tolerant. It is these less tolerant (and rarer) taxa that make the best indicators of good water quality. In order to adequately monitor the ecological health of rivers and streams, the California Department of Fish and Wildlife needs information on these rare species, so they have developed an extensive but repeatable sampling protocol to facilitate the collection of such taxa. The protocol allows the department to generate a reasonable snapshot of the state's annual water quality and monitor changes over time, all through the use of bioindicator species.

Although log-normal species-abundance distributions are the norm, there are real variations in this distribution that often reflect interactions between species and their environments. For example, Wilco Verberk (2011) makes a comparison between the plant species-abundance relationship for salt marshes and freshwater wetlands (**Figure 11.14**). Most of the plants present in a salt marsh are from only one or two plant species. In contrast, in a freshwater wetland, the individuals are spread more evenly across plant species, and there are more total species represented than in the salt marsh (Figure 11.14).

This difference in species-abundance relationships may be a product of the challenging physical conditions plants experience in salt marshes compared to freshwater wetlands. Many salt-marsh plants must contend with twice-daily submersion in salt water due to tidal inundation. The species that can survive regular saltwater submersion have evolved fascinating and complex physiological mechanisms to desalinate the water they take up to achieve their necessary water

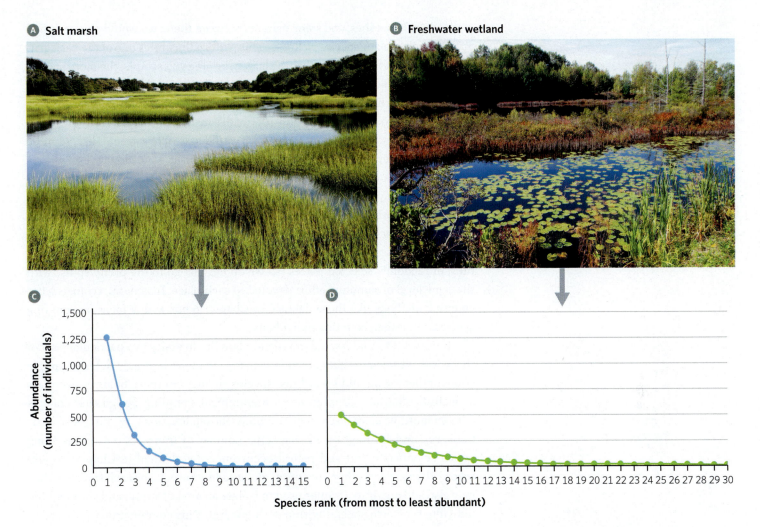

Figure 11.14 Hypothetical species-abundance distributions for wetland plants illustrate the differences in species abundances for Ⓐ a salt marsh and Ⓑ a freshwater wetland. Ⓒ In the salt marsh, the 15 species present have very unequal distributions, whereas Ⓓ in the freshwater wetland, we see a relatively more even distribution of individuals across more than 30 species (Verberk 2011).

balance (Chapter 6). Even the plants that grow on high enough ground to experience only occasional tidal inundation are nearly constantly subjected to saltwater spray and strong winds. These winds require a plant to have a substantial rooting system to combat physical forces that are regularly working to pry them out of the ground. All the physically challenging conditions filter out most plant species, leaving only a few that can exist in a salt marsh and even fewer that manage to thrive and become common (Chapters 5 and 6).

In contrast, the plants in a freshwater wetland experience slow-moving, nonsaline water with little variation in water levels and much less wind. This set of environmental conditions is less physiologically challenging. In addition, the stability of the environment and the slow-moving water allow fine sediment particles to settle out, thereby creating a range of substrate topographies and habitats. These hyperlocal variations in environmental conditions in a freshwater wetland provide a wide variety of niche dimensions (Chapter 5). Having more niche dimensions facilitates the evolution of specialist species and keeps any particular species from dominating the entire wetland.

Species-Area Relationship

From the preceding comparison, we can see that small-scale variation at each wetland location contributes to the diversity of the entire habitat. If we were to sample a much larger area, such as the entire Chesapeake Bay, which has freshwater wetlands, salt marshes, and some habitats between them, we would expect to find an even higher plant species richness. This pattern illustrates one of the basic patterns of biodiversity, namely, that the larger the area you sample, the more species you find. This deceptively simple axiom is called the **species-area relationship**, and it seems to hold across all of ecology. In this section, we explore why this relationship is so universal and how it provides insight into biodiversity patterns.

Let's start by again imagining how we would assemble a species list. Suppose we want to determine all the plant species present in a meadow. We might try walking through the whole field and recording each species seen. Using this process, it is unlikely that we will sample every spot in the field, but we can perhaps cover a good portion of the field and find some of the rare species. The problem with this approach is that we may only record flashy species or species that are tall or that otherwise stand out. We need a sampling design that forces us to look carefully, in randomly selected places (so we do not just go to the pretty spots) with the same level of sampling effort devoted to each place. In general, ecologists use counting methods like those detailed in Chapter 6 but with a focus on counting species rather than counting individuals.

If we want to understand grassland plant biodiversity in a particular part of the world, we need to think about how many meadows or grassy fields to sample and whether we should record just species richness in these fields or evenness as well. Ecologists will often use a hierarchical sampling design that collects data on multiple spatial scales at randomly chosen locations. Within those locations, researchers might walk a specific number of transects (i.e., paths along which scientists count and record occurrences, as defined in Chapter 6) and record all the species they encounter along each transect. For each transect, they may also add quadrats (frames used to isolate an area of study) and count all the individuals of each species within each quadrat. This combination of censusing (counting individuals) and surveying (counting species) provides information on both richness and evenness while covering a lot of area. This technique produces information across spatial scales (fields, transects, and quadrats) as well as an unbiased sampling effort.

To illustrate, let's look at data for meadow plant species from Massachusetts collected by Martha Hoopes and Erin Coates-Connor (unpublished data, 2009). They collected survey data from three sites, with three fields at each site and three 50 m (164 ft.) transects in each field, giving them 27 transects in total. They also employed square quadrats on the ground, each measuring 4 m² (43 sq. ft.), to delineate sample areas within which they identified the species of each individual plant and then recorded the percent cover, or portion of the quadrat occupied by each species. They placed five of these quadrats along each transect, for a total of 135 quadrats across the 27 transects (**Figure 11.15A**).

The sampled fields differed in their water availability and the amount of disturbance they experienced, but the data showed many of the same species across quadrats and sites. Yet some species were fairly rare and only occurred in a single site, field, transect, or quadrat. If we use these data to create a cumulative species list for one site and its 45 quadrats, we can see the common species were observed quickly in the beginning of their sampling efforts, leading to an initial steep rise in the number of species tallied from the first few quadrats they sampled (**Figure 11.15B**). Once the most common species were added to the list, though, the rate of increase

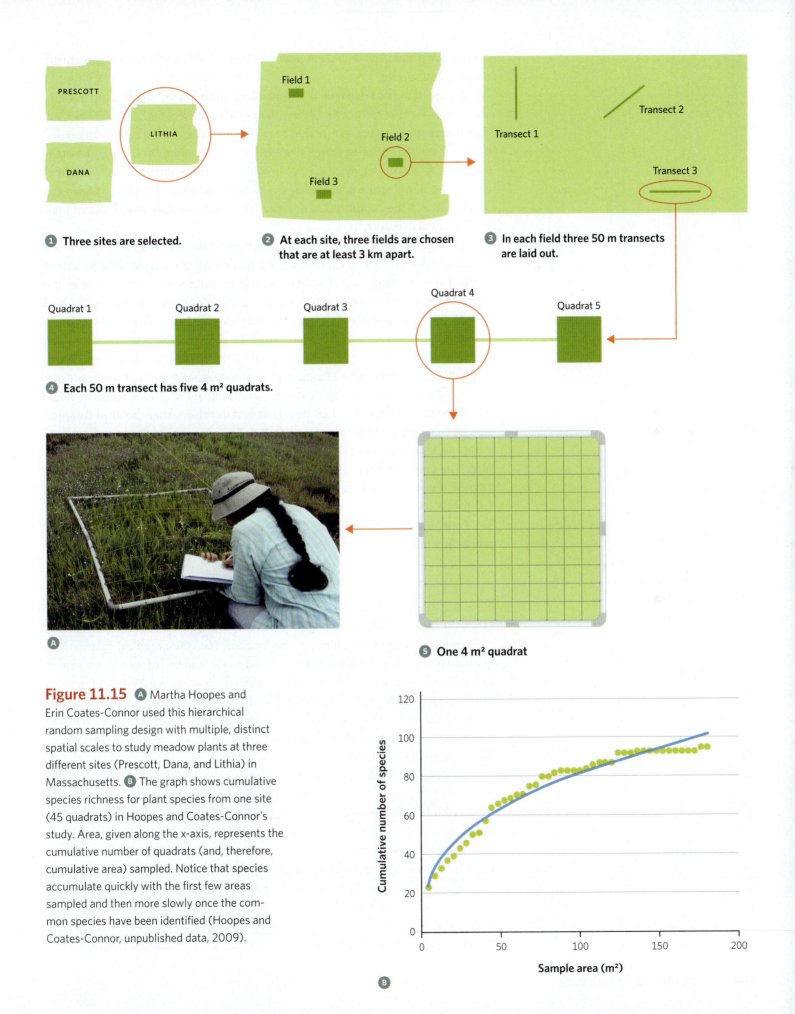

① Three sites are selected.

② At each site, three fields are chosen that are at least 3 km apart.

③ In each field three 50 m transects are laid out.

④ Each 50 m transect has five 4 m² quadrats.

⑤ One 4 m² quadrat

Figure 11.15 Ⓐ Martha Hoopes and Erin Coates-Connor used this hierarchical random sampling design with multiple, distinct spatial scales to study meadow plants at three different sites (Prescott, Dana, and Lithia) in Massachusetts. Ⓑ The graph shows cumulative species richness for plant species from one site (45 quadrats) in Hoopes and Coates-Connor's study. Area, given along the x-axis, represents the cumulative number of quadrats (and, therefore, cumulative area) sampled. Notice that species accumulate quickly with the first few areas sampled and then more slowly once the common species have been identified (Hoopes and Coates-Connor, unpublished data, 2009).

slowed down and kept slowing as new quadrats added only rare and increasingly hard-to-find species.

This pattern is an example of the standard species-area relationship. Hoopes and Coates-Connor did not find every plant species present in these meadows, but the missing species were so rare that much greater effort would have been needed to find them. On the other hand, if we expanded the graph to include the other two sites, we would see more species because natural variation in environmental conditions across space makes each site different enough that some species can only survive in the conditions at one site. These new species would add to their total species richness tally.

The species-area relationship, however, informs more than just how much sampling effort it takes to capture a site's species richness. Enter Frank Preston again. Preston (1962) was the first person to mathematically represent the species-area relationship in ecology. He published data on the number of bird species found across islands of various sizes in the East Indies (New Guinea, Java, Philippines; **Figure 11.16A** and **Figure 11.16C**) and the West Indies (Cuba, Jamaica, Bahamas; **Figure 11.16B** and **Figure 11.16D**). The data he collated clearly show the same pattern that Hoopes and Coates-Connor found in their plant surveys in Massachusetts meadows. As the total sampled area increases, the number of encountered species rises steeply at first but slows after the most common species have been added to the species list. The cumulative number of encountered species continues to slow as only rare species are added as more area is sampled. The same pattern has been found when considering plants in the Galápagos Islands, vertebrates on islands in Lake Michigan, aquatic insects in streams across California, and many other taxa in many other places.

Preston's innovation was to fit a mathematical model to such data using a power function:

$$S = cA^z \tag{11.1}$$

Here A is the cumulative spatial area from which species richness samples are taken; S is the cumulative number of species found within that area (i.e., species richness); z describes the shape of the function (i.e., the graphical relationship between S and A); and c is a constant related to the number of species we expect to find in one unit of sampled area. This equation captures our basic understanding from the meadow plant and island bird examples that more species will be encountered and counted as more area is sampled. However, the rate at which new species are encountered with an additional unit of area sampled will slow according to the exponent z.

Let's explore z a little bit more. To start, imagine what this equation would mean if z were equal to one. In that case, the cumulative number of species (S) would simply be the amount of area (A) times the number of species we expect to see in any average unit of area or in a single, average sample (c). In other words, species richness would increase linearly with area. But we know that we will encounter many of the same species in each new sample or area and that we will see fewer new species in later samples. That means that z must be less than one because the number of new species decreases as our cumulative area (or number of samples) gets bigger. The bigger z is, the faster species accumulate with area. If z is small, though, then the species-area curve is less steep, and species accumulate more slowly.

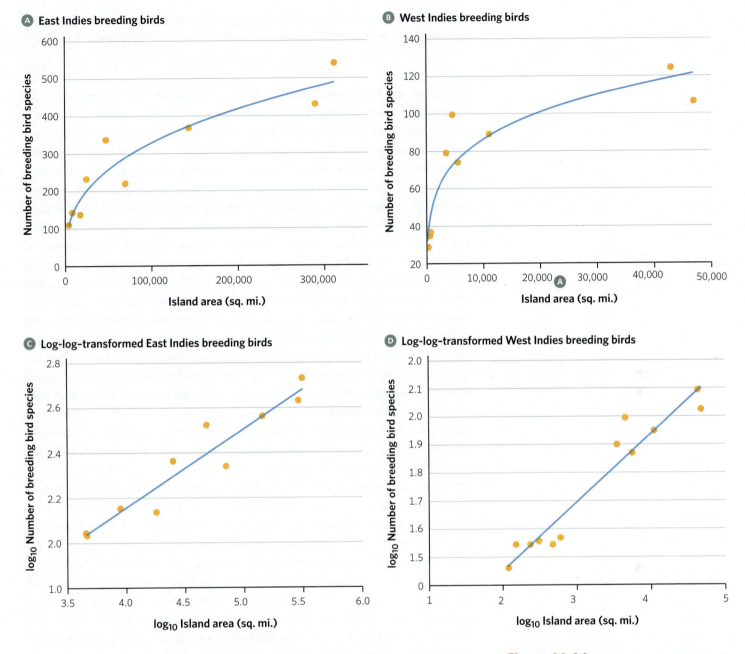

C Log-log-transformed East Indies breeding birds

D Log-log-transformed West Indies breeding birds

Figure 11.16 Preston (1962) determined species-area relationships for the number of breeding bird species on islands of different sizes in **A** the East Indies and **B** the West Indies. The lines on these two graphs are fitted power functions. **C**, **D** These two graphs represent the base-10 log-log–transformations of the graphs in A and B, respectively, with fitted linear regression lines.

This equation works well whether we want to look at the way the number of species accumulates as we sample more individual areas (Figure 11.15B) or compare the number of species we find in different-sized areas (as in Figure 11.16). When we apply equation 11.1 to the Massachusetts meadow data (the solid line in Figure 11.15B), and the bird data from the East and West Indies (the solid lines in Figures 11.16A and 11.16B), the power function described by Preston fits each of the sets of data fairly well.

Despite this nice fit of the power function to the data, our brains have a much easier time interpreting linear graphs. Basically, it would be easier to see the role of c and z in the relationship between species richness and area if we could look at that relationship on a linear scale. Luckily for us, we can log-transform both our area data (x-axis) and species richness data (y-axis) and produce a straight line. This is not magic but is just a feature of taking the logarithm of both sides of any equation with an exponent. The *log-log transformation* shifts exponential functions to linear functions. If we take the log of both sides of Equation 11.1, we get

$$\log S = \log c + z \log A \tag{11.2}$$

As you can see, Equation 11.2 has the format of a straight line ($y = mx + b$), and the equation shows us exactly how to transform the data (by taking the log of A and the log of S and plotting them on the x- and y-axes, respectively). Doing this to Preston's data on island birds produces Figures 11.16C and 11.16D, which now have a linear form. Notice that we did not specify the base of the logarithmic function. We used $\log_{10}$ to build these figures, but the function will become a line with any log base, and ecologists can choose the specific log function that best fits their data.

As promised, log-log transforming the species-area relationship makes the z and c terms in the power function a bit easier to understand. In the transformed version "$\log c$" is the intercept point of the line on the y-axis. That intercept is going to occur when $A = 1$ ($A^0 = 1$, $\log 1 = 0$). This fits with our previous statement that c tells us, on average, how many species are in one unit of area sampled; it just doesn't tell us if the species in this added area were previously found in other areas or if they are new. The value of c is the y-intercept because the first area sampled can *only* include unique species. In the transformed equation, z is the slope of the straight line described by the relationship. The z-value thus tells us how the number of *new* species added to our total changes as area gets bigger. A low z-value indicates that the cumulative number of unique species increases slowly as we increase the area sampled. On the other hand, a high z-value indicates that the cumulative number of unique species increases quite rapidly as we increase the area sampled.

After many ecologists fitted Preston's species-area function (log-log) to their data, patterns began to emerge in these z-values. For example, island z-values are consistently larger than continental z-values. Also, in their review of 794 published species-area relationships, Stina Drakare and colleagues (2006) observed that locations at lower latitudes and taxa with larger body sizes tended to produce larger z-values than higher latitude locations and smaller-sized taxa. These patterns in z-values suggest that, although you always get more species with increasing sampling area, the rate of accumulation varies according to ecological and evolutionary factors. In particular, any ecological or evolutionary factor that increases spatial differences and habitat heterogeneity will probably add new niches with new area and thus will result in higher z-values.

Alpha, Beta, and Gamma Diversity

But how do spatial differences and habitat heterogeneity affect the relationship between local and regional species richness? One of the pioneers in studying the spatial aspects of biodiversity was Robert Harding Whittaker, a distinguished Cornell University plant ecologist whose published research between the early 1950s and 1981 (the last ones appearing posthumously) informed much of our understanding of spatial patterns of diversity. Whittaker recognized that biodiversity values can change as you move across spatial scales from local to regional. His major contribution to this field was formally dividing spatial diversity metrics into three components (Whittaker 1972).

The first diversity component, called **gamma diversity (γ)**, is the species richness across large regional areas like a landscape or a biome. If we think about the factors that contribute to this large-scale regional diversity, then a reasonable starting point to consider is how many fairly distinct smaller areas, or habitats, can be found within it. From the species-area curves, we know that it is unlikely that regional richness would be a straight summation of richness values across each of these local habitats. You can imagine a scenario in which regional richness fairly closely matches the richness of any one local habitat, and other scenarios in which it may be much greater. The interesting question is, when should we expect one outcome versus the other?

If you ponder for a moment, it makes sense that some regions may be made up of a collection of local areas that are similar in habitat and thus in species composition (i.e., homogeneous). Other regions may be made up of locations representing lots of different habitats and thus lots of different species (i.e., heterogeneous). The region with homogeneous habitats should have a regional species richness that is nearly the same as the species richness of any single location within that region. The region with heterogeneous habitats, though, will have a regional richness much higher than the richness of any single location. In heterogeneous regions, each new locality has novel species and, therefore, adds more to the regional richness tally than it would in a homogeneous region.

These connections between local richness and habitat differences among sites within a region summarize the two remaining diversity components that Whittaker identified as contributing to gamma diversity. The average number of species found *within* small, local-scale samples (habitats) in a region is **alpha diversity (α)**, and **beta diversity (β)** highlights the heterogeneity in species composition *among* localities (habitats).

Whittaker combined these elements as

$$\gamma = \beta \times \alpha \tag{11.3A}$$

which can also be written as

$$\beta = \frac{\gamma}{\alpha} \tag{11.3B}$$

Equation 11.3A recognizes that both α and β contribute to γ. Equation 11.3B acknowledges that it is generally much easier to empirically obtain γ and α (simply by going into nature and counting species at sites), and to then use them to calculate β.

These spatial measures of diversity can be very useful in discussions about how local diversity (α) and heterogeneity in species distributions (β) contribute to regional diversity (γ). However, we should point out that this topic gets quite complicated, and there are several different formulas for examining this relationship. We are using the simplest and most common, but readers should be aware that the definition of β can change with other formulations.

As an example of how Whittaker's simple formula captures spatial scale differences in diversity and species distribution, imagine you are studying plant species richness on two different island groups. Let's call them the Empire Islands and the Alliance Islands (for *Star Wars* fans). Each island group contains a set of species, which we represent in **Figure 11.17** with letters of the alphabet, with each letter representing a distinct species (e.g., species A, species B, and so forth). From a careful look at this figure, we can see that the local species richness (*S*) differs among islands and that not all islands share the same species. Just from glancing

Figure 11.17 We applied Robert Whittaker's diversity measures to two separate imaginary island groups, the Empire Islands and the Alliance Islands. Green and tan shapes represent individual islands, and letters represent unique plant species. Notice that the Empire Islands have higher average local diversity (α) but lower heterogeneity or turnover of species among islands (β). This example suggests that lots of heterogeneity and high β among small local areas can lead to higher total or regional richness (γ), even when average local diversity (α) is quite low.

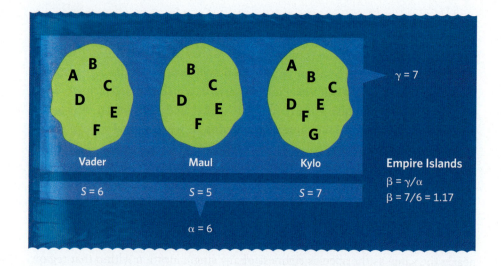

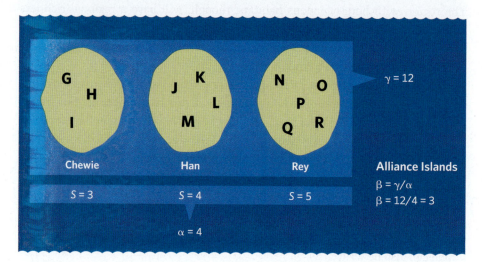

at the letters, we can see that Kylo Island of the Empire group has the highest local plant species richness ($S_{Kylo} = 7$), but we have to average the local species richness for each island group to see that the Empire Islands on average have higher local species richness ($\alpha_{Empire} = 6$) than the Alliance Islands ($\alpha_{Alliance} = 4$).

From our previous discussion, it would be easy to assume that the Empire Islands must also have higher regional species richness (γ), but Equation 11.3A makes it clear that we have to think about the contributions of local areas, or α, and the uniqueness of species at different islands, or β. When we calculate γ across the island groups, the Alliance Island group has the highest regional species richness ($\gamma_{Alliance} = 12$). It is not always the case that β is more important than α. In this case, though, the Alliance Islands have higher γ because each Alliance island holds more unique species ($\beta_{Alliance} = 3$) than the islands in the Empire group, which tend to hold the same species ($\beta_{Empire} = 1.17$).

If you worked for an international conservation group and your goal was to protect the most plant species on small islands, which of Whittaker's metrics would be useful for deciding how to spend money on island protection? In our hypothetical example, you could suggest that Kylo in the Empire Islands is conservation-worthy because it holds the highest number of plant species of all the islands. If there were enough resources to protect an entire island group, γ metrics

Shannon-Wiener Diversity Index

Our discussion of abundance and richness illustrates why evenness is an important component of biodiversity, but evenness is rarely examined by itself. Instead, ecologists typically combine measures of richness and evenness into a single *diversity index*. Although there are many different indices, the most commonly used is the **Shannon-Wiener diversity index (H')**, which was developed by Claude Shannon in 1948 on the basis of previous work by Norbert Wiener (Shannon 1948). Wiener was a philosopher and professor of mathematics at the Massachusetts Institute of Technology. His work on communication and cybernetics formed the basis of much of Shannon's later work as well as much of the world's recent advances in artificial intelligence. Shannon was a mathematician, electrical engineer, and cryptographer who worked at Bell Labs during World War II. His work on probability during this time provided a foundation for information theory, and his research was critical in the development of modern electronic communications. You can thank him for the insights that led to some of our most common and integrated technologies. Yet in terms of ecology, it is his diversity index that is most remembered.

The Shannon-Wiener index combines calculations of both evenness and richness, producing a single measure, H'.

$$H' = -\Sigma[p_i \ln(p_i)] \qquad (11.4)$$

In this equation p_i is the proportion of the whole community represented by individuals in species i. To get that proportion we can divide the number of individuals in species i by the number of individuals in the community or sample. To illustrate how to calculate Shannon-Wiener diversity, we use hypothetical data for a freshwater fish community sampled along a 10-meter section of San Pablo Creek in California (Table 11.2).

Table 11.2 Fish Collected in a 10-Meter Section of San Pablo Creek

Fish species	Number of individuals	p_i	$p_i \ln(p_i)$
Chinook salmon (*Oncorhynchus tshawytscha*)	3	3/32 = 0.09375	−0.22192
California roach (*Hesperoleucus symmetricus*)	15	15/32 = 0.46875	−0.35516
Sacramento sucker (*Catostomus occidentalis*)	1	1/32 = 0.03125	−0.10830
Riffle sculpin (*Cottus gulosus*)	1	1/32 = 0.03125	−0.10830
Pacific lamprey (*Entosphenus tridentatus*)	10	10/32 = 0.31250	−0.36348
Sacramento pikeminnow (*Ptychocheilus grandis*)	2	2/32 = 0.06250	−0.17329
Total	**32**		**H' = 1.33045**

would instead suggest protecting the Alliance Islands, as they contain more plant species overall. If we wanted to protect endemic species, defined as those species found on one island and no others, we would definitely get more from protecting the Alliance Islands because each island has fairly unique species.

Ecologists have been very interested in the way α, β, and γ change across geographic space or through time, particularly in the wake of anthropogenic disturbance to ecosystems. Not surprisingly, there are no consistent patterns in these diversity components across all taxa and locations, but calculations of all three components do provide interesting insights about biological diversity. For example, Richard Hofmann and colleagues studied long-term changes in diversity by examining the diversity of benthic marine fossils from the Cambrian explosion (541 million years ago) through to the present (Hofmann et al. 2019). They found that β was important in the very early stages of species evolution and diversification, but that α contributed more to overall γ once γ levels were no longer extremely low (**Figure 11.18**).

If we focus more on changes to diversity patterns in the Anthropocene, we can explore the way α and β contribute to regional γ as humans introduce invasive species to new locations. Agnieszka Nobis explored how the presence of invasive plant species affects the diversity of native plants in 140 plots, measuring 1 km² (0.4 sq. mi.) each, across ten transects in the San River valley in southeast Poland. The riparian areas (meaning the strip of terrestrial and aquatic habitats associated with a river or stream) are extremely diverse, with more than 750 native plant species, 32 of which are considered at risk of extinction (i.e., threatened). The valley is also loaded with invasive species, with at least 47 species of invasive plants recorded at the time of the study (Nobis et al. 2016). Nobis and her colleagues found a general pattern of higher α and lower β and γ in the presence of high densities of invasive species (**Figure 11.19**).

It may be surprising that the α of threatened species increased as invasive species richness increased, but we can understand that result a little better in the context of the β and γ effects. A dig through the Nobis data indicates that plots with lots of native plant species, including those threatened with extinction, also have lots of invasive species (Figure 11.19). This is a common pattern in the Anthropocene, as locations rich in nutrients or other critical resources are equally as good for supporting invasive species as they are native species. This pattern is sometimes called "the rich getting richer," indicating that sites rich in native species also tend to support the establishment of many invasive species.

This change in α, however, is contrasted with fairly large drops in β and γ when more invasive plants are present (Figure 11.19). How can α increase while β and γ decrease? The answer can be found by taking a close look at Equation 11.3. If nearly the same set of species are found in all plots, β will be very low and consequently γ will be low, even if each plot has high α. In the Nobis data, it is the invasive plants that are typically found across all plots; in contrast, the plant

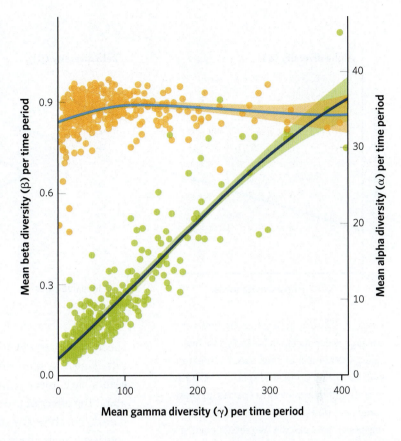

Figure 11.18 Hofmann and colleagues (2019) studied the contributions of α (green dots and dark blue line) and β (orange dots and light blue line) to γ in marine benthic communities over the past 541 million years (Phanerozoic era). Note that γ increases proportionally to α (dark blue line), whereas it does not increase with increasing β (light blue line) except at very low γ levels. This suggests that β has a stronger influence in diversifying communities, but α contributes more at highest diversity levels. Shaded areas represent the 95% confidence intervals (the area within which we are 95% confident that the mean lies) for the fitted curves.

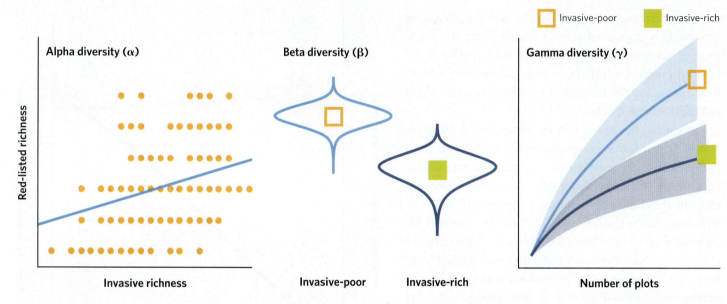

Alpha diversity (α)

Red-listed richness

Invasive richness

Beta diversity (β)

Invasive-poor Invasive-rich

□ Invasive-poor   ■ Invasive-rich

Gamma diversity (γ)

Number of plots

Figure 11.19 In the study by Nobis and colleagues conducted in Poland's San River valley, the number of plant species threatened with extinction (red-listed species) increased in plots with many invasive species, as measured by α (left graph), with the line showing a regression between the two sets of plant species. The middle figure contrasts β between plots with few invasive plant species and plots with many invasive plant species; the bulging shape stretches outward to show higher densities of points, and the central square in each bulging shape marks the mean richness for each category. The right graph shows how γ increased, as the number of plots increased, for plots with few invasive species and plots with many invasive species; the shaded areas show the 95% confidence intervals for the relationships (Nobis et al. 2016).

species threatened with extinction are found in very specific locations and habitats. Thus, the set of sites with more invasive species as compared to threatened plants have lower β and γ (Figure 11.19). One final insight comes from thinking about the effect of the invasive species on the native species. Although the Nobis data do not show any effects over time, it is possible that the increase in threatened species at high levels of invasive species is specifically due to negative effects of the invasive species. A greater number of invasive species leads to more negative effects and may lead to a greater number of native species becoming threatened—an anthropogenic effect of humans moving species around!

Drops in β with invasion—even if the other two metrics rise—are worrisome to ecologists. They suggest that the list of species that originally occupied the various sites changes from being biologically distinct to becoming very similar or homogeneous across sites. This increase in the taxonomic similarity of sites through time has been called **biotic homogenization** (McKinney and Lockwood 1999). An equivalent in everyday life is the replacement of locally owned coffee shops by chain stores that offer the exact same cup of coffee, no matter where you are in the world. The overall number of coffee shops increases when chain stores arrive, but some of the unique and interesting local coffee shops eventually close, leaving the same homogeneous set of options in every city. Across a whole state, province, or country, the effect is to reduce the number of unique coffee shops. In terms of biodiversity, ecologists have demonstrated the same homogenization pattern across many taxonomic groups (e.g., plants, fish, birds) with cosmopolitan non-native species playing the role of chain coffee shops around the world.

We clearly still have a lot to learn about how biodiversity responds to human-caused alterations to ecosystems (see Chapter 15). However, the empirical evidence we have at hand strongly suggests that humans are having an increasingly large effect on planetary biodiversity and that this effect is likely to continue to increase in the future. As we can see from exploring Whittaker's classic ecological insights about diversity, simple metrics such as species richness are insufficient to fully represent the effects these changes have on biodiversity, and instead we must employ our full arsenal of diversity metrics.

11.4 Moving beyond Species Richness

If simple measures of species richness are not enough, what other options exist? From the preceding discussions of evenness and spatial measures, it is probably clear that simply counting up the number of species provides only a coarse view of biodiversity. In particular, such tallies consider all species as equivalent, whether they are rare or common, or found only in one geographic location or worldwide.

To demonstrate why this assumption can be problematic, consider two bird species: the common mallard (*Anas platyrhynchos*) and Spix's macaw (*Cyanopsitta spixii*). Are these two species equivalent "elements" of global biodiversity? Before you answer, it may be useful to know that mallards can be found paddling around ponds in nearly every city on Earth, with the global population estimated at nearly 20 million individuals. In contrast, Spix's macaws are tree-nesting parrots native to the dry forest habitat in Bahia, Brazil. They are now rarely found outside of captivity, and their total global population hovers around 100 individuals (Figure 11.20).

Although it is unlikely that you think the two birds are equivalent, it may be difficult to explain exactly why considering them equivalent is a bad idea. It is true that one is more common than the other, and their abundance and distribution affect their evenness. But there are other ways in which the contributions toward diversity for these two species amount to more than just a tally mark to be counted. It would be useful if we could capture the effects of evenness on diversity and perhaps even think about what a species does and how rare it is in an evolutionary sense. Spix's macaws and mallards definitely do not do the same things in an ecological sense, nor do they have the same evolutionary history, despite their both being birds. Different evolutionary histories allow different evolutionary futures. Ecologists have devised ways to examine diversity that capture some of this important information that lies beyond species richness.

Figure 11.20 Ⓐ The common mallard duck (*Anas platyrhynchos*) and Ⓑ Spix's macaw (*Cyanopsitta spixii*).

Ⓐ

Ⓑ

To start we need to translate these data into the terms in the Shannon-Wiener index. For each species we first calculate p_i. For Chinook salmon p_i is 3/32 since there are three individuals of this species out of the total of 32 individuals tallied across the entire sample. For California roach, p_i is 15/32, and so on for all species in the sample. We then take the natural logarithm (ln) of the p_i for each species (not shown in Table 11.2) and multiply p_i by ln (p_i) for each species. Notice that all of these products are negative because the log of any number less than one is always negative. Then we add these values for all species together and take the negative of that sum to find that, for San Pablo Creek, $H' = 1.33045$.

It is important to notice that calculating H' includes aspects of both species richness and species evenness. Most people are more aware of the evenness because of the proportional abundance calculation, and that is indeed where evenness comes in. Species richness is a little more hidden because p_i is the only variable. Notice, however, that richness enters in the summation. We carry out the operation for *each species* and add all those quantities together. If we added an additional species, it would increase H'.

Now that we have H' for San Pablo Creek, what can this number tell us? Perhaps unexpectedly, the Shannon-Wiener index in itself is not terribly informative. Its value lies principally in comparing the biodiversity among locations. For example, we can compare the H' for San Pablo Creek to the H' for another creek, the nearby Wildcat Creek. If we calculated the H' for Wildcat to be 1.88865 (we provide the calculations soon), we could say, based on the difference, that Wildcat Creek is more diverse, but can we say why or how? Think about how richness and evenness might differ between the two creeks. Once you've given this some thought, look at **Table 11.3** to see the species abundance distribution and the H' calculation for Wildcat.

Table 11.3 Fish Collected in a 10-Meter Section of Wildcat Creek

Fish species	Number of individuals	p_i	p_i ln (p_i)
Chinook salmon (*Oncorhynchus tshawytscha*)	9	9/45 = 0.20000	−0.32189
California roach (*Hesperoleucus symmetricus*)	5	5/45 = 0.11111	−0.24414
Sacramento sucker (*Catostomus occidentalis*)	4	4/45 = 0.08888	−0.21514
Riffle sculpin (*Cottus gulosus*)	6	6/45 = 0.13333	−0.26865
Pacific lamprey (*Entosphenus tridentatus*)	7	7/45 = 0.15555	−0.28945
Sacramento pikeminnow (*Ptychocheilus grandis*)	4	4/45 = 0.08888	−0.21514
Three-spined stickleback (*Gasterosteus aculeatus*)	10	10/45 = 0.22222	−0.33424
Total	**45**		**$H' = 1.88865$**

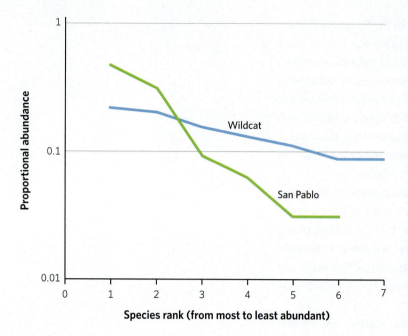

Figure 11.21 This graph shows rank abundance curves for hypothetical fish communities in Wildcat Creek and San Pablo Creek, California. The rank abundance on the x-axis lists each species in order from the most to the least common. These types of graphs do not include species names on the x-axis because different species can occupy different rank slots in different creeks or local areas. A steeper curve indicates bigger differences in abundance from the most common to the least common species, and a longer line indicates higher species richness.

Because the Shannon-Wiener index is calculated using two pieces of information, species richness and evenness, the index can increase if either or both of these aspects of diversity increases. In the hypothetical sample from Wildcat Creek, six species had fairly similar abundances (Table 11.3). Compared to San Pablo Creek, the H' for Wildcat rose because of higher species richness from the addition of a single species, the three-spined stickleback, combined with a substantial increase in evenness among the other species.

Were your intuitions correct about the reasons for the higher diversity index in Wildcat Creek before you looked at Table 11.3? Do not feel dismayed if not. There was, in fact, no way to know whether richness, evenness, or both had changed. We can intellectually understand that both species richness and species evenness play a role in the higher H' for Wildcat Creek by comparing the calculations in both Table 11.2 and 11.3, but we cannot guess the factors just from looking at the H' values. It is, therefore, often helpful to pair H' comparisons with rank abundance graphs.

Figure 11.21 shows *rank abundance curves* for both creeks. Notice that these graphs are a little different from some of our previous abundance graphs (Figures 11.10 and 11.12). First, rank abundance curves have species rank on the x-axis. Second, they rank species from most common to least common, like in Figures 11.11 and 11.14. Notice that the rankings lead to different species order for the two streams, but that does not matter because Figure 11.21 does not list species on the x-axis, just rank. Finally, notice that these graphs use log-transformed proportional abundance on the y-axis. The log transformation allows rare species to show up much better than they do in Figure 11.14. In rank abundance graphs, the length of each line indicates the species richness for each creek, and the steepness or shallowness of the curve indicates evenness. A steeper drop at the start of the curve suggests that the most common species is much more abundant than other species. We can see that the line for Wildcat Creek is both longer and shallower than the line for San Pablo Creek, indicating both higher species richness and higher evenness.

The numbers in Tables 11.2 and 11.3 are hypothetical, but the two creeks are real and were included in a more comprehensive study examining the consequences of urbanization on stream fishes in Northern California. Wildcat Creek runs through a protected recreation area, whereas San Pablo Creek flows through a relatively urban area. In general, Shannon-Wiener index values are lower for stream environments with significant human disturbance and pollution. We mentioned earlier that both urbanization and pollution tend to produce environmental conditions that only a few species of aquatic organisms find suitable. These tolerant species can reach very high abundances in these conditions, whereas some less tolerant species that coexist with them are often rare or even absent in the presence of such anthropogenic effects.

Seven Forms of Rarity

It is clear from the preceding examples that "rarity" often plays a role in determining overall diversity. We pointed this out in the section on species-area relationships, and we can see that rarity affects biodiversity measures such as the Shannon-Wiener diversity index. One question that may have popped into your mind while reading these sections, though, is whether species that are rare in one habitat are rare everywhere. Or, alternatively, are locally rare species ever found anywhere else? If so, are rare species simply restricted by having a constrained ecological niche? Can generalist species appear rare at their range boundaries where environmental conditions might not be ideal? Posing these questions highlights the insufficiency of labeling species as "rare" simply because they are not found in high abundance in one habitat. Although we have used the term *rare* to mean "low abundance" several times in this chapter, there is clearly more to rarity than this usage suggests.

Deborah Rabinowitz, a Cornell University plant ecologist, published a highly influential book chapter that categorized the notion of rarity into seven different forms (1981). She recognized that researchers had previously used the term *rare* to describe species that were present in low abundances, were found in very few geographic locations, or were highly specialized to one or a few habitats. Rabinowitz wanted to explore how all these definitions might intersect or combine. For simplicity, she created categories or "bins" to describe common and uncommon species based on their local population abundance, geographic distribution, and habitat specificity. For local population abundance, she labeled the bins "large somewhere" or "small everywhere"; for geographic distribution, she labeled them "wide" or "restricted"; and for habitat specificity, "broad" or "narrow." For each of these ways of looking at species distribution, the "wider" or "bigger" bin should hold the common species, and the "narrower" or "smaller" bin should hold the rarer species. Rabinowitz put these three different approaches to studying the rarity and commonness of species into a $2 \times 2 \times 2$ table that produced eight different combinations. Only one of these combinations describes species that are always common (and, hence, not considered rare); all seven of the other combinations describe species that are rare in at least one of the three categories (Figure 11.22A).

Species that have a wide geographic range, broad habitat specificity, and are highly abundant somewhere are clearly what ecologists would refer to as common. There are many species that fit this description, including lambsquarters (*Chenopodium album*), the weedy-plant example suggested by Rabinowitz, but also such species as the European starling (*Sturnus vulgaris*), the common dandelion (*Taraxacum officinale*), and the brown rat (*Rattus norvegicus*). All of these species are widely distributed and extremely (even annoyingly) common. But *all* the other categories in the Rabinowitz framework include some way of being rare, thus producing seven different forms of rarity.

This insight that there are different types of rarity is the most important take-home message from Rabinowitz's work. In what follows, we explore each of the seven forms in a little more detail, but ecologists often stop at this larger insight. One reason is that the point of Rabinowitz's work was to get ecologists to think more clearly about how species can be rare. Seven separate categories do

A The seven forms of rarity defined by three axes (distribution, habitat, and population)

Geographic distribution	Wide		Restricted	
Habitat specificity	Broad	Narrow	Broad	Narrow
Population abundance—large somewhere	Common	Predictable	Endemic (unlikely)	Endemic (narrow)
Population abundance—small everywhere	Sparse	Predictable (sparse)	Endemic (nonexistent)	Endemic

B Distribution of seven forms of rarity among 160 species in the British flora

Geographic distribution	Wide		Restricted	
Habitat specificity	Broad	Narrow	Broad	Narrow
Population abundance—large somewhere	36%	44%	4%	9%
Population abundance—small everywhere	1%	4%	0%	2%

C Distribution of seven forms of rarity among 40 species of freshwater fishes in Trinidad and Tobago

Geographic distribution	Wide		Restricted	
Habitat specificity	Broad	Narrow	Broad	Narrow
Population abundance—large somewhere	29%	13%	3%	16%
Population abundance—small everywhere	13%	13%	0%	13%

Figure 11.22 Rabinowitz's seven forms of rarity. **A** Only species with wide geographic distribution, broad habitat specificity, and large population abundance somewhere are actually common (green box). All of the other seven combinations of geographic distribution, habitat specificity, and population abundance describe a type of rarity (blue boxes). The terms in the boxes attempt to describe the type of species that would fit the criteria. **B** The percentage of the British flora (Rabinowitz et al. 1986) and **C** the percentage of fishes from Trinidad and Tobago that fit into each category (Phillip 1998).

that. Secondly, different taxa (plants, mammals, fungi) variously fit into these categories, and the scope or scale of reference strongly affects the number of species that slot into each category. Lastly, notice that Rabinowitz took three continuous variables (abundance, geographic distribution or range size, and habitat specificity) and divided them each into two opposite categories. If she had included the entire scale from small to large for each category, then the data would have to be presented in a complex three-dimensional figure instead of a simple 2 × 2 × 2 table. That figure would be very hard to create and its details would depend tremendously on the particular taxa under consideration. Rabinowitz's simplification was, therefore, appropriate and useful for the basic insight that there are a number of ways to be rare. Nonetheless, there are also many intermediate conditions that do not really slot into the eight categories. With these caveats, let's explore the rarity categories in a little more detail.

We can start with the three categories of rarity that have wide geographic distribution—"sparse," "predictable," and "predictable (sparse)" (Figure 11.22A). Species with broad habitat specificity (like common species) but with small population abundances are sometimes called sparse. These are species that many people have seen, but experts would be shocked to ever find them constituting more than 15% of the total number of individuals at a single site. Barn owls (*Tyto alba*) are an excellent example. They are often the first owls that people see and have a worldwide distribution, but it is still always a surprise and delight to see them. Rabinowitz describes these sparse species as "those, which when one wants to show the species to a visitor, one can never locate a specimen!" (Rabinowitz 1981).

So-called predictable species are found over a wide geographic range but have narrow habitat specificity and can be either locally abundant or more scarce—"predictable" or "predictable (sparse)," respectively, as in Figure 11.22A. Good examples are plants that specialize in occupying the harsh environments along coastlines, such as red mangrove (*Rhizophora mangle*), smooth cordgrass (*Spartina alterniflora*), and common eelgrass (*Zostera marina*). These species are only found in very particular environments, but they are common along coastlines in several parts of the world and can be very abundant there.

Rabinowitz tried to assess how plant species in the United Kingdom were distributed among the eight categories ("common" plus the seven forms of rarity) by asking local botanists to fill out a survey in which they placed individual plant species into appropriate categories (Rabinowitz et al. 1986). Of course, these traits vary continuously in real life, but the botanists were asked to slot each species of plant into the category it most closely fit. The majority of these plants ended up being labeled as either "common" or "predictable" (**Figure 11.22B**). A similar exercise with freshwater fishes from the islands of Trinidad and Tobago led to fairly comparable results (**Figure 11.22C**).

If we shift focus to restricted geographic distributions, any species with a restricted geographic range can be considered rare because it is only found in one or a few locations. All such species could be labeled as **endemic species**; that is, they can be found in a restricted set of locations and nowhere else. When we add abundance and habitat specificity on top of a restricted geographic range, we divide the group of endemics into four subsets (Figure 11.22A).

Restricted-range species that can be locally quite abundant but with very narrow habitat specificity are often called *narrow endemics*. These species have fascinated ecologists for centuries because it is unclear whether they evolved as narrow endemics or are relict populations of a formerly widely distributed species. Examples include the Florida nutmeg (*Torreya taxifolia*), which is a tree in the yew family that is only found in the limestone bluffs of northern Florida and southwestern Georgia, and the Hawaiian stilt (*Himantopus mexicanus knudseni*), a wading bird found exclusively in the few freshwater and brackish wetlands on the main Hawaiian islands. Narrow endemic species are often highly threatened with extinction.

The two categories with restricted geographic ranges but broad habitat specificity are hard to populate with actual species. In other words, it is unusual for species to be rare in this way. The relationship between geographic range size and habitat or niche breadth is variable, but the current consensus is that range size generally increases in concert with niche breadth (Slayter et al. 2013). As a result, these two types of rarity are unlikely to be very prevalent in nature. If we make our scale quite large, though, examples of species with the occasional large population, restricted geographic range, and broad habitat specificity could include many species on the Australian continent. Wombats (mammal species in the family Vombatidae), for example, are endemic to Australia but use a wide range of habitats.

Coming up with examples of species with naturally restricted geographic ranges, broad habitat specificity, and small population abundances is almost impossible. Why this is so is still an active area of debate, but at least one anthropogenic explanation makes logical sense. In this era of global travel, humans tend to disperse species around the globe. Many species with broad habitat specificity are now found in a wide geographic range, far from their native range.

In some ways it is not surprising that we have identified four different ways of being an endemic species. As ecologists have accumulated data on global species distributions, it has become evident that most species on the planet have restricted geographic ranges (**Figure 11.23**). This observation has some important implications for understanding global patterns in biodiversity, which are addressed later in this chapter. The categorization of rarity, combined with observations of which species have or are likely to become extinct, makes it clear that endemic specialists are particularly vulnerable to the many alterations humans have made to the Earth system (Chapter 15).

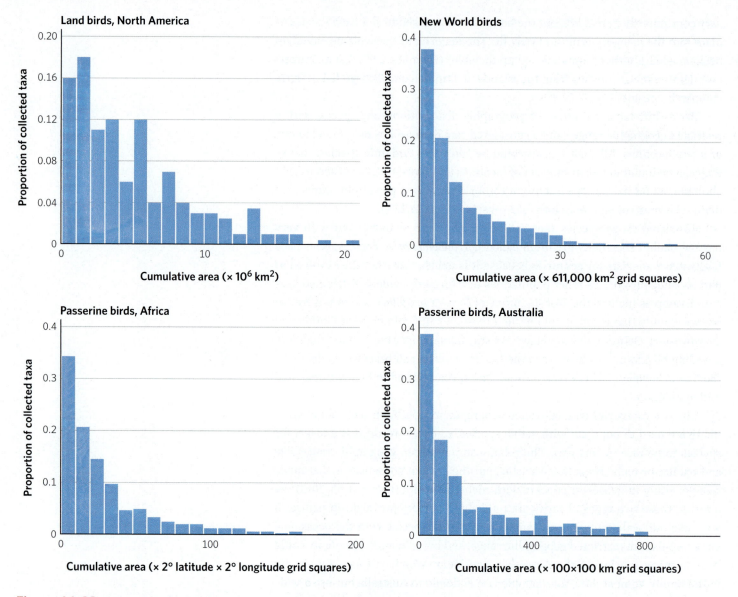

Figure 11.23 These graphs of species range size for different geographic groups of birds from four different studies demonstrate that most birds in each location have small range sizes. Notice also that the four graphs have different ways they measure range sizes (the x-axis labels). This highlights the difficulties of making comparisons between studies, but the observed pattern is consistent (Gaston 1996).

Functional Diversity

Exploring rarity is one way to examine how species differ from one another, but there are many other ways to explore how differences can contribute to biodiversity. One of the three components of the UN definition of biodiversity is the relationship between species and the ecosystems they create. A central element in this is a consideration of what species "do" in an ecosystem. For example, some species are carnivores while others are herbivores, and some plant species transpire lots of water to the atmosphere, whereas other species store respired water for long periods of time. These differences in species' traits influence how they combine to form communities (Chapter 10 and Chapter 12) and ecosystems (Chapter 14).

Functional traits are characteristics of species that describe their ecological roles in their community, including how they interact with each other and the environment. In light of the UN definition of biodiversity, therefore, functional traits can be an important element of biodiversity. These traits are often tightly coupled to the species' evolutionary trajectory (Chapter 4) and life history characteristics (Chapter 5). This aspect of biodiversity is important for understanding how species and communities provide humans with necessary ecosystem services, such as water filtration, carbon sequestration, and nitrogen availability. Functional traits also provide a more nuanced way to measure how an ecological community may be trophically structured, or more broadly, how species contribute to overall food web complexity (Chapter 10).

As with species diversity, there are several ways to measure functional trait diversity. In all cases, the information that feeds into these measurements is species' trait values. For example, ecologists have examined functional feeding traits in bird communities. Individual bird species can be categorized as *frugivores* (fruit eating), *insectivores* (insect eating), *granivores* (grain eating), *nectivores* (nectar eating), mixed (more than one category), or *raptors* (vertebrate predators) (Coetzee and Chown 2016). For plants, ecologists have defined functional traits of species in a number of ways, including their morphology (e.g., degree of stem woodiness, leaf characteristics, or plant height), how they contribute to forest canopy architecture, or according to their abilities to fix or process nutrients (Vandewalle et al. 2010).

Functional traits can be measured on continuous scales or placed into categories. For example, if we consider plant-canopy height as a functional trait, then individual species in a community can have values that vary continuously along this trait axis, or individual species can be grouped into categories, such as overstory, subcanopy, and ground cover. Such categories have an inherent order to them, based on a measurable trait. For example, average ground-cover height will always have a smaller value than average subcanopy height. On the other hand, if we define a functional trait on the basis of categories with no inherent order, such as fruit type or flower color, then the trait "values" mean less. Although it is possible to assign a number to such categories, the number is more like an identifier than a value. Using categorical traits allows only a broad differentiation among functional groups and produces a situation in which all species within a category are assumed to play the exact same functional role. This is not always the most reasonable biological assumption but is often one that is tolerated out of necessity.

Once all species under consideration are assigned a trait value or level, the information needs to be incorporated into a biodiversity metric. One of the simplest is **functional richness**, which is simply the number of different functional traits present in a community. In terms of the diversity of functional roles in a community, functional richness is much like species richness. The only problem is that the "distances" among functional traits are less defined or distinct than between species because each researcher can individually define their own set of species functional traits. Thus, a researcher who more finely partitions functional roles may find a higher functional richness in a community than a researcher who uses broader divisions in the same community.

Yet when assessing a location's biodiversity, perhaps it is not the *number* of functional traits that matters, but instead it is the relative difference in trait values across the community. Such a metric is a way of measuring the overall niche space (see Chapter 5) that is occupied by a suite of co-occurring species. We might consider a community with species that have very different functional trait values as more diverse than a community with several species all showing only slightly different functional traits, even if there are a lot of species.

This difference in perspective may be more easily visualized using continuous functional traits. For example, if we focus again on plant height, the functional

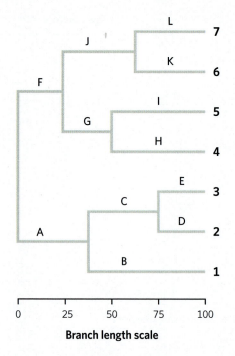

Figure 11.24 Petchey and Gaston (2002) proposed the use of functional diversity (FD) dendrograms, such as the one shown here, to map the differences between functional traits among species. The sum total of branch lengths is a measure of community FD. So, for the community of seven species represented here (numbered 1–7), the FD is the total length of branches A–L. The scale used to measure branch length is determined by the researcher. If species 6 and 7 go extinct, the FD remaining becomes the sum total length of only branches A–I. The vertical lines do not contribute to FD.

diversity in a tropical rainforest is much greater than the functional diversity in a tropical savanna. The array of different plant heights and levels is much larger in the tropical rainforest than in the savanna, where most species are close to the same height. As a result, a tropical rainforest could be said to occupy significantly more functional trait space than a savanna.

In order to visualize this aspect of functional trait diversity, Owen Petchey and Kevin Gaston (2002) suggested using a **functional diversity dendrogram**, which maps the differences between traits (**Figure 11.24**). If you add up the branch lengths in this dendrogram it produces an estimate of community **functional diversity (FD)**. Higher functional richness increases FD if functional richness is spread across divergent traits rather than concentrated in closely clumped traits.

It is also possible to weight the branch lengths by the number of individual species with each trait. In that case, FD also goes up if the proportion of individuals within each functional group is more even, which is called **functional evenness**. Ecologists are gathering experimental and analytical evidence that FD can be a useful tool for detecting the effects of species loss on ecosystem function. Even with a subjective scale, FD allows us to compare communities or local areas and identify which communities have a wider array of ecological functions. Such information could be important in identifying communities that are more resilient to anthropogenic change or disturbance.

A fascinating example of measuring the effect of humans on functional diversity was documented by Alison Boyer and Walter Jetz, who tracked which functional traits were lost when the endemic birds of Pacific islands were driven to extinction by the arrival of Native Pacific and European human colonists. Bird species tend to show high levels of both endemism and habitat specialization across the islands in the Pacific Ocean (think of the Hawaiian stilt mentioned earlier). Unfortunately, over the past 3,500 years, the colonization of these islands by Pacific Islanders and then by Europeans likely drove several hundred bird species to extinction. Boyer and Jetz's data (2014) indicate that with every bird extinction there was an associated loss of functional diversity and richness, leaving a functionally **depauperate** (meaning poor in diversity) bird community across many islands (**Figure 11.25** and **Figure 11.26**). It is interesting to note that larger circles in

Figure 11.25 Boyer and Jetz (2014) examined and reported changes in the functional diversity (FD) of Pacific island bird communities. Larger circles indicate larger original FD. Warmer colors (red and orange) indicate greater FD loss; cooler colors (green and blue) indicate less loss.

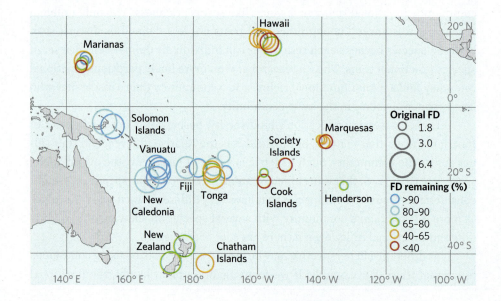

Figure 11.25, indicating higher initial FD, also tend to be cooler colors (green and blues), indicating that they lost a smaller proportion of their overall diversity.

As discussed in Chapter 10, certain species and perhaps functional groups may play unusually important roles in ecological communities. In particular, the loss of ground-foraging birds on a number of islands left disproportionately high numbers of insectivore and frugivore species. The ground-foraging species likely played a role in structuring soil and leaf-litter invertebrate communities, which in turn probably reshaped membership in these communities. Ground-foraging birds also play a role in dispersing seeds, especially those dropped on the forest floor. Changes like these to soils, nutrient cycling, and seed dispersal would likely have negatively affected the fitness of some of the native trees and shrubs and, therefore, affected the whole community. It is also notable that functional evenness also decreased with the extinction of bird species, leaving wide functional gaps between the surviving species. Finally, and perhaps not surprisingly, Boyer and Jetz's data indicate that the extinct birds tended to be functional specialists, leaving mostly generalists around today. This result has interesting ties to our discussion of both rarity and beta diversity.

Phylogenetic Diversity

In Chapter 4, we explored how a single population of one species begins to diverge into two species lineages, each evolving independently of the other. By considering much longer timescales than we did in Chapter 4, we can build on this understanding and demonstrate evolutionary relationships among a whole suite of species and quantitatively depict these relationships in a **phylogeny**.

Biologists often refer to phylogenies as "trees" because their shape is one of continual branching out from a single trunk or root. In order to interpret these "trees," it is useful to note that they can be oriented in many more ways than real trees. The "root" may be at the top, bottom, or side, basically whatever orientation allows the figure to show the "branches" most clearly. In any phylogeny there are two evolutionary pieces of information being depicted that create this tree-like shape: **nodes** and **branches** (Figure 11.27). Points of branching within the phylogeny are called nodes, and the branches connect the nodes. The single branch and node at the base of the phylogeny is called the **root** (just like in our tree analogy). The other nodes represent an unknown hypothetical common ancestor to all the taxa that originate from that node. A group of species that share a single common ancestor node is called a **clade**.

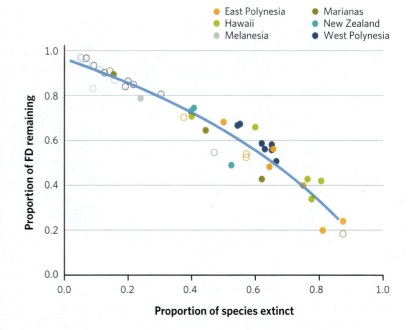

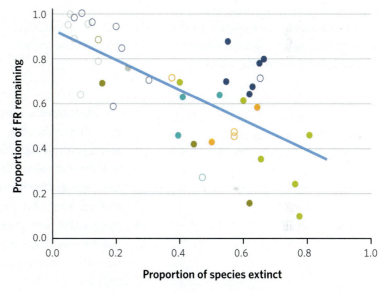

Figure 11.26 These graphs show the loss of functional diversity (FD) and functional richness (FR) due to species extinctions in 44 Pacific island bird communities from 3,500 years ago to the present. Filled points indicate the best sampled islands (Boyer and Jetz 2014).

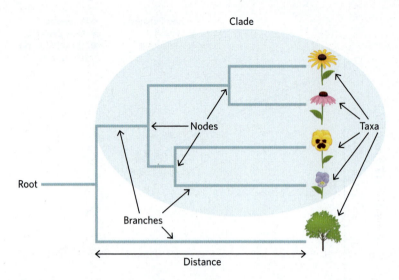

Figure 11.27 This "tree" illustrates the common terminology used in phylogenies (Winter et al. 2013). Distance in a phylogeny is measured as either time (e.g., millions of years) or number of phenotypic differences. Phylogenetic distance is, thus, the sum of all branch lengths along the distance axis (the horizontal branch lengths), similar to the calculation of functional diversity (FD). As with FD, we do not include branches perpendicular to the distance axis (i.e., the vertical branches).

A tree can be drawn so that the length of the branches (distance) is proportional to the evolutionary divergence between nodes. Usually, this divergence is measured on a timescale, but sometimes it is measured on a scale of the number of phenotypic changes that have occurred. Longer branches represent either a longer time period since divergence or more phenotypic changes since divergence from the preceding node, depending on the type of scale. This depiction is similar to showing trait difference with different branch lengths on a functional diversity dendrogram (except those branch lengths always measure difference, not time).

Building on the ideas we explored with FD dendrograms, we can use a phylogenetic tree to measure and compare the evolutionary aspect of biodiversity. A commonly used metric for this purpose is **phylogenetic diversity (PD)**, originally developed by Daniel Faith (1992). PD simply adds up all the branch lengths within a phylogeny to create a single number that can be compared across locations or through time. A set of species that are evolutionarily very different from one another will have a higher PD score than a set that contains species that are closely related.

Phylogenetic diversity quantitatively "measures" each species by how much evolutionary history it represents relative to the other species under consideration, just the way we measured differences between species in their functional trait diversity using dendrograms. For example, lemurs (superfamily Lemuroidea) are a group of primates that are only found on the island of Madagascar off the eastern coast of Africa. There are 99 living lemur species. At least 17 other species went extinct within the last 2,000 years, most likely during the time of initial human colonization of their island home. The evolutionary isolation and varied habitats of Madagascar created optimal conditions for the generation of ecologically and morphologically different lemur species.

When we depict this variability on a phylogeny as in **Figure 11.28**, we gain an understanding of how much evolutionary history is captured in this one group. Closely related nonlemur primates are shown with gray branches at the top of the phylogeny (e.g., lorises, New World monkeys), and all lemurs are depicted with blue branches. To simplify the presentation in Figure 11.28, most of the branches in this phylogeny represent sets of taxa that are closely related to each other but evolutionarily distinct from the taxa listed on other branches. This simplification allows us to avoid listing each lemur species, as well as all the New World monkeys and close relatives. For example, the branch labeled *sportive lemurs* actually contains 17 distinct but closely related species. In this phylogeny, the time since taxa diverged is recorded in millions of years, measured along the horizontal length of a branch and indicated on the calibration timeline given at the bottom of the figure.

One way to assess general features of biodiversity from a phylogeny is to count the species represented at each branch to get species richness. Of course, this approach will not work with our simplified version in Figure 11.28, but James Herrera and Liliana Dávalos (2016) included 87 living and 14 extinct lemurs in their more complete phylogeny. Two key insights that come from Figure 11.28, however, are that any subset of the lemur phylogeny contains less diversity than the whole tree, and that clades with longer branches represent higher PD than clades with shorter branches. For example, the clade with the sportive lemurs and the mouse lemurs has higher PD than the clade with the pachylemurs and bamboo lemurs.

On the basis of this phylogeny, which lemur taxa do you think have the longest overall evolutionary history, and which have the shortest? For the longest, we have a tie between the aye-aye and an extinct lemur, *Daubentonia robustus*. These two taxa are closely related to each other but are connected to all other lemurs through a single branch that represents about 45 million years of evolutionary history. In contrast, several lemurs connect to the rest of the phylogeny via relatively short branch lengths, representing species that diverged from their **sister species** (closest evolutionary relatives) much more recently.

Notable in this regard is the clade of four groups—*Archaeoindris*, *Babakotia*, *Mesopropithecus*, and *Palaeopropithecus*—located at the bottom of Figure 11.28, all of whose members are now extinct. This clade includes the sloth lemurs and other species that are relatively large-bodied but are only known through fossil remains. Unlike many other fossils (e.g., dinosaurs), these fossil lemur species have not been extinct that long, probably for less than a century. These extinctions resulted in the loss of a substantial amount of the evolutionary history from the lemur phylogeny.

Similarly to the way Boyer and Jetz explored declines in functional diversity as species became extinct on Pacific islands as a result of human activities, Matt Davis and colleagues (2018) recorded how mammalian extinctions worldwide are expected to reduce phylogenetic diversity. Mammals have been driven extinct via the actions of humans for thousands of years, starting with the hunting of large mammals in prehistoric times 130,000 years ago. Similar hunting appears to be what led to the recent demise of several lemur species. The loss of mammals to extinction continues today, but at a much higher rate (Chapter 15). Davis and colleagues built a mammalian phylogenetic tree and simply "lopped off" branches for species that have already gone extinct or are likely to go extinct within the next 100 years (based on their status as threatened or endangered). They then recorded how much evolutionary history, as measured by lost PD, was gone due to these current and future extinctions. They also compared the loss of PD from species extinction

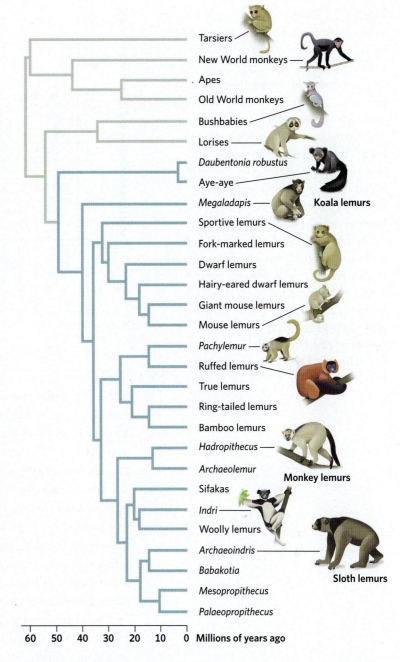

Figure 11.28 This phylogeny showing the evolutionary relationships of nearly all known lemur taxa (blue branches) is based on molecular and morphological data from Herrera and Dávalos (2016).

Figure 11.29 Each taxonomic group of mammals contributes differently (absolute value on left and percentage on right) to the overall phylogenetic diversity (PD) for this subset of mammals. The total height of each bar outline represents the contribution to PD before the influence of humans 130,000 years ago, and the height of the colored portion of each bar represents the current or near-future expected PD contribution. Gray portions of bars represent species already lost to extinction; colors represent the relationship between loss of species richness and loss of PD. Warmer colors (reds) suggest a dispro-portionately large PD loss with the loss of each species, and cooler colors (blues) represent a smaller-than-expected loss of PD with the extinction of a species. The two completely gray bars represent groups that have lost all species to extinction, with associated total loss of the evolutionary history contained in that group.

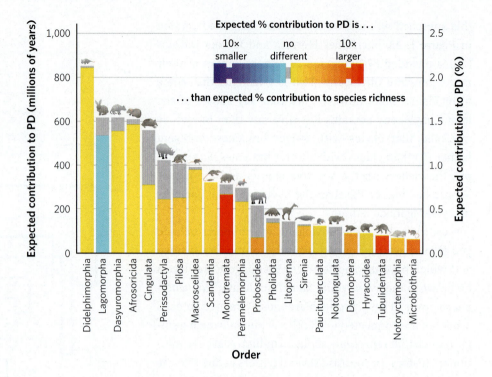

to the decrease in species richness these extinctions caused. This comparison provides a way to highlight that the loss of one species is not the same as the loss of another in terms of evolutionary history; instead, some extinctions represent the destruction of a relatively large chunk of mammalian evolutionary history, and others very little (**Figure 11.29**).

For example, because the species in the order Lagomorpha (rabbits and hares) have recently diverged from a common ancestor, these species are connected to one another by short branch lengths. According to Davis and colleagues' work, the overall contribution to mammalian phylogenetic diversity from lagomorphs is low, so species extinctions from this group do not change phylogenetic diversity much from that found before human impact. Other groups, such as the endemic South American ungulates Litopterna and Notoungulata, were completely wiped out by early humans, resulting in a total loss of the PD they represent (zero remaining PD; Figure 11.29). The prehistoric or recent losses of species in other evolutionary lineages, such as armadillos and anteaters (order Cingulata), have resulted in less than 50% of their initial PD remaining today. Overall, Davis and colleagues show that human-linked prehistoric extinctions resulted in a loss of two billion years of evolutionary history, and recent extinctions will result in an additional loss of 500 million years of evolutionary history.

11.5 Global Geographic Patterns in Biodiversity

Now that you understand how geographic range and habitat specificity drive where we find species, can you predict patterns of species diversity for the entire globe? Are all species spread evenly across the Earth, or do some places have

Figure 11.30 The three-dimensional spatial complexity of tropical rainforests leads to a much higher density of species than we see anywhere else, except perhaps in coral reefs (which cover much less of the globe). This is plainly visible in the Ecuadorian rainforest pictured here.

higher densities of species than others? The answer, as you might expect (thinking back on Chapter 4), is that the Earth's biodiversity has a lot of spatial patterning to it. But now that you have a broader way of thinking about diversity beyond just tallies of species, you may also wonder if spatial patterns in species richness are mirrored in functional and phylogenetic diversity.

The largest concentration of terrestrial biodiversity, as measured by species richness, occurs in the tropical rainforests of the world. This fact is not particularly shocking once you see how crammed with species even a small patch of rainforest can be (**Figure 11.30**). Tropical rainforests house nearly half of all the terrestrial animal and plant species on the planet. A single hectare of uncut tropical rainforest may contain 42,000 different species of insects, 300 species of trees, and more than 1,500 species of land plants. The only other ecosystem where we regularly see as many species in a single place is the tropical coral reef system. Sometimes referred to as the rainforests of the sea, coral reefs occupy less than 0.1% of the world's ocean area yet are home to nearly 25% of the world's marine species.

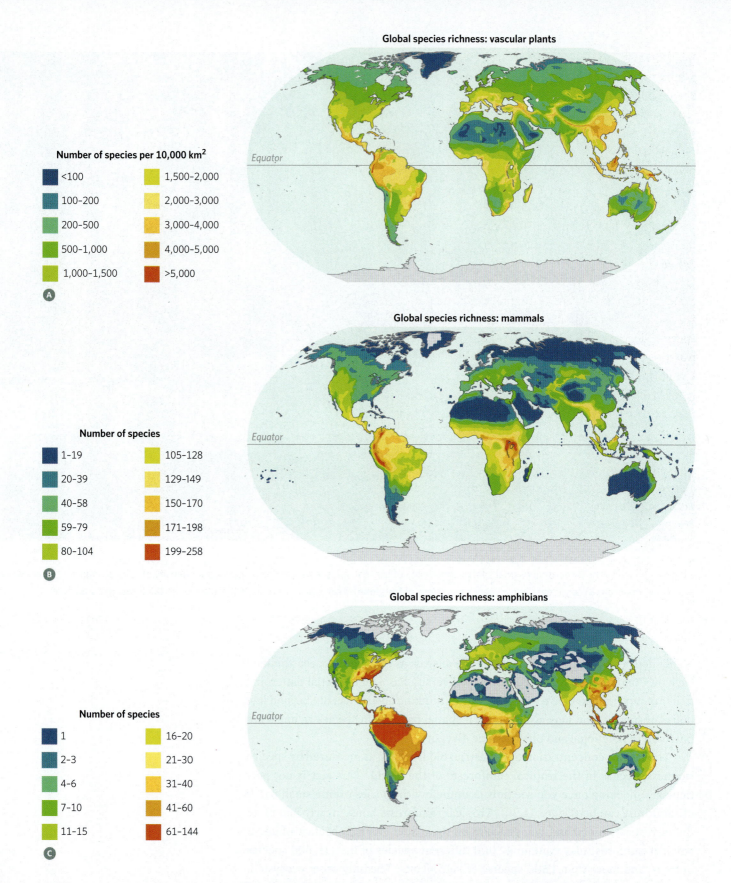

Global species richness: vascular plants

Number of species per 10,000 km²

- ■ <100
- ■ 100–200
- ■ 200–500
- ■ 500–1,000
- ■ 1,000–1,500
- ■ 1,500–2,000
- ■ 2,000–3,000
- ■ 3,000–4,000
- ■ 4,000–5,000
- ■ >5,000

A

Global species richness: mammals

Number of species

- ■ 1–19
- ■ 20–39
- ■ 40–58
- ■ 59–79
- ■ 80–104
- ■ 105–128
- ■ 129–149
- ■ 150–170
- ■ 171–198
- ■ 199–258

B

Global species richness: amphibians

Number of species

- ■ 1
- ■ 2–3
- ■ 4–6
- ■ 7–10
- ■ 11–15
- ■ 16–20
- ■ 21–30
- ■ 31–40
- ■ 41–60
- ■ 61–144

C

Figure 11.31 Maps of global species richness for Ⓐ vascular plants (Barthlott et al. 2016), Ⓑ mammals (Davies and Buckley 2011), and Ⓒ amphibians (International Union for Conservation of Nature 2008). Warm colors (reds) indicate many species; oranges and yellows indicate intermediate numbers of species; and cool colors (greens and blues) indicate very low species richness. Gray indicates extremely low species richness.

Together these two types of ecosystems encompass a huge fraction of the world's total diversity, and both are found at or near the equator. This brings us to the most obvious spatial pattern in species richness: as one moves from the poles toward the equator, species richness increases dramatically. This trend is apparent across nearly every well-studied taxonomic group, including vascular plants, mammals, and amphibians (**Figure 11.31**). In fact, we often find this same pattern when we look at more narrow taxonomic groups or smaller spatial scales, such as the species of dragonflies found across Africa (**Figure 11.32**).

Yet even a casual look at Figures 11.31 and 11.32 reveals that, although the overall latitudinal trend is clear, there are spatial anomalies. There are distinct locations where species richness is relatively high, called **biodiversity hotspots**, which are not located in the tropics. For example, one of the more species-rich areas globally for amphibians is in the southeastern United States on the coastal plain coming out of the Appalachian Mountains (North and South Carolina, Georgia, and Alabama). This area lies at about 34° north latitude, well outside anything we would call tropical, yet it is home to more than 46 species of amphibians. At approximately the same latitude in Los Angeles, California, only 8–10 amphibian species are present. We can see similar breaks in this pattern within the vascular plants. For example, the Cape Floristic Region at the tip of South Africa houses an incredibly high number of plant species, given its more southerly location outside the tropics (**Figure 11.33**).

Identifying these patterns is interesting, but it raises the question of why most of the Earth's plant and vertebrate species are located near the equator. Is there something inherently special about the tropical latitudes, or is it just a matter of a simple observation that the temperature is mild year-round, which creates few physiological challenges to temperature regulation or seasonality of resource supplies? If this reasoning holds, what leads to hotspots outside the tropics? Ecologists have proposed many hypotheses (including some involving temperature) to explain these phenomena, but we describe only a few here because no single explanation or theory works to explain the patterns completely.

One of the simplest explanations is that the tropics are species-rich because these areas have a large amount of primary productivity per unit area due to more direct and consistent solar energy (Chapter 3). More photosynthesis produces more biomass, which provides more ecological resources upon which higher trophic levels can feed. More plant biomass may also translate into more plant species and more ways consumer species can feed (i.e., more ecological niches and more functional diversity) and thus higher rates of evolutionary diversification. On the surface this seems like a plausible, fairly simple, and entirely logical explanation for the latitudinal pattern. However, many habitats around the planet are highly

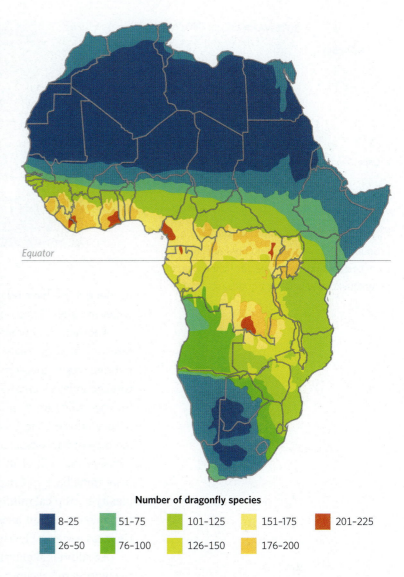

Equator

Number of dragonfly species

8-25	51-75	101-125	151-175	201-225
26-50	76-100	126-150	176-200	

Figure 11.32 As seen on this species richness map for African dragonflies, the equatorial tropical regions of the continent hold the highest number of species, which mirrors the same global pattern observed across several taxonomic groups (Clausnitzer et al. 2012).

Figure 11.33 The Cape Floristic Region in South Africa has a large number of endemic plant species and higher species richness than other locations at similar latitudes.

productive but have relatively low species richness, such as seagrass beds and saltwater marshes. These exceptions to the rule call this explanation into question.

Another, somewhat-related hypothesis is that tropical habitats such as rainforests and coral reefs have more structural complexity (e.g., height diversity, microhabitats) than other habitats. The presence of three-dimensional architectural complexity creates more physical niches and more ecological opportunities for specialization. Unfortunately, this explanation does not really fit the pattern because there is no gradient in the heterogeneity or complexity of the physical features of habitats as one moves from the poles to the equator. We can actually envision the lack of fit between structural complexity and latitudinal gradients if we think back to Chapter 3 and our discussion of biomes; high structural complexity in tropical rainforests declines as we go into tropical savannas and deserts but increases again as we move into temperate and then boreal forests. Species richness patterns clearly do not follow suit.

An additional plausible hypothesis focuses on time. Perhaps "older" regions of the biosphere have more diversity because of the "extra" time for the evolutionary machinery to produce new species. Tropical areas have not undergone glaciation, so the current unglaciated flora and fauna have had more time to evolve into the great variety of species we see in the tropics today. The slow march of glaciers over a landscape essentially resets the evolutionary and ecological clock back to zero (see ecological succession in Chapter 13), and any set of species that occupies these places after glacial retreat has to colonize them anew.

Unfortunately, this hypothesis also breaks down on close examination. For example, consider the species richness of the deep ocean (the abyssopelagic zone), which is an extremely old ecosystem that has maintained environmental constancy throughout hundreds of millions of years (e.g., the dark waters are always 4°C, or 39°F). Despite their evolutionary and ecological "age," locations in the deep ocean do not appear nearly as diverse as some locations on land. Plus, there is evidence that while tropical areas may not have undergone the same cooling phases as higher latitudes, many tropical areas have experienced extended, extreme droughts of such intensity that the entire biome would have changed. In that case, evolution would not have been acting in one direction for a long period, and the evolution of diversity would have had to "start over" in the tropics as well.

Do Functional and Phylogenetic Diversity Follow the Same Global Pattern?

When we measure biodiversity in terms of evolutionary history or functional traits, we often obtain a different perspective than if we only consider taxonomic species richness. This is also very true for global geographic patterns in biodiversity.

For example, if we focus on the global diversity of mammals, we can compare gradients in species richness to gradients in phylogenetic diversity (**Figure 11.34**). For species richness, we see that more species of mammals reside in tropical Africa than at lower or higher latitudes on that continent, and a similar

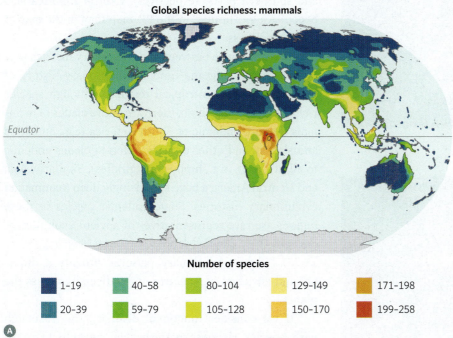

Global species richness: mammals

Number of species

■ 1–19	■ 40–58	■ 80–104
■ 20–39	■ 59–79	■ 105–128

□ 129–149	■ 171–198
■ 150–170	■ 199–258

A

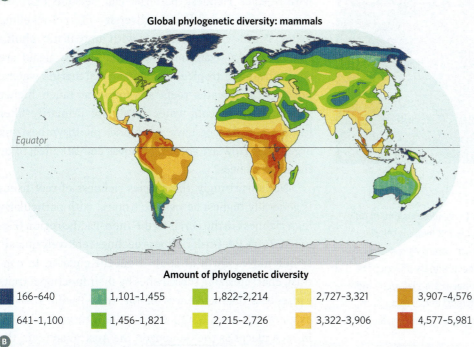

Global phylogenetic diversity: mammals

Amount of phylogenetic diversity

■ 166–640	■ 1,101–1,455	■ 1,822–2,214
■ 641–1,100	■ 1,456–1,821	■ 2,215–2,726

□ 2,727–3,321	■ 3,907–4,576
■ 3,322–3,906	■ 4,577–5,981

B

Figure 11.34 These maps compare **A** global mammal species richness with **B** global mammalian phylogenetic diversity (PD). Note that peak diversity occurs in the tropics for both measures, but that the relative importance of Asia, Central America, and the southern portions and west coast of Africa increase when we look at PD (Davies and Buckley 2011).

Figure 11.35 **Ⓐ** The clouded leopard (*Neofelis nebulosa*) is one of only two species in the genus *Neofelis*. **Ⓑ** The Sunda clouded leopard (*N. diardi*) was considered a subspecies until 2006. These two extremely rare species are the only living examples of their ancient evolutionary lineage. The genus is considered the link between the Pantherinae (cats that roar) and the Felinae (cats that purr because of their bony hyoid) and is thought to have diverged from the *Panthera* genus (e.g., lions, tigers, panthers, leopards, jaguars) more than six million years ago.

pattern appears for mammals in tropical Central and South America. Notably, relatively few species live in the Asian tropics compared to other tropical regions. When we weight each mammal species by how much evolutionary history it represents, this global biodiversity pattern shifts and expands. The concentration of PD remains in the tropics but now includes a "hotspot" of mammalian evolutionary diversity in Asia. In particular, the map expands PD concentrations to include nearly all of Africa and South America, except the extreme northern and southern parts, respectively. Central America also becomes a lot "hotter" in the PD map than in the map of species richness.

The difference between these diversity metrics may be due to the long and unique evolutionary history of mammals in the nontropical regions. Although species richness may not be high, there are entire clades of mammals that are endemic to South America (e.g., anteaters and sloths), and other entirely different clades endemic to all of Africa (e.g., hippos and giraffes). Similarly, Asia and Central America have surprisingly deep mammalian evolutionary histories. Some endemic species in these regions are rare representatives of ancient evolutionary lineages (e.g., clouded leopards [genus *Neofelis*]; **Figure 11.35**). These evolutionary patterns provide a different and enriched perspective on biodiversity across the planet.

If we consider functional diversity, the mismatch with species richness patterns can become extreme. Consider for example, the diversity of reef-dwelling fishes. Warm-water coral reefs with mutualistic, photosynthesizing zooxanthellae (i.e., mutualistic algae) are concentrated in tropical and subtropical marine waters; however, cold-water reefs are found at much higher latitudes. These cold-water corals are primarily filter feeders and do not have symbiotic, photosynthesizing organisms. How are species richness and FD affected by these differences?

Not surprisingly, the species richness of reef fishes is generally highest across the tropics, with particularly high richness in the corals of the Indo-Pacific region (**Figure 11.36A**). Reef-fish richness declines relatively quickly as one moves north or south from the equator. In contrast, characterizing these fishes by their functional traits causes the distinctiveness of the Indo-Pacific reefs to evaporate (**Figure 11.36B**). Most of the FD is instead found in such places as the Galápagos, the Australian reefs, the

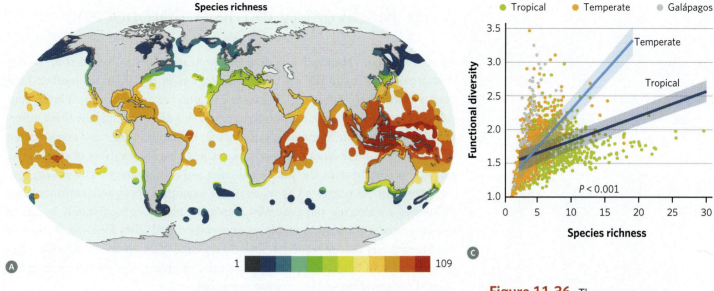

Species richness

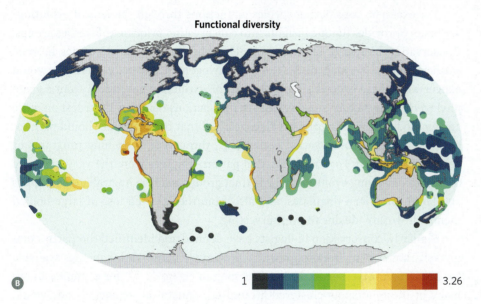

Functional diversity

A

B

C

Figure 11.36 These maps compare **A** global species richness and **B** global functional diversity (FD) in reef fishes (Stuart-Smith et al. 2013). Notice the shift in peak diversity from the waters off the east coasts of continents for species richness to waters off the west coasts of continents for FD. Peaks in FD also seem concentrated at slightly higher latitudes than peaks in species richness. **C** FD increases in reef fishes as species richness increases, but interesting regional variability in the pattern suggests that different forces may act at higher latitudes (Stuart-Smith et al. 2013).

western equatorial African coast, and within the nearshore waters of the Arabian and Caribbean Seas. Although FD also drops off quickly as one moves north and south in latitude, there are pockets of fishes with unique functional traits at surprisingly northerly or southerly locations (e.g., Southern California, United States; Baja California, Mexico; and the southwestern parts of Africa).

In general, the FD of reef fishes increases as species richness increases (**Figure 11.36C**). However, the rate at which FD increases relative to species richness varies tremendously across locations with temperate locations increasing at a faster rate. These trends suggest that, as you add species to a tally, some of the added species introduce new functional traits, but many are functionally redundant with a species that was already included. The degree of redundancy, however, can be quite low in some instances, such as with reef fishes in the Galápagos. In this case, nearly every newly added species brings with it a new functional trait.

11.6 Biodiversity in the Anthropocene

An appreciation of the Earth's biodiversity is one of the primary aspects of ecology that draws students to the field. Many ecologists, at one point or another, have been awestruck by the range of species encountered in a location. Throughout this chapter, we have explored how to measure this diversity and how these different measures provide fascinating insights into the maintenance and conservation of biodiversity. Biodiversity, however, is not just about the variety of species you see; it also encompasses the functional roles and evolutionary history these species represent. As suggested by many of the examples presented, anthropogenic influences are altering and generally decreasing planetary biodiversity by any of the measures we choose to employ. Conservation biologists focus on the loss of biodiversity and how to stem that loss. To that end, they use every definition and measurement of biodiversity presented in the chapter to decide where to invest limited conservation and restoration resources.

For example, considering species extinctions through the lens of evolutionary history provides unique insights into conservation priorities of species groups. Conservation biologists have documented that, in general, as species become extinct, there is a much higher loss of phylogenetic diversity than one would expect were these species simply chosen for extinction at random from a given phylogeny. This pattern suggests that human actions tend to fall hardest on more ancient and distinct evolutionary lineages, and that these groups should perhaps be prioritized for conservation action (e.g., through antipoaching programs or habitat protection). For mammals, this pattern emerges from the very high rate of extinction among evolutionarily distinct groups, such as Australian marsupials and South American ungulates, and also from the targeted loss of large-bodied mammals worldwide due to overharvesting.

Similarly, conservation biologists have defined and identified the Earth's biodiversity hotspots in order to prioritize the limited conservation monies available for protection (Myers et al. 2000). The map in **Figure 11.37**, for example, shows many of these hotspots clustered around the equatorial regions, as one would expect. But it also shows other hotspots located at much higher latitudes (e.g., Central Chile, New Zealand, and the Caucasus). Such hotspot maps have become the basis for governmental and agency decisions about where to install national parks and marine protected areas.

Finally, even the species-area relationship, as old and well-worn as it is, provides much-needed insight into the very modern and widespread problem of how to balance land-use decisions, habitat loss, and species extinctions. For example, when humans reduce the original expanse of a forest by converting land into residential housing or agricultural plantations, conservation biologists ask how many forest-specialist species might go extinct as a result. A simple calculation using the species-area equation (Equation 11.1) can help answer that question and point to the amount of forest conversion that will lead to substantial biodiversity declines. We can see this pattern in Figures 11.15B, 11.16A, and 11.16B. If we follow the curves in these figures backward, toward smaller areas, we can see that species richness drops off slowly at first and then sharply. Conservation biologists definitely want to avoid the sharp species richness drop-offs that come with large land conversions.

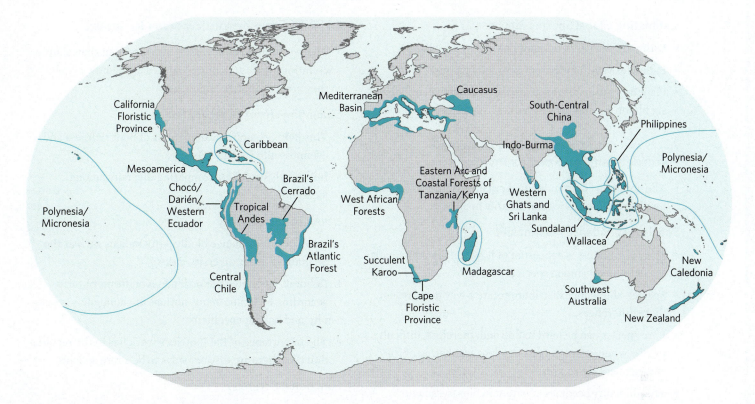

Figure 11.37 Conservation biologists have identified 25 plant and vertebrate biodiversity hotspots around the world, shown in blue on this map. Included locations met criteria for housing a significant percentage of the world's vascular plant and vertebrate diversity.

The anthropogenic forces that threaten biodiversity are many, and they interact in ways that challenge even the best-laid plans to stem the loss of species to extinction. We will revisit many of the ideas described here in Chapter 15, combining them with other tools that emerge from an ecological education, to address the fate of ecosystems in the Anthropocene.

SUMMARY

11.1 What Is Biodiversity?

- The term *biodiversity* conjures images of verdant rainforests and coral reefs.

- The number of species on Earth (i.e., species richness) is only one aspect of biodiversity.

- A full definition of biodiversity includes all living organisms, the communities they form, and the evolutionary diversity they contain.

11.2 Global Species Richness

- It is hard to answer the question of how many species are on the planet.

Formal Species Descriptions

- Approximately two million species have been described by scientists.

- This does not accurately reflect the actual number of species on Earth for several reasons:

  — Many species have multiple names (synonymy).

  — Some species, known as cryptic species, do not exhibit morphological differences from each other and get "hidden" under one species name.

  — Many species are small and hard to find.

  — Exploration of habitats is not carried out evenly over the planet.

  — There are few taxonomists with the knowledge to describe new species.

Estimating Global Species Richness

- Estimates are based on sampling and extrapolation.
- Estimates vary for different taxonomic groups.
- Planetary estimates range anywhere from five million to six billion species on the Earth.

11.3 Effects of Abundance, Area, and Heterogeneity on Species Richness

- The ability to count individual organisms influences the numbers of species we can identify in an area.

Species-Abundance Relationship

- Species evenness is a measure of how equally individuals are distributed among species.
- Species-abundance distributions are a way to represent these differences.
- Rare species can be hard to find and, therefore, difficult to tally when attempting a complete accounting of species richness for an area.
- This difficulty becomes apparent in bioassessment protocols.
- Different environments can also produce differences in abundance, further confusing the issue.

Species-Area Relationship

- The larger the area you sample, the more species you find.
- Tiered sampling protocols help with sampling issues.
- As the size of an island increases, so does the number of species on the island.
- Frank Preston fit this species-area relationship to a power function that holds true across taxa and habitats.

Alpha, Beta, and Gamma Diversity

- Robert Whittaker examined biodiversity changes across spatial scales and formulated three distinct measures to highlight the differences.
  - Alpha diversity (α) is the average number of species within a small region.
  - Beta diversity (β) is turnover or the heterogeneity among locations.
  - Gamma diversity (γ) is species richness across a large regional landscape.
- All three measures are related mathematically.
- Invasive species can affect these three measures and produce biotic homogenization of local communities.

11.4 Moving beyond Species Richness

- There is more to biodiversity measurement than simply counting species (richness) or measuring how evenly they are distributed (evenness).

Shannon-Wiener Diversity Index

- This index combines calculations of both richness and evenness into a single measure.
- It is mostly useful for comparing diversity among locations.

Seven Forms of Rarity

- Rare species influence biodiversity indices such as the Shannon-Wiener diversity index.
- Deborah Rabinowitz described seven forms of rarity according to dichotomous notions of abundance, geography, and habitat specificity.
- The importance of Rabinowitz's work lies in the recognition that there are distinct ways to be a rare species.

Functional Diversity

- An alternative measure of diversity comes from looking at the functional roles that species play in communities.
- These can be described by the investigator and are variously measured and enumerated.
- Functional richness counts the number of different functional roles occupied in a community and allows comparisons among locations and communities.
- Functional diversity (FD) measures the range of traits in a community by examining how different traits are.
- Creating an FD dendrogram for a location allows you to measure FD by summing the branch lengths on the dendrogram.
- Locations with large numbers of human-caused extinctions have lost significant FD.

Phylogenetic Diversity

- It is also possible to measure diversity in an evolutionary or phylogenetic sense, using phylogenetic trees.
- Phylogenetic trees contain information on evolutionary history.
- Phylogenetic diversity (PD) can be measured by summing the lengths of branches on a phylogenetic tree.
- In the modern era, some locations have lost significant PD due to human-caused extinctions.

11.5 Global Geographic Patterns in Biodiversity

- Earth's species are spread over the planet in distinct patterns.
- In general, more species are found near the equator than near the poles.
- However, there are hotspots of diversity that buck this trend.
- Many hypotheses for the latitudinal pattern have been proposed, but none seems to completely explain the pattern.

Do Functional and Phylogenetic Diversity Follow the Same Global Pattern?

- There is some concordance between these measures, but FD and PD also show different and more pronounced patterns.

11.6 Biodiversity in the Anthropocene

- Anthropogenic influences are severely decreasing planetary biodiversity.
- Identifying biodiversity hotspots can help ecologists prioritize conservation and protection efforts.
- Species-area relationships can be used to make land-use decisions and reduce species extinctions.

KEY TERMS

alpha diversity (α)
beta diversity (β)
biodiversity
biodiversity hotspot
biotic homogenization
branch
clade
cryptic species
depauperate

endemic species
functional diversity (FD)
functional diversity dendrogram
functional evenness
functional richness
functional trait
gamma diversity (γ)
node
phylogenetic diversity (PD)

phylogeny
root
Shannon-Wiener diversity index (H')
sister species
species abundance
species evenness
species richness (S)
species-area relationship
synonymy

CONCEPTUAL QUESTIONS

1. Using what you learned about biomes in Chapter 3, and about physiology and niches in Chapter 5, explain why we expect to see higher z-values in the species-area curves at lower latitudes.

2. Figure 11.11 (cumulative number of individuals found in the 15 most common species starting with the most common and moving toward the 15th most common) and Figure 11.16 (cumulative number of species found as area sampled increases) look somewhat similar but represent different things. Explain the differences, and explain why they look similar.

3. This chapter mentions that urbanization and pollution can lead to lower oxygen levels and alter the densities and types of species we find in stream habitats. Examine Tables 11.2 and 11.3, and divide the species into two groups—those with higher tolerance of pollution and those with lower tolerance. Base your determinations solely on the numbers in the two tables (assume San Pablo Creek is urban and Wildcat Creek is protected). Explain your criteria.

4. Explain why greater functional diversity may mean that an environment has more niches.

5. Create a comprehensive definition of biodiversity by trying to include all the different measures discussed in the text. Compare your definition to the UN definition and discuss how the two definitions would be applied by a conservation organization versus a political organization.

6. What do you think is the third most biodiverse habitat on the planet after tropical rainforests and coral reefs? Explain your answer. Can you find any evidence to support your idea?

7. Discuss the observation of latitudinal gradients in biodiversity along with the potential reasons for the pattern. Which of the reasons hold the most value in your opinion? Describe the kinds of evidence or patterns you would look for to support or refute the different lines of reasoning.

QUANTITATIVE QUESTIONS

1. The numbers we gave for Terry Erwin's estimate of global species richness are rounded off.
   a. Starting with 163 specialist beetles on one tree and 50,000 different tree species, how many specialist beetles are there globally?
   b. If beetles make up 40% of all insect species, how many insects are there globally?
   c. If insects make up two-thirds of the world's species, how many species would Erwin calculate for overall global species richness?
   d. How reasonable do these calculations seem to you? Even if the calculations do not seem reasonable for accurate assessments, under what circumstances might such approaches be useful?

2. The data for graphs B and D in Figure 11.16 are provided in the following table.
   a. Using a spreadsheet, create a graph that looks like Figure 11.16B.
   b. Log-transform the S (species richness) and A (area) values from the table, and create a graph like Figure 11.16D.
   c. Using the graph that you created in 2a, change your axes to a log scale (this operation is fairly straightforward in spreadsheet software programs). Your new graph should look just like your graph from 2b but with different numbers along the axes (e.g., 3.0 may appear on the x-axis in 2b, but would be 1,000 in 2c). Explain why.

Island	Area in square miles	Number of breeding bird species
Cuba	43,000	124
Isle of Pines	11,000	89
Hispaniola	47,000	106
Jamaica	4,470	99
Puerto Rico	3,435	79
Bahamas	5,450	74
Virgin Islands	465	35
Guadalupe	600	37
Dominica	304	36
St. Lucia	233	35
St. Vincent	150	35
Grenada	120	29

3. Using the data from Tables 11.2 and 11.3, calculate α, β, and γ for the two streams. Assume that each stream is a local area and that the region is made of just these two streams.

4. The following table lists the number of individuals of different tree species found in plots on the south and north sides of a mountain range in western Massachusetts.
   a. Calculate S for each side of the mountain.
   b. Which side do you think has a higher H'?
   c. Calculate H' for each side of the mountain.

d. Which do you think contributes more to the difference in H', species richness or species evenness?

e. Create a rank abundance curve like the one in Figure 11.21, with a line for each side.

f. Looking at the rank abundance curve, how would you now answer the question about the way species richness and evenness affect H'?

Species	South	North
Red oak (*Quercus rubra*)	19	4
White oak (*Q. alba*)	11	1
Chestnut oak (*Q. prinus*)	20	0
American beech (*Fagus grandifolia*)	13	19
American chestnut (*Castanea dentata*)	2	0
Black birch (*Betula lenta*)	8	18
Paper birch (*B. papyrifera*)	0	4
Ironwood (*Carpinus caroliniana*)	0	5
Hornbeam (*Ostrya virginiana*)	46	6
Red maple (*Acer rubrum*)	30	1
Sugar maple (*A. saccharum*)	29	19
Striped maple (*A. pensylvanicum*)	0	4
Shagbark hickory (*Carya ovata*)	7	3
White ash (*Fraxinus americana*)	3	0
Eastern hemlock (*Tsuga canadensis*)	2	87
White pine (*Pinus strobus*)	12	0

5. Explain how the graphs in Figure 11.23, which show species range sizes for different groups of birds, fit with Rabinowitz's ideas about rarity.

6. Look at Figure 11.36C, which plots species richness against functional diversity for reef fishes, and explain the difference in slope in the lines for the temperate and tropical zones. Think about the different densities and richness of species in temperate latitudes and tropical latitudes and how much more space might be required to increase species richness at higher latitudes. In light of what you learned in Chapters 3, 4, and 5, how might such differences in the amount of space available at different latitudes affect the variety of habitats and functional adaptations at higher latitudes?

SUGGESTED READING

Davies, T. J. and L. B. Buckley. 2011. Phylogenetic diversity as a window into the evolutionary and biogeographic histories of present-day richness gradients. *Philosophical Transactions of the Royal Society B* 366(1576): 2414–2425.

Gaston, K. J. 1996. Species-range-size distributions: Patterns, mechanisms and implications. *Trends in Ecology and Evolution* 11(5): 197–201.

Larsen, B. B., E. C. Miller, M. K. Rhodes, and J. J. Wiens. 2017. Inordinate fondness multiplied and redistributed: The number of species on Earth and the new pie of life. *Quarterly Review of Biology* 92(3): 229–265.

Petchey, O. L. and K. J. Gaston. 2002. Functional diversity (FD), species richness and community diversity. *Ecology Letters* 5(3): 402–411.

Preston, F. W. 1962. The canonical distribution of commonness and rarity: Part I. *Ecology* 43(2): 185–215.

Mexican free-tailed bats (*Tadarida brasiliensis*) swarm out at dusk from the Davis Ranch Blowout Cave in Blanco County, Texas. Movement such as this is a common occurrence for many organisms.

chapter 12

SPATIAL DYNAMICS

12.1 Nobody Stays in One Place

Although we can look around and see all sorts of organisms that appear to stay in one place—such as most plants and fungi and many invertebrates—movement is an important aspect in the life history of almost all organisms. Even those that are reliably and predictably in the same place every day probably moved there at one point and then settled in to stay. Others move every day and possibly every hour or minute. Although we discussed some of the effects of movement on evolution and physiology in Chapters 4 and 5, we have largely ignored the way movement alters population dynamics and interactions between species. Let's start exploring these spatial dynamics by thinking about humans.

In most human societies, kids live with their parents, and sometimes in larger familial groups, until they reach an age of independence. The age of independence may vary based on resources and culture, but it usually coincides at least somewhat with sexual maturity and the desire to start a family and perhaps produce some children of their own. At that point, newly independent humans will "move out" to a different country, city, house, or bedroom. The shove toward independence may reflect societal norms, a desire for children to experience the world, financial pressures, or a variety of other drivers. Whatever the source, the push toward independence on the part of the parents is often matched by a desire for independence in the newly "adult children." Many of us find that sticking around our parents' house for too long can cramp our style in all sorts of ways, not least of which may be our ability to find suitable mates.

LEARNING OBJECTIVES

- Explain the differences between migration and dispersal using examples.

- Explain how ecologists can add movement to the exponential and logistic growth models.

- Distinguish the features that separate a population from a metapopulation.

- Explain as fully as possible the Levins metapopulation model.

- Assess how a metapopulation approach to population dynamics would affect the ecological understanding and conservation of populations in the real world.

- Identify three assumptions of the Levins metapopulation model that are difficult to meet in the real world.

- Explain how a source-sink model differs from the Levins metapopulation model.

- Restate the equilibrium theory of island biogeography proposed by MacArthur and Wilson.

- Apply the ideas of the MacArthur and Wilson theory to the protection of species.

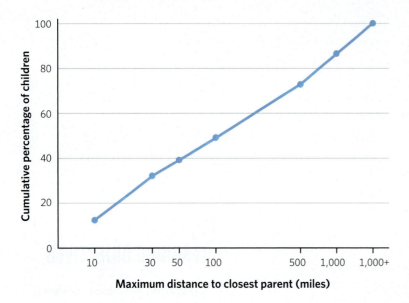

Figure 12.1 The distance between parents and their adult children (University of Michigan with funding from the National Institute on Aging 2010). Notice that the x-axis is a log scale. Although about half of the adult children respondents live within 100 miles of a parent, some adult children are quite far from their parents.

We are not alone in this transition, as most organisms move away from their birth sites by the time they reach reproductive maturity. Some, like dispersing seeds and many marine invertebrates, make their moves right at "birth," while others do it at maturity. If we look hard enough, we can find examples of species with departures at almost any point between birth and maturity. So, what does all this moving do, and does it matter when organisms move, how they move, and how far they move? The basic answer is yes.

Let's start with how far they move and focus on humans again. We know that physical proximity for extended families varies with culture. Some families stay closely connected, with several generations living in one house. But it is also relatively common in some cultures for adult children to move to different cities or even different countries. With such a wide range of movement distances, we may not gain much understanding from looking at averages. Instead, we can try to characterize how common it is to see short-, medium-, or long-distance moves.

For example, the Institute for Social Research at the University of Michigan explores physical distance between parents and their adult children in the Health and Retirement Study (sponsored by the National Institute on Aging), an ongoing study of older Americans. According to this study, half of all adult US children in 2004 lived within 100 miles (160 km) of a parent's residence, while only 13% lived farther than 1,000 miles (1,667 km) from their parent (**Figure 12.1**).

Can we use this figure to generalize about all humans, though? Clearly not. In fact, the pattern varies even within the United States. If we divide the 2004 data by geographic region, proximity between adult children and their parents was much closer in states east of the Mississippi River than in states west of the Mississippi River. Do you think we would see differences based on a person's income, gender, or religion? Would the same patterns be present in data from 1904 instead of 2004? How about if we had data comparing the same relationships between India and the United States? Human demographers spend a lot of time thinking about how all these factors may alter human movements because such dynamics have huge implications for social, cultural, and economic welfare.

Ecologists think about the spatial movements of nonhuman species for similar reasons. The spatial patterns of organisms' movements underlie many ecological and evolutionary factors, such as reproductive success, population density, and survival, as well as population extinction risk, genetic diversity, and speciation rates. With a role in all of these factors for individual species, it is probably not surprising that movement also influences large-scale biodiversity patterns. But what aspects of movement are important to measure?

From the human example, we know that we cannot capture the important features of movement with a single number such as an average, because we can see that the data are very skewed. We also know that movement may differ based on age, location, and environmental conditions. We can envision further complications if we notice that the human example only looked at one type of movement, **dispersal**, which involves individual movements. But what if we also consider the effects of coordinated movements of whole populations or segments of populations? Think about nomadic human groups—whole communities moving together, often in predictable ways. We have fewer nomadic societies than we used to, but this form of movement, called **migration**, is very common among nonhuman species.

Because we could add a spatial component to almost every aspect of population and community ecology, this chapter does not have a single organizing model. Instead, we define various forms of movement within an ecological context and focus on the timing of movement, the number of individuals that move, the predictability of movement, and the distribution of movement distances. With an awareness of these concepts, we can examine how movement enriches our previous understanding of population growth and spread. Using insights from spatial population and species interaction models, we can explore two new types of ecological models that directly incorporate spatial movement: metapopulations and island biogeography. Eventually, we demonstrate how incorporating space and movement can transform the way we think about core ecological principles.

12.2 Why Do Organisms Move?

With rapid technological advances, our ability to measure movement has increased vastly over the last few decades. However, capturing movement and measuring it are still difficult, time-consuming, and expensive undertakings. Ecologists, therefore, have focused on why we need movement information in order to prioritize data collection. We can identify four essential questions that ecologists seek to answer with movement data: (1) How do organisms move (tied to physiology and anatomy)? (2) Why do they move (tied to life history and fitness)? (3) Where and when do they move (tied to navigation)? and (4) How do external factors interact with and affect the how, why, where, and when of movement (Nathan et al. 2008)? In this chapter, we focus primarily on the second, third, and fourth questions, but note that some aspects of the first question were discussed in Chapter 5.

When thinking about movement, one important point to keep in mind is that organisms generally move to improve their individual fitness, and any hypothesis we propose to explain or predict movement needs to take fitness into account. How does movement improve evolutionary fitness? To tackle this question, we start by focusing on the differences between migration and dispersal.

Migration versus Dispersal

Although informal language will often use *migration* and *dispersal* interchangeably, ecologists attach important differences to these terms. Migration involves movement by many individuals at approximately the same time from one habitat to another habitat, often at a predictable life stage or time of the year. Migration is also typically a round-trip event, although the trip can sometimes be spread over more than one generation. Dispersal, on the other hand, is usually a solo event whereby an individual moves to a location that is not known or identified beforehand, and it does not normally include a return ticket. Although these forms of movement seem very different, the impetus can still be boiled down to a drive to improve individual fitness.

Coast range newts (*Taricha torosa*) in California illustrate the distinction between dispersal and migration (**Figure 12.2**). Young individuals *disperse* from their natal ponds (meaning the ponds in which they emerged from eggs) to upland, nonbreeding habitat after they mature into adults. This solitary dispersal event takes place early in their lives. Yet as adults, individuals all make annual trips to

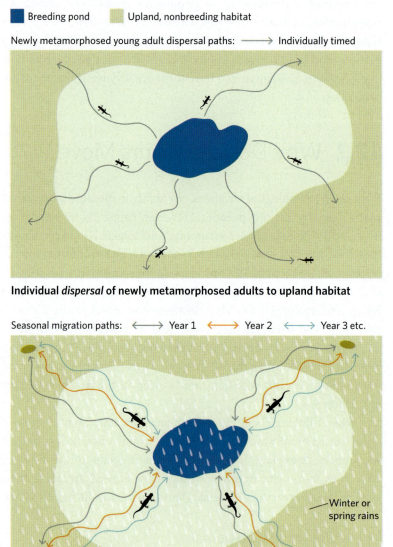

Individual *dispersal* of newly metamorphosed adults to upland habitat

Seasonal *migration* paths of many individuals to and from breeding pond at the same time every year

Figure 12.2 Ⓐ A coast range newt (*Taricha torosa*). Ⓑ The movement behavior of coast range newts exemplifies the distinction between dispersal and migration. Young individuals *disperse* by themselves from their natal ponds to upland, nonbreeding habitat after they mature into adults. As adults, however, they make a seasonal breeding *migration* from their summer dry-season burrows to ponds and then return to the upland after breeding.

a breeding pond or stream during the breeding season. This *migration* movement involves many breeding individuals moving in the same season and then returning to the upland habitat where they live when not breeding. The seasonal migration from upland burrows during the dry summer season down to breeding sites in ponds and streams during the wet winter or spring seasons can lead to surprisingly high abundances of newts crossing roads at the same time. Once their eggs are deposited, they migrate back upslope to hide in summer burrows and retreats.

To clarify the two types of movement, in Figure 12.2 we have shown dispersal from one pond to upland areas, and migration from burrows to a pond and back again. Ponds can be both breeding locations for adults and natal ponds for juveniles, and adults from all upland areas will migrate annually to breed. There are many more types of dispersal and migration, as discussed next, but this example underscores the distinction between individual dispersal movements that are often in one direction to an unpredictable location and migration movements that are typically seasonal, round-trip, and undertaken by many or all individuals in a population at the same time.

Migration   Migrating animals provide some of the most spectacular visual events in the natural world. Just watching nature shows or your social media feed will introduce you to migration and how it affects individual survival and reproduction. Perhaps you have seen a video showing massive herds of wildebeest (*Connochaetes* spp.) migrating across the Serengeti between their grazing lands and their calving grounds in an annual cycle (**Figure 12.3A**). Or you could visit one of the online sites that track the movements of humpback whales (*Megaptera novaeangliae*). One of the 14 known breeding groups of humpbacks migrates from feeding grounds in the cold waters of the North Atlantic to the warm waters of the Caribbean, where they birth and care for their young (**Figure 12.3B**). Or you may have noticed Canada geese (*Branta canadensis*) flying south in their V formation for the winter or north again in the spring (**Figure 12.3C**). In each migration example, the individuals move en masse, going to fairly predictable locations at fairly predictable times of the year and eventually returning to their starting habitat.

Nature documentaries that depict these movements often feature a dramatic scene in which a migrating juvenile dies from predation or physiological stress. These scenes create drama for the viewing audience, but they also graphically illustrate that migration poses substantial risks to the traveling individuals. Ecologists have carefully documented these risks across several species and have found that migrating individuals suffer high mortality rates relative to nonmigrating individuals or species. Yet this finding seems to contradict our evolutionary insight about increased fitness. Why risk dying simply to get to a new location?

Unfortunately, "Because everyone else is doing it" is not a good enough answer, yet it does illuminate an interesting part of the puzzle. Migratory movements impose higher risks, but "everyone" in a particular age class or cohort in a migrating species takes on these risks. This universality suggests that the reasons for migrating must be compelling.

Migratory behavior likely evolved as a way to balance reproductive and survival trade-offs (see Chapters 4 and 5) for two distinct locations and potentially for two different life stages. The ecological conditions provided by one location may enhance reproduction during some periods but be suboptimal for survival

Figure 12.3 Ⓐ Wildebeest (*Connochaetes* spp.) in Maasai Mara, Kenya, make annual migrations around East Africa. Ⓑ Humpback whales (*Megaptera novaeangliae*) from different numbered pods make long-distance migrations (indicated by dashed lines) across the oceans of the world. Notice that pod 1 moves from its feeding areas (green shapes) in the cold waters of the North Atlantic to the warm Caribbean waters that are its birthing grounds (blue area 1) (NOAA Fisheries 2016).

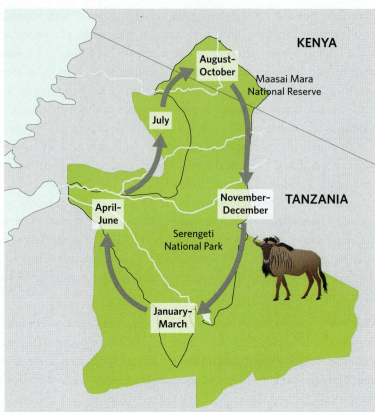

Ⓐ

Ⓑ

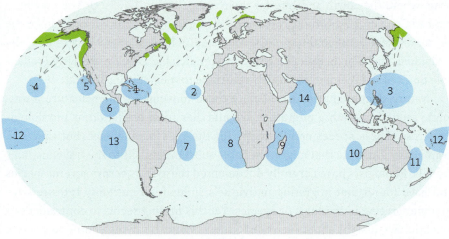

1 West Indies	6 Central America	11 East Australia
2 Cape Verde/Northwest Africa	7 Brazil	12 Oceania
3 Western North Pacific	8 Gabon/Southwest Africa	13 Southeastern Pacific
4 Hawaii	9 Southeast Africa/Madagascar	14 Arabian Sea
5 Mexico	10 West Australia	

Figure 12.3 (continued) 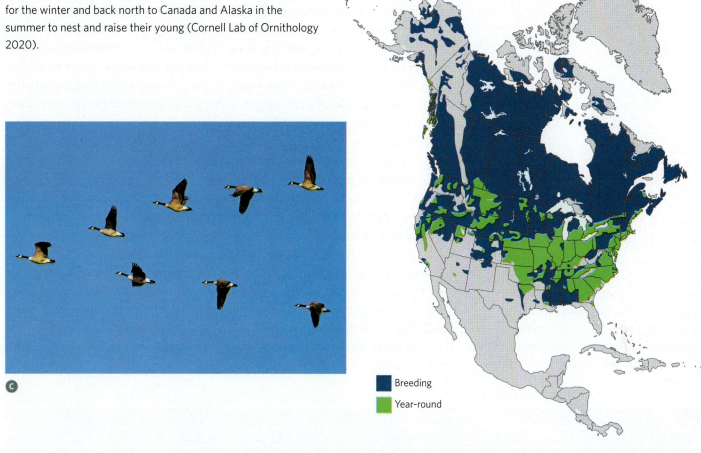 C Canada geese (*Branta canadensis*) make long migrations to the southern United States for the winter and back north to Canada and Alaska in the summer to nest and raise their young (Cornell Lab of Ornithology 2020).

C

■ Breeding
■ Year-round

during other periods. Likewise, a location may be conducive to survival but provide lousy conditions for reproduction. This type of split is particularly common in species in which adults and juveniles have drastically different habitat requirements or have very different survival abilities. If juveniles are extremely vulnerable, then it may make sense to go to a location with less food but also less danger while juveniles mature. For this strategy to work, the difference in reproduction or survival between the two locations must be large relative to the costs imposed by the migratory journey itself. Therefore, we should expect to see substantial fitness differences in one location versus the other for at least some age or stage classes.

Going back to the example of humpback whales, individuals of this species spend most of the year in cold polar waters because the habitat is amazingly productive, providing tons of marine amphipods year-round for the whales to eat. We also know that whales can physiologically tolerate the cold of Arctic winters because individuals have been seen in far northern waters in the dead of winter (Corkeron and Connor 1999). In fact, their fat stores make these cooler temperatures preferable. Nonetheless, during the coldest parts of the year, female humpback whales migrate thousands of miles south to give birth to their young in the warm waters at lower latitudes (Figure 12.3B). They make this journey because it offers a number of survival benefits for juvenile humpbacks.

First, the more southerly waters are warmer and potentially less energetically challenging for young whales. Recall from Chapter 5 that many individual organisms maintain an internal temperature balance. Whales are mammals and maintain a relatively high internal temperature. Adult whales have plenty of fat stores to assist this process, but juveniles have a higher ratio of surface area to volume and less fat and are therefore more likely to lose internal heat to the surrounding water. This difference makes the warm, southerly waters physiologically better for the juvenile whales even though the water is a little warmer than preferred and requires more energetic output for the mothers.

Perhaps more importantly, though, the warmer waters have lower densities of orcas (also known as killer whales, *Orcinus orca*; Figure 12.4), which actively target young whales as prey (Corkeron and Connor 1999). Although orcas can live almost anywhere, their densities are significantly lower in tropical areas because—like adult humpback whales—their layer of fatty insulation makes it more stressful for adult orcas to exert themselves in warm water than in cold water. An orca chasing scarce prey in warm water would likely be foraging suboptimally with respect to energy use (see Chapter 5), particularly when there are plenty of high-density prey available at higher latitudes. As a result, in the warm Caribbean Sea, humpback whales can tend to their vulnerable calves in habitats that mimic the warm conditions inside the pregnant mother and are almost completely free from the threat posed by predatory orcas.

Figure 12.4 Orcas, also known as killer whales (*Orcinus orca*), travel and hunt in pods. As a group, such pods are very effective predators that can kill much larger whales. Orcas have been known to kill many types of whales, including humpback whales, and are a common mortality source for young humpbacks as they migrate to summer feeding grounds in the Arctic from the warm waters where they were born.

The same issues that keep orcas away also make the calving grounds unsuitable for year-round use by humpback whales. The low density of predators in warm ocean waters is matched by a low density of food, so each year humpback whales move back to more productive, food-rich, high-latitude waters with their young in tow. Along the way, young humpback whales grow larger and, therefore, become increasingly protected from both cold waters and orca predation.

Humpback whales, wildebeest, and Canada geese make these seasonal back-and-forth movements as migrations, but migratory movements do not always follow a strict yearly movement pattern. For example, some species migrate only once in their lives, typically spending their adult nonbreeding period far from their site of birth, and then returning to their natal habitats (the habitats of their birth) to breed. Among the many examples of organisms that exhibit this type of migratory pattern are sockeye salmon (*Oncorhynchus nerka*; **Figure 12.5A**). These fish migrate from their natal inland streams to the Pacific Ocean, where they spend two to three years growing to enormous sizes while feeding in the ocean. They then return to the freshwater rivers of Canada and Alaska to mate and deposit their eggs. These fish follow this migration route (**Figure 12.5B**) only once in their lifetime because the adults die soon after breeding. Nonetheless, this movement is considered migration because it happens at predictable times, with all individuals of the species following the same predictable routes.

In rare situations, a species can complete a round-trip migration path over multiple generations, as is the case with some African dragonfly species and the better-known monarch butterfly (*Danaus plexippus*). Monarch butterflies follow a variety of migration routes (**Figure 12.6**). Some populations migrate twice in one lifetime, whereas others complete one round trip over multiple generations. In either case, the route and timing are quite predictable and lead to mass movements of butterflies to overwinter (i.e., spend the winter) in either central Mexico or a location in California.

Finally, there is a whole suite of species that engage in group movements to breed but that do not go to predictable locations or return to their original location. This type of movement starts to blur the line between migration and dispersal, but the fact that most individuals in the population undertake the trip together helps to push these movements into the migration category. Many insect species migrate this way in response to reduced fitness in years when massive numbers of larvae emerge simultaneously. As an example, when

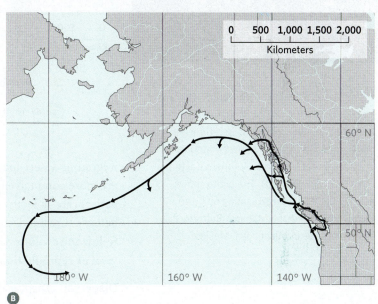

Figure 12.5 Ⓐ Sockeye salmon (*Oncorhynchus nerka*) on a spawning run in the Adams River in British Columbia, Canada. Ⓑ The generalized Pacific Ocean migration route taken by outgoing sockeye from the northwestern United States and western Canada (Beacham et al. 2014).

Monarch butterfly: fall and spring migration

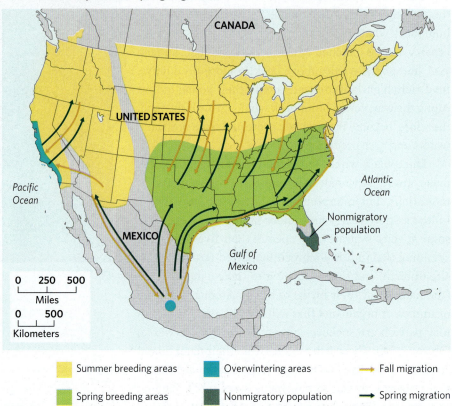

| Summer breeding areas | Overwintering areas | → Fall migration |
| Spring breeding areas | Nonmigratory population | → Spring migration |

Figure 12.6 Monarch butterflies (*Danaus plexippus*) follow a variety of migration routes from their breeding locations in central and northern North America to overwintering areas in central Mexico or Southern California (Monarch Watch 2010). The photo shows monarch butterflies aggregating on Cerro Pelón mountain in the Monarch Butterfly Biosphere Reserve in Michoacán, Mexico.

migrating desert locusts (*Schistocerca gregaria*; **Figure 12.7**) find locations that are more conducive to breeding, they stay for a few generations until conditions deteriorate as locust population density increases. When individual fitness dips below a threshold, the mass movement happens again. These movements clearly follow the evolutionary "rule" that large-scale movements serve to increase survival or reproduction (Holland et al. 2006).

Dispersal   Dispersal is movement by individuals and generally does not involve predictable routes or a return to a starting point (see Figure 12.2). Such movement is always driven by fitness gains, but the improvements may not be so predictable that they force mass movement at regular times or to certain locations. In addition, any increase in fitness may be very specific to individuals rather than to the entire species, population, or age class. There are many reasons for individual dispersal, but the most common drivers include reducing the probability of genetic inbreeding, escaping high-density locales, and going in search of scarce resources. The propensity to disperse can vary a great deal among individuals of the same species. In some species, one gender is much more likely to disperse than the other, and scientists are finding increasing evidence that an individual's genetic makeup can affect its tendency to disperse larger distances or more frequently than others in the population.

Ecologists have defined several types of dispersal, but one key differentiating factor among them is whether an individual moves from its place of birth, called **natal dispersal**, or from one breeding location to another, called **breeding dispersal**. Nearly every individual organism, no matter the species, undergoes some amount of natal dispersal, even if it is very slight. The human example from the beginning of the chapter and the newt-dispersal example from Figure 12.2 are both cases of natal dispersal because they involve moving away from the place (and perhaps even the community) of birth to find an adult settling location. The newts also migrate in the wet season to breed, but the initial movement from their natal pond is natal dispersal.

Natal dispersal can happen at different stages in the life cycle for different species. It occurs in trees when seeds are blown away from the mother tree, in oysters when larvae are carried by ocean currents away from their spawning reef, and in lions when parents impose physical or social pressure on their youngsters to leave the pride. The ubiquity with which natal dispersal occurs indicates a strong need for young individuals to find locations that maximize their own survival and reproduction and remove them from competition with their parents.

Breeding dispersal, on the other hand, is found only in animals that can move under their own power, and it represents movements that adult individuals make to improve their reproductive or breeding success. Birds exhibit many instances of breeding dispersal. For example, Jeffrey Hoover (2003) tracked the breeding success and movements of hundreds of prothonotary warblers (*Protonotaria citrea*) living in the swamp forests of southern Illinois (**Figure 12.8**). Prothonotary warblers migrate to the Caribbean and northern South America during the winter of each year and return to southeastern North America to breed each spring and summer. Their breeding dispersal takes place once they reach the breeding grounds and involves individual movements to find a specific nesting-territory location.

During Hoover's herculean field research, he captured and placed colored bands on the legs of mating warblers so he could document which birds returned to the same nesting location every year. He also recorded the breeding success of these tracked individuals. Hoover's hard work provided clear evidence that the warblers were less likely to return to the previous year's nesting territory if their nest in the previous season failed (Figure 12.8). Warblers that chose to look for new nesting territories were choosing breeding dispersal over staying in a previous site.

Prothonotary warblers can produce multiple broods (separate clutches of eggs) per breeding season, which means Hoover had a large range of potential reproductive output over which he could quantify nesting success for a pair of warblers. He found that nearly 100% of males and females that successfully reared two broods in one season returned to the same nesting site (but not necessarily the same mate) the next year; that is, they did not disperse. The return percentage was still fairly high if the individual managed to successfully rear only one brood. However, an individual (male or female) that did not have a single successful nest in a particular year was much less

Figure 12.7 Desert locusts (*Schistocerca gregaria*) move in massive numbers across North Africa.

Figure 12.8 Jeffrey Hoover tracked territory fidelity (percentage returning to the previous year's nesting site) relative to the prior year's breeding success (i.e., number of successful broods) in prothonotary warblers. Note the high return rate to the same nesting location if the previous year's broods were successful, and the low return rate, and thus high breeding dispersal rate, after experiencing no successful broods the previous year (Hoover 2003). Inset: A prothonotary warbler (*Protonotaria citrea*) perches on a tree.

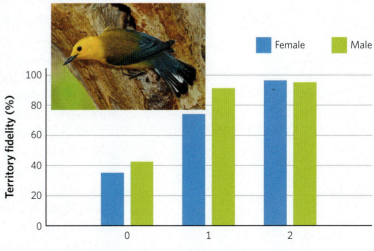

likely to return to the site the next year and presumably dispersed in search of a new nesting location, where it might have more success the next time.

Effects of Migration versus Dispersal on Geographic Ranges   The differences between migration and dispersal cause these two types of movement to have disparate effects on both population dynamics and **geographic range**, the spatial distribution over which individuals of a species can be found. In general, migration leads to widespread effects on entire populations, with fitness improvements (or disadvantages) throughout the population. Dispersal, on the other hand, leads to very individualized effects that alter the success of particular genotypes and phenotypes.

Both migrating and dispersing individuals can alter the geographic range of the species as well as the within-range distribution of individuals. Yet the small changes inside a species' distribution are more likely driven by dispersal than migration. Let's take the breeding dispersal in prothonotary warblers as an example. Birds that disperse to find more successful nesting spots within the breeding grounds may stretch the range of locations to which annual migrations occur. Over time, this breeding dispersal may shift annual breeding migration toward more northern areas, but that shift will come in small increments from individual choices in breeding dispersal rather than a large-scale shift in the migration patterns of the whole species.

This example illustrates the importance of three of the questions mentioned at the start of this section: Why do organisms move? Where and when do they move? How do environmental factors affect that movement? For migration, the answers are tied to the fitness differences that migrating organisms experience between the locations connected by migration. Understanding the evolutionary migratory drive for one individual may allow us to answer these questions for most individuals in the species. For dispersal, though, the answers are going to vary much more among individuals. Although evolutionary drivers may be similar for all individuals in a species or population, those drivers will lead to different dispersal locations and distances, depending on individual genotypes and phenotypes.

Spread and Distribution: The Effects of Movement Distance

Variation in dispersal distances can have big implications for how populations grow and spread out across space. Short-distance movements can reinforce local dynamics and contribute to the population and species interactions that we discussed in Chapters 6–9. The few individuals that disperse over really long distances and then survive to adulthood are the ones that make the biggest difference in terms of the geographic range and distribution of a species.

We know from a variety of dispersal studies that these long-distance dispersers initiate new populations by essentially colonizing new habitats. These nascent populations may grow and eventually produce their own dispersing juveniles, which then disperse and spread the species even further. The importance of long-distance dispersal to species distributions has placed a premium on documenting and modeling the consequences of rare, long-distance dispersal events. Long-distance movements can precipitate encounters with new environments, resources, prey, predators, competitors, and mutualists. A new species in an environment can even alter the way the whole ecosystem works, as with invasive species such as beavers in Tierra del Fuego (Chapter 6) and with the examples of invasional meltdown discussed in Chapter 9.

Characterizing Dispersal Distances: Dispersal Kernels Given all these effects, a key ecological question is, How far do individuals move? In fact, we not only want to know how far they move, but we also want to be able to characterize the distribution of dispersal distances for different individuals. In order to understand why, let's look at the density of seeds found at different distances from a single "parent" tree of heaven (*Ailanthus altissima*) along an urban corridor in Berlin, Germany (**Figure 12.9**; Kowarik and von der Lippe 2011). In Figure 12.9, we can see a range of seed dispersal distances. With the exception of the unexpectedly high number of seeds found at 40 m (131 ft.), the density of dispersed seeds seems to increase as we move away from the mother tree until we hit about 140 m (459 ft.), at which point it decreases and continues to decrease until it peters out around 450 m (1,476 ft.) from the mother tree. If natal dispersal is driven by a need to move offspring away from competition for resources with a parent, then this graph suggests that it works pretty well for this tree species.

At the same time, we can see that no single distance would adequately describe where we expect to find members of the next generation of tree-of-heaven individuals within Berlin. Even more important, the average, or mean, distance traveled by a seed would not tell us very much at all about how far trees of heaven might spread in a generation. Although most seeds traveled 100–200 m (328–656 ft.) from the mother tree, some seeds blew almost half a kilometer. It is these longest-distance dispersers that are most likely to drive the spread of the species to new locations. Ecologists in Berlin are interested in such spread because tree of heaven is considered a noxious, invasive species in Germany.

It is also worth examining the anomalous concentration of tree of heaven seeds at 40 m. This large spike was caused by a bus shelter located 40 m downwind of the mother tree, which served as an effective windbreak and, therefore, as a barrier to seed dispersal. The presence of the bus shelter highlights the important role that local environmental variations and serendipity can play in dispersal and movement.

We often call a graph that shows a distribution of dispersal distances, like that in Figure 12.9, a **dispersal kernel**. A dispersal kernel, in general, tends to peak some distance from the parent and have a long tail representing a very small number of very-long-distance dispersal events. Distribution graphs for wind-borne or waterborne seeds often display the approximate shape of the graph in Figure 12.9, with an increase toward a peak some distance from the parent and then a decrease, plus anomalous high or low concentrations caused by environmental features. Not surprisingly, capturing enough data to get a realistic picture of a dispersal kernel is difficult, partially because densities can get low in the long tail of the distribution. Marine invertebrates, for example, sometimes have an even more stretched dispersal

Figure 12.9 The tree of heaven (*Ailanthus altissima*, inset) grows well in urban settings. The graph shows the pattern of seed dispersal produced by wind from a single tree of heaven in Berlin, Germany. The concentration of seeds at 40 m (131 ft.) from the mother tree comes from seeds that were stopped by a bus shelter. Although the average dispersal distance is likely around 100 m (328 ft.), some seeds traveled much farther (Kowarik and von der Lippe 2011).

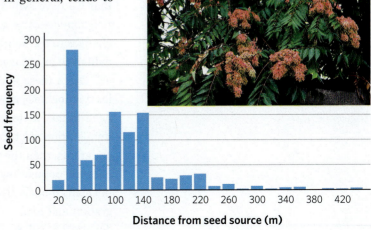

kernel than the one displayed in Figure 12.9, because marine currents can move organisms extremely long distances in one direction. Marine dispersal kernels also tend to strongly reflect the environment because marine invertebrates can only settle in certain locations as dictated by their physical shape and highly localized water-movement patterns.

12.3 Population Growth Model with Dispersal

As soon as we recognize that individuals move around a landscape either after they are born or between breeding bouts, we have to consider ecological linkages across these locations in space. In Chapter 6, we simplified our population models by ignoring the influence of immigration and emigration on population abundance and dynamics. Now that we understand a bit more about how individuals move around a landscape, we can return to population models and make them more realistic and perhaps more useful by adding in immigration and emigration. If we do this, our simplest discrete-time population growth model (Equation 6.3) becomes

$$N_{t+1} = N_t + B_t - D_t + I_t - E_t \qquad (12.1)$$

where t represents time, and $t + 1$ represents the next time step; N is population abundance; B is total births; D is total deaths; I is total number of immigrants entering the population; and E is total number of emigrants leaving the population.

In Chapter 6, we quickly left this equation behind to explore a model in which we could predict total births (B_t) and total deaths (D_t) based solely on per capita birth and death rates multiplied by the population abundance. In Equation 12.1, a similar approach might work for outgoing dispersers (E_t) if emigration is entirely dependent on population abundance. Yet the preceding discussion suggests that dispersal away from a site is often dependent on features of the local environment or the fitness benefits of other nearby environments. Also, it should be clear that it is not possible to predict the number of individuals that enter a population (I_t) by multiplying the current population density by a per capita rate because immigrants come from *outside* the population being modeled. Notice that we have switched to using the term *population density* instead of *population abundance* as we did in previous chapters when we focused on the way that changes to a growing population affected fitness by affecting the density of individuals in one place.

Given these issues, instead of trying to build a model of population dynamics that includes movement from the basic population growth model, perhaps we need to take a step back and think more broadly about the emigration (E_t) and immigration (I_t) terms. First, if the number of immigrants equals the number of emigrants, then the two terms cancel each other out and should have little to no effect on population dynamics. In fact, if we consider the more realistic (and complicated) models from Chapter 6, we may decide that low levels of both immigration and emigration are unlikely to make a difference unless the population density is small. In small populations, losing individuals to emigration could head the population toward extinction, and gaining individuals from immigration could mean that the population will approach the carrying capacity faster. For populations closer to the carrying capacity, dispersal in or out may not be noticeable if the population density stays around the carrying capacity. Lastly, and perhaps most importantly, we can recognize that immigration and emigration are only possible if individuals move between populations spread across a landscape.

12.4 Dynamics and Movement between Populations: Metapopulations

Once we start looking at dispersal and population dynamics this way, we realize that the interesting question is not how a single population changes in abundance or density when we include spatial elements, but rather how dispersal can link a network of populations across a landscape (**Figure 12.10**). A **metapopulation** is a set of populations of a single species spread across a landscape and linked by dispersal. To distinguish these investigations from ones focused on the dynamics of a single population, we call each of the populations in the metapopulation a **subpopulation** to make it clear that we are focusing on the dynamics of the larger metapopulation. Finally, the subpopulation occupies a particular chunk of suitable habitat called a **habitat patch**.

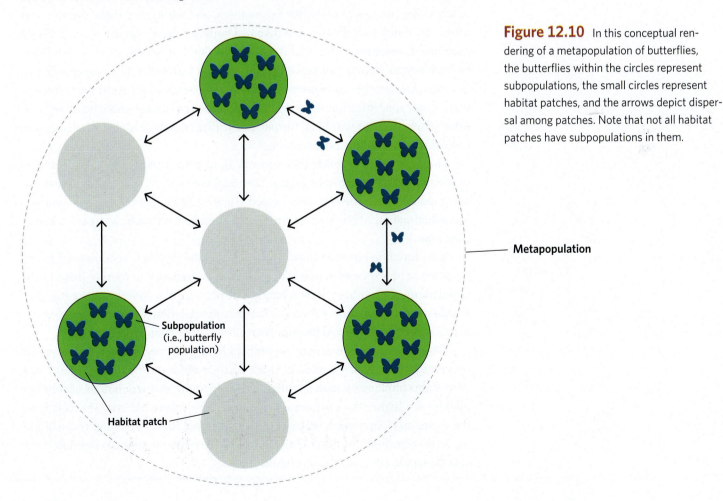

Figure 12.10 In this conceptual rendering of a metapopulation of butterflies, the butterflies within the circles represent subpopulations, the small circles represent habitat patches, and the arrows depict dispersal among patches. Note that not all habitat patches have subpopulations in them.

Metapopulation

Subpopulation (i.e., butterfly population)

Habitat patch

To explore the dynamics of a metapopulation, we need to link subpopulations and habitat patches together by dispersal. That means individuals of one subpopulation leave their natal patch and move to another habitat patch. If that patch already has a subpopulation on it, then the added individuals may have little effect, as previously mentioned. If the new patch is unoccupied, though, then the dispersing individuals may survive and breed and create a new subpopulation in that habitat patch (Figure 12.10).

If the individuals dispersing into a habitat patch make very little difference to the dynamics of the subpopulation already occupying that patch, we can ignore the dynamics within each subpopulation and treat each patch as though it has only two possible states: no subpopulation or a subpopulation that is at carrying capacity. This simplification implies that successful dispersal between patches happens so infrequently that times between bouts of immigration are long enough for subpopulations to either grow to their carrying capacity or decline to extinction. This simplification allows us to explore the dynamics of this metapopulation without becoming overwhelmed by the task. It is somewhat the opposite of the simplification in the growth model from Chapter 6, because instead of only considering births and deaths, while leaving out immigration and emigration, in this case we do the opposite; we leave out births and deaths and *only* consider immigration and emigration.

When focusing on metapopulation dynamics, we need to make an additional simplification. In a real landscape, an individual patch will be closer to some neighboring patches and farther from others, and we expect these distances to affect the ability of individuals to disperse between these patches. As previously mentioned, sometimes dispersal is assisted by wind or water currents, and these environmental factors can make distant patches just as likely to receive immigrants as near patches. In order to focus on the essential elements of the metapopulation, our simplification supposes that immigrants from any patch can reach any other patch. In other words, all the patches will be equally accessible to dispersing individuals from all the other patches.

It may help to visualize this network if, in your mind, you take the configuration of habitat patches in Figure 12.10 and mentally "light up" some that are occupied and keep the rest unoccupied (**Figure 12.11**). You can further imagine individual patches "blinking on and off," depending on their occupation status over time.

This simplified picture allows us to focus on the factors that contribute to the persistence of a species across a region within a network of loosely linked subpopulations. Each subpopulation may go extinct over time, but the entire network should persist much longer than any single subpopulation because extinct habitat patches can be recolonized by immigrants from other subpopulations.

In order to assess whether a metapopulation will persist on a landscape or eventually go extinct, we need to think further about our "blinking lights." Are more lights blinking off than are blinking on? If so, the extinction rates of subpopulations are higher than the colonization rates for empty patches. If this is true, the entire metapopulation will eventually go extinct. But if more lights are blinking on than are blinking off, colonization rates are higher than the extinction rates, and the whole metapopulation should persist.

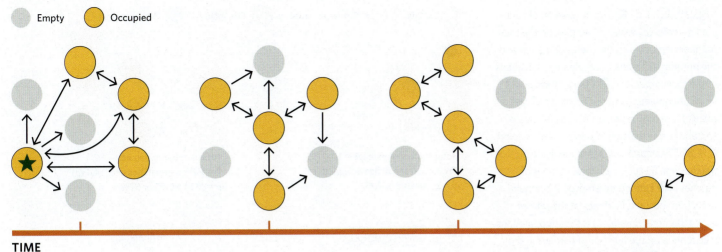

Legend: ⬤ Empty (gray) ⬤ Occupied (yellow)

TIME

Figure 12.11 This conceptual figure of a metapopulation represents variation in the occupancy of different habitat patches through time. Filled circles represent occupied habitat patches. Unfilled circles represent unoccupied patches. Arrows represent dispersal. We represent all possible dispersal arrows for only the left-most occupied patch in the first time step (labeled with a star). Dispersers can move to any patch from any occupied patch. Dispersal to an unoccupied patch can lead to a newly occupied patch in the next time step. Random extinction leads to the loss of the subpopulation on occupied patches. Note that, at different times, various patches are occupied or unoccupied, thereby producing a pattern of individual lights that are randomly "blinking" over time.

It is also important to question whether the lights are all blinking on together and off together, or if they are more random, with individual lights blinking on or off completely out of synchronization with other lights. A completely synchronized system would not really be a metapopulation because the subpopulations would be so tightly linked by dispersal that all of them would really be acting as one big population. Representing this conceptual model with equations may allow us to gain additional insights.

The Levins Metapopulation Model

In 1969, Richard Levins, a mathematical ecologist with an agriculture degree and experience as a farmer, devised a model to describe the population dynamics of insects across different agricultural fields (Levins 1969). His goal was to provide practical insight into how to manage pest insects, presumably so crop production could remain high. Eradicating pests in one field is not useful if the field can be recolonized by individuals dispersing from adjacent untreated fields. Although Levins was originally interested in eradicating the entire network of insect subpopulations, he later generalized his model to explore metapopulation persistence (Levins 1970). His model was explored and expanded by later ecologists, spawning the whole subfield of metapopulation biology. The generalized version of his original model is now commonly called the **Levins metapopulation model**, and it is a relatively simple place to start thinking about metapopulations and how changes in dispersal can influence our perception of ecological processes.

The Levins metapopulation model keeps track of the proportion of patches with subpopulations (i.e., occupied patches) (p) and how that proportion changes over time (t). The proportion of occupied patches (p) ranges from 0 to 1 (**Figure 12.12A**), where $p = 1$ means that all the patches are occupied, and $p = 0$ means that the metapopulation is extinct because there are no occupied patches. The Levins model assumes that colonization by dispersing individuals leads directly to a filled patch—what we call a subpopulation. The easiest way to envision colonization (c) is to imagine that each occupied patch sends out colonists that successfully reach

Figure 12.12 **(A)** The proportion of occupied patches (p) varies depending on the total number of patches in the metapopulation and the number of patches occupied by individuals of the focal organism. **(B)** Colonization adds new subpopulations based on the colonization rate (c), the proportion of patches occupied (p), and the proportion of patches unoccupied ($1 - p$). The actual number of newly occupied patches will be $cp(1 - p)$(total number of patches). **(C)** Extinction removes subpopulations from occupied patches at the rate ep. Again, the total number of lost subpopulations will depend on the number of patches in the population. **(D)** At equilibrium, we will see the same number of occupied patches from one time step to the next, but the actual patches that are occupied may have changed.

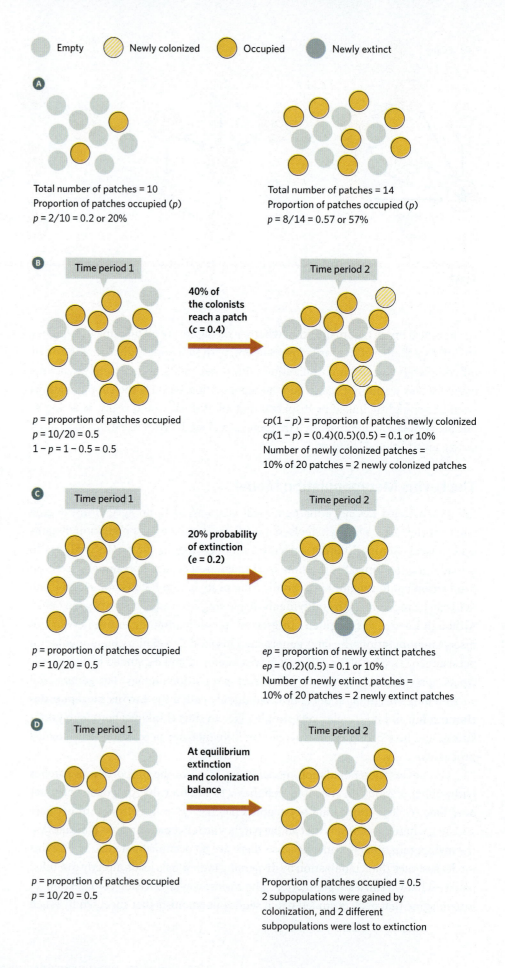

Empty   Newly colonized   Occupied   Newly extinct

(A)

Total number of patches = 10
Proportion of patches occupied (p)
p = 2/10 = 0.2 or 20%

Total number of patches = 14
Proportion of patches occupied (p)
p = 8/14 = 0.57 or 57%

(B)

Time period 1

40% of the colonists reach a patch ($c = 0.4$)

Time period 2

p = proportion of patches occupied
p = 10/20 = 0.5
$1 - p = 1 - 0.5 = 0.5$

$cp(1 - p)$ = proportion of patches newly colonized
$cp(1 - p)$ = (0.4)(0.5)(0.5) = 0.1 or 10%
Number of newly colonized patches =
10% of 20 patches = 2 newly colonized patches

(C)

Time period 1

20% probability of extinction ($e = 0.2$)

Time period 2

p = proportion of patches occupied
p = 10/20 = 0.5

ep = proportion of newly extinct patches
ep = (0.2)(0.5) = 0.1 or 10%
Number of newly extinct patches =
10% of 20 patches = 2 newly extinct patches

(D)

Time period 1

At equilibrium extinction and colonization balance

Time period 2

p = proportion of patches occupied
p = 10/20 = 0.5

Proportion of patches occupied = 0.5
2 subpopulations were gained by colonization, and 2 different subpopulations were lost to extinction

a constant fraction of the patches in the metapopulation. If p is the proportion of the patches that are full, then $1 - p$ must be the proportion of patches that are empty. If all patches are equally accessible, then the colonists bump into empty patches in proportion to the density of these patches on the landscape $(1 - p)$. When colonists encounter occupied patches, they have no effect because those patches are already filled (**Figure 12.12B**).

We also assume that an occupied patch can become unoccupied, meaning the subpopulation occupying that patch goes extinct from one time period to the next and that this will happen at a constant probability, denoted as e (for extinction). This constant probability of extinction for a single subpopulation is mathematically the same as saying that a constant fraction of subpopulations (occupied patches) goes extinct in a time step (**Figure 12.12C**). Because c and e can both be proportions, they also range in value from 0 to 1. A high value for colonization (c) indicates a network with lots of dispersing individuals and thus a high probability of patch colonization. A high value for extinction (e) indicates that subpopulations will have a high tendency to go extinct.

We have one final bit of logical housekeeping to note. We need to assume that each subpopulation goes extinct independently of how many other subpopulations there are (i.e., how many other patches are occupied). In other words, a subpopulation will go extinct solely on the basis of its own internal dynamics. This caveat is built into our statement that e is constant. On the other hand, if colonists come from occupied patches, then the probability that a patch becomes colonized *depends on* how many other patches are occupied. The more occupied patches that exist within the network, the more dispersing individuals there are on the landscape to bump into unoccupied patches (Figure 12.12B). We put all this together in Equation 12.2.

$$\frac{dp}{dt} = cp(1 - p) - ep \tag{12.2}$$

From Chapter 6 you may recognize that this model represents continuous-time dynamics. Here we are tracking the change in the proportion of occupied patches (dp) over tiny slices of time (dt).

Patches "blink on" according to the first term on the right-hand side of the equation, $cp(1 - p)$. This term says that p occupied patches send colonists out to c other patches. These colonists create new subpopulations only if they bump into unoccupied patches, $(1 - p)$. The term $cp(1 - p)$ is an example of the law of mass action, like predators and prey bumping into each other based on their densities in an area (Figure 12.12B). In order to get any newly occupied patches, all the factors in this term must be nonzero. Ecologically this means that, if there are no patches occupied now $(p = 0)$, then no colonization can happen in the future because there are no occupied patches to produce dispersers in the system. If all the patches are occupied now $(1 - p = 0)$, then again none can be colonized in the next time step because there is nothing left to "blink on." Finally, if patches send out no colonists $(c$ is zero), then no patch will be colonized because there is no chance of colonization.

We represent patches "blinking off" between time steps in this equation using the second term on the right-hand side of the equation, ep (Figure 12.12C). To make sense of this term, it is useful to consider what it means when either of these variables equals zero. No subpopulation can go extinct if all patches are already empty ($p = 0$), and no subpopulations will go extinct if the probability of extinction is zero ($e = 0$).

We keep track of the change in the proportion of occupied patches through time (dp/dt) by subtracting the proportion that went extinct (blinked off) from those that were colonized (blinked on). We can solve for the equilibrium solution to this equation by setting dp/dt equal to zero, meaning that the occupied fraction of patches is not changing. Once we do this, we can rearrange Equation 12.2 and solve for the equilibrium fraction of occupied patches (p^*), which is $1 - \frac{e}{c}$. We leave the algebra to you in the Quantitative Questions at the end of the chapter. If you are stumped, look at Chapter 6 to see how we solved for similar equilibria in single population growth models.

Let's review what all this means and see how it fits the conceptual model that we came up with in the previous section. First, the equilibrium fraction of occupied patches will only be greater than zero if $c > e$ (i.e., colonization is greater than extinction), which fits our previous predictions. If the equilibrium fraction of patches is zero or less than zero, we will not have a surviving metapopulation. This outcome may remind you of the discussion of per capita birth and death rates in population growth from Chapter 6. Just as populations can go extinct if deaths exceed births, entire metapopulations can go extinct if patch extinction probability is higher than patch colonization ($e > c$).

Even more interesting is that we also have a stable proportion of occupied patches, sort of like a stable population density. We did not predict this outcome in the conceptual model, but it makes sense. In Chapter 6, we noticed that we could have both deaths and births in a given time step without changing the population density if the number of births is made up for by the number of deaths. The same is true here. If colonizations balance extinctions, then subpopulations can go extinct without changing the proportion of occupied patches (**Figure 12.12D**). Even with $c > e$, we will not head to a completely occupied network because the subpopulations on some fraction of the occupied patches will always go extinct, and the new colonizations can only make up for those extinctions to the level set by $1 - \frac{e}{c}$.

It is important to notice that even when an equilibrium point is reached, no one set of particular patches always stays occupied. Instead, a random but constant proportion of patches is occupied through time (Figure 12.12D). Our blinking network is still blinking away at equilibrium, but from one moment to the next, the *proportion* of occupied patches will be the same. This means that, even though subpopulations on individual patches go extinct over time, the entire metapopulation occupancy rate stays constant, and the metapopulation sticks around.

Just like the population growth models presented in Chapter 6, this metapopulation model should be used with caution, as it has some fairly unrealistic assumptions built into it. Nevertheless, it provides interesting insights that have clear implications for real-world ecological scenarios. In particular, this model helps ecologists consider the types of factors that influence population persistence across a landscape and the number of occupied habitat patches. As expected from the conceptual model, reducing the colonization rate can have dire consequences whether the reduction occurs in the colonization term or in the number of occupied patches in the network. If the colonization probability drops below the extinction probability for each subpopulation on an occupied habitat patch, the entire metapopulation will go extinct.

On the other hand, if the colonization rate is so high that each patch sends dispersers to every other patch, then we have a completely linked system. In this case, the patches will no longer blink on and off independently. Instead, the lights will all go on and stay on together, and they may also all go off together.

A less intuitive outcome is that, if unoccupied patches are somehow removed from the metapopulation, then the equilibrium number of occupied patches will decrease. This unexpected result comes from the fact that the equilibrium is based on the proportion of the entire metapopulation network that is occupied. If, for example, $c/e = 0.5$, then we expect half the patches to be occupied at equilibrium. If the total number of patches is reduced, then the number of patches occupied at equilibrium will also be reduced.

Let's see how such loss works by considering the impact of anthropogenic land development on a metapopulation. Suppose a land developer in the midwestern United States wants to build houses in a county with a mixture of forested areas and grasslands but must avoid building in habitats suitable for the Karner blue butterfly (*Lycaeides melissa samuelensis*; **Figure 12.13**). The Karner blue is an endangered butterfly found in grasslands that contain perennial lupine (*Lupinus perennis*), an essential plant for its caterpillar's growth. Anthropogenic habitat loss and changes to the natural pattern and frequency of fires (fire is essential for grassland maintenance in many areas) have contributed to the Karner blue's decline.

To avoid negatively affecting the endangered Karner blue, the developer pays an environmental consulting company to survey grassland habitats for the butterfly and its caterpillars during a single spring season when caterpillars are present. They find that 10 out of 20 grassland habitat patches across the county were unoccupied at the time of the survey. From our understanding of metapopulations, we may conclude that $c/e = 0.5$ and that empty patches might be occupied in the following year. The developer, however, may interpret the presence of the empty patches very differently. These empty locations may be considered unsuitable as Karner blue habitat, and the developer may, therefore, argue that they can build on those sites without harming the butterfly's regional abundance. If this logic allows the developer to move forward, then those 10 unoccupied patches would shift from butterfly habitat to houses for humans.

Unfortunately, this would have a negative effect on the butterflies because developing the unoccupied patches would reduce the metapopulation network to just 10 of the original 20 *suitable* habitat patches. Over time, the newly reduced metapopulation would equilibrate so that half the patches are occupied (i.e., $c/e = 0.5$). This occupancy level would now translate to only five occupied patches in any given year instead of 10 if the original 20 habitat patches had been protected. Clearly, this population reduction would be a major loss for the Karner blue.

Real-World Implications of the Levins Model

What are the real-world implications of this model? The first and most obvious is that the Levins model demonstrates that regional persistence of a metapopulation can happen even in the face of frequent individual subpopulation extinction, as long as the patches are linked by dispersal. Second, the importance of dispersal

Figure 12.13 Some butterflies, like this Karner blue (*Lycaeides melissa samuelensis*), require specific habitat types and even specific plants for larval caterpillar development. Such habitats and plants are often patchily distributed across a landscape in such a way that each patch can support a subpopulation in a butterfly metapopulation.

to metapopulation persistence can completely change how we environmentally manage areas, as was the case with the Karner blue butterfly. Rather than managing individual separate populations or subpopulations of a rare species, managers can focus on regional persistence of the species by ensuring that individuals can disperse between subpopulations. This insight has become a critical tenet of conservation science and has led to the implementation of such features as wildlife corridors, amphibian tunnels, and connecting habitat strips between parcels of protected forest.

Nonetheless, we know that the real world is in a constant state of change and that sometimes colonization rates may decrease in real metapopulations. For example, we have evidence that dispersal distances can vary according to the presence or absence of a dispersal mutualist. What would happen to a metapopulation if the dispersal mutualist were lost from the ecosystem?

One interesting example comes from Amazonian rainforests, where several tree species rely on seed dispersal by a fruit-eating fish, the tambaqui (*Colossoma macropomum*; **Figure 12.14**). These fish swim among the submerged trees during annual flood events and routinely deposit fruit seeds (in their feces) 300–500 m (984–1,640 ft.) from their source, with long-distance dispersal events reaching more than 5 km (3.1 mi.)! This frugivorous fish is better at long-distance dispersal than most birds, monkeys, or other seed mutualists in the Amazonian rainforest. Unfortunately, the tambaqui is also commonly captured by local fisheries and is currently overfished in many locations across its geographic range (Anderson et al. 2011).

This overfishing pressure reduces seed dispersal and colonization and could negatively affect the metapopulation structure of several tree and vine species. Because the Amazon rainforest is so diverse, individuals of particular tree species tend to be spaced far apart, with potentially only one or two individuals in a single hectare. If dispersal and colonization rates decline to below the rate at which we lose individuals in each hectare, entire species of trees could disappear from the Amazon. In other words, human fishing could cause the extinction of a tree species from the Amazonian floodplain forest without any loss of habitat. This example illustrates how anthropogenically caused extinctions can occur for reasons that are much more subtle than one would initially imagine.

We commented earlier on the importance of empty patches in the Levins metapopulation model, and we want to emphasize that the model implies that empty patches should be a real-world phenomenon. If they do not exist, then the model really does not offer much insight into real-world metapopulation dynamics. As with the Karner blue butterfly example, empty patches are unoccupied not because they are "bad" habitat for the species, but because they have not yet been colonized by dispersing individuals. If individuals of the species colonized such patches, they would have sufficient reproduction and survival rates to produce a growing subpopulation there. If this is a common feature of natural systems, then ecologists have to be very careful about defining "good" and "bad" habitat solely based on where they find individuals. The subset of current locations for a species is unlikely to represent the full set of locations where that species could happily survive. This possibility makes a strong argument for protecting more habitat than is currently occupied by a species.

Figure 12.14 The tambaqui (*Colossoma macropomum*) is a fruit-eating fish found in the Amazon rainforests. During seasonal floods, this species can disperse seeds from Amazonian trees by eating their fruit and transporting the seeds in its gut (and potentially in its feces) up to 5 km (3.1 mi.) away.

Interestingly, this exact argument has come under fire in several high-profile conservation legal cases in which ecologists argued against habitat destruction in order to protect rare species. One of the most notable cases in the 1980s and 1990s involved the spotted owl (*Strix occidentalis*) in the western United States (**Figure 12.15**). The spotted owl successfully nests only among widely spaced, large trees that are found in old-growth forests. These same forests are of great value to the logging industry. Young owls disperse and establish mating territories in the old-growth forest but will use second-growth forest if no other habitat is available. Lamentably, the young, inexperienced owls generally do not succeed at breeding in these less desired locations because the surrounding forest generally has low densities of their prey species.

Conservation biologists argued in the associated legal cases that they could apply metapopulation models to the owls and treat each mating pair as a subpopulation. Therefore, potential old-growth mating sites that did not house a mating pair were in fact not poor habitat that could be logged. Instead, empty mating sites represented a suitable habitat patch whose mating pair had aged and died but where a new pair might happily reproduce in the future. Ecologists contended that maintaining the unoccupied nesting trees was essential for protecting the hunting and breeding territories for juvenile owls. Although the metapopulation concept helped in listing the owl in 1990 under the US Endangered Species Act, spotted-owl populations continue to be a subject of disagreement between conservation managers and loggers, cattle grazers, and developers, whose activities affect forest habitat for the owls.

A final real-world insight from the Levins metapopulation model is that it may be possible for colonization rates (*c*) to be so high that the extinction probability of the whole metapopulation network rises instead of falls. Typically, in a metapopulation, when the colonization rate is above the patch extinction rate but low enough that individuals rarely move between patches, the entire metapopulation has a high probability of regional persistence. Under these conditions, each subpopulation experiences different local risks, which means it is very unlikely that the entire network will experience a single event that leads to simultaneous extinction. Instead, risk is spread among multiple subpopulations that operate almost completely independently. That independence is important for maintaining the regional network of subpopulations so that at any given time at least one patch is occupied.

However, if colonization rates are very high, then the subpopulations are closely linked and mostly *not* independent. Although the Levins metapopulation model assumes a constant extinction probability, it also assumes that extinction in each subpopulation is random and not linked to simultaneous extinction in another subpopulation. When subpopulations are very closely linked, the factors that lead to extinction for one subpopulation may lead to extinction for all the subpopulations. As we mentioned before, this would be like all the lights blinking on together, but also blinking *off* together (**Figure 12.16**).

Figure 12.15 This young spotted owl (*Strix occidentalis*) has captured a rodent in a Canadian forest. Spotted owls have been used by ecologists as an example of a metapopulation. For the spotted owl, ecologists considered one breeding pair and their juvenile offspring as a "subpopulation" and defined habitat patches by old-growth forest habitat large enough to support nesting and hunting territory for a breeding pair.

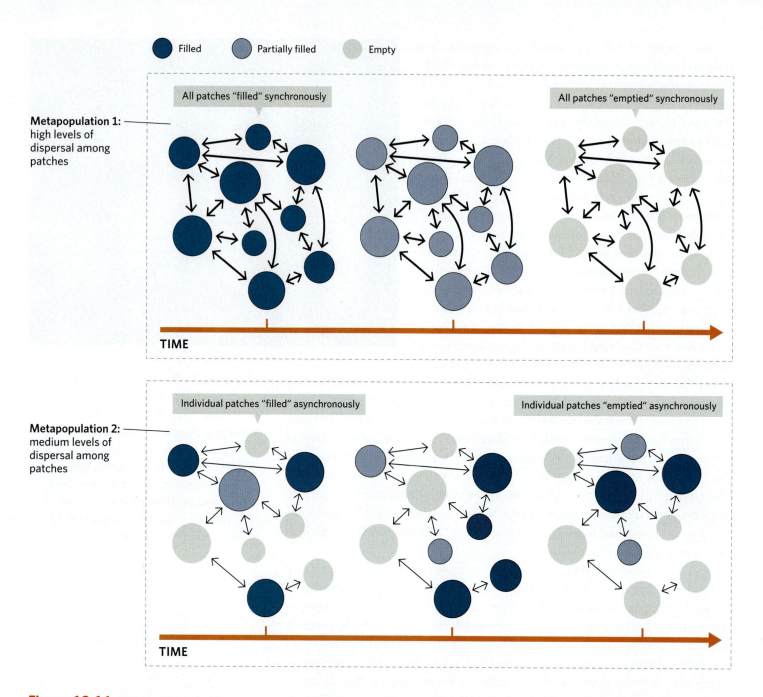

Figure 12.16 Each row shows a metapopulation through time. The top row has higher dispersal (indicated by thicker arrows), leading to subpopulations with strong links and more-synchronous population dynamics. In the lower row, there is less dispersal (thinner arrows) between subpopulations, leading to asynchronous dynamics and a higher probability of persistence for the whole metapopulation. Note the intermediate stage (light blue patches) between patches that have a subpopulation at carrying capacity (dark blue patches) and patches with no subpopulation (gray patches) (Hoopes et al. 2005).

The Levins model stipulates that the extinction rate comes from random, external factors, which could be weather or other more catastrophic disturbances, such as floods or fires. If patches are close enough in space that their subpopulations are completely linked via dispersal, then it may not make sense to assume that the subpopulations experience different weather or different catastrophes. In that case, the whole metapopulation may experience the same cold-weather spell, hurricane-induced flood, or drought-induced fire, making simultaneous extinction possible across all the subpopulations. Another reasonable scenario for too much connection could arise if the focal species has a predator or disease that can decimate whole subpopulations. If the habitat allows easy dispersal for the predator or pathogen, then it is possible to drive all subpopulations extinct quickly relative to colonization rates. For predators that operate at different scales from their prey, this scenario is more common than one might imagine.

12.5 Beyond Metapopulations to More Realistic Spatial Dynamics

We can see that the Levins metapopulation model offers some insights that are useful for understanding how dispersal can link populations and affect regional dynamics for a species. Yet when ecologists try to apply the Levins model to real situations, they sometimes find a mismatch between the model's simplifying assumptions and the real world. In particular, the Levins model assumes that subpopulations regularly go extinct, or "blink off" in the terminology of our visual analogy. This assumption may be fine when considering insect populations across various agricultural fields, as these species do vary tremendously across space and time. However, in most situations, subpopulations do not regularly switch from being present to absent (blink on and off) over short time periods.

This mismatch is partially due to our concept of time. The Levins model takes a very long view of the world. The view is so long, in fact, that population dynamics happen instantaneously in comparison with the temporal rate of dispersal. Ecologists rarely have the luxury of viewing the world that way, and we also may not have data to explore such long-term dynamics. Even more importantly, though, it may not make sense to assume that metapopulation dynamics stay constant over long periods of time. For example, is it reasonable to assume that dispersal will stay the same across a region where habitats and weather patterns are changing due to anthropogenic climate change?

In addition, it may not make sense to assume that the local dynamics of a particular subpopulation do not matter. In previous chapters, almost all the discussions of population dynamics and species interactions point to the importance of local conditions: the presence of a mutualist or predator, the density of resources, and local disturbances. But what happens if some patches are inherently higher quality than others and are more conducive to maintaining high birth rates, low death rates, and a steady supply of dispersing subadults? Also, what if some patches send dispersers more frequently to closer patches than to distant patches? All of these scenarios seem biologically reasonable. In fact, does it seem reasonable to ever assume that all metapopulation patches are equally accessible or of equal quality? Not really.

Does this mean that the whole concept of metapopulations is useless for real-world situations? Probably not. Most ecologists consider the metapopulation concept to be an important component of our current ecological understanding. Although the Levins model is simple, it illustrates the way dispersal can increase regional persistence when it links a network of subpopulations just enough to maintain asynchronous dynamics (Figure 12.16). Once we have that insight, we can tweak the model to add reality as we need it.

Releasing the entire metapopulation concept from a strict interpretation of the Levins model provides better and more nuanced insights for real-world applications. Metapopulation dynamics increasingly apply to how we protect and manage species. Human activities commonly chop up habitats into smaller and smaller remnant chunks. Even if a species did not exist as a metapopulation under natural conditions, it may appear as one if the habitat is fragmented into small

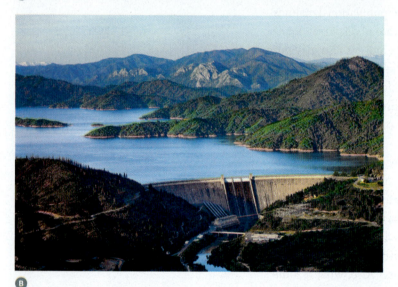

Figure 12.17 Examples of fragmented habitats. Ⓐ Forest fragments dot the agricultural fields in the foothills of the Austrian Alps. The agricultural lands create a matrix of uninhabitable area between forest fragments (patches), and thus can create metapopulations for species that rely on the forest habitat. Ⓑ Dams block linkages among rivers and streams around the world. Here we can see Shasta Dam, which fragmented the Sacramento River in Northern California by creating Shasta Lake and cutting off the movement of aquatic organisms between sections of the river.

blocks scattered across a human-altered landscape. There are so many possible examples of this type of **fragmentation** that it is hard to capture them all here. Examples include the anthropogenic division of forests to allow for agricultural production, or the compartmentalization of aquatic ecosystems after the placement of dams on large and small rivers (**Figure 12.17**).

Applying metapopulation theory to help address these human-caused problems requires us to think carefully about how dispersal and subpopulation extinctions function in real environments. This is an active and growing field of interest in ecology, with findings that are as varied as the systems in which ecologists apply the models. However, there are two elaborations on the Levins model that provide general ecological insights. The first involves including subpopulation dynamics, and the second involves recognizing that not all patches are equally likely to be colonized. We explore these elaborations next.

Quality Matters: Source-Sink Dynamics

First, we consider how our understanding of metapopulations changes if we move beyond simply calling a habitat patch "occupied" or "unoccupied." Instead let's follow the internal dynamics of each subpopulation and consider how this alters our understanding of metapopulations. In our initial explorations of the Levins metapopulation model, we assumed that every subpopulation had a constant and equal extinction probability. As we mentioned, though, this assumption is unlikely to be true in the real world.

For simplicity's sake, let's consider only two populations, one that is growing in abundance (with $\lambda > 1$) and another that is declining (with $\lambda < 1$). Remember from Chapter 6 that λ, or the finite per capita rate of population growth, describes each individual's contribution to the population abundance in the next year. If $\lambda = 1$, the population abundance stays the same. We can link our two populations via dispersal and refer to them as subpopulations of a very simple metapopulation. Because of the population dynamics within each patch, only the subpopulation in which $\lambda > 1$ will reach its carrying capacity and, therefore, produce "excess" individuals that can disperse. Individuals in this growing population may experience intense intraspecific competition for increasingly scarce resources or for breeding territories. One way to ease that competition is for individuals to move in search of unused resources or territories. Because this metapopulation only has two patches, these dispersing individuals can only move from one patch to the other. For simplicity, let's imagine that this dispersal is natal dispersal and that individuals do not move again once they have dispersed.

The other patch, in contrast, sends out no dispersers because its subpopulation never comes close to reaching carrying capacity (i.e., $\lambda < 1$). Such a scenario might arise if this patch lacks breeding habitat or essential nutrients or contains competitors or predators with which the species cannot coexist, thus making it a "low-quality" patch. As discussed in Chapter 5, organisms can be surprisingly good at determining the quality of foraging habitat, but they still may find themselves in locations that provide poor conditions for population growth. This situation could occur if the factors that reduce habitat quality are undetectable to the individuals residing there, as is the case when habitats are altered by humans in ways that do not mimic natural disturbance, such as with toxin inputs. Alternatively, sometimes organisms that do not direct their own dispersal (such as seeds or marine planktonic larvae) may randomly land in poor habitat. Finally, even organisms that are capable of actively choosing their dispersal pathways may have little choice but to accept low-quality territories or habitats if they are juveniles or subdominant adults that cannot behaviorally compete with older, more experienced adults.

If the hypothetical low-quality patch never produces dispersers, then we see unidirectional dispersal from the other (high-quality) patch toward the low-quality patch in these two subpopulations (**Figure 12.18**). In this case, the patch that produces dispersers is called a **source** because it provides all the dispersing individuals that arrive and then settle in the other patch.

Let's now consider the internal dynamics of the low-quality patch. At this location, population death rates exceed birth rates (Chapter 6). Without input and dispersal of individuals from the high-quality patch, this subpopulation would gradually decline and go extinct. However, because a source patch is sending over dispersing individuals, this subpopulation is continually being bolstered. Unfortunately, the newly arriving individuals are consigned to a fate of low reproduction and perhaps early death because they settled in a lousy patch, just like the individuals born in that patch. This poor-quality patch, therefore, absorbs all the individuals that the other subpopulation sends it and is referred to as a **sink**. Although deaths outnumber births in the sink, as long as the sink is continually being refilled with dispersing individuals from the source, its subpopulation will not go extinct. Metapopulations that are driven by the differences in quality between source patches and sink patches show what are called **source-sink dynamics**.

If source-sink dynamics occur in real-world situations, then we should see two patterns. First, some habitats that are continually occupied by a species may not be high-quality habitats (i.e., ones that allow survival and reproduction to balance deaths). Instead, we may have sink subpopulations supported by a continuous trickle of individuals from one or more source habitats (**Figure 12.19**). Once again, we see from considering metapopulation dynamics that the presence or absence of individuals of a species in a particular location does not automatically tell us if a habitat there is "good" or "bad." However, in this case, the insight is the reverse of

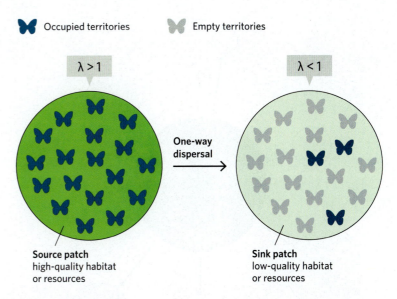

Figure 12.18 In this simple two-patch metapopulation model of source-sink dynamics, the circles represent habitat patches, with the darker green color indicating high-quality habitat. Blue butterfly symbols represent occupied territories within each patch, and gray symbols are empty territories. The low-quality habitat will have empty territories that may be enticing to dispersing subadults that cannot obtain a territory in the high-quality patch. Lambda (λ) is the finite per capita rate of population increase described in Chapter 6.

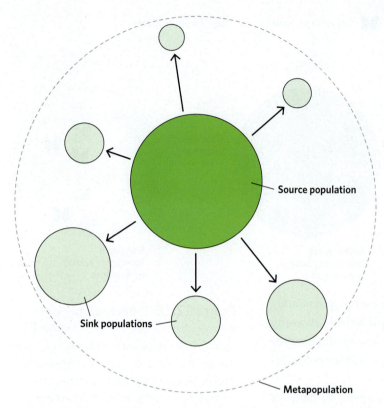

Source population

Sink populations

Metapopulation

Figure 12.19 In this source-sink meta-population representation, the large source population shows positive population growth and thereby produces dispersers (arrows). These individuals move to other locations that ostensibly fit their needs but do not allow population increases. These sink populations only receive individuals and never produce dispersers. Loss of the sink habitat patches will have very little effect on metapopulation dynamics, but loss of the source would lead to loss of the entire metapopulation.

what we found earlier when considering a classic Levins metapopulation model. There the absence of individuals did not necessarily indicate poor-quality habitat, and here the presence of individuals does not necessarily indicate high-quality habitat.

Second, in a metapopulation that includes several sinks (Figure 12.19), the loss of a single patch can have enormous impacts on the persistence of the whole network. Consider the difference between losing a sink patch versus losing a source patch. The sink patch contributes no dispersing individuals to the network. The loss of a sink patch still represents a loss of individuals from the entire metapopulation, but that loss will mean very little to the persistence of the metapopulation or every other subpopulation. In stark contrast, if a habitat patch supports a source subpopulation, then its loss will mean a reduction in overall metapopulation abundance *and* the loss of dispersers. This loss of dispersers has a twofold effect because individuals who disperse from source patches may subsidize sink subpopulations, or they may colonize new source patches. If the lost source patch was the only source in the metapopulation, then the entire metapopulation will go extinct with the loss of one habitat patch. The existence of source-sink metapopulation dynamics, therefore, suggests that even minimal habitat loss can lead to massive reductions in species' regional persistence if the lost habitat acts as a source of dispersers for other patches.

Distance Matters: Patch Size and Connectivity

Now that it is clear that differences in habitat quality can have big effects on regional metapopulation dynamics, let's consider what happens if we include the effects of real-world dispersal distances. Up to this point, distance has been more implied than real in the metapopulation models. Spatial models like the Levins metapopulation model are called *spatially implicit models* because they acknowledge the existence and effects of spatial separations without ever explicitly measuring those separations. In fact, spatially implicit models do not recognize any effect of distance on colonization rates. Separation, whether near or far, is all the same. The fact that the Levins model uses implicit space is clear from the fact that all patches are equally accessible.

As we mentioned earlier, it is rare in the real world for all patches to be equally accessible, regardless of how close they are to each other. Most critically, the number of individuals that move long distances is usually very small (see Figure 12.9). Thus, a habitat patch that is far from all other patches is unlikely to be reached by colonizing dispersers very often. In contrast, a patch that lies close to other patches, and particularly near big patches or high-quality patches, should receive many dispersers.

It is possible to create a *spatially explicit model* by removing the assumption made in the Levins model that all patches are equally likely to be colonized. To do this, the model must acknowledge that spatial separations can differ between patches and that these differences are likely to alter colonization rates at each patch. Of course, the model must also take into account the actual distances between patches and provide a means for figuring out how to include the effects of these distances. There are metapopulation models, called incidence function models, that account for distances explicitly, but they are too complicated for use in this book, and the insights they provide tend to be very specific to each habitat and focal organism.

A Mainland-Island Model A simpler way to explore the effects of distance is to investigate movement from one big mainland to an array of islands that lie at different distances from the mainland (**Figure 12.20**). This **mainland-island model** allows us to examine how distance and patch size influence regional dynamics. In this conceptual model, the large mainland habitat holds many individuals, has a population that is unlikely to go extinct, and acts as a source for dispersing individuals out to the islands. Island populations can go extinct, like subpopulations on patches in the Levins metapopulation model, but they do not produce dispersers. If a subpopulation on an island goes extinct, the island can only be recolonized by dispersal from the mainland. Dispersal to each island is determined by the island's distance from the mainland. Assuming that no dispersers come from islands may seem artificial, but it allows us to focus on the effect of distance from the mainland. Plus, dispersal may be so rare from islands in comparison with the mainland that this assumption allows us to examine the effects of the vast majority of dispersal events.

If the islands that are far from the mainland lose their subpopulations to random extinction, they could be unoccupied for long periods of time. Recolonization of these far islands would require rare long-distance dispersal. This result differs strongly from what we found with the Levins population model, which assumes that all "islands" (patches) are equally likely to be occupied, and to be colonized if unoccupied. The low frequency of colonization of distant islands suggests that the Levins metapopulation model may overestimate the proportion of the metapopulation patches that are occupied at any one time, because the Levins model allows distant patches to be colonized too easily and frequently.

The mainland-island model also highlights the importance of environmental context and barriers in dispersal. It should be fairly obvious that, at certain times, dispersal across a landscape is dependent on the conditions that dispersing individuals experience en route to their destination. In other words, the habitat conditions between the mainland and islands, or between Levins's habitat patches, are very likely to affect dispersal success. Up to this point, we have not discussed the **matrix**, or surrounding environment, in which metapopulation patches are embedded. We know that patches contain habitat that is suitable for

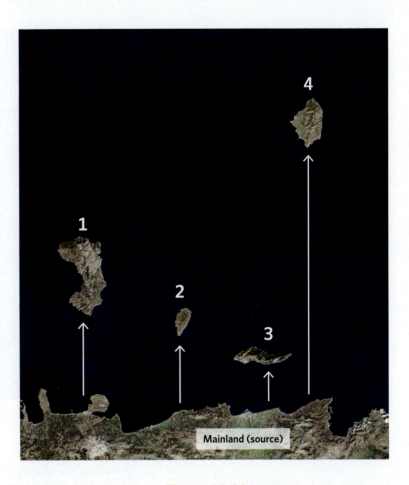

Figure 12.20 This simple mainland-island model includes four oceanic islands of various sizes and distances from the mainland. Arrow lengths indicate dispersal distances necessary for colonization of each island. Note that island 4 is much harder to colonize from the mainland than any of the other islands.

Figure 12.21 For a mountain lion, it is much easier to move long distances through Santa Monica National Recreation Area than it is through urban Los Angeles, whose lights can be seen in the background.

subpopulations (or almost suitable, in the case of sink patches). By extension, therefore, the matrix must be composed of habitat that is not suitable, possibly even inhospitable or dangerous for individuals. In the mainland-island model, an aquatic matrix serves that purpose clearly for terrestrial island organisms. This concept can work in terrestrial environments as well, though, if the matrix falls outside an organism's physiological tolerance or is a completely anthropogenic habitat—such as asphalt or a crowded city. If the habitat in the matrix creates conditions that impede movement, then the distance between patches becomes even more relevant. Distance, therefore, may be relative in the context of the matrix habitat, because physical distances will be experienced differently by populations based on the quality of the matrix habitat. For example, it would be much harder for a young, dispersing mountain lion to successfully move 10 km (6.2 mi.) through downtown Los Angeles than to move the same distance through a nearby forested wilderness area (**Figure 12.21**).

Ecologists use the term *functional dispersal distance* to distinguish between the simple straight-line distance between two points in space and the distance experienced with real-world hurdles, barriers, and dangers. Understanding functional dispersal distances requires a deep appreciation of the biology, ecology, and natural history of the organism being studied. How individuals move in response to perceived predation threats, their ability to travel over hostile terrain, and even how their sensory systems respond to stark shifts in environmental conditions, such as light, can determine functional dispersal distances. For sessile plants and animals that disperse via physical processes, such as wind and water currents, functional dispersal is dictated more by the dynamics of the currents than the straight-line distances between two points.

12.6 Multispecies Spatial Models

Adding a spatial dimension to ecological dynamics clearly offers more realism, but it also creates more complications. So far, we have only looked at these effects when considering the dynamics of one species. What if we think about the effects of movement on communities or multispecies assemblages? Not surprisingly, multispecies spatial models become even more complicated. Yet, we can gain some interesting insights if we ignore explicit interactions between species and instead focus on the effects of movement on species diversity. We can start by thinking about multiple species in the mainland-island model.

Island Biogeography

If we add multiple species to the simple mainland-island model, we can explore the way dispersal and colonization influence patterns in overall species richness across locations. For example, as you may recall from the chapter on biodiversity (Chapter 11), one of the few seemingly invariant patterns in ecology is that the number of species in an area rises as the size of that area increases. We modeled this relationship between habitat area and number of species using a simple power function, $S = cA^z$, where A is the cumulative spatial area of the sampled locations, S is the cumulative number of species (species richness) in these samples, c is a constant, and z is an exponent less than one. This relationship suggests that species richness accumulates quickly at first and then more and more slowly as the sampled area increases in size. As mentioned in Chapter 11, larger areas are also more likely to have more habitats, which may support a wider array of species (Figure 12.22). In addition, larger areas allow larger populations, which are less likely to go extinct (Chapter 6). The rate at which we add new species with newly

(A) (B)

Figure 12.22 Larger areas in general are more likely to have a greater diversity of habitat types than small areas. We can see this difference when comparing oceanic islands. (A) Tiny islands have limited diversity of habitat, whereas (B) larger islands (e.g., the island of Kauai, Hawaii) have a much greater range of habitats available.

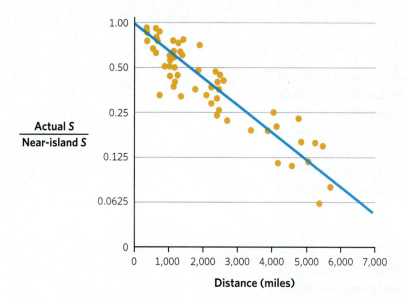

Figure 12.23 Relative species richness (*S*) on tropical southwest Pacific islands decreases as distance from mainland New Guinea increases. For each point, the y-axis value represents the ratio of the actual number of species on that island to the number of species found on the island nearest to mainland New Guinea. Notice that the y-axis is a logarithmic scale (log base 2) because each marked fraction is twice the size of the previous fraction (Diamond 1972).

sampled areas decreases as the area gets bigger because the samples usually add more individuals of the same species rather than adding new species. In Chapter 11, we showed a good deal of evidence to support this pattern, particularly from island archipelagos.

Yet Chapter 11 left out an important feature, which our newfound understanding of dispersal brings to the fore. The overall area of an island cannot be the only factor that influences species richness on that island. The ability of organisms to disperse to an island must also influence the species found there. Remoteness or distance plays a role because it affects dispersal success. Based on the mainland-island model, we can imagine that the number of species on two oceanic islands with roughly the same area will differ if the islands lie at different distances from the mainland (Figure 12.20). The mainland-island model suggests that a close island is likely to have more species because it is much easier for individuals of almost any taxon to get there from the mainland. More remote islands are harder to reach, meaning fewer species have the potential to colonize them.

A classic example of this effect of island isolation comes from Jared Diamond's work (1972) on bird species richness across various oceanic islands near the large Pacific island of New Guinea. When he plotted how far smaller islands were from mainland New Guinea versus a measure of the number of species on the islands, he found that species richness declined sharply with distance from the mainland (**Figure 12.23**). Although Figure 12.23 looks like a linear relationship, notice that Diamond's y-axis has a logarithmic scale, which means that the richness of the farthest islands falls off much more rapidly than the richness of near islands. In general, lots of species on the New Guinea mainland can reach near islands via dispersal, but very few can make it to the farthest islands.

The pattern that Diamond found fits a theoretical idea developed in the late 1960s by Robert MacArthur and E. O. Wilson, who used both the area and distance effects that we have been discussing to understand patterns in species richness across islands. They suggested that two opposing ecological processes, colonization and extinction, both operated in real time. The number of species on islands was determined by the balance of these two opposing forces. Their resulting **equilibrium theory of island biogeography** (MacArthur and Wilson [1967] 2001) was able to account for many of the species richness patterns actually observed on islands and has become one of the more enduring and often-cited theories in the field of ecology. The basic premise of the model is that any given island has an equilibrium number of species resulting from a balance between colonization and extinction. This pattern is similar to the equilibrium proportion of sites occupied in the Levins metapopulation model discussed earlier.

To begin, we have to imagine a mainland with a permanent source pool of species. Any of these mainland species can potentially colonize islands if dispersing individuals can reach those islands. Two key assumptions in this model are that

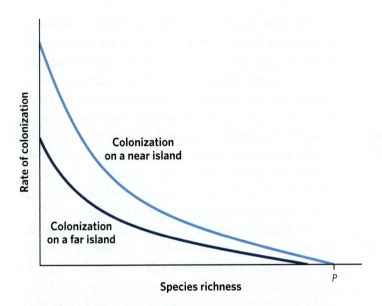

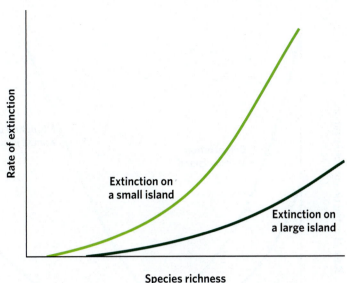

Figure 12.24 As species richness increases, the rate of novel colonizations decreases in near and far islands. Near islands, though, have higher colonization rates (notice that the curve for near islands sits above the curve for far islands) and can acquire a greater total number of species (notice where each curve intersects with the x-axis) because they are more accessible to species with lower dispersal capacity. Although the mainland contains a finite number of species (*P*), not all of those species may be able to reach the far islands.

Figure 12.25 As species richness increases, extinction rates also increase, but these increases are different for large islands and small islands. On all islands, having more species leads to more competition, which increases extinction rates. On small islands, low resources, small population abundance, and fewer habitat types cause extinction to start at lower species richness (notice where each curve starts on the x-axis) and increase faster than on large islands (notice that the curve for small islands sits above the curve for large islands).

(1) the mainland contains a finite number of species (*P*), and (2) the time frame is too short to allow any possibility of speciation on the islands (i.e., one species cannot evolve into two or more species). Colonization of each island is, therefore, determined exclusively by dispersal of individuals from the mainland. More dispersing individuals will be able to reach nearer islands than farther islands, meaning more species will colonize near rather than far islands. Over time, as species richness builds up on an island, colonization rates should decrease because fewer and fewer new species (i.e., ones not already present on the island) will be arriving (**Figure 12.24**).

On the other hand, extinction should be driven primarily by forces that occur on the island, and species on small islands are considered more likely to go extinct for at least three reasons. First, small islands will have fewer resources, leading to smaller populations. Small populations are inherently more prone to extinction than large populations (Chapter 6). Second, small islands should have fewer habitats, which means that each species is less likely to find a habitat in which it is a superior competitor and can survive as more species arrive. Third, as species richness on an island increases, extinction rates should also increase, not only because there are more species to go extinct but also because the presence of more species leads to more competition for limited resources and less habitat or territory for each species (**Figure 12.25**).

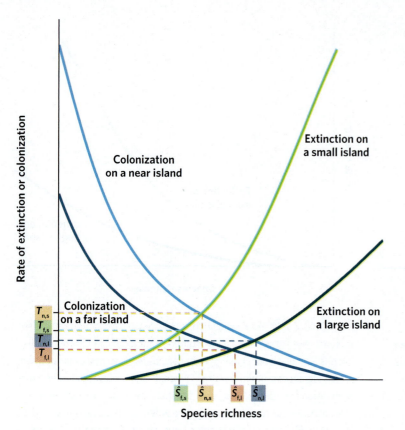

Figure 12.26 Combining colonization and extinction curves produces four distinct $\hat{S}$ values where the curves cross. Notice that $\hat{S}$ for near, large islands ($\hat{S}_{n,l}$) is much higher than $\hat{S}$ for far, small islands ($\hat{S}_{f,s}$). It is much harder to predict which of the intermediately diverse islands will have higher $\hat{S}$; it depends on the actual shapes of the curves. The T values on the y-axis represent species turnover (the number of disappearing species replaced by arriving species per unit time), with the subscripts the same as for the $\hat{S}$ values. $\hat{S}$ = equilibrium species richness; T = turnover rate; n = near; f = far; s = small; l = large.

In order to determine the equilibrium number of species on an island (i.e., the island's equilibrium species richness), we can put the extinction and colonization curves together (**Figure 12.26**). These combined curves tell us that the equilibrium species richness ($\hat{S}$) for an island depends on the colonization rate, the extinction rate, and the size of the source pool of colonizing species on the mainland. $\hat{S}$ for an island is the point on the x-axis where the colonization curve crosses the extinction curve, meaning that gain from newly colonizing species balances the loss from species going extinct.

In Figure 12.26, we depict $\hat{S}$ for four islands: an island far away from the mainland and small in size ($\hat{S}_{f,s}$), an island near the mainland and small in size ($\hat{S}_{n,s}$), an island far away from the mainland and large in size ($\hat{S}_{f,l}$), and an island near the mainland and large in size ($\hat{S}_{n,l}$). The large and near island has the highest $\hat{S}$ relative to the other three, as it receives lots of dispersing individuals, representing lots of species and lots of colonization, and it is large enough to experience little extinction. In contrast, the small island far from the mainland has the lowest $\hat{S}$ relative to the others because it receives fewer dispersing individuals, meaning less colonization, and its small size increases the risk of extinction.

These $\hat{S}$ values do not mean that any given island keeps the same suite of species consistently over time. Instead, newly colonizing species replace newly extinct species to keep the same $\hat{S}$ with different member species. We see a similar *dynamic equilibrium* in the changing occupancy of patches in a metapopulation at equilibrium (Figure 12.12D) or in the changing individuals at carrying capacity in a population (Chapter 6). Newly colonized patches replace newly unoccupied ones at equilibrium in the metapopulation, and newly born individuals exactly replace newly dead individuals to create a constant population abundance at the carrying capacity. The number of newly colonizing species that replace newly extinct species per unit of time is called the **turnover rate** (T). We can identify this rate on the y-axis in Figure 12.26 at the point where the colonization and extinction curves cross. Turnover is an interesting and critical property of this model, as it implies there is continuous colonization and extinction of species on islands. The relative order of the four islands in Figure 12.26 in terms of T is different from their order for $\hat{S}$. Turnover rate is highest for islands near the mainland but small in size and lowest for large islands that are far from the mainland. Since T is a function of high extinction *and* high colonization, it makes sense that islands that are small *and* near the mainland have the highest T.

One of the first tests of MacArthur and Wilson's equilibrium theory of island biogeography was carried out by Daniel Simberloff and E. O. Wilson (1969, 1970) through an ambitious large-scale experiment that examined insect diversity on small mangrove islands off the coast of southern Florida (**Figure 12.27**). The researchers chose eight roughly circular mangrove islands ranging from 11 m to 18 m (36 ft. to 59 ft.) in diameter. The distance of the islands from the "mainland" (a nearby but very large mangrove island) varied from 2 m to more than 1,000 m (6.6 ft. to more than 3,280 ft.). Using large tents to cover each island entirely, Simberloff and Wilson hired pest control contractors to pump insecticide into the enclosures and kill all the insects and invertebrates on the study islands. After removing all the insects from the island (a process called *defaunation*, meaning all animals were removed), the researchers returned to each island repeatedly over time to observe the recolonization process. They were curious to see if each island's insect community returned to the predefaunation level of species richness. Interestingly, the recolonization by arthropods was remarkably rapid. Within two years, most of the islands had returned to their approximate original species richness values, and the islands as a group seemed to approach a stable equilibrium species richness (**Figure 12.28**). In addition, as the equilibrium theory of island biogeography would suggest, species richness was highest on the nearest island, lowest on the farthest island, and intermediate on the islands that were at middle distances from the mainland.

One of the most interesting features of this study was that it supported the turnover in species predicted by the equilibrium theory of island biogeography. Although Simberloff and Wilson found a steady number of species at each census, the actual species on each island changed between censuses (**Table 12.1**). In other words, the equilibrium species richness on each island was dynamic. In fact, looking at each island individually, we see there was never more than a 55% overlap in the species found on a particular island in the censuses taken before defaunation, one year after defaunation, and two years after defaunation. This study demonstrates that the equilibrium theory of island biogeography captures some very real features of the effects of distance and island size on community richness and turnover.

Despite this initial supportive experimental evidence, in the 50 years since its publication, MacArthur and Wilson's theory has come under substantial criticism (Lomolino 2000). One significant issue is that the theory ignores the critical role of interspecific interactions beyond competition, such as predation and mutualism, in shaping and assembling ecological communities through time. It also assumes that the presence of more species leads to more extinction. It is possible, though, that more

Figure 12.27 This small mangrove island in Florida is similar to the ones chosen by Simberloff and Wilson to test the equilibrium theory of island biogeography.

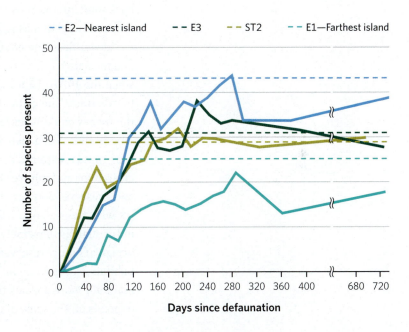

Figure 12.28 Simberloff and Wilson (1970) recorded this pattern of recolonization and equilibration of arthropod assemblages on mangrove islands after the islands were sprayed with pesticide to remove all insects. E2 is the nearest island to the mainland; E1 is the most distant; and E3 and ST2 are at intermediate distances. Dashed lines indicate the predefaunation species richness for each island. Solid lines of the same color indicate the changing species richness after defaunation. The islands equilibrated at or near the species richness level from the start of the experiment.

Table 12.1 Comparisons of Simberloff and Wilson's Pre- and Postdefaunation Censuses of Arthropods on Four Mangrove Islands

Island name	Comparison of the community before defaunation to one year later		Comparison of the community before defaunation to two years later		Comparison of the year-one community to the year-two community	
	Number of species in common	Percentage in common	Number of species in common	Percentage in common	Number of species in common	Percentage in common
E1	2	6.9	5	19.2	7	38.8
E2	10	18.5	13	25.5	16	37.2
E3	8	20.0	7	20.0	16	51.6
ST2	11	29.7	17	54.8	12	35.3

E1 was the island farthest from the mainland. E2 was the island nearest to the mainland. Islands E3 and ST2 were at intermediate distances from the mainland (Simberloff and Wilson 1970).

species could increase the potential for facilitation (Chapter 9) or different food web configurations (Chapter 10) that might support a higher richness of species rather than encourage extinction. Lots of research since the late 1960s has shown that the structure and dynamics of ecological communities are influenced by feedbacks and interactions among species and ecosystem components, all of which are not represented in MacArthur and Wilson's theory.

Regardless, the equilibrium theory of island biogeography has played a large and formative role in the fields of ecology, biogeography, and conservation biology since the 1970s. In terms of conservation, the importance of the theory is hard to overstate. The critical conservation application came with the realization that the theory could be extended to other kinds of "islandlike" habitats, not just oceanic islands. Such "islands" could be patches of coniferous forest on mountaintops, ponds scattered across a landscape, or fragments of once contiguous habitat cleared for agriculture (Figure 12.17). If a goal of conservation science is to protect and preserve biodiversity, having a model to examine how species richness varies with the size of a habitat patch and the distance from a source of dispersing individuals is extremely useful. The theory has been used as a rationale for the design of nature reserves and protected areas, and it suggests that larger protected habitats closer to a "mainland" habitat source of species will both contain and maintain more species than small, isolated nature reserves. Of course, this result depends on the scale of dispersal of the protected species.

Moving Further Afield

So far, we have almost completely ignored multispecies interactions on spatial patterns in populations and communities. What if we wanted to explore spatial versions of the competition or predator-prey models from Chapters 7 and 8, or investigate the dynamics of entire communities that are spatially distributed (called **metacommunities**)? Adding spatial elements into some of the models

Figure 12.29 Spotted Lake in British Columbia, Canada, dries substantially every summer, leaving mineral deposits that create solid separations between remaining pools of water. The network of naturally segregated pondlike habitats creates a perfect environment for an aquatic organism metacommunity.

from previous chapters makes them much more complicated but also leads to significantly more interesting dynamics. This is a rich area of research that is growing rapidly as new technology improves our ability to explore complex models and track organism movements across space. The topic of metacommunities is too large to cover here in any detail, but in a nutshell, metacommunities are a community-level version of metapopulations, with communities of species distributed across a region on patches of habitat (**Figure 12.29**). Metacommunity dynamics are more complicated than metapopulation dynamics because they include spatial dynamics and interactions among species. If you are interested in this idea, we recommend looking into the growing literature on the subject.

Clearly, spatial interactions can and do play an incredibly important role in the dynamics of single species and the assembly of ecological communities. In this chapter, we barely scratched the surface of the use of spatial models in ecology. Every model in this book could have multiple different pathways to address spatial effects. The simplest models are spatially implicit, like the Levins metapopulation model, and explore the overall effects of movement. These models tend to offer generalizations that expand our fundamental understanding of how species and communities exist across a landscape. As we saw with metapopulation models, some of these models can guide biodiversity conservation efforts across a wide array of habitats and situations. But these spatially implicit models always have assumptions that violate many of the critical aspects of dispersal or ignore the specific behavior of focal species, and these models rarely fit a specific habitat or system well.

Models that are more spatially explicit tend to be tailored to specific ecosystems and organisms. They give detailed, local information, but their predictions may not produce general insights for the larger field of ecology. We have not spent time on more detailed approaches here, but the field of spatial ecology is enormous, and we encourage you to both dive in and disperse.

SUMMARY

12.1 Nobody Stays in One Place

- Human children generally move away from their parents.
- Many organisms also move away from their birthplace.
- Dispersal occurs when individuals move on their own.
- Migration occurs when whole communities or populations move together.

12.2 Why Do Organisms Move?

- Scientific advances allow ecologists to measure movement more accurately than in the past.
- Organisms generally move to improve their evolutionary fitness.

Migration versus Dispersal

- **Migration**
  - Migrations can present a stunning visual display.
  - Migration likely evolved to balance reproduction and survival trade-offs.
- **Dispersal**
  - Dispersal is driven by an individual's impulse to improve its fitness.
  - Ecologists distinguish between natal dispersal and breeding dispersal.
- **Effects of Migration versus Dispersal on Geographic Ranges**
  - The process of migration produces fitness benefits for the species involved.
  - Dispersal benefits accrue uniquely to the moving individuals.

Spread and Distribution: The Effects of Movement Distance

- Dispersal distances can vary among individuals in a population, and this has significant implications for population growth and spread.
- **Characterizing Dispersal Distances: Dispersal Kernels**
  - Dispersal kernels are graphs that plot the distribution of dispersal distances and allow ecologists to better visualize how far individuals move.

12.3 Population Growth Model with Dispersal

- We can add movement (i.e., immigration and emigration) to the population growth models from Chapter 6.
- It is difficult to get actual values for immigration and emigration from the natural world.
- Ecologists have built new models that involve only immigration and emigration.

12.4 Dynamics and Movement between Populations: Metapopulations

- A metapopulation is a linked network of discrete subpopulations occupying habitat patches that are spread out across a landscape.
- These patches are linked via dispersal.
- For simplicity, scientists assume each patch is either occupied or unoccupied, and that individual patches continually get colonized and go extinct (i.e., become unoccupied again) over time.
- This produces a "blinking lights" effect in a set of patches spread over space.

The Levins Metapopulation Model

- The Levins model incorporates the proportion of occupied patches, the rate of colonization for empty patches, and the extinction probability for patch subpopulations into a continuous-time model that keeps track of the change in the proportion of occupied patches through time.
- The model predicts that, at equilibrium, a random but constant proportion of patches will be occupied through time producing a "blinking lights" phenomenon.
- Under this model, even though individual patches go extinct through time, colonization among patches allows the entire network to persist.

Real-World Implications of the Levins Model

- Regional persistence of a metapopulation can occur even with frequent patch extinctions.
- The dispersal rate among patches is vital to this process.
- In the real world, dispersal and colonization are often variable and can, therefore, have big effects on metapopulation dynamics.

12.5 Beyond Metapopulations to More Realistic Spatial Dynamics

- Several critical assumptions built into the Levins metapopulation model are hard to meet in real-world situations—namely, the assumptions of equal size among patches, equal dispersal among patches, and instantaneous responses.

- Yet the Levins model has been important in stimulating thought among scientists.

- More recent work has suggested a continuum of situations involving patch connectivity, with metapopulation dynamics being an outcome when movement levels are moderate.

- Ideas of metapopulation dynamics have been applied to conservation efforts and rare species management in both the terrestrial and marine environments.

Quality Matters: Source-Sink Dynamics

- If we take into account internal patch-population dynamics, we find that certain patches may produce more dispersers, whereas others may produce none.

- This leads to an asymmetry among patches: some become sources of colonists, and others become "sinks" where colonists go and eventually die.

- These source-sink dynamics are likely common in both natural and human-altered situations.

Distance Matters: Patch Size and Connectivity

- In the real world, not all patches are the same size, nor are they equidistant; therefore, differential dispersal occurs among patches.

- **A Mainland-Island Model**

  - A mainland-island model explores how dispersal distance affects colonization and persistence of island populations.

— The habitat around the islands (i.e., the matrix) is not habitable, and the quality of the matrix can alter the dispersal rate for individuals and species.

12.6 Multispecies Spatial Models

Island Biogeography

- Dispersal can influence an area's overall species richness.

- Building on the species-area relationship from Chapter 10, we can add the effect of distance on an area's species richness.

- Robert MacArthur and Edward O. Wilson proposed the equilibrium theory of island biogeography in 1967; it includes effects of island area and island distance on an island's species richness.

- The theory predicts an equilibrium number of species on an island resulting from two forces: immigration and extinction.

- The theory also predicts that the exact membership of the community should change through time, but that the overall species richness should remain relatively fixed.

- The theory has been tested in many situations over the years and has received various levels of support.

- MacArthur and Wilson's theory has played a critical role in conservation biology through its recognition that habitat fragments can act as islands.

Moving Further Afield

- New technologies make it possible to add spatial context to the multispecies models from Chapters 7 and 8.

- Community-level versions of metapopulations are called metacommunities and involve multispecies assemblages of organisms on habitat patches distributed in a matrix across a landscape.

KEY TERMS

breeding dispersal
dispersal
dispersal kernel
equilibrium theory of island
   biogeography
fragmentation
geographic range

habitat patch
Levins metapopulation model
mainland-island model
matrix
metacommunity
metapopulation
migration

natal dispersal
sink
source
source-sink dynamics
subpopulation
turnover rate

CONCEPTUAL QUESTIONS

1. Name an organism in your area that disperses and one that migrates. Explain why the movement of one is dispersal and that of the other is migration.

2. What is a metapopulation and how does its probability of extinction compare with that of an independent population? Describe some of the ways metapopulations may be biologically and ecologically reasonable.

3. Which would you expect to survive longer (and why): several small populations connected by dispersal or a single large population? Explain your answer.

4. Describe factors that may play a role in the low persistence of populations in small habitat patches.

5. True metapopulations according to the Levins metapopulation model may be rare. Explain why, and describe more realistic ways that a species may be distributed and linked across a landscape.

6. The Levins metapopulation model has been applied to various species of pond-breeding amphibians by treating the ponds as patches across a terrestrial landscape. Given that many amphibians only breed in the ponds for a few months out of the year, can you think of any reasons why this might not be the best model for these amphibians?

7. Use the concepts of metapopulations and source-sink dynamics to explain why we might find high-quality habitat with no individuals or low-quality habitat with lots of individuals.

8. What is a dynamic equilibrium? Explain why the equilibrium species richness ($\hat{S}$) in the equilibrium theory of island biogeography is a dynamic equilibrium.

QUANTITATIVE QUESTIONS

1. Examine Figures 12.1 and 12.9. Are these figures showing the same kind of information for different species? If so, explain why the figures look so different. If not, explain why not.

2. In the chapter, we state that long-distance dispersal suggests that mean or average dispersal distances may not be useful numbers for thinking about how fast a species might spread.

   a. Can you explain why average dispersal distances would not be useful in this situation?

   b. Explain the importance of long-distance dispersal.

3. As we mentioned in the text, collecting dispersal data gets harder and harder at larger distances from a source, partially because it is very hard to detect rare dispersers. This effort is easier, though, if the dispersers can only move in a line (such as along a stream). If dispersers can radiate out in all directions from a source, then ecologists must sample larger areas as they move away from the source. This effect can be important in designing dispersal studies. Can you explain why it is necessary to sample increased area at increased distances from the source? (Hint: Think about the difference between the radius and the area of a circle, and relate those to the area that an ecologist must survey at each distance.)

4. From Equation 12.2, derive the equilibrium occupancy for a metapopulation ($p^* = 1 - \frac{e}{c}$). Explain what this equilibrium means.

5. In the mainland-island model, distant islands remain unoccupied for much longer periods than any patches in the Levins model with similar dispersal. Can you explain why? Think about the way that dispersers are produced in each model.

6. Jared Diamond examined bird species richness on the Channel Islands off the coast of California and found the results shown in the table here (Diamond 1975). Do these results match the expected results based on MacArthur and Wilson's equilibrium theory of island biogeography? Explain your answer.

Island	Area (mi²)	Distance from mainland (mi)	Species
Santa Catalina	75	20	30
Santa Rosa	84	27	14
Santa Cruz	96	19	36

SUGGESTED READING

Hoopes, M. F., R. D. Holt, and M. Holyoak. 2005. The effects of spatial processes on two species interactions. Pp. 35–67 in M. Holyoak, M. A Leibold, and R. D. Holt, eds. *Metacommunities: Spatial Dynamics and Ecological Communities*. Chicago: University of Chicago Press.

Levins, R. 1969. Some demographic and genetic consequences of environmental heterogeneity for biological control. *Bulletin of the Entomological Society of America* 15(3): 237–240.

MacArthur, R. H. and E. O. Wilson. (1967) 2001. *The Theory of Island Biogeography*. Reprinted with a new preface by E. O. Wilson. Princeton Landmarks in Biology. Princeton, NJ: Princeton University Press.

Nathan, R., W. M. Getz, E. Revilla, M. Holyoak, R. Kadmon, D. Saltz, and P. E. Smouse. 2008. A movement ecology paradigm for unifying organismal movement research. *Proceedings of the National Academy of Sciences* 105(49): 19052–19059.

Simberloff, D. S. and E. O. Wilson. 1970. Experimental zoogeography of islands: A two-year record of colonization. *Ecology* 51(5): 934–937.

A May 2018 eruption on Mt. Kilauea sent a river of hot lava through forests and fields on Hawaii Island. The lava flows created 3.54 km² (1.37 sq. mi.) of new land in the ocean and permanently transformed 35 km² (13.5 sq. mi.) of the terrestrial habitat.

chapter 13

COMMUNITIES THROUGH TIME

LEARNING OBJECTIVES

- Distinguish between Clements's and Gleason's versions of the structure of ecological communities.

- Compare the processes of primary and secondary succession.

- Evaluate plant successional patterns in comparison to animal successional patterns.

- Identify how chance events and the biotic community influence successional patterns and processes.

- Explain the three contrasting models of succession proposed by Connell and Slatyer.

- Justify the idea that there are rules that govern how communities are assembled.

- Explain the concept of disturbance, and describe how it plays a role in community dynamics.

- Assess the intermediate disturbance hypothesis in terms of community assembly and membership.

- Compare the ideas of steady state and disturbance ecology in terms of community dynamics.

13.1 Nature Is Dynamic

Before Darwin and the publication of his *On the Origin of Species* (1859), the natural world and all its biodiversity were generally considered to be in a steady state, static and unchanging (Benson 2000). Little thought, appreciation, or understanding was given to the dynamic relationships between organisms and the environment. Although Darwin is mainly remembered for the theory of natural selection (see Chapter 4), he was also one of the founders of the science of ecology. Much of his work examined the way organisms interact with their environments, and his explorations led to some of the earliest recorded ecological experiments (Hector and Hooper 2002).

One of the longest-lasting impacts of Darwin's work, beyond evolutionary theory, is the recognition that the natural world is dynamic—that change is the rule rather than the exception. This observation seems fairly obvious and rather prosaic to us today, but in 1859, the idea was transformative and, frankly, revolutionary. It led to our current understanding that not only do species change through time, but the communities and ecosystems made up of species are also in a constant state of flux.

Although we are no longer surprised by the idea of a dynamic natural world, there are still many situations in which we expect predictability, stability, and "natural balance" in ecological systems. How is it possible to expect predictability and stability while also knowing that living systems are constantly in flux? This chapter explores these ideas and contradictions with a particular focus on the ways ecological communities change through time. We begin by examining predictable changes and then show how additional factors can push communities away from a steady state toward more dynamic situations.

13.2 The Nature of Communities

Defining Community

In previous chapters, we tackled many of the important elements and building blocks of community dynamics, including interspecific interactions (competition, exploitation, and mutualism in Chapters 7, 8, and 9, respectively), energy flow and trophic relationships (Chapter 10), biodiversity (Chapter 11), and colonization and extinction (Chapter 12), but we have not really explored how to *identify* communities.

In earlier chapters, when introducing the topic, we defined a **community** as a group of interacting species in the same location. That definition still holds true for our exploration of the temporal dynamics of communities. However, sometimes the term *community* is used in a narrower sense to refer to a particular taxonomic group, such as the bird community of Quito, Ecuador; the insect community in the province of Ontario, Canada; or the plant community in the prefecture of Chiba, Japan. In general, in this chapter we use the term *community* in a broader sense, one that includes all the species in a particular location and how they change through time, but on occasion we may use it to refer to a particular taxonomic group.

Although the preceding broad definition may be fine for the purposes of defining the word on an exam, it is not especially helpful to ecologists trying to identify a community in the real world. The question of how to identify communities may sound overly philosophical, but it is a serious and practical concern that leads to a host of important ecological questions. For example, is it possible to stand in one community and wholly differentiate it from other nearby communities (**Figure 13.1**)? Is it feasible to see or measure discrete boundaries delineating communities? Are there rules or patterns that govern how communities change through time? Questions like these have confronted and challenged ecologists since Darwin's time.

We examined similar sets of questions in Chapters 4 and 6 when defining *species* and *populations*. As we discovered then, there are theoretical and real-world practicalities to consider. Although it is important to grapple with these complexities, at times we can work around or even ignore some of them in particular contexts. Still, confronting these issues allows us to recognize temporal trends and dynamics, and—frankly—that is where some of the fun lies.

Clements versus Gleason

One of the great early debates in this field of study focused on the fundamental nature of ecological communities. The argument originated with two plant ecologists, Frederic Clements and Henry Gleason. Clements received a doctorate in botany in 1898 and became a professor at the University of Nebraska in 1904. While there, he conducted research on the grasslands, coniferous forests, and

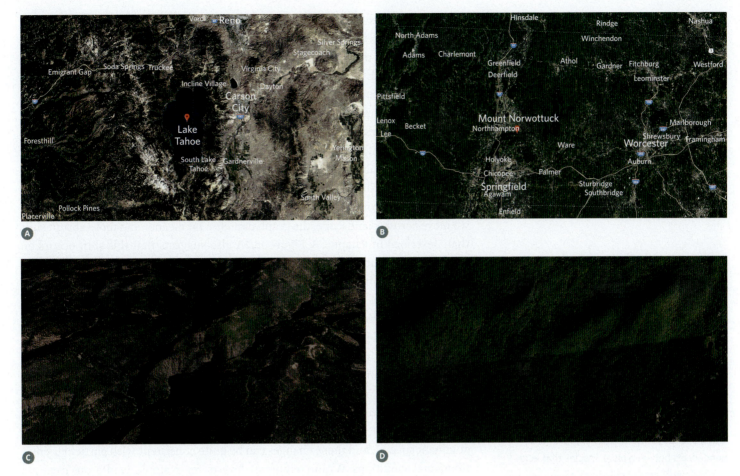

prairies of Nebraska and Colorado. He also published the first American textbook on ecology in 1905.

In his research, Clements observed vegetation assemblages in the midwestern United States and saw that suites of plant species repeatedly seemed to co-occur across the landscape. In other words, when he walked through the landscape, he found that certain species—let's call them A, B, C, and D—were always present together. He also noticed that over long periods of time, membership in these suites of species, or communities, tended to converge on a predictable and permanent group of species that he called a **climax community**. Clements viewed the climax community as the endpoint of a process called *succession*, which we explore in greater detail later in the chapter.

These and other observations led Clements to think of communities as coordinated, holistic entities composed of interdependent parts. He likened plant communities to *supraorganisms*, with individual species playing the roles of different organs within a body. This analogy suggested that every species in a climax community is as vital to the functioning of the community as different organs and systems are to the functioning of a body.

We mentioned in Chapter 3 that soils and climate can determine the dominant vegetative patterns in biomes, but Clements's theory went further. He suggested that the species in a particular community worked together over time to facilitate a shift in community composition toward the climax version of that community, and that the climax, when reached, would be stable for thousands of years.

Figure 13.1 Is it possible to stand in one community and distinguish it from other nearby communities? **A** Along the Sierra Nevada mountains between California and Nevada, a satellite photo shows distinct differences in plant communities as temperature and water availability change. **B** From a similar altitudinal view, the only observable community differences in the Mount Holyoke Range in Massachusetts are attributable to rivers and the presence of cities and towns on the landscape. **C** Returning to the Sierra Nevada and zooming in closer on the western slopes in California, we can see distinct transitions between woodland and grassland communities. **D** In the Mount Holyoke Range, though, no clear distinctions are visible between communities, not even when we move in closer and include both the northern and southern slopes of the mountains.

Clements's climax communities were coevolved, clearly identifiable, and largely unchanging in their species composition, given the climate and soils found in a particular part of the world.

We can sketch a figure depicting Clements's view of the world (**Figure 13.2A**), showing an environmental gradient on the x-axis and species abundance on the y-axis. An environmental gradient is a change in abiotic conditions across distance in a landscape; this change could be a shift from wet to dry soils, for example, or from warm to cool temperatures. In Figure 13.2A, each colored line represents the distribution and abundance of a specific plant species along this gradient (similar to resource utilization curves in Chapter 5). Although this graph fits with the idea of biomes from Chapter 3 by suggesting that plant communities change as we move across a landscape, the graph depicts a far cleaner and more abrupt shift than described in Chapter 3. Figure 13.2A also suggests that this shift occurs on much smaller spatial scales than in biomes. Clements proposed that all species within a community responded in the same way to environmental gradients. This Clementsian view implies that there should be sharp boundaries between communities, with little overlap in plant species distributions among communities.

Clements's ideas about the dynamics of plant communities were appealingly tidy, and they fit reasonably well with general observations. For example, it is apparent that plant communities do, in fact, shift in composition across climate or soil gradients. You can see these changes as you drive across a landscape in which the plant communities shift from forest to grassland over a rainfall gradient (see Chapter 3). Even within a biome, though, we often see shifts in associated dominant species, and those changes can be quite consistent with particular environmental conditions. Within temperate forests, for example, coniferous trees are more prevalent in wetter or cooler habitats, and deciduous trees are more prevalent in warmer or drier habitats.

Other evidence in support of Clements's views came from looking at plant communities that developed after farmland abandonment. In the last half of the nineteenth century and the first half of the twentieth century, when Clements was making his observations, much of the United States was moving away from

Figure 13.2 Frederic Clements and Henry Gleason held contrasting views of how species in plant communities change across an environmental gradient. In both graphs, each colored line represents a different species of plant. **A** Clements's view posited that coevolution and close interactions create interdependencies between plant species that lead to predictable sets of co-occurring species and distinct and separate plant communities with fairly sharp boundaries dividing them. **B** Gleason's individualistic concept of plant associations suggests that species respond to environmental factors according to their own unique physiology. Thus, species do not consistently co-occur.

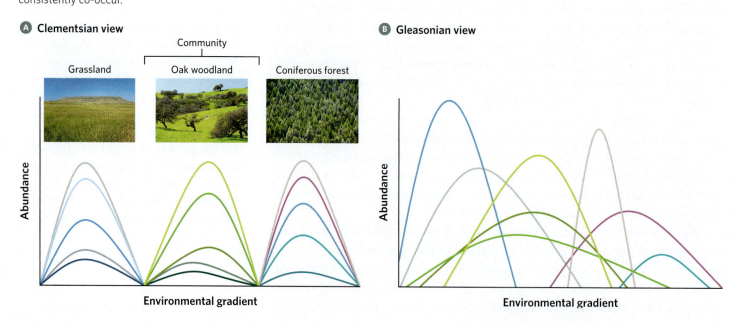

A Clementsian view

Community

Grassland · Oak woodland · Coniferous forest

Abundance

Environmental gradient

B Gleasonian view

Abundance

Environmental gradient

a strongly agrarian culture. Many farms still dotted the nation's landscape (particularly in Nebraska, where Clements grew up), but the peak in farmland coverage had passed, and many former agricultural fields lay fallow as people moved to cities and into industrial work. Ecologists often observed that abandoned fields on these empty farms shifted over time in fairly predictable ways from barren plowed fields to mature forests or grasslands. These simple observations fit quite well with Clements's theory.

In fact, Clements used former pastures of different ages—"age" being the length of time since the last plowing—to illustrate his vision of a climax community. Ecologists can collect data about the makeup of a community at different times, and the set of data from all these time points is called a **chronosequence**, because the changes in the community through time can be seen in a sequence of "snapshots." These chronosequence snapshots can come from one place at different times or from several places that are in different stages of the same process, such as abandoned fields with a range of years since their last plowing. For Clements and his adherents, these observations clearly illustrated how the plant communities progressed along a trajectory from empty fields to a forest or grassland climax community. This pattern can still be seen today near abandoned farm fields or even abandoned city lots. The regularity in the way plant communities change through time demonstrates that, in some fundamental ways, Clements got a few important features of community dynamics right.

Yet, not all ecologists saw eye to eye with Clements. One vociferous opponent was Henry Gleason, who received his doctorate from Columbia University in 1906 and became a professor of botany at the University of Michigan in 1910. In his early years, Gleason's views and research agenda were heavily influenced by Clements's ideas. Yet by 1918, Gleason began expressing considerable doubt about Clements's holistic view of plant communities, and in 1926 he published a paper describing an alternative that he called the *individualistic concept of plant associations* (Gleason 1926).

Gleason noted that every individual plant species has a unique set of physiological requirements and adaptations. (You may notice that this sounds similar to the concept of the niche we discussed in Chapter 5.) In addition, Gleason recognized that seed dispersal and germination are largely a matter of chance in that, given the right set of conditions, a seed can germinate and grow in many different locations. This combination of ideas and observations led him to suggest that Clements's ideas were nonsense. He completely rejected the idea that communities were supraorganisms consisting of coevolved and coadapted plant species working together to move toward a climax.

A visual depiction of Gleason's view of plant communities might look something like **Figure 13.2B**. Here, again, each colored line represents a single plant species distributed along an environmental gradient. Because Gleason suggested that each individual species has its own unique set of requirements, his species do not necessarily covary with each other. His view implies that every species is independent of the others, and each responds to local environmental conditions differently than the others.

The ensuing debate between proponents of Clementsian and Gleasonian views of ecological communities raged quite fiercely for several decades. In hindsight, the persistence of the debate is somewhat surprising because ecologists could have tested the predictions of both ideas early on by making some simple observations. For example, researchers could have collected data on plant species along an environmental gradient and compared real-world patterns to both hypothesized patterns. Such obtainable data should have settled the debate—in Gleason's favor. In fact, it is easy to demonstrate that such data produce Gleasonian patterns, which fit with topics discussed previously in this book about individual responses to the environment, natural selection, and the niche.

Why, then, did the debate rage for so long? And why did so many ecologists cling to Clementsian views until the 1950s? The issue lay in figuring out how to reconcile individualistic associations with the clearly repeatable patterns that Clements saw, such as the chronosequences of abandoned farmland. In fact, how can we reconcile this individualistic view with the evidence provided by our own eyes, which can often detect clear patterns in plant community composition through time?

The quick answer lies in the loose associations among species that arise from biological interactions and responses to the environment. It is not necessary for species to work together like a Clementsian supraorganism in order to coexist and produce predictable patterns in compositional change through time. The fact that individual species have their own physiological limits and niches does not mean that they cannot overlap in many of their requirements.

If you walk through a forest, you might see many of the same species repeatedly as you hike along a path. Many of those species might appear together in a nearby forest, as well, because species with similar requirements that can coexist tend to establish themselves in similar places. This pattern occurs not because the species are working together holistically in a Clementsian way but because they are responding to the same environmental conditions or resource levels. An analogy may help to illustrate. We often see flowers blooming at the same time of year that people start eating more ice cream. The flowers do not cause ice cream consumption (we are pretty sure about this), and eating ice cream does not cause flowers to bloom. Both the flowering plants and the humans are instead responding to an underlying change in temperature.

Although Gleason's ideas eventually became the standard interpretation among ecologists, Clements's ideas about how communities change through time still work remarkably well. His observations were critical to identifying associations between species, defining communities, and noticing the way that communities change. In the section on succession in this chapter, we discuss how this predictability is possible, even though we know species are not working together to facilitate a climax community.

Community Boundaries

Two of the main take-home points from the debate between Clements and Gleason are that ecological communities are not set-in-stone physical entities and that, at times, communities can be somewhat hard to delineate. To make progress in identifying communities, we need to tackle two related definitional issues. First, can we define the repeatable assemblages that Clements saw and label them as different communities? And second, can we define specific boundaries between communities?

As to the first issue, consider what happens if we define forest community A by the presence of species 1, 2, 3, 4, and 5. If we come across a forested area that has species 1 through 4 but not 5, should we consider it the same community as community A or a different one? In practice, ecologists generally do not use such strict criteria when determining communities (hence the fairly broad definition of the term *community* given earlier). Ecologists may note the dominant plant species that they expect in a particular community, but they do not expect the entire suite of species to repeat from one place to another. Instead, ecologists often talk about *community assemblages* (or associations), or they define a community in fairly broad strokes (e.g., a New Jersey pine barren community or a California chaparral community). These associations focus on dominant species or on a suite of adaptations, leaving the door open for localized differences in specific plant or animal distributions and abundances.

Regarding the second issue, ecologists generally expect to see gradual shifts in species as they move along an environmental gradient. These changes can make it difficult to draw a clear boundary between two communities. Of course, sudden changes in the abiotic conditions can at times lead to clear-cut boundaries, such as that between an alpine stream community and a nearby mountain meadow. One is either in the stream and wet or outside the stream in the meadow and fairly dry. Within the stream itself, though, there may be no clear boundary between the alpine portion of the water and the warmer lowland stream. In addition, the meadow itself may shift quite abruptly to forest in certain areas, but in others it may slowly grade into forest; such a transitional area is sometimes called an ecotone.

As with defining populations, delineating boundaries is rarely as problematic as it initially seems. Ecologists are often interested in a particular community because of a conservation or management issue, in which case the boundary of an ecological reserve or the spatial scale of a managed site effectively determines the community. Therefore, in practice, a community is often determined by the ecological question at hand. For example, if the goal of a project is to protect seasonal wetland communities, ecologists can use the presence of certain critical plants or animals as criteria to include or exclude particular sites from consideration. Alternatively, they can decide that the presence of standing water on a certain calendar date is the criterion that delineates the community. Choosing between these options may depend on the programmatic or institutional requirements of the agencies or organizations involved in the project. Either way, ecologists are often given some practical guidance on how or why to delineate communities.

13.3 Succession

Predictable Change

If community membership is dynamic through time, but some changes in species composition are predictable, is it possible to explain why this predictability exists? To a certain extent, yes. Communities change in fairly predictable ways over time as they go from bare rock or lifeless water to locations filled with interacting species. The process of that change is called ecological **succession** because a series of associated species succeed each other through time. We call each sequence of

Successional change on Hawaiian Islands

Bare substrate   Initial colonizers   Small shrubs   Tropical forest

Time

Figure 13.3 Successional change on the Hawaiian Islands is relatively easy to observe. Starting with bare lava substrate, a site is first colonized by mosses and lichens, then ferns and small grasses. After decades, a site develops into scrublands and then into dense tropical forests after a century or more.

Figure 13.4 The process of primary succession requires bare mineral substrate that contains no living organisms. Two places where this bare substrate occurs are on **A** recent lava flows and **B** ground exposed by glacial retreat.

A

B

a plant community a **sere**, and each sere is distinguished by its dominant plant species. Typically, the timeline of changes begins with the initial colonization of a new habitat or follows a disturbance event, such as a fire, flood, or bulldozing.

It is relatively easy to find evidence of succession if you know what to look for, and it is a particularly conspicuous feature of newly created sites that start barren of all noticeable life. For example, if you lived on the volcanic island of Hawaii (the biggest island in the Hawaiian chain), you could easily observe over time how recently cooled lava with no living species on it slowly accumulates organic material and begins to support mosses, lichens, and a few small plants, such as ferns and grasses (**Figure 13.3**). Given more time, these early species give way to small shrubs; and after many years, the old lava field becomes home to a complement of vegetation typical for the island. The study of succession has traditionally focused on how plant communities change through time, but it is easy to recognize that plant successional processes correspond with changes to the animal, fungal, bacterial, and protist communities present as well.

Primary and Secondary Succession

The process of **primary succession** occurs on substrates containing no organisms and no organic material, meaning that the community assembles completely from scratch. Natural locations for primary succession are rare but include such geologic features as sites of recent lava flows and exposed land after glacial retreat (**Figure 13.4**). Anthropogenically initiated sites of primary succession are much more common. For example, ecologists have studied the colonization of abandoned urban sites from which a covering of asphalt or cement has been removed as a form of primary succession. The revegetation of industrial waste sites may also qualify as primary succession; for example, primary succession can occur on mine tailings, which are the piled ore waste from mining activities. Despite a mudlike consistency, the substrate in these sites is completely mineralized soil containing no living organisms.

During primary succession, the only way for a community to form is for organisms to disperse to the site from elsewhere, sometimes after traveling long distances. Typically, the initial stage of succession is rather slow, as the first colonists must not only arrive but also essentially transform the environment, allowing it to hold resources that support life. The physical and chemical weathering of rocks allows the abraded material to hold water, perhaps providing a location for the first colonists. Over time, as the colonist species' own organic matter accumulates and breaks down in the presence of microorganisms, an organic soil layer is slowly formed. In general, the initial colonizing species in the first sere are short-lived, small-stature organisms, such as mosses, lichens, ferns, and grasses, with relatively modest resource requirements.

The first few seres in primary succession alter the environment in ways that are essential for the survival of species in later seres. Once the initial colonizers establish themselves in a location, the changes that they induce tend to provide a toehold for the establishment of other organisms, once soils have formed. This facilitating process is what led Clements to think that the species in a community behaved like a supraorganism, with all the species working together toward a final climax community.

Secondary succession is similar, but it occurs on already established soils following a disturbance to the previous habitat. In this case, the disturbance resets the system to a previous early successional state, in which the location is not entirely devoid of life or organic matter (**Figure 13.5**). For example, if a large, intense forest fire burns through an area of deciduous forest, it may kill off all the aboveground plants and cause animals to move from the area, but it still leaves organic soil and

Figure 13.5 Many natural and anthropogenic factors can remove vegetation and initiate secondary succession, including Ⓐ forest fires, Ⓑ hurricane damage to mangrove forests, Ⓒ bulldozing areas for mining, and Ⓓ clear-cutting during timber harvests.

soil organisms intact. Even if the fire is intense enough to burn through the top few inches of organic soil, the next layer of soil is generally unharmed and contains viable plant seeds and often viable roots. These lower soil layers also harbor animals and microorganisms, including microbes, mycorrhizae, earthworms, beetles, and sometimes vertebrates such as lizards, voles, or gophers. With this base of species and organic material, the initial stages of secondary succession can proceed much faster than the first stages of primary succession because of the intact nature of the resident soil community.

Both types of succession are most easily observed along a spatial chronosequence that shows snapshots of processes occurring over extremely long periods of time by looking at locations in different stages of the same process. Lava flows of different ages, or ground exposed by melting glaciers, are good examples. Secondary succession is more common and more easily observable, even without a chronosequence, because significant pieces of the process can be seen within a human lifetime or even a couple of decades.

Life History and Succession

One consistent pattern shared by primary and secondary succession is that the species inhabiting early successional stages have different life history strategies from those that inhabit late successional stages. Compare, for example, grasses and hickory trees (**Figure 13.6**). Many early successional grasses are small annual plants that live for one year, whereas hickory trees are large and tend to live for decades or even hundreds of years. If we make a list of traits exhibited by early successional species and another for late successional taxa, patterns emerge that take us back to an understanding of how populations change through time (Chapter 6).

Early successional plant species tend to be small and have high rates of reproduction, low rates of survival, short generation times, rapid development, and

Figure 13.6 We can see the differences between early and late successional species by comparing an early successional species such as Ⓐ little barley (*Hordeum pusillum*) and a late successional species such as Ⓑ mockernut hickory (*Carya tomentosa*).

r-selected species   *K*-selected species   *r*-selected species   *K*-selected species

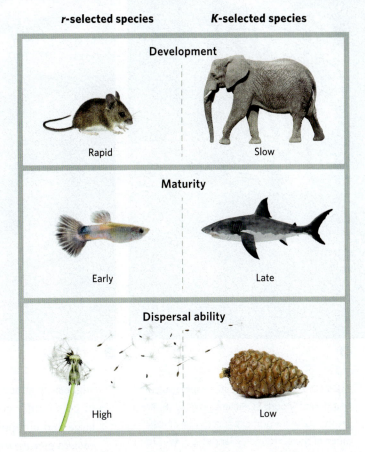

Figure 13.7 Comparison of *r*-selected and *K*-selected species and their associated life history traits.

early maturity. These traits lead to high per capita rates of population growth and the ability to colonize new sites quickly and produce offspring before being out-competed by later-arriving species (**Figure 13.7**). Species that exhibit these traits are often called **_r_-selected species**. Note that the *r* in "*r*-selected" refers to r_{max}, the intrinsic per capita rate of population growth in the logistic population growth model from Chapter 6. These plants are also sometimes called weedy species, as another way to highlight their traits. Common *r*-selected plants include annual grasses, mosses, ferns, liverworts, sedges, and many of the small flowering plants seen along roadsides. This same suite of *r*-selected traits can also be recognized in animal species. For example, mice, mosquitos, cockroaches, or rabbits would all be considered *r*-selected.

Late successional species share an entirely different suite of life history traits. They tend to be large and have low rates of reproduction, high rates of survival, long generation times, slow development, and late maturity (Figure 13.7). These types of organisms are often called **_K_-selected species**. The *K* in "*K*-selected" refers to the carrying capacity term in the logistic growth model from Chapter 6,

Figure 13.8 Ⓐ The catastrophic volcanic eruption of Mount St. Helens on May 18, 1980, set the stage for primary succession across much of the surrounding landscape. Ⓑ After the 1980 eruption, the landscape was bare. The green bits in the foreground of this photograph may look like signs of life, but those are dead trees, and behind them lie hundreds of square miles covered with ash and lava.

suggesting a more stable population size through time. Late successional changes in community composition are slow to materialize, and species that can grow large and compete for resources in these circumstances tend to dominate. It is easy to picture trees as the most common *K*-selected plant species, but in nonforest biomes *K*-selected plants can be other large or long-lived species, such as cacti, thorny shrubs, or even perennial grasses. Many animals also exhibit *K*-selected traits, including elephants, whales, humans, cicadas, and tortoises, to name a few. Although the ideas of *r*- and *K*-selection are often a little too broad for applied uses, classifying any organism (plant or animal) according to these two life history categories remains useful as a tool of comparison, particularly when studying general patterns of ecological succession.

Examples of Succession

Primary Succession on Mount St. Helens   Mount St. Helens is located in southwestern Washington State, about 50 miles northeast of Portland, Oregon, and is one of several volcanoes in the Cascade Range of the Pacific Northwest of the United States. It is a composite volcano and, as such, tends to erupt explosively, as compared to shield volcanoes, which have less explosive and slower-moving eruptions (like those in Hawaii and elsewhere). On March 20, 1980, an earthquake of magnitude 4.2 on the Richter scale was the first indication that Mount St. Helens was awakening from a 123-year dormant period. Earthquake activity increased during the following weeks as magma continued to move into the shallow crust beneath the volcano. Then on May 18, at about 20 seconds after 8:32 a.m. local time, a magnitude 5.1 earthquake struck 1.6 km (1 mi.) beneath Mount St. Helens, causing its unstable north flank to collapse, triggering a rapid and violent explosion (**Figure 13.8A**).

   This cataclysmic eruption resulted in widespread devastation and the loss of 57 human lives, including that of a volcanologist studying the mountain. The volcanic eruption released 24 megatons of thermal energy, ejected more than 2.79 km³ (0.67 cu. mi.) of volcanic material, and reduced the mountain's height by 400 m (1,312 ft.). The ecological devastation following this eruption was staggering (**Figure 13.8B**), as it completely obliterated more than 600 km² (232 sq. mi.) of coniferous forests. Not surprisingly, in the following days and weeks, ecologists were drawn to the newly devastated landscape. Researchers wanted to document and study changes on the landscape to better understand how natural forces contribute to the successional processes that build ecosystems from scratch.

One of the first scientists to set foot on the new ground was a University of Washington plant ecologist named Roger del Moral. The work of del Moral and his lab over the four decades since the eruption has produced one of the longest continuous records of primary succession in the world and has resulted in nearly 100 scientific papers. Del Moral arrived at what would become permanent field sites a little more than a month after the eruption and found severely blasted landscapes that, in many areas, contained no living organisms. By repeatedly collecting ecological data from the same sites every year, del Moral and colleagues documented primary ecological succession and community recovery over time and got a firsthand look at how bare substrate morphs into various plant communities over time (**Figure 13.9**).

Figure 13.9 These photos, taken by Roger del Moral and collaborators at the same spot over the years since the 1980 eruption of Mount St. Helens, provide visual evidence of primary succession.

1982

1991

1999

2008

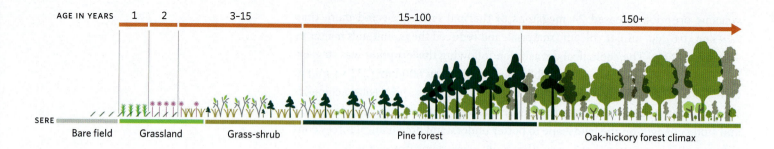

SERE

Bare field   Grassland   Grass-shrub   Pine forest   Oak-hickory forest climax

Figure 13.10 In the process of secondary succession in the Piedmont region of North Carolina, a bare (but not entirely lifeless) field changes to grassland, then scrubland, then pine forest, and eventually into a mature forest of oak and hickory (Odum 1971).

CRABGRASS   HORSEWEED   ASTER   BROOM SEDGE   SHRUBS   PINE   HARDWOOD UNDERSTORY   OAK   HICKORY

KEY

Secondary Succession in North Carolina's Piedmont Forests   The pattern of community change associated with secondary succession varies depending on site location and the stage at which the successional sequence begins. We use as an example the secondary succession following the abandonment of agricultural fields in the forests of the Piedmont region of North Carolina (**Figure 13.10**). It is important to note that the secondary successional process in the Piedmont forests takes about 150 years. Because of this timescale, it is impossible for any one person to witness the entire process or study it from start to finish at one site. Thus, ecologists must turn to alternative methods, such as chronosequences, to study the entire process. Nonetheless, succession of abandoned farmland into North Carolina Piedmont forest is one of our best-understood examples of secondary succession because of the careful work of Catherine Keever (1950), who documented these successional processes in North Carolina starting in 1947.

Thanks to Keever's efforts, we have a clear understanding of the process, which progresses as follows: Initially, crabgrass (*Digitaria sanguinalis*) and other grasses colonize abandoned fields. Within one or two years of abandonment, horseweed (*Erigeron canadensis*), asters (*Symphyotrichum* spp., and often *S. pilosum* in Keever's work), and broom sedge (*Andropogon virginicus*) colonize, creating a meadow community. Shrubs colonize the site next, followed by pines, which create a closed-canopy pine forest within about 15 years of abandonment. The pines only last approximately one human lifetime, with dying pine trees being replaced by oaks and hickories (**Figure 13.11**). Within 150 years, a mature oak-and-hickory forest dominates the former farm field. This plant community will persist indefinitely or until a disturbance resets the site to an earlier successional state.

It is worth noting that Keever's later retrospective work includes a discussion of the ideas of Gleason and Clements from the perspective of someone who was originally taught Clements's theories but eventually found her own research supporting Gleasonian ideas (Keever 1983).

Animal Succession in the Piedmont   Though plant species are clearly the drivers of successional processes, animal communities also undergo compositional changes that can be considered succession. **Figure 13.12** depicts some of the breeding passerine bird species found in different successional communities in

Figure 13.11 Pine trees are slowly being replaced by oaks and hickories in this picture of a Piedmont forest in North Carolina.

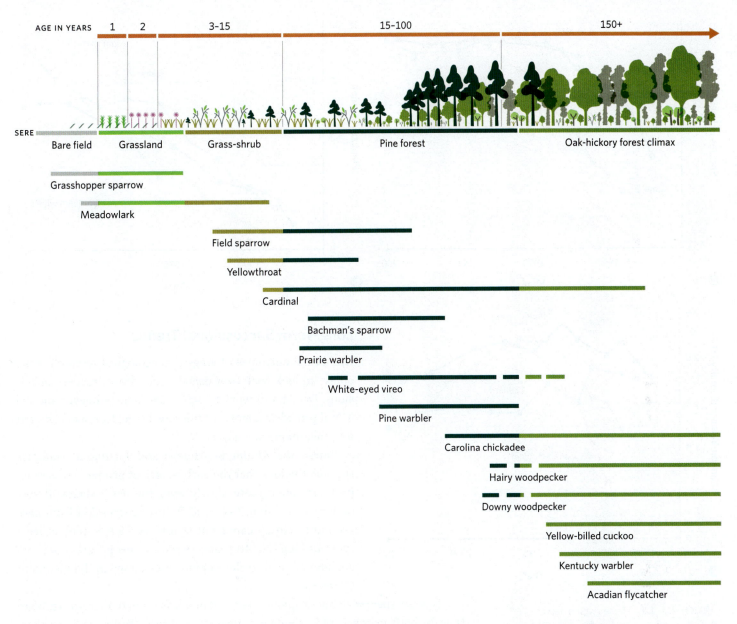

AGE IN YEARS 1 2 3-15 15-100 150+

SERE Bare field Grassland Grass-shrub Pine forest Oak-hickory forest climax

Grasshopper sparrow

Meadowlark

Field sparrow

Yellowthroat

Cardinal

Bachman's sparrow

Prairie warbler

White-eyed vireo

Pine warbler

Carolina chickadee

Hairy woodpecker

Downy woodpecker

Yellow-billed cuckoo

Kentucky warbler

Acadian flycatcher

the Piedmont region. Notice that shifts in the bird community composition over time seem to exhibit patterns similar to those in the plant community. Although the transition is gradual, there is little to no overlap in the bird species found in the earliest successional stages and the final successional stages.

This connection between the plant and bird communities is mediated through each bird species' habitat requirements. Some species, including grasshopper sparrows (*Ammodramus savannarum*), require sparsely vegetated grasslands for nesting, and such sites are typical of successional seres at one to three years after farmland abandonment. Other bird species, such as the Kentucky warbler (*Geothlypis formosa*), prefer deep, shaded woods with dense thickets, and bottomlands near creeks or ravines in upland deciduous forests. These types of forest develop some 100 or more years after farmland abandonment. For other organisms—such as amphibians, insects, or fungi—we also see the species composition change in a fairly predictable manner as a successional timeline proceeds, with the speed and alterations in diversity changing across different organisms.

Figure 13.12 The shifts in breeding-bird distribution across the different successional stages in the Piedmont forests of North Carolina exhibit patterns similar to those in the plant community. Note that there is little to no overlap in the bird species present in the earliest successional stages and the final successional stages (Odum 1971).

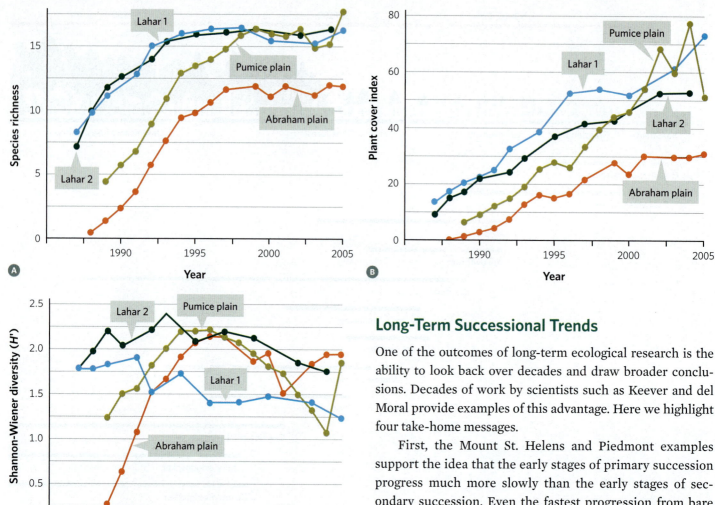

Long-Term Successional Trends

One of the outcomes of long-term ecological research is the ability to look back over decades and draw broader conclusions. Decades of work by scientists such as Keever and del Moral provide examples of this advantage. Here we highlight four take-home messages.

First, the Mount St. Helens and Piedmont examples support the idea that the early stages of primary succession progress much more slowly than the early stages of secondary succession. Even the fastest progression from bare ground to closed canopy at Mount St. Helens took at least twice as long (30–40 years) as the slowest progression from abandoned field to closed pine-tree canopy in the Piedmont (15 years).

Figure 13.13 These graphs depict Ⓐ plant species richness, Ⓑ plant cover, and Ⓒ Shannon-Wiener diversity (*H'*) since the 1980 eruption of Mount St. Helens across four sampling locations within the blast zone. In all graphs, each line represents one sampling plot that was surveyed yearly from the 1980s to 2005. Both species richness and plant cover increase over the time period, but *H'* levels off or decreases after an initial increase in all four sampling locations within the blast zone (del Moral, n.d.).

Second, species richness, plant cover, and biomass seem to increase through time in both primary and secondary succession, although diversity—as measured by the Shannon-Wiener diversity index (*H'*; Chapter 11)—may not. Recall that the Shannon-Wiener index is a diversity measure that takes into account both species richness and species evenness. For example, regardless of how much vegetation was left at Mount St. Helens study sites immediately after the eruption, all the sites increased in plant species richness through time (**Figure 13.13A**). That pattern also holds for plant cover, which is a measure of how much space plants occupy (**Figure 13.13B**). During the first few years, there was close to zero plant cover at Mount St. Helens, but by the end of 2005 some sites had nearly 80% coverage. Interestingly, following a rapid early colonization, Shannon-Wiener diversity seemed to level off quite quickly (**Figure 13.13C**). This trend suggests that community evenness declined as richness increased. Nonetheless, as you can see from the pictures in Figure 13.9, the successional process overall moved the plant communities along a path toward more ground cover and vegetation. Based on these data, del Moral and colleagues have estimated that it may be another 75–150 years before many of the Mount St. Helens sites ecologically resemble the pre-eruption forest community.

A third lesson is that chance events are significant factors in determining the specifics of the successional process. On Mount St. Helens, the timing of the blast directed the successional processes that followed. If the volcano had blown later in the year, the loss of existing vegetation would have been far worse, and the successional process even slower. However, because the volcano erupted in May and many higher-elevation forests still had snow on the ground, a fraction of the existing vegetation survived under an insulating and protective blanket of snow. Additionally, the summer following the blast (summer of 1981) turned out to be very hot and dry, and this likely restricted the process of initial seedling establishment by creating relatively hostile germination conditions. Although the snow allowed some species to survive and speed up the process, the dry conditions slowed the process, and different combinations of these factors in conjunction with the physiology of colonizing species led to different rates of succession around the blast zone. In the Piedmont example of secondary succession, it was exactly this type of random variation that led Keever to reflect back on her career and note vegetation patterns that fit Gleason's predictions more than those of Clements.

We see this serendipity not only in time but also across space. Environmental variability present across landscapes can play a key role in determining the pace and outcome of succession. Particularly during early successional stages, the dispersal ability of plants is a key limitation on community membership. Distance effectively filters the potential suite of plant species that are able to colonize a new area due to some species' limited dispersal capabilities (Chapter 12). On Mount St. Helens, the initial colonists of the severely denuded study sites were the few species with light, buoyant, wind-dispersed seeds (**Figure 13.14**) that managed to ride the air currents from locations fairly far away. In contrast, at denuded sites immediately adjacent to intact plant communities, seeds came from plants in those nearby intact communities. These differences led to a greater diversity of plant species in the less-isolated sites and produced slow-moving colonization fronts along all the intact vegetation boundaries.

In some cases, long-distance dispersal of plant seeds on Mount St. Helens occurred through the movement of larger animals, such as birds, elk, and coyotes (Wood and del Moral 2000). Such mutualistic seed dispersal allowed poorly dispersing plants to "jump over" the slow colonization front in a few locations. These animal-assisted jumps created little oases of more mature vegetation in otherwise bare areas within the blast zone. These patches of more complex plant communities began to expand in size, producing a *spatial mosaic* of plant communities across the formerly barren landscape (**Figure 13.15**). Observations like these demonstrate that there is no single linear path to succession. Sites adjacent to each other may reach successional stages at different rates and may end up with communities that differ in subtle ways in terms of species composition.

Finally, one of the most important findings from del Moral and colleagues' research is that the process of succession is generated through a complex set of biotic interactions. Recall that early Clementsian succession theories suggested

Figure 13.14 After the eruption of Mount St. Helens, aerially dispersed fireweed seeds (*Chamerion angustifolium*) played a big role in colonizing bare sites within the blast zone. The long, fuzzy structure on each seed, called a pappus, allowed the seeds to float in the air and move long distances with the wind from unaffected areas into the blast zone.

Figure 13.15 By 2006, the successional patterns of plant community recovery at Mount St. Helens created a spatial mosaic of "islands" of plant communities, with different species and densities occurring across these colonized patches.

Figure 13.16 Prairie lupine (*Lupinus lepidus*) was a critical early colonizer in the Mount St. Helens blast zone because of its ability to increase soil nitrogen levels. A mutualism with symbiotic bacteria allows lupine to capture atmospheric nitrogen and convert it to nitrogen that is available to plants. This species, therefore, provided a critical and scarce nutrient resource to the volcanic ash soils.

that abiotic factors drove changes in species composition and that species facilitated each other's colonization of a site, with these two forces moving the community inexorably toward a climax state. Gleason's individualistic theory, on the other hand, suggests that repeated community assemblages arise because the constituent species share similar abiotic requirements. But what about biotic interactions such as competition or predation? We know from Chapters 7–11 that species in close proximity will interact in a variety of ways that affect abundance, survival, coexistence, and community composition. How do these factors determine successional patterns in community composition?

Research on Mount St. Helens suggests that the introduction of organic material, such as seeds, pollen, dead insects, and spiders, into study sites was crucial for determining plant community membership. Some early-colonizing plants, such as prairie lupine (*Lupinus lepidus*) (**Figure 13.16**), increased the available soil nitrogen (Chapter 14), which in turn increased survival for later successional plant species. The presence of these early colonizers also provided "safe sites" with water, nutrients, and shade for the establishment of other plant species. Additionally, small burrowing mammals and their entrance and exit tunnels became important islands for plant establishment by aerating the soil and providing nutrients in the form of feces, which are rich in nitrogen.

Given the importance of competition and predation (Chapters 7 and 8), it should be no surprise that negative ecological interactions also play important roles in succession. Adding such interactions to our understanding of succession makes the ecological story more complicated than Clements's or Gleason's tidy theories. Is it possible to combine abiotic effects and these positive and negative biotic effects to come up with a theory or model of succession that organizes all these factors and allows ecologists to predict future community states?

Modeling Succession

Conceptual Models of Succession Our understanding of succession has increased tremendously since the 1920s and 1930s. In fact, a number of conceptual models for this process have been proposed since then. In 1977, Joseph Connell and Ralph Slatyer published an influential paper on succession that examined ecological mechanisms encompassing a range of relevant factors and outcomes. Connell and Slatyer wrestled particularly with why ecologists often observe a fairly predictable order of replacement of species at a site after a disturbance. To help explain this process, they suggested three separate conceptual models that could each produce species-by-species replacement: facilitation, inhibition, and tolerance (Connell and Slatyer 1977).

The idea behind the **facilitation model of succession** is relatively simple and fits well with a Clementsian view. The facilitation model posits that earlier species alter the environment in ways that enhance the establishment and growth of later successional species. Of course, these early species do not alter the environment *because* they are trying to facilitate later species. Instead, the alterations improve the fitness of early colonizers by improving their own access to resources or changing their exposure to abiotic elements. Along the way, though, the changes also improve conditions for other species, eventually paving the way for competitive dominance by later taxa and potential exclusion of the early colonizers.

Let's think of how this works in a practical sense using the Mount St. Helens example. The blast zone is initially colonized by small *r*-selected species, as described earlier. These mosses, grasses, and ferns have characteristics that make them good early colonizers. Once in place, these early species begin to transform the denuded landscape by providing soil-building organic matter (**Figure 13.17**), lifting water from deep underground via their roots, or creating shade oases in an open, sunbaked habitat. These alterations to the physical environment facilitate the establishment of later successional plants, such as shrubs. Notice that these alterations are not intended to assist later species but are either an effect of just being there (providing shade) and dying there (supplying organic material) or of improving individual fitness (drawing water from underground crevices). These unintentional changes to the environment by plants in early seres allow shrubs to outcompete the early species for available resources such as water and sunlight. The shrubs, in turn, further modify the environment and pave the way for the establishment of tree seedlings by adding more organic material. Eventually, a final, or climax, community is reached (Figure 13.17), chock full of *K*-selected species, and the environmental transformations cease until a disturbance resets the successional clock.

The **inhibition model of succession** works in almost the exact opposite way. Again, remember that this is a conceptual model to explain observed patterns, and the model does not suggest that organisms are choosing a path or following rules set by the model. In the inhibition model, early successional species actively inhibit and prevent the establishment and growth of species that arrive later. This suppression is periodically broken when an environmental stress or disturbance alters the competitive hierarchy, at which point new, later-stage species are able to successfully colonize and take over a site. In the Piedmont example of secondary succession, horseweed is an *r*-selected species that establishes in the community soon after farm abandonment. Its tiny seeds with feathery structures that assist long-distance dispersal (like the fireweed seeds in Figure 13.14) arrive early and take advantage of empty ground. Each horseweed plant grows first as a flat circle of leaves called a rosette (like the leaves at the bottom of a dandelion) and then sends up a tall stalk with a multitude of small flowers that produce thousands of seeds. The rapid production of seeds and new plants and the consumption of space by rosettes prevent or slow later plant species from colonizing the site. In similar examples, inhibition may involve **allelopathy**, the release of toxins by plants to impede the growth of other plants.

In an extreme example of inhibition, the community remains in one successional stage until a disturbance removes the inhibiting dominant species and allows a different wave of plant species to colonize (**Figure 13.18**). In the

Facilitation model of succession

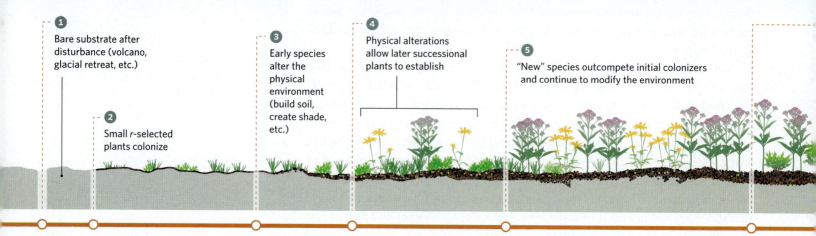

① Bare substrate after disturbance (volcano, glacial retreat, etc.)

② Small *r*-selected plants colonize

③ Early species alter the physical environment (build soil, create shade, etc.)

④ Physical alterations allow later successional plants to establish

⑤ "New" species outcompete initial colonizers and continue to modify the environment

TIME

Figure 13.17 Under the facilitation model of succession, plants that are early colonizers grow and provide organic matter that helps break down rock and build soil. The new soil holds water and nutrients that allow later species to germinate, grow, and eventually replace the early species. The newer plants add more organic matter and more nutrients and may provide nursery areas for shrubs. The facilitated shrubs eventually outcompete smaller plants through shading, add more organic material and nutrients, and often change the soil pH. All of these changes encourage the germination and growth of trees that outcompete the shrubs and form a climax community.

Inhibition model of succession

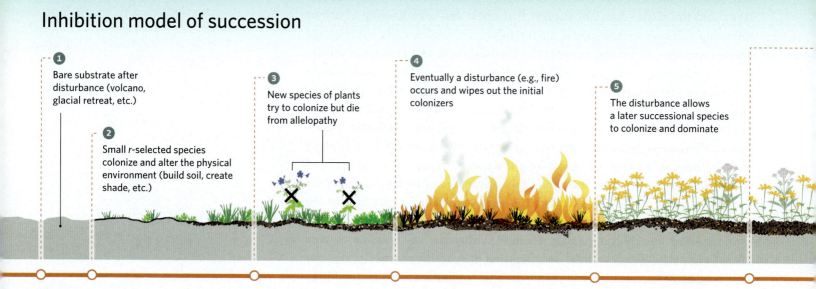

① Bare substrate after disturbance (volcano, glacial retreat, etc.)

② Small *r*-selected species colonize and alter the physical environment (build soil, create shade, etc.)

③ New species of plants try to colonize but die from allelopathy

④ Eventually a disturbance (e.g., fire) occurs and wipes out the initial colonizers

⑤ The disturbance allows a later successional species to colonize and dominate

TIME

Figure 13.18 If inhibition were dominant throughout the successional process, we might see the stages shown in this example. At each stage, a disturbance, such as fire or grazing, is necessary to remove the dominant early-colonizing species and allow other species to colonize or perhaps eventually to outcompete the early colonizers.

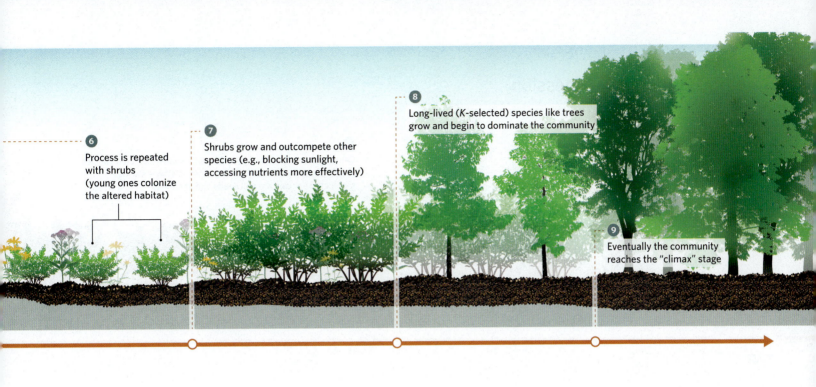

6 Process is repeated with shrubs (young ones colonize the altered habitat)

7 Shrubs grow and outcompete other species (e.g., blocking sunlight, accessing nutrients more effectively)

8 Long-lived (*K*-selected) species like trees grow and begin to dominate the community

9 Eventually the community reaches the "climax" stage

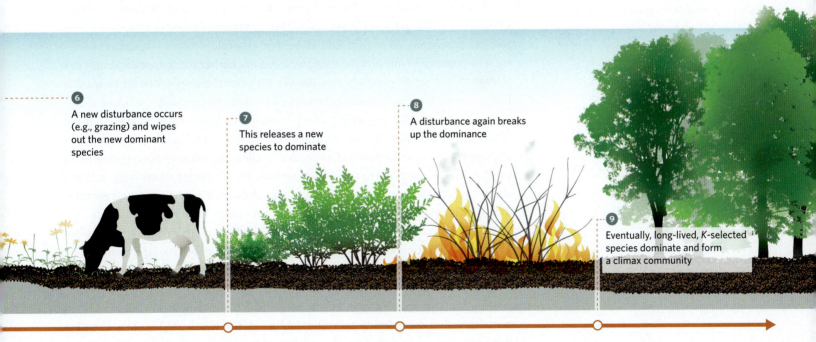

6 A new disturbance occurs (e.g., grazing) and wipes out the new dominant species

7 This releases a new species to dominate

8 A disturbance again breaks up the dominance

9 Eventually, long-lived, *K*-selected species dominate and form a climax community

Tolerance model of succession

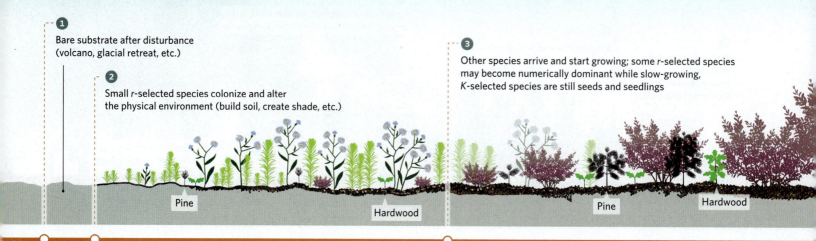

1 Bare substrate after disturbance (volcano, glacial retreat, etc.)

2 Small *r*-selected species colonize and alter the physical environment (build soil, create shade, etc.)

3 Other species arrive and start growing; some *r*-selected species may become numerically dominant while slow-growing, *K*-selected species are still seeds and seedlings

Pine

Hardwood

Pine

Hardwood

TIME

Figure 13.19 Under the tolerance model of succession, species arriving later in the process are neither helped nor hindered by earlier colonizers. Because later *K*-selected species grow slowly, *r*-selected species dominate early successional stages. Eventually the *K*-selected species outcompete the *r*-selected species and dominate the community.

inhibition model, this cycle of inhibition and release from inhibition continues until a final climax community is reached. Under this model, the final community of *K*-selected plants has life history characteristics that allow it to exclude other species at the site, and it takes a significant disturbance to reset the playing field and begin the successional process again.

Inhibition generally relies on periodic stresses or disturbances in order to progress to later successional stages, but sometimes other species can play a role in overcoming ecological inhibition. For example, when a dominant species such as horseweed inhibits the establishment of another plant species, it may not inhibit all the other plant species. In the Piedmont example, broom sedge can help to break the dominance of horseweed because broom sedge releases allelopathic chemicals that negatively affect horseweed. Asters take advantage of the newly opened ecological spaces vacated by horseweed faster than broom sedge can, mostly because the seeds of asters are smaller and colonize open spaces more rapidly. The interactions in these early seres in the Piedmont forest suggest that a single successional model does not have to be the only force structuring a community. In this case, horseweed inhibits, and broom sedge facilitates.

Finally, Connell and Slatyer suggested a **tolerance model of succession**, in which early successional species have little to no effect on later stages. The early *r*-selected colonists arrive first and dominate as the *K*-selected species are also arriving (**Figure 13.19**). The *K*-selected species grow slowly, but they are larger and longer-lived and eventually block the sunlight for *r*-selected species. Once their population is well-established, the *K*-selected species can hinder early taxa from dominating again, often by altering the soil or shading out smaller plants. Because the late-stage species take a much longer time to mature, it takes a long time for them to dominate the site after they colonize. The tolerance model, therefore, reflects the inherent differences in life histories among species in terms of growth, size, time to maturity, and related competitive ability.

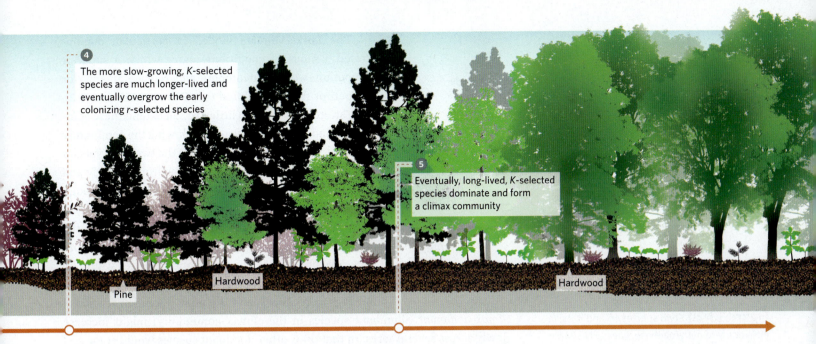

The more slow-growing, *K*-selected species are much longer-lived and eventually overgrow the early colonizing *r*-selected species

Eventually, long-lived, *K*-selected species dominate and form a climax community

Pine

Hardwood

Hardwood

The tolerance model has found quite a bit of support from recent research on the late-successional dynamics of forested systems.

Connell and Slatyer proposed these models to move beyond the Clementsian model of succession, which is based solely on facilitation. The three models offer a wider range of interactions that might be at play in succession, and there is no reason to assume that only one of these models regulates succession through all seres. All three of Connell and Slatyer's models have been experimentally tested many times since 1977. The accumulated evidence suggests that, although one model might work quite well to explain the change in dominance from one sere to the next, long-term successional processes rarely conform to any single model. These processes instead often exhibit aspects of two or more of the models simultaneously. In general, ecologists have found quite a bit of evidence for facilitation in the early stages of primary succession, although it can appear at other stages of the process as well. There is also plenty of evidence for inhibition, but it does not seem to be restricted to, or most common at, a particular successional stage, as suggested by Connell and Slatyer. Good evidence for the tolerance model is found in the later stages of succession in both tropical and temperate forests, where late successional species seem unimpeded and often unfacilitated by species in earlier seres.

Quantitative Models of Succession   At this point, it may be clear that succession is not a simple process but is instead the result of a series of complex processes involving life history characteristics and interspecific interactions. Connell and Slatyer's models for this process are conceptual, but there is a clear ecological basis for them in the theories we presented in Chapter 9 (mutualisms) and Chapter 7 (competition). Based on either the Tilman resource-ratio model or the Lotka-Volterra model of competition (Chapter 7), we should expect that, over time, species will outcompete and replace their inferior competitors.

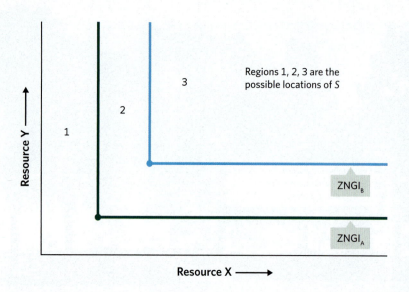

Figure 13.20 Two species (A and B) have different zero net growth isoclines (ZNGIs) for two resources (X and Y). The ZNGIs create three distinct areas on the graph (regions 1, 2, and 3), and the potential for coexistence depends on which region contains the environmental resource supply point (S). In region 1, resource levels are too low to support either species, and both go extinct if S lies in this region. In region 2, the resource levels are too low to support species B, and only species A can survive. In region 3, there are enough resources to support both species, but consumption by species A draws the resource levels down into region 2 and eventually to an equilibrium point along the ZNGI for species A. At this point, species A outcompetes species B because it can use resources at a lower level than species B. Therefore, in this scenario, species A competes better for both resources and always outcompetes species B.

The Lotka-Volterra model, in particular, demonstrated a range of conditions under which a superior competitor would outcompete and replace an inferior competitor that arrived first (see Figures 7.14B and 7.15). If we stretch our minds a little, we can extend these competition models to succession by allowing the community to include more than two species. If the two focal species are, in fact, the two dominant species from an early and a later sere, then we can also envision stepping through a series of Lotka-Volterra models, with superior competitors in one sere becoming the inferior competitors in the next sere. In this progression through successively more superior competitors, the final climax species would be the most superior. The individuals of the climax species might mature slowly, and the population density of the *climax dominant* (dominant species in the climax community) might increase slowly across several seres. In that time, other dominant species would come and go, and the climax dominant would tolerate those species as it moved toward its own dominant sere.

We can also understand succession through the resource-ratio hypothesis (Tilman 1977). It offers a mechanistic model that allows us to explore a range of resource levels and observe under what resource conditions a superior competitor outcompetes an inferior competitor (Chapter 7 and **Figure 13.20**). The scenario in Figure 13.20 includes just two species and assumes that one of them competes better for all resources. We can imagine, though, stacking such a model with a zero net growth isocline (ZNGI) for the dominant species of each sere, so that we move through a series of species and end up at the ZNGI of the climax dominant. Again, that sort of model fits well with the tolerance model, in which superior competitors eventually outcompete inferior competitors without the use of facilitation or inhibition. This insight from the Tilman model does not offer much beyond what we gained from the Lotka-Volterra model.

But what if resource levels change over time, as they do in succession? In that case, it makes sense that dominance could shift, and new superior competitors could replace previously dominant species. That situation is illustrated by **Figure 13.21**, where species A is the superior competitor when resource levels fall in regions 2 and 3, but species B is the superior competitor when resource levels fall in regions 5 and 6 (see Chapter 7 for more details). In the Mount St. Helens example, water availability, soil nutrient retention, and nitrogen content all change throughout the different successional stages. Those sorts of changes could shift the ecosystem resource levels from regions 2 and 3 of Figure 13.21 toward regions 5 and 6. This resource change through time could lead to a shift in the balance of competition between two (or more) species.

As the conditions and resources change with new species membership and new layers of organic matter in the soil, we can imagine community dominance shifting from one species to another through a series of species until we reach the climax community state. At that point, the community resource levels may be in region 4 of Figure 13.21, where the final set of dominant species coexists (see Chapter 7). In fact, at each sere along the way, the community may be populated by a set of species that can coexist under either resource-ratio or Lotka-Volterra competition until conditions change or a new species arrives.

What about some of the more complicated aspects of succession, though, such as the effects of history or the order of arrival of species? Both Keever's and del Moral's findings suggest that different sites proceed at slightly different paces through succession, depending on their proximity to seed sources and the initial environmental conditions immediately following volcanic eruption or farmland abandonment. In fact, the inhibition model of succession is an acknowledgment of the effects of order of arrival. Given this connection, can we explain successional inhibition with our models of competition?

The simple answer is yes. Remember that both the Lotka-Volterra and resource-ratio models of competition predicted that species interactions may lead to unstable coexistence (Chapter 7). Both models suggest that, in such a situation, the species that arrives first can exclude later-colonizing species. This outcome fits well with Connell and Slatyer's inhibition model.

In general, when an early-arriving species inhibits or facilitates a later-arriving species, ecologists call this a **priority effect**. Priority effects occur when the order of species' arrivals influences the final community membership. All three successional models suggest that priority effects can be quite important.

13.4 Community Assembly

Assembly Rules

As communities develop through time, priority effects can have profound consequences on species membership. Although the successional models work well to explain community transitions after they happen, small variations along the path of succession are not predictable with Connell and Slatyer's three models. To explain these differences, ecologists have turned to the study of **assembly rules**. These "rules" are more like guiding principles that explore how the timing of species arrival or the initial suite of colonizing species can determine the species composition of the final community.

If priority effects can contribute to the final makeup of a community, as models and empirical work suggest, it makes sense to ask whether there is any logic, order, or pattern to the assembly or construction of communities in nature. Ecologists have thought about the rules of community assembly since Jared Diamond first proposed them in the mid-1970s (Diamond 1975). The general concept is that nature may follow a set of "rules" that favor certain combinations of species or particular orders of arrival that lead to particular community memberships. For example, in a community that contains species A, B, and C, it may be that this set of species can only occur if A arrives at the site before species B, and that species B must be present before species C can colonize. This order of arrival would constitute a simple assembly rule that leads to a particular set of species coexisting in a community.

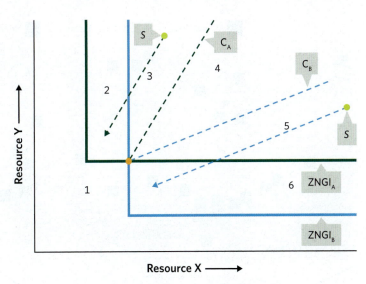

Figure 13.21 Sometimes the ZNGIs for two species overlap, meaning one species is better at exploiting resource X and the other is better at exploiting resource Y. The consumption vectors (C_A and C_B) suggest that species A is better at exploiting resource Y and species B is better at exploiting resource X (look at the slopes and notice which resource each species takes up faster), which creates six regions on the graph. To explore survival when the resource supply point (S) is in regions 3, 4, and 5, we need to think about how consumption will alter resource levels in each region. The light green dots each represent a sample S in regions 3 and 5, and the dashed arrow from the S shows the way one species can pull resource levels outside the "survivable envelope" for the other species. These arrows demonstrate that species A wins in region 3, and species B wins in region 5. In region 4, the two species can coexist, and resource availability should stabilize at the orange intersection point.

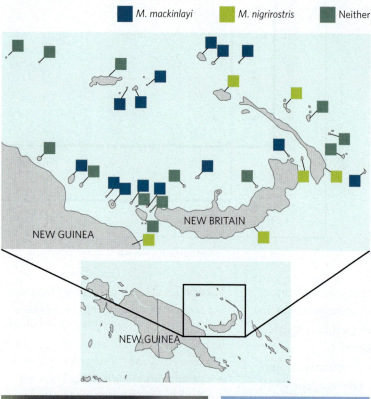

M. mackinlayi M. nigrirostris Neither

NEW BRITAIN

NEW GUINEA

NEW GUINEA

NEW GUINEA

M. nigrirostris

M. mackinlayi

Figure 13.22 Jared Diamond found that each of the New Guinean islands he explored had one of two cuckoo-dove species present (*Macropygia nigrirostris* or *M. mackinlayi*) or neither species present, but never both species. He likened this pattern to a checkerboard (Diamond 1975).

Initially, ecologists suggested that any data that demonstrated a nonrandom geographic pattern of how often two (or more) species co-occur provided evidence of an assembly rule in nature. Species *co-occurrence patterns* can range anywhere from two species always co-occurring no matter where we look across a landscape to two species never co-occurring. Diamond looked for such co-occurrence patterns among the avian fauna on New Guinean islands and found that some bird pairs never co-occurred on the same island. **Figure 13.22** shows the nonoverlapping distribution pattern for two species of native doves, the bar-tailed cuckoo-dove (*Macropygia nigrirostris*) and Mackinlay's cuckoo-dove (*Macropygia mackinlayi*). Notice that islands have either one or the other cuckoo-dove species, but never both. He called the resulting co-occurrence pattern a checkerboard, for its resemblance to the alternating colored squares of the popular game board.

Diamond suggested that the only way a checkerboard distribution pattern could arise was if there were underlying "rules" that prohibited the two species from coexisting on the same island. This assumption drew quick criticism from other ecologists. Some argued that such extreme patterns are rare in nature, but most of the early discussion about assembly rules centered on how to demonstrate statistically that co-occurrence patterns are truly nonrandom.

Demonstrating nonrandom patterns can be close to impossible, given the number of potential species combinations and the order in which the species can arrive through time. The total number of unique pathways to assemble an *n*-species community can be determined by taking the factorial of the number of species in the community (*n*!, which means *n* multiplied by every integer smaller than *n*). For example, there are 720 unique ways to assemble a simple six-species community (6! = 6 × 5 × 4 × 3 × 2 × 1 = 720). Even a modest increase from six to eight species in a community produces a substantial increase from 720 to 40,320 unique assembly routes. Most natural communities have significantly more species, which means that exploring even a moderately diverse community of 50 species requires considering a whopping 3.04×10^{64} possible assembly pathways. With that many combinations, it becomes impossible to determine whether a co-occurrence pattern we observe in nature, like the birds across the islands of New Guinea, represents the outcome of a nonrandom order of community assembly. Even if we are sure that the final outcome pattern seems nonrandom, there are so many assembly routes that may create that pattern that it is almost impossible to explore them all and identify general insights or rules about the ways that communities assemble.

The arguments around co-occurrence patterns and assembly rules eventually abated with a general consensus that simply finding patterns is not enough to demonstrate assembly rules. Yet the concept of rules that guide the assembly of communities was not abandoned. Logically, it seems there must be some assembly rules in nature, partly because we already know some exist. For example, a predator cannot colonize an area without appropriate prey being present first. Similarly, a species involved in an obligate mutualism cannot successfully colonize a community unless its mutualist partner is also present. And finally, for any species, all vital resources must be present at a location for that species to colonize and create a self-perpetuating population. We also know that other rules likely exist because, when ecologists arbitrarily choose species and/or randomly combine them in either models or small-scale communities, the results generally fail to produce a persistent suite of species (Weiher and Keddy 1999).

So instead of abandoning the idea of assembly rules altogether, ecologists turned to experiments. Experiments on community assembly might seem impossible if we only consider communities composed of organisms that we see and notice every day. How would a researcher experimentally order, and then reorder, the arrival of plants, herbivores, and carnivores across Piedmont forest communities for appropriate replication and follow them for long enough to see community assembly outcomes? Clearly, the logistics, timescale, and replication of such experiments are prohibitive.

Nonetheless, there are some approaches that offer meaningful insight. For example, ecologists can focus on plants and associated soil organisms and pollinators. Similar approaches have been used successfully in restoration ecology (see next section) but still pose significant experimental challenges. Fortunately, not all ecological communities are composed of large organisms. Ecologists have at their disposal a whole world of primary producers, herbivores, and predators that are microscopic, living in water, soils, or even hydrothermal vents. These communities move through dozens, hundreds, or even thousands of generations relatively quickly and can be manipulated in a laboratory setting. This approach offers exciting opportunities to experimentally explore community assembly rules.

Some of the more influential studies of assembly rules occurred at the University of Tennessee laboratory of James Drake. In order to investigate the effect of assembly dynamics on community composition, Drake (1991) constructed experimental aquatic communities within flasks kept under controlled laboratory conditions and then varied the sequences of species additions into these flask communities. His communities were still rather small (fewer than 15 species), which allowed him to run a series of replicated experiments in which he kept environmental conditions exactly the same but altered the sequence of species' colonization events until each species in his species pool had at least a chance to establish within each community.

For his experimental communities, Drake created two sizes of replicated environments, small (250 mL flasks) and large (40 L aquaria). To populate these environments, he chose a group of six species of aquatic organisms for the small system and 15 for the large (**Figure 13.23**). This group of species included primary producers, large and small herbivores, and some predatory zooplankton. Although fairly tiny, these organisms offered the same trophic levels that are found in meadows with plants, herbivores, and carnivores. Drake's treatments manipulated the sequential introductions of the species into the experimental communities over weeks. For example, the six different small-flask treatments were each replicated

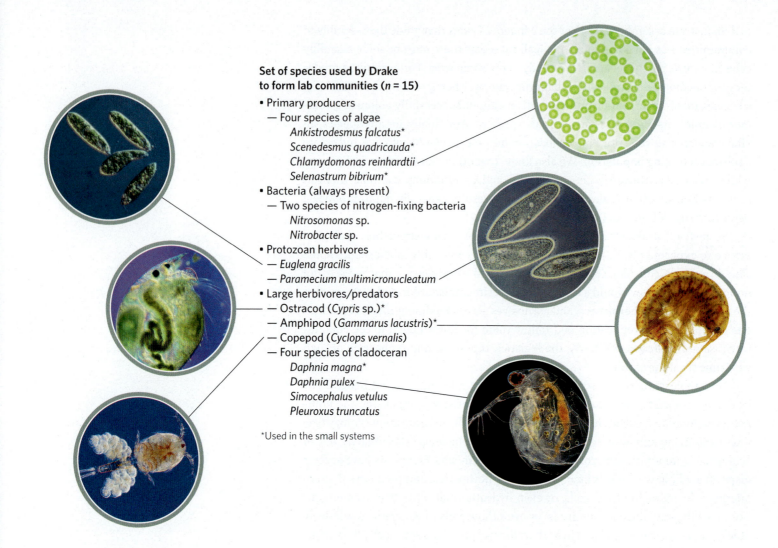

Set of species used by Drake to form lab communities (n = 15)

- Primary producers
  - Four species of algae
    - *Ankistrodesmus falcatus**
    - *Scenedesmus quadricauda**
    - *Chlamydomonas reinhardtii*
    - *Selenastrum bibrium**
- Bacteria (always present)
  - Two species of nitrogen-fixing bacteria
    - *Nitrosomonas* sp.
    - *Nitrobacter* sp.
- Protozoan herbivores
  - *Euglena gracilis*
  - *Paramecium multimicronucleatum*
- Large herbivores/predators
  - Ostracod (*Cypris* sp.)*
  - Amphipod (*Gammarus lacustris*)*
  - Copepod (*Cyclops vernalis*)
  - Four species of cladoceran
    - *Daphnia magna**
    - *Daphnia pulex*
    - *Simocephalus vetulus*
    - *Pleuroxus truncatus*

*Used in the small systems

Figure 13.23 James Drake (1991) used an elaborate and exacting experimental design to examine community-assembly patterns in a series of lab-based aquatic communities. This is a list of the species used to populate his experimental communities.

three times and followed a predetermined pattern of introducing three algal species over the span of a few weeks before adding any primary consumers. In the large aquaria, he followed a similar approach that involved all four algal taxa, but he introduced them over a longer period of time.

Not surprisingly, Drake's experimental setup demanded some herculean efforts to overcome the various methodological challenges associated with creating species-pure containers. For example, he needed to develop a synthetic freshwater medium that contained the right ratios of nutrients to sustain aquatic life for an extended period of time (more than one year), which required making a mixture of growth media used for both algal and zooplankton taxa. It was also challenging to raise populations of single species of plankton and then add them to the experiment without contaminating the flask with "hitchhiker" species, such as the food that the focal species was eating. In addition, he had to standardize the reproductive state of the consumers being added to the treatments. If half of the consumer population was effectively pregnant, he could inadvertently introduce 25 or 50 individuals when trying to introduce only 10.

Despite these and other challenges, Drake was able to complete the experiments and produce some interesting results. For example, when looking at the three algal species in the small-flask experiments, he found that one species

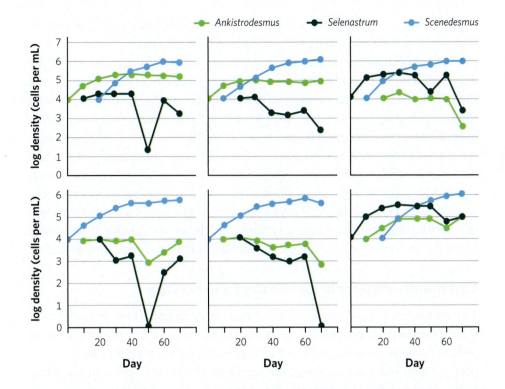

Figure 13.24 Drake plotted the population-growth trajectories of the primary producers (species of algae) in the small-flask community assembly experiment. Each panel represents a different sequence of invasion for the small flask experiment. Notice that *Scenedesmus quadricauda* population numbers dominate at the end of 60 days, regardless of the order in which the species were introduced (Drake 1991).

(*Scenedesmus quadricauda*) always dominated the community, regardless of when it was added to the flask environments (**Figure 13.24**), meaning there was no assembly rule for dominance. The abundance and even survival of the other two species, though, depended on the order in which they were allowed to colonize (Figure 13.24). This second result suggested there was an assembly rule that dictated the density and coexistence of the competitively inferior species.

Yet when Drake examined the results from the larger systems, with four species of algae instead of three, the hard-and-fast "rules" observed in the small-flask environments fell apart. In these larger aquaria systems, *S. quadricauda* did not consistently competitively dominate. In fact, in the larger experimental ecosystems, Drake found no particular dominance by any of the algal taxa. The only somewhat-repeatable pattern across treatments was that, in the majority, the first algal species added became the competitively dominant one, suggesting a kind of priority effect and assembly rule rolled into one.

Drake's results for the consumer species were also complex. The major insights were as follows:

- Widely different communities were produced solely through alterations in the sequence of species' colonizations.

- Variability among treatment replicates was low, indicating that whatever process of community assembly was happening was relatively repeatable.

- Some sequences of colonization led to community compositions that prevented any further species colonization.

- At least for the algal species, species dominance was due to interspecific competition.

Drake's exploration of assembly rules ushered in an era of experimental community ecology that persists to this day. However, the focus solely on identifying consistent assembly rules per se has waned. In its place has emerged a quest to understand the various ecological mechanisms behind priority effects, the influence of environmental variation on community assembly trajectories, and the ways in which species' life history traits influence community assembly beyond the simple r- and K-selection insights mentioned previously. Ecologists pursuing this line of inquiry have recently considered how the contemporaneous evolution of species' traits influences community assembly. This research involves not only creating experimental communities in the laboratory but also experimentally evolving the species themselves. Such undertakings are far too complex to summarize here, but interested readers might explore the fascinating research emerging from Tadashi Fukami's lab at Stanford University. The ecological understanding of communities has come a long way from the debates of Clements and Gleason, as has the awareness of how community membership may change through time.

Community Assembly and Restoration Ecology

The discussion of how communities change through time leads us back to the topic of restoration ecology (Chapter 9). Though there are many situations in which ecologists cannot offer clear-cut theories to predict changes or even explain past changes, restoration ecology has significantly advanced our understanding of how the process of succession, interactions between dominant species, and assembly rules might work on local scales. Restoration ecology is like a giant series of ecological experiments to see if we can re-create communities that provide key ecosystem functions and services (see Chapter 14). In some cases, restoration efforts encourage succession on highly degraded sites, such as abandoned mines or abandoned agricultural lands (**Figure 13.25**). In other situations, restoration involves creating habitats de novo (e.g., as in wetland mitigation) by using succession to build a community from scratch. Either way, ecological restoration generally involves the application of successional ideas.

Left on their own, most abandoned sites will eventually acquire an assemblage of both plant and animal species. Management choices informed by ecological science help to ensure that the resulting community approaches a version of the mature natural community that originally existed at the site. An interesting example can be found in the work of Jason Andras, Katherine Ballantine, and

Figure 13.25 The foreground shows a 15-year-old restored riverside (riparian) forest that is now part of the Sacramento River National Wildlife Refuge, California. This restoration project is one of the largest and most ambitious riparian restoration projects in the United States. Before the native vegetation was restored, the land was used for agriculture.

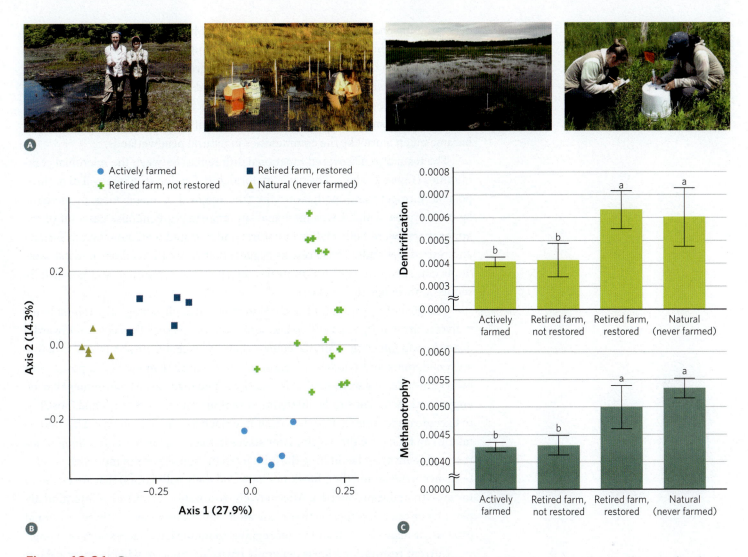

Figure 13.26 Ⓐ Jason Andras, Katherine Ballantine, and their collaborators explored the effect of restoration on the microbiomes of wetland soils. Ⓑ Their data showed that the prokaryotic soil communities in active cranberry farms and natural peat wetlands were very different from each other; that soil communities from unrestored, retired farms were very similar to those of active farms; and that the soil communites of restored farms eventually became much more like those of natural peat wetlands. Ⓒ The team also found differences in how the microbial communities functioned. In each graph, the y-axis represents the fraction of genes found in the soil biota that code for a function that reduces an anthropogenic effect. Bars with different letters (a vs. b) have statistically significant differences between their means and error bars represent ±1 standard error. These results suggest that restoration of microbial communities can speed up succession and potentially reduce anthropogenic influences (Andras et al. 2020).

their collaborators (2020; **Figure 13.26A**), who explored soil microbial communities and ecosytem functions across four types of wetland ecosystems in Massachusetts: active cranberry farms, retired cranberry farms with no restoration treatments, retired cranberry farms that received restoration treatments, and natural peat wetlands that were never farmed. They used genetic sequencing techniques to identify more than 151,000 distinct soil microbial taxa. (Note that these taxa are not necessarily species, but they are genetically distinct units, and researchers can use them to look at diversity and similarity in much the same way that we discussed in Chapter 11.) Statistical analysis then allowed them to plot sites based on similarities and differences among these soil microbial communities (**Figure 13.26B**). Each axis in the graph shown represents a composite measure of community characteristics. The clumping of points from different types of wetlands suggests that the microbial communities differed across treatments. They found that soil microbial communities in active cranberry farms and natural peat wetlands were very different from each other, and that retired farms with no restoration treatments were very similar to active farms. After just four years of restoration treatments, however, the microbial soil communities in restored farms became much more like the communities in natural peat wetlands.

The team also discovered functional differences between the microbial communities (**Figure 13.26C**). The capacity for denitrification (the removal of anthropogenic nitrogen) and methanotrophy (the removal or metabolism of methane by prokaryotes, which keeps methane—an important greenhouse gas—out of the atmosphere) were both higher in natural and restored wetlands than in farmed and retired wetlands. These results suggest that restoration efforts in these wetlands can speed up succession in the microbial communities and potentially reduce anthropogenic effects.

The goal of restoration projects is to make sites physiologically tolerable for a diverse array of species and to accelerate the processes of succession toward a desired endpoint community. Of course, there is much debate about how to define such endpoints and whether a "natural" or "original" community is possible or even knowable. Regardless, what is clear is that changes to an environment can be engineered by humans to facilitate important or critical organisms and, possibly, to exclude organisms that are less desirable, such as invasive species. This undertaking can then pave the way for later stages in a successional process. Sometimes restoration involves facilitating the joint introduction of pairs of mutualist species (e.g., mycorrhizae and their hosts, or plants and their pollinators) that are essential for growth and reproduction. Alternatively, restoration can involve intentionally planting or releasing species that come after inhibition stages of the successional process, or managing soil and water regimes to mimic later successional systems.

Current restoration efforts generally try to incorporate what is known about successional pathways and local environmental heterogeneity into the planning stages. They also play an important role in providing large-scale experimental opportunities to study the process of community development and succession in the field. In addition, a firm knowledge of community ecology is used to assess the success of restoration projects, and we encourage anyone interested to pursue their studies in this growing and fascinating field of ecology. Some scientists suggest that the future of conservation work, and perhaps the future of the planet, belongs to the study of restoration science, where we can put much of our accumulated ecological knowledge into positive action.

13.5 Disturbance Dynamics

At the start of this chapter, we said we would explore different elements of change in communities. Although we have made it clear that natural systems are not static and that stochastic or serendipitous elements play a role in community dynamics, we have also focused largely on predictable patterns and trends, inadvertently emphasizing steady-state dynamics. It should come as no surprise, though, that natural community dynamics are much more complicated. In order to delve into this complexity, we need to restart the discussion somewhat by shifting our focus to changes that disrupt the status quo and alter the predictable path of succession. To do that, we need to think about predictability.

Change in environmental conditions occurs on hourly, daily, yearly, and longer timescales. Many of these changes are predictable and happen frequently enough that they become part of the background environment to which an organism is evolutionarily adapted. For example, the predictable daily patterns of light availability and tidal cycles, and the predictable seasonal patterns of temperature, daylight hours, precipitation, and water flow, all create evolutionary forces to which species adapt. All organisms have a suite of evolutionary adaptations attuned to these predictable variations in the environmental conditions they regularly and predictably encounter. Even if change in environmental conditions is common, some changes lie outside the adaptive reach of many species, especially if they occur in an unpredictable fashion or if each event affects only local portions of a species' range.

Why, you may wonder, are we talking about species-level effects of environmental change in a chapter about changing communities? Well, we can understand how environmental factors alter community composition by examining both the predictability of these factors through time and the adaptability of species in a community to these changes.

At this point, it is important to define **disturbance**, a fairly simple concept that is surprisingly difficult to define in a single sentence. A complete definition needs to encompass both the factors that cause changes and the changes themselves, and perhaps even the processes associated with such changes. Ideas about disturbance and its effects on community composition have a long and varied history in ecology. Some of the earliest ecologists (including Darwin) wrote about disturbance in terms of ecological communities. Clements even discussed the role of disturbance in the process of forest succession, although he generally dismissed it as a relatively unimportant step on the inevitable path to realizing a forest climax community. This dismissal of disturbance as present but not particularly important was typical of the early decades in ecology. Not until the 1970s and 1980s did interest in the role of disturbance take off (McIntosh 1985).

Definitions of Disturbance

Many ecologists have attempted to define the concept of ecological disturbance, often relying on a particular ecosystem or taxon to help focus their definition. The similarities and differences are obvious among the following most commonly cited definitions:

- A process removing or damaging biomass (Grime 1979)

- A discrete or punctuated killing, displacement, or damaging of one or more individuals (or colonies) that directly or indirectly creates an opportunity for new individuals (or colonies) to be established (Sousa 1984)

- Any relatively discrete event in time that disrupts ecosystem, community, or population structure and changes resource pools, substrate availability, or the physical environment (White and Pickett 1985)

- Any process that alters the birth and/or death rates of individuals in a community by killing or affecting resources, natural enemies, or competitors in ways that alter survival and fecundity (Petraitis et al. 1989)

Figure 13.27 Is a hurricane a disturbance event or a process? The answer depends on who is experiencing it. For a long-lived animal or plant, one hurricane may be a single event. In contrast, for a short-lived organism, the hurricane is experienced as more of a process, with periods of high winds, heavy rains, and tidal surges.

- A change in community structure caused by factors external to the system of interest (Pickett et al. 1989)

- An event or process that causes the mortality of some but not all of the biological organisms making up the community in an ecosystem or the larger metacommunity (Newman 2019)

Some of these definitions focus on disturbance as a process, whereas others suggest that disturbance is an event. This discrepancy is mostly due to the temporal scale under consideration. For example, is a hurricane an event or a process (**Figure 13.27**)? The answer depends on the perspective from which the hurricane is viewed. To a long-lived tree or animal, the hurricane may be one disrupting event among many throughout its lifetime. To a short-lived insect, though, a hurricane may be a process involving high winds and rain that lead to unprecedented levels of dispersal and potential bodily injury. How about floods or fires? Are they events or processes?

Clearly, the timescale matters when defining a disturbance. For an organism that lives much of its life during the disturbance, it is a process. For one that lives for considerable time before or after the disturbance (or both), it is generally an event. How we characterize a flood event, for example, may depend on the focal species. Whatever the case, disturbance is generally linked to some type of ecological disruption, which then leads to an open opportunity, available resources, or a vacant area in a community.

The complex nature of ecological disturbance makes a cogent description of all disturbances quite difficult. Yet natural systems may have a signature **disturbance regime**, which, if described well, allows us to compare and measure disturbances through time or across space (Sousa 1984; White and Pickett 1985). The most common features of a disturbance regime are the following:

- Distribution: the size and spatial extent of the disturbed area, including the relationship to geographic and environmental gradients

- Magnitude: the intensity (physical force of the event) and severity (the impact on the organism, community, or ecosystem involved) of the disruptive force

- Frequency: the mean number of disturbances per unit time; sometimes synonymous with the probability of disturbance

- Predictability: the regularity of the disturbance event through time

- Return interval: the mean time between disturbances, or the inverse of the frequency

- Synergism: the effects of the disturbance on the occurrence of other disturbances, meaning one disturbance affects the frequency of another (as in the way the frequency of drought can affect the frequency of fires)

These disturbance metrics highlight the challenging and complex nature of ecological disturbance, reminding us that making generalizations and predictions may be inherently difficult (Lockwood et al. 2013).

In our discussion of succession, we mentioned disturbances a few times without giving much detail, but now it should be clear that disturbances can and do play an important role in succession. When explaining the difference between primary and secondary succession, we pointed to different temporal scales of disturbance. Some disturbances, such as farming, leave behind organic matter and species that can then start the process of secondary succession. Other disturbances, such as volcanic eruptions, result in the loss of most, if not all, organic matter and species, and thus start the process of primary succession. In this context, a disturbance essentially halts or resets successional processes.

On the other hand, and perhaps confusingly, we can also see a role for disturbance in moving succession forward, particularly when thinking about the inhibition model of succession. A disturbance might remove a competitively dominant species, thus facilitating the colonization success of species arriving later. Are there particular features of disturbances that might affect succession and community change in particular ways?

Types of Disturbance

Ecologists have suggested various schemes for categorizing types of disturbance. One simple dichotomy identifies disturbance as either *natural* or *human-caused*. Natural disturbances are ones without anthropogenic roots; examples include hurricanes, floods, landslides, fires, and even the removal of biomass due to the movement and grazing of large native herbivores. Human-caused disturbances, on the other hand, have their origins in human activity.

This dichotomy is somewhat arbitrary because many human disturbances can mimic natural disturbances. Think, for example, of the release of water from a dam, the harvesting of trees from a forest, a human-ignited fire, or the mowing of a field. All these disturbance events can mimic a natural disturbance if they match the magnitude and frequency of natural events. The practice of using prescribed fires for forest management (**Figure 13.28**), for example, is predicated on the idea that humans can start fires that mimic natural events in terms of their timing, intensity, and extent. But unlike natural forest fires, prescribed fires can be controlled through safeguards, thus minimizing the risk they pose to infrastructure or human lives. Naturally occurring fires do not send out text messages to coordinate their arrival with the schedules of nearby firefighting units. Nor do they start in precise, carefully selected locations to facilitate directing them away from houses or urban areas. Prescribed fires and other human-caused disturbances that mimic natural disturbance regimes can be critical to maintaining community composition in a more natural state.

Figure 13.28 A fire scientist initiates a prescribed burn in a Florida longleaf pine forest to aid in the management of the community and the regeneration of the longleaf pine trees.

(A)

(B)

Figure 13.29 These photos are of a grasshopper outbreak in Kansas in the 1930s and the aftermath. (A) A swarm of voracious, flying herbivores of this density can decimate vegetation, and the damage was extensive. (B) The pictured field was not harvested by farmers but eaten to the ground by the pests.

Nonetheless, some human-caused disturbances are fundamentally different from natural disturbances because of their type, frequency, duration, or magnitude. For example, the release of synthetic toxic chemicals or nuclear radiation into the environment has no natural analogue. Human-caused fires, even prescribed fires, can occur in locations, at times of the year, or at temporal frequencies that fall outside the natural fire regime for a community. And there are few natural equivalents to pouring asphalt or concrete over large areas of the ground or spilling millions of barrels of oil into the sea.

Sometimes, an anthropogenic disturbance can involve the *removal* of a natural seasonal event. For example, all flowing (lotic) water systems have what is called a **characteristic flow regime** for each particular geographic region (also referred to as a natural hydrologic regime; Poff et al. 1997). The flow regime for a region often includes periods of seasonal flooding, which can be eliminated by humans through the construction of dams or other artificial structures. In these lotic systems, the normal bouts of flooding can themselves act as disturbance by clearing species and organic material from locations within the stream channel. Also, any alteration to the natural flow regime can disturb the system because it changes the return interval, magnitude, or frequency of flooding toward regimes that the native species have not evolved to withstand. Flooding rivers often deposit nutrients on adjacent land, and the loss of flooding can fundamentally alter the productivity of adjacent terrestrial areas.

A key distinction, in terms of ecological outcomes, is that human-caused disturbances that mimic natural events tend to cause few changes in community composition, whereas human-caused events that fall outside the bounds of natural disturbance events—whether they lead to more or fewer disturbances, more- or less-intense disturbances, or longer or shorter disturbances (in terms of time)—can have profound effects on community composition.

Another simple dichotomy ecologists use to characterize disturbances is that of *abiotic* versus *biotic* mechanisms. Abiotic mechanisms are features of weather, geology, or climate. Biotic mechanisms, as we have said before, are factors driven by other species. Both types of disturbance can create a change in community composition, as both a drought (abiotic) and a trampling herd of buffalo (biotic) are able to alter grassland vegetation. Often, abiotic disturbances are easier to imagine, and we have already mentioned fires, floods, and hurricanes. One powerful example of biotic disturbance is grasshopper outbreaks; when grasshopper population abundances explode, these pests can consume vast amounts of vegetation and cause extensive damage (**Figure 13.29**). Similarly, the introduction of novel, non-native predators to a formerly predator-free ecosystem can have massive effects on species composition. A 2016 study found that non-native mammalian predators have led to the extinction of 142 species worldwide (Doherty et al. 2016), which is clearly a disturbance for those individual communities. Finally, biotic disturbances can be brought about by ecosystem engineers, which are species that alter their physical environment in ways that can lead to major changes in the composition of the community they inhabit (Chapter 10). The non-native beavers in South America discussed in Chapter 6 are an excellent example of biotic disturbance agents.

A third dichotomy for categorizing types of disturbances is *exogenous* versus *endogenous* (Simberloff and Von Holle 1999). **Exogenous disturbances** are those that come from outside the system (e.g., asteroid impacts or volcanic eruptions) and generally lie beyond the evolutionary range that the system has experienced in the past. **Endogenous disturbances** have a historical or evolutionary presence in the system and produce change from within (e.g., successional forces or predation). Unfortunately, these distinctions are often hard to make in natural systems and become even more problematic when we consider factors such as the introduction of non-native species. For example, is the extreme level of predation by the non-native Nile perch (*Lates niloticus*; **Figure 13.30**) in Africa's Lake Victoria an exogenous disturbance (due to the human introduction of this non-native species) or an endogenous disturbance (due to predation)? This dichotomy is perhaps better thought of as a continuum, with endogenous and exogenous as the two endpoints.

Table 13.1 lists a few of the more common forms of large disturbances (both biotic and abiotic) and indicates some typical biomes and aquatic biological zones where they commonly occur. The table is woefully incomplete, however, as there are potentially as many types of disturbance as there are organisms that can be affected by the disturbance. This diversity of disturbance types and effects makes it hard to develop theories or make generalizations about how disturbance alters communities.

Figure 13.30 Nile perch (*Lates niloticus*) were released by humans into a number of lakes in central Africa, including Lake Victoria, in the 1950s in the hopes of increasing the food supply. Unfortunately, these huge, voracious, non-native predators quickly decimated the native fish fauna, altering the entire aquatic food web in the process.

Table 13.1 Types of Ecological Disturbances and Where They Commonly Occur

Disturbance		Terrestrial Biomes and Aquatic Biological Zones
Abiotic	Drought	Forest, grassland, scrubland, desert, and savanna biomes; freshwater biological zones
	Fire	Terrestrial biomes of all types
	Flood	Freshwater, riparian, and nearshore biological zones
	Hurricane	Nearshore aquatic biological zones; tropical and low-latitude temperate terrestrial biomes
	Ice storm	Temperate deciduous forests and boreal forests
	Landslide	Terrestrial biomes with steep topography
	Salinity change	Aquatic biological zones of all types
	Wind	Forest biomes of all types
Biotic	Agriculture	Terrestrial biomes of all types
	Burrowing	Terrestrial biomes of all types
	Dam	Freshwater biological zones
	Disease	All
	Grazing, mowing	Terrestrial biomes of all types
	Human-caused	All
	Insect outbreak	Terrestrial biomes of all types
	Predation	All
	Toxins, pollution	All

Adapted from White and Pickett 1985.

The Intermediate Disturbance Hypothesis

Despite a strong research focus on disturbance dynamics over the past 50 years, the field of ecology does not offer many theories about disturbance except for one that garnered much attention in the late 1970s and 1980s. The **intermediate disturbance hypothesis** suggests that the highest levels of species diversity (measured either as species richness or using the Shannon-Wiener diversity index) are found in ecosystems that have intermediate levels of disturbance (Connell 1978).

The logic follows a similar thread to that described earlier regarding r-selected and K-selected species. A disturbance creates an opportunity (e.g., a clear patch of soil or other available resources) and allows species that can take advantage of the opportunity to establish themselves and grow in population abundance. The hypothesis suggests that systems that experience very infrequent or mild disturbances will likely be controlled by competitive exclusion from a few highly competitive taxa (often K-selected species) and thus have low overall species diversity. At the other extreme, systems with very frequent or intense disturbances will likely exclude all but the species most adapted to disturbance (often r-selected species). In the middle, at intermediate levels, there is enough disturbance to keep strong competitors from excluding weak competitors, but not so much disruption that species requiring a relatively stable environment are excluded.

In other words, under the intermediate disturbance hypothesis, disturbance acts as a reset button on community dynamics. If the button is rarely pushed or pushed too often, the habitat will favor fewer species than if the button is pushed at a frequency in the midrange (**Figure 13.31**).

Evidence for the intermediate disturbance hypothesis originally came from small-scale experiments in marine intertidal systems (such as Jane Lubchenco's research on New England tide pool communities—see Chapter 1) and some larger experiments in grassland communities. Intuitively, the hypothesis makes sense. We can imagine that most of the species in highly disturbed sites would be r-selected species, and that the species in rarely disturbed sites would largely be K-selected. And in the middle, between these extremes, it seems logical that the forces favoring strong competitors and the forces favoring good colonizers would "even out" to keep both r- and K-selected species in the community.

Nonetheless, the hypothesis is not without its problems (Fox 2013). For example, how do we define "intermediate"? Do both the frequency and intensity of disturbance have to be in some middle range for it to be considered intermediate? Additionally, how do we define "high-frequency"? It seems possible to set an arbitrarily high level and thereby omit some even larger historical disturbance events. Finally, how can we quantify the spatial component of disturbance effects? Disturbance effects vary with spatial scale; therefore, we would also have to define intermediate levels of disturbance for a particular scale of interest.

A review of the scientific literature identified more than 100 diversity-disturbance relationships from around the globe and found that the diversity peak predicted by the intermediate disturbance hypothesis

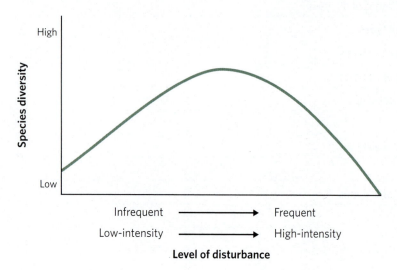

Figure 13.31 The intermediate disturbance hypothesis suggests that disturbances of intermediate frequency and intensity promote species diversity by preventing competitive exclusion. When a system experiences frequent or intense disturbances, all but a few of the most disturbance-adapted species will be excluded. With very infrequent or low-intensity disturbances, superior competitors may outcompete inferior competitors.

occurred in less than 20% of cases (Mackey and Currie 2001); clearly, the most common outcome is not the one predicted by the hypothesis. Additionally, the theory is fairly general in its scope and ignores trophic levels and more complex interactions (Fox 2013).

Some modern ecologists have labeled the intermediate disturbance hypothesis a *zombie theory* that has limited support and should be dead but manages to lurch on, causing difficulties as it spreads among new ecological thinkers. A possible middle ground might be to think of the intermediate disturbance hypothesis as interesting and a potentially useful factor in determining diversity. Any theory that can explain nearly 20% of the outcomes in a science as complicated as ecology is probably worth understanding. It may offer the most useful insights in well-studied systems where the disturbances are understood. Neither the intermediate disturbance hypothesis nor any other single ecological idea is likely to work perfectly when trying to predict diversity at a range of scales, communities, and disturbance types.

Another difficulty with the intermediate disturbance hypothesis is that it perpetuates the idea that disturbances change communities in gentle, predictable ways that maintain some dynamic equilibrium. Given all the ways we have struggled to define disturbance or come to any general conclusions about it, perhaps we should ask whether disturbance can alter the way we think about community change through time in much more fundamental ways. Do ecological disturbances mean that community change is much less predictable than we have suggested?

13.6 Beyond the Steady State

One of the pedagogical threads throughout this book is that models, whether conceptual or quantitative, can help us understand how ecological systems work. Our simplest models leave out many details but still help us to identify patterns. On the other hand, very complex models often hit limits in their applicability. At this point in the chapter, we are hitting a limit, bringing into question many of our previous generalizations and insights. Given the ubiquity and importance of disturbance, predictable change and stable community endpoints may not be the norm. How then can we rationalize our understanding of how important disturbance is in ecological systems with all the steady-state models discussed in the previous chapters?

This chapter deals with change, and an important feature of the dynamic nature of communities is that they rarely have a chance to reach an equilibrium, or steady state, in the natural world. The idea of a steady state in nature is directly related to the idea of a *balance of nature*, a traditional concept in Western thought dating back to antiquity (DeAngelis and Waterhouse 1987). Many of the earliest natural historians and Western culture at large embraced the premise that nature is always in some form of equilibrium. Certainly, we can see this point of view in Clements's concept of a supraorganism and in the enthusiasm with which ecologists embraced his view and failed to recognize data that refuted it.

As ecology formalized into a distinct branch of biology, the idea of a balance of nature infiltrated most of the initial efforts to model ecological systems.

Consequently, the foundational models of early community dynamics assumed communities would eventually settle into an equilibrium state. This perspective has been referred to as the **equilibrium view** in ecology (McIntosh 1985). It has always had a fair number of skeptics, whose dissatisfaction with the idea of equilibrium hinges on whether it is valid to define the existence of an equilibrium state at all.

We have relied on this equilibrium view quite frequently in previous chapters. For example, Chapters 6–9 explored the equilibrium outcomes of two-species models. When studying the natural world, it is perfectly reasonable to think about equilibrium outcomes of models and to explore steady-state endpoints. These outcomes tell us what to expect from interactions if they continue to work according to how they were conceptualized and parameterized. But it is not reasonable to assume that such steady states are the only outcomes, or even the most likely outcomes, for populations and communities in the real world.

In Chapter 2, we discussed *stochastic models*, which are similar to equilibrium models except they include elements of random variation. Many ecologists today use stochastic models to incorporate disturbances, but the results are so varied that it is hard to generalize about what random disturbances do to population growth, competition, predation, or mutualism, let alone all of them together. Nonetheless, we can extract some insights. Some understanding comes from thinking about disturbances that alter population growth or species interactions, especially if we consider human actions as disturbances.

For example, if we think about a growing population, repeated disturbances that remove individuals (e.g., fishing or hunting) could keep the population from ever reaching its carrying capacity (**Figure 13.32**). Similarly, disturbances such as altered water flows in a river due to dam construction could easily shift the balance between two competing species. A superior competitor might never exclude an inferior competitor if such a disturbance occurs frequently. Such *nonequilibrium dynamics* might be extremely common in the Anthropocene, particularly for processes or interactions that take multiple generations to materialize. In fact, this is the premise of the intermediate disturbance hypothesis, but as we have pointed out, there is no reason to assume that such disturbances would lead to another steady state.

What about during succession? Repeated disturbances during succession may have a similar effect and prevent communities from ever reaching a climax assemblage, or they could even shift successional processes sufficiently to lead to a different type of climax community. Both scenarios appear to be much more likely in the Anthropocene as climate change and other human actions alter the disturbance regimes to which many species are adapted. Natural disturbance regimes may also limit species' membership in communities by determining which species can colonize or survive.

For example, increased temperatures and associated extreme fire seasons in the Australian Alps led to burns across 85% of this region from 2002 to 2014

Figure 13.32 Population growth with repeated disturbances (at each arrow) setting the population back to a lower density (blue line). The dark green line shows the trajectory of logistic population growth without disturbances. K = carrying capacity.

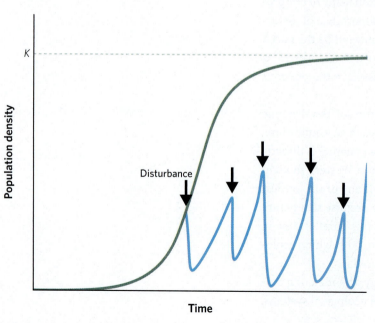

(Bowman et al. 2014; **Figure 13.33**). Alpine ash (*Eucalyptus delegatensis*) is a dominant and endemic tree species in these communities. Once these trees reach maturity, at around 20 years of age, their thick bark protects them from low-intensity fires. The species is, in fact, fire-reliant because the trees' seeds only germinate after experiencing a fire, and young trees grow prolifically in fire-cleared areas. These traits evolved from generations of living in a fire-prone environment. However, although immature trees grow well in the aftermath of a fire, they are also easily killed by repeated, frequent burns. David Bowman and colleagues (2014) showed that, when these Australian forests experienced two fires within the span of 6–10 years, alpine-ash population density declined, and other, nonendemic tree species began to dominate. Over the long term, climate-induced changes in the frequency of wildfires in Australia could result in the complete replacement of alpine-ash forests by other types of forest communities.

Fire regimes are just one of the natural disturbances that have been substantially altered by climate change, land use change, invasive species, and other hallmarks of the Anthropocene. We should, therefore, expect nonequilibrium dynamics to become more common in the future, a topic we return to in Chapter 15. Such redirected communities may seem like an exception rather than the norm, but there are plenty of other examples. If you think back to Chapter 3, you may remember that fire was a factor that helped distinguish between grassland and forested biomes in temperate and tropical systems. In a similar manner, wave action and changes in salinity help to determine where salt marshes exist around the world.

Clearly, timescale is an issue. Succession is an extremely long process, and our ability to explore successional stages over centuries tends to be concentrated on a few well-followed examples. Frequent disturbances may define the "normal" abiotic conditions in an area, so that we actually expect the community to be "in sync" with the disturbance. Plus, if we expand the timescale of succession to centuries, is there really any reason to expect that a single climax community will remain in perpetuity? That seems crazy, frankly. The one absolutely reliable constant in all of nature is that things change, and recognizing this natural trend toward change is an important lesson for all ecologists.

Figure 13.33 Ⓐ In the Australian Alps bioregion, alpine ash (*Eucalyptus delegatensis*) is endemic and the dominant tree species. Ⓑ Recent changes to fire frequency, however, threaten this species because fires in quick succession kill immature trees. In this picture from January 2020, firefighters in Alpine National Park, Victoria, Australia, take measures to control a bush fire.

SUMMARY

13.1 Nature Is Dynamic

- The natural world is constantly changing.

- As a result, ecological communities do not remain static through time.

13.2 The Nature of Communities

Defining Community

- A community is a group of interacting species in the same location.

Clements versus Gleason

- Frederic Clements

  - In the early 1900s, Clements suggested that communities of plants changed predictably through time and that long-term stable groups of species should be called climax communities.

  - He called the process leading to a climax community *succession*.

  - He likened a climax community to a functioning organism (a supraorganism) in which all species present were interdependent and working in synchrony.

  - Evidence for this worldview came from looking at plant communities of different ages and observing this widespread pattern.

  - Clements's ideas were embraced by the early crop of ecologists.

- Henry Gleason

  - As a contemporary, Gleason vehemently disagreed with Clements and argued that each plant species responded in an individualistic manner to the environment around it.

  - He rejected the idea of holistic, coevolved plant communities and instead suggested that each individual species finds a place to grow according to its physiological requirements.

- A debate between these two early ecologists and their supporters ensued, with the data eventually supporting Gleason's ideas.

Community Boundaries

- Out of this debate came the understanding that communities are not static entities unchanging through time and are difficult to cleanly define.

13.3 Succession

Predictable Change

- Ecological communities change in predictable ways through time, and the process is called succession.

Primary and Secondary Succession

- Primary succession

  - In primary succession, the successional trajectory begins on a bare substrate void of life (e.g., lava flows or glacial till).

  - Early stages progress slowly as the bare area is colonized by small but hearty plant species that build an environment (i.e., create soil).

  - These changes allow other species to establish, eventually leading to other successional stages over time.

- Secondary succession

  - Secondary succession occurs after a disturbance on an already established habitat.

  - This process is much faster than primary succession because typically a bank of living organisms is already present (i.e., microbes, invertebrates, seeds, roots, and so on).

Life History and Succession

- Early successional species tend to be *r*-selected; they are small and have high reproduction rates, low survival rates, and short generation times.

- Late successional species tend to be *K*-selected; they are large and have low reproduction rates, high survival rates, and long generation times.

Examples of Succession

- **Primary Succession on Mount St. Helens**

  - The eruption of Mount St. Helens in 1980, and the resulting extensive devastation, provided a unique glimpse into the process of primary succession.

  - Roger del Moral and colleagues have studied this successional process for more than 40 years.

- **Secondary Succession in North Carolina's Piedmont Forests**

  - Catherine Keever and others studied secondary successional processes in the forests of North Carolina's Piedmont region over four decades.

- **Animal Succession in the Piedmont**
  - Animals and animal communities respond to the process of plant succession in predictable ways.

Long-Term Successional Trends

- Succession is a very slow process, particularly primary succession.
- Many important community traits (e.g., species richness and biomass) increase through time as succession progresses.
- Chance events and serendipity play a significant role in the overall successional process in terms of the speed of succession and spatial patterning.
- Biotic components play an important role in the successional process.

Modeling Succession

- **Conceptual Models of Succession**
  - Connell and Slatyer suggested three contrasting conceptual models for succession: facilitation, inhibition, and tolerance.
  - In the facilitation model, early species alter the environment in positive ways that facilitate the establishment of later species.
  - In the inhibition model, early species prevent later species from establishing. Periodic disturbances break these periods of dominance and allow later taxa to colonize.
  - In the tolerance model, early stages have no real effect on later stages, and patterning through time is the result of different life history traits, such as longevity and growth rate.
- **Quantitative Models of Succession**
  - Both the Lotka-Volterra and resource-ratio models of competition from Chapter 7 can be used to examine successional processes.

13.4 Community Assembly

Assembly Rules

- Species can assemble or combine in a multitude of ways to form communities. Are there rules that guide or govern the construction of communities?
- Initially, ecologists looked for nonrandom patterns of occurrence to demonstrate assembly rules, but this approach was abandoned because of the vast variety of potential assembly combinations that can create these patterns.

- Yet, some assembly rules clearly must exist. For example, a predator cannot successfully colonize an area if its prey is absent.
- James Drake investigated rules for freshwater community assembly through a series of challenging laboratory experiments.
- The results were mixed but suggested that there are inherent limits to how and which species can assemble to form a community, and that history and chance make a difference in the final outcome.

Community Assembly and Restoration Ecology

- Repairing or re-creating damaged communities is an active area of research among ecologists that relies heavily on ideas of succession and an understanding of disturbance dynamics.
- Some scientists believe that the future of applied ecology, and perhaps the future of the planet, lies in the direction of restoration ecology.

13.5 Disturbance Dynamics

- Change is a natural and integral part of ecological communities. Some changes are predictable, and some are not. Unpredictable change is sometimes called disturbance.
- The concept of disturbance includes the factors that cause change, the changes themselves, and the processes associated with such changes.

Definitions of Disturbance

- Ecologists have proposed many different and often contrasting definitions of ecological disturbance.
- Some definitions focus on disturbance as a process, whereas others cast it as an event.
- This distinction may have to do with the temporal scale and the organisms for which a disturbance is being defined.
- Describing the critical aspects of all disturbances is challenging but commonly involves identifying the following disturbance features: distribution, magnitude, frequency, predictability, return interval, and synergism.

Types of Disturbance

- Various schemes for categorizing disturbances exist (natural vs. human-caused, abiotic vs. biotic, exogenous vs. endogenous).
- A disturbance can sometimes result from the removal of an already existing disruptive force.

The Intermediate Disturbance Hypothesis

- This hypothesis from the 1970s suggests that species diversity is highest in communities that experience intermediate levels of disturbance.

- Very strong or very frequent disturbances prevent many species from existing in a community, whereas infrequent or weak disturbances often allow a few competitively superior species to dominate.

- Although this hypothesis was initially viewed as one of the central elements of community ecology, it is simplistic, and less than 20% of examined communities fit the predicted patterns.

- Its ability to explain even a fraction of patterns, however, may mean that the intermediate disturbance hypothesis contains important insights into the relationship between disturbance and diversity.

13.6 Beyond the Steady State

- Most of the models presented so far in this book have not included elements of disturbance, despite the important role that change plays in systems.

- Underlying this apparent oversight is an old ecological idea that the world is in equilibrium, or a steady state, or that there is a balance of nature.

- Most of the early ecological models were based on this concept.

- Yet it is possible to include elements of change, dynamism, and disturbance into these equilibrium models by developing stochastic models.

- When stochastic forces are taken into account, most of the models show different and interesting dynamics.

KEY TERMS

allelopathy
assembly rule
characteristic flow regime
chronosequence
climax community
community
disturbance
disturbance regime

endogenous disturbance
equilibrium view
exogenous disturbance
facilitation model of succession
inhibition model of succession
intermediate disturbance hypothesis
K-selected species

primary succession
priority effect
r-selected species
secondary succession
sere
succession
tolerance model of succession

CONCEPTUAL QUESTIONS

1. Define the successional process and discuss the various types of succession using examples from your local environment when possible.

2. Compare and contrast Gleason's and Clements's views on communities and the successional process. In what ways did Clements get things right?

3. Discuss the differences between the three models proposed by Connell and Slatyer to explain the successional process.

4. Describe how competition could be a factor in the facilitation model of succession.

5. Describe how natural selection could produce tolerance, inhibition, or facilitation in succession.

6. How important do you expect seed dispersal to be in primary succession versus secondary succession? How do you think biomass and diversity may change through time with succession? Explain your answers.

7. Explain some of Roger del Moral and colleagues' findings on primary succession from their research on Mount St. Helens.

8. Define and explain the six features that help describe disturbances.

9. Describe a restoration project that would be possible in your area. Be as specific as you can about how you would design the project given what you know about succession.

QUANTITATIVE QUESTIONS

1. We reused some figures from Chapter 7 in this chapter to suggest that competition models play a role in Connell and Slatyer's models of succession. Revisit those figures, and try to explain how each one fits with the succession models.

2. People often talk about "100-year floods" as disturbances, which suggests that we expect to get one such flood in any 100-year period. The flood could come in the very first year of that period, the last year, or any year in between. Other types of disturbances come with extreme regularity, though. Some cicadas have distinct cycles built into their life history and appear in explosive densities at set intervals, eating and decimating plant communities. Explain the difference between the probability or frequency of the 100-year flood and the periodical nature of the cicada outbreaks. Which one is going to be more predictable?

3. Figure 13.32 offers a glimpse into the way disturbance might alter population dynamics. Can you create a similar graph to describe elements of communities that change through time and how repeated disturbances might alter such elements?

4. Lubchenco's snail experiments (from Chapter 1) were offered as an example of experimental work that tests (and supports) the intermediate disturbance hypothesis.
   a. Describe how Lubchenco's findings fit our expectations of the intermediate disturbance hypothesis, and draw a graph like Figure 13.31 based on Lubchenco's work.
   b. Describe how Lubchenco's findings differ from, or do not fit, the intermediate disturbance hypothesis.

5. Community composition seems like a very descriptive aspect of ecology, but describing community differences is one of the most quantitative aspects of ecology, particularly plant ecology. Explain how change through time complicates our estimates of diversity from Chapter 11.

SUGGESTED READING

Connell, J. H. and R. O. Slatyer. 1977. Mechanisms of succession in natural communities and their role in community stability and organization. *American Naturalist* 111(982): 1119–1144.

Diamond, J. M. 1975. Assembly of species communities. Pp. 342–444 in M. Cody and J. Diamond, eds. *Ecology and Evolution of Communities*. Cambridge, MA: Harvard University Press.

Drake, J. A. 1991. Community-assembly mechanics and the structure of an experimental species ensemble. *American Naturalist* 137(1): 1–26.

Keever, C. 1983. A retrospective view of old field succession after 35 years. *American Midland Naturalist* 110(2): 397–404.

Pickett, S. T. A., J. Kolasa, J. J. Armesto, and S. L. Collins. 1989. The ecological concept of disturbance and its expression at various hierarchical levels. *Oikos* 54(2): 129–136.

In this Massachusetts ecosystem, the organisms, lake, and forest exchange carbon, nitrogen, phosphorus, and water through respiration, photosynthesis, falling leaves and sediments, nutrient uptake, and the movement of wind and water.

chapter 14

NUTRIENT CYCLING AND ECOSYSTEM SERVICES

LEARNING OBJECTIVES

- Explain the concept of an ecosystem and how ecosystems differ from communities.

- Describe the steps in the terrestrial carbon cycle, using mangrove forests as an example.

- Distinguish between the slow and fast carbon cycles.

- Explain the effects of anthropogenic forces on the global carbon cycle.

- Distinguish between the atmospheric and terrestrial portions of the nitrogen cycle.

- Outline the improvements to human survival and decreases in biodiversity from anthropogenic alterations to the nitrogen cycle.

- Compare and contrast the global phosphorus, nitrogen, and carbon cycles.

- Explain how the water cycle links to other biogeochemical cycles.

- Explain what ecosystem services are, and provide examples.

14.1 Ecosystems Are Inclusive

If you have ever had the pleasure of sitting by a body of water—whether a stream, a lake, the ocean, or even a tiny pond—or relaxing beside a tree, then perhaps you also know the experience of letting your mind drift off in the warmth of the sunshine, thinking of nothing in particular for a few moments. Sometimes, in that relaxed space, it is possible to take in the whole environment around you: the water, the trees, the grasses, and countless other organisms living and interacting there. All of these living parts of your surroundings, as well as the nonliving parts, are intimately connected to each other through the natural features of the place, such as sunlight, water, oxygen, and carbon dioxide. When you daydreamed and observed the world around you in this way, you were contemplating and perhaps even experiencing your surroundings as an *ecosystem*—an integrated entity made

Figure 14.1 A closed ecosphere is on display at the American Museum of Natural History in New York City. It contains bacteria, algae, and tiny shrimp, was sealed in 1999, and has not been opened since. In a sense, it is a tiny, self-sustaining, individual ecosystem.

of organic and inorganic parts. An ecosystem is both dynamic and responsive, as we will demonstrate in this chapter.

In previous chapters, we explored interactions among species (Chapters 7–10) that led to dynamic collections of species (Chapter 11) across space (Chapter 12) and through time (Chapter 13). Here we consider exchanges between the biotic and abiotic environments. This focus hearkens back to our understanding of how the physical environment affects an individual's physiology and defines a species' niche (Chapter 5). In fact, you can think of ecosystem ecology as physiological ecology written across an entire ecological community. If a community includes all the biotic aspects (organisms) of a location, then an ecosystem is the community plus all the abiotic components. We have used the word *ecosystem* throughout this book, with the implicit understanding that you would read it as representing a larger "whole." In this chapter, we formalize this concept so that we can scientifically explore how collections of species integrate with the environment to influence whole-Earth processes. Through their use and dependence on water, carbon, nitrogen, and phosphorus (as well as many other molecules and elements), living organisms create myriad dynamic connections with their environment (**Figure 14.1**). We can see this interconnectedness if we follow these essential elements through the ecosystem.

It is often within the context of studying ecosystems that students begin to consider how ecological principles inform some of the more profound changes humans have wreaked on our Earth, such as global warming, ocean acidification, eutrophication, and global freshwater scarcity. One of the more challenging aspects of such broadscale thinking is that it requires an ecological understanding of **nutrient cycles**, meaning the ways that molecules cycle through the living and nonliving world. In essence, ecosystem ecology requires you to think big by analyzing the very small.

Using an impressive and growing set of tools, ecologists can trace the movement of life's critical elements and molecules from the atmosphere or water into the tissues of living individuals and back out to the environment once these individuals die. Dead, excreted, or decomposed tissue eventually returns these molecules from the living world back to the atmosphere, water, or soil, where they may combine with other molecules as they travel across parts of the Earth's system but are never lost. For this reason, ecosystem ecologists describe these processes as cyclical. To explore the exchange between biotic and abiotic portions of the biosphere, ecologists build **nutrient budgets** to keep track of the concentrations of specific types of molecules as they move through and among the living and nonliving components of an ecosystem.

Ecosystem ecologists often focus on budgets for carbon, nitrogen, phosphorus, and water, but it is possible to build an ecosystem budget for any molecule, provided the measurement tools are available. Indeed, biogeochemists do this for a variety of molecules, some of which interface strongly with the living world and others of which do not. For ecological scientists, these budgets are of primary interest when they affect evolutionary processes, influence the numbers or types of species in an ecosystem, or directly impact global climate. These connections to the living world are why carbon, nitrogen, phosphorus, and water are the molecules most commonly studied. All four are critical to sustaining biological life on the planet, and all are covered in this chapter.

First, we concentrate on building a budget for carbon in a single ecosystem (a mangrove forest) and then consider the carbon budget for the whole Earth. This initial and deep exploration of carbon is useful for two reasons. For one, life as we know it on this planet is carbon-based, so understanding the process by which carbon enters and leaves living systems is essential. Second, as detailed in Chapter 3, the massive increase in carbon emissions over the last two centuries is responsible for a variety of substantive changes to the Earth's climate. These changes have clear consequences for humans and biodiversity. As a citizen of this planet, you will have to make decisions about how to respond and adapt to such changes during your lifetime. To do this well, it is important to learn everything you can about carbon.

In the latter part of the chapter, we consider the nitrogen, phosphorus, and water cycles, skipping over the nitty-gritty details of their budgets in favor of describing why these elements are critical to life, how the cycles have been altered by human actions, and what these changes mean for biodiversity. We end the chapter by flipping the story a bit, and instead of asking how changes in these cycles affect biodiversity, we ask how biodiversity affects these cycles.

14.2 Budget Basics

A standard bit of life advice for adults, young and old, is to create a budget for personal finances and stick to it. Creating a financial budget requires keeping track of all the money coming into your bank account, including from jobs, student loans, lottery windfalls, and coins found on the sidewalk. It is useful to have a set time frame for your budget, often monthly, that coincides with paychecks or the calendar month. A true budget doesn't just follow monthly income but also tracks money that flows out each month. An accurate accounting would include the basics, such as paying rent and buying food, but also the smaller one-time purchases, such as a movie ticket, a pack of gum, and the hopeful lottery ticket. In the end, a good financial budget accounts for every single dollar that cycles through your life in the span of one month. If you do this budgeting exercise, you may gain some interesting insights, particularly about the ways in which you are either saving or overspending your money.

One of the primary methods of ecosystem ecology is to follow molecules in more or less the same way that financial budgets follow money. An **ecosystem budget** is depicted as a series of **ecosystem pools** (sometimes called stocks or reservoirs) and **ecosystem flows** (sometimes also referred to as fluxes) (**Figure 14.2**). Pools are parts of an ecosystem where material (e.g., water, carbon, nitrogen, phosphorus) collects and is stored. These pools are measured in basic units, such as grams per hectare. Flows represent the movement of material from one part of the ecosystem to another. Like all flows, including the flow of

Figure 14.2 This simple depiction of how carbon cycles among autotrophs, heterotrophs, and the atmosphere offers an example of an ecosystem budget. We demonstrate more detailed budgets later, so view this one solely as a way to orient yourself. The shapes outlined in gray indicate pools where the focal molecules may build up or be stored, and arrows represent the flow of material between pools.

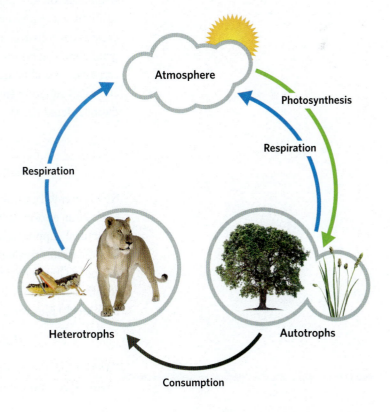

money out of your wallet, these are measured as a temporal rate of change, such as number of grams lost per hectare per year. Flow arrows point in the direction that material is moving. The flow back and forth between pools is often referred to as an **ecosystem loop**.

A Carbon Primer

To make all this budgeting slightly more real, let's work through an example of a carbon budget for mangrove forests. We already mentioned why concentrating on carbon as an example of ecosystem processes is important, but a bit more background on carbon and its role in life will help you understand why we care about how it moves through an ecosystem.

Nearly all structures that make up a living organism are built with carbon atoms. Proteins, carbohydrates, lipids, and nucleic acids all have carbon atoms as their base; this base bonds with some combination of hydrogen, oxygen, nitrogen, or phosphorus. Carbon-based molecules used by living organisms are called **organic carbon**, and they have bonds with both hydrogen and oxygen. Carbon-based molecules are also readily found outside of living organisms; examples include carbon monoxide (CO), carbon dioxide (CO_2), and methane (CH_4). These molecules, referred to as **inorganic carbon**, contain either oxygen or hydrogen, but not both.

The Earth is full of inorganic carbon in its various forms. The trick for building living organisms is to take available inorganic carbon and biologically "fix" it into organic carbon molecules. In this instance, *fix* means to convert into a stable organic molecule or form. **Carbon fixation** happens principally via photosynthesis, although some bacteria and archaea can also fix carbon via chemosynthesis. As you may recall from Chapter 5, photosynthesis involves a chemical reaction in which carbon dioxide from the atmosphere and water combine with energy from sunlight to create a carbohydrate (e.g., glucose) and oxygen. Chemosynthesis also creates carbohydrates from carbon dioxide and water but generally uses energy from chemical bonds to power the process. Chemosynthesis is less common and can take several forms, but we can represent the photosynthetic reaction using a chemical equation (noting, of course, that the process is much more complicated than depicted here):

$$6CO_2 + 6H_2O \xrightarrow{\text{sunlight}} C_6H_{12}O_6 + 6O_2 \tag{14.1}$$

$$\text{Carbon dioxide} + \text{Water} \xrightarrow{\text{sunlight}} \text{Carbohydrate (glucose)} + \text{Oxygen}$$

Newly created carbohydrates are great sources of energy, but they can also be used to make an enormous variety of other organic molecules for use in everything from structural support to roots. Plants can chemically *reduce* the carbon structures in carbohydrates through an interaction with oxygen. This process, which was mentioned in Chapter 5, is necessary any time an organism uses energy. This reduction reaction creates ATP (adenosine triphosphate, the energy storage molecule for life's work) plus water and carbon dioxide.

$$C_6H_{12}O_6 + 6O_2 \xrightarrow{\text{yields}} 6CO_2 + 6H_2O + \text{Energy (ATP)} \tag{14.2}$$

Can you see how an organism that uses energy is basically doing the reverse of the chemical reaction of photosynthesis? In fact, we can create a generalized equation that shows how photosynthesis and respiration (i.e., the biotic use of energy) are complements of each other, using energy in one direction and releasing energy in the other direction.

$$6CO_2 + 6H_2O \xrightleftharpoons[\text{yields } E]{\text{uses sunlight}} C_6H_{12}O_6 + 6O_2 \qquad (14.3)$$

When carbon goes from the left- to the right-hand side in Equation 14.3, it is fixing or storing energy and is moving from an inorganic to organic state in the process of carbon fixation. When carbon goes from the right- to the left-hand side in this equation, it is releasing energy and is moving from an organic to inorganic state in a process called *respiration* (Figure 14.2).

As discussed in Chapter 5, all the organisms on Earth that are able to fix carbon through photosynthesis are called autotrophs. Any organism that gets its carbon from eating an autotroph or any other organism is called a heterotroph. From Chapters 8 and 10, it is clear that heterotrophs come in a variety of forms, including herbivores, carnivores, omnivores, and detritivores. All living organisms— autotrophs or heterotrophs—respire, thereby giving off carbon dioxide and water as they use the energy released from breaking the chemical bonds of organic carbon. In most ecosystems, the amount of carbon released in respiration by the heterotrophic organisms is minimal, except in the case of detritivores. Detritivores do a lot of heavy lifting in terms of converting carbon from an organic to an inorganic form in most ecosystems on Earth. Because of this disparity, ecosystem budgets include detritivores but generally ignore the carbon processed by other types of heterotrophs.

Much of this information is a review of material from Chapter 5 and from biology classes you may have taken, but next we will detail how every part of an ecosystem fixes and respires carbon to derive the sum amount of carbon held within or emitted from a whole ecosystem. Should you get lost in the details ahead, come back to Equation 14.3 and the terms defined in this section.

The Mangrove Forest Carbon Budget

We examine the carbon budget for mangrove forests in part because these ecosystems store a lot of carbon relative to the area of the Earth they occupy, which makes them relevant when considering carbon flows and the global climate (as we will do later in the chapter). They are also fairly simple ecosystems. Unlike terrestrial forests, mangrove forests—which are nearshore aquatic biological zones (Chapter 3)—have no real understory and are, therefore, limited in the number of plant species they can support. If we understand how carbon moves through mangrove trees and through the soil and water, we will have a fairly complete picture of the carbon cycle for mangrove forests. Truthfully, though, the main reason we decided to use them for our example is that they are supremely cool ecosystems.

Recall from Chapter 3 that mangrove forests are restricted to the coastal margins of continents and islands in the tropics and subtropics (**Figure 14.3A**). Because these areas serve as the transition zone between the open ocean and terrestrial areas, mangrove trees experience constant wave action on a good day. On a bad day, they face the high winds and heavy waves of tropical storms, tsunamis, and hurricanes. All this wave energy means that mangrove trees lose leaves and branches but also trap bits of material tossed at them by waves. The crazy interwoven pattern of mangrove root systems (**Figure 14.3B**) serves to slow the flow of

A

B

Figure 14.3 Ⓐ The roots of a mangrove forest along the coast of Mali experience inundation and exposure with tidal fluctuations. Ⓑ Beneath the water's surface, the tangled mangrove roots serve to slow down wave action and allow silt to settle. They also trap carbon and many plant nutrients.

ocean water as it comes in with the tides. As the water slows, fine silt particles settle out of the water and around mangrove roots. If the mangrove forest is situated at the mouth of a large river, the mangrove root system captures much of the large and small sediments coming from upriver terrestrial ecosystems as well. These sediments are often filled with organic and inorganic carbon, as well as nitrogen and phosphorus and other essential nutrients.

Living in the intertidal zone means that mangrove roots are completely submerged in salt water at least twice a day as the tides rise and fall. In the low-elevation reaches of mangrove forests, tree roots are nearly always underwater, even if that water is sometimes quite shallow at low tide. This submersion helps produce *anoxic* (oxygenless) conditions for substantial periods each day, especially below about 50–100 cm (20–39 in.) from the soil surface. Oxygen-starved sedimental soils favor microorganisms that produce methane and that restrict the ability of mangrove roots to gather nutrients.

Figure 14.4 is a carbon budget for mangrove forests produced by Daniel Alongi (2014), in which he brings together research from across a variety of mangrove forests worldwide to create a general understanding of the carbon budget of these ecosystems. We will work through all the pools and flows of this budget so that they make sense. Along the way, we will define important terms used in carbon accounting. It is important to recognize at the outset that Alongi was not able to make all the numbers in the budget add up exactly, particularly when dealing with belowground processes. Part of this difficulty results from trying to account for what happens underground within sedimental soils, which is extraordinarily difficult to do given the limitations of our current technology. The budget presented in Figure 14.4 is the best estimate he was able to make at the time and is quite impressive, given the complexity of the ecosystem and the limited ability of scientists to track what is happening in some carbon pools (e.g., sediments and soils).

Let's take the carbon pools first. Figure 14.4 shows eight labeled pools (i.e., A through H) that represent the amount of carbon that mangrove trees (A) fix through photosynthesis, (B) have left over from photosynthesis after respiration, (C) put into the wood of the trunk and branches, (D) put into roots, and (E) drop as leaf or wood litter. The other three pools represent (F) the carbon from litter that is buried under sediments and remains in the soil, (G) the amount that is fixed by photosynthetic algae that grow on the surface of the soil or on the mangrove roots, and (H) the carbon created by the mangrove trees but dropped as litter and consumed by soil bacteria. Although we do not need to understand every detail of these pools in order to understand a general carbon cycle, they represent common carbon pools in many ecosystems and introduce terms that are useful as we examine the global carbon budget.

Notice that each oval has a number that represents the amount of carbon estimated to reside in the respective pool. Usually, these measurements are calculated as grams of carbon per year. There are a lot of atoms of carbon even in a single leaf of a mangrove tree. If we weigh all the atoms of carbon across all mangrove trees present worldwide, we are talking about tons of carbon. In Figure 14.4, the numbers within each oval represent teragrams (Tg) of carbon stored per year for all mangrove trees worldwide, which is abbreviated as Tg C y^{-1}. One teragram is equivalent to 10^{12} grams, or one million metric tons.

Figure 14.4 Daniel Alongi (2014) developed this carbon budget for the major pathways of carbon flow through a mangrove forest. Arrows indicate carbon flows, with solid arrows showing well-established and understood flows and dashed arrows showing the best estimates based on the literature. Green boxes indicate the amount of carbon flowing out to the environment. All numerical values are in teragrams of carbon stored per year (Tg C y^{-1}). The budget assumes a global mangrove area of 138,000 km^2 (53,282 sq. mi.). Ovals represent carbon pools. Letter labels (A–H) correspond with values named in the text. DIC = dissolved inorganic carbon; DOC = dissolved organic carbon; GPP = gross primary production; NPP = net primary production; POC = particulate organic carbon.

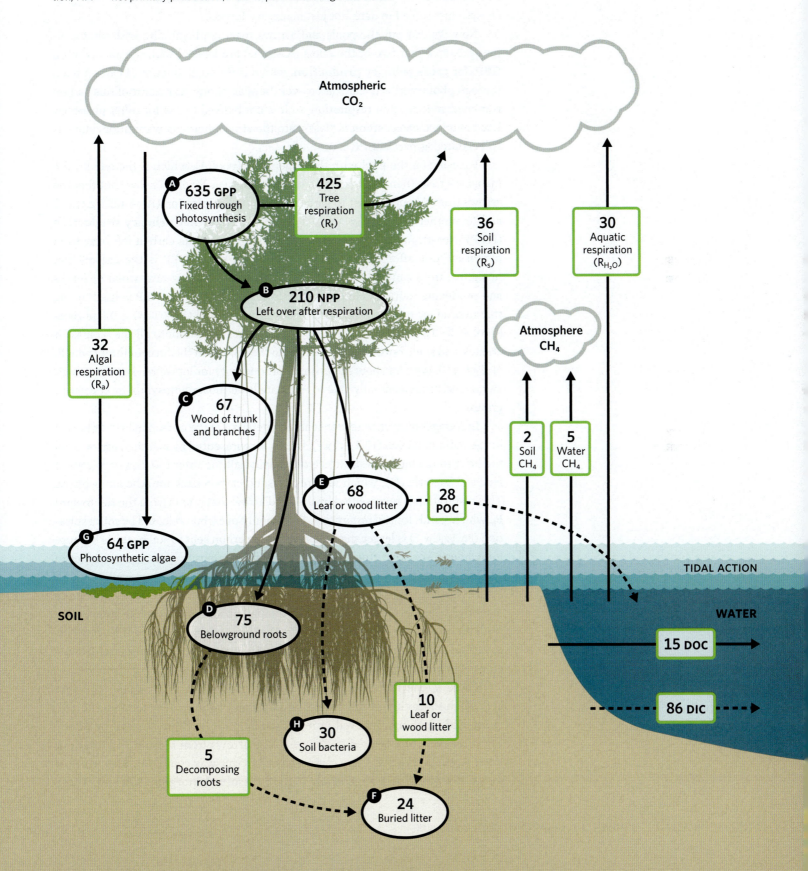

Now let's look at the arrows in Figure 14.4, which represent the directional flows of carbon through the system. Flows that are well documented in the literature are represented by solid black arrows. Dashed arrows represent flows for which Alongi had only best estimates. If you look at the placement of solid and dashed arrows, you can see that scientists know more about aboveground carbon flows than belowground movement. This disparity is common across nearly all ecosystems studied to date, not just mangrove forests.

Now we can put the pools and arrows together, beginning with photosynthesis. Notice that two of the pools (A and G) are labeled with the abbreviation **GPP**, for **gross primary production**, which is the total amount of carbon fixed through photosynthesis, calculated over the span of one year. Some of this carbon is immediately used in respiration, so it is not banked to use for other purposes. Used energy means carbon is "lost" into the atmosphere via respiration, but it is still counted as part of GPP.

Focusing on the pool with the larger GPP (pool A), which is the one for the mangrove trees themselves, we can see that the trees fix 635 Tg C y^{-1} but that the arrows from this pool point to carbon lost to tree respiration (R_t = 425 Tg C y^{-1}) and to another 210 Tg C y^{-1} that flows to pool B, or **net primary production (NPP)**. Net primary production is the amount of organic carbon left over from photosynthesis after respiration. In a practical sense, NPP is the carbon that mangrove trees can use for increasing biomass by making new wood or leaves and producing roots. In fact, in Figure 14.4, we can see how NPP is used by the tree to produce new aboveground woody growth (pool C, 67 Tg C y^{-1}) and roots (pool D, 75 Tg C y^{-1}). Of course, no tree manages to hold onto all the biomass it creates, and some portion of NPP drops to the ground in the form of litter (pool E, 68 Tg C y^{-1}). We care about net primary productivity for lots of reasons in ecology, and scientists are generally able to calculate it for most photosynthetic species or groups.

In Alongi's budget, we see that about one-sixth of the carbon in litter is buried in the sedimental soil (10 Tg C y^{-1}), which is represented by a dashed arrow from the litter to the burial oval. Notice that about half the litter (30 Tg C y^{-1}) goes to feed soil bacteria (H), which respire and send carbon back into the atmosphere. This respiration is represented by the solid arrow that points from the soil toward R_s, which stands for *soil respiration*. The amount of carbon released by soil respiration (36 Tg C y^{-1}) is larger than the amount of carbon sent to soil bacteria because it includes respiration from soil bacteria and decomposition of organic matter in the soil. There is also some loss of carbon from the soil to the atmosphere in the form of *methane* (CH_4), produced from anoxic microorganisms in the sediment, as mentioned earlier. A very small amount of methane is also lost to the atmosphere from the water around mangrove roots. We do not want to ignore this source of inorganic carbon, though, because methane is a strong greenhouse gas with enormous effects on global climate.

There is one more way that mangrove ecosystems lose carbon, and that is through the action of the tides. The rising and falling tidewaters tend to pull water, and anything dissolved in it or floating on it, out into the nearby ocean. This process is represented in Figure 14.4 as arrows from the soil and litter out into the water. The mangrove carbon moved in this way comes in three principal forms: **dissolved organic carbon (DOC)**, **dissolved inorganic carbon (DIC)**, and **particulate organic carbon (POC)**. As suggested by the "dissolved" part of their name, DOC (solid arrow) and DIC (dashed arrow) are such tiny bits of carbon that they are effectively invisible in the water column. You can think of POC (dashed arrow from the litter to the water) as the visible leaves, stems, branches, and roots that the tides detach from mangrove trees and carry out into the ocean. There is a fair amount of POC in mangrove forests because of the constant physical action of waves against the trees. These pieces of tree litter, as well as the roots in the water, contribute to aquatic respiration (R_{H_2O}), which sends 30 Tg C y^{-1} into the atmosphere.

The flows of DIC, DOC, and POC point to the way the carbon budget of mangrove forests is connected to the ocean. The arrows pointing outward from the sediment and soil into the water represent a source of carbon for marine organisms that is not generated within the marine food web itself. We call movements of carbon from one ecosystem to another *trophic subsidies* (Chapter 10). Although it is easy to think that the bits of tree in POC would be the largest subsidies, DIC is the biggest input from mangrove ecosystems to ocean ecosystems, as you can see from the number associated with the DIC arrow (86 Tg C y^{-1}). Although not explicitly shown in Figure 14.4, mangrove forests that grow near the mouths of large rivers receive a fair amount of carbon as DIC, DOC, or POC from terrestrial ecosystems and river water upstream. So, mangrove forests also pass along to the ocean carbon subsidies that come from inland watersheds.

Ecosystem ecologists have shown that such subsidy connections between ecosystems are quite common and include many other critical nutrients, including nitrogen and phosphorus. Indeed, you can think of global circulation as an effective way of connecting and subsidizing whole biomes in terms of water and the various elements that may be carried on or in global air and ocean currents (Chapter 3).

Mangrove Forests as Carbon Sinks   Documenting and calibrating all these pools and flows is quite difficult for a number of technical and logistical reasons, even for a relatively simple ecosystem like a mangrove forest. A good portion of ecosystem ecology research involves trying to find ways to make these measurements and then compare them across ecosystem types and situations. The numbers themselves are not static and change according to the available carbon in the atmosphere, the availability of water, and the availability of other elements, such as nitrogen. Nevertheless, once these budgets are created for several ecosystems, we can ask some interesting ecological questions at the global scale.

For example, does an ecosystem, such as a mangrove forest, fix and store more carbon than it releases via respiration or exports? If a system fixes and stores more carbon than it releases, we call it a *sink ecosystem*, meaning that carbon goes in and stays. If a system releases more carbon than it fixes and stores, it is called a *source ecosystem*. These terms match up with the source-sink population dynamics terminology discussed in Chapter 12. In fact, the population dynamics language came from the use of the terms in physiology and ecosystem ecology. Sources export, and sinks absorb.

Mangrove forests, when intact, are carbon sinks. In the mangrove forest carbon budget, the flows of carbon into the mangrove (fixation) and out of the mangrove (respiration and movement into the ocean) do not balance. Mangrove forests tend to fix and store more carbon than they release. We can see this imbalance in Figure 14.4 because the amount of carbon fixed adds up to 699 Tg C y^{-1}, about 40 Tg C y^{-1} more than the total respiration, leaving about 40 Tg C y^{-1} in storage.

$$
\begin{array}{rr}
\text{GPP from mangrove trees:} & 635 \text{ Tg} \\
\text{GPP from algae:} & +64 \text{ Tg} \\
\hline
\text{Total carbon fixed:} & 699 \text{ Tg}
\end{array}
$$

POC:	28 Tg			
DOC:	15 Tg			
DIC:	86 Tg			
R_{H_2O}:	30 Tg	Total carbon fixed:	699 Tg
CH_4 from the water:	5 Tg	Total carbon released:	$-$ 659 Tg
R_t:	425 Tg	Total carbon stored:	40 Tg
R_s:	36 Tg		
CH_4 from the soil:	2 Tg		
R_a:	$+$ 32 Tg		
Total carbon released:	659 Tg		

Although we know that some of Alongi's numbers are best guesses, the budget suggests that, overall, mangrove forests add to the storage of carbon and remove carbon from the atmosphere.

The sink suggested by Alongi's mangrove carbon budget is supported by other studies. On a per hectare basis, mangrove forests store more carbon than other tropical and subtropical ecosystems, including terrestrial forests, and most of that carbon is stored in the sediment (Pendleton et al. 2012). Even though they pack a lot of carbon storage potential into a small spatial area, mangroves do not make a huge dent in the global carbon budget because they account for only a small percentage of the total landmass of the planet (**Figure 14.5**). Thus, ecosystems that cover much larger areas of the Earth's surface will contribute more heavily to the movement and fate of carbon worldwide. Mangrove forests provide additional benefits beyond carbon storage, such as reducing terrestrial erosion during storms

Figure 14.5 This map of global mangrove forest distribution shows modeled patterns of aboveground biomass per unit area (Hutchinson et al. 2014).

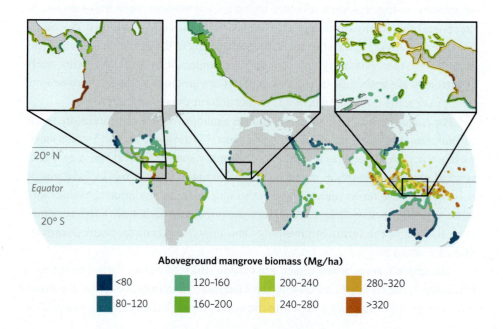

Aboveground mangrove biomass (Mg/ha)

<80	120–160
80–120	160–200

200–240   280–320
240–280   >320

and hurricanes. At the end of this chapter, we explore ways that functioning eco-systems provide benefits to the health of the planet and offer services that improve human well-being.

Ecosystem budgets like Alongi's mangrove carbon budget also allow us to ask how long an ecological sink lasts, and when and under what circumstances a sink can become a source. How long does carbon stay fixed and stored in a par-ticular carbon pool? The carbon that is fixed and put toward the growth of leaves clearly does not stay in leaves for long, especially in deciduous forests. Leaves eventually drop and become leaf litter, and detritivores access carbon from that litter for their growth, reproduction, and respiration. Each pool in an ecosys-tem carbon budget has a unique time frame for how long a single carbon atom is likely to stay in that pool. The rate at which carbon enters and then leaves a carbon pool is called the *turnover rate*. Some carbon pools (like leaf-litter pools) turn over quickly, whereas others can store carbon for centuries. If we can iden-tify a timescale on which small pools release fixed carbon, can we do the same for whole ecosystems? Even in sink ecosystems, major disturbances such as fires can release stored carbon. How often will this happen, and what happens to all the released carbon?

Even for simple systems like mangrove forests, the answer can be compli-cated. Mangrove trees tend to grow quickly toward their maximum size and then stay that size for up to 80 years before they die. Their extensive root systems make them resistant to wind and water disturbance, and the near-constant presence of water makes them less prone to burning. Mangrove forests, once established, do not experience much in the way of succession (Chapter 13). Although the litter from mangroves may create short-lived pools, even the leaves are not deciduous, and the turnover rate for the largest carbon pools in Figure 14.4 is on the order of many decades.

The exception to this pattern is in the sediments underneath mangrove trees. Because sediments from both rivers and the ocean are trapped within the complex root networks characteristic of mangrove forests, litter from trees and upstream sources is quickly buried. Interestingly, only a relatively thin layer at the top of the sediments under a mangrove forest is inhabited by detritivores. If the leaf material is buried quickly below this bioactive top layer, it never gets processed by detriti-vores and can be stored in these sediments for centuries.

This process enables mangrove soils to hold much more carbon than other ecosystems. We see similar effects in some other aquatic ecosystems, such as peat bogs. This buried carbon is removed from circulation in the global carbon cycle for a relatively long time (centuries), though not for nearly as long as carbon that is fos-silized as oil or gas (see the discussion of the slow carbon cycle in the next section). Nevertheless, mangrove forests are considered to be **sequestering ecosystems** because of the long-term storage of carbon in their soils. Sequestering ecosystems are globally important for countering the anthropogenic releases of carbon into the atmosphere, primarily from the burning of fossil fuels.

Finally, with a budget, we can ask what happens when humans reconfigure, develop, or destroy ecosystems. If we bulldoze or chop down one acre of man-grove forest, a series of effects kicks in. Not surprisingly, we completely alter the budgeting when we alter the ecosystem, and this change in budgeting helps explain why sinks sometimes become sources. If human actions remove or destroy photosynthesizing organisms, the ecosystem's ability to draw carbon out of the atmosphere and fix it in biological forms decreases. In addition, killing organisms causes more carbon to be released into the atmosphere. Often, we speed up this release by burning vegetative or woody materials, but microorganisms also decompose dead, carbon-laden material that we leave behind. In addition, when human actions disturb the soil, carbon sequestration may be reversed, "unsealing" centuries-old carbon and allowing it to flow into the atmosphere as well. The destruction of ecosystems effectively converts all the stored carbon into outward-facing arrows (in the imagery of our budget schematic, Figure 14.4), emitting it to the atmosphere or exporting it to other ecosystems.

Given that the soils of mangrove forests store centuries-old carbon, their destruction adds significantly to the carbon that reaches the atmosphere. Mangroves also sit at the interface of terrestrial and marine systems. Alongi (2014) suggests that, if we lump them with marine systems, mangrove forests compose 10%–15% of the ocean's carbon sequestration ability. Unfortunately, the world's mangrove forests are being converted to urban developments or aquaculture farms (particularly shrimp farms) at a rate of 1%–10% per year (Pendleton et al. 2012). These new land uses certainly have their own carbon budgets, but they do not store carbon for long periods of time, and many are net-carbon sources instead of sinks. Linwood Pendleton and colleagues (2012) estimated that each hectare of mangrove forest destruction has the potential to add 1,500 megagrams of carbon dioxide to the atmosphere (1 megagram [Mg] = 1,000,000 grams). Mangrove forests are not alone in this regard. Peat bogs, boreal forests, and other deep-soiled or wet ecosystems are also carbon sinks. The loss of ecosystems that tend to store carbon for long periods of time is increasing in the Anthropocene and impeding efforts to reduce carbon concentrations in the atmosphere.

14.3 The Global Carbon Cycle

Mangrove forests are but one ecosystem among many on Earth. Each ecosystem has its own carbon budget, although only a fraction of these budgets has been well characterized by scientists. Still, there is enough information to scale budgets up in order to estimate the flow of carbon through the entire Earth system. At the global scale, we sum carbon pools and flows within the atmosphere, land, and ocean to understand the location and time frames of carbon storage and the processes that move carbon between storage pools (**Figure 14.6**).

Slow Carbon Cycle

The vast majority of the carbon on Earth is stored in an inorganic form within the Earth's solid parts—the geosphere—with most of the surface carbon stored in limestone. This carbon can be exchanged with the other components of the global system through several geological processes, including volcanic eruptions and weathering. Carbon movements into and out of the geosphere have dictated much of Earth's climate over the millennia, including producing and maintaining the relatively warm and hospitable climate that allowed life to evolve on the planet in the first place. However, this type of carbon movement is very slow, leaving it

Figure 14.6 In this simplified diagram of the Earth's carbon cycle, the numbers represent approximate metric gigatons (Gt) of carbon (Riebeek 2011). The only portions of the slow carbon cycle that are depicted are those that contribute to flows. Green boxes represent natural flows; red arrows, boxes, and text indicate human contributions; and black ovals indicate stored carbon. The red arrows with red boxes indicate the major extraction-and-release pathways for anthropogenic carbon. The numbers for long-term, belowground carbon storage are approximate and do not necessarily add up. You may be used to thinking about atmospheric carbon in terms of concentrations of molecules in the air (measured in parts per million, or ppm). Concentration does not work well for the other pools and flows, but the approximately 800 Gt of carbon in Earth's atmosphere translated to an average global atmospheric concentration of 418.94 ppm of carbon in July 2021 (Scripps Institution of Oceanography and NOAA Global Monitoring Laboratory 2021).

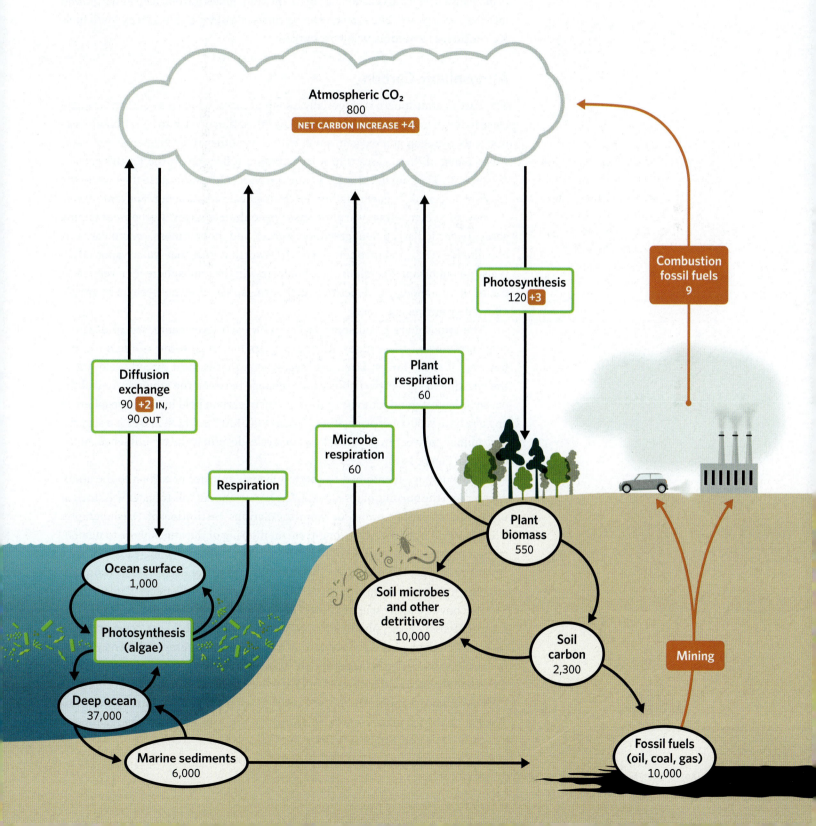

largely irrelevant to discussions of how carbon influences ecological processes. This movement of carbon from the Earth's rocks is often referred to as the **slow carbon cycle**, leaving all other carbon pools and exchanges in the **fast carbon cycle**.

Fossil carbon makes up an extremely small but important organic part of the geosphere. However, humans extract and burn this fossil carbon, or *fossil fuel*, that exists in solid (coal), liquid (oil), or gas (natural gas) phases. The burning of fossil fuels converts organic carbon from millions of years ago into inorganic carbon, principally carbon dioxide but also methane, and moves it from deep underground into the atmosphere. The extremely long timescale of formation is why fossil fuels are considered a nonrenewable resource.

Atmospheric Carbon

The Earth's atmosphere is essentially a soup of gases, which we call "air," trapped close to the planet's surface by gravity. Any inhalation of air into your lungs brings in mostly nitrogen and oxygen, along with a few other trace elements and water vapor. Some of these gases trap solar radiation reflecting off the Earth's surface (Chapter 3). They act like the glass panes on a greenhouse because they warm the Earth by letting light energy in but not letting heat radiation out. For that reason, we call these blanketing, warming gases **greenhouse gases**. Carbon-containing gases are particularly potent greenhouse gases, and the two most common are carbon dioxide (CO_2) and methane (CH_4). It is worth noting that water vapor (H_2O) is also an important greenhouse gas, but it stays in the atmosphere for a very short period of time and its residence time in the atmosphere does not seem to be greatly affected by human activities.

The atmosphere has distinct layers, with each layer composed of different concentrations of these gases. All these layers matter in terms of life on Earth; however, it is the lowest layer, the *troposphere*, that is most directly relevant to the question of carbon availability for exchange between the atmosphere and the ocean or land. A good estimate of the amount of carbon held in the troposphere at any one time is 800 metric gigatons (Gt), or 800,000 Tg, which makes it a surprisingly minor "player" in the global carbon cycle, despite its heavy influence on life on Earth (Figure 14.6).

Previously, we mentioned the anthropogenic transfer of carbon to the atmosphere from the burning of fossil fuels. The importance of this transfer was unclear until an enterprising professor from the Scripps Institution of Oceanography, Charles David Keeling, set up long-term monitoring of the atmosphere to demonstrate trends in how much carbon it held. Keeling followed some early evidence from Guy Stewart Callendar, who had tied changes in the Earth's surface air temperature to changes in atmospheric CO_2 concentrations. Callendar's data were too local to convince many scientists that CO_2 was increasing through time, so Keeling set up infrared spectrophotometers to measure atmospheric gas concentrations in three locations in the continental United States (Monterey, California, where the Scripps Institution is based, and the states of Washington and Arizona). There was some suggestion that these readings were affected by local urban centers, so Keeling deployed sensors in 1958 at a geophysical laboratory on top of Mauna Loa volcano on the island of Hawaii. Instruments have continued to take readings at regular intervals at this location ever since (**Figure 14.7**).

The location of Keeling's instrument is essential to the insight it has provided over the years. Mauna Loa is high enough in elevation that it sits above the tree line, thus reducing the influence of respiration and photosynthesis from the island's trees on the CO_2 readings taken there. Mauna Loa is also located in the most remote island archipelago in the world (the Hawaiian Islands), thus reducing the effect of industrial activity from cities on CO_2 readings. The instrument is situated high enough that it avoids effects from the thermal inversion layer and contamination from volcanic vents, as well. These attributes make the record kept by Keeling's instrument one of our best sources for long-term estimates of how much CO_2 has been added to the atmosphere in the last six decades.

The record from the Mauna Loa monitors, called the **Keeling curve**, is stunning for two reasons. First, a close look at the curve reveals that the line has a constant wave pattern to it (Figure 14.7). This waviness is due to the seasonal flux of CO_2 in the atmosphere that results from Northern Hemisphere forests' fixing massive amounts of carbon in the spring (as they grow and leaf out) and then releasing this carbon in the autumn (when growth stops and leaves fall and decompose).

The second reason the Keeling curve is so remarkable is that it provides clear evidence that CO_2 concentrations in the atmosphere have risen every year since 1958, and that concentrations have been rising at an accelerating rate in recent decades (Figure 14.7). It is now quite evident from other records and experiments that the increase seen in the Keeling curve is attributable in large part to the massive amounts of CO_2 that are moved from the geosphere to the atmosphere by the burning of fossil fuels. Fossil fuels, combined with other human sources of carbon release (e.g., from mining limestone for cement, and the effects of land use changes that move long-stored carbon into the atmosphere), have put 9 Gt of carbon into the atmosphere each year. These inputs move carbon from the slow carbon cycle into the fast carbon cycle (Figure 14.6), and these 9 Gt are the primary driver of global warming from greenhouse gases.

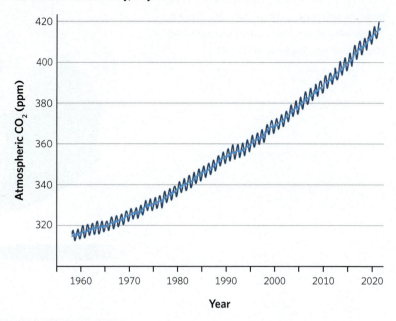

Mauna Loa Observatory, July 2021

Figure 14.7 Atmospheric carbon dioxide concentrations have been measured from the observatory on Mauna Loa volcano on the island of Hawaii since 1958, producing a record known as the Keeling curve. As of July 2021, atmospheric carbon was 418.94 ppm (Scripps Institution of Oceanography and NOAA Global Monitoring Laboratory 2021).

Oceanic Carbon

The ocean is a major player in both the fast and slow carbon cycles. Carbon flows into the ocean from terrestrial sources via rainwater and streams. These inflows contain inorganic carbon from the geosphere and organic matter from organisms that live on the continent (e.g., tree leaves and freshwater microorganisms; Figure 14.6). However, the vast majority of the carbon in the ocean comes from interactions between its surface waters and the atmosphere.

Carbon, mostly in the form of CO_2, is exchanged freely between the atmosphere and the ocean's surface waters so that, in general, about 90 Gt of carbon are exchanged each year (Figure 14.6). This exchange of carbon is balanced at a large scale and represents a *fast loop* in the global carbon cycle, serving as neither a source nor a sink. But this loop can be shifted away from balance. The

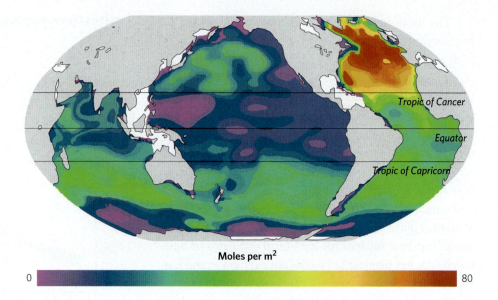

Figure 14.8 Oceanic storage of anthropogenic carbon varies with latitude, and there is a large area of very high carbon storage in the North Atlantic (Sabine et al. 2004).

Moles per m²

0   80

ocean becomes a carbon sink when atmospheric CO_2 dissolves in marine surface waters and is converted into dissolved inorganic carbon, mostly bicarbonate ions (HCO_3^-). The surface waters of the ocean tend to hold this form of carbon for relatively long periods of time (decades). In general, ocean surface waters hold about 1,000 Gt of carbon.

Surface-water carbon storage is not evenly distributed around the globe, though. Colder waters convert more atmospheric CO_2 into bicarbonate ions than warmer waters. Thus, at the Earth's poles, the ocean takes in more carbon from the atmosphere than it releases. This cold, dense water sinks and is then moved toward the equator via ocean circulation patterns (Chapter 3). When these waters warm and return to the ocean surface near the equator, all of the dissolved inorganic carbon they hold moves back into the atmosphere. This *thermohaline pump* is responsible for very large-scale patterns in carbon movement and, in particular, serves to store inorganic carbon in the deep ocean over relatively long timescales. Even with the movement by ocean currents, surface waters in high latitudes hold more carbon than in the tropics, with the waters in the North Atlantic holding 23% of oceanic CO_2, despite representing only 15% of global ocean surface area (**Figure 14.8**).

Another way that the deep oceanic waters serve as a global carbon store is through a *biological pump*, which converts inorganic carbon into organic carbon via photosynthesis. Collectively, phytoplankton are the most common primary producers in the open ocean, and they gather much of the carbon they need for photosynthesis from dissolved inorganic forms of carbon in the waters around them. Just as in terrestrial systems (like the mangrove forests), the total amount of oceanic carbon fixed by phytoplankton is called gross primary production, or GPP. This carbon is used by phytoplankton for growth and respiration, and the amount ultimately kept within phytoplankton is referred to as net primary production, or NPP.

When phytoplankton die, their organic carbon either breaks down in the surface waters or sinks to the ocean floor, where it may be deposited as sediment. The creatures that eat phytoplankton are also (obviously) made of carbon, and much

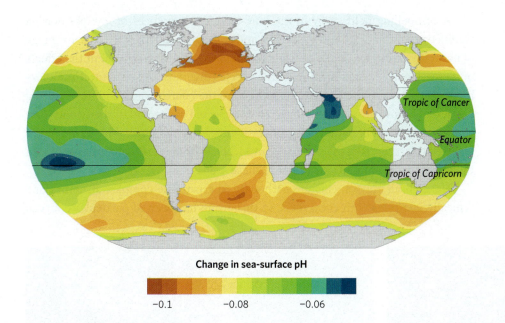

Change in sea-surface pH

−0.1   −0.08   −0.06

Figure 14.9 Change in the sea-surface pH from the 1700s to the 1990s. Warmer colors indicate a greater shift toward acidity (i.e., decrease in pH) (Plumbago 2009).

of their carbon also sinks into the deep ocean when they die. All of this sinking organic carbon will eventually reach very deep ocean waters, staying there in dissolved form or deposited as marine sediment (Figure 14.6). This store of carbon will only reach the surface again through oceanic circulation patterns, upwelling, or a massive geologic event (e.g., an eruption) that returns these oceanic sediments to the surface.

The dual action of the thermohaline and biological pumps tends to capture about 2 Gt of anthropogenic carbon per year from the atmosphere and store it in deep ocean waters or deposit it as marine sediment. Globally, on average, deep ocean waters store 37,000 Gt of carbon, and marine sediments store an additional 6,000 Gt (Figure 14.6).

There is one final bit of ocean carbon chemistry that is important for us to consider from an ecological point of view. When CO_2 from the atmosphere dissolves in ocean surface waters and forms bicarbonate ions, hydrogen ions are also released, which decreases the water's pH and makes the ocean more acidic. Over the last two centuries, the global ocean has dropped in pH from an average of 8.25 to 8.14 because of high rates of CO_2 emissions into the atmosphere and the subsequent dissolution of this CO_2 in the ocean waters. Recall that the pH scale is logarithmic, so even a slight drop in pH represents a substantial change toward more acidic conditions in the ocean. Despite this global trend, there is sizable variation in acidification across the Earth's oceans (**Figure 14.9**).

Ocean acidification has a number of harmful effects on marine life, including creating conditions that are physiologically stressful for many marine species. One of the more significant impacts is a weakening of carbonate-based shells and support systems in many mollusks, snails, corals, crabs, zooplankton, and krill. In all of these species, dissolved calcium from marine waters chemically combines with carbonate ions to create calcium carbonate ($CaCO_3$) for the organisms' hard parts. The decreased pH of the ocean indicates the presence of more hydrogen ions in the water, which react with carbonate ions to form more bicarbonate ions. This chemical reaction results in less available carbonate and a dissolving of the constructed shells of these organisms (**Figure 14.10**).

Ⓐ

Ⓑ

Figure 14.10 Pteropods are small, free-swimming snails that are an important food source for many marine organisms. These photos show what happens to a pteropod's shell in seawater that is too acidic: Ⓐ a live pteropod from the Southern Ocean, where acidity is not too high; Ⓑ a pteropod shell from a region where the water is more acidic.

A compelling field study of the long-term consequences of low-pH marine waters comes from Katharina Fabricius and colleagues (2011), who took advantage of naturally occurring shallow-water volcanic CO_2 seeps to document the effects of increasing ocean acidification on the species diversity and structural complexity of coral reefs. These seeps bubble CO_2 into the nearby water column, creating a gradient of low pH (7.8, similar to levels expected at the end of the twenty-first century as a result of climate change) near the seeps and normal background pH levels (8.1) farther from the seeps (**Figure 14.11**). Fabricius and colleagues explored the natural coral communities close to these seeps.

Hard corals rely on calcium carbonate to create the complex structure one typically identifies with coral reefs, so this group has been of particular concern

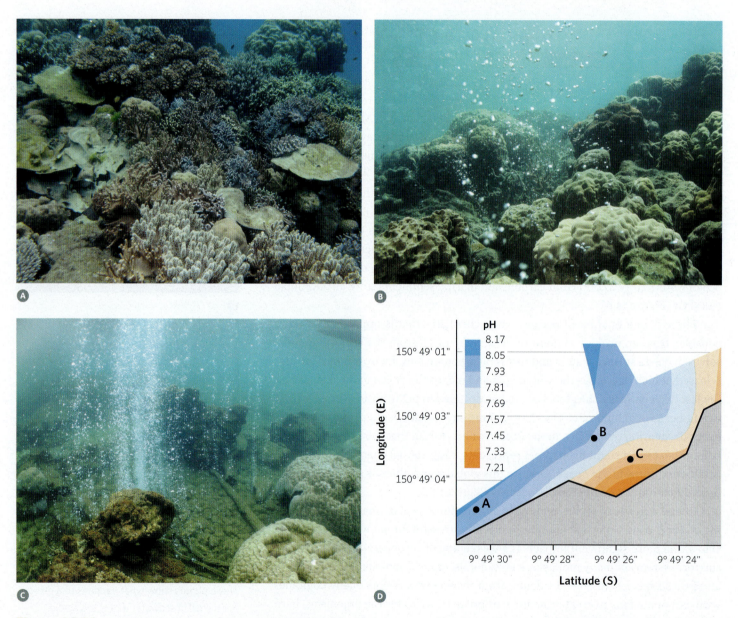

Figure 14.11 The long-term effects of ocean acidification on coral reef development can be physically observed in Milne Bay, Papua New Guinea. At this location, decreasing pH has caused a decrease in biodiversity and structural complexity, as represented in these photographs and labeled on the graph (Fabricius et al. 2011): **A** a site with low CO_2 levels (pH 8.1); **B** a site with moderate CO_2 levels (pH 7.8–8.0); and **C** a site with high CO_2 levels (pH <7.7). **D** The graph shows the gradient of oceanic pH created by natural CO_2 seeps (Fabricius et al. 2011).

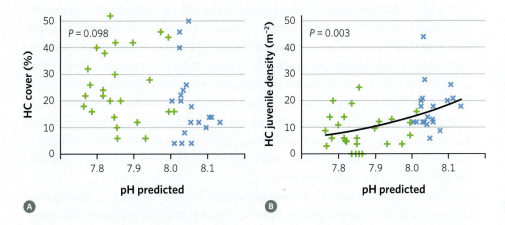

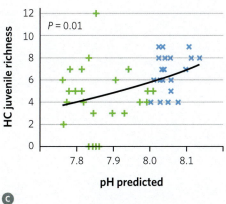

with increasing ocean acidification. Fabricius and her fellow researchers found that hard corals covered about the same amount of area across pH levels (**Figure 14.12**), but that there was a significant decline in the species density and richness of juvenile hard corals at the low-pH end, near the seeps (Figure 14.12). In addition, a form of macroalgae that uses calcium to create hard structures—crustose coralline algae—decreased sevenfold in density and species richness from the low-CO_2, high-pH area to the high-CO_2, low-pH area. Together, these changes to the species composition of coral reefs in conditions of low pH suggest a loss of structural complexity, which reduces the habitat availability and quality for coral reef fishes and invertebrates.

Of particular note in Fabricius's study was the major reduction in the species richness and spatial cover of juvenile corals of any species (Figure 14.12). The absence of young coral is a strong indicator that these corals may not survive in a future low-pH ocean because they cannot replace aging individuals by recruiting young individuals into the population (Chapter 6). One reason for low juvenile recruitment is the increasing presence in low-pH conditions of noncalcareous macroalgae and seagrass that reduce available space for coral juveniles to settle. Crustose coralline algae create settlement sites for juvenile hard corals, and their replacement by soft macroalgae and seagrass creates a further reduction in coral regeneration. A low juvenile recruitment rate in lower-pH conditions suggests that, with increasingly acidic ocean conditions in the future, corals will have a hard time recovering from disturbance events such as heavy storms (Fabricius et al. 2011).

The final worrisome insight from the Fabricius study is that the researchers found absolutely no corals around the seeps when the pH fell below 7.7. This observation suggests the existence of a threshold pH level below which corals of any kind simply cannot survive. Current predictions for ocean acidification suggest that, as long as carbon emissions continue to accelerate, seawater pH will drop by 0.14 to 0.35 units globally, resulting in an average global ocean pH of 7.9 to 7.7 by 2100. The work of Fabricius and colleagues paints an unsettling picture for the Earth's coral reefs over the next century.

Terrestrial Carbon

As we have detailed, photosynthesis moves carbon from the atmosphere into living organisms in terrestrial ecosystems. Budgets can help us follow this carbon as

Figure 14.12 **A** Fabricius and colleagues (2011) found that the percentage of hard coral (HC) spatial cover did not change much between sites with low pH (green symbols) and sites with high pH (blue symbols), but the **B** juvenile density and **C** juvenile species richness both significantly decreased at low pH. The black lines indicate a log-linear fit to the data. Also presented at the top of each figure is the statistical significance (i.e., the p-value). The pH along the x-axis is predicted rather than measured because the study used a variety of seawater measurements to arrive at an average pH level.

it flows back into the atmosphere, gathers in the soil, or drifts into the ocean. We have seen how this works for mangrove forests, but all other terrestrial ecosystems have their own budgets. If we combine all of these budgets and consider the role of terrestrial ecosystems as a whole in the global carbon budget, then an estimated 120 Gt of carbon flow from the atmosphere into terrestrial plants per year, with an additional 3 Gt estimated to come from human-produced carbon sources (Figure 14.6). About half of the carbon fixed by photosynthesis globally goes back into the atmosphere through respiration of primary producers. If we consider all plants as stores of carbon, at least until they die, then terrestrial ecosystems hold about 550 Gt of carbon (Figure 14.6).

The big carbon player in terrestrial systems is soil. Soil contains large quantities of organic and inorganic carbon, most of which is derived from the organisms that once lived in it or on it but also with a bit of input from the geosphere (rock weathering, volcanic ash). This carbon can stay in soil storage pools anywhere from a few hours to centuries, depending on a host of factors, including the species of detritivores (bacteria, fungi) present; the availability of water, nitrogen, and oxygen; and soil temperature. Global estimates suggest that about 60 Gt of carbon per year move from the soil into the atmosphere via *microbial respiration* or as inorganic carbon gases that are produced through microbial decomposition of organic soil carbon. Another 2,300 Gt of carbon are stored in soils. If we count the organic carbon that is trapped in deep sediments as fossils, we can add another 10,000 Gt of long-term storage (Figure 14.6).

Human Influences on the Global Carbon Cycle

In Figure 14.6, the red numbers indicate human-derived inputs of inorganic carbon into the cycle. Notice that humans produce or emit about 9 Gt of inorganic carbon per year, but only a portion of that carbon stays in the atmosphere. Estimates vary, but anywhere from a third to 45% of the carbon stays in the atmosphere ($3–4$ Gt y^{-1}), with about a third entering the oceans ($2–3$ Gt y^{-1}), and the remaining third (3 Gt y^{-1}) entering the terrestrial fixed-carbon pool from photosynthesis. Clearly, the global carbon cycle is complex, but it may seem surprising that such relatively small additions of carbon from human sources can have such a big influence on the global climate. Remember, though, that we are talking about annual inputs. Anyone who has tried to balance their own financial budget knows that a small but steady imbalance can produce big problems if that imbalance leads to a steady depletion of a savings account. Similarly, the steady shift of carbon from stored fossil reserves into the atmosphere and our oceans can lead to a change in the way the global system works. In addition, think about the size of some of the fast loops. The human-derived alterations to these loops are in the percentage range of many annual salary increases. Imagine, though, if your salary stayed the same but your costs increased by 2%–3% every year. Such a situation would not be sustainable for long before becoming disastrous financially. We consider the implications of a rapidly changing climate for biodiversity in much more detail in Chapter 15. For now, it is only important that you recognize the importance of the carbon cycle in regulating the Earth's climate, including the key role that such organisms as trees, phytoplankton, and soil bacteria play in this regulation.

14.4 The Nitrogen Cycle

Forms of Nitrogen

Carbon is but one of the elements critical to life on Earth. After carbon, nitrogen probably plays the largest role in most ecosystem dynamics. Nitrogen is essential for the production of proteins and nucleic acids, making this element vital to all life on the planet.

Exploring the nitrogen cycle is a little more challenging than diving into the carbon cycle, which may seem impossible after having just tracked how tons of carbon move through trees, oceans, and the atmosphere. Carbon comes in many forms, most of which interact in some way with living tissues or organisms, but the essential interactions are fairly easy to encapsulate by talking about photosynthesis, carbohydrates, and respiration. Nitrogen, however, is primarily found in an unreactive form, as *inert* nitrogen gas (N_2), but it is also present in several reactive forms that are biologically available. These reactive forms are sometimes divided into **inorganic nitrogen** (not derived from living tissues) and **organic nitrogen** (coming from living or formerly living tissues). The main biologically available inorganic forms are ammonia (NH_3), ammonium (NH_4^+), and nitrate (NO_3^-), and the organic forms include the full range of compounds that organisms make using nitrogen, such as DNA, amino acids, and proteins. Each of these reactive forms of nitrogen has its own particular set of characteristics and interactions.

To complicate matters, although nitrogen is necessary for life, many of its reactive forms are toxic to living organisms, so individuals are constantly expelling nitrogen and then looking for more. Perhaps that process is not that different from how we expel CO_2 yet breathe it in again to get more oxygen, but the process of breathing is one that we all understand. Most people are less familiar with all the relevant forms of nitrogen. Therefore, it is good to start by explaining why N_2, NH_4^+, and NO_3^- are so important and then use that information to build a general nitrogen cycle for the entire planet.

Nitrogen is the most abundant chemical element in the air we breathe. Each time we inhale, we breathe in approximately 78% nitrogen and only about 21% oxygen. Obviously, everything else is even less common than these two elements. The vast majority of the gaseous atmospheric nitrogen exists in its elemental form as N_2. Unfortunately, this abundance of N_2 is particularly difficult for organisms to use because of a triple covalent bond that links the two nitrogen atoms together. This bond is what makes atmospheric nitrogen inert, or nonreactive.

You may or may not remember from chemistry that covalent bonds are fairly strong, and a triple covalent bond is extremely stable. Converting N_2 gas into biologically useable forms is, therefore, an energetically difficult task. This energy barrier means that atmospheric nitrogen does not readily react with any other elements around it. Although nitrogen forms the backbone of our DNA and proteins and is essential to life, our bodies (and the bodies of most living things) cannot use even a single molecule of the abundant N_2 we take in with every breath. After breathing it in, we breathe it right back out again.

How do living things get access to this most abundant but inaccessible element if it is so difficult to use? Well, one way is through lightning strikes. Heating

N_2 gas to extremely high temperatures causes it to react with atmospheric oxygen, creating nitrogen oxides, such as nitrite (NO_2^-) and nitrate (NO_3^-), which are biologically available. This process happens naturally around the planet when lightning heats the air or soil, meaning lightning is partially responsible for life on Earth.

Waiting for lightning to strike would be a pretty inefficient way to support life, though, so luckily there is a second natural route to biologically available nitrogen. Some bacteria and archaea have evolved the ability to convert atmospheric N_2 into ammonia (NH_3) in the presence of an enzyme, **nitrogenase**. These nitrogen-fixing organisms, called *diazotrophs*, are widespread and are the only organisms that can grow using only atmospheric N_2. We call this process of converting N_2 into a biologically available form **biological nitrogen fixation**, and it is an *anaerobic* process, meaning a chemical reaction that can only occur if there is no oxygen present. Earth is an oxygen-rich planet, making anoxic conditions difficult to achieve in most places, so the bacteria must fix nitrogen in isolated situations. One potential fixation space is in specialized structures within diazotrophs called *heterocysts*, where oxygen is excluded.

Nitrogen fixation is not a common characteristic of life on Earth, even within the bacteria and archaea groups. **Cyanobacteria**, rhizobia, bacteria of the genus *Frankia*, and some taxa of archaea can all convert N_2 to reactive forms of biologically available nitrogen. Luckily for the rest of the organisms on Earth, these diazotrophs are widely distributed around the planet in both terrestrial and aquatic systems. Some diazotrophs are free-living, but others have formed mutualistic relationships with terrestrial plants through which they exchange nitrogen for carbohydrates (Chapter 9). There are even a few extremely rare associations between diazotrophs and animals, such as termites.

Among plants, such mutualistic associations occur only in a few families, but they are particularly common between legumes (peas, beans, alfalfa, and locust trees) and bacteria in the genus *Rhizobium*. About two dozen other plant genera, including alders and bayberries, have developed similar nitrogen-fixing associations with *Frankia*. Plants with nitrogen-fixing associations generally have anoxic nodules in their root systems to house the symbiotic bacteria. In these nodules, the symbiotic bacteria can control oxygen levels in order to fix atmospheric nitrogen in anaerobic conditions. Plants with these symbioses act as nitrogen fixers in early successional stages, as mentioned in Chapter 13.

Ammonium (NH_4^+), the positive ion of ammonia (NH_3), is one of two primary forms of nitrogen taken up by plants. In aqueous solutions, particularly those with a pH below 8, ammonia (NH_3) from biological nitrogen fixation breaks down into ions and creates ammonium (NH_4^+) and hydroxide (OH^-). In the environment, NH_4^+ is a short-lived and not particularly mobile molecule. It quickly forms other compounds because its associated positive charge is attracted to the common negative charges in soils. Ammonium is toxic to most heterotrophs, and *ammonium toxicity* is even possible for plants because NH_4^+ is not easily stored.

Nitrate (NO_3^-) is the form of inorganic nitrogen that is most easily assimilated by plants and microbes. It does not become toxic to plants and is found in abundance in green, leafy plant tissues. Unlike NH_4^+, NO_3^- readily moves through soils and can enter groundwater. Nitrates are commonly used as agricultural fertilizers because of their high solubility.

Cycling Nitrogen

So far, we have been looking at different forms of nitrogen, but now we move to understanding the basics of how nitrogen cycles around the planet (**Figure 14.13**), which is complex in terms of both its chemistry and ecology. We have tried to simplify the process presented here but realize there are many details that could be added.

Let's start with information that we already know: breaking the bonds of atmospheric nitrogen (N_2) is a tough job. Lightning and nitrogen-fixing bacteria are the only two natural processes that accomplish this nitrogen fixation. With only two avenues for creating reactive nitrogen, nitrogen often limits ecosystem productivity (without human intervention). Both natural pathways of nitrogen fixation create forms of nitrogen that are fairly easily converted into ammonium or nitrate, which are greedily taken up by plants and microbes and converted into organic forms of nitrogen, such as amino acids, proteins, and urea. This process of turning inorganic nitrogenous compounds into organic ones is generally called *nitrogen assimilation*.

As plants grow, they produce a whole suite of other nitrogen-containing compounds. Remember that nitrogen is in all DNA and proteins, so it is present in all living tissues. Of course, once an ecosystem has plants, it can support a full community of herbivores, omnivores, carnivores, and detritivores, which get their nitrogen from the organic forms in plant tissues or the tissues of other organisms that eat plants. Nitrogen fixation and assimilation are, therefore, critical to our understanding of food webs, trophic interactions (Chapter 10), and physiological ecology (Chapter 5).

Nitrogen from living organisms is recycled back into the environment via either the death and decomposition of individuals or the production of waste products (Figure 14.13). This movement of nitrogen from living organisms into the environment happens constantly because nitrogen is a common by-product of digestion. One reason we urinate is to expel excess water, but the more pressing concern is to expel the toxic nitrogenous wastes produced by metabolism. In fact, sometimes, animals can barely spare the water for urine but must still get rid of the ammonia and ammonium; you can review the discussion of the kangaroo rat in Chapter 5, which provides a good example of how one organism evolved ways to balance these competing physiological needs. Once urine reaches the soil or water, it quickly deionizes, leading to free-floating ammonium. This source of ammonium is toxic to most animals, but it is useable by plants and many microbes, and thus the nitrogen moves in a closed loop between photosynthetic organisms (phytoplankton or plants), animals, and the microbial food webs of soils and waters (Figure 14.13).

The human component of the nitrogen cycle (red arrows in Figure 14.13) is another important part of this global system. When humans invented a way to produce biologically available nitrogen through industrial means and applied it to agriculture, the discovery permanently altered the planet's nitrogen cycle and the associated ecosystems. Humans produce several million kilograms of nitrate annually and apply it as agricultural fertilizer worldwide. Much of this applied nitrogen either runs off the land and into streams, lakes, and oceans or enters the groundwater. Production of cheap forms of nitrogen fertilizer for agriculture has

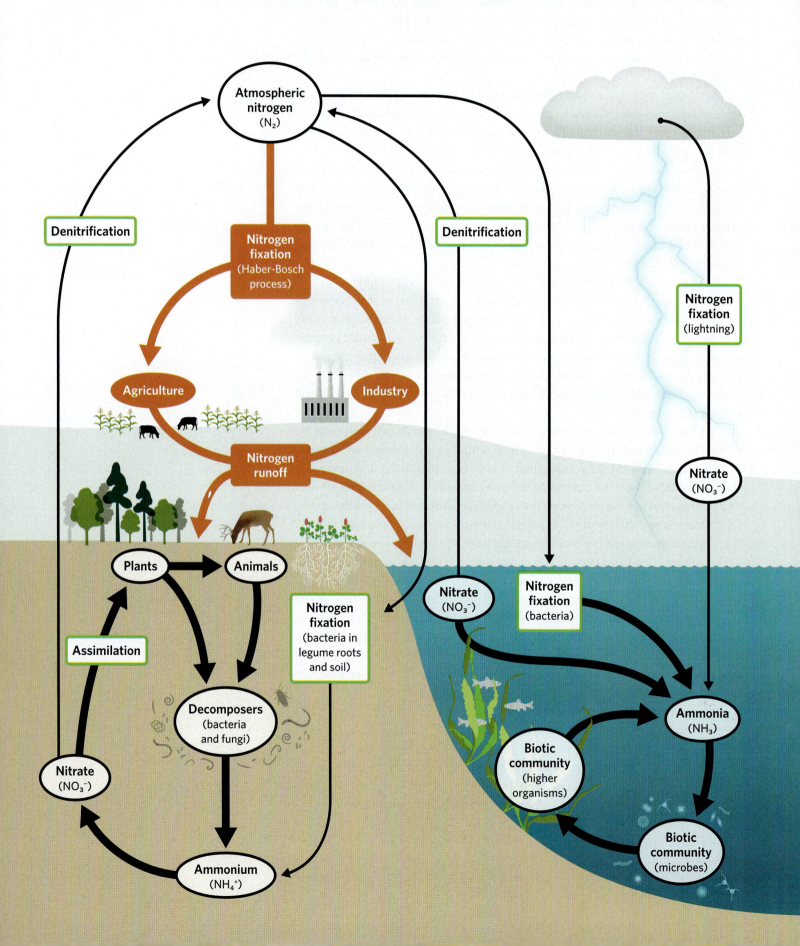

Figure 14.13 In this simple diagram of the nitrogen cycle, the thickness of arrows corresponds to the relative amount of nitrogen moved, with thicker arrows indicating more movement. Red arrows indicate movement of anthropogenic nitrogen.

been a blessing because it has increased food production, but it has also been a curse because of the massive increase of nitrogen flowing into natural ecosystems.

In order to close the full loop so that nitrogen is converted from reactive forms back into atmospheric N_2, we have to consider another microbial process called **denitrification** (Figure 14.13). In denitrification, a different set of specialized bacteria chemically reduce NO_3^- to N_2 through a series of intermediate steps. Because these bacteria are pulling oxygen away from the nitrogen, denitrification typically occurs in low-oxygen settings, such as the anoxic soils of wetlands or the oxygen-poor waters of the deep ocean. Human environmental engineers have created systems to favor the growth of denitrifying bacteria in order to denitrify human wastewater and stormwater runoff before mixing such water with natural systems.

Two things should be clear from this overview of the nitrogen cycle. First, microbial communities in the soil and oceanic waters are essential to sustaining life on Earth. Most living creatures cannot take advantage of the abundant nitrogen in the atmosphere, even though they all need it to survive and reproduce. Microbial food webs serve as a critical bridge for moving the various forms of nitrogen through the cycle, ultimately making nitrogen available for all other life forms and returning that nitrogen back to the atmosphere. Second, we should expect to find plenty of evidence that the availability of reactive and thus useable nitrogen is a central limiting factor to where species occur, how large they may grow, how long they may live, and what species they may eat as prey. We explore several of these factors next.

Nitrogen in Food Webs

Recall that carbon derived from mangrove trees flows with tidal waters out into the open ocean, thereby providing coastal and deeper-ocean food webs with a source of "outside," or *exogenous*, carbon. The term *exogenous* here means coming from outside the ecosystem, as it did when we discussed forms of disturbance in Chapter 13 and external inputs to food webs in Chapter 10. Nitrogen moves between ecosystems similarly to the way carbon does, so that nitrogen from mangrove forests provides a reactive, useable form of nitrogen for marine species in the open ocean. Similarly, riverine floodplains can receive a hefty dose of nitrogen from upstream sources. High-water flooding events often wash away soils in streambanks and lead to dead vegetation floating downstream. As this water slows in velocity, it deposits reactive nitrogen on floodplain soils and along riverbanks. This process explains why floodplains make such attractive soil for growing crops. The relatively abundant nitrogen in these soils leads to large yields that make up for the risk of crop loss from floods (Figure 14.14).

Animal movements also provide nitrogen subsidies (Chapter 10) between ecosystems. For example, salmon grow to adulthood in the open ocean, where they feed on marine-derived sources of nitrogen (Chapter 12). When salmon are ready to breed, however, they swim miles up into freshwater rivers until they reach their natal

Figure 14.14 Flooded banks of the River Eden in an agricultural region of the United Kingdom. Floodplains tend to have high levels of naturally available nitrogen, which lead to high crop yields.

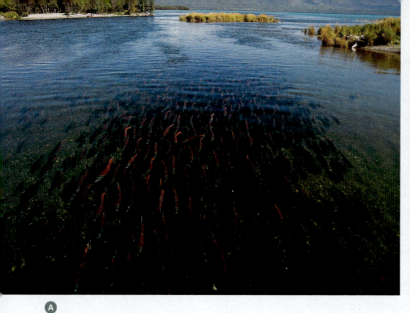

(A)

(B)

Figure 14.15 Animal movements provide nitrogen subsidies. (A) Migrating sockeye salmon (*Oncorhynchus nerka*) in Alaska's Brooks River routinely number between 200,000 and 400,000 individual fish per year. When these fish spawn and die, they contribute their carcasses and nutrients to the Alaskan forest food webs. (B) Grizzly bears (*Ursus arctos*) prey on sockeye salmon returning to spawn. Bears foraging on salmon move into the adjacent forest ecosystem to defecate, depositing marine-derived nitrogen in the terrestrial habitat.

Figure 14.16 Salmon carcasses provide a significant "fertilizer" effect to the rivers and associated forests where they spawn. Chichagof Island in Alaska has two rivers with significant salmon runs, the Indian and the Kadashan. Research has shown a threefold increase in tree growth within 25 m of the rivers' spawning sites compared to the reference sites. No tree growth differences were detected beyond 25 m from the river when comparing the spawning and reference sites. Error bars represent the standard error of the mean (Helfield and Naiman 2001).

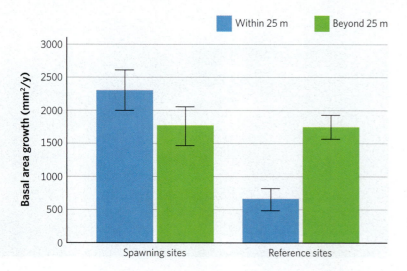

spawning grounds in headwater streams (**Figure 14.15A**). During this migration, they are entirely focused on breeding, and they do not eat. Most Pacific salmon die after spawning, and their carcasses provide gigantic marine-derived nutrient buffets for bears, local fish, aquatic invertebrates, and even algae that live in the stream year-round (**Figure 14.15B**). In headwater streams with salmon runs, up to 40% of the useable nitrogen is derived from marine sources that found their way into these locations via spawning salmon.

Salmon carcasses subsidize food webs outside of the stream as well. Detailed fieldwork done by James Helfield and Robert Naiman (2001) in southern Alaska shows that, along streams with salmon runs, Sitka spruce trees (*Picea sitchensis*) use marine nitrogen to enhance their growth rates (**Figure 14.16**). Some of this nitrogen may be coming to the spruce from belowground root interactions with the stream, but lots of nitrogen travels from the stream into terrestrial environments because of organisms that move between the aquatic and forest habitats. Animals, such as bears and wolves, feed on salmon in the river but deposit their waste (feces, empty carcasses) deep in the forest. This marine-derived nitrogen can release forest trees from severe nitrogen limitation and allow increased growth in biomass. In turn, the growth of healthy spruce forest communities helps to produce the cool, clear, shaded water habitat needed by the salmon for successful spawning. It is fascinating to see such interconnected loops of food and nutrients in natural ecosystems.

At times, marine-derived nitrogen subsidies can include surprising links to our daily lives. A study by Joseph Merz and Peter Moyle (2006) was able to track salmon-derived nutrients from the Mokelumne River,

California, into the native streamside vegetation (as was the case in Alaska), as well as into cultivated wine grapes growing adjacent to the river. In fact, Merz and Moyle showed that 18%–25% of the nitrogen in the grape leaves was derived from marine sources when compared with reference sites without salmon carcasses (**Figure 14.17**). More nitrogen results in higher grape production in these riverside vineyards, and perhaps better wine. You may be able to partially thank salmon if you consume a tasty glass of California wine.

Anthropogenic Nitrogen

Nitrogen is so often limiting in ecosystems that it restricted the population size of *Homo sapiens* for millennia. High human population densities demand large quantities of food that are difficult to acquire by hunting and gathering alone. The invention of agriculture was a huge step in the development of complex human societies. Agricultural production, however, was severely limited by the availability of nitrogen. It was simply not possible to grow large amounts of vegetables, fruits, or grains with the low or average nitrogen levels found naturally in soils around the world.

Many early cultures had discovered that legumes seemed to "feed" the soil, but in the 1880s scientists identified the microbial symbionts (rhizobia) that fix atmospheric nitrogen and substantially increase the availability of reactive nitrogen in the soils around legume roots. Thus began an extended effort to manipulate the growth of legumes within agricultural settings to enable a step-up in crop production. This process often involved planting legumes alongside focal crops or rotating between legumes and other crops each year. Although the process allowed agriculture to continue in one field over many years without depleting soil nitrogen, it also required planting legumes that were not always needed or wanted.

Another natural source of useable nitrogen was seabird guano (droppings). The nineteenth century had a thriving maritime industry in harvesting guano from oceanic islands (believe it or not), hauling it onto a ship, and then selling it to farmers as a way to increase crop production (**Figure 14.18**). Of course, chicken guano also worked, and many farmers raised chickens not only for their eggs but also for their fertilizer production. It takes a lot of chickens to produce enough guano to cover a field, though.

The biggest revolution in agricultural production came when two German scientists perfected an industrial approach to fixing atmospheric nitrogen. In the early 1900s, Fritz Haber and Carl Bosch developed a process that takes atmospheric N_2 and runs it over a metal

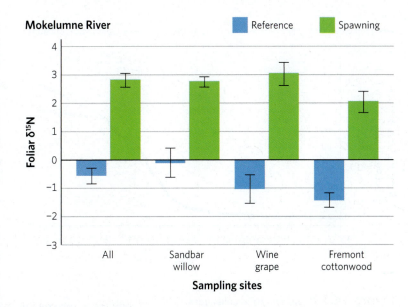

Figure 14.17 Because salmon returning to their freshwater spawning grounds have consumed most of their food out in the ocean, their tissues have a very distinct nitrogen isotope ratio ($\delta^{15}N$) when compared to terrestrial plants and animals. This marine-derived isotopic signature from the salmon tissue was detected in the streamside sandbar willows, wine grapes, and cottonwood trees growing near the spawning sites on the Mokelumne River in California. Plant tissues from reference sites above where salmon can spawn did not show the marine-derived nitrogen signature. Bars show mean values of $\delta^{15}N$, and error bars are ± 1 standard error (Merz and Moyle 2006).

Figure 14.18 Seabird nesting sites can become covered in guano, as seen here on one of the Farne Islands, offshore from the village of Seahouses in Northumberland, United Kingdom.

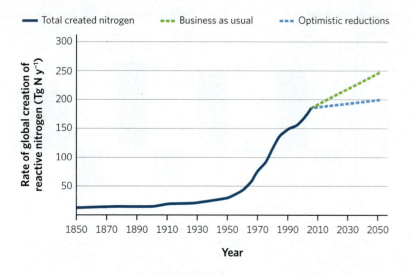

Figure 14.19 Galloway and colleagues plotted the global creation of reactive nitrogen since the invention of the Haber-Bosch process. The solid line represents historical data, and the dashed lines are projections based on no change (business as usual) or on best-case possibilities for reductions in human-created reactive nitrogen (Galloway et al. 2014).

catalyst with hydrogen under extreme pressure and at high temperatures to produce ammonia (NH_3). The high temperature and metal catalyst help to loosen the strong triple covalent bond in N_2, and the pressure helps to increase exposure and efficiency. The **Haber-Bosch process** was originally important for creating explosives in World War I, but it also revolutionized agricultural production by allowing the synthesis of a reactive form of nitrogenous fertilizer that did not require long oceanic guano-harvesting voyages, rotating crops, or handling chicken droppings. Since World War I, the Haber-Bosch process has provided the needed reactive nitrogen for farmers to grow high-yielding crops at a fast rotation.

The ability to synthesize nitrogen was a huge boon for food production; however, it has had significant consequences for species and ecosystems far removed from agriculture. In general, crops take up only about a third of the nitrogen applied to agricultural fields. The rest of the applied nitrogen leaches into the soil or is washed off fields and into nearby streams and rivers during rainstorms. Today, synthetic nitrogen applications in agriculture generally exceed 120 Tg N y^{-1}, which nearly doubles the amount of nitrogen available in terrestrial systems globally.

Fertilizers are not the only human-derived sources of reactive nitrogen. Our vehicles mimic the effects of lightning and produce nitrogen oxides (NO_x). As we drive, fossil fuel combustion allows our engines to achieve very high temperatures, and the heat and pressure cause atmospheric N_2 and oxygen to combine. Estimates differ, but this process adds at least an additional 25 Tg N y^{-1} to the atmosphere, although in some countries this has been addressed to some extent with environmental legislation. With these two sources of human-derived reactive nitrogen, it is clear that humans have drastically altered the availability of nitrogen over the last century.

James Galloway and colleagues (2014) estimated the total amount of reactive nitrogen created by biological and industrial processes since 1850 and predicted the ongoing production through 2050 (**Figure 14.19**). The increase in available nitrogen that occurred between the 1950s and the 1990s is remarkable, showing a rise of more than 800%. The solid line in Figure 14.19 represents empirically derived estimates of human-derived available nitrogen, and the dotted lines represent future scenarios based on alternative environmental policies. Why are these social and political changes so important? In a world where nitrogen has limited the plant and phytoplankton growth in natural ecosystems for millions of years, abundant reactive nitrogen leads to profound ecological changes. Next, we detail some of the effects of human increases in reactive nitrogen.

Ecological Effects of Anthropogenic Nitrogen

One of the first steps to understanding how human-derived reactive nitrogen is changing ecosystems and ecological interactions is to focus on the nitrogen budget. It is difficult to pull out just anthropogenic sources, but William Schlesinger (2009) used previously collected data and analyses to estimate the fate of anthropogenically derived reactive nitrogen on an annual basis. Because most of this reactive nitrogen comes from applied fertilizers, he focused specifically on global land-surface inputs and final storage pools of those inputs. He started with the previously published estimate that industrial production creates about 150 Tg y^{-1} of reactive nitrogen, with about 125 Tg N y^{-1} coming from fertilizer production and 25 Tg N y^{-1} coming from combustion engines (**Table 14.1**). An additional 20 Tg N y^{-1} comes from human manipulations of nitrogen-fixing mutualisms with plants, such as soybean production and crop rotation with legumes. These last 20 Tg N y^{-1} lead to organic nitrogen that is generally incorporated into biotic systems through human consumption. Although these 20 Tg may contribute to the inorganic reactive nitrogen after excretion, that contribution was harder to isolate and was less likely to have large global effects, so he focused on the industrial production of reactive nitrogen.

Schlesinger used local estimates, taken from various locations around the globe, of leaching and movement of fertilizer, nitrogen levels in groundwater and streams, and rates of denitrification to explain the fate of the anthropogenic inputs. After he excluded the input from biological nitrogen fixation, he was only able to trace about 83% of the remaining 150 Tg N y^{-1} of reactive nitrogen that humans put into the global system (he identified the fates of 124 Tg N y^{-1} of the 150 Tg N y^{-1} of human inputs, excluding the input from biological fixation in Table 14.1). The largest surface portion (roughly 23%) flows off the land into streams and rivers and eventually makes its way into estuaries and, potentially, into the ocean (river flow fate in Table 14.1). These fluvial inputs contribute to eutrophication (Chapter 10) of local lakes and ponds and coastal systems. An

Table 14.1 Estimated Budget of Nitrogen on the Global Land Surface

	Preindustrial	Anthropogenic	Total
Inputs			
Biological nitrogen fixation	120	20	140
Lightning	5	0	5
Industrial nitrogen fixation	0	125	125
Fossil fuel combustion	0	25	25
Totals	125	170	295
Fates			
Biosphere uptake	0	9	9
Stream and river flow	27	35	62
Groundwater	0	15	15
Denitrification	92	17	109
Atmospheric transport to oceans	6	48	54
Totals	125	124	249

All numbers are in Tg N y^{-1}. The details of how the fates of the nitrogen are estimated can be found in Schlesinger (2009).

Figure 14.20 The Mississippi River (which looks like a big green root in this satellite image) drains the central part of North America, transporting vast quantities of sediment and depositing them into the Gulf of Mexico. The sediment (the cloudy and dark-green feathery areas in the image) contains topsoil, decaying plants, and nutrients such as nitrogen. Much of this deposited material acts as fertilizer for the phytoplankton living in the gulf's waters. Each summer, the phytoplankton population explodes, fueled by the long summer days and the nutrients supplied by the Mississippi. The resulting combination of respiration and decay creates a hypoxic (low-oxygen) or anoxic (zero-oxygen) dead zone, killing zooplankton, crustaceans, mollusks, and fish.

Mississippi River

Gulf of Mexico

additional 10% ends up in groundwater, where it can remain for tens to thousands of years. Such nitrogen in our groundwater has the potential to create toxic levels of nitrate in drinking water. An astonishing 32% ends up in the atmosphere, where global winds and currents move it around the world and may deposit it again across a range of biomes and ecosystems. He could trace less than 20% to "natural" sinks, such as immediate denitrification (approximately 11%) by soil microbes and uptake in forest biomass (around 6%).

Although Schlesinger's work is an impressive feat of ecological sleuthing, discovering the fate of the missing industrial reactive nitrogen could be incredibly important for understanding global ecosystem processes and identifying future ecosystems with heavy impacts. For example, the massive increases in reactive nitrogen in coastal waters have produced explosive growth in phytoplankton. These tiny photosynthesizing algae are naturally nitrogen-limited, and their populations explode when they encounter enormous nitrogen inputs from agricultural runoff in coastal waters (**Figure 14.20**). Although we know that photosynthesis produces oxygen, it is a two-part process, with oxygen appearing in the second half. The bursts of photosynthesis by massive numbers of algae cause oxygen to be released into the atmosphere but also lead to massive oxygen reductions in the water column as the algae respire and then decay. The zooplankton, fish, and other consumer species cannot survive in such low-oxygen environments and die in multitudes. These deaths produce huge marine *dead zones*, where there are few living organisms other than algae (**Figure 14.21**).

As a whole, forests are a key nitrogen pool worldwide, as they take in about one-third of the human-derived reactive nitrogen deposited on land. This number may appear to contradict Schlesinger's number, but remember that he was looking at immediate fates. Atmospheric and fluvial movement of reactive nitrogen lead to additional opportunities for uptake in forest systems, but not always with positive effects. Ecologists have shown that the influx of reactive nitrogen into forests can increase the growth rate of trees, resulting in greater tree biomass and higher net primary productivity. Generally speaking, coniferous forests retain more human-derived nitrogen than deciduous forests. At first glance, this may be encouraging news because forests also serve as long-term storage of atmospheric carbon. Given the rise in CO_2, any increase in carbon storage must be helpful, and it does appear that increases in nitrogen availability allow some forest communities to take up more carbon. Nonetheless, digging a bit deeper into forest ecosystem ecology reveals a less rosy picture.

A central player in forest carbon storage comes not from the sheer biomass of wood in trees, but from the forest soil microbiome. As discussed in Chapter 9, mycorrhizal fungi form mutualistic associations with tree roots, enhancing trees' ability to access nutrients and water. There are two types of mycorrhizal fungi: **arbuscular mycorrhizae (AM)** and **ectomycor-rhizae (EM)**. AM fungi primarily use inorganic forms of nitrogen (NO_3^-, NH_4^+), which are typically produced by human actions. EM fungi, however, rely heavily on large-molecule forms of organic nitrogen, such as those found in proteins, peptides, and other complex organic molecules from living organisms. The association of EM fungi with organic nitrogen means that EM-dominated soils tend to hold organic carbon for long periods.

Colin Averill and his colleagues (2018) made two predictions based on these observations. First, they suggested that deposition of anthropogenic inorganic nitrogen into forest ecosystems should favor AM-associated trees at the expense of EM-associated trees. Second, they predicted that declines in the occurrence of EM-dominated soils would reduce the ability of these forest soils to store carbon. They tested their predictions using data from the US Forest Service's annual survey of tree composition conducted across the United States. They used previous data on the associations between tree species and the two types of mycorrhizae to create AM- or EM-associated forest types. Averill and colleagues added to the tree survey information on the mycorrhizal composition of associated soils and compared changes in these ecological variables across a gradient of inorganic nitrogen deposition rates (**Figure 14.22**).

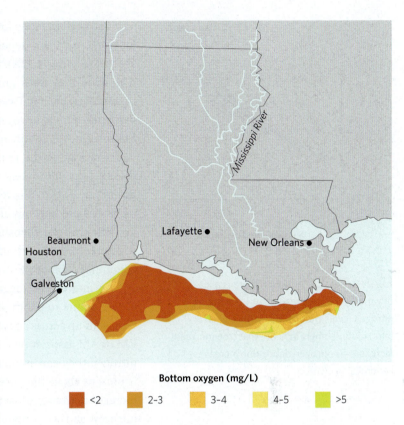

Bottom oxygen (mg/L)

| ■ <2 | ■ 2-3 | ■ 3-4 | ■ 4-5 | ■ >5 |

Figure 14.21 This color-coded map shows oxygen levels in the benthic waters of the Gulf of Mexico in 2017 off the coast of Louisiana. The darkest colors indicate oxygen levels below two parts per million, which is considered hypoxic, producing a dead zone (Sherman and Montgomery 2020).

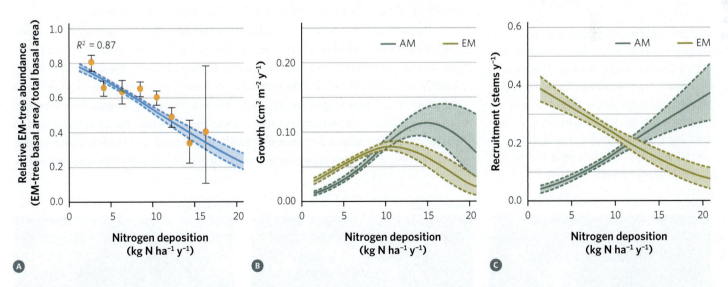

Figure 14.22 Averill and colleagues studied changes in the tree communities associated with different mycorrhizae across various levels of inorganic nitrogen deposition. Increased nitrogen can lead to the following shifts in trees with arbuscular mycorrhizal (AM) and ectomycorrhizal (EM) associations: **A** decreasing abundance of EM-associated trees; **B** faster growth in AM-associated trees than in EM-associated trees; and **C** a decrease in recruitment for EM-associated trees but an increase in recruitment for AM-associated trees (Averill et al. 2018).

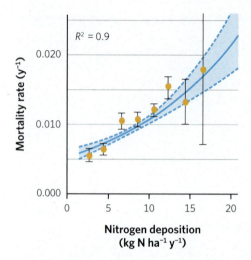

$R^2 = 0.9$

Figure 14.23 Although there is a lot of variability in tree responses to changing nitrogen levels (notice the increasing error bars), data strongly indicate that tree mortality increases as nitrogen deposition increases (Averill et al. 2018).

Figure 14.24 The chronic addition of nitrogen to forests affects forest structure by making the trees visibly stressed. This effect can be seen in pictures from a 15-year nitrogen-addition experiment at the Harvard Forest Long Term Ecological Research (LTER) program site in Massachusetts. Note that in the high nitrogen plots the trees show significantly reduced leaf area (Magill et al. 2004).

The patterns they found confirmed their predictions. Sites with high rates of inorganic nitrogen deposition had a reduced abundance of EM-associated tree species (Figure 14.22A). They also found that, as nitrogen deposition increased, AM-associated trees grew faster and recruited new offspring at higher rates than EM-associated trees (Figures 14.22B and 14.22C). Their results suggest that increased human nitrogen deposition is shifting forest soil mycorrhizal communities toward AM dominance and triggering a shift in tree community composition.

A change in the mycorrhizal soil biota also has an impact on the ability of these ecosystems to sequester carbon. AM-dominated soils do not hold carbon for as long as EM-dominated soils. The soils in EM-associated forests tend to hold more carbon, but only if nitrogen deposition is low. Although tree growth goes up with nitrogen deposition, the new growth does not always increase the ability of forests to store carbon because the nitrogen can shift above- and belowground community composition away from carbon-holding species assemblages.

An interesting nugget of information from the analysis by Averill and team (2018) points to another ecological effect. They found that, as nitrogen deposition went up across sites, the mortality rate of tree species increased, regardless of EM or AM association (**Figure 14.23**). Plants have an upper limit to the amount of nitrogen they can assimilate. If nitrogen, particularly in the form of ammonium, is present above this threshold, it can become toxic and reduce plant growth and increase mortality. Most plants do not have physiological systems in place to store nitrogen and will only take up as much nitrogen from the soil as they can use. When plants cease nitrogen uptake, excess nitrogen in the soil can form various salts that reduce the ability of plants to access water. These nitrogenous salts can also damage plant tissues. More subtly, when nitrogen is plentiful, other nutrients such as phosphorus, magnesium, and potassium become limiting for plants. If these other nutrients are not available to support plant growth, the plant will begin to show negative effects, such as reduced leaf production or reduced ability to ward off herbivores via secondary chemical defenses (Chapter 8).

We have other evidence that suggests a strong pattern to these effects. Ecologists have conducted many nitrogen addition experiments in forests worldwide to investigate what happens when nitrogen goes from being limited to being overly available. Trees' initial response to increases in nitrogen availability is to grow faster and larger, as we might expect. However, continuous nitrogen additions lead to excess soil nitrogen, which leaches into the groundwater and nearby streams. When nitrogen hits really high levels, the trees become visibly stressed (**Figure 14.24**), resulting in increased mortality and reduced reproductive

Control

Low nitrogen

High nitrogen

output, just as Averill and colleagues (2018) found with current background levels of human-derived nitrogen deposition.

At the community level, there is consistent evidence that plant species richness decreases as nitrogen availability increases. A study by Carly Stevens and colleagues (2004) related plant species richness across grasslands in Great Britain to human-derived inorganic nitrogen deposition rates, which vary from under 12 kg y^{-1} in some locations to more than 25 kg y^{-1} in others (**Figure 14.25A**). They found that their four-meter-square plots on average lost one grassland plant species with each 2.5 kg of nitrogen deposition per year (**Figure 14.25B**). At the mean rate at which human-derived reactive nitrogen was deposited across Europe in the early 2000s (17 kg y^{-1}), this result predicted a 23% loss of grassland plant species per year.

Similar results have now been documented for other nitrogen-limited groups, such as lichens and bryophytes, and for other ecosystem types around the world. Species are lost from communities at high nitrogen levels because they suffer the toxic effects of excess nitrogen or are outcompeted by nitrophilic (nitrogen-loving) species. Of course, for many species, decline is driven by both mechanisms in combination. If we think back to Schlesinger's (2009) estimates of atmospheric movement of deposited nitrogen, we can imagine that these effects on biodiversity are very widespread.

14.5 The Phosphorus Cycle

The final nutrient that we follow through an ecosystem budget is phosphorus (P). We focus on nitrogen, carbon, and phosphorus because they are the most commonly used essential nutrients, and their availability affects all life on Earth. Phosphorus is critical in the formation of ATP (the energy currency of cells), DNA, RNA, and the phospholipid layer in cell membranes. It is, therefore, a central player in the movement of energy and molecules through all organisms, as well as a component of their genome and protein coding.

As with carbon and nitrogen, we can create a budget for the movement of phosphorus through a system, but this budget is a little different. Unlike carbon and nitrogen, phosphorus is almost never present in a gaseous phase, so its cycle does not really involve an atmospheric pool. Most of the Earth's phosphorus is stored in rocks and marine sediments. The geologic uplift of marine sediment due to plate tectonics moves deep-marine phosphorus back into terrestrial systems (**Figure 14.26**), where it is weathered over time. Because this process moves at the pace of plate tectonics, it may take hundreds of thousands of years for a single ion

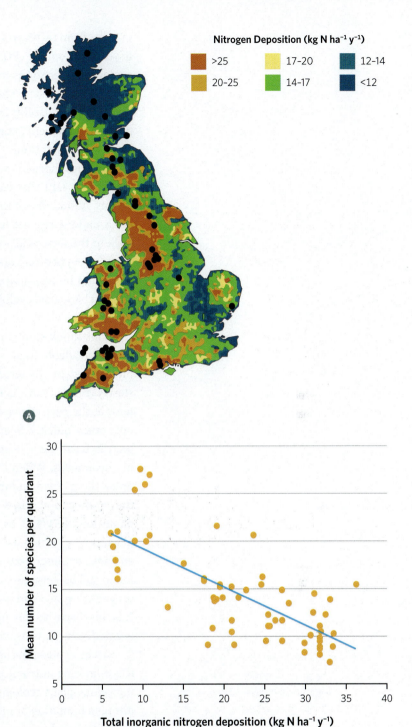

Figure 14.25 **A** Stevens and colleagues (2004) related grassland species richness to this spatial pattern in anthropogenic inorganic nitrogen deposition across Great Britain. Their field sites are indicated by black dots. **B** The group found that the grassland plant species richness across Great Britain decreased by one species for each 2.5 kg increase in annual nitrogen deposition.

to move from resting on the ocean floor to becoming bioavailable. Over thousands of years, phosphate (PO_4^{3-}, an ionic form of phosphorus) is eroded from rocks and geological formations into soils and water, where it becomes available to plants, fungi, and microbes. There is no special microbial aspect to this cycle to increase the rate of availability, but once phosphorus makes it into living organisms like plants, the nutrient is tightly and quickly (over days) cycled through other organisms by consumption.

As phosphorus is weathered and eroded from rocks, it can form a number of different salts that wash out of soils and move downstream to enter lakes and oceans. About 90% of this terrestrially derived phosphorus is taken up by aquatic biota, and the rest settles into oceanic sediments to begin the long, slow journey back to the terrestrial system (Figure 14.26). The bulk movement of phosphorus can be thought of as a one-way transport from rocks to soil and water to eventually ending up in deep-marine sediments. Yet along this pathway there is significant and fast biological cycling of phosphorus among all types of living organisms (Figure 14.26).

This quick summary may make it sound as though phosphorus should be readily available in all systems, but two issues limit its availability. First, phosphorus returns to terrestrial environments from marine sediments *extremely* slowly (remember it is moving out of marine sediments and into biological systems at the pace of plate tectonics), so if it washes out of a system locally, it may not return quickly. Second, plants are the primary organisms that move phosphorus from the soil into food webs, and plants are picky about their sources of phosphorus. When living organisms containing phosphorus die, detritivores break them down and make their phosphorus available to new organisms. Here the complication arises. Plants take up phosphorus primarily as inorganic phosphate ions (HPO_4^{2-} or $H_2PO_4^-$), but decomposing organisms release organic phosphorus, which is not very accessible to plants. Plants need particular enzymes to mineralize organic phosphorus and make it accessible for their uses. Fungi and bacteria are much better at acquiring phosphorus than plants are, and they take up two and nine times more phosphorus, respectively, than do plants. It is this disparity that helps to explain why mycorrhizae can form such interesting trade mutualisms with plants involving phosphorus (Chapter 9).

Because phosphorus can wash out of soils quickly but tends to return slowly, it is a limiting nutrient just as often as nitrogen is. In fact, discussions are ongoing among plant ecologists about when and where we should expect to see more nitrogen limitation or more phosphorus limitation. Of course, such scarcity limits agriculture and drives human ingenuity, so it should not be surprising that humans have discovered ways to increase our access to phosphorus by mining it from rock. Tectonic uplift has returned phosphorus to the terrestrial environment in fairly specific locations on Earth, and the world's primary phosphate mining locations are in China, Morocco, and the United States.

As with the carbon and nitrogen cycles, human influences on the phosphorus cycle can have big effects on ecosystems. Because phosphorus is a significant component of all agricultural fertilizers and many detergents, it finds its way into most streams and waters. Our fertilizers, therefore, contribute to anthropogenic

Figure 14.26 This diagram represents the global phosphorus cycle. Outside of biological systems, phosphorus is stored primarily in rocks and ocean sediments. Once it weathers out of rocks and enters plants, it passes through food webs by consumption (biological cycling) and then flows in water to the ocean (arrows), where it eventually settles into ocean sediments. From these sediments, phosphorus only reenters the terrestrial system with the slow movement of tectonic plates. Note that there is no atmospheric or gaseous pool of phosphorus. Geologic processes are in green boxes, and anthropogenic processes are shown in red.

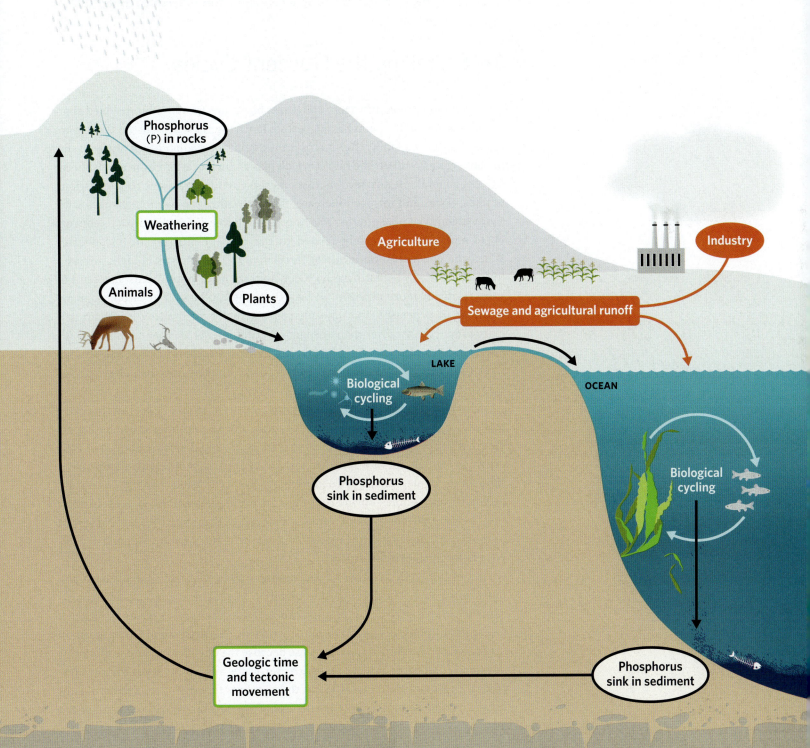

eutrophication by adding both reactive nitrogen and phosphorus to natural systems. William Schlesinger and Emily Bernhardt (2013) calculated that the phosphorus levels in streams could be as much as 300 times higher than preindustrial levels. Although this increase speeds up access and cycling of a naturally rare nutrient, it also shifts the balance of available nutrients and encourages the growth of algae to the detriment of other species. In addition, anthropogenically added phosphorus comes primarily from mined sources, and environmental geologists have begun to worry that we could deplete these sources in the next few centuries. Of course, by then many of our ecosystems may be very different (Chapter 15).

14.6 Linking the Nutrient Cycles

The cycles of carbon, nitrogen, and phosphorus are far more intertwined than our discussion to this point has suggested. In Chapter 5, we discussed the importance of water availability for photosynthesis, but the availability of nitrogen and phosphorus also affects the rate at which plants can fix carbon from atmospheric CO_2. In general, photosynthesis rates increase within leaves as concentrations of phosphorus and nitrogen increase. Similarly, nitrogen fixation changes with the availability of phosphorus. The rate at which free-living diazotrophs convert N_2 to ammonia is inversely related to the ratio of nitrogen to phosphorus. When the concentration of nitrogen relative to phosphorus is high, nitrogen fixation slows. Soils that have high concentrations of phosphorus also tend to store relatively more nitrogen. In fact, the same can be said of carbon in many locations. It is possible that phosphorus is actually *the* limiting factor for primary productivity rather than nitrogen. It is definitely true that increased availability of both nitrogen and phosphorus increases carbon fixation and storage rates (although, as we discussed earlier, these changes can also alter species composition).

These interlinking effects are not only apparent in the uptake of inorganic molecules. Increased concentrations of nitrogen and phosphorus can be important to the speed at which organic nutrients cycle. You may have noticed in your everyday life that the rate of decomposition varies across different organic materials. For example, banana peels in compost disappear faster than dry leaves, and deciduous leaves break down faster than the needles of coniferous trees. In addition, the rate of decomposition varies with location. Fruit piled in a bowl may start to rot faster than fruit sitting on a stone counter, and leaves in a wet pile decompose faster than leaves in dry areas.

All of these effects occur as decomposers work to acquire the nitrogen and phosphorus found in decomposing material. Higher nitrogen and phosphorus content can speed the decomposition process, whereas higher carbon-to-nitrogen or carbon-to-phosphorus ratios can slow down decomposition. Decomposition also speeds up as temperatures increase because the chemical reactions that drive decomposition occur faster. Finally, greater water availability leads to faster terrestrial decomposition rates. Clearly, the nutrients that cycle through both natural and human-dominated ecosystems create complex biological connections that determine the fate of all species on Earth, including our own. This brings us to a non-nutrient molecule that is perhaps most critical for life as we know it: fresh water.

14.7 The Water Cycle

We know from Chapter 3 that most of the Earth's water is found in the oceans, leaving only a small fraction that is useable in terrestrial and freshwater aquatic ecosystems. Only about 3% of the Earth's water is fresh enough to drink or to be used by most plants, and nearly 70% of this fresh water is locked in glaciers and the polar ice caps. Nevertheless, the cycling of this tiny fraction of the Earth's water between the atmosphere, landforms, and oceans is critical to life on this planet.

As mentioned in the brief coverage of the **hydrologic cycle**, or water cycle, in Chapter 3, the cycle is driven by precipitation, evaporation, and transpiration (**Figure 14.27**). Precipitation happens as either rainfall or snow, and evaporation occurs when temperatures are high enough to turn surface water into vapor that moves into the atmosphere. *Transpiration* is the movement of water by plants out of the soil via roots, up into leaves, and eventually out into the atmosphere as water vapor. Transpired water is lost to the atmosphere as a by-product of photosynthesis (see earlier in this chapter and in Chapter 5). Figure 14.27 shows the pools and flows of water through the global water cycle, including a budget of how much water is typically transported between terrestrial, oceanic, and lake or riverine systems.

Both freshwater storage and the movement of water are essential for most terrestrial life. The seasonality and spatial patterns of precipitation and evaporation play a large role in the location of the global biomes described in Chapter 3. Indeed, animals and plants have evolved to match their growth and reproductive cycles to the seasonal and spatial patterns of precipitation, including features like the spectacular migratory movements of herbivores within grasslands (Chapters 3 and 12). Human society also depends heavily on water in lakes, rivers, and belowground aquifers for drinking water, crop irrigation, and energy generation.

Recent theoretical and field-derived research shows that the planet's hydrologic cycle has intensified, meaning that the rate of evaporation and transpiration has increased, thereby increasing the global average amount of precipitation. This intensification of the water cycle is due to an increase in the global average surface temperature, and we know rising temperatures are largely attributable to human-derived increases in greenhouse gas concentrations. Higher planetary temperatures lead to faster evaporation and transpiration with subsequent changes in precipitation patterns across time and location.

Because of the actions of the global climate system (Chapter 3), water often will not come down anywhere close to where evaporation or transpiration put it into the atmosphere. Scientists now understand that the intensification of the water cycle will tend to make the wetter areas of the Earth wetter and the drier

Figure 14.27 This diagram depicts the global hydrologic cycle. Pools of water are labeled with ovals containing their respective values in units of km³. Water flows or movement are indicated by arrows, with values given in units of km³/y. Rivers are a significant storage pool for fresh water but serve primarily as flows on the scale of the global water cycle. (Data are from Trenberth et al. 2007.)

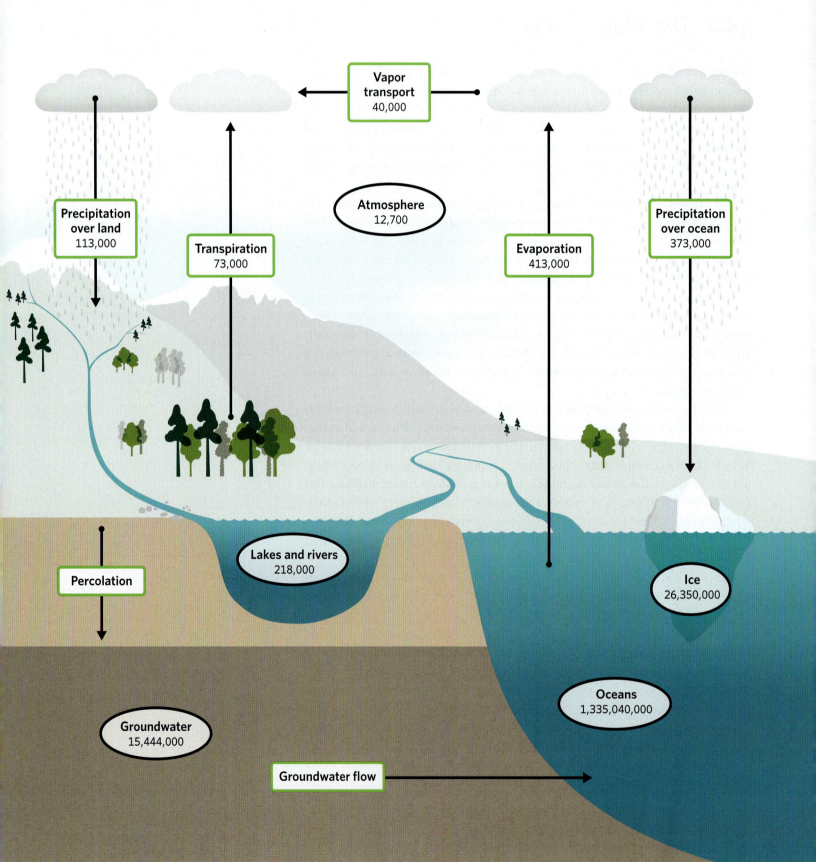

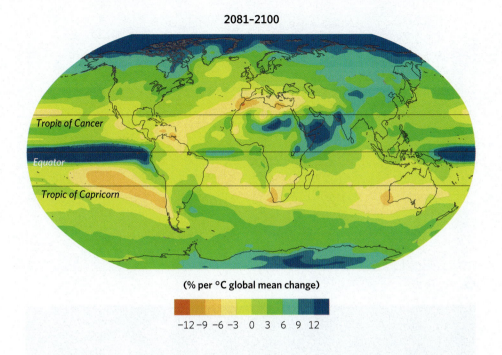

2081–2100

Tropic of Cancer

Equator

Tropic of Capricorn

(% per °C global mean change)

−12 −9 −6 −3 0 3 6 9 12

Figure 14.28 Scientists have been able to predict large-scale changes in precipitation (percentage change for each degree Celsius change in temperature) for the period 2081–2100 relative to 1986–2005. Note that predictability from this model is uneven across the globe, and some geographic locations are difficult to assess, but we generally expect to see precipitation decreasing in some of the areas of the globe that are already quite dry (Intergovernmental Panel on Climate Change 2013).

parts drier (**Figure 14.28**). These outcomes are bad news for humans because we structure our lives around predictable rainfall patterns—but so do most other species. These changes in patterns of water availability mean that an enormous variety of species across a wide range of locations will be physiologically stressed. Many may disappear from ecosystems or become globally extinct.

Perhaps the most obvious place to look for such ecological changes is in the more arid biomes, such as deserts or tropical dry forests. In these locations, plants and animals have evolved to survive the natural periods of water scarcity and the high evaporation rates associated with warm dry air and to take advantage of the highly seasonal pulses of precipitation, often called *monsoons*. The evolutionary specialization of these species makes them especially susceptible to even slight increases in surface temperatures, delays in the onset of seasonal rains, or the shortening of the rainy season.

Craig Allen and colleagues (2010) systematically searched for evidence of increasing forest tree mortality as surface air temperatures have risen and drought conditions have become more frequent, both products of climate change. They found increasing documentation of severe tree species mortality, with no region of the Earth left untouched (**Figure 14.29A**). However, tree mortality was particularly severe in historically dry biomes, such as those of eastern Australia (**Figure 14.29B**).

In a follow-up investigation, A. Park Williams and others (2010) explored the relationship between increasing drought and forest ecology through detailed analysis of tree growth in the forests of the southwestern United States. This region is characterized by arid forests, which are often found at higher elevations. Williams and colleagues used the incredible storehouse of data on tree growth captured in the International Tree-Ring Data Bank (ITRDB) to calculate the incremental growth of trees in southwestern forests and relate growth patterns to precipitation and air temperatures across the same time frame.

Figure 14.29 ⒶThe dots represent locations around the globe with documented tree mortality events related to climatic stress from drought and high temperatures (Allen et al. 2010). ⒷThe numbers on this satellite map of Australia and New Zealand indicate locations where drought-induced tree mortality occurred in the past. The photo on the right shows a 2007 die-off of mulga trees (*Acacia aneura*), the dominant tree species across large areas of semiarid Australia. The photo on the left shows a 1996 die-off of yellow-branched ironbark trees (*Eucalyptus xanthoclada*) in Queensland (Allen et al. 2010).

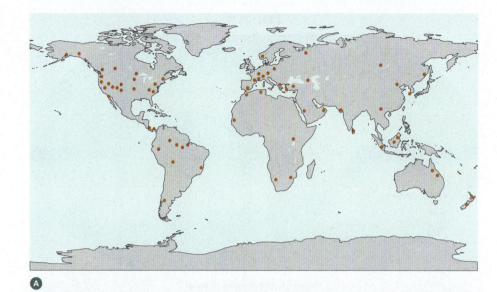

Ⓐ

Ⓑ

In many temperate-zone trees, the amount of new growth in a year is "recorded" as a distinct ring in the wood itself (**Figure 14.30A**). If you grew up near a temperate forest, you may have had a chance to count tree rings in a cross-section of tree trunk, where the number of rings translates into the age of that tree when it was chopped down. What you may not have realized is that the width of each of those rings tells you how well that tree grew in a single year, with thick rings indicating excellent growth conditions and narrow rings indicating challenging conditions. Drilling into a tree and taking out a narrow core of wood from the trunk can tell you the age of the tree and how the environmental conditions for its growth have varied throughout its lifetime (**Figure 14.30B**). It is worth noting that taking a core does not kill the tree, because the drill removes only a tiny fraction of the tree's living biomass (**Figure 14.30C**). Every time foresters or ecologists take a tree core, they have the option of sending it to the ITRDB for long-term data storage.

Figure 14.30 **A** Varying tree-ring widths indicate different amounts of growth from year to year. **B** A scientist uses an instrument called an increment borer to take a core sample from a tree. **C** The resulting core is only a tiny portion of the tree's biomass, but it shows the tree's yearly growth rings, which can be counted and measured.

Williams and team analyzed tree-ring data from the ITRDB for trees in the southwestern United States. When they compared the ring-width index with historical records of annual precipitation and maximum yearly temperatures, they found that ring width decreased substantially with either prolonged drought or high average air temperatures. The forests of the southwestern United States are dominated by three species: pinyon pine (*Pinus edulis*), ponderosa pine (*Pinus ponderosa*), and Douglas fir (*Pseudotsuga menziesii*). Williams and colleagues found these species to be especially sensitive in their yearly growth patterns to the amount of annual winter precipitation. To add worse news to bad, decreases in annual precipitation and increased temperatures in these forests also led to more frequent and intense wildfires, as well as outbreaks of destructive bark beetles, both of which increased tree mortality (Williams et al. 2010).

Changes in the water cycle influence more than just the drier biomes of the world. Tropical rainforests, such as those in the Amazon basin of South America (Chapter 3), are some of the wettest places on Earth. Nevertheless, these locations experience seasonal pulses in rainfall that are big enough to cause all the major tributaries of the Amazon to overflow their banks during the wet season. The ecological effects of this flooding pattern were highlighted in Chapter 12, when we discussed an interesting mutualism in which river fish disperse the seeds of trees over unexpectedly long distances during these monsoon floods.

The monsoons in the Amazon are driven by the local-scale interactions between the rainforest trees and the atmosphere, and by weather patterns affected by sea surface temperatures in the Atlantic and Pacific Oceans. Climate change has increased sea surface temperatures, and massive deforestation has altered the connection between forests and the atmospheric recycling of water. The end result is a clear intensification of the Amazonian water cycle, with reduced rainfall during the annual dry season in the eastern and southern parts of the Amazon basin and increased rainfall during the wet season in the northern basin (Marengo and Espinoza 2016). Other than an increase in large fires during the drier periods, little is known about the ecological response of the many thousands of species that make their homes in these forests.

One exception has come from a detailed study by Cristhiana Röpke and others (2017) that documented shifts in fish assemblages within a lake in the central Amazonian floodplain, near Manaus, Brazil (**Figure 14.31A**). These ecologists provided evidence that the annual variability in precipitation in this area increased markedly after 2005, with more periods of low rainfall. Through the use of long-term fish assemblage records in this lake, they showed that this shift in the water cycle coincided with a sharp drop in fish species richness and evenness (**Figure 14.31B**). They also documented distinct changes in the taxonomic composition, life history, and trophic structure of the fish community after 2005 (**Figure 14.32**).

The mechanisms behind these changes are not fully understood, but Röpke and colleagues suggested that some species dropped out of the community after 2005 because of altered or decreased connections between water bodies. Floods in this area provide aquatic connections between lakes and create opportunities for dispersal between lakes during the monsoon season. As discussed in Chapter 12, this kind of dispersal can lead to metapopulation dynamics and more persistent local and regional populations. With drier conditions, these flooding events were no longer common, thus severely limiting fish dispersal.

Another possible mechanism was an increase in harsh environmental conditions or predation. As lake levels drop and water temperatures rise, available oxygen declines. Low oxygen levels cause physiological stress for fish inhabiting these waters. The reduction in lake area during dry conditions also increases the density of fish per unit volume of water, which increases the chances that predators are able to find their prey. Although there is no documentation of these changes, such general responses could also have produced the changes in fish communities. No matter the mechanism, the ecological outcome was a massive shift in the freshwater fish fauna of this Amazonian lake. Given that the Amazon is the most diverse area on Earth for freshwater fishes, the changing water cycle has raised concern among conservation biologists.

We can even see the effects of the change in precipitation patterns in the ocean. Evaporation of ocean water into the atmosphere leaves salt behind. When ocean waters experience low rainfall (drought), they become noticeably saltier, and when they experience high rainfall (floods), they become less salty. Global measurements of ocean water salinity taken between

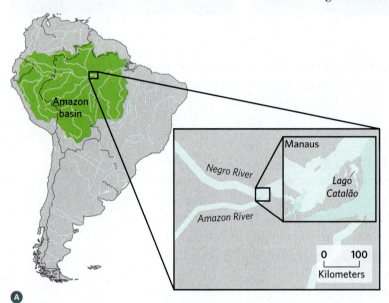

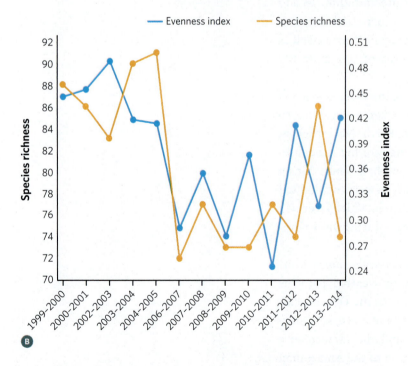

Figure 14.31 Ⓐ Seen here is the location of the seasonal lake (Lago Catalão) studied by Röpke and colleagues in the Amazon. Ⓑ Fish species richness and fish species evenness in this lake both decreased after drought periods in 2005–2006 (Röpke et al. 2017).

Figure 14.32 Röpke and colleagues recorded changes in the taxonomic composition, life history, and trophic structure in a food web for fishes found in an Amazonian lake following alterations to the water cycle due to increased drought periods after 2005–2006. The y-axes represent a composite variable derived from several variables related to the topic (i.e., taxonomy, life history, or trophic position). Solid lines represent linear regression models and dashed lines represent confidence intervals. Notice that each panel shows a distinctly different pattern after 2005–2006 than before (Röpke et al. 2017). PCA = principal components analysis; PCoA = principal coordinates analysis. The "1" on each y-axis label (e.g., PCA 1) indicates that this composite variable explained the most variation in the relevant topic.

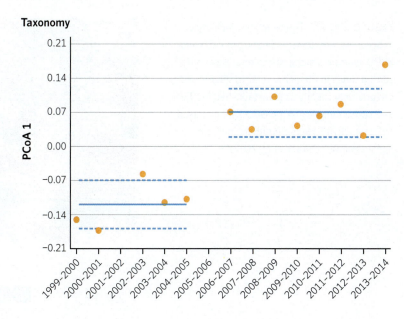

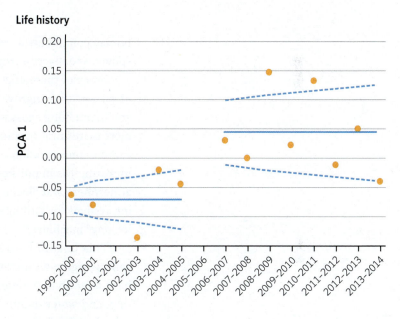

1950 and 2000 show that huge swaths of the ocean have changed substantially in their salinity (**Figure 14.33**). These changes seem to intensify the alteration of precipitation events on land, making the droughts experienced in places like the Amazon last even longer than predicted.

14.8 Ecosystem Services

If we take a step back from all the intricate details of these global cycles, we can see the critical role that functioning ecosystems play in providing enough fresh water, carbon, nitrogen, and phosphorus for all species, including humans. The delivery of these and other critical components by natural ecosystems is collectively referred to as **ecosystem services**, or *nature's contribution to people.*

Wetlands, for example, are amazing contributors to the cycles of carbon, nitrogen, and water. It should not surprise anyone that wetlands store fresh water, but they also clean that water (and make it safer for human consumption) by reducing the levels of nitrogen (and other nutrients) in the water through both denitrification and uptake of nitrogen by wetland plants. Growing wetland plants also remove carbon from the atmosphere. Because of the wet conditions and nutrient-rich materials, wetland plants decompose fairly quickly, leading to available sources of energy for other organisms. The saturated soils also allow slow seepage of clean water into groundwater aquifers for long-term storage. If we convert wetlands into parking lots, we reduce drinking-water stores, increase the risk of flooding in adjacent

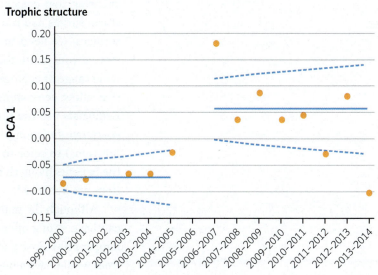

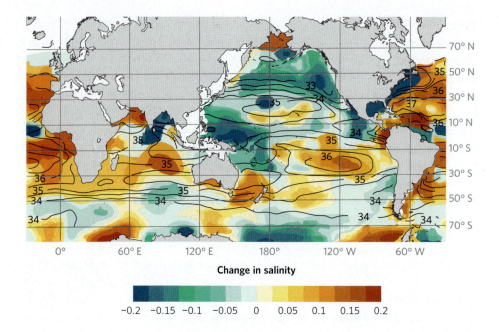

Figure 14.33 This map shows changes in ocean salinity observed between 1950 and 2000. Numbers along the color scale and the contour lines in the figure refer to the PSS-78, the practical salinity scale established in 1978 as a consistent way to measure salinity based on conductivity. Higher numbers indicate higher salinity (Durack et al. 2012).

human communities, reduce carbon uptake, release stored carbon into the atmosphere, and divert nitrogen-rich runoff into streams, lakes, and oceans.

Many of these effects are similar to the ones explored earlier in the context of the loss of mangrove forests. It is instructive to note, however, that the loss of any functioning ecosystem is likely to reduce carbon uptake and simultaneously increase respiration and decomposition, which all ends up increasing atmospheric carbon concentrations. If we wish to ensure that the Earth does not continue to warm, it is in our best interest to keep as much carbon as possible out of the atmosphere and "locked" in plants or soils instead. Similarly, plants offer amazing systems for pulling excess nutrients out of water, acting as food- and carbon-locking "machines."

This linkage between functioning ecosystems and human well-being is a central argument for the preservation of intact natural ecosystems. Researchers have not only calculated the economic benefits of carbon storage, food production, and water quality; they have also made equally compelling arguments for ensuring that coastal dunes and salt marshes remain intact as buffers against storm surges and hurricane-force winds. If we want to protect such ecosystem services, we need to understand how the organisms and processes in an ecosystem contribute to **ecosystem function**. Ecosystem function encompasses a wide range of ecological outcomes and processes and includes the mechanisms that allow ecosystems to continue to cycle nutrients and support organisms. Ecologists use the term primarily to refer to functionality (hence the singular *ecosystem function*), but occasionally they use it in the plural form (*ecosystem functions*) to refer to specific outcomes and processes. Ecosystem services are the benefits living things enjoy from the rate and scale at which ecosystem functions are delivered.

Although the mangrove example made it clear that specific species or genera can have strong effects on ecosystem functions, we have not yet explored what happens if we lose particular species from an ecosystem. Clearly, losing mangrove trees would have a huge effect in mangroves, but what about losing other species?

Connections between biodiversity and ecosystem function have been hotly debated over several decades, spawning a large body of literature on the subject. Ecologists have run a plethora of laboratory and field experiments in which they have manipulated the number of species across treatments and measured the subsequent differences in nitrogen removal or fixation, carbon uptake, water cycling, and other ecosystem functions.

In 2012, Bradley Cardinale and 16 colleagues collated more than 600 of these studies to assess what we know and do not know about the relationship between biodiversity and ecosystem function. They explored a huge range of types of biodiversity including genes, species, and functional groups (Chapter 11). Their results strongly indicated that the loss of biodiversity "reduces the efficiency by which ecological communities capture biologically essential resources, produce biomass, decompose and recycle biologically essential nutrients" (Cardinale et al. 2012). They also found strong evidence that ecosystems that are more diverse provide these services more consistently through time. Their analysis, however, also suggested that this relationship is nonlinear (**Figure 14.34**). Small losses in biodiversity may only minimally decrease ecosystem function, but decreases in function accelerate as biodiversity losses increase. We can see this pattern in Figure 14.34 if we "slide down" the slope from the right of the figure to the left.

Unfortunately, for ecosystem managers, there is no clear threshold at which the biodiversity of a system produces maximized functionality or where biodiversity losses suddenly have oversized effects. Part of the reason for this lack of threshold comes from the range of ecosystems, functions, and participating species. It is clear, though, that not all species contribute equally to this relationship. Species with particular traits play a key role in the delivery of ecosystem functions, and the loss of these species has an especially big impact on the rate at which ecosystem processes occur. A lot of the uncertainty seen in the biodiversity-functionality relationship depicted in Figure 14.34 and the location of the inflection point in that relationship (the point on the line where ecosystem function suddenly levels off) is driven by which functional traits are lost and in what order. This connection between species and order of loss makes sense given our discussions of community assembly in Chapter 13 and keystone species in Chapter 10. Finally, when the lost species represent more than one trophic level, ecosystem function decreases more than if the same number of species disappeared from a single trophic level. This result echoes our discussion of functional diversity in Chapter 11.

Notably, all of the experiments reviewed by Cardinale and colleagues explored the role that biodiversity plays in the rate at which ecosystem processes occur, but they do not often explicitly tie these processes to the benefits they provide human societies. Although there is solid evidence for an overall positive effect of biodiversity on ecosystem function, connecting this effect to the provision of services and then calculating the economic or social value of these services is extremely difficult. There are many links in this chain, and we should perhaps expect that the relationship will be less clear as a result. Nonetheless, it

Figure 14.34 This simple graph captures the conceptual relationship between biodiversity and ecosystem function. The green line shows the average change across all combinations of genes, species, or traits. The light green band represents a 95% confidence interval, meaning a range around that average where 95% of the simulations lay, indicating a level of uncertainty around the estimate represented by the line.

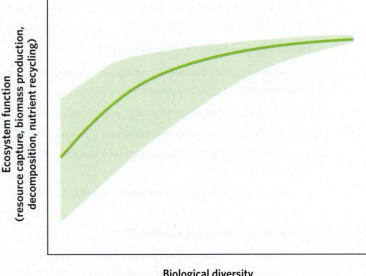

Ecosystem function
(resource capture, biomass production, decomposition, nutrient recycling)

Biological diversity
(variation in genes, species, functional traits)

is obvious that ecosystem functions often benefit human society, and that ecosystems produce these outcomes more efficiently and cheaply than technological or engineered solutions. Thus, as Cardinale and others have suggested, these linkages are vital to our understanding of how ecosystem health affects human health, and thus are well worth the effort to research and explore.

SUMMARY

14.1 Ecosystems Are Inclusive

- An ecosystem is all the living organisms and the nonliving components in a web of interconnection at a certain location.

- Ecologists trace critical chemical nutrients through ecosystems in terms of chemical cycles and nutrient budgets, typically focusing on carbon, nitrogen, phosphorus, and water.

14.2 Budget Basics

- An ecosystem budget is an accounting of a nutrient's movement through and within a system and is typically depicted as a series of pools and flows.

A Carbon Primer

- Carbon is critical for life and is found as either organic or inorganic carbon.

- Carbon fixation occurs when chemosynthesizers (bacteria and archaea) or photosynthesizers (bacteria, archaea, and plants) take inorganic carbon and convert it to organic carbon.

- This process provides the carbon building blocks for the rest of the Earth's trophic systems.

The Mangrove Forest Carbon Budget

- Mangrove forests are nearshore aquatic biological zones that sit at the coastal margins in tropical locations.

- Their intertidal location produces anoxic soil conditions.

- Carbon is fixed by mangrove trees and flows into the soil, and at the same time it is released into the surrounding water as leaf litter.

- Mangrove forests export carbon as a subsidy to the nearby marine ecosystems.

- **Mangrove Forests as Carbon Sinks**

  — Mangrove forests produce more carbon than they export to other systems; therefore, a mangrove forest is a net carbon sink.

— The excess carbon is stored within the system for a long period of time.

14.3 The Global Carbon Cycle

Slow Carbon Cycle

- The vast majority of carbon on Earth is locked away in rocks, as fossil carbon. This carbon eventually, but very slowly, cycles to the living components over millennia.

Atmospheric Carbon

- The atmosphere contains very little carbon, but the forms of carbon that dominate (carbon dioxide and methane) are potent greenhouse gases.

- The burning of fossil fuels transfers carbon from the geosphere to the atmosphere.

- The Keeling curve shows that there has been a steady increase in atmospheric CO_2 since the 1950s.

Oceanic Carbon

- The oceans contain vast amounts of carbon in the form of dissolved CO_2.

- Oceanic carbon levels vary around the globe and are influenced by temperature and the movement of water.

- Oceanic carbon also affects the pH of ocean waters, in that higher levels of dissolved CO_2 create water that is more acidic.

- More-acidic ocean waters make it difficult for many marine organisms, particularly corals, to live.

Terrestrial Carbon

- Most of the mobile terrestrial carbon is located in soils in the form of organic matter.

Human Influences on the Global Carbon Cycle

- Humans have a fairly large influence on the entire carbon cycle at many stages of the cycle.

14.4 The Nitrogen Cycle

Forms of Nitrogen

- Nitrogen comes in a number of forms: inert nitrogen (N_2 gas); reactive forms, such as ammonia (NH_3), ammonium (NH_4^+), and nitrate (NO_3^-); and the myriad organic forms found in living organisms.

- The atmosphere consists mostly of inert nitrogen gas, which, because of its triple covalent bond, is inaccessible to most organisms.

- A few bacteria and archaea can "fix" N_2 into more biologically available forms using the enzyme nitrogenase, although the fixation process is, by necessity, anaerobic.

Cycling Nitrogen

- The nitrogen cycle is more complicated than the carbon cycle, and a complete budget is best understood from a figure (Figure 14.13).

- Microbial communities are important for nitrogen cycling.

- Humans have influenced global nitrogen cycling through the application of agricultural fertilizers.

Nitrogen in Food Webs

- Supplies of nitrogen are sometimes moved among systems by natural means, thereby providing needed resources.

- This happens, for example, when animals move from place to place, when rivers transport nitrogen downstream, and when organisms such as salmon deposit nitrogen from their carcasses when they die.

Anthropogenic Nitrogen

- Humans have been adding nitrogen to agricultural fields to boost production for hundreds of years.

- In the 1900s, scientists found a way to industrially fix atmospheric nitrogen and spawned the birth of cheap chemical fertilizers.

- Large-scale application of nitrogen fertilizer around the world is having a significant impact on many systems.

Ecological Effects of Anthropogenic Nitrogen

- Nitrogen fertilizers are often applied in excess, with some amount of the nitrogen running off the application site.

- Excess nitrogen causes algae outbreaks in aquatic systems, which in turn can strip a body of water of all of the available oxygen.

- This creates huge aquatic dead zones where anoxic waters prevent the growth and reproduction of marine and aquatic organisms.

- For terrestrial plants, extra nitrogen is a boon, because it boosts production and carbon sequestration.

- However, extra nitrogen can have a negative impact on the microbial soil community and cause a loss of species richness.

14.5 The Phosphorus Cycle

- Phosphorus plays a key role in all life on Earth, but the global phosphorus cycle is much slower and more linear than the carbon and nitrogen cycles because it has no gaseous component.

- Phosphorus generally moves from rocks into soils via weathering, and then into terrestrial plant communities, from which it cycles through the biological community.

- Alternatively, weathered phosphorus can move through water into aquatic systems, where it is taken up by organisms and again tightly cycled.

- Eventually, all phosphorus ends up in deep-marine sediments and remains there for millions of years until it is resurfaced by geologic activity.

- Activities such as mining and industrial production have greatly added to the cycling of phosphorus, causing ecological disruption.

14.6 Linking the Nutrient Cycles

- All three nutrient cycles—carbon, nitrogen, and phosphorus—are tightly coupled and interact, affecting the availability and speed of the cycles.

14.7 The Water Cycle

- Life as we know it depends on the availability of fresh water.

- Human-induced climate change has intensified the global hydrologic cycle, disrupting many aspects of the planet's climate and weather patterns.

14.8 Ecosystem Services

- All global cycles contributed to the formation of the intricate ecosystems around the planet.

- These ecosystems provide large-scale ecosystem services that humans often rely on, such as clean water and fresh air.

- Understanding the links between these services and human well-being is an important lesson that we, as a species, must learn.

KEY TERMS

ammonium (NH_4^+)

arbuscular mycorrhizae (AM)

biological nitrogen fixation

carbon fixation

cyanobacteria

denitrification

dissolved inorganic carbon (DIC)

dissolved organic carbon (DOC)

ecosystem budget

ecosystem flow

ecosystem function

ecosystem loop

ecosystem pool

ecosystem services

ectomycorrhizae (EM)

fast carbon cycle

greenhouse gas

gross primary production (GPP)

Haber-Bosch process

hydrologic cycle

inorganic carbon

inorganic nitrogen

Keeling curve

net primary production (NPP)

nitrate (NO_3^-)

nitrogenase

nutrient budget

nutrient cycle

ocean acidification

organic carbon

organic nitrogen

particulate organic carbon (POC)

sequestering ecosystem

slow carbon cycle

CONCEPTUAL QUESTIONS

1. Given that nutrient cycles are inherently connected, describe some of the ways the hydrologic cycle can affect the global carbon cycle.

2. Following the reasoning for why mangroves are carbon sinks, suggest a biome or vegetation formation that you think might be a carbon source. Explain why, and then do some research to see if you can find ecological support for your suggestion.

3. Describe how an atom of carbon would move from the atmosphere through portions of the carbon cycle to end up in a bar of chocolate. If you eat that chocolate, how will that same carbon atom get back into the atmosphere?

4. Using your knowledge of ecology, come up with some ways to permanently (or semipermanently) remove CO_2 from the atmosphere and store it for long periods

of time. Discuss the issues that make such solutions difficult.

5. If you were a plant growing in a relatively pristine location and you could choose to have either nitrogen or phosphorus be the limiting nutrient in your environment, which of these two elements would you choose? Explain the reasons for your choice. (Hint: It is useful to say at the outset what type of environment you are in.)

6. Describe some of the ways the hydrologic cycle can affect the global nitrogen cycle.

7. Describe an ecosystem service that affects the area where you live. Tie this service to at least one of the global cycles we discussed (i.e., carbon, nitrogen, phosphorus, or water).

QUANTITATIVE QUESTIONS

1. Although we go through Figure 14.4 (the carbon budget for mangrove forests) in more detail than any other figure, we do not point out all the ways in which the figure reveals uncertainty in the estimated sizes of carbon pools and flows.

   a. Identify the places where we cannot see all the sources for a particular pool or flow.

   b. Explain whether those uncertainties affect our assessment of mangrove forests as carbon sinks.

2. Revisit Figure 14.6, and explain how all the anthropogenic effects (in red text) balance. It may help to start by explaining why the overall anthropogenic effect on atmospheric carbon is not as big as the contribution from fossil fuel combustion.

3. The shutdown associated with the COVID-19 pandemic in 2020 started between January and March in much of the Northern Hemisphere and led to decreases in travel and fossil fuel combustion. How might such a decrease show up in the Keeling curve (Figure 14.7)? Explain how the timing of the oscillation in the Keeling curve might make such a decrease hard to see.

4. In Figure 14.13 (the nitrogen cycle), we did not give exact numbers but showed the relative size of nitrogen flows by the width of the arrows. On the basis of those arrow widths, assess the role of the Haber-Bosch process in the global nitrogen cycle.

5. Many of the graphs and figures for this chapter have wide bands of uncertainty around the mean value. Explain why that may be. (Hint: Think about variability in responses or in data sources.)

6. Figure 14.22 shows several changes for two different groups—arbuscular mycorrhizae (AM)-associated trees and ectomycorrhizae (EM)-associated trees. Interpreting such graphs can be difficult but is an important quantitative skill to have in the field of ecology and in many aspects of modern life.

   a. Explain how Figure 14.22B suggests that growth decreases for all examined trees, both AM-associated and EM-associated, at high levels of nitrogen deposition.

   b. If growth decreases at high nitrogen levels for trees with both types of mycorrhizal association, why do increases in nitrogen deposition lead to more growth of AM-associated trees?

   c. Use another panel (A or C) of Figure 14.22 to explain why AM-associated trees replace EM-associated trees as nitrogen deposition increases.

7. Figure 14.34 explores the relationship between biodiversity and ecosystem functions. If you were going to redraw this figure to explore the relationship between functional diversity (x-axis) and ecosystem function (y-axis), following the current Figure 14.34, how do you think the figure might change? Specify whether you expect a positive relationship or any other changes to the slope, and whether you expect a linear or nonlinear relationship.

SUGGESTED READING

Alongi, D. M. 2014. Carbon cycling and storage in mangrove forests. *Annual Review of Marine Science* 6: 195–219.

Cardinale, B. J., J. E. Duffy, A. Gonzalez, D. U. Hooper, C. Perrings, P. Venail, A. Narwani, et al. 2012. Biodiversity loss and its impact on humanity. *Nature* 486: 59–67.

Intergovernmental Panel on Climate Change. 2013. Figure 12.41 in T. F. Stocker, D. Qin, G.-K. Plattner, M. Tignor, S. K. Allen, J. Boschung, A. Nauels, Y. Xia, V. Bex, and P. M. Midgley, eds. *Climate Change 2013: The Physical Science Basis. Contribution of Working Group I to the Fifth Assessment Report of the Intergovernmental Panel on Climate Change.* Cambridge: Cambridge University Press. https://www.ipcc.ch/report/ar5/wg1/long-term-climate-change-projections-commitments-and-irreversibility/. Accessed September 16, 2021.

Schlesinger, W. H. and E. S. Bernhardt. 2013. *Biogeochemistry: An Analysis of Global Change,* 3rd ed. Waltham, MA: Academic Press.

Williams, A. P., C. D. Allen, C. I. Millar, T. W. Swetnam, J. Michaelsen, C. J. Still, and S. W. Leavitt. 2010. Forest responses to increasing aridity and warmth in the southwestern United States. *Proceedings of the National Academy of Sciences* 107(50): 21289–212294.

The Anthropocene is a time of great planetary upheaval, largely because of the influence of human beings. The effects of human population growth and habitation can be seen across all landscapes and biomes, and they influence all species on the planet.

chapter 15

ANTHROPOCENE ECOLOGY

LEARNING OBJECTIVES

- Explain the idea of the Anthropocene, and assess the evidence for its existence.

- Summarize the ways that climate change is affecting biodiversity.

- Explain how phenological studies have been used to examine the effects of climate change.

- Explain why the rate of extinction is higher in the Anthropocene than in the past, and summarize how scientists have been able to demonstrate this difference.

- Describe ways in which the loss of a top predator can influence an ecological system.

- Identify the social and political factors that appear to influence global rates of species invasion.

- Explain the impacts that invasive species can have on ecosystems.

- Define the term *anthrome*, and assess its utility.

- Identify ways that ecological concepts can be applied to diverse fields outside of the sciences.

- Identify ways to apply ideas from this book to reverse the rate of human-induced climate change or loss of biodiversity.

15.1 The Long History of Planetary Change Meets Human Acceleration

The world today is not the same as it was yesterday. Some version of this sentiment has probably been uttered by human beings since they first started using language. We understand intellectually that, as time marches forward, the natural world changes around us in a multitude of ways. Yet recognizing that the natural world is not static and that humans have contributed to some of the changes taking place around us is a deep and profound philosophical switch in our cognitive processing. Some people find it hard to accept that human actions can influence the entire biosphere, but the scientific and climatologic evidence—not to mention the evidence visible right outside our windows—of significant anthropogenic effects on our planetary system is growing. The somewhat lofty goal of this chapter is to synthesize some of what we know about ecology in the Anthropocene and provide markers, signposts, and perhaps directions for how the application of ecological principles plays a role in determining the future quality of life on this planet.

Most ecological science was conceived and tested within a framework that supposed a relatively stable Earth system. This framework included ideas about variability but assumed that many of these variations were random and nondirectional. In the grand scheme of the Earth's history, humans have been around for barely the blink of an eye (**Figure 15.1**). Before the evolution of *Homo sapiens*, the planet experienced many intense climatic shifts that eventually led to

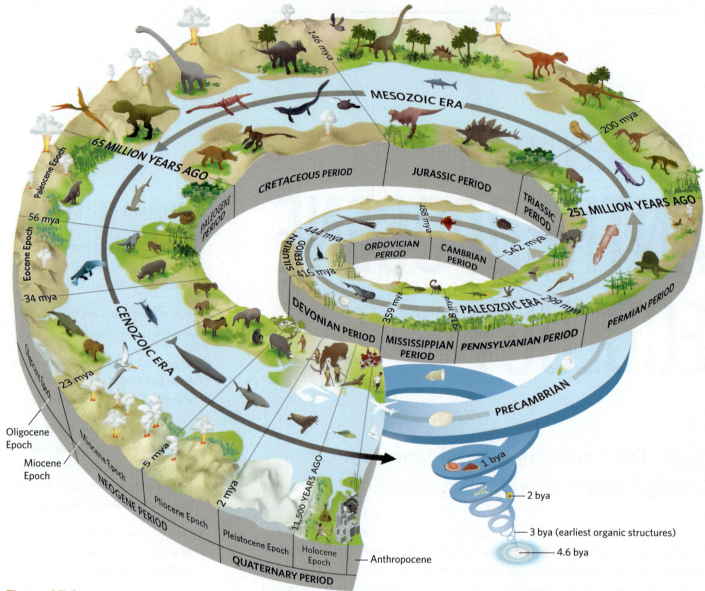

Figure 15.1 A timeline of the evolution of life on Earth, with approximate dates and evolutionary events from the formation of the Earth to the current Anthropocene epoch given in millions of years ago (mya) and billions of years ago (bya).

an abundance of living organisms. For example, in the first few billion years of its existence, Earth underwent massive changes to its land, waters, and atmosphere. Around 4.6 billion years ago, when the planet first formed, it was so molten that it had little to no atmosphere. As the Earth cooled, droplets of methane hung in clouds and slowly created a blanket over the Earth, keeping the planet warm despite the fact that the Sun produced only about 70% of the energy it produces today. This initial methane-rich atmosphere would have been toxic to most organisms alive today, but cyanobacteria evolved and flourished in the oceans during this time.

The emergence of photosynthesizing cyanobacteria about 2.5 billion years ago led to a slow atmospheric change toward a more oxygen-rich atmosphere. But even as recently as 600 million years ago, oxygen levels were only about one-fifth of their current levels. For most organisms alive at that time, though, oxygen was toxic. The buildup of oxygen in the atmosphere led to the extinction of most of these organisms except for the few species that retreated into anoxic (oxygen-free) environments or evolved to survive in the new oxygen-rich world. Thus, the accumulation of oxygen from photosynthesis continued and eventually led to our current atmosphere, which is dominated by nitrogen and oxygen.

Since the oxygen-rich atmosphere developed, the Earth has experienced temperature cycles, driven by changes in its position relative to the Sun, that have led to periods of glaciation every 40,000–100,000 years. Compared to the pace of change over the first three to four billion years of Earth's history, these glacial cycles are rapid, but to humans they are imperceptible. Even to ecologists and evolutionary biologists who study long-term dynamics, these cyclical changes appear as climatic predictability overlaid on *very* long-term climate trends.

Since the end of the last glaciation (about 10,000 years ago), the planet's sea and air surface temperatures have varied somewhat over time but have generally hovered around a long-term average value, overlaid by local seasonality. During this stable period, some species have arisen via evolutionary speciation or have colonized new locations, while others have gone extinct from natural events.

The appearance of humans definitely altered these rhythms for some species, but in our first few thousand years on the planet, humans were not a dominant species. Human impacts intensified with the use of fire, the domestication of animals and plants, and cultural changes that increased hunting efficiency. The pace of these changes increased drastically around the time of the Industrial Revolution. Since then, human activities, consumption habits, and waste materials have altered the Earth's atmosphere, weather, biomes, and biodiversity. What has changed in the past 50–150 years to cause these alterations?

In this chapter, we revisit the suggestion that our current era is the *Anthropocene* (Chapter 1), and we explore the implications of such far-reaching human impacts in more detail, now that you have a better understanding of ecological principles. In earlier chapters, for example, we explored numerous examples of how humans have produced significant changes in the Earth's biosphere, including changes in evolutionary patterns, individual organisms' physiology and behavior, community interactions, food web structures, species' phenology, ecological succession, and biogeochemical cycles. In this chapter, we pull several of these topics together into ecological "stories" about how human actions have come to dominate many of the functions of the Earth's systems. Together these stories illustrate that the natural world is rapidly shifting toward an unknown but significantly altered future state. The ever-present influence of people on biodiversity means that the ecologists of today must confront new and unexpected challenges when studying the natural world, particularly with respect to the social, cultural, and political importance of their work.

We start this exploration by reviewing the concept of the Anthropocene epoch and then synthesize ideas from previous chapters to explore how climate change affects where species will be found in the future, the significant loss of biodiversity through extinctions, the widespread impacts of invasive species, and the replacement of biomes with human-created anthromes. We conclude the chapter and book by discussing how ecological principles provide a blueprint for realizing a more ecologically sustainable future.

15.2 Is the Anthropocene Real?

In Chapter 1, we introduced you to Paul Crutzen and Eugene Stoermer, who wrote a short but highly influential article in 2000 in which they highlighted the large number of biological and geological changes to the Earth caused by humans. They stated, "It seems to us more than appropriate to emphasize the central role of mankind in geology and ecology by proposing to use the term 'anthropocene' for the current geological epoch" (Crutzen and Stoermer 2000). To suggest that the entire planet has shifted into a new geological epoch, one that is clearly influenced by human activities, was a bold assertion. When geologists of the future look at the sediments, rocks, and fossils they dig up from today, will they be able to clearly and distinctly see a global shift from the Holocene to the Anthropocene? If so, their geological data would point toward an Earth profoundly altered by people (**Figure 15.2**). In the years since Crutzen and Stoermer began making their case for the Anthropocene, supporting evidence for their idea has accumulated. The International Commission on Stratigraphy convened the Anthropocene Working Group in 2009 to assess the evidence for such a distinction. This group, composed of climate scientists, ecologists, archaeologists, historians, and oceanographers, identified global signals in the atmosphere and ice cores that are markedly distinct from the rest of the Holocene. A large portion of the difference comes from the presence of radioactive carbon and plutonium from the use and testing of nuclear weapons in the mid-twentieth century (Waters et al. 2016). We can further detect the transition into the Anthropocene epoch from the global deposition of pollution from plastics, concrete, and black carbon (fossil fuel ash) since the mid-twentieth century (**Figure 15.3**).

Given these massive anthropogenic effects on the Earth, ecologists clearly need to consider the role humans play in altering planetary ecology. With this consideration in mind, we, as a species, may take one of two paths. We may decide that the world is already so damaged by humans that further destruction is inevitable and unimportant. If the world has no untouched or "natural" places left, then what is to stop society from just deciding to turn the world into a huge parking lot? Alternatively, perhaps our impacts create a moral and self-preserving imperative for biodiversity conservation and restoration. Wilderness and natural communities are still relatively common and certainly worth conserving. In that vein, recognition of the Anthropocene drives us to examine human coexistence with biodiversity and functioning ecosystems.

In practice, a good amount of current ecological research considers anthropogenically driven global trends. Today a majority of ecological research appreciates the need to move away from the stable steady-state models and theories of the past. Scientists are increasingly emphasizing ideas about interactions, feedbacks, and thresholds in their research and models. Ecologists

Figure 15.2 As evidenced by this sediment core from western Greenland, glacial retreat due to climate warming has resulted in a transition from glacial and preglacial sediment deposition to nonglacial organic matter, effectively demarcating the onset of the Anthropocene (Waters et al. 2016).

Live moss

Organic sediment

Glacial sediment

often explicitly consider the role that humans directly and indirectly play in shaping ecological processes. In the sections that follow, we highlight ecological principles discussed in earlier chapters that are critical to understanding ecology in the Anthropocene.

15.3 Species' Response to Climate Change

Rapid global shifts in climate are some of the most obvious markers of the Anthropocene. In Chapter 3, we explained many of the processes that are altering the climate system and the outcomes of these processes (e.g., increased average surface temperature, increased ocean temperatures, increased ocean acidification, and a rise in sea levels). In several chapters, we touched on the ecological effects of rapid climate change, such as the extinction rates of iguanas in tropical dry forests (Chapter 5) and reduced shell calcification in marine organisms (Chapter 14), but we did not consider the effects of rapid climate change on species' geographic ranges. To do so we must synthesize several topics from previous chapters.

Geographic Range Changes

Much of what we know about how species alter their geographic ranges comes from our understanding of the ecological niche. Recall from Chapter 5 that a species' fundamental niche is the set of factors that determine individual fitness and where an organism can survive and reproduce. These multidimensional niche factors include the abiotic characteristics of a location, such as the amount and timing of rainfall, the frequency of days with very high (or very low) air or water temperatures, or the presence of high (or low) soil nitrogen. In addition, the fundamental niche also involves biotic elements, such as the presence (or absence) of competitors, predators, prey, or mutualists.

Climate change can degrade the local environmental characteristics so that a habitat no longer provides one or more critical niche factors for a species. Paradoxically, climate change can also improve local characteristics to better fit a species' niche, leading to higher fitness for individuals. But given the complexity of the ecological niche and the multitude of factors involved in a species' niche, how do we identify the range of niche components that determine why, under the current climate, a species exists in certain locations? And then, how do we predict shifts in niche factors as the climate warms, sea levels rise, weather patterns transform, and oceans acidify? What happens when one species responds to altered niche conditions, and the response brings it into contact with entirely new

Figure 15.3 **A** This graph shows the increase in the production of plastic, concrete, and black carbon (fossil fuel ash) over the last 150 years and the spike in radioactive plutonium (Pu) and carbon (^{14}C) fallout from nuclear weapons during the 1940s–1960s (Waters et al. 2016). PBq = petabecquerel; Tg = teragram. **B** Here, we see the longer-range signals of nitrate (NO_3^-), carbon dioxide (CO_2), and methane (CH_4) in ice cores since the beginning of the Holocene (Waters et al. 2016). ppb = parts per billion; ppm = parts per million.

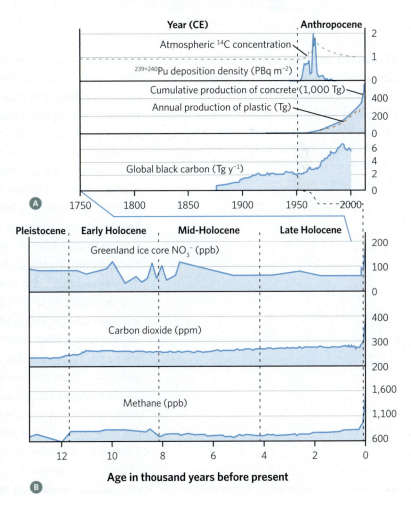

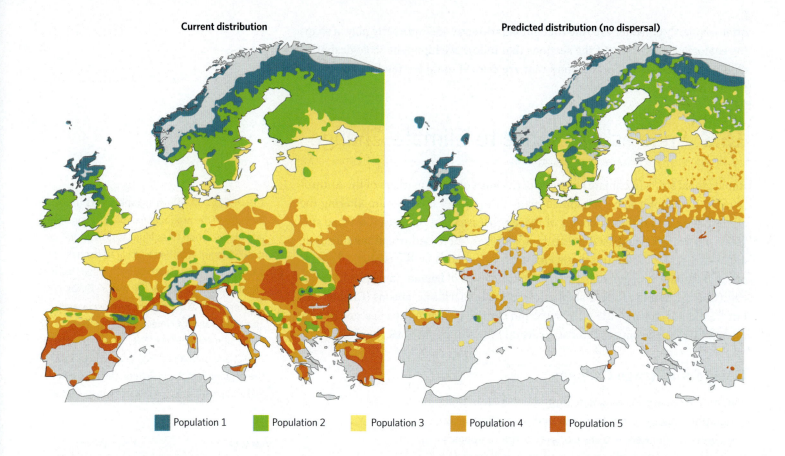

Population 1 Population 2 Population 3 Population 4 Population 5

Figure 15.4 The results of a computer simulation model show the change in distribution of five populations of a hypothetical plant species occurring across Europe given a 3.6°C (6.5°F) increase in temperature. When individuals cannot disperse, we see that most populations decline and some may not survive (Valladares et al. 2014).

predators, competitors, prey, or mutualists? Is it possible for a species to evolve fast enough to shift its ecological niche to suit a future climate? Forecasting such alterations and then planning conservation approaches to protect biodiversity in the face of a rapidly changing climate are two of the principal challenges facing modern-day ecologists.

As a first attempt to address the problem, ecologists often combine a map of a species' current distribution with predictions of future climate variables (e.g., precipitation and temperature). From this, they can forecast how well the future climate will match with a species' niche requirements across a defined geographic space. Advances in technology and data collection have produced accurate fine-scale data on air temperatures, water temperatures, and hourly rainfall amounts for literally millions of locations around the globe. Using these types of data, researchers can search for correspondence with individual species' niche requirements. **Figure 15.4** illustrates a simulation experiment for a hypothetical widespread species of plant showing how individual populations of that plant might change their distributions across Europe given a 3.6°C (6.5°F) increase in temperature. Notice that a species is not homogeneous—populations of the same species perform differently with climate change.

Not surprisingly, one of the earliest insights from analyses like these was that many species are not likely to persist in their current locations and will need to disperse to new sites where conditions allow their survival and reproduction. Therefore, in order to explore the larger biodiversity implications of climate change, we also need to consider a species' dispersal ability (Lenoir and Svenning 2015). Combining the ideas of niche flexibility and dispersal ability allows scientists to develop a more nuanced understanding of the ways in which species' geographic ranges are likely to respond to climate change (Table 15.1).

Extinction is likely if a species does not disperse easily and has low evolutionary fitness in the future climate. Species with broad niches may be able to persist in their current locations as the local climate changes, but they may experience a sharp drop in abundance or have a population crash as the quality of the local habitat declines. At the other extreme are scenarios in which species can disperse easily and have flexible enough niche requirements that changes in environmental factors will not drastically reduce individual fitness. Species with narrow niches but high dispersal capacity will simply move their ranges, expanding into newly suitable locations in some directions but dying out in places where climate change makes conditions unsuitable. Some of the more adaptive species or species with a wide range of habitable conditions will persist where they are and also expand their geographic ranges to incorporate new locations in which climate change has created habitable conditions.

At this point, it is not clear which of these outcomes (extinction, population decline, shifted range, or expanded range) is more likely for most species, but we can identify certain species that fit each scenario. Endemic species with low dispersal capacity are most likely to fall into the category of shorter-term extinction. One example may be the stinking cedar tree (*Torreya taxifolia*), an endemic species restricted to a very small area at the border of Florida and Georgia (Figure 15.5). Although there are organisms that eat its seeds, they do not seem to disperse the seeds, and some ecologists have suggested that the tree's dispersal agent may have gone extinct sometime in the past. The tree cannot move to the cooler environment to the north without a dispersal mutualist. The population has been steadily shrinking since it was first placed on the federal endangered species list in 1984 because of a variety of threatening factors, including climate change.

Small, short-lived species that are not endemic but have narrow temperature requirements and low movement rates are also likely to suffer population declines in all or part of their current geographic ranges. For example, global lizard diversity seems to be decreasing with climate change (Sinervo et al. 2010). As discussed in Chapter 5, lizards are ectotherms that use conditions in different microenvironments to regulate their internal temperature. Juveniles of many lizard species are particularly susceptible to changes in temperature, and most lizards are not great at dispersing long distances or traversing areas with different thermal habitats. For all of these reasons, we are seeing a global decrease in lizard diversity as appropriate habitats shrink, and populations crash or retract and then dwindle away.

Some species are more likely to expand their range, particularly very mobile, large-bodied, or long-lived species. Large-bodied organisms can often buffer themselves

Table 15.1 Conceptual Outcomes for Species and Their Potential Range Shifts Based on Species Movement Rates and Species Persistence Rates

Species persistence rate (niche flexibility)	Species movement rate (dispersal ability)	
	Low	High
Low	Species extinction	Species range shift in location
High	Population crash	Species range expansion

Adapted from Lenoir and Svenning 2015.

Figure 15.5 The endangered stinking cedar tree (*Torreya taxifolia*), found in southern Georgia and northern Florida, has low dispersal ability and a fairly narrow thermal niche.

from temperature shifts better than small species can (surface-to-volume ratios decrease with organism size, meaning large-bodied organisms have relatively less surface area through which they can lose or gain heat). In addition, long-lived organisms may have adults that survive in place well after the environment has shifted in ways that are completely inhospitable to juveniles. For example, large-bodied birds may survive well with changes in their habitat, and trees whose seeds are dispersed by birds may expand their distribution, if older individuals survive for decades in original habitats and juveniles thrive in new territories at higher latitudes. We see similar patterns in marine invertebrates and other small ectotherms whose adult individuals tend to die in areas closer to the equator, but whose juveniles can move long distances on currents and shift their distribution toward the poles.

Even mobile species with broad, flexible niches may encounter barriers that limit their ability to escape a changing climate. These species may shift their ranges but still experience declines in population abundance. For example, many butterfly species in Europe are moving poleward to escape warming, but their movement often seems restricted by mountains. They can disperse relatively easily, but eventually their range shift may be compromised by encountering altitudinal gradients they cannot cross. Freshwater aquatic organisms that disperse well in streams, ponds, or lakes frequently encounter similar barriers to movement between water bodies. A lack of aquatic connections may limit the ability of a species to shift its range in response to a changing climate.

Combining dispersal ability and environmental factors to forecast changes in species' ranges is a clear improvement over forecasting using environmental niche requirements alone, but it is still an enormous simplification. Everything you have learned about ecology up to this point should suggest that this approach cannot adequately describe how the world's biodiversity will respond to the altered climate of the future. For example, we should expect that, if each species in a community is responding individually to climate shifts, the resulting changes will produce a plethora of novel ecological interactions as time goes on. Even those species that manage to persist in their current range will encounter new competitors, predators, prey, and mutualists as the other species around them change their geographic ranges to adjust to a novel climate. These new interspecific interactions will have at least as large an effect as any abiotic changes.

Key Ecological Mechanisms and Adaptive Modeling

Mark Urban and colleagues (2016) confronted this dilemma by pointing out that more nuanced and accurate ecological models include ecological mechanisms beyond dispersal ability and niche flexibility. They identified six key ecological mechanisms that would improve predictions of biological responses to climate change: evolution, environment and land use, physiology, demography, dispersal, and species interactions (Figure 15.6). As an ecologically literate student, you will likely recognize in Figure 15.6 how and why the details of each of these mechanisms would be relevant to future work, but can you imagine how we might put together such models? Note that in the chapter on spatial dynamics (Chapter 12), we did not include explicit spatial models for multispecies interactions. The reason we omitted them is that these models can get *very* complicated (Figure 15.7). Nonetheless, the approach outlined in Figure 15.7 offers a framework for how researchers might incorporate these ecological mechanisms into modeling approaches.

1
Evolution
Using genetic models to examine anticipated adaptive responses

2
Environment
Using geographic data to predict land use changes

3
Physiology
Predicting physiological responses using mass balance and energy equations

4
Demography
Predicting population dynamics using climate-specific models

5
Dispersal
Using climate-influenced dispersal behavior to predict spatial distribution

6
Species interactions
Predicting novel communities using species interaction models

Aedes aegypti
Yellow fever mosquito

Map of land use in eastern Pennsylvania

Lates niloticus
Nile perch

Sequoiadendron giganteum
Giant redwood

Danaus plexippus
Monarch butterfly

Gecarcoidea natalis
Christmas Island red crab
Anoplolepis gracilipes
Yellow crazy ant

Figure 15.6 The inclusion of these six key ecological mechanisms in models should improve ecologists' predictions of how species will respond to climate change. Models incorporating the identified mechanisms have helped predict (1) the adaptation of disease-harboring mosquitoes; (2) future habitats and land use; (3) the impacts and spread of invasive species (such as Nile perch) based on their physiology; (4) redwood tree population responses to the warming of their habitats; (5) climate-dependent dispersal in butterflies; and (6) unpredictable interactions between native and non-native species (Urban et al. 2016).

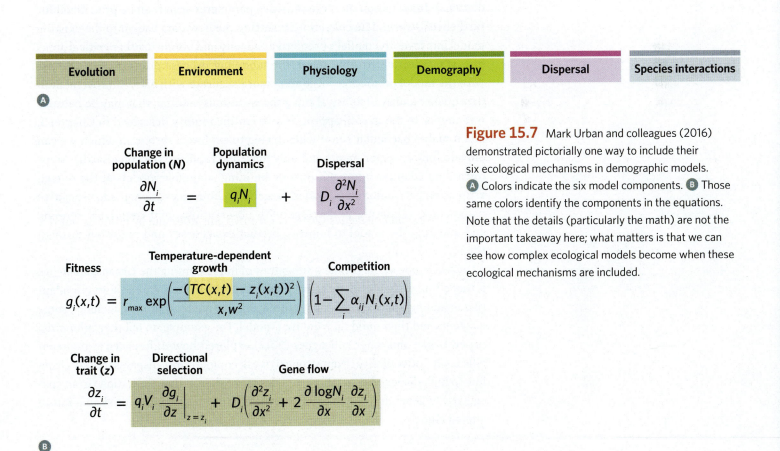

Figure 15.7 Mark Urban and colleagues (2016) demonstrated pictorially one way to include their six ecological mechanisms in demographic models. **A** Colors indicate the six model components. **B** Those same colors identify the components in the equations. Note that the details (particularly the math) are not the important takeaway here; what matters is that we can see how complex ecological models become when these ecological mechanisms are included.

Evolution   Environment   Physiology   Demography   Dispersal   Species interactions

(A)

Change in population (N)   **Population dynamics**   **Dispersal**

$$\frac{\partial N_i}{\partial t} = q_i N_i + D_i \frac{\partial^2 N_i}{\partial x^2}$$

Fitness   **Temperature-dependent growth**   **Competition**

$$g_i(x,t) = r_{max} \exp\left(\frac{-(TC(x,t) - z_i(x,t))^2}{x, w^2}\right)\left(1 - \sum_i \alpha_{ij} N_i(x,t)\right)$$

Change in trait (z)   **Directional selection**   **Gene flow**

$$\frac{\partial z_i}{\partial t} = q_i V_i \left.\frac{\partial g_i}{\partial z}\right|_{z = z_i} + D_i\left(\frac{\partial^2 z_i}{\partial x^2} + 2\frac{\partial \log N_i}{\partial x}\frac{\partial z_i}{\partial x}\right)$$

(B)

Urban and his colleagues, using an approach that incorporated elements of Figure 15.7, explored which mechanisms might determine species survival in a future climate. They found that, at the most basic level, the millions of available data points on abiotic conditions around the globe do not match up with the available data on the relevant biological features of the world's species. Thus, even for the most well-studied species, Urban and team found that scientists do not have the specific biological data to use this modeling framework. For example, for well-studied species, we may have fairly good information on such variables as population abundance or life history characteristics, but we lack information on the parameters that make those same variables change. The models of population growth presented in Chapter 6 predicted population abundance based on per capita birth and death rates, but those rates change with habitat quality. It is much easier to collect data on population abundance than on the birth and death rates that lead to changes in population abundance, but we need information on these rates *and* how they change when the climate is altered.

Urban and colleagues, therefore, recommended that future-forecasting models embrace an **adaptive modeling** approach. Working within this framework, scientists admit at the outset that their initial model is not going to be fully accurate and tightly predictive. By using such models to make predictions and then testing these predictions against real-world data, researchers can identify information gaps that, when filled, will allow them to create better predictions. Researchers can also identify which model parameters are particularly sensitive (i.e., the parameters that, when slightly altered, drastically modify the predictions). Information on these sensitive parameters can then be prioritized for further real-world data collection. Inserting the new data back into the existing model improves the model's predictive capacity and points to other critical areas for data collection. In this sense, the modeling process progresses iteratively, with the model predictions becoming more accurate with each improvement in data quality and availability. If this process sounds familiar, that may be because it is similar to the overall approach to scientific inquiry described in Chapter 1, and it makes particular sense when trying to explore a system in which we can only do limited experiments and only have one replicate (i.e., one Earth). Steps toward an adaptive approach include building adaptable models at the outset, estimating parameter sensitivities, targeting better measurements for sensitive parameters, validating projections with observations, and iteratively refining and updating the model to improve predictive accuracy and precision through time (**Figure 15.8**).

Clearly, one of the primary difficulties of implementing the Urban framework is that all six of their ecological mechanisms often interact. We can start our adaptive modeling approach, though, with a smaller model that only includes some elements and then build up from that model. For example, in another influential paper, Urban and other colleagues (2012) explored how differences in interspecific competitive ability (from strong to weak competitors), dispersal ability (from low to high dispersal capability), and degree of thermal specialization (from specialized to generalized niches) interacted to influence extinction risk in a future altered climate.

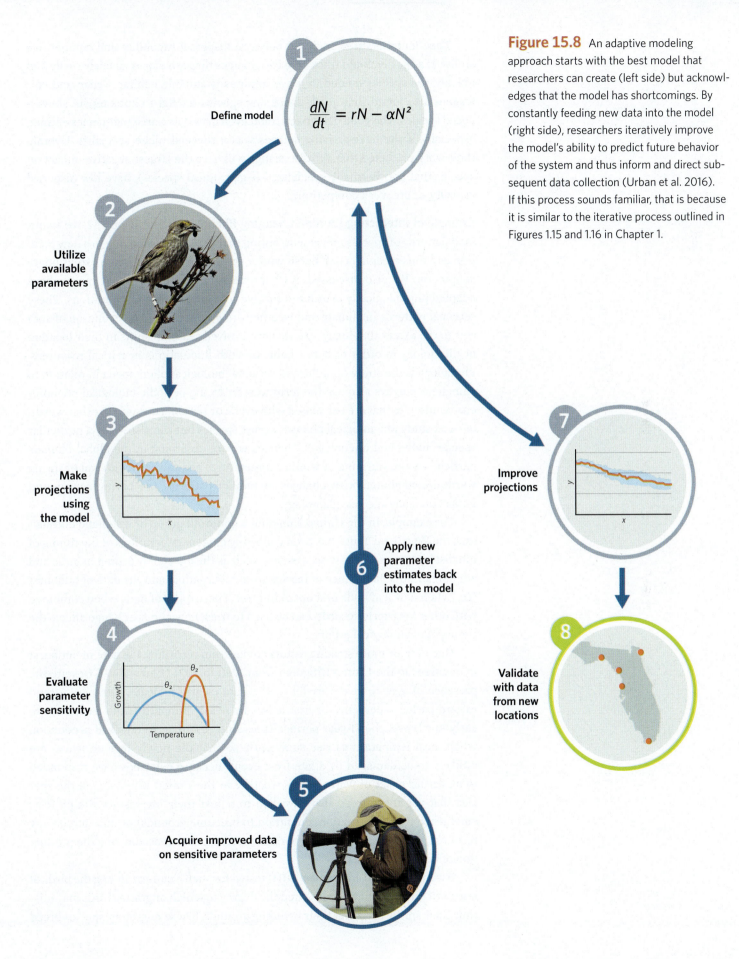

1 Define model

$$\frac{dN}{dt} = rN - \alpha N^2$$

2 Utilize available parameters

3 Make projections using the model

4 Evaluate parameter sensitivity

5 Acquire improved data on sensitive parameters

6 Apply new parameter estimates back into the model

7 Improve projections

8 Validate with data from new locations

Figure 15.8 An adaptive modeling approach starts with the best model that researchers can create (left side) but acknowledges that the model has shortcomings. By constantly feeding new data into the model (right side), researchers iteratively improve the model's ability to predict future behavior of the system and thus inform and direct subsequent data collection (Urban et al. 2016). If this process sounds familiar, that is because it is similar to the iterative process outlined in Figures 1.15 and 1.16 in Chapter 1.

They found a sharp trade-off between dispersal capability and competitive ability. If a species could track climatic changes through dispersal fairly easily and quickly, the species tended to move into newly suitable habitat. Urban and colleagues also found that quick-dispersing species tended to outcompete slower-paced species. The good dispersers could even drive later-arriving species extinct, especially if the later-arriving species was a thermal niche specialist. Overall, their work predicted that climate change will have the largest negative impact on species that are thermally specialized (e.g., tropical species), have low dispersal capacity, or are weak competitors.

Example of Interacting Factors: Changing Phenology   In Chapter 12, we examined patterns of species movement, noting that temporarily moving to a new location and thus escaping cold, harsh winter conditions was a common evolutionary adaptation. We also discussed in Chapter 5 some of the ways that species have adapted physiologically in order to survive cold or dry seasonal conditions. These seasonal patterns in climate and weather drive the ecology and evolution of species to the extent that many species have evolved seasonal shifts in their location or physiology in order to better coincide with fluctuations in critical resources. Phenology is the study of cyclic and seasonal biological phenomena in relation to climate; ecologists also use the term to refer to the periodic biological phenomena themselves that are correlated with cyclic or seasonal events. (In other words, one can study phenological changes across species but can also study a particular species' individual phenology.) Changes we are observing in the global climate, particularly the warming of winter temperatures and the earlier arrival of spring warm-up, combined with changes in rainfall patterns, are having a profound impact on many species' phenology.

For example, in the United Kingdom a nonprofit group of woodland enthusiasts, the Woodland Trust, has been collecting citizen observations of the timing of life history events in forest ecosystems, such as the onset of breeding in birds and amphibians, the emergence of insects from hibernation, and the date of budburst (i.e., when new leaf buds first open) in trees. These types of data, when combined with other long-term records, can be used to track how climate change affects the timing of phenological events.

One of their more striking results comes from recording the date of budburst in oak trees in the United Kingdom from 1950 to 2021 (**Figure 15.9**). Despite the presence of year-to-year variability, the data clearly indicate a dramatic trend toward earlier emergence of first leaves by nearly a month (from early May to early April) over the 71-year period. Trees shift resources to new leaf production when local temperatures rise sufficiently so that the newly formed leaves are unlikely to be damaged by a late frost event. Because oak trees have responded to an earlier onset of warm spring weather in the United Kingdom, all the species that use these trees also have had to adjust their life history timing (e.g., emergence from hibernation, migration to breeding grounds) or miss out on any resource pulses (i.e., short-lived bursts of resources) that accompany the production of new oak leaves.

Nicola Saino and colleagues (2011) tested for such a pattern among the birds of northern Europe. They documented the degree to which migrating birds may mistime their spring arrival at their breeding grounds. These migratory species breed

Figure 15.9 The Woodland Trust has been recording budburst dates for oak trees in the United Kingdom since 1950 (Woodland Trust 2021). Although there is a lot of variability from year to year, the overall trend is a shift toward an earlier calendar date for the first fully opened leaves.

in the varied habitats of northern Europe but overwinter in southern Europe or North Africa. When they return to their breeding grounds, they confront several possibilities. If they arrive too early at the breeding grounds, they run the risk of encountering cold spring events and poor food supplies. Their food resources are generally insects or seeds, which are themselves trying to time their own emergence to spring warm-up conditions (as with the UK oaks and their budburst). If the birds arrive too late at the breeding grounds, they face poor odds of securing mates or good territories, or they may be too late to take advantage of early season food pulses. Either manner of mistiming their northern spring arrival has significant fitness consequences. There is widespread evidence that migratory birds have shifted their spring arrival to earlier in the year in order to stay in sync with the earlier spring warm-up, which is likely to optimize the birds' fitness.

The question Saino and colleagues addressed was whether the observed shift in arrival date was sufficient to maintain these bird species' populations through time. The results of their assessment suggest that many migratory bird species, especially those that traverse the long distances from sub-Saharan Africa to northern Europe, have not sufficiently changed their spring arrival time to fully coincide with the timing of peak fitness opportunities in the breeding grounds. The species that show the largest mismatches between their phenology and the onset of spring in their breeding locations also show the largest population declines between 1959 and 2009 (Saino et al. 2011). Even with a shift in their migration phenology, they have not been able to escape negative demographic consequences of climate change.

A central challenge to anyone studying ecological processes in the Anthropocene is how to integrate the various ecological mechanisms outlined by Urban and colleagues (2016) with the anticipated changes to the Earth's climate. The examples we have given highlight promising avenues of research and shed some light on how we may expect biodiversity to fare in the altered climate of the future. There is a definite need for future experiments to explore the mechanisms behind range collapse (or range expansion) and phenological shifts, among many other possible outcomes. Ecologists also need observations that document species interactions, physiology, and population demography. These complementary approaches, often based on the conceptual and quantitative models detailed in previous chapters, will play key roles in producing ecological science that is relevant to the changing climate.

15.4 Extinction in the Anthropocene

Accelerated Extinctions

One of the hallmarks of the Anthropocene is the vastly accelerated rate at which species are going extinct relative to what we would expect in the absence of human influence. The evidence for this acceleration comes mostly from our understanding of vertebrate extinctions. Most people have heard the story about the end of the dinosaurs, but through the ages millions of other vertebrate taxa have gone extinct as a result of events or processes that have nothing to do with humans. Using the unusually rich paleontological record of natural mammal extinctions, and a few inferences about the fossil record itself, scientists estimate that we would expect to see 1.8 mammals go extinct per 100 mammalian species every 10,000 years in the absence of human actions (Ceballos et al. 2015). Although it may be hard to do the mental math to figure out how many mammal species this adds up to, it is important to recognize that this is not a very high rate of extinction. The fossil record for mammals is reasonably complete, but rates for other groups are much harder to estimate; thus, the mammalian extinction rate is regularly used as a surrogate for "background" natural extinction rates for other vertebrate groups.

Gerardo Ceballos and his colleagues (2015) compared the mammalian background estimate of the rate of vertebrate extinction to the observed rate of vertebrate extinctions during and just before the Anthropocene (post-1900). First, they rounded the background extinction rate to two species (instead of 1.8) per 100 species every 10,000 years, which means their background rate estimate should be a little high. This sort of conservative estimate acknowledges that there might be some error in the comparisons, and it keeps us from sounding the alarm

Figure 15.10 This graph shows the cumulative number of vertebrate species recorded as either "extinct" (no reasonable doubt that the last individual has died) or "extinct in the wild" (known only to survive in captivity) by the International Union for Conservation of Nature (IUCN) as a percentage of the number of species evaluated. The dotted line represents the number of extinctions expected under a constant standard background rate (Ceballos et al. 2015).

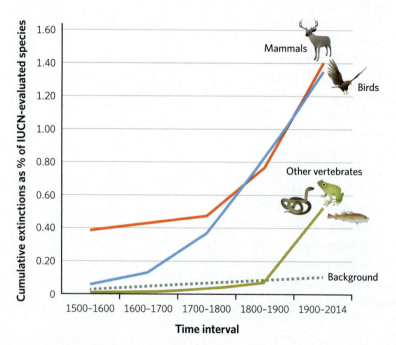

unnecessarily. They calculated that, if this rate of natural extinction were happening today, we would expect around nine species of vertebrates to have gone extinct between 1900 and 2014 (114 years) in the absence of human actions. What we see instead is very different.

If we only count those species that have most certainly gone extinct since 1900 and do not include those that are poised on the brink of extinction, we have witnessed 198 vertebrate extinctions over this 114-year period. This observation breaks down into 35 mammal, 57 bird, 8 reptile, 32 amphibian, and 66 fish species extinctions since 1900 (**Figure 15.10**). This number of lost species suggests that 189 more vertebrate species have gone extinct since 1900 than the natural background rate would predict.

Another way of viewing these data is to estimate how many years it would have taken for the observed species extinctions to occur without human influences. For this calculation, Ceballos and colleagues included the 198 definite extinctions already mentioned as well as a less conservative measure that included species that are "extinct in the wild" (known only to exist in captivity) or "presumed extinct" (no individuals have been observed anywhere in at least a decade; **Figure 15.11**). Using the mammalian background extinction rate as a guide for each group, it would have naturally taken 900–11,000 years to witness the number of vertebrate extinctions the Earth has experienced over the 114 years from 1900 to 2014.

Species extinctions are one of the most obvious effects on biodiversity in the Anthropocene. All extinctions result from severe and widespread population declines that eventually allow the last individual of a species to die. If we track only the population declines and not just the species extinctions, do we see equally dire prospects for biodiversity in the Anthropocene?

Rodolfo Dirzo and his colleagues (2014) synthesized available information on population declines in all kinds of fauna (not just vertebrates) and found widespread evidence that the Earth is rapidly losing its animal species, a process known as **defaunation**. For example, using standardized records for insect populations worldwide, Dirzo and colleagues showed that 33% of documented insect populations are declining globally (**Figure 15.12**). They found substantial variation among insect orders: for example, most species of crickets and grasshoppers (order Orthoptera) showed severe evidence of population decline, whereas populations of dragonflies (order Odonata) appeared to be in better shape, with the majority having fairly stable populations. Notably, very few insect populations anywhere showed increasing trends (Figure 15.12). Regrettably, only a small number of insect species had enough data to show any population trends for this analysis. As a result, Figure 15.12 is based on only about 1% of all insects known today. Even with these limited data, the trends suggest that the loss of insects globally is an underappreciated impact that rivals the loss of vertebrates. We clearly need a better way to track insect populations over large spatial scales.

Figure 15.11 This bar graph shows the number of years it would take for the observed vertebrate species extinctions of the past 114 years to occur based on the estimated background extinction rate. The "very conservative" rate includes only confirmed extinctions, as discussed in the text and in Figure 15.10. The "conservative" rate includes additional species that are now either extinct in the wild or presumed extinct (Ceballos et al. 2015).

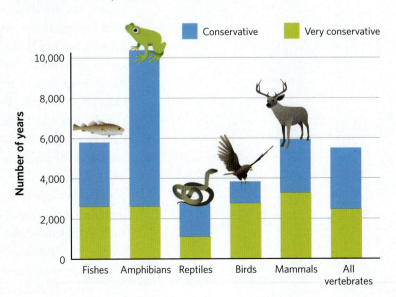

Figure 15.12 Rodolfo Dirzo and colleagues synthesized the available data from the International Union for Conservation of Nature (IUCN) to determine the percentage of insect species across various taxonomic orders that have documented changes in population abundance (Dirzo et al. 2014). Overall, 33% of IUCN-evaluated insects are declining in population abundance. The percentage of species with declines is indicated in black and the percentage of species increasing is indicated in blue.

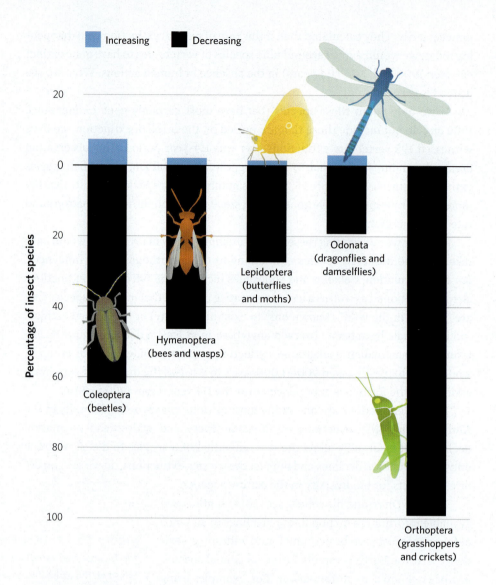

Effects of Extinction

We are losing organisms and losing them faster in the Anthropocene than ever before, but are there ways to identify patterns that can guide management and help ecologists develop strategies to address these losses? Ecologists have produced a few key insights in this regard. First, we know that the loss of animal species is nonrandom within a food web. Large-bodied top predators are much more likely to go extinct than other species, including smaller-bodied predators. For example, Francesco Ferretti and colleagues (2010) collated data on global shark populations and found widespread evidence that sharks have been less abundant since the 1950s. Pelagic sharks—large predatory sharks that live in the open ocean—seem particularly at risk of extinction. Data for pelagic sharks show only one species, the tiger shark in the eastern United States, as having an increasing population since the 1970s (**Figure 15.13**). The recent global causes of shark loss include overfishing of sharks for food and sport, the demand for shark products such as their fins, and the unintentional capture and killing of sharks as bycatch within other fisheries.

Pelagic sharks represent some of the most iconic oceanic predators, including great white (*Carcharodon carcharias*), mako (*Isurus* spp.), and thresher (*Alopias*

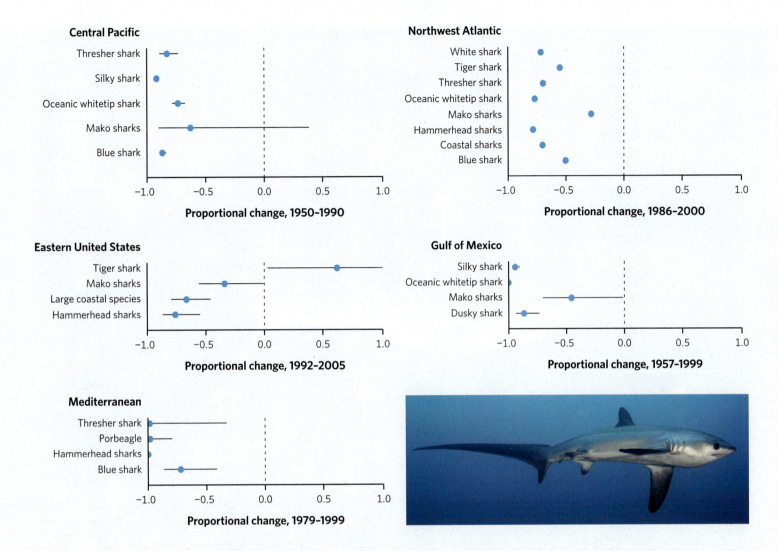

Central Pacific

Proportional change, 1950–1990

Northwest Atlantic

Proportional change, 1986–2000

Eastern United States

Proportional change, 1992–2005

Gulf of Mexico

Proportional change, 1957–1999

Mediterranean

Proportional change, 1979–1999

spp.) sharks (Figure 15.13). These large-bodied sharks migrate over large expanses of ocean, have few (if any) nonhuman predators, and mostly eat smaller fish but can also prey on nearshore coastal marine animals, such as seals and sea turtles. A central question in marine ecology and conservation is, therefore, what happens to marine ecosystems when they lose their top predators?

A second insight from the extinction data is that the loss or reduction in large vertebrates affects other species. In the preceding example, Ferretti and colleagues (2010) found that the removal of sharks from marine ecosystems leads to a trophic cascade. The prey of sharks experience less predation pressure when humans remove top predatory sharks (**Figure 15.14**). Because sharks have diverse diets that include several important midlevel predators (mesopredators), these midlevel species increase in abundance and thereby increase predation pressure on the trophic level below them—which includes fishes, seagrass, and invertebrates—leading to a trophic cascade.

Some of the species at the bottom trophic level in Figure 15.14 support important fisheries worldwide, and others are species that play critical ecosystem functions (e.g., seagrass). The plight of oceanic sharks caused by overexploitation has complex direct and indirect impacts on several valuable ecological services. Sometimes these effects are felt far away from where the actual fishing occurs, making the task of recognizing and managing the loss of ecosystem services difficult.

Figure 15.13 These graphs present estimates of relative changes in pelagic shark population abundance across various oceanic basins around the world. The dashed line down the middle of each graph indicates zero change in population abundance through time; markers (dots) to the left of the dashed line indicate declines in abundance, and markers to the right indicate increases. Lines on the left and right of the markers indicate variability in the proportional change estimates (Ferretti et al. 2010). Inset: A thresher shark (*Alopias* sp.).

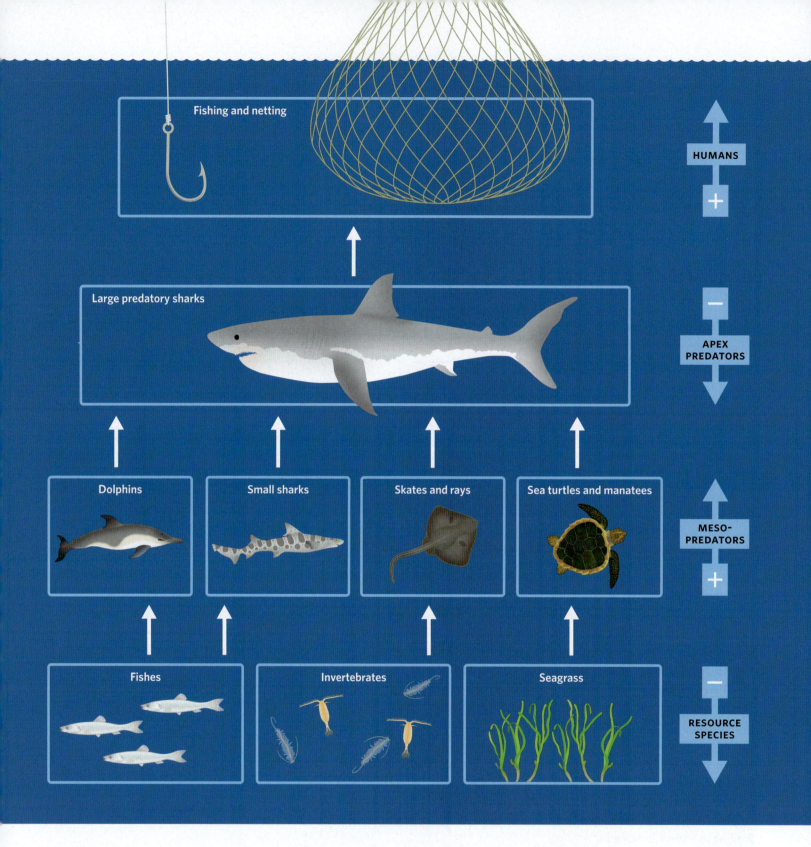

Figure 15.14 A study documented the food web effects of overexploitation of large pelagic sharks. Depicted are trophic interactions (white arrows) between humans, apex-predator and mesopredator sharks, and their prey species. Trophic arrows point in the direction of energy flow (see Chapters 5 and 10). The blue block arrows on the right represent overall abundance trends of the various functional groups across trophic levels (Ferretti et al. 2010).

More generally, ecologists are finding growing evidence that removing large-bodied top predators can shift an entire ecosystem into an **alternative stable state** (Figure 15.15). Ecosystems and the food webs within them can be fairly resilient to losses of some species or to changes in the availability of energy, thermal conditions, or harsh weather events. When minor changes occur in a food web, most of the species experience little change in abundance; however, larger changes can lead to dramatic shifts to radically novel or extremely transformed food webs or communities (Figure 15.15). These alternative states are often persistent, and the former configuration is not easily restored or recovered. These community alterations can remain even if some of the factors that caused the change in the first place are remedied or removed. These alternative stable states do not come only from changes in predation but can come from other anthropogenic or natural forces that sufficiently alter communities. The yellow crazy ant invasion on Christmas Island discussed in Chapter 9 (and the ensuing invasional meltdown) is an example of how an anthropogenic shift can lead to an alternative stable state.

A particularly well-documented and surprisingly widespread example of an alternative stable state caused by human exploitation can be observed in the rising dominance of jellyfish in many marine ecosystems. These outbreaks seem to occur when overfishing and bycatch reduce or remove fish and other jellyfish predators (Richardson et al. 2009). Jellyfish, as a group, are typical midlevel predators in nearshore marine ecosystems where they feed on fish larvae, zooplankton, and other small animals. Introducing non-native jellyfish or removing the predators that eat jellyfish can lead to population explosions, or "blooms" (Figure 15.16). These blooms are common in the coastal waters of Asia, as in the case of the outbreak of Nomura's jellyfish (*Nemopilema nomurai*) off the coast of Japan in 2009. Similarly, in 2000, the Australian spotted jellyfish (*Phyllorhiza punctata*), a native of the Pacific, was introduced to the Gulf of Mexico, where its population also exploded, reaching an estimated five million jellyfish in a 150 km² (58 sq. mi.) area.

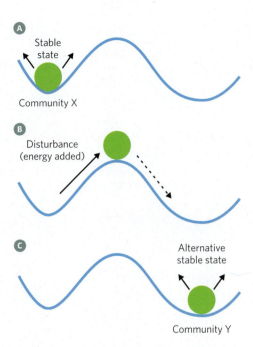

Figure 15.15 Major changes in food webs can cause ecosystems to shift to alternative stable states. Ⓐ An ecological community sits in a stable state and will generally return to this stable state after small disturbances (small solid arrows). Ⓑ A large enough disturbance (large solid arrow), however, can add sufficient ecological "momentum" to shift the community composition out of its stable state, which allows it to move toward another stable state (dashed arrow), with different species in the community. Ⓒ When this happens, the community composition stabilizes at a new, alternative stable state, and it is unlikely that the community will return to the original stable state because of the large ecological momentum required for the reversal.

Figure 15.16 A diver rises from the sea off the coast of Japan in 2009 amid a bloom of Nomura's jellyfish (*Nemopilema nomurai*). One of the world's largest jellyfish, Nomura may exceed 2 m (6.6 ft.) in width and 100 kg (220 lb.) in weight.

As we have suggested, human exploitation and the effects of climate change do not explain the entirety of the rate of species extinctions. Nonetheless, the observed acceleration of species loss suggests a large anthropogenic role in these extinction dynamics. Both the shark and jellyfish examples illustrate how complicated trophic interactions can amplify human impacts. The species we drive extinct play key roles in functioning ecosystems, and their removal can lead to effects that can spiral, amplify, and cause a permanent change to food webs and ecosystems. The jellyfish example, however, suggests another avenue of human impact. So far, we have focused on the species that humans *remove*, but what about the species that humans *add* to ecosystems, as in the case of the invasive *P. punctata* in the Gulf of Mexico?

15.5 Invasive Species

One of the ecological signatures of the Great Acceleration (Figure 1.2; Steffen et al. 2015) is the huge increase in the number of invasive species found worldwide. Invasive species influence ecological dynamics in myriad ways, as documented in many of the previous chapters. To refresh our memories, an invasive species (1) is present in a location where it did not evolve, (2) was moved there via human actions, and (3) has spread in its new location, producing negative impacts on the environment, human health, or economic systems (Lockwood et al. 2013). The more general term *non-native* (synonymous with *alien* and *nonindigenous*) denotes species that meet all but the last of these criteria. Thus, invasive species are the subset of all non-native species that have large impacts or aggressive spread.

Global Distribution of Non-native Species

To understand the role invasive species play in the Anthropocene, we must focus on the global distribution of non-native species. Regardless of whether one is

Figure 15.17 Ⓐ The cumulative number of plants from outside Europe that have been introduced into Europe has increased sharply since 1500. The dots represent the data, and the accompanying line is an exponential curve statistically fitted to the data (Lambdon et al. 2008). Ⓑ The numbers of newly introduced non-native species in Europe across various time periods reveal that all taxa show an upward trend through time (Hulme 2009).

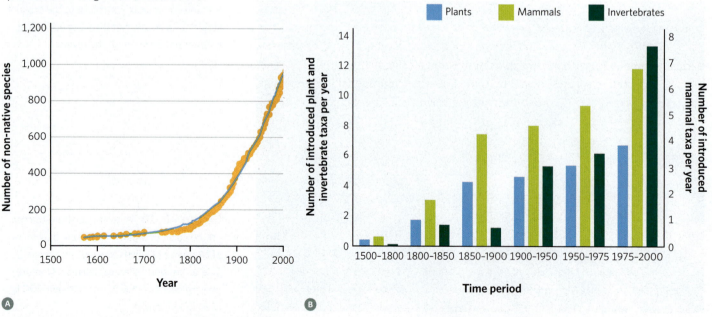

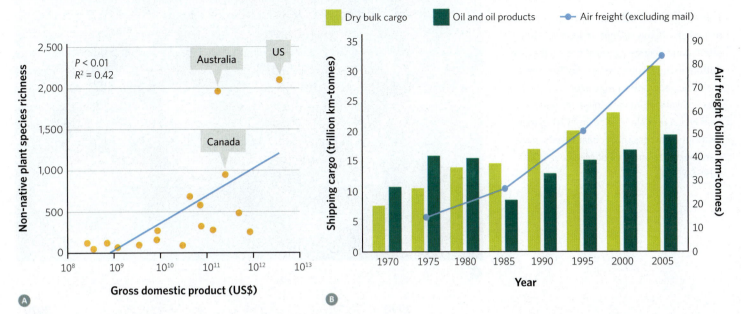

considering invasive or non-native species, the key to their role in the Anthropocene is the influence of humans in their global distribution. As illustrated in Chapter 11, very few species naturally have large geographic distributions, and most species exist in small populations within specific habitats and ecosystems. As mentioned previously, niche constraints and dispersal limitations help to keep populations small and local. Human modes of transportation, therefore, allow non-native and invasive species to overcome these natural geographic limitations.

The numbers of non-native and invasive species often show accelerating rates of accumulation as we approach the present (**Figure 15.17**). On a country-by-country basis, there is a strong correlation between the number of non-native species and gross domestic product (GDP) (**Figure 15.18A**). The GDP is an economic measure of the value of all goods and services produced by a country over a specific time period. Not surprisingly, GDP is closely linked to the volume of traded goods and services, and these trades are an efficient way for species to hitchhike from one location to another (Lockwood et al. 2013). Trade has increased dramatically since 1970, with air transportation showing astonishing growth (**Figure 15.18B**). Non-native species moved by trade are either intentionally transported (e.g., garden plants, pets) or are unintentional stowaways or contaminants in the traded goods (e.g., extra seeds in grain, marine organisms in ballast water). In other words, it seems likely that we are transporting and establishing non-native species around the planet very rapidly as a result of our fast-paced global economy. This reorganization of biodiversity via globalization is one of the clearest examples of human influence on the environment in the Anthropocene.

As described in Chapter 11, the establishment of non-native and invasive species can lead to biodiversity homogenization, which is itself a signature of the Anthropocene. But when invasive species have ecological impacts, they can do more than just alter species counts across locations, as in the jellyfish example. There are a growing number of situations in which a single invasive species creates a very large ecological footprint.

Figure 15.18 Ⓐ A positive linear relationship is evident between country-level gross domestic product (GDP, in US dollars) and non-native plant species richness for continental regions of the world. The p-value is less than 0.05, which indicates a statistically significant relationship between plant richness and GDP. The R^2 value indicates the proportion of the variance in the data that is explained by the regression line. Ⓑ The volume of global shipping cargo (dry bulk and oil) and air freight trended dramatically upward from 1970 to 2005. Note the three orders of magnitude difference in scales between the shipping-cargo axis and the air-freight axis (Hulme 2009).

Figure 15.19 Ⓐ Non-native populations of lionfish (*Pterois volitans* and *P. miles*) are often found in large aggregations on the sea bottom in the Caribbean Sea. Ⓑ The maps show the geographic range expansion of invasive *P. volitans* from their initial introduction near Ft. Lauderdale, Florida, in 1990 to their distribution across most coastal reefs in the Caribbean Sea and Gulf of Mexico in 2020 (USGS 2020).

Ⓐ

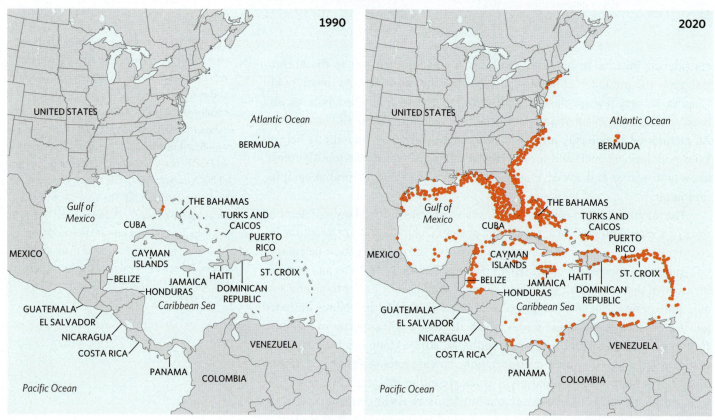

Ⓑ

Impacts of Invasive Species

Take for example the invasion of Pacific lionfish (*Pterois volitans* and *P. miles*) in the Caribbean Sea. This invasion began in the late 1980s after aquarium enthusiasts, pet shop owners, or a hurricane released a few unwanted pet lionfish into the coastal waters off of southern Florida (**Figure 15.19A**). These beautiful species are native to the Indo-Pacific. Their ornate and colorful architecture is a prime reason for their popularity among aquarium owners. Pacific lionfish also have an amazing set of evolutionary adaptations, including highly venomous spines, an extremely broad diet, rapid reproduction, and a planktonic larval stage that

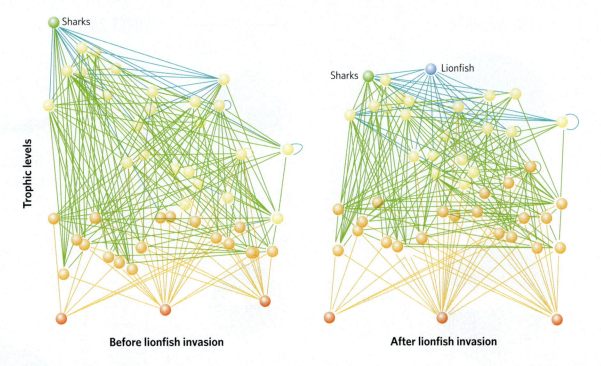

Trophic levels

Sharks

Before lionfish invasion

Sharks Lionfish

After lionfish invasion

disperses their offspring widely. With a reproductive rate of two million eggs per year from a single female, it is not surprising that a few introduced lionfish individuals had a rapid population increase and spread to coral reefs throughout the Gulf of Mexico, the Caribbean Sea, and portions of the western Atlantic Ocean (**Figure 15.19B**).

Unfortunately, the invasive lionfish populations are triggering significant trophic alterations by disrupting coral reef food webs (**Figure 15.20**; Arias-González et al. 2011). Lionfish are also competing for resources (food and space) with economically important species, such as snapper (family Lutjanidae) and grouper (subfamily Epinephelinae) (Morris 2012). Additionally, the lionfish invasion may hamper the population recovery of marine fish of conservation concern, such as the Nassau and Warsaw groupers (*Epinephelus striatus* and *E. nigritus*, respectively). These grouper populations have been overfished and are not likely to recover with the additional predation pressure imposed by lionfish.

The greatest concern with the lionfish invasion is how their predatory nature interacts with other anthropogenic stressors that affect reef ecosystems in the Anthropocene (Albins and Hixon 2013). Coral reefs in the northwestern Atlantic are already burdened from overfishing, climate change, and pollution. The additional stress of a new predatory species is likely to accelerate the degradation of entire coral reef ecosystems in unexpected ways (**Figure 15.21**). On a brighter note, lionfish are appearing on restaurant menus and dinner plates around the Caribbean and Florida, thereby increasing fishing pressure on this invasive species. The unknown future effect of human fishing pressure on lionfish is indicated by a question mark in Figure 15.21 and should be the focus of control efforts.

As is clear from the lionfish example, the types of impacts a single invasive species can produce are extremely varied. Unfortunately, most ecosystems receive not one but many invasive species that produce both individual and synergistic impacts. For example, the Colorado River has lost a large percentage of its native fish species as a result of invasive species and other anthropogenic effects,

Figure 15.20 This figure represents the food web of the Alacranes Reef ecosystem, Mexico, before and after its invasion by lionfish. The different-colored nodes represent species at different trophic levels: red nodes represent producer species at the basal level; orange nodes represent intermediate species; and yellow nodes represent primary predators at the top. Note that sharks were the top predators before the lionfish invasion; after the invasion, lionfish shared that role. Diet information for the reef organisms is based on published data (Arias-González et al. 2011).

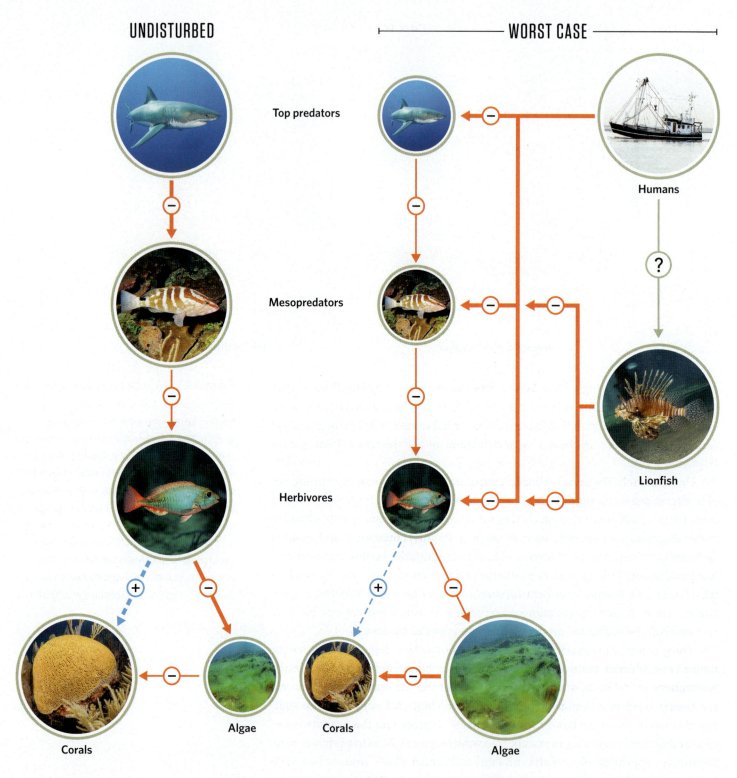

UNDISTURBED

WORST CASE

Top predators

Mesopredators

Herbivores

Humans

?

Lionfish

Corals

Algae

Corals

Algae

Figure 15.21 If we compare conceptual figures of an undisturbed Caribbean coral-reef food web and a food web with anthropogenic overfishing and invasive lionfish, we can see that the system with overfishing and lionfish moves away from corals and toward algae, with more mesopredators and fewer top predators and herbivorous fish. The size of each kind of organism represents relative abundance between the two scenarios, and the thickness of arrows indicates the relative interaction strength between organisms. Solid arrows are predation (including fishing) and competition. Dashed arrows are indirect positive effects of herbivores on reef-building corals (Albins and Hixon 2013).

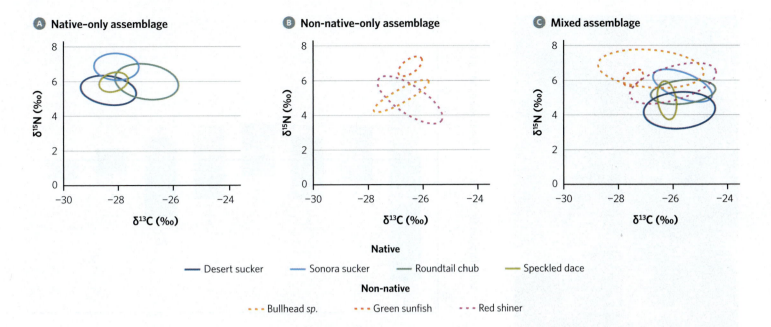

A Native-only assemblage **B** Non-native-only assemblage **C** Mixed assemblage

Native

— Desert sucker   — Sonora sucker   — Roundtail chub   — Speckled dace

Non-native

- - - Bullhead *sp.*   - - - Green sunfish   - - - Red shiner

including substantial water diversion to urban and agricultural areas. The river drains a significant portion of the southwestern United States and has formed major geological features, including the Grand Canyon. The Colorado River system was historically home to 49 species of native fishes, 80% of which are currently extinct or threatened with extinction. One of the major causes for these declines is the establishment of more than 40 non-native fish species in the watershed.

In their investigation into food web changes in the Bill Williams River (a small tributary of the Colorado River in Arizona), scientists Jane Rogosch and Julian Olden (2020) used stable-isotope relationships to examine trophic alterations within the native fish community in the presence of non-native fish species. The researchers were able to locate a section of the river where only native fishes were present, a section where only non-native fishes were present, and a section where both groups were present together. They examined the carbon-isotope and nitrogen-isotope ratios from the fish tissues in all three river sections and constructed isotopic niche "maps" for each (**Figure 15.22**).

Based on ecological theory, they predicted that both the non-native and native fishes would shift their niche space at the location where they lived together (in a mixed assemblage). Instead, they found an asymmetrical interaction; the native fishes altered their niche space in the presence of the non-native fishes, but the non-native species shifted very little in the presence of the natives and even increased their niche space in some cases. Such asymmetric competitive interactions may partly explain how non-native species come to dominate fish assemblages (Rogosch and Olden 2020).

From these aquatic examples, it is relatively easy to see shifts in species richness, niche space, and food web structure, but many terrestrial examples exist as well. Invasive species can also alter ecosystem processes. One example with which you are already familiar is invasion by ecosystem engineers, such as the North American beaver (*Castor canadensis*) in Argentina and Chile (Chapter 6). The invasive beavers fundamentally altered both water and nutrient cycling in vast stretches of both countries.

Figure 15.22 Researchers studying food web changes in the Bill Williams River in Arizona identified isotopic niche spaces occupied by **A** native-only (solid lines), **B** non-native–only (dashed lines), and **C** mixed native and non-native assemblages of fish species. Note that the native species showed altered niche spaces for both carbon and nitrogen when in the mixed assemblage, but the non-native species did not. Isotopic niche space is indicated using an ellipse area corrected for sample size. Isotopic content is expressed in δ values, which represent the relative difference, in parts per thousand (‰), between sample and conventional standards for ratios of carbon ($^{13}C/^{12}C$) and nitrogen ($^{15}N/^{14}N$) isotopes (Rogosch and Olden 2020).

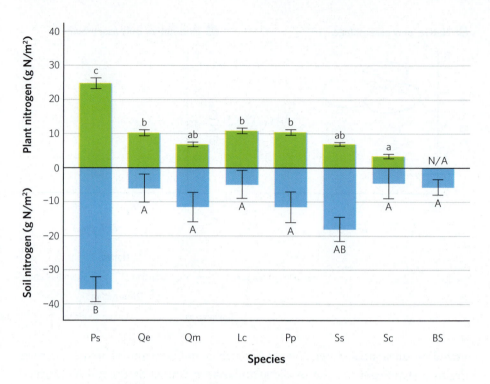

Figure 15.23 This graph compares plant nitrogen allocation in the invasive eastern white pine (*Pinus strobus*) versus six other tree species and bare soil over a number of growing seasons. *P. strobus* stores significantly more nitrogen in its tissues (upper bars) and removes significantly more nitrogen from the soil (lower bars) than any of the other species. Bars labeled with different letters (a, b, c, A, B) were found to be statistically different from one another (Laungani and Knops 2009). In order along the bottom: Ps = *Pinus strobus*; Qe = *Quercus ellipsoidalis*; Qm = *Q. macrocarpa*; Lc = *Lespedeza capitata*; Pp = *Poa pratensis*; Ss = *Schizachyrium scoparium*; Sc = *Solidago canadensis*; BS = bare soil.

Perhaps surprisingly, invasive plants can have similar impacts. For example, eastern white pines (*Pinus strobus*), which are an invasive species in multiple locations around the world, hold more nitrogen in their tissues and particularly in their photosynthesizing needles than deciduous trees. The needles stay on the trees longer and decompose more slowly than deciduous leaves. This difference allows the pines to take nitrogen out of the soil in nitrogen-limited areas and hold it in their tissues, thereby altering the productivity and nitrogen cycles in invaded ecosystems (**Figure 15.23**; Laungani and Knops 2009). Invasion by eastern white pine has had significant impacts in Hawaii and portions of Central Europe where the species is particularly invasive.

Nitrogen-fixing plants can also invade low-nutrient areas and affect the system by adding nitrogen, which may lead to a very different climax community. Humans have spread a number of these species—such as clovers and alfalfas—around the world because we use them to increase nitrogen in our agricultural systems. Many of these species are also used to stabilize low-nutrient soils that are prone to erosion. Examples such as autumn olive (*Elaeagnus umbellata*), black locust (*Robinia pseudoacacia*), and yellow bush lupine (*Lupinus arboreus*) are invasive on multiple continents, where each has a demonstrated negative effect on the biomass of native plant species. Part of the reason for this negative effect on native species is that these nitrogen-fixing plants facilitate other invasives. By adding nutrients to nutrient-poor areas, these species improve conditions for a range of non-native species that thrive in high-nutrient environments.

Finally, plants have the ability to alter a region's natural disturbance cycles. A number of invasive grasses are both more flammable and more able to resprout or regrow after a fire than native vegetation. As a result, invasive grasses have altered natural fire cycles around the world. Carla D'Antonio and Peter Vitousek (1992) first highlighted this impact and investigated the effects of invasive grasses on fire regimes. Since then, a number of studies have identified the particular effects of cheatgrass (*Bromus tectorum*), which has invaded much of the western United States and Canada between the Rocky Mountains and the Sierra Nevada. Bethany Bradley and colleagues (2018) looked at the distribution of cheatgrass and fire frequency and found that even very low amounts of cheatgrass (i.e., as low as 1% cover) significantly increased fire frequency (**Figure 15.24**). This change in the fire regime is dramatic, as it alters nutrient cycling and water availability, as well as the composition of plant and animal communities.

Percentage of location burned (y-axis: 0, 2, 4, 6, 8, 10, 12, 14, 16, 18)

a   a   b   b   b   b

Cheatgrass percentage cover (x-axis: 0, <1, 1–5, 5–10, 10–15, >15)

The Future of Invasive Species

Taken as a whole, much of the research on invasive species suggests that the habitats of the future will be different from the habitats of today as a result of these invaders spreading around the globe. Yet the future of invasions is not entirely bleak. Research by Hanno Seebens and colleagues (2017) suggests that, at least for some taxonomic groups, the rate of non-native species introductions globally seems to be slowing (**Figure 15.25**). The establishment of new non-native mammals and fishes worldwide hit a peak around the 1950s, and invasions by both types of taxa have dramatically declined since then. These declines seem to result from a reduction in the deliberate establishment of mammals as game or for fur production, and a reduction in deliberate fish-species releases for aquaculture and fisheries (Seebens et al. 2017). The rate of establishment of non-native vascular plants has also declined in this period.

Unfortunately, the rest of the news is not so positive. As indicated in Figure 15.25, for birds, insects, and mollusks, there is no evidence of a slowdown. Instead, all still show an increase in newly introduced species. These global tallies suggest that environmental legislation and international agreements to reduce invasive species introductions have not been very effective (Seebens et al. 2017). The Anthropocene will likely be characterized both evolutionarily and ecologically by this massive exchange of organisms.

We could go into many other aspects of species invasion, but one worth mentioning here is that invasive species do not affect every part of the globe equally. Regan Early and colleagues (2016) combined spatial data on airport and seaport activity with numbers of total trade imports to produce a map showing where the future threat of invasive species is highest (**Figure 15.26**). It is important to notice that the threat is concentrated in locations with high human density, wealth, and economic activity.

This research also suggests that, as a country's economy develops over time and becomes more connected through global trade, the number of invasive species arriving in that country increases. Early and colleagues examined the political and social capacity of different countries to respond to and address these invasive species issues (**Figure 15.27**). They distinguished between proactive capacity (prevention of future introductions) and reactive capacity (understanding and managing invasions). Not surprisingly, the wealthiest countries have the best chance of coping with invasive species issues in the future, while the poorest ones may be left to deal with the negative ecological and economic consequences.

Figure 15.24 Cheatgrass (*Bromus tectorum*) has invaded much of the western United States and Canada, affecting natural fire regimes. This graph plots the percentage of habitat that burned at a location (with standard deviation error bars) versus the amount of cheatgrass present at the location. Notably, fire frequency increased even when cheatgrass accounted for only 1%–5% of the vegetation. The letters at the top of the figure (a, b) indicate which treatments are statistically different, so those with the same letter are indistinguishable statistically (Bradley et al. 2018).

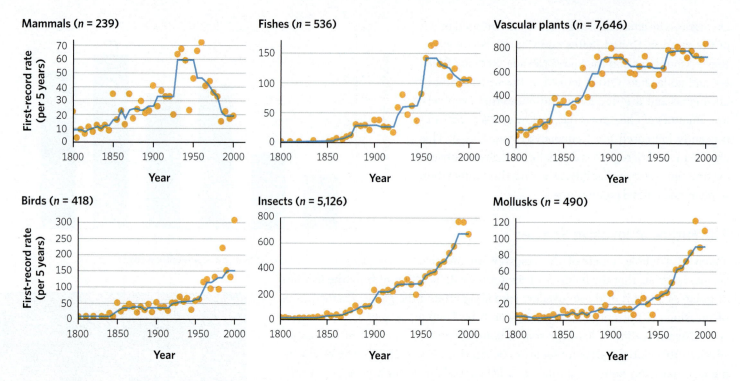

Figure 15.25 As these graphs show, the introduction of non-native species around the globe seems to have slowed for some taxa (e.g., mammals, fishes, and vascular plants) in recent decades. Yet for other taxa, there is no sign of a slowdown. The "first-record rate" is equal to the number of scientific publications within a year that reported a newly introduced species in a location (Seebens et al. 2017).

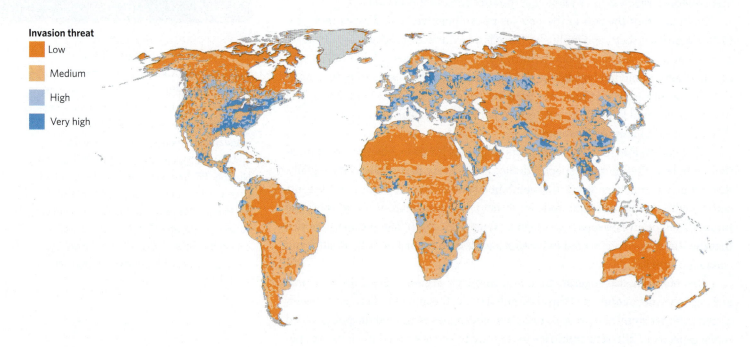

Figure 15.26 Researchers assessed the global species-invasion threat for the twenty-first century, based on airport and seaport capacity as well as total shipping imports from 2000 to 2009, and mapped the results. Note the uneven distribution of threat levels across the globe, with the largest concern existing in the most developed nations that have the largest economies and the most global travel (Early et al. 2016).

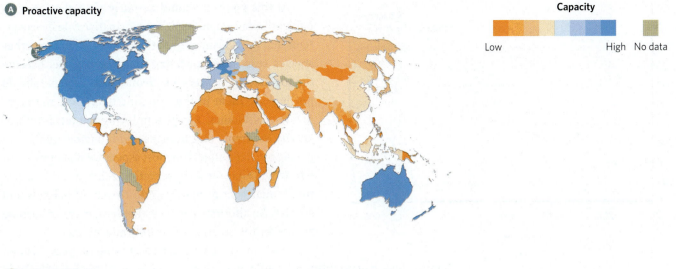

A Proactive capacity

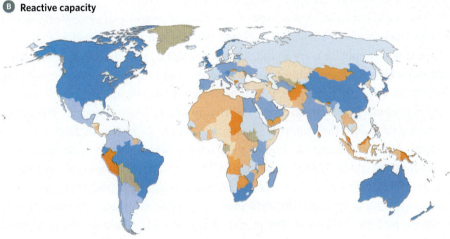

B Reactive capacity

Figure 15.27 Researchers estimated the political and social capacity of countries to respond to the threat of emerging species invasions and mapped the results (Early et al. 2016). **A** Proactive capacity measures the ability of a country to prevent the introduction of invasive species. **B** Reactive capacity estimates a country's internal knowledge regarding invasive species and the degree to which the nation has an action plan to manage invasions.

15.6 Anthromes and the Ecology of a "Used" World

In Chapter 3, we presented a classical view of how to divide the biosphere into standard categories called biomes. Biomes were defined by dominant vegetation, climate (particularly rainfall and temperature), and soils. We made similar divisions within aquatic systems where depth, oxygen availability, and salinity defined aquatic biological zones. This standard conceptualization of the biosphere assumes that only nonhuman factors determine the locations of plants, animals, and microbes on the planet. We included no consideration of the influence that biomes have on human evolution, societal complexity, or ecology. And we assumed no connection between the welfare and activity of human societies and the species that are found in each biome.

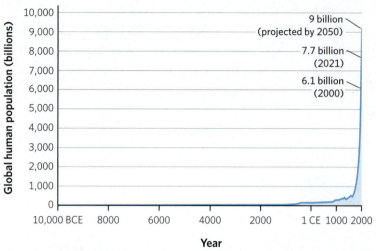

Figure 15.28 This graph of estimated human population growth since 10,000 BCE shows the incredibly sharp increase in the modern era (UNEP 2011).

At this point, it should be fairly clear that humans currently play a significant transformative role in ecosystems and across landscapes. Of course, so do many other species (e.g., nitrogen-fixing plants and algae, oceanic phytoplankton, beavers, elephants), yet we generally do not include them in the scientific definition and organization of biomes. But by employing the term *Anthropocene*, we imply that humans and human activities are having outsized effects on most aspects of planetary ecology. In addition, most conservation-science texts suggest that humans act primarily as destroyers of habitats and species. An alternative view is that humans have become agents for the creation of entirely new biomes.

Over most of the Earth's 4.6-billion-year history, humans have had no substantial influence on the structure and function of biomes. We either did not exist, as was the case for the vast majority of geological time, or our numbers were so small and localized that we had little appreciable effect in shaping where species were found. There is evidence that the ancestors of modern humans were formidable hunters whose animal harvests led to extinctions, but effects were still mostly local. Yet the evidence favoring the designation of the Anthropocene epoch strongly indicates that this is no longer the case. The global human population, for example, was approximately 7.7 billion in 2021, which is several orders of magnitude higher than what it was for most of the Holocene (which began about 12,000 years ago) (**Figure 15.28**). Humans now appropriate about one-third of the Earth's net primary productivity and move more soil and convert more nitrogen to a biologically available form than all other natural sources combined (Ellis and Ramankutty 2008). Once we embrace the concept of the Anthropocene, can scientists really continue to discuss the idea of people-free biomes? If not, what happens to the way we define biomes when human influence is explicitly considered in our classification scheme?

Erle Ellis and Navin Ramankutty (2008) have explored this question in detail, using modern and historical constructions of land use and land cover worldwide. These authors used highly detailed large-scale maps of the historical dominant vegetation found across terrestrial systems. On top of these, they overlaid information about human population density, the presence of settlements (large and small), and the extent of agriculture and rangeland in order to construct new categories of human-influenced land types they call **anthromes** (i.e., anthropogenic biomes).

Ellis and Ramankutty divided all the terrestrial landmass of the planet into small geographic "cells," each equivalent to about 86 km² (33 sq. mi.). In their first pass, they categorized all the cells that had no human settlement, agriculture, or rangeland into a division called "wildlands." They then subdivided the wildland cells into three categories: "wild forests," "sparse trees," and "barren" (the latter having no tree growth, as, for example, in the tundra or on ice sheets). They grouped the remaining, non-wildland grid cells into five divisions (dense settlements, villages, croplands, rangelands, and forested) comprising 18 anthromes and mapped them according to their geographic extent in the year 2000 (**Figure 15.29**).

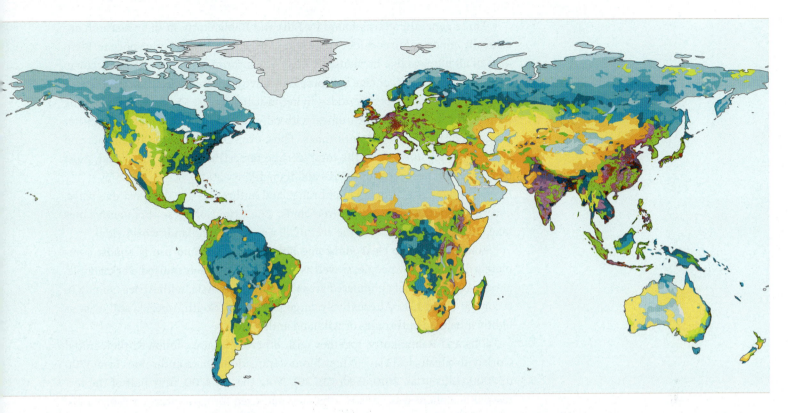

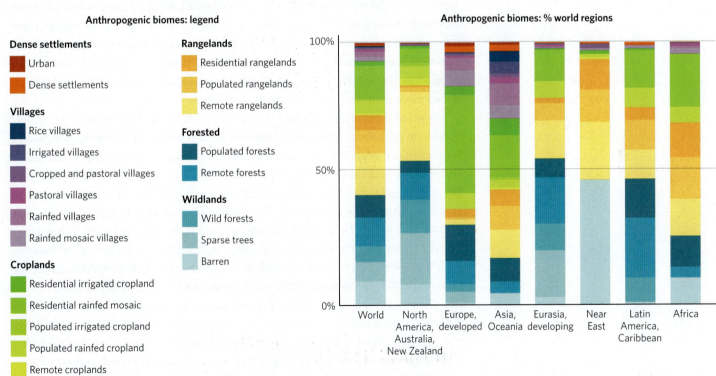

Anthropogenic biomes: legend

Dense settlements
- Urban
- Dense settlements

Villages
- Rice villages
- Irrigated villages
- Cropped and pastoral villages
- Pastoral villages
- Rainfed villages
- Rainfed mosaic villages

Croplands
- Residential irrigated cropland
- Residential rainfed mosaic
- Populated irrigated cropland
- Populated rainfed cropland
- Remote croplands

Rangelands
- Residential rangelands
- Populated rangelands
- Remote rangelands

Forested
- Populated forests
- Remote forests

Wildlands
- Wild forests
- Sparse trees
- Barren

Figure 15.29 Ellis and Ramankutty (2008) developed a complex map of anthromes—human-influenced land types—around the world. The bars on the lower right indicate the relative percentage of landmass occupied by each anthrome in different geographic regions of the world.

Three general patterns found by Ellis and Ramankutty in their research are critically important to consider when studying ecology in the Anthropocene. First, 40% of all humans live in "dense settlement" anthromes, 90% of which occur in urban areas. This percentage has increased since they published their paper, with more than half of all people today living in urban areas. Most of the remaining humans live in "village" anthromes, with only scattered numbers living in either "rangeland" or "cropland" anthromes. Thus, even though dense settlements and villages occupy only 7% of the total ice-free terrestrial area on the planet, most people live in these locations (Ellis and Ramankutty 2008).

Second, village, cropland, and rangeland anthromes are the most extensive, covering about 20% of the ice-free lands. The extent of these three anthromes taken together is larger than the area of all remaining wildlands worldwide.

Finally, there are clear anthrome hotspots, where land use is almost completely dominated by people, as well as areas that are largely unused. For example, almost 40% of densely populated areas are located in Asia, with Africa coming in second at about 13%. Alternatively, large portions of South America still exist as lightly inhabited anthromes or wildland anthromes.

Ellis and Ramankutty, together with other colleagues, followed their initial work with a historical view of how biomes transitioned into anthromes from 1700 to 2000 (Ellis et al. 2010; **Figure 15.30**). Note that by 1700, fully half of the terrestrial biosphere was already inhabited and used for agriculture or rangelands. However, by 2000 the vast majority of terrestrial lands had been converted from wildland or seminatural forests into anthromes. The rate at which biomes were converted into anthromes over this time period was heterogeneous (Figure 15.30). Boreal forests, deserts, mixed woodlands, and tundra went largely unchanged from 1700 to 2000. However, grasslands, savannas, and scrubland biomes experienced more than 80% conversion into anthromes.

A striking pattern seen in both Figure 15.29 and Figure 15.30 is the extreme heterogeneity of the anthromes, especially when compared to the biomes they replaced. As human use intensifies, terrestrial landmasses become an incredibly variable mix of cropland, seminatural forest, villages, and urban areas. The landscapes that people create tend to be an amalgam of anthrome types and often include small patches of unused lands (not devoted to urban use or agriculture). In fact, as of the year 2000, more than 60% of the remaining unused lands were embedded in a dense matrix of settlements, villages, croplands, rangelands, or seminatural anthromes. Only 40% of unused lands worldwide would still be classified as "wildlands" under Ellis and colleagues' scheme (Ellis and Ramankutty 2008; Ellis et al. 2010).

Clearly, there is a lot of information to unpack in the maps and analyses of Ellis and colleagues and other scientists who are doing this kind of research. Yet the loss of wildlands and the heterogeneous nature of unused lands embedded within human-impacted landscapes has been a topic of major interest to conservation biologists for decades. This pattern is often discussed in the context of deforestation. However, Figures 15.29 and 15.30 indicate that the loss of grasslands and savannas has been even more dramatic than the loss of forests.

Why, though, do these conversions matter? They matter because preventing species extinctions requires preserving or restoring natural habitat. When these habitats are converted to anthromes for human-dominated uses, they often

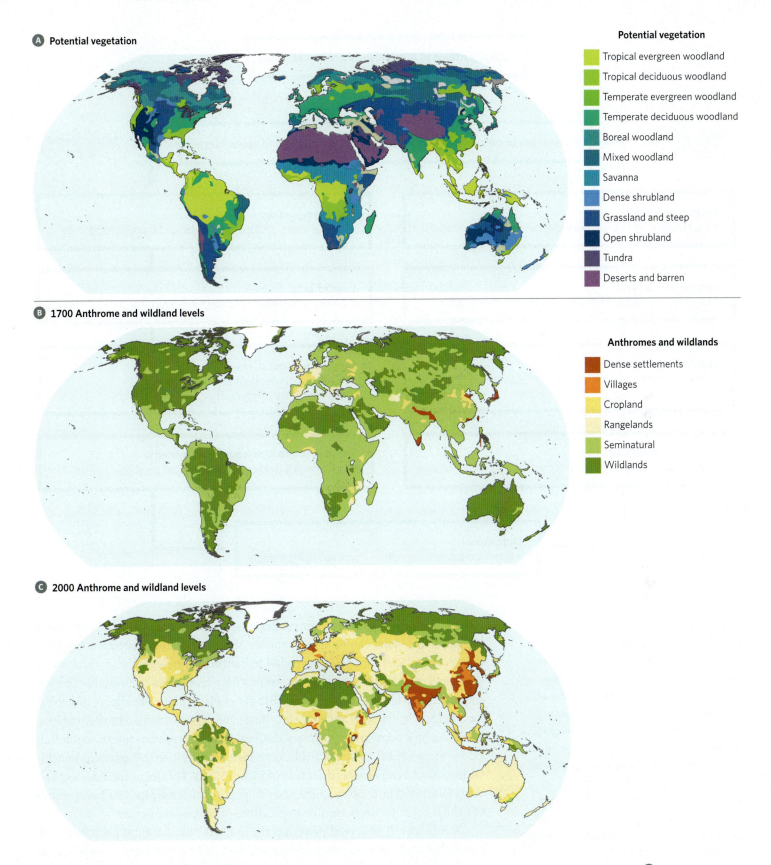

A Potential vegetation

Potential vegetation

- Tropical evergreen woodland
- Tropical deciduous woodland
- Temperate evergreen woodland
- Temperate deciduous woodland
- Boreal woodland
- Mixed woodland
- Savanna
- Dense shrubland
- Grassland and steep
- Open shrubland
- Tundra
- Deserts and barren

B 1700 Anthrome and wildland levels

Anthromes and wildlands

- Dense settlements
- Villages
- Cropland
- Rangelands
- Seminatural
- Wildlands

C 2000 Anthrome and wildland levels

Figure 15.30 The conversion of natural biomes to anthromes through time, as determined by Ellis and colleagues (2010). **A** The "potential vegetation" map indicates what naturally occurring vegetation patterns would look like around the world in the absence of human interference. **B** This map shows the extent of conversion to anthromes by the year 1700. **C** The third map indicates the extent of both natural vegetation and anthromes as of the year 2000. The shift away from shades of green indicates the vast increase of human influence on vegetation and the accelerated creation of anthromes.

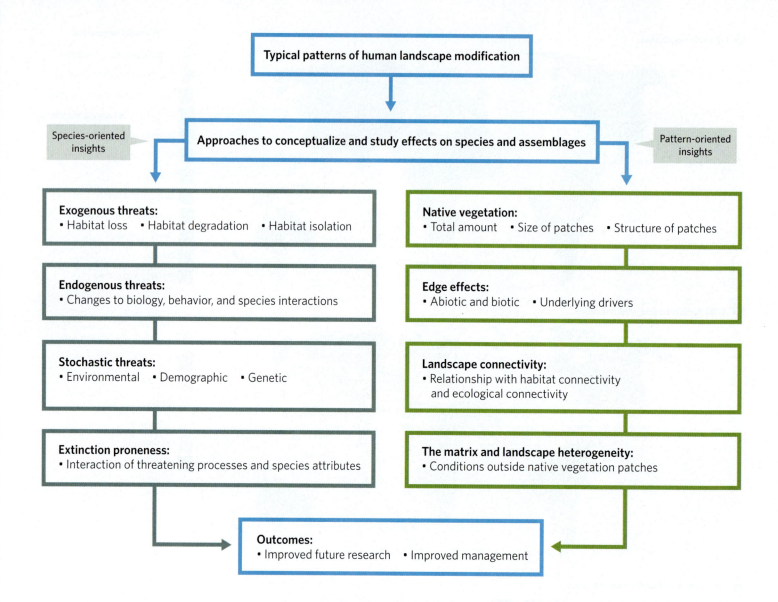

Figure 15.31 Ecologists use this type of conceptual framework to study the effects of landscape modification and habitat fragmentation on ecosystems. Because ecologists know more about certain well-studied species than they do about whole communities or ecosystems, these research efforts often focus on the effects on particular species (in which impacts are more easily detected), following the procedure on the left. Alternatively (or concurrently), ecologists can look at larger, pattern-oriented insights (landscape-scale patterns), following the procedure on the right (Fischer and Lindenmayer 2007).

lose their ability to support entire suites of species. Understanding the effects of habitat conversions can be more complex than this simple statement suggests, though. **Figure 15.31** shows a conceptual framework for how ecologists think about landscape modification. Because they rarely have landscape-, ecosystem-, or even community-level information, ecologists generally explore the effects of habitat modification by tracking and modeling how single species respond to habitat loss, including how population processes combine to drive a species to extinction (left side of Figure 15.31). When possible, ecologists also explore how patterns in habitat loss affect biodiversity (right side of Figure 15.31). This focus includes explicit consideration of how easily species can move across a landscape and how patches of habitat combine to create metapopulations or metacommunities.

A consistently observed pattern is the gradual loss of natural habitat in such a way that it initially becomes variegated, then patchy, and finally exists only as small, leftover habitat fragments. Such small patches of habitat offer only marginal fitness benefits to the remaining endemic or dependent species (**Figure 15.32**). This progression of habitat loss is clearly one reason anthromes are highly heterogeneous. What is not clear from Figure 15.32 is the fate of the land once it is

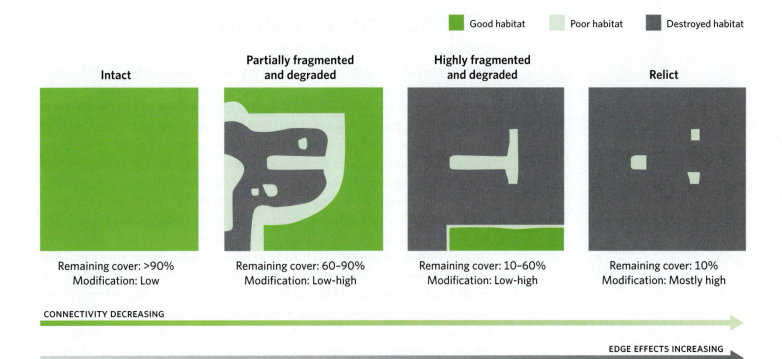

Intact	Partially fragmented and degraded	Highly fragmented and degraded	Relict

Remaining cover: >90%
Modification: Low

Remaining cover: 60–90%
Modification: Low-high

Remaining cover: 10–60%
Modification: Low-high

Remaining cover: 10%
Modification: Mostly high

CONNECTIVITY DECREASING

EDGE EFFECTS INCREASING

converted from its more natural state. In the conceptualization shown, the lost habitat is just . . . lost. It is depicted as a gray nothingness. However, what Ellis and others have shown is that these "gray spaces" are converted into human-dominated uses.

The degree to which the modified land is now unavailable to species native to the (former) wildland habitat varies tremendously. Some species truly cannot survive in habitats that have been even mildly transformed by human uses, in which case Figure 15.32 is likely a good representation of their plight. Some species, though, can exist fairly well in human-dominated habitat, and some actually thrive in such landscapes (e.g., crows, rats, cockroaches, and dandelions).

The issue then becomes one of understanding the value of wildland habitats and the various species they support. Can anthromes be reconfigured to accommodate more than a few human-tolerant species? This question has become a central theme in applied ecological research and restoration ecology and one that interfaces strongly with other fields, such as sustainable urban design.

Perhaps the bigger challenge is to recognize that we now inhabit a permanently transformed anthropogenic Earth. In other words, the human footprint has a massive influence on the natural ecosystems around us. Ellis and colleagues in particular suggest an intricate and tight association between human actions and natural ecosystems, and ecologists must concentrate on studying these links. Understanding how we humans affect ecosystems can help us develop sustainable solutions for species and ecosystem protection. To better understand the connections between humans and natural systems, Ellis and his colleagues suggest that ecological science should focus on the local and study the ecological processes where humans live (i.e., in urban areas), instead of going off "into the wild" to study the "natural" world. Similarly, Jason Kaye and his colleagues (2006) write, "Although urban ecosystems are literally 'in our backyard,' they are a frontier for ecology."

Figure 15.32 This conceptual model shows the process of landscape conversion from a fully intact habitat to one where the location is dominated by an alternative habitat. Pale green coloring indicates a partly degraded version of the original habitat, and gray indicates destroyed habitat. A *relict* habitat is a disconnected remaining piece of a previous habitat (Fischer and Lindenmayer 2007; McIntyre and Hobbs 1999).

15.7 Science for Solutions

At the start of this book, we introduced you to a famous quotation from Aldo Leopold in which he suggested that one of the "penalties" of gaining an ecological education is the inability to ever close your eyes again to the ecological wounds of the world (Leopold 1970). He also suggested that ecologists face a choice between ignoring these wounds or becoming a doctor to an unwilling patient. As he put it, "Our job [as ecologists] is to sharpen our tools and make them cut the right way" (Leopold 1992).

Perhaps at no other time in history have the "tools" of ecology been more important than today for finding solutions to the environmental crises we face as a global community. Ecologists can either sharpen their ecological tools to see more clearly how and why ecological systems are transformed by pervasive human impacts, or they can sharpen them to build the science to support solutions. Most of this chapter has described how researchers have used their tools for the former, but here we want to show how they can employ them to achieve the latter. When using ecological science to ensure persistence of natural ecosystems, ecologists must explicitly consider the connection between social and economic forces and natural ecosystems. This integration between the human and the natural world, often referred to as socioecological systems, explicitly recognizes that biodiversity is affected by humans and, in turn, affects human well-being. Tying the two together, especially in the face of global-scale crises, such as climate change, should result in improved futures for all parties involved.

The growing recognition that the effects of climate change are upon us has highlighted how natural ecosystems buffer people from the harshest of impacts, and how ecological communities can play a substantial role in reducing the rate of climate warming. Perhaps the most obvious connection is the role of ecosystems in capturing carbon dioxide from the atmosphere and holding it for long periods of time. For example, as we have mentioned, mangrove forests, and particularly their soils, are huge carbon sinks but cover a tiny fraction of the Earth's surface. Preventing the destruction of mangrove forests can, therefore, be a key factor in efforts to reduce atmospheric carbon. The same principle applies to conservation of several carbon-sink ecosystems, such as tropical rainforests, seagrass meadows, salt marshes, and peatlands. Detailed ecological knowledge of how these systems fix carbon and release it can help prioritize habitat conservation efforts.

We can take this principle one step further, though, and consider what ecological science can do to increase the ability of natural ecosystems to store carbon. These interventions can include purposeful establishment, or reestablishment, of species of trees, algae, or grasses that fix more carbon than they respire. It can also include managing habitats so that the aboveground species composition we create fosters belowground microscopic assemblages that sequester relatively large amounts of soil carbon. Knowledge of successional dynamics, along with carbon cycling, may help forecast how restored or recovering communities can serve as carbon sinks.

For example, tropical forests (both rainforests and dry forests) that are recovering from extensive logging accounted for around 30% of terrestrial lands in Central and South America in 2008 (Chazdon et al. 2016). These *second-growth forests* (forests regrowing after being harvested) represent mid-successional ecological communities that are 1–60 years old relative to their last deforestation

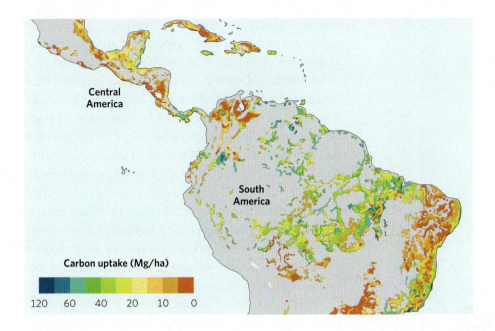

Carbon uptake (Mg/ha)

| 120 | 60 | 40 | 20 | 10 | 0 |

Figure 15.33 Chazdon and colleagues predicted the total potential sequestered carbon (in megagrams per hectare, Mg/ha) for all young and middle-aged (1–60 years) second-growth tropical forests in Central and South America for the years 2008 to 2048. The gray areas are areas with no data because they are either above 1 km (3,281 ft.) in elevation or they are savannas, rivers, lakes, old-growth forests, or urban areas (Chazdon et al. 2016).

event. Second-growth forests are full of tree species that fix more carbon than they transpire, and we can expect these forests to sequester more carbon as the size of their trees increases over time. Robin Chazdon and her colleagues (2016) combined information on the location of second-growth forests in Central and South America in 2008 with carbon-flux equations for these forests and predictions of how the forests would change in composition and growth rates to calculate how much carbon would be sequestered by second-growth forests in this region over the ensuing 40 years. Their results were incredibly encouraging (**Figure 15.33**). They predicted that, between 2008 and 2048, second-growth forests would sequester the same amount of carbon that was released by fossil fuel and industrial emissions in Central and South America and the Caribbean from 1993 to 2014 (Chazdon et al. 2016). Given the right ecological conditions, these forests naturally regenerate, but they can also be "pushed along" through intentional tree planting or restoration of environmental conditions that are conducive to tree growth.

In addition to carbon sequestration, functioning ecosystems offer other services that improve human well-being. For example, coastal wetlands reduce ocean wave energy and the depth of flooding during hurricanes. Siddharth Narayan and colleagues (2017) mapped the location of coastal wetlands across 12 US states from North Carolina to Maine and calculated that coastal wetlands substantially reduced flooding heights from Hurricane Sandy in 2012 and helped property owners avoid $625 million in damage. These positive effects of coastal wetlands did not come just from pristine habitats but were especially noticeable in areas with highly urbanized coastlines, such as New Jersey and New York.

Ecological systems, such as coastal wetlands (**Figure 15.34**), that provide protection services from climate

Figure 15.34 Coastal wetlands, such as this one in Portugal (the Castro Marim and Vila Real de Santo António Marshland Nature Reserve), serve double duty in that they protect the nearby areas from storm damage and often serve as net sinks for carbon storage. They are an example of a nature-based infrastructure.

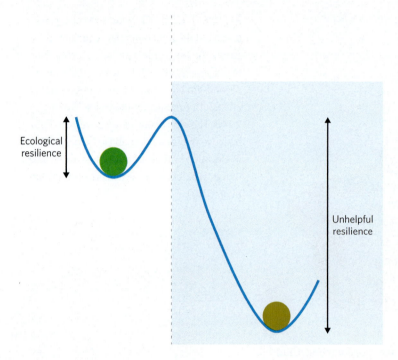

Ecological resilience

Unhelpful resilience

Figure 15.35 Alternative stable states can occur in natural systems and make it difficult to recover the natural state due to unhelpful resilience, which creates barriers to escaping the alternative stable state. Even if management can move the ecosystem toward the natural state, failing to move far enough means the ecosystem will return to the alternative stable state. Ecological science can help provide key ideas and road maps for how to recover such damaged systems (Standish et al. 2014).

change impacts are termed *nature-based solutions* or *nature-based infrastructure* (Seddon et al. 2020). Intact ecosystems often minimize the cultural, social, and economic impacts of climate change and pollution better than engineered solutions (Seddon et al. 2020). In order to take advantage of such nature-based infrastructure, though, we need more information on how species composition and functional diversity of ecological communities alter protective functions. In this regard, ecologists and ecological "tools" can play a strong role in creating systems that can help protect society from the effects of climate change. The unintended consequence of haphazard applications of nature-based solutions is the creation of monocultures of weedy or non-native species, or the replacement of natural ecosystems with less-functional anthromes.

Finally, ecological science plays a key role in managing natural ecosystems so that they are resilient to the inevitable impacts they will suffer in the Anthropocene. *Ecological resilience* and *recovery* take us back to the idea of alternative stable states. A resilient ecosystem is one in which ecological relationships persist after a disturbance event. Recovery, in this context, is the time it takes an impacted ecosystem to return to its predisturbance functionality (Standish et al. 2014). If resilience is not high, or a disturbance is long-lasting or severe, the ecosystem may switch into an alternative stable state. That stable state will have its own resilience. If the alternative stable state has lower functionality than the original ecosystem, then the resilience of the new stable state is "unhelpful" in that it keeps the ecosystem from recovering (and delivering useful ecosystem services) (**Figure 15.35**).

Even if an ecosystem is resilient and does not switch states, it may take a long time to return to its former configuration. From the viewpoint of ecologists who are concerned with ensuring the persistence of valuable ecosystems (e.g., coral reefs, tropical forests, desert springs), Rachel Standish and colleagues (2014) suggest that the salient questions here are

- How much disturbance can an ecosystem absorb before switching to another state?
- Where is the threshold associated with the switch between ecosystem states?
- Will ecosystems recover from disturbance without intervention?

Ecological theories, experiments, and observations provide the answers to these questions. In particular, basic ecological knowledge about population growth, species interactions, and dispersal can provide important information on when, and how, to implement active measures to ensure that diverse and valuable ecosystems are not lost in the Anthropocene.

Coral reefs offer an excellent example of using ecological tools for biosphere protection. In several chapters in this book (Chapters 1, 3, 6, 9, and 11), tropical coral reefs have offered examples of complex ecological interactions and biodiversity. In Chapter 14, it also became clear that coral reefs have been profoundly influenced by human societies. These compelling ecosystems are highly threatened by a variety of forces, including overfishing, habitat destruction, eutrophication of coastal waters, invasive species, and climate change. The stories are so consistently dire that you may wonder if there is any positive news about the persistence of coral reefs in the Anthropocene. These good-news stories may be few and far between, but they do exist, and they offer insight into the factors that allow reefs to persist.

For example, a heat-driven bleaching event reduced coral reef cover and extent off the northern coast of Western Australia by 91%. Within 12 years, this isolated reef had recovered to 44% of its original size. Recovery after a bleaching event is never guaranteed and often requires decades. The fast rate of recovery for this reef may be attributable to its isolation, which protects it from many other anthropogenic influences.

In comparison, coral reefs off the coast of Belize have suffered bleaching from climate change, chronic pollution, and overharvesting, but they have not recovered the way the reef in Western Australia did. Although several Belizean reefs have been set aside within marine protected areas since the early 2000s, many of these reefs have shown little sign of recovery, and some have switched to an alternative stable state in which macroalgae cover the coral structures. Although macroalgae photosynthesize, they do not form a mutualistic symbiosis with coral polyps the way that zooxanthellae do. However, Peter Mumby and colleagues (2021) have demonstrated that one avenue for increasing the resilience of these less-isolated coral reefs is to specifically protect herbivorous fishes (e.g., parrotfishes) from fishing pressure. Herbivorous parrotfishes (family Scaridae) consume or displace macroalgae growing on coral (**Figure 15.36**). The reduction in macroalgae allows zooxanthellae to recolonize the corals, allowing the reefs to recover and grow. Thus, assisting the establishment of a food web through managing fishing pressure can facilitate coral reef recovery.

When reef ecologists considered these admittedly spotty but clear success stories, they arrived at a few insights that are relevant not only to coral reefs but also to other biomes and ecosystems. First, connections between healthy ecosystems matter. Reefs have higher resilience to anthropogenic pressures when they have intact dispersal connections to another, relatively healthy reef nearby. Second, anthropogenic effects can have supersized impacts when they combine. We see this effect in coral reefs when bleaching events combine with pollution and overfishing. In general, ecosystems fare better when disturbance events are spaced out over time, allowing some recovery time between stresses. Third, indirect effects can be important, and healthy food webs may depend unexpectedly on species that we would not intuitively identify as critical to ecosystem function, as in the case of herbivorous parrotfishes that help reduce macroalgae density and decrease macroalgae cover and influence.

These ecological insights led to a variety of suggestions, including ensuring the long-term protection of at least some reefs so that they can provide a source population for the recovery of nearby damaged reefs. More

Figure 15.36 After a ban on the removal and fishing of parrotfishes (family Scaridae) on two island marine reserves in Belize, the parrotfish population rebounded. Higher densities of parrotfishes have coincided with a significant reduction in the amount of smothering macroalgae on Belizean coral reefs around two islands (Long Cay and Middle Cay) (Mumby et al. 2021).

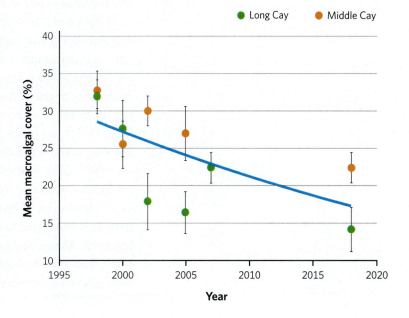

unexpectedly, ecologists point out that we need to concern ourselves with the health of our own (human) communities, because healthy human communities lead to healthier reef communities. If coastal cities and towns are themselves resilient to economic and social upheaval, then their actions are less likely to produce regular or intense disturbance events on coral reefs. When impacts are more severe or global in extent, as is the case with increasing sea surface temperatures, interventions may need to be more intensive. Here again, ecological tools play a role by informing when and how to "plant" coral for restoration or how to recognize when coral species are evolving to withstand higher water temperatures (Darling et al. 2019). The future of coral reefs is not assured, because of the many negative impacts they suffer in the Anthropocene, but harnessing our ecological tools can provide a way forward to continued coexistence with these majestic ecosystems.

15.8 But I'm Not an Ecologist . . .

The truth is that most students using this book will not become ecologists in the classical, research-based manner in which the authors were trained and practice. For many of you, working through this text may be the only time you ever directly engage with ecological science. You may instead follow pathways to other interesting careers in the health sciences, business, environmental policy, architecture, engineering, film studies, law, and many other fields. So, how do we anticipate the knowledge you gain from this class will influence your future?

We can think about this question in two ways. The first is to recognize that nature, broadly defined, provides you and your family with a huge array of services that, if lost, would be immediately noticeable as a decrease in your well-being. These ecosystem services include providing you with clean water, storm buffering, food in the form of protein or vegetables, fibers for manufacturing the everyday items you need, and a variety of other tangible resources.

Ecological systems and the science of ecology also have a profound influence on your health. From the rise in emerging infectious diseases, including COVID-19, to the decline of physical fitness and mental health, ecology has a role to play in our daily lives. Gregory Bratman and his colleagues (2015) did a study examining mood and well-being and found that people's self-reported well-being, their cognition, and their detectable brain patterns showed marked improvements after walking for 90 minutes in a natural setting. Recognizing and valuing these connections to health may be one of the more tangible benefits of this course. Those of you entering the health professions or public health fields in particular will find that ecological processes or the availability of natural ecosystems will influence your daily practice, even though this may not often be recognized by your peers.

The second way that ecology may influence your life is in how your chosen profession interfaces with sustainable ecological systems and their benefits. Certainly, those of you who pursue engineering, architecture, or landscape design will have a profound role to play in creating anthropogenic environments with ecological value. Urban design solutions can build systems that are more hospitable to nonhuman species than those that exist now, and, in particular, will increasingly seek nature-based solutions over those that are constructed.

In the end, we find it easy to argue that ecological science will permeate your daily life and profession in a multitude of ways, including many beyond our current understanding. We hope you will walk away with some of this ecological knowledge and use it well.

SUMMARY

15.1 The Long History of Planetary Change Meets Human Acceleration

- The Earth has a long history of geological and ecological change.

- Most of our ecological science was developed during the geologically recent period of relative stability.

15.2 Is the Anthropocene Real?

- Evidence suggests that the entire planet has shifted into a new geological epoch influenced by human activities; scientists are calling this epoch the Anthropocene.

- Scientists around the world are working to evaluate this concept.

- Today, it is difficult to do ecological research without considering anthropogenic trends and changes.

15.3 Species' Response to Climate Change

- Climate change is one of the most obvious markers of the Anthropocene and is altering how ecologists conduct ecological science.

- By employing new technologies, ecologists are beginning to predict how species will respond to changing climatic conditions.

Geographic Range Changes

- Much of this effort hinges on considerations of a species' niche and dispersal abilities, resulting in multiple possible outcomes.

Key Ecological Mechanisms and Adaptive Modeling

- Predictions may be improved by incorporating key ecological mechanisms such as evolution, environment and land use, physiology, demography, dispersal, and species interactions.

- Unfortunately, much of these species-specific data are not currently available.

- As a result, an adaptive modeling approach is helpful, as it allows for the inclusion of additional information and data in the future.

- **Example of Interacting Factors: Changing Phenology**

  — Studies of phenology have been useful in demonstrating the effects of climate change.

15.4 Extinction in the Anthropocene

Accelerated Extinctions

- The current rate of species extinction is much higher than in the past, so species extinctions are a hallmark of the Anthropocene.

Effects of Extinction

- The loss of species is not confined to one or a few taxonomic groups but seems to be happening across the board.

- The loss of top predators seems to have particularly large effects on other species, which sometimes kicks a system into an alternative state.

15.5 Invasive Species

- The movement of species into new areas by humans is a significant driver of ecological change in the Anthropocene.

Global Distribution of Non-native Species

- The number of non-native species is accelerating, and this acceleration is tied to nations' gross domestic product and the movement of goods and services.

Impacts of Invasive Species

- This widespread movement of species is producing a homogenization of the world's biota, another signature of the Anthropocene.

- Pacific lionfish invading the Caribbean are an example of how a single species can have drastic impacts on whole ecosystems.

  — Lionfish alter food webs.

  — They are hindering the recovery of fish species of concern.

  — The invasion compounds the effects of other stressors, such as climate change, overfishing, and pollution.

- Invasive fish species can alter trophic niche space for the whole fish community.
- Invasive plant species can alter ecosystem processes such as nitrogen cycling and disturbance regimes.

The Future of Invasive Species

- For some taxonomic groups of invaders, the rate of invasion seems to be slowing, although this is not the case for many other groups.
- Invasive species are not evenly distributed around the planet; concentrations are found in areas with lots of humans and economic activity, but areas with high economic activity are also more likely to be able to respond to invasions.

15.6 Anthromes and the Ecology of a "Used" World

- Humans have significantly altered a large fraction of the Earth's biomes.
- The resulting human-dominated landscapes can be called anthromes.
- Anthromes currently make up a significant portion of the terrestrial landscape.
- The global transition to anthromes started in the 1700s and seems to follow repeatable patterns.
- Ecologists have much to learn about these human-dominated landscapes, as they still contain a significant nonhuman presence.
- Future ecological science should examine these transformed landscapes and include specific recognition of the role that social forces play in their maintenance.

15.7 Science for Solutions

- Natural systems buffer human societies from the effects of climate change, and an understanding of ecology can help protect these buffering systems.
- Protecting mangrove forests can reduce atmospheric carbon.
- Second-growth forests sequester more carbon than they respire, and that sequestered carbon may offset local carbon released by fossil fuels.
- Coastal wetlands can provide nature-based solutions or nature-based infrastructure that minimizes the cultural, social, and economic impacts of climate change and pollution better than engineered solutions.
- Understanding disturbance patterns, resilience, and recovery processes can help scientists identify ecosystems in danger and critical factors for their protection.
- Because anthropogenic effects can combine and create oversize effects, distance from human effects may be critical for the protection of natural ecosystems.
- Where distance is not possible, we should enhance connections between healthy ecosystems and identify and support important indirect effects to increase natural ecosystem resilience.

15.8 But I'm Not an Ecologist . . .

- Many students who study ecology will not become professional ecological scientists.
- Yet the ideas, theories, models, and concepts gained from studying ecology can be applied in other aspects of life.
- Understanding ecological systems can help you be an intelligent and informed citizen of the planet.

KEY TERMS

adaptive modeling	anthrome
alternative stable state	defaunation

CONCEPTUAL QUESTIONS

1. Visit one of the many ecological footprint calculators available online, such as the US Environmental Protection Agency's carbon footprint calculator (www3.epa.gov/carbon-footprint-calculator).

   a. Follow the prompts and calculate your ecological footprint. How does yours compare to the average ones reported on the calculator?

b. Now go back and do it again, but this time answer the questions in ways that lower or minimize your footprint. Discuss the factors that had the largest influence on lowering your footprint.

c. How does the ecological footprint calculator tie into the idea of the Anthropocene?

2. Find three figures (not from this textbook) that highlight changes that support the idea that anthropogenic activity is changing ecological interactions, species distributions, or Earth systems globally. Explain your interpretation of each figure.

3. Research the change in distribution for at least one species or taxonomic group that is a result of climate change, and explain your findings. If you are having trouble finding data, there are quite a few studies on butterflies.

4. Compare the effect of the loss of sharks, as depicted in Figure 15.14, with the trophic cascade models in Chapter 10.

5. Identify an anthrome near you. Describe why you think it is a human-dominated biome.

6. How might an understanding of the Anthropocene affect urban planning?

QUANTITATIVE QUESTIONS

1. Look at Figure 15.3, and explain how this figure suggests that the Anthropocene is distinct from the Holocene.

2. Models for many of the ecological mechanisms in Figure 15.6 appear in previous chapters. Pick one of these mechanisms and describe how you might include it in a model that also explores climate change or how you might add climate change to one of the models from a previous chapter.

3. In Chapter 4, we described the flowering period and possible evolution of an invasive St. John's wort species (*Hypericum perforatum*; Figures 4.17 and 4.18) and a native St. John's wort species (*Hypericum punctatum*; Figure 4.20). Using Figure 4.19, explain the flowering-time figures (Figures 4.18 and 4.20) based on what you now know about the Anthropocene, invasions, and the effects of climate change.

4. Background extinction rates suggest that we should lose about two mammal species for every 100 species over a period of 10,000 years.

a. What percentage of species does this number suggest we should lose per 1,000 years, per 100 years, and per year?

b. Using the same rate of loss as in part a, calculate the percentage of mammal species we would lose in 114 years. At this rate, if you start with 200 mammal species in a local state or province, how many would you expect to go extinct in 114 years?

c. Figure 15.10 suggests that the actual percent loss for mammals in 114 years was 1.4%. Using that actual rate, how many of your 200 mammal species from part b would you expect to go extinct in 114 years?

5. On the basis of Figure 15.17B, which taxonomic group (plants, invertebrates, or mammals) do you think has gained the most introduced species in Europe since 1800? Explain your answer. Notice that the scales are not the same for mammals as for other groups and that the scale on the x-axis changes for the last two time periods.

SUGGESTED READING

Ceballos, G., P. R. Ehrlich, A. D. Barnosky, A. García, R. M. Pringle, and T. M. Palmer. 2015. Accelerated modern human–induced species losses: Entering the sixth mass extinction. *Science Advances* 1(5): e1400253.

Ellis, E. C. and N. Ramankutty. 2008. Putting people in the map: Anthropogenic biomes of the world. *Frontiers in Ecology and the Environment* 6(8): 439–447.

Fischer, J. and D. B. Lindenmayer. 2007. Landscape modification and habitat fragmentation: A synthesis. *Global Ecology and Biogeography* 16(3): 265–280.

Saino, N., R. Ambrosini, D. Rubolini, J. van Hardenberg, A. Provenzale, K. Huppop, O. Huppop, et al. 2011. Climate warming, ecological mismatch at arrival and population decline in migratory birds. *Proceedings of the Royal Society B* 278(1707): 835–842.

Steffen, W., W. Broadgate, L. Deutsch, O. Gaffney, and C. Ludwig. 2015. The trajectory of the Anthropocene: The Great Acceleration. *Anthropocene Review* 2(1): 81–98.

SELECTED ANSWERS

Conceptual Questions

Chapter 1

1. The field of ecology differs from some of the other biological fields (e.g., genetics, biochemistry) in that it is heavily interdisciplinary and combines elements from many areas of study, including evolution, physiology, physics, mathematics, chemistry, biochemistry, geology, meteorology, genetics, environmental science, economics, psychology, and others. It is similar to other biological fields in that its subject is living organisms.

3. The scientific method used by ecologists generally starts with some observation of the natural world from which researchers propose a tentative explanation or hypothesis to explain the observation or pattern. The observation and hypothesis together can be formulated as a conceptual model that leads to predictions that are testable. Ecologists then go out and collect data from the natural world to test or compare against the hypothesis or model. These data may come in the form of further observations, manipulative experiments, or natural experiments. Analyzing the data allows ecologists to assess the validity of the conceptual model by basically asking whether the data support the model. If the data do not support the model, then the model can lead to reevaluations, new conceptual models or hypotheses, refined data collection, and even scrapping the first hypothesis or model to propose an entirely new conceptual model or hypothesis. If the data do support the model, the model can be used to explore the system's behaviors, make predictions, propose models for alternative locations or systems, and manage ecosystems and ecological interactions. Researchers can add complexity to the model or use it as a jumping-off place for new insights and new models. In general, this process always leads to new questions and observations about the natural world, thereby initiating the process again.

5. Ecologists use the word *model* in a number of different ways. The term can refer to a conceptual model, in which the scientist assembles a theoretical construct of how the world or system of interest might work. It can also refer to a mathematical model, in which the researcher uses mathematical constructs to quantitatively represent a concept or the interaction and outcome of multiple factors (variables and parameters). A mathematical model can be analytical, from which general behaviors are clear from complete solutions, or it can be a simulation model, offering insight into a range of behaviors by exploring outcomes through multiple iterations. Of these definitions, the first is often considered the most conventional.

Chapter 3

3. The four ways we can easily modify the generalized atmospheric circulation model or make it more realistic are by (1) adding the effect of the Earth's daily spin (i.e., the Coriolis effect) on air and water circulation patterns; (2) examining the effect that topography (e.g., mountains) has on air circulation patterns (e.g., rain shadow effects); (3) incorporating the ameliorating effects of large water bodies on the climate (e.g., the continental effect); and (4) understanding the role that living plants play in the global climate through the transpiration of water vapor through their leaves.

5. Based on the information in the question, it seems reasonable that a rain shadow contributes to the aridity of the Atacama Desert. At those latitudes, the dominant winds at ground level will be moving from the east toward the west. Moist air blowing across the land from the Atlantic Ocean helps form rainforests before reaching the foot of the Andes mountain range. The air is then forced up in elevation, where it expands due to lower pressure. As it expands, it cools and can hold less moisture than warm air. This moisture comes down as significant precipitation on the east side and the top of the Andes. After this air loses much of its moisture, the dry, cold air sinks down the back or west side of the range. As it flows down the mountains, it is under more pressure, so it compresses and warms. The increased warmth also increases the water-holding capacity of the air, which allows it to draw even more water from the land surface on the west side of the mountains to form the Atacama Desert.

7. Rising sea levels associated with anthropogenic climate change are more likely to directly affect biomes that sit in close proximity to the ocean. The most affected biomes will be those that sit at the terrestrial-aquatic interface, such as mangroves, estuaries, and salt marshes. Purely terrestrial biomes that may experience large impacts include tropical rainforests, tropical dry forests, Mediterranean scrublands, temperate forests, and tundra, although a good case could be made for how rising sea levels will affect all the Earth's biomes, as a result of alterations in the atmospheric circulation patterns.

Chapter 4

1. Two organisms that morphologically seem like different species are a Great Dane and a Chihuahua. They do not look very similar, given that the Great Dane can exceed 180 lbs. (82.6 kg) and a Chihuahua seldom weighs more than 6 lbs. (2.7 kg). But under the biological species concept, they would be considered the same species because they can freely mate and produce fertile offspring. Other examples are plentiful.

3. Natural selection is the differential survival and reproduction of individuals in a population in response to biotic and abiotic factors in their environment. That means the environment selects which individuals survive and reproduce and which ones do not. Selection occurs because resources are limited and cannot support all individuals. In a world of limited resources, some individuals have traits that allow them to acquire resources more effectively than other individuals can. The individuals that acquire more resources survive to reproduce more and pass on their genes to the next generation, thereby altering allele frequencies.

5. The three mechanisms of evolutionary change other than natural selection are genetic mutation, gene flow, and genetic drift.

Genetic mutations arise most frequently from mistakes in DNA replication or sexual reproduction, and they produce a change in allele frequency at the population level by adding brand new alleles to the population, thereby producing evolutionary change. Gene flow transports alleles into a population. This influx of new alleles produces a change in allele frequency, thereby producing evolutionary change, and it may introduce novel alleles. Genetic drift is the random loss of alleles from a population. The easiest way to understand such random loss is to think of the loss that comes from the random movement or death of individuals. Genetic drift is the most likely mechanism to appreciably alter allele frequencies in small populations.

7. A third graph that would be useful for showing the full process of sexual selection on peacock tails would be one that showed how well the chicks themselves reproduced (e.g., the number of life-time offspring they produced). This graph would have the mean area of father's eyespots on the x-axis, and the y-axis would represent something like the number of offspring the father's babies themselves produced in the following generation.

9. (a) One explanation for why some of the colonial bentgrass off-spring may not have high copper tolerance is that they are the result of a genetic mutation during sexual reproduction (i.e., new alleles enter the population through mutation). A second expla-nation is that if copper tolerance is a dominant trait, then hetero-zygous parents can breed and create offspring that are homozygous for the recessive allele (no mutation required), and these offspring will not have copper tolerance. This possibility requires no change in allele frequencies. A third explanation is that if copper tolerance is a recessive trait, then all resident par-ents should be homozygous for the recessive allele (in order to have survived to reproduce), but pollen from parents outside the copper mine tailings (i.e., gene flow) may bring in dominant alleles that code for a lack of copper tolerance. The first and third options would change allele frequencies (i.e., involve evolution). (b) Colonial bentgrass is a non-native species in the United States because it originates from China but was brought to the United States for mine revegetation.

Chapter 5

1. Two major constraints of desert environments are a lack of water and high temperatures. Kangaroo rats have adapted to these con-straints by having extremely efficient kidneys to reduce water loss. In addition, they spend the majority of the daylight hours underground to avoid the heat, and they conserve water while sleeping by making a high-moisture microenvironment.

3. Optimal foraging theory is the theory that animals act to gain the most energy for the least cost when making foraging decisions. The simple optimal foraging model follows this formula:

$$P = \frac{E}{S + H}$$

where P = profitability of the prey, E = energy content of the prey, S = search time of the prey, and H = handling time of the prey. The model has been shown to work in great tits, a species of bird, in an experimental laboratory setting.

5. Because they are reptiles, lace monitors do not have any fur or feathers to help regulate their body temperature, yet the species manages to maintain a fairly even internal thermal environment. In order to regulate their temperature, they must rely on behav-ioral mechanisms. These include moving to microhabitats that either add heat (e.g., a place where they can bask in the sun) or

cooling themselves off (e.g., by moving into shade). For a graph showing this internal consistency, see Figure 5.23.

7. The ecological niche is the specific set of physical and biological conditions under which a species can live and reproduce. Any example of a three-dimensional niche would need to include three axes of environmental or biological conditions, such as temperature, light, water availability, phosphorus content, and so on.

Chapter 6

5. (a) An annual storm cycle that wipes out half the population would be density independent because it does not depend on the population density. (b) A fixed amount of food that limits popu-lation growth would be density dependent because if the popula-tion is low, the amount of food would not be limiting, and, therefore, the effects depend on population density. (c) A fixed number of shelters that increase mortality at high density is an example of density dependence because at high density more individuals die. (d) A lethal disease that increases transmission with contact is density dependent because at high population densities the transmission increases and, therefore, so does the lethality. (e) A cold spell that kills individuals in the same frac-tion across different populations would be considered density independent because the same proportion of individuals dies in both large and small populations.

7. Figure 6.23 shows a hypothetical age pyramid for humans. If all females of reproductive age (light green bars in the figure) were to instantaneously start limiting themselves to two children, the population would still continue to grow. This is because of the gray bars in the figure beneath the reproductive age individuals—all these prereproductive individuals will grow older and start having babies. Yet even if these people also only have two off-spring, there are still significantly more of them in the prerepro-ductive stage. This continued growth would proceed until all the current prereproductive individuals became reproductive. At that point, if the two-child limit were still in place, then the popula-tion would begin to stabilize.

Chapter 7

1. (a) The situation with garlic mustard is one of interference com-petition (the plant's use of chemicals points to allelopathy to remove other plant species that might also seek space and soil nutrients). (b) The scenario of lions attacking hyenas to steal their catch is interference competition. (c) In this scenario, if the lions and hyenas both prey on gazelles (which seems to be implied by part [b]), then this is an example of resource competi-tion. (d) The situation with the male frog calls is not competition. (e) The scenario of western ground squirrels and the tiger sala-manders sharing the same burrows is likely not competition unless the two species somehow interfere with each other's use of the burrows. From the description, it appears that they have partitioned the burrows in time and do not share this limiting resource at the same time of the year.

5. One positive feature of the Lotka-Volterra competition model is that it is fairly intuitive and the α and β terms make sense. Another advantage of this model is that it recreates the four situ-ations we see in nature (one species wins, the other species wins, stable coexistence, and a winner is determined by initial condi-tions, such as which species arrived first). A third positive is that the model's outcomes have been demonstrated with real organ-isms (e.g., flour beetles). But the model also has some cons. One

issue with the model is that it is based on the continuous-time growth model and, therefore, assumes that interactions play out in tiny increments of time, which implies instantaneous responses by the species; such instantaneous population responses are unrealistic. Another difficulty is that the model suggests populations grow steadily until they reach equilibrium, but the natural world is more variable than this, and factors such as carrying capacity (K) change over time. A third issue with this model is that the competition coefficients α and β are not always easy to define or measure for some competition situations. Finally, measuring K can also be quite difficult. There are many other pros and cons of this model; these are only a few.

7. (a) Character displacement involves long-term adaptation to another species' niche use and to interspecific competition, and it requires evolution (changes in gene frequencies). In this case, there is no evidence of genetic shifts. (b) Niche shift is when one species' niche use along a single dimension is altered. In this situation with the mynah and the rosellas, the term may apply in that the rosellas avoid eucalyptus trees with mynahs in them, suggesting a niche shift along this niche axis. (c) and (d) The fundamental niche is the full range of conditions (biotic and abiotic) and resources in which individuals of a species can survive and reproduce; the realized niche is the niche space that individuals of a species can access in the presence of their competitors. In this scenario, the terms apply in that the rosellas avoid eucalyptus tree hollows with mynahs in them (the niche without the eucalyptus trees is a realized niche), whereas the rosella's fundamental niche would include all tree hollows. (e) Resource partitioning is the idea that two competing species divide up resources and avoid resources used by the other species. Generally, resource partitioning requires some time, but we see the beginnings of resource partitioning in the division of trees (with rosellas avoiding eucalyptus trees).

Chapter 8

1. Predation: using another living organism for food. Herbivory: consumption of photosynthetic organisms, in whole or in parts, by an exploiter. Parasite: an organism that obtains its energy from a host organism by consuming the host's body, organs, tissues, or fluids often without killing the host. Pathogen: a parasitic organism that causes diseases. Parasitoid: an insect exploiter that attacks other insects and lays eggs or larvae inside the host insect; unlike most parasites, parasitoid larvae generally kill the host by eating it from inside out.

3. Classic predators and parasitoids perhaps best fit the Lotka-Volterra exploiter-prey model because consumption leads to death (and a decrease in the prey population). Herbivores and parasites do not always kill their hosts, so they may not fit quite as well. All of the additions (carrying capacity, functional responses, and Allee effects) add realism to the Lotka-Volterra model and make the model fit various predation scenarios better.

5. Animals use a number of generalized prey defense strategies. One general strategy is to avoid being attacked by predators, and this can be achieved in several ways. One way is to avoid being seen in the first place, through camouflage or cryptic coloration; organisms such as frogfish and zebras use this strategy. Another way is to appear large, dangerous, or noxious; this strategy is used by many insects, such as the Io moth, which reveals huge eye spots on its wings to make it resemble a large predator. An additional strategy for avoiding attack is to announce one's toxic

or noxious characteristics by displaying bright colors to potential predators; this is called aposematic coloration and exists in many organisms, such as in the poison arrow frogs of Central and South America. Avoiding predators can also be achieved by mimicking another organism that is toxic, dangerous, or noxious; this can take two forms, Müllerian mimicry (in which an organism is toxic and also looks like other toxic species so the signal is widespread) and Batesian mimicry (in which an organism is not toxic but looks like an organism that is toxic). Most bees and wasps use Müllerian mimicry in that they all share a similar black and yellow banding pattern, which announces that they can sting.

A second overall defense strategy is escape. Speed (whether by running, flying, or scurrying away) can allow prey to escape predators, and some distraction techniques can also help with escape (think about fish in schools or organisms that direct attention away from themselves).

A third overall strategy involves fighting back. This defense may involve some variation on the first strategy (e.g., announcing "dangerousness") but must also include something that makes the prey actually dangerous, such as defense compounds or spines in plants; spines, teeth, toxins, or huge size in terrestrial vertebrates; and toxins in many marine invertebrates and insects.

These three overall strategies fit generally into the primary, secondary, and tertiary defense strategy categories.

Chapter 9

1. Mutualism is an interaction between individuals of two species in which both species benefit from the interaction—for example, as between a flower and a bee. Commensalism is an interaction between individuals of two species in which one species benefits from the interaction while the other species neither benefits nor is harmed—for example, as between a remora and a shark. Both types of interaction fall under the larger category of facilitation, which is a general term for interspecific interactions that benefit at least one of the two species and neither is harmed. The two general types of mutualisms are obligate and facultative. A mutualism is obligate for a species that cannot survive or reproduce without the mutualist partner. For example, zooxanthellae cannot survive without their coral mutualists. A mutualism is facultative for a species that benefits from the interaction but can survive without it—for example, as with fruit-eating animals and fruit-producing plants. Mutualisms can be obligate for one partner and facultative for the other.

3. Ecologists have run into some issues in their efforts to model mutualisms. When they tried to use the Lotka-Volterra competition equations and just switch the sign of the effect of the other species from negative (competition) to positive (mutualism), the resulting equations gave unrealistic answers (i.e., the populations exploded to infinite abundance). A similar thing happened when ecologists tried to modify the Lotka-Volterra predator-prey equations—the resulting mutualistic populations exploded to infinity. To get around these issues, ecologists turned to other fields of study for inspiration. For example, they turned to game theory, using the prisoner's dilemma scenario to investigate how mutualisms could evolve. They also adapted economic models to show how two species could realistically share resources through specialization and trade. Both of these approaches proved useful.

5. Game theory has been applied to studying mutualisms through the use of the prisoner's dilemma scenario to investigate how

mutualisms could evolve. It is not clear from the outset that the two mutualistic partners should not cheat on each other and defect from the relationship. In fact, there is reasonable evidence to suggest that cheating and defection should be common. But using an iterative version of the prisoner's dilemma game, ecologists were able to come up with potential evolutionarily stable strategies through which both species should be able to cooperate and maintain the mutualism. One such strategy was called tit-for-tat and involved cooperation in the first instance and then mirroring the other species' strategy for all subsequent encounters.

Chapter 10

1. Trophic ecology is the study of how species are linked together by trophic (feeding) interactions. One way to represent these interactions is with a linear chain of feeding relationships (i.e., who eats whom). Although such food chains are extremely useful on occasion, they are not always the best representation of the natural world. The reasons for this discrepancy are many but include the fact that many or most organisms eat a variety of things (omnivory), that diets often shift as an organism develops (ontogenetic diet shifts), and that food chains often omit the reality that all organisms die and become part of the detrital food chain that eventually provides nutrients for plants. Consequently, a food web model of trophic interactions is often preferred to a chain model. An additional problem with both food chains and food webs is that not all interactions boil down to trophic interactions.

3. Energy is only transferred between trophic levels at the rate of 10%–20%, meaning that only 10%–20% of the energy consumed by a predator is made into predator tissue and, therefore, available if and when that predator is eaten by an organism at the next higher trophic level. In other words, 80%–90% of the consumed food energy is "lost" by the organism that does the eating. This fact of nature puts very real limits on the natural world. If all humans, for example, were to eat mainly at the primary producer level (i.e., plants), there would be much more energy available for the human population than if all humans were to eat mainly cows (the second trophic level), which have to be fed an enormous amount of primary producers (plants). This same concept helps explain why there are no top-top predators, such as a predator that only eats bald eagles. There would not be enough energy for a top-top predator because the bald eagle population is small and could not support a predator trophically above it (because of the energy loss at each trophic interaction).

5. Stable isotopes are alternative forms of atoms that generally have extra neutrons in the nucleus and do not spontaneously radioactively decay. Two common isotopes employed in ecological studies are ^{13}C and ^{15}N. The ratio of isotopic to nonisotopic forms of the atom is measured from tissue samples using a mass spectrometer. Carbon isotope ratios are conserved from prey to predator, meaning "you are what you eat." Nitrogen isotope ratios are different in that they accumulate as you move up the food chain by approximately 3.4 units. So, if a predator eats a prey that has a nitrogen isotope ratio of 1, the predator will have a nitrogen isotope ratio of 4.4. Using these two isotopes, scientists can make a graph showing the two isotope ratios for different organisms in a food web. The trophic relationship among these organisms can then be inferred from the graph, which creates a sort of food web diagram based on the isotope data. Figure 10.21 shows how this can work.

7. (a) Given the scenario described, the eagle is most likely to be a keystone species because, by definition, keystone species are species that have large impacts on food webs despite having small abundances. (b) Apparent competition occurs when two prey species appear to be competing with each other (by having reduced population abundances) but, in fact, the reduction in population abundances is caused by a shared predator. A number of trios in Figure 10.15 could represent apparent competition, but one clear example is between grasshoppers and squirrels mediated by loggerhead shrike predators. (c) Five examples of resource competition in Figure 10.15 are (i) squirrels and grasshoppers competing for creosote bush; (ii) shrikes and scorpions competing for grasshoppers; (iii) shrikes and roadrunners competing for grasshoppers; (iv) owls and kit foxes competing for scorpions; and (v) owls and eagles competing for roadrunners. (d) Removal of the golden eagle might cause a trophic cascade because it sits at the top of the food web and eats three other organisms below it (roadrunners, squirrels, and kit foxes). In a similar manner, the removal of the great horned owl may also cause a trophic cascade, as it consumes three other organisms below it (roadrunners, scorpions, and kit foxes). It is also possible that the removal of the kit fox could cause a trophic cascade, as it eats shrikes, squirrels, and scorpions.

Chapter 11

1. In the section on the species-area relationship, we discussed Preston's mathematical model ($S = cA^z$; equation 11.1) of how species richness accumulates as more and more area is sampled. Taking the log of both sides of this equation, we transform this exponential function into a linear function. In the linear version of this relationship, the z term is the slope of the straight line of the function. The z-value thus tells us how the number of new species added changes as the area sampled gets larger. We would expect larger z-values at lower latitudes (i.e., closer to the equator) because we expect to see more heterogeneity at lower latitudes. Some of this explanation is circular, but the higher solar input leads to more species and more differences in local areas, partially because of competition. In general, there are more species present at lower latitudes than at higher latitudes.

3. In order to evaluate which species are tolerant to pollution, it is useful to remember that tolerant species in disturbed or polluted environments will have larger abundances. For this reason, when we look at the two tables, it is clear that the California roach and the Pacific lamprey are the most abundant species in the disturbed San Pablo Creek, which would suggest they are the most tolerant of poor water quality. In the more protected Wildcat Creek, three-spined stickleback and Chinook salmon show large increases, suggesting that these two species are perhaps the most intolerant of disturbance or pollution. All the other species fall somewhere in between these two extremes.

Chapter 12

3. If each small population were able to survive for several generations (or long enough for the population to get to carrying capacity and stay there for a while), then we would expect that, everything else being equal, several small populations connected through dispersal (i.e., a metapopulation) would survive longer than a single large population. The probability of extinction for an entire metapopulation is lower than the probability of extinction for one single population, even if that population is large.

This is because movement or colonization among subpopulations essentially spreads the risk of global extinction among the individual patches, thereby lowering the overall probability of extinction. This generalization does not hold if each small population has a very high risk of extinction or if the small populations are so connected that they share risks and are all likely to go extinct together.

5. If we accept the assumptions of the Levins model, true metapopulations are likely to be rare. The major assumptions behind the Levins model are: (1) the patches are homogeneous in size and character (not likely to be the case in the natural world, as patch size and quality vary); (2) the probabilities of extinction and immigration are constant through time (not likely to be true in the natural world, as these probabilities will vary through time and in different environments); (3) there is no spatial structuring among the patches, or, in other words, each patch is equidistant or equally likely to be colonized by every other patch (this also cannot be the case in the natural world because more than three unique physical spaces cannot all be equidistant); and (4) all dynamics play out after a population has reached carrying capacity (meaning that extinction is very rare, which is becoming less and less true in a rapidly changing world). More realistic ways to represent patchily distributed species include models that keep track of the individuals in each patch and the quality of the patches (like source-sink models), and models that keep track of the distance between patches and make dispersal and colonization dependent on habitat quality and distance to neighboring patches (a combination of source-sink and mainland-island models; incidence function models—not mentioned in the chapter—are common ways to handle such colonization).

7. The concept of spatially structured populations (e.g., metapopulations and source-sink populations) helps us understand phenomena that wildlife managers and ecologists see in nature. For example, it is not uncommon in nature to find a high-quality habitat with no individuals in it or, alternatively, to find a poor-quality habitat with some individuals present. Metapopulation ideas help us understand these situations. With a spatially structured metapopulation, some patches will have gone extinct even if the habitat patches are of good quality. The model suggests that, over time, these empty patches will be recolonized. In terms of source-sink situations, it is common to see a relatively poor-quality habitat with individuals present because there is a nearby source population whose per capita rate of increase (λ) is greater than one and which is producing lots of dispersing individuals. These dispersers have to go somewhere, and they may end up in a poor-quality patch (i.e., a sink) even though they will not successfully reproduce there.

Chapter 13

1. Ecological succession is the process by which ecological communities change in fairly predictable ways over time. This process can start from bare rock or lifeless water and proceed to locations filled with interacting species. The process is called succession because a series of species succeed each other in the community through time. Each sequential stage of a terrestrial plant community's succession is called a sere, and each sere is generally distinguished by its dominant plant species. The timeline of changes begins with the initial colonization of a new habitat or following a disturbance event, such as a fire, flood, or bulldozing. When a location starts at zero (no species present), the successional process is called primary succession; when it starts later (some species are already present), it is called secondary succession. Examples of succession are relatively easy to spot in most locales.

3. Connell and Slatyer propose three models to explain the successional process. The first is the facilitation model, in which the initial colonizing species alter the environment and essentially enhance it, thereby facilitating later successional species to colonize the area and succeed as the next dominant species. In this model, each sere or stage of the process makes it possible for later stages to occur, eventually leading to a climax community. The inhibition model takes an essentially opposing viewpoint, positing that each stage in the succession process prevents (inhibits) later species or stages from colonizing. This inhibition is broken up periodically by some disturbance event (e.g., fire, flood, drought, or hurricane). After the disturbance, the inhibiting species are removed and the next-stage species can colonize and dominate. The third model is the tolerance model, which proposes that later successional species are neither helped nor hindered by earlier species. The tolerance model relies on the observation that most dominant species in climax communities are K-selected species that arrived on their own schedule and grew in abundance with very little delay or facilitation by early species. Their eventual dominance is due to their own life history characteristics and traits. Given enough time, such long-lived K-selected species take over and become the dominant taxa, excluding earlier (probably r-selected) species. These models can all operate in a single community at different stages of succession. Facilitation is most common in the earliest stages; inhibition can happen at almost any stage; and tolerance is most common in the species that end up in the climax community.

5. Natural selection, or the differential survival and reproduction of favorable genetic traits, could produce all three successional model types. For facilitation, individuals who improve their local environment also improve their own fitness by increasing their access to water, nutrients, and other resources. Such individuals will produce more offspring, and their alleles and traits will become more common in the population. Even if they are outcompeted by other individuals of other species, they will still be the most successful individuals of their species, and their offspring will likely be the ones who survive to disperse to a new habitat and start the process all over again.

Natural selection could produce species that use inhibition because inhibition improves individual success. Perhaps the easiest way to argue for inhibition is to think about the selective advantage of allelopathy. A plant that can produce toxic chemicals that keep other species or individuals at a distance is a plant that will have greater access to space and the resources in that space. Such an individual will survive better and reproduce better than individuals without allelopathic potential.

Finally, natural selection could produce organisms that thrive in the tolerance model of succession in the same way that natural selection would produce a highly competitive species. If individuals in a species differ in their competitive abilities in a highly competitive environment with lots of resources, then we would expect that the individuals who can best capture those resources and thrive, despite the presence of other individuals, will be the ones to survive and reproduce the most in its population. Such reproductive success will lead to more copies of the alleles of that individual in the next generation, which will lead to a higher frequency of individuals who are well adapted to thriving in dense environments.

Note that each of these arguments works in some situations but not in all. The tolerance model would not work in an environment with low resources and little organic soil. The facilitation argument would not work in an environment with lots of resources and species.

7. First, del Moral's research on primary succession at Mount St. Helens demonstrated that the early stages of primary succession progress much more slowly than the early stages of secondary succession. In other words, primary succession takes time. Second, del Moral and colleagues found that species richness, plant cover, and biomass seemed to increase through time at Mount St. Helens, but that the pattern for some of these biodiversity metrics eventually leveled off or slowed the rate of change. Third, they observed that chance events seem to play a significant role in determining how the successional process unfolds. Such chance events can be the time of year that the disturbance initiating the process happens, the weather patterns following the disturbance, or the specific features of individual species or landscapes that influence the successional process. A fourth finding of their work was that biotic interactions among species play a significant role in the process of succession, and without them the process would likely stall.

Chapter 14

1. Though there are many ways to answer this question, here are some of the more obvious ways the hydrologic cycle can affect the global carbon cycle: The hydrologic cycle clearly affects the diversity, density, and biomass of living terrestrial organisms because, when there is little water, all three of these factors decrease (e.g., think of deserts). The amount of water present will affect the amount of plant biomass, which, in turn, affects the amount of photosynthesis (carbon capture) and plant respiration (carbon release). In addition, plant respiration is linked to both the carbon cycle (carbon release back to the atmosphere) and to the hydrologic cycle, in terms of respired atmospheric water.

3. The carbon would first have to be captured from the atmosphere by a cacao tree (the plant from which chocolate is derived) in the form of carbon dioxide, which would enter the plant through its open stomata. Once inside the cacao tree, the carbon would be used in photosynthesis, converted into sugar, and used to create the fruit (pod) of the cacao plant. Humans would harvest the cacao pod that holds the atom of carbon and grind up the pod to create powder to use to make chocolate. That powder would then be combined with milk and sugar (and perhaps other ingredients) to create a chocolate bar. If you were to consume the chocolate bar, the atom of carbon would end up in one of three places: your feces, your exhaled breath, or your body. Let's assume it ends up in your feces. At this point, it would likely be consumed by decomposing bacteria or fungi. Then, the bacteria would either "respire" the carbon atom or be consumed by another organism. Eventually, the carbon atom would end up either buried in sediment or released to the atmosphere. If it is buried in sediment, it might take millions of years to be released to the atmosphere.

Quantitative Questions

Chapter 2

3. (a) Categorical. (b) Numerical. (c) Numerical. (d) Numerical. (e) Ordinal categorical. (f) Discrete numerical or ordinal categorical—when indicated by names, they are more clearly categorical. (g) Although this is a discrete numerical variable, it is harder to turn into an ordinal categorical variable, but each year could be indicated by a letter. (h) Numerical. (i) Ordinal categorical, which means it could be replaced with a discrete numerical variable. (j) Discrete numerical, but it is hard to turn into an ordinal categorical variable, although letters can work. (k) Categorical.

5. (a) The explanatory variable is categorical (e.g., low, medium, high) or numerical (measured); the response variable is numerical. (b) The explanatory variable is categorical; the response variable is categorical or numerical. (c) The explanatory variable is categorical; the response variable is categorical. (d) The explanatory variable is categorical or numerical; the response variable is categorical or numerical. (e) The explanatory variable is categorical or numerical; the response variable is categorical or numerical.

Chapter 3

1. (a) The most important aspect of this change would be that the surface of the Earth would no longer curve away from the sun at higher latitudes, which would mean that solar intensity would no longer change with latitude in the same way. Instead, the poles would be likely to receive peak solar input on the solstices, while the length of the cylinder would receive fairly consistently distributed solar input at other times. This change would probably mean that we would no longer see Hadley, Ferrel, or Polar cells, and solar intensity would vary much less with latitude. We would still have seasonality because of the tilt of the Earth, but biomes as we know them would disappear. The poles would probably be very cold or very hot, depending on the time of year. (b) Our precipitation model would completely change with this reversal. Rising warm air would not carry moisture, and it would not drop moisture at high altitudes. Instead, moisture in the air would stay there, and moisture on land might not rise with warm air (though it might rise with cold air). Moisture might concentrate in cold portions of the atmosphere (i.e., high elevations) and would probably block much of the sunlight, making the world a much colder and probably icier place. (c) If wind could go through mountains, there would be no rain shadows. It is the process of going over the mountains that leads to adiabatic cooling and rainfall on the windward side of mountains and rain shadows on the leeward side.

3. The curve in each line suggests that the rate of change (the slope) is changing. In Figure 3.48B, the meters of water are decreasing, and that decrease is getting steeper, meaning that the rate of glacier loss is increasing. In Figure 3.49, the y-axis measures change in sea level relative to an average. The fact that the change is negative in earlier years and positive in later years already lets us see that the sea level is increasing. The slight upward curve in later years suggests that the slope is increasing and that the rate of increase is getting a little bigger each year.

Chapter 4

1. (a) In the graph shown here, the blue bars represent all females, and the gray bars represent females who survived.
   The most obvious pattern is that the smallest and biggest birds did not survive, which suggests that selection is choosing the intermediate-sized birds. (b) This would be stabilizing selection.

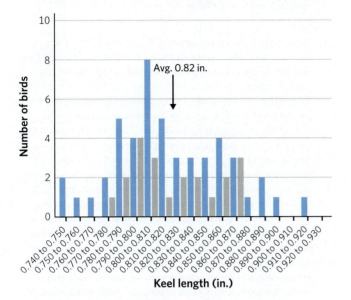

3. (a) Proportions will vary based on coin tosses. (b) Mean proportions will vary based on coin tosses. (c) The means for the 30-toss sets should be closer to each other and closer to 0.5 than the means for the 4-toss sets. This is similar to how traits with no effect on fitness should be more likely to persist in a larger population than in a smaller population (i.e., less genetic drift in large populations). (d) With 1,000 tosses (or 1,000 individuals in a population), we would expect the mean to be even closer to 0.5, and we would expect even less loss of alleles to genetic drift.

5. Stabilizing selection will not lead to any loss of genetic variation if it does not draw the frequency distribution more closely around one perfect phenotype. If, instead, the frequencies for all phenotypes stay as they are, then there will be no loss of genetic variation. The graph you draw for this should have two distributions that are close to identical overlaid on top of each other for the distribution before and after selection.

Chapter 5

1. Organisms need to balance costs with energy intake. The travel time between patches is a cost of moving away from the current patch, and organisms should avoid that cost until they have offset it with extra energy consumed.

3. Time in patches does seem to increase as travel time increases, but it is not a linear relationship. This suggests that there may be a limit to how long an organism will stay in a patch and decrease local resources, even if travel time is *very* long; but it also highlights that the relationship to energy is not the same for consuming resources as it is for traveling (i.e., travel costs and energy consumption on site are not linearly related).

5. First, each bird is unique, so we would expect that individual birds might have slightly different "foraging" (removal of mealworms from sawdust) success, which would affect their

time in patches. Using the time it takes to remove a cover as a surrogate for travel time between patches also may not be a very good substitution, because some birds may actually experience frustration and not just a time delay. It will also take birds a while to determine that there are time delays because the cover-removal time will not fit into any previous evolutionarily adapted assessment of forage time; for example, the birds will not be able to look at all the dowels and understand the time between foraging bouts the way they might in a forest.

7. It would be cool to create multiple "forests" of dowels with different distances between clumps of dowels (instead of different covers).

Chapter 6

1. (a) The cockroach population abundance in the basement is

$$N = \frac{nM}{m} = 28 \times \frac{21}{6} = 98$$

(b) It is possible that the blue paint affects the cockroaches. And three days might not be a good amount of time to let the released cockroaches mix. It is also possible that the bait attracts some cockroaches more than others (specifically, it might attract the cockroaches that were already trapped).

3. $N_0 = 25$, $t = 5$, $\lambda = 1.257$. Plugging those into Equation 6.8, we get 78.454236, which rounds down to 78 individuals.

5. Yes, this leads to the classic S-shaped curve. It may make sense to use this number. The population abundance may be too big to assume it is not experiencing density dependence.

7. (a) Yes, all four behaviors can be produced. (b) No, neither action changes the overall dynamic behavior, nor does it bump the system into a different category.

Chapter 7

1. (a) Species 1: $\frac{K_1}{\alpha} = \frac{100}{0.25} = 400$.

   Species 2: $\frac{K_2}{\beta} = \frac{50}{0.25} = 200$.

   Given these conditions, the Lotka-Volterra model would predict that the two species would coexist.

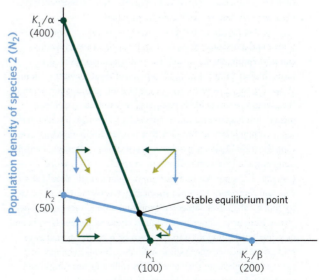

(b) Species 1: $\dfrac{K_1}{\alpha} = \dfrac{100}{0.75} = 133.3$.

Species 2: $\dfrac{K_2}{\beta} = \dfrac{50}{0.50} = 100$.

Given these conditions, the Lotka-Volterra model would predict that species 1 would win.

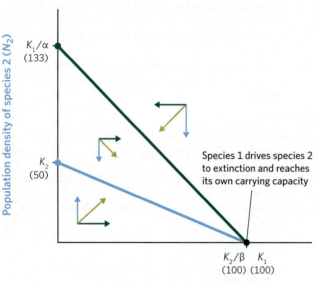

Population density of species 1 (N_1)

3. Under the Tilman model, stable coexistence can occur if each species experiences positive population growth at a lower resource level than the other species, for one of two resources (i.e., when the two species' ZNGIs intersect [Figure 7.26]). In other words, this can happen when each species is a better competitor for one of two resources. If this ZNGI situation occurs, stable coexistence *can* occur, but only if the environmental resource supply point (i.e., the natural level of resources the environment returns to) lies between the two species' consumption vectors. When these two conditions are met, the model predicts stable coexistence.

5. If the situation with the mosquitoes is as shown in the figure, then we would expect mosquito species A to feed more on birds in location 2, based on mosquito wing length.

    The two mosquito species have similar wing lengths in locations 1 and 2 because, most likely, both species are advantageously feeding on mammals (low to the ground) as well as birds (higher from the ground) whenever they encounter them. When the two mosquito species are present in the same location (i.e., sympatry), they likely compete for resources (i.e., blood meals). But because mosquito species A on average has shorter wings, when it is confronted with a competitor (sympatry), over time the individuals in species A that perform better will be those that compete less with individuals in species B (i.e., shorter-winged mosquitoes that feed more on mammals should survive and reproduce better in species A). Over time, therefore, the frequency of shorter-winged mosquitoes should increase in species A. Similarly, longer-winged individuals in species B will compete less with individuals in species A by focusing primarily on bird hosts. Because these shifts in feeding involve changes in traits or morphology, they most likely reflect genetic shifts and would be characterized as character displacement.

    If a third mosquito species arrives at location 3, one way it could coexist with the two species already present is if it had

wings of intermediate length (4–6 mm) and fed on both mammals and birds, but only mammals and birds that exist at an intermediate height (e.g., tree squirrels and low-foraging birds). That may be a very narrow niche with few resources. Alternatively, a third species could coexist if it had wings (or other traits) that allowed it to feed on an entirely different food source that is not used by species A and B (e.g., smallest mammals).

Chapter 8

5. With a Type II functional response, the number of prey eaten increases at first but eventually levels off at high prey density. This is generally due to some kind of constraint or limitation on the predator's ability to continuously eat prey (gut size, prey handling time, and so on). When this kind of limitation is imposed on the predator, it changes the prey isocline because prey can escape from predator control at higher densities. Figure 8.18 shows that the prey isocline is likely to have a positive slope. With this prey isocline and the predator isocline from Figure 8.11, the resultant trajectories lead to cycles with larger and larger amplitudes, which will result in one or both species going extinct (see Figure 8.22B).

7. (a) Based on the general pattern of population growth discussed in Chapter 6, it is possible that a new non-native population may remain small and local for a long time because the initial phases of population growth (exponential or logistic) are generally slow. This pattern can be accentuated if the non-native species has a long generation time or experiences a significant time lag in terms of reproduction. Once a larger population abundance is reached, both logistic and exponential growth models predict very rapid growth, much like what we see in nature with non-native species. (b) With an Allee effect, at low population densities, fitness declines as the population density decreases, meaning that at low population density the species may experience population decreases instead of increases. An Allee effect could keep a population from growing until enough individuals were imported to overcome the Allee effect or until local conditions changed enough to lead to a higher population growth rate.

Chapter 9

3. These isoclines can be confusing, so referring to Figure 9.11 should help. If we make an analogy to the competition isoclines, we can think of each species as decreasing in density (because the species exceeds its carrying capacity) when it sits in phase-plane space above its carrying capacity. For species 1, the area of the phase plane where the population abundance is above its carrying capacity is the area that sits to the right of the dark green isocline. In that area, we see either green shading (no coexistence) or an arrow that shows species 1 declining in population density (and species 2 increasing). Species 2's population abundance will be above its carrying capacity in the phase-plane space that sits above the blue line (where we see green shading or an arrow that shows species 2 decreasing in population density and species 1 increasing). In the space between the two isoclines, neither species has grown past its isocline (i.e., species 1 has not passed its isocline horizontally, and species 2 has not passed its isocline vertically), so both populations are still increasing (hence the arrow that shows both increasing).

Chapter 10

1. (a) Eight billion $= 8 \times 10^9$. If 1,500 calories are needed for every person and they all get their energy from the first trophic level (eating plants), then $1,500 \times (8 \times 10^9)$ calories, or 1.2×10^{13} calories, would be needed. But if everyone ate at the third trophic level (two levels higher up when only 15% of the energy from

that first trophic level is transferred to the next level), then $1,500/(0.15 \times 0.15) \times (8 \times 10^9)$ calories, or 5.3×10^{14} calories, would be needed, which is a significant increase. (b) Eating at the fifth trophic level would require 2.35×10^{16} calories at the basal level (using 15% efficiency). If only 10%–20% of energy actually transfers between trophic levels at each feeding, significantly more energy would be required at the basal trophic levels (an order of magnitude more at each succeeding trophic level) if humans were eating fish that occupied the fourth or fifth trophic level. We would need huge oceans teeming with primary producers (phytoplankton), zooplankton (primary consumers), small fish (secondary consumers), and midlevel fish (tertiary consumers) in order to sustain that kind of consumption.

3. Figure 10.16 shows the cumulative number of different scorpion prey species Polis and his researchers recorded (y-axis) as they did more and more studies (the x-axis represents the survey number). It shows that, as of survey number 200, the scorpions of the Coachella Valley were known to eat at least 100 different species of prey, which is a very broad diet. But what we do not see in the figure is any sign that the number of different prey items was leveling off. In other words, it seemed that almost every other time the researchers went out and looked at what the scorpions were eating, they found new species in their diets (i.e., they found new prey in half the samples). If the researchers had begun to approach discovering the total number of different species that the scorpions ate, the line of the graph would have started to curve and level off. Because there is no indication of leveling off, it appears that the diet of the scorpions is enormous.

5. We can see that the effects of competition are likely to lead to declines in both species (until one goes extinct or until they hit equilibrium and coexist), but the phase planes for predator and prey suggest that species will cycle. Perhaps, with apparent competition, we will see cycling that shifts to a steady decline when a second prey enters the picture. The second prey may make it so that the predator population no longer declines when the focal prey density goes down. With competitors, we should also see an increased rate of decline in the shared limiting resource. With noncompetitors that share a predator, resource densities should not decrease at a higher rate (and may even increase).

Chapter 11

1. (a) $163 \times 50,000 = 8,150,000$ specialist beetles. (b) To answer this, make a ratio

$$\frac{8,150,000}{0.4} = \frac{x}{1.0}$$

and solve for x ($x = 20,375,000$). (c) The question can be translated to

$$20,375,000 = \left(\tfrac{2}{3}\right)x$$

To solve for x, we can rearrange the equation by multiplying each side by $\tfrac{3}{2}$ to get

$$\left(\tfrac{3}{2}\right)20,375,000 = x = 30,562,500$$

(d) We make some pretty big leaps with this set of calculations. It is not at all clear that every species of tree would have 163 different specialist beetle species associated with it. This may be possible for trees in the tropical rainforest, but this number is likely much too high for trees in the temperate or boreal forests. Regardless, this kind of "back of the envelope" calculation is useful for trying to estimate a number that is very difficult to pin down. These kinds of estimations help scientists determine the upper bounds of what the number of species on the planet could be.

3. Using the data for fishes in the two streams, the values are as follows: α (i.e., the average local S) $= (6 + 7)/2 = 6.5$; γ (i.e., the total number of species found in the two streams together) $= 7$; and β (which is γ/α) $= 7/6.5 = 1.08$.

5. Rabinowitz's ideas about rarity suggest that a few species are common but that most species are rare (in one of seven ways). The graphs in Figure 11.23, which depict the range sizes for birds in different geographic regions, all show a similar pattern, namely that the majority of bird species have small range sizes and that a small proportion of the examined bird species have very large range sizes. Taken together these two ideas support each other, if we assume that birds with small range sizes are rare and that birds with large range sizes are common. Of course, this assumption does not quite work because these species could be sparsely distributed across their range. Nonetheless, that just reinforces the idea that only a very small proportion of species have large range sizes and are common or in dense populations across that whole range.

Chapter 12

1. Interestingly, the two figures actually show the same pattern. For humans in the United States, most offspring stay fairly close to their mother (50% stay within 100 miles), and a smaller number move long distances from their mother (15%–20% move more than 1,000 miles away). The same pattern is shown with the seeds from the tree of heaven in Figure 12.9. Here, most of the seeds landed within 140 m of the parent tree, and a very few made it more than 420 m from the parent tree. The figures look so different because different measures were used and different ways of determining distances moved were employed.

3. To answer to this question, think about dispersers moving from a starting point and spreading outward in concentric circles. For example, if a parent plant releases 100 seeds in any direction, but the seeds only move up to 1 m (3.3 ft.) from the parent, you would have a fairly easy time finding all those seeds, because the area they potentially moved into is relatively small (i.e., a small circle with a 1 m radius around the plant). Alternatively, if the same plant released 100 seeds but they could move up to 100 m (328 ft.) from the parent, then the seeds might occupy any spot in a circle with a 100 m radius. With this bigger circle, it would be extremely difficult to find those seeds because the area is much larger. The reason sampling has to be more intense at further distances is because the arc of the circle gets bigger further from the center. Think about 20 seeds that are in an arc that covers 90° of a circle. At a distance of 1 m from the parent plant, that arc would be easy to search. But as the arc gets further and further from the parent plant, the seeds that followed the path to get there will be more and more spread out, and the sampling protocol needs to increase proportionately to the arc in order to have an equal chance of finding the seeds. The circumference of a circle is $2\pi r$ (where r is the radius). If you want to cover 1/4 of a circle that has a 1 m radius, you only need to search a curve of $\pi/2$ m. But 100 m away, the line on which you must search needs to be 100 times as big.

5. In the mainland-island model, distant islands remain unoccupied because the distance to them affects the rate at which dispersers arrive. Larger distance means fewer dispersers. But an assumption built into the Levins model is that each patch is equally likely to be colonized, which means that distance does not affect the numbers or likelihood of dispersers.

Chapter 13

1. The two figures involved are Figure 13.20 and Figure 13.21, which both refer to Tilman's resource-ratio hypothesis of competition. In Tilman's model, two species compete for two different resources, and each species has its own physiological requirements for the two resources. The outcome of competition between the two species depends on the species' resource requirements and how much of each resource the environment supplies. In Connell and Slatyer's models of succession, there is often a pattern in which one species that is present either paves the way for another species to dominate (facilitation) or prevents another species from dominating (inhibition). The Tilman model helps us see how, as the environment changes (either by species present or by qualities that change over time), the resource supply point will vary and alter the competitive outcomes among species.

3. There are many ways to answer this question, but they all should be evaluated in light of the ecological principles presented in this text. One simple approach would be to think of a community-level property that we have talked about (e.g., biodiversity) and, placing that on the y-axis, envisioning what would happen with a pattern of repeated disturbance.

5. The fact that all communities change through time seriously complicates our estimates of biodiversity. All the ways in which we examined biodiversity in Chapter 11 (species richness, species evenness, indices of biodiversity, functional diversity, phylogenetic diversity) rested on the idea that the measure was taken or assessed at a particular location and at a particular time. But if that community or habitat is changing through time (as we know it does), any biodiversity assessment we make will be linked to a particular time. In other words, all assessments of biodiversity (no matter how detailed or sophisticated) are likely to be wrong after we make them because communities are in a constant state of flux, with successional patterns being one contributor to that variation. But the fact that successional changes generally take time means the assessments we make are not likely to be very much different in the short term.

Chapter 14

1. (a) There are several places in Figure 14.4 where we cannot see all the sources of carbon. One fairly obvious place is in the leaf and wood litter pool. This value can be influenced not only by what the trees drop but also by what is brought in by the tides. This tidal influx could contribute significant carbon from long distances and could account for large amounts on occasions (i.e., hurricanes or large storms). Another area of uncertainty lies in all the belowground processes. These are inherently difficult to study and get concise numbers from, due to the small size of the organisms involved and the challenges associated with measuring something you cannot see but that is spatially and temporally variable. (b) Despite these uncertainties, the overall assessment of mangrove forests as carbon sinks is fairly robust. This is because the variable or uncertain parts of the budget are not likely to vary enough to alter the overall pattern of carbon sequestration. In other words, it would take a huge change to alter these values enough to affect the overall result.

3. Remember that the Keeling curve shows increasing annual carbon dioxide levels from on top of Mauna Loa, and that one distinctive feature of the graph is the regular periodicity between spring and fall in the Northern Hemisphere. This cycling is due to the leafing out of the extensive forests of the Northern Hemisphere in the spring, which sequesters massive amounts of carbon dioxide, and the dropping of leaves in the fall, which then releases this carbon back into the atmosphere. The international shutdown associated with COVID-19 in the spring of 2020 led to a significant drop in global anthropogenic carbon production, as cars, buses, and planes stopped releasing carbon in their exhaust. This may have been possible to see in the Keeling curve, but it was likely masked by the annual sequestering of carbon dioxide by the trees in that spring period. In other words, this decrease in anthropogenic carbon production may be overwhelmed by the natural decrease from the trees, thereby making it hard to see in Figure 14.7.

5. Natural data tend to be quite variable due to many factors, such as annual and daily fluctuations resulting from the Earth's pathway around the Sun and its rotation on its axis. Other factors that contribute to this variability include the daily weather patterns at a location, the myriad sources of disturbance (both natural and anthropogenic) that occur at all times in all places, and the unpredictable behaviors and choices made by individual organisms. All of these contribute to the "noise" that is always present in ecological data. Ecologists and environmental scientists, therefore, often work with mean or average values, but they retain the variation by including information such as standard deviations, standard errors, and confidence intervals. This variability in data is often depicted in graphs as wider bands around the mean values.

7. The overall relationship probably wouldn't change much, regardless of the measure of biodiversity used for the x-axis. Functional diversity and phylogenetic diversity will both likely give the same nonlinear pattern of rapid increases at low diversity levels with leveling off at high diversity. This is because each of the biodiversity measures is sort of "circling" the same philosophical idea—namely, that of diversity of living organisms. The specifics of how fast the increase occurs or the height the pattern achieves before leveling off may be different for each measure, but the overall pattern is likely the same.

Chapter 15

1. Figure 15.3 shows the massive increases in nitrate, carbon dioxide, and methane (bottom three graphs) over the last 50–100 years. These increases are indicated at the far right of these graphs, where the blue lines spike up significantly as we approach the present. The top three graphs try to give some explanation for why we see these increases in atmospheric carbon. They show increases in the production of plastic, concrete, and black carbon over the past 70 years. All these changes are drastic and significant alterations to the previous thousands of years. Taken together, they strongly suggest that the current era (the Anthropocene) is distinct from the rest of the Holocene.

3. The data collected for the first flowering dates of the invasive St. John's wort in the pre-1900 period versus the 1966–1999 period (Figure 4.18) show that the plant is now flowering later into the year than it did when it first was introduced to New England. This fits with the fact that Figure 4.19 shows a steady increase in the annual average high temperature at the Harvard Forest field station in Massachusetts. As the Earth has been warming, there has been selective pressure on the invasive plant to adjust to the temperature changes, and, therefore, over time some individuals have lengthened their flowering period by more than 60 days to take advantage of this. These individuals likely leave more offspring than ones with a shorter flowering period, thereby

changing the genetic makeup of the populations. In a similar way the native spotted St. John's wort has tended to flower earlier than it did previously, again in an attempt to take advantage of the earlier onset of spring.

5. First, it's important to note that the two y-axes in Figure 15.17B are not the same. The axis for the mammals only goes from 0 to 7, whereas the one for invertebrates and plants goes from 0 to 14. So, we can eliminate the mammals, because the lower scale means they definitely do not have the most introduced species. To decide between the plants and invertebrates, we can take two different routes: (1) try to count up all species introductions for each group, or (2) estimate the difference in introductions. Either approach requires estimating the number of introductions per year across all the time periods, again keeping in mind that the time periods are different lengths. For example, the first time period is 300 years long. Using the first route, if we assume 0.5 plants were introduced per year during this time period, that means 150 species of plants were introduced during the years 1500–1800. Doing this across all the time periods for invertebrates (depending on how we interpret the bars on the graph), we arrive at approximately 900 invertebrate taxa introduced. Doing the same for plants, we get approximately 950 plant taxa introduced, suggesting there would be more plants overall.

Using the second route, we arrive at the opposite answer. Notice that there are more plant introductions than insect introductions in all but the last three time periods. Those time periods, though, only account for 100 years. It looks as though there may have been 6×25 more insects in 1975–2000, 1×25 more insects in 1950–1975, and another 1×50 insects in 1900–1950, for a total of 225 more insects in more recent years. On the other hand, it looks as though, on average, about 0.5 more plants were introduced per year from 1500 to 1850 (the first two time periods; 175 species) and about three more plant species were introduced per year from 1850 to 1900 (150 species), for a total of 325 more plant species in the first 400 years. With those estimates, we see that there were more total plant introductions.

GLOSSARY

A

ad hoc fallacy A story or narrative that explains an observed pattern or process without any proof and that is believed to be correct because it seems logical. Also known as a "just-so-story." (Ch. 1)

adaptation (1) A trait that has changed over time due to natural selection; the current phenotype helps an organism to survive and reproduce in a particular environment. (2) The process of evolution that leads to an adaptive trait becoming more common or fixed in a population. (Ch. 4)

adaptive evolution Evolution by natural selection. (Ch. 4)

adaptive modeling An approach to quantitative modeling in which the construction of the model, data input, and outputs can evolve with the changing information provided during the modeling process. (Ch. 15)

age structure A type of population structure in species such that survival and annual fecundity rates vary with age group. (Ch. 6)

Allee effect A pattern, observed in some species, in which there is a fitness decline as density declines. (Ch. 8)

allele An alternate form of a gene that has slight differences in the exact arrangement and type of nucleic acids compared to other versions. (Ch. 4)

allele frequency The number of occurrences of an allele that one would expect to find in a population of a given size. (Ch. 4)

allelopathy A biological phenomenon in which a plant releases toxins that impede the growth or reproduction of other plants. (Ch. 13)

allochthonous input Resource input into a food web from a source outside the system. (Ch. 10)

allopatric speciation Speciation between two or more spatially disjunct (i.e., not in the same place) populations. The spatial separation disrupts gene flow and makes it easier for a reproductive isolating mechanism to evolve. (Ch. 4)

allopatry The spatially disjunct existence of two or more groups—that is, the existence of groups in separate and unconnected locations. (Ch. 7)

alpha diversity (α) The average number of species found within small, local-scale samples (habitats) in a region. (Ch. 11)

alternative hypothesis A statement proposing that the focal explanatory factor or factors in an experiment do have an effect on the system of interest. (Ch. 1)

alternative stable state A novel equilibrium state for any dynamic system; in community dynamics this term may refer to succession that does not lead to the climax community; in food webs, it refers to a radically new trophic configuration of a community. (Ch. 15)

altruism A facilitative interaction in which one species or individual offers benefits to another without the expectation of a return benefit. (Ch. 9)

ammonium (NH_4^+) A form of organic nitrogen. (Ch. 14)

analytical model A model that can be solved mathematically or for which the relationships among the variables can be interpreted through mathematics. (Ch. 1)

anthrome The anthropogenic version of a biome. (Ch. 15)

Anthropocene The current geological epoch that is characterized by the large and influential effect humans have on the planet. (Ch. 1)

anthropogenic effect An impact on Earth's ecology that is caused by human actions. (Ch. 1)

aphotic zone The depths of lakes and ocean waters that receive too little light for photosynthesis. The actual depth of this zone varies among water bodies, depending on conditions in the water. (Ch. 3)

aposematic coloration A coloration pattern found in prey organisms that signals danger (e.g., poison) to a potential predator. (Ch. 8)

apparent competition An indirect, reciprocal negative effect between two species that share a predator. This (–,–) effect looks like competition, but the negative effects are produced because each species increases the density of the shared predator. (Ch. 10)

aquatic biological zone An ecological zone in a water-based environment; similar to a biome but defined by the characteristics of physical features such as light, salinity, and temperature. (Ch. 3)

arbuscular mycorrhizae (AM) A type of mycorrhizal fungus that penetrates a plant's root cells and forms a mutualism with the plant. (Ch. 14)

assembly rule A "rule" for how the timing of species arrival or the composition of the initial suite of colonizing species determines the success of subsequent colonizations and the final species composition of a community. (Ch. 13)

B

Batesian mimicry A form of interspecies mimicry in which the mimic sends a false signal to interacting species. Batesian mimicry is common in prey species that do not actually have a mechanism to kill or harm the predator (e.g., poison) but show the same signal or pattern as a species that does. (Ch. 8)

behavior The way in which individuals of a species physically interact with each other, with individuals of other species, and with their environment. (Ch. 5)

benthic zone An ocean or lake bottom. (Ch. 3)

beta diversity (β) A measure of the heterogeneity of species composition among localities (habitats); although there are many equations for beta diversity, it is most simply calculated as gamma diversity divided by alpha diversity (γ/α). (Ch. 11)

biodiversity The total number of all species living on Earth (global species richness), the evolutionary diversity represented by these species, and the diverse communities and ecosystems these species build. The term can also be applied to a locality. (Ch. 11)

biodiversity hotspot A distinct location on the planet where species richness is relatively high. (Ch. 11)

biological nitrogen fixation The process of converting N_2 into a biologically available form of nitrogen. (Ch. 14)

biological species concept The definition of a species as consisting of all individuals that can actually or potentially interbreed and produce viable, fertile offspring. (Ch. 4)

biomass Living tissue or growth, as well as storage of energy as fat, carbohydrates, or proteins. (Ch. 5)

biome A large geographic area affected by similar climatic and physical factors that lead to distinctive associations of plants and animals. (Ch. 3)

biosphere All the living organisms on Earth, including the environments in which they live. (Ch. 3)

biotic homogenization An increase in the taxonomic similarity of sites through time. (Ch. 11)

birth In the context of population dynamics, a mechanism of population increase whereby adult organisms in a population produce offspring within a given time period. (Ch. 6)

boreal forest A terrestrial biome at a high latitude that is characterized by coniferous forests and long winters during which temperatures drop below freezing, leading to permafrost. (Ch. 3)

bottom-up control Control of the energy flow in a food web by organisms at the basal trophic levels of the web (autotrophs). The autotrophs act as gatekeepers to the upward flow of energy, thereby constraining the abundance of species at the higher trophic positions. (Ch. 10)

branch A part of a phylogeny that depicts different taxa. (Ch. 11)

breeding dispersal The dispersal of adult organisms from one location to another in order to reproduce. (Ch. 12)

C

C₃ photosynthesis The most common form of photosynthesis in plants whereby the carbon capture happens at the same time and place as the Calvin-Benson cycle, which makes sugars. In this form of photosynthesis, carbon dioxide is first fixed into a molecule with three carbons, the 3-phosphoglycerate (PGA) molecule. (Ch. 5)

C₄ photosynthesis A form of photosynthesis in which carbon dioxide is first fixed into a compound containing four carbon atoms before being moved to different cells where the Calvin-Benson cycle takes place to create sugars, reducing water loss by reducing the amount of time stomata must be open to capture carbon dioxide. (Ch. 5)

Calvin-Benson cycle A process by which activated electrons from sunlight provide the energy for plants to build sugars, such as glucose, that contain energy-rich bonds. (Ch. 5)

camouflage A color pattern or morphology that allows an organism to blend in with its background and avoid being seen by potential predators. (Ch. 8)

capture rate (*f*) A term in the Lotka-Volterra predator-prey model that represents the combination of the capture success and eating mode of an organism. (Ch. 8)

carbon fixation The conversion of inorganic carbon to an organic state. The most common forms of carbon fixation involve chemosynthesis and photosynthesis. (Ch. 5, 14)

carnivore An organism that eats animal tissue. (Ch. 8)

carrying capacity (*K*) The number of individuals in a population that the resources in a habitat can sustain. (Ch. 6)

character displacement An adaptive shift in a population-level mean trait or phenotype for a trait that is critical to resource acquisition or other competitive interactions. The shift is driven by selective pressures from competition and reduces the strength of competitive interactions between two or more species. (Ch. 7)

characteristic flow regime The seasonal water flow pattern for a lotic system (e.g., a river) in a given region; often characterized by the timing of seasonal flooding. (Ch. 13)

chromosome A tight coil of DNA, which is a double-stranded molecule of bonded nucleic acids. (Ch. 4)

chronosequence A set of communities across a landscape that have different successional ages but similar environmental conditions. (Ch. 13)

clade A group of species that share a single common ancestor node on a phylogeny. (Ch. 11)

climate An expected pattern in the physical factors that make up the world (e.g., precipitation, heat, wind). (Ch. 3)

climate diagram A type of graph that combines data on a region's annual patterns of precipitation and environmental temperature. (Ch. 3)

climax community A predictable and permanent assemblage of species that occurs in a location after a long period of succession. (Ch. 13)

coevolution Evolution by natural selection through reciprocal selective pressures between two or more closely interacting species. (Ch. 8)

cohort A group of individuals of a species that are all born at the same time. (Ch. 6)

commensalism An interaction between individuals of two species that benefits one individual but neither benefits nor harms the other individual. (Ch. 9)

community A group of interacting species in the same location. (Ch. 13)

community ecology The study of ecological interactions between individuals of two or more species. (Ch. 7)

competition coefficient The numeric conversion term in the Lotka-Volterra competition equations that translates the relative effect of an individual of a competing species into units of individuals of the focal species. (Ch. 7)

competitive exclusion principle The principle that when two species are too similar in their resource use, they cannot coexist. (Ch. 7)

conceptual model A theoretical construct that specifies how various components of a system fit in relation to each other. (Ch. 1)

consumption vector In a graphical representation of the Tilman model of interspecific competition, a line representing the rate at which individuals of a species consume available resources. (Ch. 7)

continental effect An increase in the range of seasonal average temperatures with distance from the regulating climatic effects of any large body of water. Continental effects lead to much hotter summers and colder winters inland than along a coast. (Ch. 3)

continuous numerical variable A data type whose possible values are numbers that can take on any value, including fractions and decimals (e.g., acorn length). (Ch. 2)

continuous-time model A model that keeps track of all the little changes in the focal variable that accumulate through time; such models treat the temporal component of the model in a nondiscrete manner using calculus. The exponential and logistic population growth models, Lotka-Volterra competition and predation models, and metapopulation models are all continuous-time models. (Ch. 6)

control An unmanipulated unit of an experiment; generally used for comparison to detect the effect of a treatment in a manipulated unit of an experiment. (Ch. 1)

coral reef An aquatic biological zone found in warm, shallow, tropical waters that is dominated by coral animals, other invertebrates, and fishes. (Ch. 3)

Coriolis effect The deflection of air and wind patterns in bands around the globe caused by the Earth's daily rotation on its axis. (Ch. 3)

Crassulacean acid metabolism (CAM) A form of photosynthesis in which there is a temporal separation between when a plant takes up carbon dioxide and when it makes sugars. With CAM, the plant opens its stomata only at night, when evaporation is lowest, and takes in as much carbon dioxide as possible. During the day, the plant uses the stored carbon in the Calvin-Benson cycle. (Ch. 5)

crypsis The ability of an organism to conceal itself. (Ch. 8)

cryptic species Two or more species "hidden" under a single species name. (Ch. 11)

cyanobacteria A photosynthetic form of bacteria that can convert N_2 into a reactive form of biologically available nitrogen. (Ch. 14)

D

damped oscillation Population dynamics that cycle with decreasing amplitude through time. (Ch. 8)

data Individual facts, statistical results, or items of information that are often (but not always) numerical. (Ch. 1)

data literacy The ability to collect, analyze, make sense of, and communicate the meaning of quantitative and qualitative data. (Ch. 2)

death In the context of population dynamics, a mechanism of population decrease whereby individual organisms in a population die within a given time period. (Ch. 6)

decomposer A bacterium, fungus, or protist that feeds on dead plant and animal parts (detritus) as its source of energetic nutrition. (Ch. 10)

defaunation The loss of animals from the biosphere. (Ch. 15)

definitive host An organism that supports the adult form of a parasite. (Ch. 8)

denitrification A process in which bacteria chemically reduce NO_3^- to N_2 through a series of intermediate steps. (Ch. 14)

density dependence A phenomenon in which individuals in a population reproduce less and have lower probabilities of survival within a given time period as population density increases. These reductions in

reproduction and survival will slow the population growth through time. (Ch. 6)

density-independent factor An event or feature that influences resource levels or influences the mortality and reproduction of individuals in a population in ways that do not depend on the density of the population. (Ch. 6)

depauperate *Of a location*: poor, or low, in biological diversity. (Ch. 11)

desert A terrestrial biome that receives very little precipitation year-round, has low primary productivity, and where evaporation exceeds precipitation. (Ch. 3)

deterministic model A mathematical model that will always produce the same results if the starting conditions are the same. (Ch. 2)

detritivore Any organism that feeds on dead plant and animal parts (detritus) as its primary source of energetic nutrition. (Ch. 8, 10)

detritus Dead plant or animal material. (Ch. 10)

directional selection A shift in the trait frequency distribution of an organism whereby the mean trait value tends to shift through time toward one or the other end of the original distribution; a form of natural selection. (Ch. 4)

discrete numerical variable A data type whose possible values are numbers but which can only occur in whole numbers, not fractions or decimals (e.g., the number of ladybugs on a leaf). (Ch. 2)

discrete-time density-dependent population growth A type of population growth in which time is considered in chunks (e.g., years, months, days, or seconds) and in which the population shows density dependence. (Ch. 6)

discrete-time population growth model A population growth model in which the calculation of population abundance or density is done once after each full time step (e.g., year, month, day, or second); in other words, time is considered discrete (coming in "chunks"), not continuous. (Ch. 6)

dispersal Movement by individuals that generally does not involve predictable routes. (Ch. 12)

dispersal kernel A mathematical or graphical representation of a distribution of dispersal distances for a given population. (Ch. 12)

disruptive selection A shift in a trait frequency distribution whereby the extremes of the frequency distribution become more

common over time; a form of natural selection. (Ch. 4)

dissolved inorganic carbon (DIC) Any inorganic form of carbon dissolved in water. (Ch. 14)

dissolved organic carbon (DOC) Any organic form of carbon dissolved in water. (Ch. 14)

disturbance Multiple ecological definitions, each of which highlights factors that cause changes in local biomass or community composition, the changes themselves, or both; sometimes includes the processes associated with such changes. (Ch. 13)

disturbance regime An overview of disturbance in a location that includes a description or quantification of multiple factors, such as distribution, frequency, magnitude, predictability, return interval, or synergism. (Ch. 13)

dominant (allele) An allele that codes for its trait even if there is only one copy of the allele at a genetic locus. (Ch. 4)

E

ecology The study of the interactions that determine the distribution and abundance of organisms. (Ch. 1)

economic model of trade A model of mutualistic interactions that assumes that two independent groups differ in their abilities to access resources in ways that make it advantageous for each species to trade some resources. (Ch. 9)

ecosystem A group of interacting organisms (biotic components) along with their physical environment (abiotic components). (Ch. 1)

ecosystem budget A quantitative depiction of the transformation and pathway of nutrients and water through an ecological system. (Ch. 14)

ecosystem engineer A species that creates, modifies, maintains, or destroys a physical habitat. (Ch. 10)

ecosystem flow The net movement of nutrients or water across a particular area over a set time period. (Ch. 14)

ecosystem function A set of ecological outcomes and processes that results from the intricate interplay among the species that make up an ecosystem and that allows the system to continue to cycle nutrients and support living organisms. (Ch. 14)

ecosystem loop The flow of nutrients or water back and forth between ecosystem pools. (Ch. 14)

ecosystem pool   A store of nutrients or water within an ecosystem. (Ch. 14)

ecosystem services   The delivery of ecosystem functions and other critical components by natural ecosystems in ways that benefit humans. (Ch. 14)

ecotype   A population of a species with locally adapted traits. (Ch. 4)

ectomycorrhizae (EM)   A type of mycorrhizal fungus that covers the outside of a plant's root cells, forming a mutualism with the plant. (Ch. 14)

ectotherm   An organism that regulates its internal temperature (i.e., thermoregulates) using only external mechanisms, such as moving into the shade or sun. (Ch. 5)

emigration   A mechanism of population decrease whereby individuals leave a population through dispersal. (Ch. 6)

endemic species   A species found only in a restricted set of locations or habitat types and nowhere else. (Ch. 11)

endogenous   Produced from within a system. (Ch. 10)

endogenous disturbance   A disturbance event that has a historical or evolutionary presence in the system and produces change from within the system. (Ch. 13)

endotherm   An organism that regulates its internal temperature (i.e., thermoregulates) using both external and internal mechanisms. (Ch. 5)

energetic profitability   The ratio of energy gained from eating a prey item to the energy costs associated with acquiring and eating that prey item. (Ch. 5)

energy flow   The direction in which nutrients and resources are moving up or down a food chain. (Ch. 10)

equilibrium theory of island biogeography   An ecological theory suggesting that the number of species on an island can be determined by a balance of colonization and extinction, where there is turnover in the species present in the community over time. (Ch. 12)

equilibrium view   A historical idea about the natural world that suggests all communities eventually settle into an equilibrium or steady state. (Ch. 13)

estuary   An aquatic biological zone that sits at the juncture of a river (or rivers) and an ocean and that has a distinct salinity gradient from fresh to salt water. (Ch. 3)

eutrophic   *Of a body of water (e.g., a lake):* supporting an abundance of basal autotrophic species and having an overall high biomass of organisms per unit area. (Ch. 10)

evolution   A change in gene frequency in a population over time. The four mechanisms of evolution are mutation, gene flow, genetic drift, and natural selection. (Ch. 4)

evolutionary arms race   A constant cycle of trait evolution in which two coevolved species respond to each other's adaptations over time. (Ch. 8)

exclusion   The restriction or removal of individuals of one species from a location by another species; a possible long-term outcome of competition or predation. (Ch. 7)

exogenous   Produced from outside a system. (Ch. 10)

exogenous disturbance   A disturbance event that comes from outside the system (e.g., an asteroid impact or volcanic eruption) and generally lies beyond the evolutionary range that the system has experienced in the past. (Ch. 13)

explanatory variable   A variable that affects, changes, or explains variation in a response variable. In an experimental study, an explanatory variable is often manipulated by the researcher. Also sometimes called the independent, or predictor, variable. (Ch. 2)

exploitation   An interaction between individuals of two species in which the individuals of one species increase in fitness by consuming individuals (or parts of individuals) of another species. Individuals of the consumed species experience a decrease in abundance or fitness. (Ch. 8)

exploiter conversion factor (*c*)   A term in the Lotka-Volterra predator-prey model that translates consumed prey into an expected number of offspring for a focal species. (Ch. 8)

exploiter-prey cycle   A complex cycle between predator (exploiter) and prey in which the density or abundance of each species alters the density or abundance of the other species; predator density increases as prey density increases, but prey density decreases as predator density increases. These reciprocal reactions lead to increasing and decreasing densities of each species with the prey species leading the cycle, and predators following 1/4 cycle behind. (Ch. 8)

exponential population growth   Population growth dictated by the equation $dN/dt = rN$ or $N_t = e^{rt}$. The exponent on the right side of the equation leads to a pattern of population growth in which abundance increases slowly at first but then accelerates through time, with no eventual limit to the growth. The continuous time formulation suggests that reproduction is possible throughout the year. (Ch. 6)

extinction   The situation when no more individuals of a population or species are present in a location or on Earth; a possible long-term outcome of competition or predation. (Ch. 7)

extraneous factor   An aspect of an ecological system that is not of interest or of importance to the researcher designing an experiment. (Ch. 1)

F

facilitation   Interactions that benefit at least one species and have either a positive or net zero effect on the other species. (Ch. 9)

facilitation model of succession   A conceptual model positing that earlier-arriving species alter the environment in ways that enhance the establishment and growth of later successional species. (Ch. 13)

facultative mutualism   A mutualistic interaction that improves the fitness of one of the participating species but is not necessary for that species' continued existence. (Ch. 9)

fast carbon cycle   The movement of carbon through life forms. (Ch. 14)

fecundity   The number of young produced per individual in a population over a given time period. (Ch. 6)

Ferrel cell   A cyclic atmospheric pattern of airflow between 30° and 60° latitude with one Ferrel cell in the northern hemisphere and one in the south. (Ch. 3)

finite per capita rate of population growth (λ)   A term in the discrete-time population growth equation that represents the combined per capita birth and survival rate. (Ch. 6)

first law of thermodynamics   The law stating that energy cannot be created or destroyed. Accordingly, the amount of energy taken in by an individual organism must be equal to the amount that it uses plus the amount that it loses. (Ch. 5)

fitness   The genetic contribution that an individual makes to future generations. Traits that increase survival and reproduction increase an organism's fitness, so long as the organism's offspring also survive to reproduce. (Ch. 4)

focal factor   An important or critical aspect at play in an ecological system that scientists focus on (alter or manipulate) when designing an experiment or constructing a model. (Ch. 1)

food chain A hierarchical linear set of interacting species depicting trophic interactions. (Ch. 10)

food web A graphical representation of the complex energetic feeding relationships among species that coexist in a community. (Ch. 10)

foundation species A species that—in proportion to its biomass—facilitates the presence of many other species, thus providing the foundation or infrastructure for a whole community or ecosystem. (Ch. 9)

fragmentation A landscape pattern whereby a single contiguous habitat is divided into small scattered blocks embedded within human-dominated habitats. (Ch. 12)

frequency distribution A graph showing the number of observed occurrences across the possible values of a variable. (Ch. 2)

functional diversity (FD) A measure of the diversity of species' characteristics or functional roles of species within a community; the sum of the branch lengths of a functional diversity dendrogram. (Ch. 11)

functional diversity dendrogram A two-dimensional visualization of the relationships among species in terms of their functional traits; uses root, nodes, and branch lengths in a fashion analogous to a phylogenetic tree. (Ch. 11)

functional evenness A measure of functional diversity that represents the proportion of individuals within each functional group across a functional dendrogram. (Ch. 11)

functional response The number of prey eaten per exploiter per unit of time. (Ch. 8)

functional richness The number of different functional traits present in a community. (Ch. 11)

functional trait Any characteristic of a species that describes an aspect of the species' ecological role in its community, including how that species interacts with other species and with the environment. (Ch. 11)

fundamental niche The full range of conditions (biotic and abiotic) and resources within which individuals of a species can survive and reproduce. (Ch. 7)

G

game theory The study of conflict and cooperation between intelligent, rational decision-makers. (Ch. 9)

gamma diversity (γ) The species richness across a region made up of smaller localities; gamma diversity is a product of alpha (α) and beta (β) diversity. (Ch. 11)

gene A sequence of deoxyribonucleic acid (DNA) that contains the code for a biological molecule with a particular physiological or behavioral function. (Ch. 4)

gene flow The movement of alleles from one population to another via the movement of organisms or their gametes across space; one of the four mechanisms of evolution (mutation, gene flow, genetic drift, and natural selection). (Ch. 4)

generalism The ability of an organism to use multiple species, resources, or habitats for a particular function. In mutualisms, generalism means at least one partner is able to swap and use a different species as its mutualistic partner. (Ch. 9)

generalist A species or organism that is able to use an array of species, resources, or habitats to survive. In competition and exploitative or trophic interactions, a generalist is a species that rarely refuses available food items and whose diet tends to include a diverse array of food types. In mutualisms, a generalist is a species whose mutualistic partner can be a range of species. (Ch. 8)

genetic drift The removal of genes from a population over time through random mating between individuals or by the random death of individuals; one of the four mechanisms of evolution (mutation, gene flow, genetic drift, and natural selection). (Ch. 4)

genetic mutation Errors in the replication of DNA within a cell that create slight differences in nucleotide sequences, leading to novel alleles; one of the four mechanisms of evolution (mutation, gene flow, genetic drift, and natural selection). (Ch. 4)

genotype A specific combination of alleles in an individual at a particular genetic locus. A genotype interacts with the environment to create a specific phenotype. (Ch. 4)

genus A level in the taxonomic hierarchy; the first part of the Latin binomial (scientific name) that identifies a species. (Ch. 4)

geographic isolating mechanism A physical barrier that prevents organism dispersal and limits gene flow. (Ch. 4)

geographic range The spatial distribution over which individuals of a species can be found. (Ch. 12)

geometric growth equation The discrete-time version of the basic population growth model. (Ch. 6)

geometric population growth Population growth dictated by the geometric growth equation and leading to a pattern in which abundance increases slowly at first but then accelerates through time, with no eventual limit to the growth. The discrete time nature of the model assumes that reproduction occurs in one distinct time period or season in a time step or year. (Ch. 6)

ghost of competition past The idea that past competition between two or more species may have altered resource use and interactions sufficiently that the species no longer compete; the only lingering signs of their previous competition are non-overlapping but similar resource use. (Ch. 7)

greenhouse effect The retention of heat by the Earth's atmosphere. The atmosphere allows sunlight energy to enter and then retains the longer-wavelength heat energy, thereby warming the planet. This effect is responsible for a climate that supports life on Earth; however, the tons of anthropogenically produced gases, such as carbon dioxide and methane, contribute to accelerating global warming, which is altering global ecological systems. (Ch. 3)

greenhouse gas A gas in the atmosphere that warms the Earth by letting light energy in but prevents reflected heat radiation from leaving the atmosphere. (Ch. 14)

gross primary production (GPP) The total amount of carbon an organism, or collection of organisms, fixes through photosynthesis; typically calculated over the span of one year. (Ch. 14)

H

Haber-Bosch process An anthropogenic process that takes atmospheric N_2 and runs it over a metal catalyst with hydrogen under extreme pressure and at high temperatures to produce ammonia (NH_3). (Ch. 14)

habitat patch A parcel or chunk of suitable habitat within a landscape. (Ch. 12)

Hadley cell A cyclic atmospheric pattern of airflow caused by the solar heating of air at the equator and the cooling and descending of air at 30° latitude. There are two Hadley cells; one from the equator to 30° north latitude and the other from the equator to 30° south latitude. (Ch. 3)

handling costs The amount of energy needed by a predator to capture, manipulate, and consume its prey. (Ch. 5)

herbivore An organism that feeds on individuals, or parts of individuals, of a primary producer species. (Ch. 8)

heritable *Of a characteristic (phenotype)*: transmissible from parent to offspring via the genetic inheritance of alleles. (Ch. 4)

heterozygous Having two different alleles for a particular gene. (Ch. 4)

homeotherm An organism that uses metabolic heat, a range of muscle motions (e.g., shivering, panting), and changes in blood circulation to keep its internal temperature within a narrow range; generally mammals and birds. (Ch. 5)

homozygous Having two copies of the same allele for a particular gene. (Ch. 4)

host An organism that harbors a parasite or pathogen. (Ch. 8)

hydrologic cycle The net movement of water through the Earth system; driven by precipitation, evaporation, and transpiration. (Ch. 14)

hypothesis A tentative or proposed explanation about the world, made on the basis of limited evidence, that is a starting point for further, typically scientific, investigation. (Ch. 1)

hypothesis testing A method of statistical or scientific inquiry used to determine whether an assumption or explanation applies to a group or sample. (Ch. 1)

I

immigration A mechanism of population increase whereby individuals enter a population through dispersal. (Ch. 6)

indirect effect The overall effect of one species on a second species when both species interact with a third species or with a shared resource; the third species or shared resource mediates the interaction in such a way that the effect may even change in sign or direction. (Ch. 10)

inhibition model of succession A conceptual model positing that early successional species actively inhibit and prevent the establishment and growth of species that arrive later. (Ch. 13)

inorganic carbon Carbon-based molecules that are not a part of living or formerly living material. (Ch. 14)

inorganic nitrogen Nitrogen that is not derived from living or formerly living tissues. (Ch. 14)

instantaneous per capita rate of population growth (r) A term in a population growth model that represents the instantaneous per capita birth rate minus the instantaneous per capita death rate. (Ch. 6)

interference competition Competition between individuals of two species that is characterized by direct interaction through aggressive behavior, behavioral displays, or the release of toxins to increase access to a limiting resource. (Ch. 7)

intermediate disturbance hypothesis The hypothesis that the highest levels of species diversity (measured either as species richness or via the Shannon-Wiener diversity index) are found in ecosystems that have middle, or intermediate, levels of disturbance. (Ch. 13)

intermediate host An organism that supports the nonadult form of a parasite. (Ch. 8)

interspecific competition Competition between individuals of different species for limiting resources that are needed for reproduction or survival. (Ch. 7)

intraspecific competition Competition between individuals of the same species for a limiting resource that is critical for reproduction or survival. (Ch. 6)

intrinsic rate of increase A population's maximum per capita growth rate in a particular habitat. (Ch. 6)

invasional meltdown An ecological phenomenon in which the presence of one invasive species can facilitate the establishment of many other invasive species. (Ch. 9)

invasive species A species introduced to a new geographic range by humans that then spreads rapidly in the new environment and potentially has significant impacts on native ecosystems and species. (Ch. 8)

isocline A line representing all points where a species has zero population growth on a phase-plane graph. (Ch. 7)

iteration The repetition of a mathematical model multiple times; sometimes the outcome from one pass through the model is then used as the starting point for the next pass through the model. (Ch. 2)

iterative prisoner's dilemma (IPD) A game theory model in which the two individuals participating in the prisoner's dilemma "game" have multiple opportunities to interact through time. (Ch. 9)

iteroparous *Of an organism*: reproducing multiple times throughout its life. (Ch. 5)

K

Keeling curve The long-term record of atmospheric carbon dioxide collected from an observatory on Mauna Loa, Hawaii; initiated by Charles Keeling in the late 1950s. (Ch. 14)

keystone mutualist A mutualist whose interactions with mutualist partners have a positive effect on community structure, function, and diversity far out of proportion to the mutualist's relative biomass in the community. (Ch. 10)

keystone species A species that has a disproportionately large effect on community structure, function, and diversity relative to its biomass or abundance. (Ch. 10)

K-selected species Species that tend to be large in body size and have low rates of reproduction, high rates of survival, long generation times, slow development, and late maturity. (Ch. 13)

L

Levins metapopulation model A model that keeps track of the proportion of habitat patches occupied by subpopulations and how that proportion changes over time. (Ch. 12)

life history The temporal sequence of events that determines survival and reproduction from an individual's birth until its death. (Ch. 6)

logistic equation An equation that includes a carrying-capacity term that reflects the effects of limiting resources on population growth rates. The logistic equation depicts limits to population growth that draw the population toward the carrying capacity. (Ch. 6)

logistic growth model A population growth model in which changes in abundance or density are governed by the logistic equation. (Ch. 6)

lotic *Of a water ecosystem*: flowing, as a river or stream. (Ch. 10)

M

macroparasite A parasite that is large enough to be seen with the naked eye (e.g., a mosquito). (Ch. 8)

mainland-island model A model that depicts the effects of organism movement from one big mainland patch to an array of islands that lie at different distances from the mainland. This model allows for the examination of how distance and patch size influence regional metapopulation dynamics. (Ch. 12)

mangrove forest An aquatic biological zone that is found at the terrestrial and marine interface in the tropics and subtropics and is dominated by mangrove trees. (Ch. 3)

manipulative experiment A scientific experiment in which the researcher intentionally alters one or more conditions in order to examine hypotheses. (Ch. 1)

mark-recapture model A model that uses observational data from tagged individuals, who are later recaptured, to estimate population abundance. (Ch. 6)

mathematical model A conceptual model that is translated into the language of mathematics. (Ch. 1)

matrix The environment, or habitat, that surrounds metapopulation patches. (Ch. 12)

Mediterranean scrubland A terrestrial biome in which rainfall occurs in the cool winters, while the summers tend to be very warm and dry. Found on the west coasts of continents, next to cool oceanic currents. (Ch. 3)

metacommunity A collection of communities on habitat patches that are distributed across a landscape and linked via dispersal. (Ch. 12)

metapopulation A set of populations of a single species on habitat patches spread across a landscape and linked by dispersal. (Ch. 12)

microbiota The collection of microbes that live in a particular site or location, as within the human digestive system. (Ch. 9)

microparasite A parasite that is tiny and often occurs intracellularly (e.g., a virus). (Ch. 8)

migration The coordinated long-range movements of whole populations or segments of populations. (Ch. 12)

mimicry An adaptation that benefits individuals that look very similar to individuals in another species; a common form of mimicry is as a type of secondary defense strategy used by prey species with very similar appearances to other species that advertise their toxicity (or that they are venomous) through coloration and patterning. (Ch. 8)

morphology An organism's appearance (from *morpho*, which means "shape"). (Ch. 4)

Müllerian mimicry A form of mimicry in which there is a selective advantage for both the mimic and the model to closely resemble each other, thereby gaining protection by reinforcing a cue sent by their similar trait. (Ch. 8)

mutualism A type of interaction between individuals of two species in which both individuals experience a benefit. (Ch. 9)

N

natal dispersal The movement of organisms from their place of birth or hatching to another location. (Ch. 12)

natural selection The differential survival and reproduction of individuals in a population in response to biotic and abiotic factors in their environment acting on heritable variation in traits of individuals; one of the four mechanisms that produces evolutionary change (mutation, gene flow, genetic drift, and natural selection). (Ch. 4)

naturalist A person who studies or observes natural organisms or phenomena in detail. (Ch. 1)

n-dimensional hypervolume G. Evelyn Hutchinson's definition of a species' ecological niche, in which *n* indicates the number of dimensions, or different environmental factors, that determine where an organism can survive. (Ch. 5)

net primary production (NPP) The amount of organic carbon left over from photosynthesis after respiration; generally calculated for a single plant or a collection of plants. (Ch. 14)

niche The specific set of abiotic and biotic conditions under which an organism can live and reproduce. (Ch. 5)

niche shift An alteration in the way individuals of a species use resources in the presence of a competitor species. (Ch. 7)

niche space A specific region in a multidimensional setting constrained or defined by environmental factors that affect one or more species' fitness; we may use niche space to explore the physiological requirements of a single species or to compare resource use for multiple species. (Ch. 5)

nitrate (NO₃⁻) The form of inorganic nitrogen that is most easily assimilated by plants and microbes. (Ch. 14)

nitrogenase An enzyme present in some bacteria and archaea that allows them to convert atmospheric N_2 into ammonia (NH_3). (Ch. 14)

node A point of branching within a phylogeny. (Ch. 11)

nominal categorical variable A data type associated with no inherent numerical value and that is defined by names (e.g., blue, green, yellow). (Ch. 2)

non-native species A population founded by individuals transported by humans and released well outside their native range. (Ch. 4)

normal distribution A continuous, symmetrical, bell-shaped frequency distribution in which the proportion of values found at particular distances from the central value is highly predictable. Often used in statistical analyses and also known as a Gaussian distribution. (Ch. 2)

null hypothesis A statement proposing that the focal explanatory factor or factors in an experiment do not have an effect on the system of interest. (Ch. 1)

numerical variable A data type whose values are numbers. (Ch. 2)

nutrient budget A quantitative record-keeping of the concentrations of specific molecules as they move through and among the living and nonliving components of an ecosystem. (Ch. 14)

nutrient cycle The way that molecules cycle through the living and nonliving world. (Ch. 14)

O

obligate mutualism A mutualistic interaction that is necessary for the survival or reproduction of at least one of the participating species. (Ch. 9)

ocean acidification The drop in the ocean water's pH caused by the high rate of carbon dioxide emissions into the atmosphere and the subsequent dissolution of this carbon dioxide in ocean waters. (Ch. 14)

oceanic zone An aquatic biological zone that consists of open ocean. (Ch. 3)

oligotrophic *Of a body of water (e.g., a lake)*: supporting few primary producers and thus having a low overall biomass of organisms per unit area. (Ch. 10)

omnivore An organism that acquires energy by consuming across multiple trophic levels. (Ch. 10)

optimal foraging theory An ecological theory positing that an individual animal acts to gain the most energy for the least cost when making its foraging decisions, with the overall goal of maximizing its evolutionary fitness. (Ch. 5)

optimization The maximization of a benefit relative to costs. In a foraging context, the maximization of energy acquisition or energetic performance of an organism relative to the costs to that organism of acquiring that energy. (Ch. 5)

ordinal categorical variable A data type associated with no inherent numerical value but that does have a natural order (e.g., small, medium, large). (Ch. 2)

organic carbon Carbon-based molecules used by living organisms. (Ch. 14)

organic nitrogen Nitrogen-based molecules used by living organisms. (Ch. 14)

organism A single individual of any type of living creature on the planet, including all plants, animals, and microbes. (Ch. 1)

P

parameter An element in a mathematical model that—unlike a variable—is not affected by the model; often represented by a lowercase Greek or Roman letter. In many ecological models, parameters vary for different species or locations but do not vary over time. (Ch. 2, 6)

parasite An organism that obtains its energy from a host. (Ch. 8)

parasitoid An insect exploiter species that attacks other insects (hosts) and lays its eggs or larvae inside the host; the larvae eventually develop, consume, and kill the host insect. (Ch. 8)

particulate organic carbon (POC) Carbon that is not dissolved in water and thus can be filtered or otherwise removed from the water. (Ch. 14)

pathogen An exploiter species that causes disease in its hosts. (Ch. 8)

pelagic zone The open water of an ocean or lake. (Ch. 3)

per capita rate A number that provides an expectation of what to expect for each individual, or "head" (*capita* is the Latin word for "head"), in a population. (Ch. 6)

per capita survival rate The number of individuals in a population that survive from one time period to the next, expressed as a fraction of the total number of individuals in the population at that time. (Ch. 6)

periphyton An alga in a shallow-water zone that attaches to substrate. (Ch. 10)

phase-plane graph A graphical representation of two population densities simultaneously, with the density of one species given along one axis and the density of the other species given along the other axis. Used to explore outcomes of interspecific competition and exploiter-prey interactions. (Ch. 7)

phenology The cyclic and seasonal events in the life of a plant or animal; also the study of cyclic and seasonal biological phenomena in relation to climate. (Ch. 4)

phenotype The specific observable physical, developmental, or behavioral trait of an organism; determined by the organism's genotype. (Ch. 4)

photic zone The portion of lakes and ocean waters where light is sufficient for

photosynthesis. The actual depth of this zone varies among water bodies, depending on conditions in the water. (Ch. 3)

photorespiration The physiological process in plants whereby oxygen (O_2) is combined with sugars and carbon dioxide (CO_2) is released. It takes place during photosynthesis. (Ch. 5)

photosynthesis The synthesis of carbohydrates from CO_2 and water, using sunlight as an energy source; photosynthesis is an autotrophic form of energy capture used by plants and some prokaryotes. (Ch. 5)

phylogenetic diversity (PD) A measure of evolutionary diversity within a community; the branch lengths within a phylogeny are summed to create a single number that can be compared across locations or through time. (Ch. 11)

phylogeny A quantitative graphical depiction of evolutionary relationships among a suite of species. (Ch. 11)

physiology The study of bodily functions and cellular mechanisms at work within living organisms. (Ch. 5)

poikilotherm An organism that does not regulate its internal temperature but instead allows its internal body temperature to follow external temperatures. (Ch. 5)

Polar cell A cyclic atmospheric pattern of airflow between 60° and 90° latitude; one cell is in the northern hemisphere and a second in the southern hemisphere. (Ch. 3)

population (1) In the context of sampling and statistics, the largest group of things under study. (2) In the context of evolution, a group of individuals of the same species that can mate together. (3) In the context of ecology, a group of individuals of the same species that is spatially distinct from other groups of individuals of the same species. (Ch. 2, 4, 6)

population abundance The number of individuals in a population; sometimes also called population size. (Ch. 6)

population density (N) The number of individuals in a population within a defined spatial area. Because all studied populations are in a defined spatial area population density is generally synonymous with population abundance, and both are represented by N in most ecological models. (Ch. 6)

predator An organism that kills and consumes prey. (Ch. 8)

predator-mediated coexistence A state of coexistence among prey species in a food web that is facilitated, indirectly, by a keystone or other predator that selectively

consumes large numbers of a competitively dominant prey species, thus allowing other species to survive and increase the diversity of the community. (Ch. 10)

prey An organism that is killed and consumed by a predator. (Ch. 8)

primary consumer An organism that mainly eats plants or other autotrophs. (Ch. 10)

primary producer An autotroph (self-feeder) that produces its energy from inorganic sources rather than by consuming other biological organisms. (Ch. 10)

primary productivity The synthesis of organic material through the process of photosynthesis or chemosynthesis; this term refers most commonly to production of biomass by plants through photosynthesis. (Ch. 3)

primary succession The temporal sequence of changing communities that begins on substrates containing no organisms and no organic material, meaning that the community assembles completely from scratch. (Ch. 13)

principle of allocation The principle stating that, because all individuals have limited access to energy, energy allocated to one of life's necessary physiological functions reduces the amount that can be allocated to other such functions. (Ch. 5)

priority effect An ecological phenomenon in which the order of species' arrivals influences the final community membership. (Ch. 13)

prisoner's dilemma A game theory model that describes the decisions of two prisoners weighing their options while being interrogated by police for a crime they committed together. (Ch. 9)

p-value A statistical term describing the probability of collecting observed data if the null hypothesis is true; in practice this often means the probability of getting the statistical results of an experiment by random chance. (Ch. 2)

Q

quantitative reasoning The application of math or mathematical thinking to measured or counted information in order to gain a deeper understanding of the topic. (Ch. 2)

R

R^* A threshold resource level in the Tilman model of competition above which survival and reproduction increase as resource availability increases; a point of zero net population growth for an individual species. (Ch. 7)

rain shadow A climatic pattern whereby mountain ranges affect air movement, causing air masses to lose or deposit moisture on the "front" side (windward side) of mountains and create very dry conditions on the "back" side (leeward side) of mountains. (Ch. 3)

realized niche In the context of interspecific competition, the niche space that individuals of a species can access in the presence of their competitors. (Ch. 7)

recessive (allele) An allele that codes for its trait only if there are two copies of the allele at a genetic locus. (Ch. 4)

reciprocal altruism A form of self-reinforcing reward for cooperation between two individuals. (Ch. 9)

recruitment A combination of factors that leads to new individuals entering the next age class, used most frequently to refer to plants entering a reproductive age class. (Ch. 8)

Red Queen hypothesis The evolutionary hypothesis that all species must constantly evolve in order to maintain their fitness in the face of a changing environment full of other evolving species. (Ch. 8)

relative fitness An individual's contribution to the gene pool of future generations relative to the contribution from other individuals in the same population. (Ch. 4)

renewal vector In a graphical representation of the Tilman model of interspecific competition, a line that represents the rate at which resources enter a system. (Ch. 7)

reproductive isolating mechanism A mechanism that separates two evolutionary lineages, making it so they can no longer interbreed; all speciation requires a reproductive isolating mechanism. (Ch. 4)

resource Anything that can be used up and for which organisms might compete. (Ch. 5)

resource competition Competition between two species for a shared resource; by consuming or exploiting that resource, individuals of each species make it less available for individuals of the other species; also called exploitative competition. (Ch. 7)

resource partitioning The splitting of limited resources among individuals of two competing species. (Ch. 7)

resource supply point (S) The steady amount of resources in a system that is naturally resupplied by the environment in the Tilman model of competition. (Ch. 7)

resource utilization curve A graph that shows how an organism uses a particular resource in its environment. (Ch. 5)

resource-ratio hypothesis An ecological hypothesis that states that the coexistence of competitor species depends on the ratio of availability of two critical resources in the Tilman model of competition. (Ch. 7)

respiration At the cellular level, a metabolic process that breaks large molecules into smaller ones and releases carbon, energy, and water in the process. (Ch. 5)

response variable A variable that is affected by, or changes as the result of, the effects of an explanatory variable. In an experimental study, this is the outcome that is measured following manipulation of the explanatory variable. Also sometimes called the dependent variable. (Ch. 2)

restoration ecology The recovery of degraded or damaged ecosystems through active human intervention, which is generally termed *management*. (Ch. 9)

resultant vector A line in a Lotka-Volterra competition or predation model phase-plane graph that indicates the direction in which the population abundances of the two species of interest will move over time. (Ch. 7)

rocky substrate An aquatic biological zone that is found along the rocky shoreline of terrestrial systems with cool to cold water. (Ch. 3)

root The single branch and node at the base of a phylogeny. (Ch. 11)

root-to-shoot ratio A comparison of the biomass of a plant's roots to the biomass of its aboveground material, such as leaves, stems, and flowers (i.e., its shoots). (Ch. 5)

r-selected species Species that tend to be small in body size and have high rates of reproduction and dispersal, low rates of survival, short generation times, rapid development, and early maturity. (Ch. 13)

S

sample A subset of individuals from a larger group used to characterize or make inferences about the larger group. (Ch. 2)

sandy bottom An aquatic biological zone dominated by a sandy substrate that is extremely dynamic and is constantly moving due to tides and wave action. (Ch. 3)

scientific name The genus and specific epithet that provide a name for a species. (Ch. 4)

seagrass bed An aquatic biological zone that spans the terrestrial and marine interface and is dominated by aquatic herbaceous plants called seagrasses. (Ch. 3)

search costs The energetic costs a predator expends finding its prey. (Ch. 5)

secondary consumer An organism that mainly eats animals that are primary consumers. (Ch. 10)

secondary metabolite An organic chemical that is typically not involved in a plant's basic growth and development but which can be essential to its survival. (Ch. 8)

secondary succession The temporal sequence of changing communities that occurs on already established soils following a disturbance to the previous habitat. (Ch. 13)

semelparous *Of an organism*: reproducing just once in a lifetime. (Ch. 5)

sequestering ecosystem An ecosystem that stores more carbon than it releases, and in which carbon is generally stored for relatively long periods of time. (Ch. 14)

sere A plant community at a particular point in the process of succession. (Ch. 13)

sexual selection Evolutionary selection for traits that increase mating success rather than survival; a form of natural selection. (Ch. 4)

Shannon-Wiener diversity index (H') A quantitative combination of both species evenness and richness that produces a single diversity measure. (Ch. 11)

simulation model A mathematical or conceptual model that is run multiple times to explore outcomes with different parameters or configurations. (Ch. 1)

sink A habitat patch with a resident subpopulation in which organism deaths outnumber births, but that is refilled with dispersing individuals from a source population. (Ch. 12)

sister species The closest evolutionary relative of a species depicted in a phylogeny. (Ch. 11)

slow carbon cycle The movement of carbon through the Earth's atmosphere, geosphere, and oceans. (Ch. 14)

source A habitat patch with a resident subpopulation in which organism births outnumber deaths and which, therefore, continually produces dispersing individuals that move to other habitat patches. (Ch. 12)

source-sink dynamics Spatial population dynamics that are governed by the differences in quality between source patches and sink patches and dispersal between them. (Ch. 12)

specialist   An organism that uses a single species, resource, or habitat to survive. In competition and exploitative or trophic interactions, a specialist is a species with a narrow diet that includes only a few food items. In mutualisms, a specialist is a species with a single mutualistic partner species. (Ch. 8)

speciation   The process by which two species arise from one common ancestral species. (Ch. 4)

species abundance   A tally of the number of individuals of each species found in a sample or community. (Ch. 11)

species evenness   A measure of how equally individuals are distributed among species in a sample. (Ch. 11)

species richness (*S*)   The number of species present at a given location. (Ch. 11)

species-area relationship   A power function that represents the way species richness accumulates with area sampled; species richness accumulates quickly with initial samples and then more slowly after common species have been found. (Ch. 11)

stabilizing selection   A change in a trait frequency distribution whereby the mean trait value does not change through time, but the variation away from that mean decreases over time; a form of natural selection. (Ch. 4)

stable coexistence   A stable equilibrium point where two species coexist at unchanging densities. Coexisting species may be competitors, predator and prey, or mutualists. (Ch. 7)

stable equilibrium   An unchanging value, or point, that a dynamic system moves toward and returns to after a disturbance. (Ch. 6)

stable isotope   A variant of an element having differing numbers of neutrons in the nucleus and which does not radioactively decay over time. (Ch. 10)

stable stage distribution   A population distribution in which the proportions of individuals in each stage class remain constant over time. (Ch. 6)

stage structure   A type of population structure in species such that survival and annual fecundity rates vary with life stage. (Ch. 6)

stage-structured population matrix model   A population growth model that assigns different survival and reproduction probabilities at different stage classes in a matrix; population growth can be modeled by

iteratively multiplying the vector of densities in each stage class by this matrix. (Ch. 6)

standard deviation   A measure of the average distance that all data points in a distribution lie from the mean value for that distribution. (Ch. 2)

statistical significance   A measure of how unlikely it is that a result is due to random chance; scientists choose specific levels of significance below which they accept that a result is not random. (Ch. 2)

stochastic model   A type of mathematical model that includes a random element in the parameters of the model, thereby allowing the results of the model to vary. (Ch. 2)

stoma   (*plural*, stomata) A small pore on the surface of a plant leaf that allows gases to diffuse in and out of the leaf. (Ch. 5)

subpopulation   A population within a metapopulation. (Ch. 12)

succession   The series of changes in a community through time at a particular location that occur in a fairly predictable way as the location goes from bare rock or lifeless water to being filled with interacting species. (Ch. 13)

symbiotic relationship   An interaction between individuals of two species that is characterized by close physical contact, often with integrated bodies and physiology. (Ch. 9)

sympatric speciation   Speciation that occurs when two (or more) population lineages occupy the same physical location but experience no gene flow between them. (Ch. 4)

sympatry   The coexistence of individuals of two species in the same spatial location. (Ch. 7)

synonymy   A formal way of saying that often many names are used to refer to the same species. (Ch. 11)

T

taiga   An alternative name for the boreal forest biome. (Ch. 3)

taxon   The singular form of a general term (*plural*, taxa) referring to biological groups at a variety of different evolutionary organizational levels from subspecies up to orders. (Ch. 4)

temperate forest   A terrestrial biome that is dominated by deciduous trees and winter temperatures that drop below freezing. (Ch. 3)

temperate grassland   A terrestrial biome that occurs between 30° and 55° north and south latitudes and where the predominant vegetation is grasses or small shrubs. (Ch. 3)

tertiary consumer   An organism that eats secondary consumers; sometimes called a top predator. (Ch. 10)

time lag   The time it takes for an effect to become apparent; used to describe the phenomenon in which the effects of density dependence on the rate of population growth are not accounted for until the next time period in a population growth model. (Ch. 6)

tit-for-tat   A game theory strategy that dictates that player A should always cooperate on the first move of the game but thereafter copy player B's most recent move. (Ch. 9)

tolerance model of succession   A conceptual model positing that early successional species have little to no effect on later-arriving species and that the pattern of succession is the result of life history characteristics of the species present. (Ch. 13)

top predator   An organism that eats secondary consumers; sometimes called a tertiary consumer. (Ch. 10)

top-down control   Control of the energy flow in a food web by species at the upper trophic levels. By eating organisms at lower trophic levels, the species at the upper levels control the biomass and abundance of the lower species and thus exert a strong influence on energy flows down the food chain. (Ch. 10)

trade-off   A balancing of factors all of which are not attainable at the same time; a giving up of one thing in return for another. In species, having one trait may preclude having another, and resources are often limited; this combination of limitations leads organisms to make trade-offs between resources in order to provide themselves with the highest possible chance of surviving or reproducing. (Ch. 5)

transpiration   The release of water vapor by plants. (Ch. 3)

treatment   An aspect of an ecological system that has been manipulated by an experimenter in the design of a scientific experiment. (Ch. 1)

trophic cascade   A pattern of alternating effects on abundance and biomass that moves down a food web (i.e., across trophic levels). (Ch. 10)

trophic interaction   An interaction between species in which the individuals of one species eat the individuals of the other. (Ch. 10)

trophic level   A feeding or eating level within a food chain or food web. (Ch. 10)

trophic pyramid   A representation of the flow of trophic energy up a food chain. (Ch. 10)

trophic specialist   An organism that feeds only from one trophic level. (Ch. 10)

tropical dry forest   A terrestrial biome that is warm to hot year-round and has substantial precipitation that is concentrated in a few months. (Ch. 3)

tropical rainforest   A terrestrial biome found at or near the equator, where the temperature is uniformly warm year-round and large amounts of precipitation fall year-round. (Ch. 3)

tropical savanna   A terrestrial biome near the equator that is characterized by large, open grasslands, seasonal rainfall, and frequent fires. (Ch. 3)

tundra   A terrestrial biome at very high latitudes (higher than 65° north or south) in which the cold winters are so extreme and the summers so short that a portion of the soil remains frozen year-round. Most vegetation consists of shrubs, mosses, and lichens. (Ch. 3)

turnover rate   In the theory of island biogeography, the number of newly colonizing species that replace newly extinct species per unit of time. (Ch. 12)

U

unstable coexistence   An outcome of a competition model that leads to an unstable equilibrium that contains both species; because the equilibrium is unstable, the two species in reality do not coexist, and the species that arrives first generally excludes the other species. (Ch. 7)

V

variable   A changeable value within a mathematical equation. (Ch. 2)

W

weather   The observed day-to-day variation in physical factors that make up the world. (Ch. 3)

X

xeric   *Of an environment*: having only a small amount of water; dry. (Ch. 5)

Z

zero net growth isocline (ZNGI)   A line in the Tilman model of competition formed by all the R^* values along a resource gradient where one would expect to see zero population growth through time. (Ch. 7)

REFERENCES

Chapter 1

Corlett, R. T. 2015. The Anthropocene concept in ecology and conservation. *Trends in Ecology and Evolution* 30(1): 36–41.

Crone, E. E. and J. L. Gehring. 1998. Population viability of *Rorippa columbiae*: Multiple models and spatial trend data. *Conservation Biology* 12(5): 1054–1065.

Crutzen, P. J. and E. F. Stoermer. 2000. The "anthropocene." *Global Change Newsletter* 41: 17–18.

Dickinson, J. L. and W. D. Koenig. 2003. Desperately seeking similarity. *Science* 300(5627): 1887–1889.

Forbes, E. 1843. *Report on the Mollusca and Radiata of the Aegean Sea: And on Their Distribution, Considered as Bearing on Geology*. London.

Hilborn, R. and M. Mangel. 1997. *The Ecological Detective: Confronting Models with Data*. Monographs in Population Biology 28. Princeton, NJ: Princeton University Press.

Krebs, C. J., S. Boutin, R. Boonstra, A. R. E. Sinclair, J. N. M. Smith, M. R. T. Dale, K. Martin, and R. Turkington. 1995. Impact of food and predation on the snowshoe hare cycle. *Science* 269(5227): 1112–1115.

Leopold, A. 1949. *A Sand County Almanac*. New York: Oxford University Press.

Lubchenco, J. 1978. Plant species diversity in a marine intertidal community: Importance of herbivore food preference and algal competitive abilities. *American Naturalist* 112(983): 23–39.

Sinervo, B. and C. M. Lively. 1996. The rock–paper–scissors game and the evolution of alternative male strategies. *Nature* 380(6571): 240–243.

Steffen, W., W. Broadgate, L. Deutsch, O. Gaffney, and C. Ludwig. 2015. The trajectory of the Anthropocene: The Great Acceleration. *Anthropocene Review* 2(1): 81–98.

Zalasiewicz, J., C. N. Waters, C. P. Summerhayes, A. P. Wolfe, A. D. Barnosky, A. Cearreta, P. Crutzen, et al. 2017. The Working Group on the Anthropocene: Summary of evidence and interim recommendations. *Anthropocene* 19(September): 55–60.

Chapter 2

Box, G. E. P. 1979. Robustness in the strategy of scientific model building. Pp. 201–236 in R. L. Launer and G. N. Wilkinson, eds. *Robustness in Statistics*. New York: Academic Press.

Burkley, M., J. Parker, S. P. Stermer, and E. Burkley. 2010. Trait beliefs that make women vulnerable to math disengagement. *Personality and Individual Differences* 48(2): 234–238.

Ford, A. 2009. *Modeling the Environment*, 2nd ed. Washington, DC: Island Press.

Peters, M. P., A. M. Prasad, S. N. Matthews, and L. R. Iverson. 2020. Climate change tree atlas, version 4. U.S. Forest Service, Northern Research Station and Northern Institute of Applied Climate Science, Delaware, OH. https://www.fs.fed.us/nrs/atlas/tree/318. Accessed October 19, 2021.

UNESCO Institute for Statistics. 2021. Literacy. http://uis.unesco.org/en/glossary-term/literacy. Accessed April 18, 2021.

Chapter 3

Brown, P. T. 2013. Unforced variability and the global warming slow down. *Patrick T. Brown, PhD* (blog). https://patricktbrown.org/2013/07/10/unforced-variability-and-the-global-warming-slow-down/. July 10, 2013. Accessed April 1, 2021.

DeLeon-Rodriguez, N., T. L. Lathem, L. M. Rodriguez-R, J. M. Barazesh, B. E. Anderson, A. J. Beyersdorf, L. D. Ziemba, M. Bergin, A. Nenes, and K. T. Konstantinidis. 2013. Microbiome of the upper troposphere: Species composition and prevalence, effects of tropical storms, and atmospheric implications. *Proceedings of the National Academy of Sciences* 110(7): 2575–2580.

Hansen, J. E. and M. Sato. 2011. Earth's climate history: Implications for tomorrow. National Aeronautics and Space Administration Goddard Institute for Space Studies. https://www.giss.nasa.gov/research/briefs/2011_hansen_15/. July 2011. Accessed April 1, 2021.

Haurwitz, B. and J. M. Austin. 1944. *Climatology*. New York: McGraw-Hill.

IPCC (Intergovernmental Panel on Climate Change). 2007. Contribution of Working Group I to the Fourth Assessment Report of the Intergovernmental Panel on Climate Change. Box TS.5 Figure 1, p. 53, in S. Solomon, D. Qin, M. Manning, Z. Chen, M. Marquis, K. B. Averyt, M. Tignor, and H. L. Miller, eds. *Climate Change 2007: The Physical Science Basis*. Cambridge, UK, and New York: Cambridge University Press.

Lieth, H., J. Berlekamp, S. Fuest, and S. Riediger. 1999. *Climate Diagram World Atlas*. Leiden, Netherlands: Backhuys.

Lindsey, R. 2020. Climate change: Glacier mass balance. NOAA Climate.gov. https://www.climate.gov/news-features/understanding-climate/climate-change-glacier-mass-balance. February 14, 2020. Accessed April 1, 2021.

Lindsey, R. 2021. Climate change: Global sea level. NOAA Climate.gov. https://www.climate.gov/news-features/understanding-climate/climate-change-global-sea-level. January 25, 2021. Accessed April 1, 2021.

Marshak S. and R. Rauber. 2020. *Earth Science*, 2nd ed. New York: W. W. Norton.

Munroe, R. n.d. Lakes and oceans (comic). xkcd (website). https://xkcd.com/1040/. Accessed April 1, 2021.

NASA (National Aeronautics and Space Administration). 2010. If Earth has warmed and cooled throughout history, what makes scientists think that humans are causing global warming now? *Climate Q&A* (blog). NASA Earth Observatory. https://earthobservatory.nasa.gov/blogs/climateqa/if-earth-has-warmed-and-cooled-throughout-history-what-makes-scientists-think-that-humans-are-causing-global-warming-now/. May 4, 2010. Accessed May 5, 2021.

NASA. 2021a. Carbon dioxide. NASA Global Climate Change and Global Warming: Vital Signs of the Planet. https://climate.nasa.gov/vital-signs/carbon-dioxide/. Accessed April 1, 2021.

NASA. 2021b. Ice sheets. NASA Global Climate Change and Global Warming: Vital Signs of the Planet. https://climate.nasa.gov/vital-signs/ice-sheets/. Accessed April 1, 2021.

National Drought Mitigation Center. 2006. Archive: climographs. University of Nebraska-Lincoln. https://drought.unl.edu/archive/climographs. Accessed April 1, 2021.

Newman, A. 2002. *Tropical Rainforest: Our Most Valuable and Endangered Habitat with a Blueprint for Its Survival into the Third Millennium*, rev. ed. New York: Checkmark Books.

NOAA National Centers for Environmental Information. 2020. Climate at a Glance: Global Time Series. https://www.ncdc.noaa.gov/cag/. August 2020. Accessed September 4, 2020.

Nunoura, T., Y. Takai, M. Hirai, S. Shimamura, A. Makabe, O. Koide, T. Kikuchi, et al. 2015. Hadal biosphere: Insight into the microbial ecosystem in the deepest ocean on Earth. *Proceedings of the National Academy of Sciences* 112(11): E1230–E1236.

Roper, L. D. 2016. Precipitation rate versus latitude and longitude. Genealogy Web Page of L. David Roper. http://www.roperld.com/science/PrecipLatitude_Longitude.htm. Last updated April 6, 2016. Accessed April 1, 2021.

SAGE (Center for Sustainability and the Global Environment, University of Wisconsin–Madison) and CRU (Climate Research Unit, University of East Anglia). n.d. Atlas of the Biosphere: Annual total precipitation [1960–1990]. Center for Sustainability and the Global Environment, University of Wisconsin–Madison (website). https://nelson.wisc.edu/sage/data-and-models/atlas/maps.php?datasetid=34&includerelatedlinks=1&dataset=34. Accessed April 1, 2021.

Vose, R. S., D. R. Easterling, K. E. Kunkel, A. N. LeGrande, and M. F. Wehner. 2017. Temperature changes in the United States. Pp. 185–206 in D. J. Wuebbles, D. W. Fahey, K. A. Hibbard, D. J. Dokken, B. C. Stewart, and T. K. Maycock, eds. *Climate Science Special Report: Fourth National Climate Assessment*, vol. 1. Washington, DC: U.S. Global Change Research Program. DOI: 10.7930/J0N29V45.

Walter, H. and H. Lieth. 1961–1967. *Klimadiagramm-Weltatlas* [Climate diagram world atlas]. Jena, Germany: Fischer Verlag.

Zepner, L., P. Karrasch, F. Wiemann, and L. Bernard. 2021. ClimateCharts.net—an interactive climate analysis web platform. *International Journal of Digital Earth* 14(3): 338–356. DOI: 10.1080/17538947.2020.1829112.

Chapter 4

American Phytopathological Society. 2021. Population genetics of plant pathogens: Gene and genotype flow. https://www.apsnet.org/edcenter/disimpactmngmnt/topc/PopGenetics/Pages/GeneGenotypeFlow.aspx. Accessed April 8, 2021.

Baker, M. R., N. W. Kendall, T. A. Branch, D. E. Schindler, and T. P. Quinn. 2011. Selection due to nonretention mortality in gillnet fisheries for salmon. *Evolutionary Applications* 4(3): 429–443. DOI: 10.1111/j.1752-4571.2010.00154.x.

Bergstrom, C. T. and L. A. Dugatkin. 2016. *Evolution*, 2nd ed. New York: W. W. Norton.

Boose, E. and E. Gould. 2019. Harvard Forest climate data since 1964. Harvard Forest Data Archive: HF300.

Darwin, C. 1859. *On the Origin of Species*. London: John Murray.

Darwin, C. 1871. *The Descent of Man, and Selection in Relation to Sex*. London: John Murray.

Kingston, T. and S. J. Rossiter. 2004. Harmonic-hopping in Wallacea's bats. *Nature* 429: 654–657.

Lockwood, J. L., M. F. Hoopes, and M. P. Marchetti. 2013. *Invasion Ecology*, 2nd ed. Chichester, West Sussex, UK: Wiley-Blackwell.

Petrie, M. 1994. Improved growth and survival of offspring of peacocks with more elaborate trains. *Nature* 371: 598–599.

Pigeon, G., M. Festa-Bianchet, D. W. Coltman, and F. Pelletier. 2016. Intense selective hunting leads to artificial evolution in horn size. *Evolutionary Applications* 9(4): 521–530.

Robson, N. K. B. 2002. Studies in the genus *Hypericum* L. (Guttiferae) 4(2). Section 9. *Hypericum* sensu lato (part 2): subsection 1. *Hypericum* series 1. *Hypericum*. *Bulletin of the Natural History Museum: Botany* 32(2): 61–123. DOI: 10.1017/S096804460200004X.

Rudh, A., B. Rogell, and J. Höglund. 2007. Non-gradual variation in colour morphs of the strawberry poison frog *Dendrobates pumilio*: Genetic and geographical isolation suggest a role for selection in maintaining polymorphism. *Molecular Ecology* 16(20): 4284–4294.

Shapiro, M. D., Z. Kronenberg, C. Li, E. T. Domyan, H. Pan, M. Campbell, H. Tan, et al. 2013. Genomic diversity and evolution of the head crest in the rock pigeon. *Science* 339(6123): 1063–1067. DOI: 10.1126/science.1230422.

Chapter 5

Berec, M., V. Krivan, and L. Berec. 2003. Are great tits (*Parus major*) really optimal foragers? *Canadian Journal of Zoology* 81(5): 780–788.

Charnov, E. L. 1976. Optimal foraging, the marginal value theorem. *Theoretical Population Biology* 9(2): 129–136.

Cowie, R. J. 1977. Optimal foraging in great tits (*Parus major*). *Nature* 268(5616): 137–139.

Delworth, T. L., A. J. Broccoli, A. Rosati, R. J. Stouffer, V. Balaji, J. A. Beesley, W. F. Cooke, et al. 2006. GFDL's CM2 global coupled climate models. Part I: Formulation and simulation characteristics. *Journal of Climate* 19(5): 643–674.

Ehleringer, J. R. and O. Björkman. 1978. Pubescence and leaf spectral characteristics in a desert shrub, *Encelia farinosa*. *Oecologia* 36(2): 151–162.

Ehleringer, J. R. and H. A. Mooney. 1978. Leaf hairs: Effects on physiological activity and adaptive value to a desert shrub. *Oecologia* 37(2): 183–200.

Huey, R. B. and M. Slatkin. 1976. Cost and benefits of lizard thermoregulation. *Quarterly Review of Biology* 51(3): 363–384.

Krebs, J. R., J. T. Erichsen, M. I. Webber, and E. L. Charnov. 1977. Optimal prey selection in the great tit (*Parus major*). *Animal Behaviour* 25(1): 30–38.

Lidicker, W. Z. 1960. *An Analysis of Intraspecific Variation in the Kangaroo Rat* Dipodomys merriami. Berkeley: University of California Press.

Reich, P. B. 2002. Root-shoot relations: Optimality in acclimation and adaptation or the "Emperor's New Clothes." Pp. 205–220 in Y. Waisel, A. Eshel, and U. Kafkafi, eds. *Plant Roots: The Hidden Half*, 3rd ed. New York: CRC Press.

Schoener, T. W. 2009. Ecological niche. Pp. 26–51 in S. A. Levin, ed. *Princeton Guide to Ecology*. Princeton, NJ: Princeton University Press.

Sih, A. and B. Christensen. 2001. Optimal diet theory: When does it work, and when and why does it fail? *Animal Behaviour* 61(2): 379–390.

Still, C. J., J. A. Berry, G. J. Collatz, and R. S. DeFries. 2003. Global distributions of C_3 and C_4 vegetation: Carbon cycle implications. *Global Biogeochemical Cycles* 17(1): 6-1–6-14. DOI: 10.1029/2001GB001807.

Tomlinson, P. T. and P. D. Anderson. 1998. Ontogeny affects response of northern red oak seedlings to elevated CO_2 and water stress: II. Recent photosynthate distribution and growth. *New Phytologist* 140(3): 493–504.

Tracy, R. L. and G. E. Walsberg. 2000. Prevalence of cutaneous evaporation in Merriam's kangaroo rat and its adaptive variation at the subspecific level. *Journal of Experimental Biology* 203(4): 773–781.

Tracy, R. L. and G. E. Walsberg. 2002. Kangaroo rats revisited: Re-evaluating a classic case of desert survival. *Oecologia* 133(4): 449–457.

Valenzuela-Ceballos, S., G. Castañeda, T. Rioja-Paradela, A. Carrillo-Reyes, and E. Bastiaans. 2015. Variation in the thermal ecology of an endemic iguana from Mexico reduces its vulnerability to global warming. *Journal of Thermal Biology* 48(February): 56–64.

Chapter 6

Crouse, D. T., L. B. Crowder, and H. Caswell. 1987. A stage-based population model for loggerhead sea turtles and implications for conservation. *Ecology* 68(5): 1412–1423.

Crowder, L. B., D. T. Crouse, S. S. Heppell, and T. H. Martin. 1994. Predicting the impact of turtle excluder devices on loggerhead sea turtle populations. *Ecological Applications* 4(3): 437–445.

Frazer, N. B. 1983. Survivorship of adult female loggerhead sea turtles, *Caretta caretta*, nesting on Little Cumberland Island, Georgia, USA. *Herpetologica* 39(4): 436–447.

Lefkovitch, L. P. 1965. The study of population growth in organisms grouped by stages. *Biometrics* 21(1): 1–18.

Parkes, J. P., J. Paulson, C. J. Donlan, and K. Campbell. 2008. *Control of North American Beavers in Tierra del Fuego: Feasibility of Eradication and Alternative Management Options*. Landcare Research Contract Report: LC0708/084. Prepared for Comité Binacional para la Estrategia de Erradicación de Castores de Patagonia Austral. Lincoln, New Zealand: Landcare Research New Zealand.

Stewart, K. M., R. T. Bowyer, B. L. Dick, B. K. Johnson, and J. G. Kie. 2005. Density-dependent effects on physical condition and reproduction in North American elk: An experimental test. *Oecologia* 143: 85–93.

Wang, Y. and E. Liang. 2016. Age and size of Smith firs at treeline in Tibet 1700–2013. Harvard Forest Data Archive: HF265. DOI: 10.6073/pasta/0252fdf7264012dffc2ac0fa4ad41983.

Chapter 7

Fausch, K. D., S. Nakano, and K. Ishigaki. 1994. Distribution of two congeneric charrs in streams of Hokkaido Island, Japan: Considering multiple factors across scales. *Oecologia* 100(1/2): 1–12.

Gause, G. F. (1934) 2019. *The Struggle for Existence: A Classic of Mathematical Biology and Ecology*. Mineola, NY: Dover. First published by Williams and Wilkins, Baltimore.

Hardin, G. 1960. The competitive exclusion principle. *Science* 131(3409): 1292–1297.

Marchetti, M. P. 1999. An experimental study of competition between the native Sacramento perch (*Archoplites interruptus*) and introduced bluegill (*Lepomis macrochirus*). *Biological Invasions* 1(1): 55–65.

McIntosh, R. P. 1985. *The Background of Ecology: Concept and Theory*. Cambridge: Cambridge University Press.

Park, T. 1948. Interspecies competition in populations of *Tribolium confusum* Duval and *Tribolium castaneum* Herbst. *Ecological Monographs* 18(2): 265–307.

Pfennig, D. W. and P. J. Murphy. 2000. Character displacement in polyphonic tadpoles. *Evolution* 54(5): 1738–1749.

Pfennig, D. W., A. M. Rice, and R. A. Martin. 2006. Ecological opportunity and phenotypic plasticity interact to promote character displacement and species coexistence. *Ecology* 87(3): 769–779.

Sax, D. F., J. J. Stachowicz, J. H. Brown, J. F. Bruno, M. N. Dawson, S. D. Gaines, R. K. Grosberg, et al. 2007. Ecological and evolutionary insights from species invasions. *Trends in Ecology and Evolution* 22(9): 465–471.

Schluter, D., T. D. Price, and P. R. Grant. 1985. Ecological character displacement in Darwin's finches. *Science* 227(4690): 1056–1059.

Taniguchi, Y. and S. Nakano. 2000. Condition-specific competition: Implications for the altitudinal distribution of stream fishes. *Ecology* 81(7): 2027–2039.

Tilman, D. 1977. Resource competition between planktonic algae: An experimental and theoretical approach. *Ecology* 58(2): 338–348.

Tilman, D. 1980. Resources: A graphical-mechanistic approach to competition and predation. *American Naturalist* 116(3): 362–393.

Weiner, J. 2014. *The Beak of the Finch: A Story of Evolution in Our Time*. New York: Vintage.

Chapter 8

Allmann, S. and I. T. Baldwin. 2010. Insects betray themselves in nature to predators by rapid isomerization of green leaf volatiles. *Science* 329(5995): 1075–1078.

Barański, M., D. Średnicka-Tober., N. Volakakis, C. Seal, R. Sanderson, G. B. Stewart, C. Benbrook, et al. 2014. Higher antioxidant and lower cadmium concentrations and lower incidence of pesticide residues in organically grown crops: A systematic literature review and meta-analyses. *British Journal of Nutrition* 112(5): 794–811.

Brodie, E. D., III and E. D. Brodie Jr. 1999. Predator-prey arms races: Asymmetrical selection on predators and prey may be reduced when prey are dangerous. *BioScience* 49(7): 557–568.

Brommer, J. E., H. Pietiäinen, K. Ahola, P. Karell, T. Karstinen, and H. Kolunen. 2010. The return of vole cycle in southern Finland refutes the generality of the loss of cycles through "climatic forcing." *Global Change Biology* 16(2): 577–586.

Caro, T. 2016. *Zebra Stripes*. Chicago: University of Chicago Press.

Denno, R. F. and D. Lewis. 2009. Predator-prey interactions. Pp. 202–212 in S. A. Levin, ed. *The Princeton Guide to Ecology*. Princeton, NJ: Princeton University Press.

Forrister, D. L., M.-J. Endara, G. C. Younkin, P. D. Coley, and T. A. Kursar. 2019. Herbivores as drivers of negative density dependence in tropical forest saplings. *Science* 363(6432): 1213–1216.

Gause, G. F. (1934) 2019. *The Struggle for Existence: A Classic of Mathematical Biology and Ecology*. Mineola, NY: Dover. First published by Williams and Wilkins, Baltimore.

Gotelli, N. J. 1995. *A Primer of Ecology*. Sunderland, MA: Sinauer Associates.

Graham, I. M. and X. Lambin. 2002. The impact of weasel predation on cyclic field-vole survival: The specialist predator hypothesis contradicted. *Journal of Animal Ecology* 71(6): 946–956.

Greene, H. W. and R. W. McDiarmid. 1981. Coral snake mimicry: Does it occur? *Science* 213(4513): 1207–1212.

Groom, M. J. 1998. Allee effects limit population viability of an annual plant. *American Naturalist* 151(6): 487–496.

Holling, C. S. 1959. The components of predation as revealed by a study of small-mammal predation of the European sawfly. *Canadian Entomologist* 91(5): 293–320.

Knapp, R. A., K. R. Matthews, and O. Sarnelle. 2001. Resistance and resilience of alpine lake fauna to fish introductions. *Ecological Monographs* 71(3): 401–421.

Korpimäki, E., L. Oksanen, T. Oksanen, K. Norrdahl, and P. B. Banks. 2005. Vole cycles and predation in temperate and boreal zones of Europe. *Journal of Animal Ecology* 74(6): 1150–1159.

Lockwood, J. L., M. F. Hoopes, and M. P. Marchetti. 2013. *Invasion Ecology*, 2nd ed. Chichester, West Sussex, UK: Wiley-Blackwell.

MacLulich, D. A. 1937. Fluctuations in the numbers of the varying hare (*Lepus americanus*). *University of Toronto Studies Biological Series* 43: 1–136.

Pfennig, D. W., W. R. Harcombe, and K. S. Pfennig. 2001. Frequency-dependent Batesian mimicry. *Nature* 410(6826): 323–323.

Sherratt, T. N. 2008. The evolution of Müllerian mimicry. *Naturwissenschaften* 95(8): 681–695.

Spencer, K. A. 1972. *Diptera: Agromyzidae*. Handbooks for the Identification of British Insects, vol. 10, pt. 5(g). London: Royal Entomological Society of London.

Chapter 9

Axelrod, R. and W. D. Hamilton. 1981. The evolution of cooperation. *Science* 211(4489): 1390–1396.

Bronstein, J. L. 1994. Our current understanding of mutualism. *Quarterly Review of Biology* 69(1): 31–51.

De Fine Licht, H. H., M. Schiøtt, A. Rogowska-Wrzesinska, S. Nygaard, P. Roepstorff, and J. J. Boomsma. 2013. Laccase detoxification mediates the nutritional alliance between leaf-cutting ants and fungus-garden symbionts. *Proceedings of the National Academy of Sciences* 110(2): 583–587.

Green, P. T., D. J. O'Dowd, K. L. Abbott, M. Jeffery, K. Retallick, and R. Mac Nally. 2011. Invasional meltdown: Invader–invader mutualism facilitates a secondary invasion. *Ecology* 92(9): 1758–1768.

Hoeksema, J. D. and M. W. Schwartz. 2003. Expanding comparative-advantage biological market models: Contingency of mutualism on partners' resource requirements and acquisition trade-offs. *Proceedings of the Royal Society of London Series B* 270(1518): 913–919.

Koenig, J. E., A. Spor, N. Scalfone, A. D. Fricker, J. Stombaugh, R. Knight, L. T. Angenent, and R. E. Ley. 2011. Succession of microbial consortia in the developing infant gut microbiome. *Proceedings of the National Academy of Sciences* 108(Supplement 1): 4578–4585.

Menz, M. H., R. D. Phillips, R. Winfree, C. Kremen, M. A. Aizen, S. D. Johnson, and K. W. Dixon. 2011. Reconnecting plants and pollinators: Challenges in the restoration of pollination mutualisms. *Trends in Plant Science* 16(1): 4–12.

Remy, W., T. N. Taylor, H. Hass, H. Kerp. 1994. Four hundred-million-year-old vesicular arbuscular mycorrhizae. *Proceedings of the National Academy of Sciences* 91(25): 11841–11843.

Schwartz, M. W. and J. D. Hoeksema. 1998. Specialization and resource trade: Biological markets as a model of mutualisms. *Ecology* 79(3): 1029–1038.

Simberloff, D. and B. Von Holle. 1999. Positive interactions of nonindigenous species: Invasional meltdown? *Biological Invasions* 1(1): 21–32.

Smits, L. P., K. E. Bouter, W. M. de Vos, T. J. Borody, and M. Nieuwdorp. 2013. Therapeutic potential of fecal microbiota transplantation. *Gastroenterology*, 145(5): 946–953.

Stachowicz, J. J. 2001. Mutualism, facilitation, and the structure of ecological communities. *Bioscience* 51(3): 235–246.

Trivers, R. L. 1971. The evolution of reciprocal altruism. *Quarterly Review of Biology* 46(1): 35–57.

Van Tussenbroek, B. I., N. Villamil, J. Márquez-Guzmán, R. Wong, L. V. Monroy-Velázquez, and V. Solis-Weiss. 2016. Experimental evidence of pollination in marine flowers by invertebrate fauna. *Nature Communications* 7, article 12980.

Von Holle, B. 2011. Invasional meltdown. Pp. 360–364 in D. Simberloff and M. Rejmanek, eds. *Encyclopedia of Biological Invasions*. Berkeley: University of California Press.

Chapter 10

Beschta, R. L. and W. J. Ripple. 2009. Large predators and trophic cascades in terrestrial ecosystems of the western United States. *Biological Conservation* 142(11): 2401–2414.

Beschta, R. L. and W. J. Ripple. 2016. Riparian vegetation recovery in Yellowstone: The first two decades after wolf reintroduction. *Biological Conservation* 198: 93–103.

Bolnick, D. I., P. Amarasekare, M. S. Araújo, R. Bürger, J. M. Levine, M. Novak, V. H. W. Rudolf, S. J. Schreiber, M. C. Urban, and D. A. Vasseur. 2011. Why intraspecific trait variation matters in community ecology. *Trends in Ecology and Evolution* 26(4): 183–192.

Dunne, J. A., K. D. Lafferty, A. P. Dobson, R. F. Hechinger, A. M. Kuris, N. D. Martinez, J. P. McLaughlin, et al. 2013. Parasites affect food web structure primarily through increased diversity and complexity. *PLoS Biology* 11(6): e1001579.

Estes, J. A. and J. F. Palmisano. 1974. Sea otters: Their role in structuring nearshore communities. *Science* 185(4156): 1058–1060.

Fretwell, S. D. 1977. The regulation of plant communities by food chains exploiting them. *Perspectives in Biology and Medicine* 20(2): 169–185.

Fretwell, S. D. 1987. Food chain dynamics: The central theory of ecology? *Oikos* 50(3): 291–301.

Fry, B. 2006. *Stable Isotope Ecology*. New York: Springer.

Ginzburg, L. R. and H. R. Akçakaya. 1992. Consequences of ratio-dependent predation for steady-state properties of ecosystems. *Ecology* 73(5): 1536–1543.

Hairston, N. G., F. E. Smith, and L. B. Slobodkin. 1960. Community structure, population control, and competition. *American Naturalist* 94(879): 421–425.

Lavigne, D. M. 1996. Ecological interactions between marine mammals, commercial fisheries and their prey: Unraveling the tangled web. Pp. 59–71 in W. A. Montevecchi, ed. *Studies of High-Latitude Seabirds. 4. Trophic Relationships and Energetics of Endotherms in Cold Ocean Systems.* Canadian Special Publications on Fisheries Aquatic Science. Ottawa: Canadian Wildlife Service.

Layhee, M., M. P. Marchetti, S. Chandra, T. Engstrom, and D. Pickard. 2014. Impacts of aquatic invasive species and land use on stream food webs in Kaua'i, Hawai'i. *Pacific Conservation Biology* 20(3): 252–271.

Paine, R. T. 1969. A note on trophic complexity and community stability. *American Naturalist* 103(929): 91–93.

Pfennig, D. W., A. M. Rice, and R. A. Martin. 2006. Ecological opportunity and phenotypic plasticity interact to promote character displacement and species coexistence. *Ecology* 87(3): 769–779.

Pianka, E. R. 1999. *Evolutionary Ecology*, 6th ed. San Francisco: Benjamin Cummings.

Pimentel, D. and M. Pimentel. 2003. Sustainability of meat-based and plant-based diets and the environment. *American Journal of Clinical Nutrition* 78(3): 660S–663S.

Polis, G. A. 1991. Complex trophic interactions in deserts: An empirical critique of food-web theory. *American Naturalist* 138(1): 123–155.

Polis, G. A. and S. D. Hurd. 1995. Extraordinarily high spider densities on islands: Flow of energy from the marine to terrestrial food webs and the absence of predation. *Proceedings of the National Academy of Sciences* 92(10): 4382–4386.

Polis, G. A. and S. D. Hurd. 1996. Linking marine and terrestrial food webs: Allochthonous input from the ocean supports high secondary productivity on small islands and coastal land communities. *American Naturalist* 147(3): 396–423.

Power, M. E. 1992. Top-down and bottom-up forces in food webs: Do plants have primacy? *Ecology* 73(3): 733–746.

Power, M. E., D. Tilman, J. A. Estes, B. A. Menge, W. J. Bond, L. S. Mills, G. Daily, J. C. Castilla, J. Lubchenco, and R. T. Paine. 1996. Challenges in the quest for keystones: Identifying keystone species is difficult—but essential to understanding how loss of species will affect ecosystems. *BioScience* 46(8): 609–620.

Ripple, W. J. and R. L. Beschta. 2012. Trophic cascades in Yellowstone: The first 15 years after wolf reintroduction. *Biological Conservation* 145(1): 205–213.

Strauss, S. Y. 1991. Indirect effects in community ecology: Their definition, study and importance. *Trends in Ecology and Evolution* 6(7): 206–210.

Vander Zanden, M. J., S. Chandra, B. C. Allen, J. E. Reuter, and C. R. Goldman. 2003. Historical food web structure and restoration of native aquatic communities in the Lake Tahoe (California–Nevada) basin. *Ecosystems* 6: 274–288.

Wootton, J. T. 1994. The nature and consequences of indirect effects in ecological communities. *Annual Review of Ecology and Systematics* 25(1): 443–466.

Chapter 11

Barthlott, W., M. D. Rafiqpoor, and W. R. Erdelen. 2016. Bionics and biodiversity—bio-inspired technical innovation for a sustainable future. Pp. 11–55 in J. Knippers, K. Nickel, and T. Speck, eds. *Biomimetic Research for Architecture and Building Construction. Biologically-Inspired Systems*, Vol. 8. Springer, Cham.

Benigno, G. M. 2011. Invertebrate drift in neighboring perennial and seasonal tributaries of the Sacramento River. Master's thesis, California State University, Chico.

Boyer, A. G. and W. Jetz. 2014. Extinctions and the loss of ecological function in island bird communities. *Global Ecology and Biogeography* 23(6): 679–688.

Chapman, A. D. 2009. *Numbers of Living Species in Australia and the World.* Canberra, Australia: Australian Biodiversity Information Service.

Clausnitzer, V., K. D. B. Dijkstra, R. Koch, J. P. Boudot, W. R. Darwall, J. Kipping, B. Samraoui, M. J. Samways, J. P. Simaika, and F. Suhling. 2012. Focus on African freshwaters: Hotspots of dragonfly diversity and conservation concern. *Frontiers in Ecology and the Environment* 10(3): 129–134.

Coetzee, B. W. T. and S. L. Chown. 2016. Land-use change promotes avian diversity at the expense of species with unique traits. *Ecology and Evolution* 6(21): 7610–7622.

Costello, M. J., R. M. May, and N. E. Stork. 2013. Can we name Earth's species before they go extinct? *Science* 339(6118): 413–416.

Davies, T. J. and L. B. Buckley. 2011. Phylogenetic diversity as a window into the evolutionary and biogeographic histories of present-day richness gradients. *Philosophical Transactions of the Royal Society B* 366(1576): 2414–2425.

Davis, M., S. Faurby, and J.-C. Svenning. 2018. Mammal diversity will take millions of years to recover from the current biodiversity crisis. *Proceedings of the National Academy of Sciences* 115(44): 11262–11267.

Drakare, S., J. J. Lennon, and H. Hillebrand. 2006. The imprint of the geographical, evolutionary and ecological context on species–area relationships. *Ecology Letters* 9(2): 215–227.

Erwin, T. L. 1982. Tropical forests: Their richness in Coleoptera and other arthropod species. *Coleopterists Bulletin* 36(1): 74–75.

Faith, D. P. 1992. Conservation evaluation and phylogenetic diversity. *Biological Conservation* 61(1): 1–10.

Fontaine, B., K. van Achterberg, M. A. Alonso-Zarazaga, R. Araujo, M. Asche, H. Aspöck, U. Aspöck, et al. 2012. New species in the Old World: Europe as a frontier in biodiversity exploration, a test bed for 21st century taxonomy. *PLoS One* 7(5): e36881.

Gaston, K. J. 1996. Species-range-size distributions: Patterns, mechanisms and implications. *Trends in Ecology and Evolution* 11(5): 197–201.

Hebert, P. D. N., E. H. Penton, J. M. Burns, D. H. Janzen, and W. Hallwachs. 2004. Ten species in one: DNA barcoding reveals cryptic species in the Neotropical skipper butterfly *Astraptes fulgerator*. *Proceedings of the National Academy of Sciences* 101(41): 14812–14817.

Hererra, J. P. and L. M. Dávalos. 2016. Phylogeny and divergence times of lemurs inferred with recent and ancient fossils in the tree. *Systematic Biology* 65(5): 772–791.

Hofmann, R., M. Tietje, and M. Aberhan. 2019. Diversity partitioning in Phanerozoic benthic marine communities. *Proceedings of the National Academy of Sciences* 116(1): 79–83.

Illies, J. 1971. Emergenz 1969 im Breitenbach: Schlitzer produktionsbiologische Studien. *Archiv für Hydrobiologie* 69(1): 14–59.

International Union for Conservation of Nature, Conservation International, and NatureServe. 2008. *The IUCN Amphibians Initiative: A Record of the 2001–2008 Amphibian Assessment Efforts for the IUCN Red List.* https://www.iucn-amphibians.org/wp-content/uploads/2019/03/Amphibians-Initiative-2008-webcontent-Downloaded-27Nov2018-1.pdf. Accessed July 29, 2021.

Larsen, B. B., E. C. Miller, M. K. Rhodes, and J. J. Wiens. 2017. Inordinate fondness multiplied and redistributed: The number of species on Earth and the new pie of life. *Quarterly Review of Biology* 92(3): 229–265.

May, R. M. 1988. How many species are there on Earth? *Science* 241(4872): 1441–1449.

McKinney, M. L. and J. L. Lockwood. 1999. Biotic homogenization: A few winners replacing many losers in the next mass extinction. *Trends in Ecology and Evolution* 14(11): 450–453.

Mora, C., D. P. Tittensor, S. Adl, A. G. Simpson, and B. Worm. 2011. How many species are there on Earth and in the ocean? *PLoS Biology* 9(8): e1001127.

Myers, N., R. A. Mittermeier, C. G. Mittermeier, G. A. Da Fonseca, and J. Kent. 2000. Biodiversity hotspots for conservation priorities. *Nature* 403(6772): 853–858.

Nobis, A., M. Zmihorski, and D. Kotowska. 2016. Linking the diversity of native flora to land cover heterogeneity and plant invasions in a river valley. *Biological Conservation* 203: 17–24.

Petchey, O. L. and K. J. Gaston. 2002. Functional diversity (FD), species richness and community diversity. *Ecology Letters* 5(3): 402–411.

Phillip, D. A. T., 1998. Biodiversity of freshwater fishes of Trinidad and Tobago, West Indies. Doctoral dissertation, University of St. Andrews.

Preston, F. W. 1962. The canonical distribution of commonness and rarity: Part I. *Ecology* 43(2): 185–215.

Purvis, A. and A. Hector. 2000. Getting the measure of biodiversity. *Nature* 405(6783): 212–219.

Rabinowitz, D. 1981. Seven forms of rarity. Pp. 205–217 in H. Synge, ed. *The Biological Aspects of Rare Plant Conservation*. New York: John Wiley and Sons.

Rabinowitz, D., S. Cairns, and T. Dillon. 1986. Seven forms of rarity and their frequency in the flora of the British Isles. Pp. 182–204 in M. E. Soulé, ed. *Conservation Biology: The Science of Scarcity and Diversity*. Sunderland, MA: Sinauer Associates, Inc.

Scheffers, B. R., L. N. Joppa, S. L. Pimm, and W. F. Laurance. 2012. What we know and don't know about Earth's missing biodiversity. *Trends in Ecology and Evolution* 27(9): 501–510.

Shannon, C. E. 1948. A mathematical theory of communication. *Bell System Technical Journal* 27(3): 379–423.

Slayter, R. A., M. Hirst, and J. P. Sexton. 2013. Niche breadth predicts geographical range size: A general pattern. *Ecology Letters* 16: 1104–1114.

Stuart-Smith, R. D., A. E. Bates, J. S. Lefcheck, J. E. Duffy, S. C. Baker, R. J. Thomson, J. F. Stuart-Smith, et al. 2013. Integrating abundance and functional traits reveals new global hotspots of fish diversity. *Nature* 501(7468): 539.

United Nations. 1992. Convention on biodiversity. https://www.cbd.int/doc/legal/cbd-en.pdf. Accessed September 7, 2021.

Vandewalle, M., F. de Bello, M. P. Berg, T. Bolger, S. Dolédec, F. Dubs, C. K. Feld, et al. 2010. Functional traits as indicators of biodiversity response to land use changes across ecosystems and organisms. *Biodiversity and Conservation* 19: 2921–2947.

Verberk, W. C. E. P. 2011. Explaining general patterns in species abundance and distributions. *Nature Education Knowledge* 3(10): 38.

Whittaker, R. H. 1972. Evolution and measurement of species diversity. *Taxon* 21(2/3): 213–251.

Winter, M., V. Devictor, and O. Schweiger. 2013. Phylogenetic diversity and nature conservation: Where are we? *Trends in Ecology and Evolution* 28(4): 199–204.

Chapter 12

Anderson, J. T., T. Nuttle, J. S. Saldaña Rojas, T. H. Pendergast, and A. S. Flecker. 2011. Extremely long-distance seed dispersal by an overfished Amazonian frugivore. *Proceedings of the Royal Society B* 278(1723): 33329–33335.

Beacham, T. D., R. J. Beamish, J. R. Candy, C. Wallace, S. Tucker, J. H. Moss, and M. Trudel. 2014. Stock-specific migration pathways of juvenile sockeye salmon in British Columbia waters and in the Gulf of Alaska. *Transactions of the American Fisheries Society* 143(6): 1386–1403.

Corkeron, P. J. and R. C. Connor. 1999. Why do baleen whales migrate? *Marine Mammal Science* 15(4): 1228–1245.

Cornell Lab of Ornithology. 2020. Canada goose range map. eBird. https://ebird.org/science/status-and-trends/cangoo/range-map. Accessed January 21, 2022.

Diamond, J. M. 1972. Biogeographic kinetics: Estimation of relaxation times for avifaunas of southwest Pacific islands. *Proceedings of the National Academy of Sciences* 69(11): 3199–3203.

Diamond, J. M. 1975. Assembly of species communities. Pp. 342–444 in M. L. Cody and J. M. Diamond, eds. *Ecology and Evolution of Communities*. Cambridge, MA: Harvard University Press.

Holland, R. A., M. Wikelski, and D. S. Wilcove. 2006. How and why do insects migrate? *Science* 313(5788): 794–796.

Hoopes, M. F., R. D. Holt, and M. Holyoak. 2005. The effects of spatial processes on two species interactions. Pp. 35–67 in M. Holyoak, M. A. Leibold, and R. D. Holt, eds. *Metacommunities: Spatial Dynamics and Ecological Communities*. Chicago: University of Chicago Press.

Hoover, J. P. 2003. Decision rules for site fidelity in a migratory bird, the prothonotary warbler. *Ecology* 84(2): 416–430.

Kowarik, I. and M. von der Lippe. 2011. Secondary wind dispersal enhances long-distance dispersal of an invasive species in urban road corridors. *NeoBiota* 9: 49–70.

Levins, R. 1969. Some demographic and genetic consequences of environmental heterogeneity for biological control. *Bulletin of the Entomological Society of America* 15(3): 237–240.

Levins, R. 1970. Extinction. Pp. 75–107 in M. Gerstenhaber, ed. *Some Mathematical Questions in Biology*. Lectures on Mathematics in the Life Sciences 2. Providence, RI: American Mathematical Society.

Lomolino, M. 2000. A call for a new paradigm of island biogeography. *Global Ecology and Biogeography* 9(1): 1–6.

MacArthur, R. H. and E. O. Wilson. (1967) 2001. *The Theory of Island Biogeography*. Reprinted with a new preface by E. O. Wilson. Princeton Landmarks in Biology. Princeton, NJ: Princeton University Press.

Monarch Watch. 2010. Two-way monarch migration map. *Monarch Watch* (blog). https://monarchwatch.org/blog/2010/05/13/two-way-monarch-migration-map/. May 13, 2010. Accessed January 20, 2022.

Nathan, R., W. M. Getz, E. Revilla, M. Holyoak, R. Kadmon, D. Saltz, and P. E. Smouse. 2008. A movement ecology paradigm for unifying organismal movement research. *Proceedings of the National Academy of Sciences* 105(49): 19052–19059.

NOAA (National Oceanic and Atmospheric Administration) Fisheries. 2016. Humpback whale distinct population segments identification map. https://www.fisheries.noaa.gov/resource/map/humpback-whale-distinct-population-segments-identification-map. Accessed August 4, 2021.

Simberloff, D. S. and E. O. Wilson. 1969. Experimental zoogeography of islands: The colonization of empty islands. *Ecology* 50(2): 278–296.

Simberloff, D. S. and E. O. Wilson. 1970. Experimental zoogeography of islands: A two-year record of colonization. *Ecology* 51(5): 934–937.

University of Michigan with funding from the National Institute on Aging. 2010. Health and retirement study. Cross-Wave Child Proximity Public Use Dataset: grant number NIA U01AG009740. Ann Arbor, MI.

Chapter 13

Andras, J. P., W. Rodriguez-Reillo, A. Truchon, J. Blanchard, E. Pierce, and K. Ballantine. 2020. Rewilding the small stuff: The effect of ecological restoration on prokaryotic communities of peatland soils. *FEMS Microbiology Ecology* 96(10): fiaa144. DOI: 10.1093/femsec/fiaa144.

Benson, K. R. 2000. The emergence of ecology from natural history. *Endeavour* 24(2): 59–62.

Bowman, D. M. J. S., B. P. Murphy, D. L. J. Neyland, G. J. Williamson, and L. D. Prior. 2014. Abrupt fire regime change may close landscape-wide loss of mature obligate seeder forests. *Global Change Biology* 20(3): 1008–1015.

Connell, J. H. 1978. Diversity in tropical rain forests and coral reefs. *Science* 199(4335): 1302–1310.

Connell, J. H. and R. O. Slatyer. 1977. Mechanisms of succession in natural communities and their role in community stability and organization. *American Naturalist* 111(982): 1119–1144.

Darwin, C. 1859. *On the Origin of Species*. London: John Murray.

DeAngelis, D. L. and J. C. Waterhouse. 1987. Equilibrium and nonequilibrium concepts in ecological models. *Ecological Monographs* 57(1): 1–21.

del Moral, R. n.d. Grids. Plant Ecology Lab in the Department of Biology [University of Washington]: Mount St. Helens Research Program (website). http://faculty.washington.edu/moral/pdf/2005Grids.pdf. Accessed August 24, 2021.

Diamond, J. M. 1975. Assembly of species communities. Pp. 342–444 in M. Cody and J. Diamond, eds. *Ecology and Evolution of Communities*. Cambridge, MA: Harvard University Press.

Doherty, T. S., A. S. Glen, D. G. Nimmo, E. G. Ritchie, and C. R. Dickman. 2016. Invasive predators and global biodiversity loss. *Proceedings of the National Academy of Sciences* 113(40): 11261–11265.

Drake, J. A. 1991. Community-assembly mechanics and the structure of an experimental species ensemble. *American Naturalist* 137(1): 1–26.

Fox, J. W. 2013. The intermediate disturbance hypothesis should be abandoned. *Trends in Ecology and Evolution* 28(2): 86–92.

Gleason, H. A. 1926. The individualistic concept of the plant association. *Bulletin of the Torrey Botanical Club* 53(1): 7–26.

Grime, J. P. 1979. *Plant Strategies and Vegetation Processes*. Chichester, UK: John Wiley and Sons.

Hector, A. and R. Hooper. 2002. Darwin and the first ecological experiment. *Science* 295(5555): 639–640.

Keever, C. 1950. Causes of succession on old fields of the Piedmont, North Carolina. *Ecological Monographs* 20(3): 229–250.

Keever, C. 1983. A retrospective view of old field succession after 35 years. *American Midland Naturalist* 110(2): 397–404.

Lockwood, J. L., M. F. Hoopes, and M. P. Marchetti. 2013. *Invasion Ecology*, 2nd ed. Chichester, West Sussex, UK: Wiley-Blackwell.

Mackey, R. L. and D. J. Currie. 2001. The diversity–disturbance relationship: Is it generally strong and peaked? *Ecology* 82(12): 3479–3492.

McIntosh, R. P. 1985. *The Background of Ecology: Concept and Theory*. Cambridge: Cambridge University Press.

Newman, E. A. 2019. Disturbance ecology in the Anthropocene. *Frontiers in Ecology and Evolution* 7:147.

Odum, E. P. 1971. *Fundamentals of Ecology*, 3rd ed. Philadelphia: Saunders.

Petraitis, P. S., R. E. Latham, and R. A. Niesenbaum. 1989. The maintenance of species diversity by disturbance. *Quarterly Review of Biology* 64(4): 393–418.

Pickett, S. T. A., J. Kolasa, J. J. Armesto, and S. L. Collins. 1989. The ecological concept of disturbance and its expression at various hierarchical levels. *Oikos* 54(2): 129–136.

Poff, N. L., J. D. Allan, M. B. Bain, J. R. Karr, K. L. Prestegaard, B. D. Richter, R. E. Sparks, and J. C. Stromberg. 1997. The natural flow regime. *Bioscience* 47(11): 769–784.

Simberloff, D. and B. Von Holle. 1999. Positive interactions of nonindigenous species: Invasional meltdown? *Biological Invasions* 1: 21–32.

Sousa, W. P. 1984. The role of disturbance in natural communities. *Annual Review of Ecology and Systematics* 15: 353–391.

Tilman, D. 1977. Resource competition between planktonic algae: An experimental and theoretical approach. *Ecology* 58(2): 338–348.

Weiher, E. and P. Keddy, eds. 1999. *Ecological Assembly Rules: Perspectives, Advances, Retreats*. Cambridge: Cambridge University Press.

White, P. S. and S. T. A. Pickett. 1985. Natural disturbance and patch dynamics: An introduction. Pp. 3–13 in S. T. A. Pickett and P. S. White, eds. *The Ecology of Natural Disturbance and Patch Dynamics*. New York: Academic Press.

Wood, D. M. and R. del Moral. 2000. Seed rain during early primary succession on Mount St. Helens, Washington. *Madroño* 47(1): 1–9.

Chapter 14

Allen, C. D., A. K. Macalady, H. Chenchouni, D. Bachelet, N. McDowell, M. Vennetier, T. Kitzberger, et al. 2010. A global review of drought and heat-induced tree mortality reveals emerging climate change risks for forests. *Forest Ecology and Management* 259(4): 660–684.

Alongi, D. M. 2014. Carbon cycling and storage in mangrove forests. *Annual Review of Marine Science* 6: 195–219.

Averill, C., M. C. Dietze, and J. M. Bhatnagar. 2018. Continental-scale nitrogen pollution is shifting forest mycorrhizal associations and soil carbon stocks. *Global Change Biology* 24(10): 4544–4553.

Cardinale, B. J., J. E. Duffy, A. Gonzalez, D. U. Hooper, C. Perrings, P. Venail, A. Narwani, et al. 2012. Biodiversity loss and its impact on humanity. *Nature* 486: 59–67.

Durack, P. J., S. E. Wijffels, and R. J. Matear. 2012. Ocean salinities reveal strong global water cycle intensification during 1950 to 2000. *Science* 336(6080): 455–458.

Fabricius, K. E., C. Langdon, S. Uthicke, C. Humphrey, S. Noonan, G. De'ath, R. Okazaki, N. Muehllehner, M. S. Glas, and J. M. Lough. 2011. Losers and winners in coral reefs acclimatized to elevated carbon dioxide concentrations. *Nature Climate Change* 1: 165–169.

Galloway, J. N., W. Winiwarter, A. Leip, A. M. Leach, A. Bleeker, and J. W. Erisman. 2014. Nitrogen footprints: Past, present and future. *Environmental Research Letters* 9(11): 115003.

Helfield, J. M. and R. J. Naiman. 2001. Effects of salmon-derived nitrogen on riparian forest growth and implications for stream productivity. *Ecology* 82(9): 2403–2409.

Hutchison, J., A. Manica, R. Swetnam, A. Balmford, and M. Spalding. 2014. Predicting global patterns in mangrove forest biomass. *Conservation Letters* 7(3): 233–240.

Intergovernmental Panel on Climate Change. 2013. Figure 12.41 in T. F. Stocker, D. Qin, G.-K. Plattner, M. Tignor, S. K. Allen, J. Boschung, A. Nauels, Y. Xia, V. Bex, and P. M. Midgley, eds. *Climate Change 2013: The Physical Science Basis. Contribution of Working Group I to the Fifth Assessment Report of the Intergovernmental Panel on Climate Change.* Cambridge: Cambridge University Press. https://www.ipcc.ch/report/ar5/wg1/long-term-climate-change-projections-commitments-and-irreversibility/. Accessed September 16, 2021.

Magill, A. H., J. D. Aber, W. S. Currie, K. J. Nadelhoffer, M. F. Martin, W. H. McDowell, J. M. Melillo, and P. Steudler. 2004. Ecosystem response to 15 years of chronic nitrogen additions at the Harvard Forest LTER, Massachusetts, USA. *Forest Ecology and Management* 196(1): 7–28.

Marengo, J. A. and J. C. Espinoza. 2016. Extreme seasonal droughts and floods in Amazonia: Causes, trends and impacts. *International Journal of Climatology* 36(3): 1033–1050.

Merz, J. E. and P. B. Moyle. 2006. Salmon, wildlife, and wine: Marine-derived nutrients in human-dominated ecosystems of central California. *Ecological Applications* 16(3): 999–1009.

Pendleton, L., D. C. Donato, B. C. Murray, S. Crooks, W. A. Jenkins, S. Sifleet, C. Craft, et al. 2012. Estimating global "blue carbon" emissions from conversion and degradation of vegetated coastal ecosystems. *PLoS One* 7(9): e43542.

Plumbago. 2009. Estimated change in seawater pH caused by human-created CO_2 between the 1700s and the 1990s, from the Global Ocean Data Analysis Project (GLODAP) and the World Ocean Atlas. Wikipedia. https://en.wikipedia.org/wiki/Ocean_acidification#/media/File:WOA05_GLODAP_del_pH_AYool.png. Accessed September 14, 2021.

Riebeek, H. 2011. The carbon cycle. NASA Earth Observatory. https://earthobservatory.nasa.gov/features/CarbonCycle. Published June 16, 2011. Accessed October 13, 2021.

Röpke, C. P., S. Amadio, J. Zuanon, E. J. G. Ferreira, C. Pereria de Deus, T. H. S. Pires, and K. O. Winemiller. 2017. Simultaneous abrupt shifts in hydrology and fish assemblage structure in a floodplain lake in the central Amazon. *Scientific Reports* 7: 40170.

Sabine, C. L., R. A. Feely, N. Gruber, R. M. Key, K. Lee, J. L. Bullister, R. Wanninkhof, et al. 2004. The oceanic sink for anthropogenic CO_2. *Science* 305(5682): 367–371.

Schlesinger, W. H. 2009. On the fate of anthropogenic nitrogen. *Proceedings of the National Academy of Science* 106(1): 203–208.

Schlesinger, W. H. and E. S. Bernhardt. 2013. *Biogeochemistry: An Analysis of Global Change*, 3rd ed. Waltham, MA: Academic Press.

Scripps Institution of Oceanography and NOAA Global Monitoring Laboratory. 2021. Atmospheric CO_2 at Mauna Loa Observatory. NOAA Global Monitoring Laboratory. https://gml.noaa.gov/webdata/ccgg/trends/co2_data_mlo.pdf. Accessed July 2021.

Sherman, D. J. and D. R. Montgomery. 2020. *Environmental Science and Sustainability*. New York: W. W. Norton.

Stevens, C. J., N. B. Dise, J. O. Mountford, and D. J. Gowing. 2004. Impact of nitrogen deposition on the species richness of grasslands. *Science* 303(5665): 1876–1879.

Trenberth, K. E., L. Smith, T. Qian, A. Dai, and J. Fasullo. 2007. Estimates of global water budget and its annual cycle using observational and model data. *Journal of Hydrometeorology* 8: 758–769.

Williams, A. P., C. D. Allen, C. I. Millar, T. W. Swetnam, J. Michaelsen, C. J. Still, and S. W. Leavitt. 2010. Forest responses to increasing aridity and warmth in the southwestern United States. *Proceedings of the National Academy of Sciences* 107(50): 21289–212294.

Chapter 15

Albins, M. A. and M. A. Hixon. 2013. Worst case scenario: Potential long-term effects of invasive predatory lionfish (*Pterois volitans*) on Atlantic and Caribbean coral-reef communities. *Environmental Biology of Fishes* 96(10–11): 1151–1157.

Arias-González, J. E., C. González-Gándara, J. L. Cabrera, and V. Christensen. 2011. Predicted impact of the invasive lionfish *Pterois volitans* on the food web of a Caribbean coral reef. *Environmental Research* 111(7): 917–925.

Bradley, B. A., C. A. Curtis, E. J. Fusco, J. T. Abatzoglou, J. K. Balch, S. Dadashi, and M. N. Tuanmu. 2018. Cheatgrass (*Bromus tectorum*) distribution in the Intermountain Western United States and its relationship to fire frequency, seasonality, and ignitions. *Biological Invasions* 20(6): 1493–1506.

Bratman, G., J. Hamilton, K. Hahn, G. Daily, and J. Gross. 2015. Nature experience reduces rumination and subgenual prefrontal cortex activation. *Proceedings of the National Academy of Sciences* 112(28): 8567–8572. DOI: 10.1073/pnas.1510459112.

Ceballos, G., P. R. Ehrlich, A. D. Barnosky, A. García, R. M. Pringle, and T. M. Palmer. 2015. Accelerated modern human–induced species losses: Entering the sixth mass extinction. *Science Advances* 1(5): e1400253.

Chazdon, R. L., E. N. Broadbent, D. M. Rozendaal, F. Bongers, A. M. A. Zambrano, T. M. Aide, P. Balvanera, et al. 2016. Carbon sequestration potential of second-growth forest regeneration in the Latin American tropics. *Science Advances* 2(5): e1501639.

Crutzen, P. J. and E. F. Stoermer. 2000. The "anthropocene." *Global Change Newsletter* 41: 17–18.

D'Antonio, C. and P. Vitousek. 1992. Biological invasions by exotic grasses, the grass/fire cycle, and global change. *Annual Review of Ecology and Systematics* 23: 63–88.

Darling, E. S., T. R. McClanahan, J. Maina, G. G. Gurney, N. A. Graham, F. Januchowski-Hartley, J. E. Cinner, et al. 2019. Social–environmental drivers inform strategic management of coral reefs in the Anthropocene. *Nature Ecology and Evolution* 3(9): 1341–1350.

Dirzo, R., H. S. Young, M. Galetti, G. Ceballos, N. B. J. Isaac, and B. Collen. 2014. Defaunation in the Anthropocene. *Science* 345(6195): 401–406.

Early, R., B. A. Bradley, J. S. Dukes, J. J. Lawler, J. D. Olden, D. M. Blumenthal, P. Gonzalez, et al. 2016. Global threats from invasive alien species in the twenty-first century and national response capacities. *Nature Communications* 7: 12485.

Ellis, E. C., K. K. Goldewijk, S. Siebert, D. Lightman, and N. Ramankutty. 2010. Anthropogenic transformation of the biomes, 1700 to 2000. *Global Ecology and Biogeography* 19(5): 589–606.

Ellis, E. C. and N. Ramankutty. 2008. Putting people in the map: Anthropogenic biomes of the world. *Frontiers in Ecology and the Environment* 6(8): 439–447.

Ferretti, F., B. Worm, G. L. Britten, M. R. Heithaus, and H. K. Lotze. 2010. Patterns and ecosystem consequences of shark declines in the ocean. *Ecology Letters* 13: 1055–1071.

Fischer, J. and D. B. Lindenmayer. 2007. Landscape modification and habitat fragmentation: A synthesis. *Global Ecology and Biogeography* 16(3): 265–280.

Hulme, P. E. 2009. Trade, transport and trouble: Managing invasive species pathways in an era of globalization. *Journal of Applied Ecology* 46(1): 10–18.

Kaye, J. P., P. M. Groffman, N. B. Grimm, L. A. Baker, and R. V. Pouyat. 2006. A distinct urban biogeochemistry? *Trends in Ecology and Evolution* 21(4): 192–199.

Lambdon, P. W., P. Pyšek, C. Basnou, M. Arianoutsou, F. Essl, M. Hejda, V. Jarošík, et al. 2008. Alien flora of Europe: Species diversity, temporal trends, geographical patterns and research needs. *Preslia* 80(2): 101–149.

Laungani, R. and J. M. H. Knops. 2009. Species-driven changes in nutrient-cycling can provide a mechanism for plant invasions. *Proceedings of the National Academy of Sciences* 106(30): 12400–12405.

Lenoir, J. and J.-C. Svenning. 2015. Climate-related range shifts—a global multidimensional synthesis and new research directions. *Ecography* 38: 15–28.

Leopold, A. 1970. *A Sand County Almanac.* New York: Ballantine.

Leopold, A. 1992. *The River of the Mother of God: And Other Essays by Aldo Leopold.* Edited by S. L. Flader and J. B. Callicott. Madison: University of Wisconsin Press.

Lockwood, J. L., M. F. Hoopes, and M. P. Marchetti. 2013. *Invasion Ecology,* 2nd ed. Chichester, West Sussex, UK: Wiley-Blackwell.

McIntyre, S. and R. Hobbs. 1999. A framework for conceptualizing human effects on landscapes and its relevance to management and research models. *Conservation Biology* 13(6): 1282–1292.

Morris, J. A., Jr., ed. 2012. *Invasive Lionfish: A Guide to Control and Management.* Gulf and Caribbean Fisheries Institute Special Publication Series 1. Marathon, FL: Gulf and Caribbean Fisheries Institute.

Mumby, P. J., R. S. Steneck, G. Roff, and V. J. Paul. 2021. Marine reserves, fisheries ban, and 20 years of positive change in a coral reef ecosystem. *Conservation Biology.* DOI: 10.1111/cobi.13738.

Narayan, S., M. W. Beck, P. Wilson, C. J. Thomas, A. Guerrero, C. C. Shepard, B. G. Reguero, G. Franco, J. C. Ingram, and D. Trespalacios. 2017. The value of coastal wetlands for flood damage reduction in the northeastern USA. *Scientific Reports* 7(1): 1–12.

Richardson, A. J., A. Bakun, G. C. Hays, and M. J. Gibbons. 2009. The jellyfish joyride: Causes, consequences and management responses to a more gelatinous future. *Trends in Ecology and Evolution* 24(6): 312–322.

Rogosch, J. S. and J. D. Olden. 2020. Invaders induce coordinated isotopic niche shifts in native fish species. *Canadian Journal of Fisheries and Aquatic Sciences* 77(8): 1348–1358.

Saino, N., R. Ambrosini, D. Rubolini, J. van Hardenberg, A. Provenzale, K. Huppop, O. Huppop, et al. 2011. Climate warming, ecological mismatch at arrival and population decline in migratory birds. *Proceedings of the Royal Society B* 278(1707): 835–842.

Seddon, N., A. Chausson, P. Berry, C. A. Girardin, A. Smith, and B. Turner. 2020. Understanding the value and limits of nature-based solutions to climate change and other global challenges. *Philosophical Transactions of the Royal Society B* 375(1794): 20190120.

Seebens, H., T. M. Blackburn, E. E. Dyer, P. Genovesi, P. E. Hulme, J. M. Jeschke, S. Pagad, et al. 2017. No saturation in the accumulation of alien species worldwide. *Nature Communications* 8: 14435.

Sinervo, B., F. Méndez-de-la-Cruz, D. B. Miles, B. Heulin, E. Bastiaans, M. Villagrán-Santa Cruz, R. Lara-Resendiz, et al. 2010. Erosion of lizard diversity by climate change and altered thermal niches. *Science* 328(5980): 894–899.

Standish, R. J., R. J. Hobbs, M. M. Mayfield, B. T. Bestelmeyer, K. N. Suding, L. L. Battaglia, V. Eviner, et al. 2014. Resilience in ecology: Abstraction, distraction, or where the action is? *Biological Conservation* 177: 43–51.

Steffen, W., W. Broadgate, L. Deutsch, O. Gaffney, and C. Ludwig. 2015. The trajectory of the Anthropocene: The Great Acceleration. *Anthropocene Review* 2(1): 81–98.

UNEP (United Nations Environment Programme). 2011. One small planet, seven billion people by year's end and 10.1 billion by century's end. UNEP Global Environmental Alert Service. https://na.unep.net/geas/getUNEPPageWithArticleIDScript.php?article_id=71. Published June 2011. Accessed September 17, 2021.

Urban, M. C., G. Bocedi, A. P. Hendry, J. B. Mihoub, G. Pe'er, A. Singer, J. R. Bridle, et al. 2016. Improving the forecast for biodiversity under climate change. *Science* 353(6304): aad8466.

Urban, M.C., J. J. Tewksburn, and K. S. Sheldon. 2012. On a collision course: Competition and dispersal differences create no-analogue communities and cause extinctions during climate change. *Proceedings of the Royal Society B* 279: 2072–2080.

USGS (US Geological Survey). 2020. Lionfish distribution, geographic spread, biology, and ecology. USGS. https://www.usgs.gov/centers/wetland-and-aquatic-research-center-warc/science/lionfish-distribution-geographic-spread?qt-science_center_objects=7#qt-science_center_objects. Accessed September 24, 2021.

Valladares F., S. Matesanz, F. Guilhaumon, M. B. Araújo, L. Balaguer, M. Benito-Garzón, W. Cornwell, et al. 2014. The effects of phenotypic plasticity and local adaptation on forecasts of species range shifts under climate change. *Ecology Letters* 17(11): 1351–1364.

Waters, C.N., J. Zalasiewicz, C. Summerhayes, A. D. Barnosky, C. Poirier, A. Gałuszka, A. Cearreta, M. Edgeworth, and E. C. Ellis. 2016. The Anthropocene is functionally and stratigraphically distinct from the-Holocene. *Science* 351(6269): aad2622.

Woodland Trust. 2021. Important phenologists. Nature's Calendar. https://naturescalendar.woodlandtrust.org.uk/what-we-record-and-why/why-we-record/important-phenologists/. Accessed September 22, 2021.

CREDITS

Chapter 1

Photos: Chapter Opener: toddy0011/Stockimo/Alamy Stock Photo; Figure 1.3A: Reproduced with permission from John van Wyhe ed. 2002. The Complete Work of Charles Darwin Online (http://darwin-online.org.uk/); Figure 1.3B: BL/Robana Picture Library/Age Fotostock; Figure 1.5: National Science Foundation funded NEON project, courtesy of Battelle; Figure 1.6: Rudyard Kipling/Alamy Stock Photo; Figure 1.9A: Derrick Alderman/Alamy Stock Photo; Figure 1.9B: David Chapman/Alamy Stock Photo; Figure 1.9C: Imagebroker/Alamy Stock Photo; Figure 1.9D: Premaphotos/Alamy Stock Photo; Figure 1.10 (left): Fritz Mueller Visuals; Figure 1.10 (right): Jeffrey Lepore/Science Source; Figure 1.12: © 2013 Justy L. Grinter; Figure 1.13A: Hafidziqram/Shutterstock; Figure 1.13B: Courtesy of Barry Sinervo; Figure 1.14A: Luca Iaconi-Stewart; Figure 1.14B: Feng Yu/Alamy Stock Photo.

Other: Figure 1.2: Republished with permission of SAGE, from Will Steffen et al., "The Trajectory of the Anthropocene: The Great Acceleration," *The Anthropocene Review* 2(1): 81–98 (2015). Permission conveyed through Copyright Clearance Center, Inc.; Figure 1.12: Republished with permission of Wiley, from Elizabeth E. Crone and Janet L. Gehring, "Population Viability of *Rorippa columbiae*: Multiple Models and Spatial Trend Data," *Conservation Biology* 12(5): 1054–1065 (2008). Permission conveyed through Copyright Clearance Center, Inc.; Figure 1.13B: From Janis L. Dickinson and Walter D. Koenig, "Desperately Seeking Similarity," *Science* 300(5627): 1887–1889 (2003). Reprinted with permission from AAAS.

Chapter 2

Photos: Chapter Opener: RGB Ventures/SuperStock/Alamy Stock Photo; Figure 2.1A: Courtesy of Michael P. Marchetti; Figure 2.1B: BIOSPHOTO/Alamy Stock Photo; Figure 2.1C: Jeffrey Isaac Greenberg 2+/Alamy Stock Photo; Figure 2.1D: Courtesy of Michael P. Marchetti; Figure 2.1E: Courtesy of Michael P. Marchetti; Figure 2.2: Tom Salyer/Alamy Stock Photo; Figure 2.3: Courtesy of Michael P. Marchetti; Figure 2.4: Nature Photographers Ltd/Alamy Stock Photo; Figure 2.6: Ian Thraves/Alamy Stock Photo; Figure 2.9: Anton Sorokin/Alamy Stock Photo, Andreas Kettenburg, Chuck Brown/Science Source, Creeping Things/Shutterstock, Will Flaxington, Tammy Lim, and Jeremiah Easter; Figure 2.10: Mike Truchon/Shutterstock, Life on White/Alamy Stock Photo, and Oleksandr Zheltobriukh/Alamy Stock Photo; Figure 2.12: Courtesy of Michael P. Marchetti; Figure 2.13A: Courtesy of CID Bio-Science; Figure 2.13B: Jit Lim/Alamy Stock Photo.

Other: Figure 2.8: From Murray Adams, "Design for six sigma: A potent supplement to QbD," Pharmamanufacturing.com. Reused with permission of Pharma Manufacturing; Figure 2.9: Adapted from Devitt, T.J., Baird, S.J. and Moritz, C. "Asymmetric reproductive isolation between terminal forms of the salamander ring species *Ensatina eschscholtzii* revealed by fine-scale genetic analysis of a hybrid zone," *BMC Evolutionary Biology* 11: 245 (2011). https://doi.org/10.1186/1471-2148-11-245. This is an open access article distributed under the terms of the Creative Commons Attribution License (http://creativecommons.org/licenses/by/2.0), which permits unrestricted use, distribution, and reproduction in any medium, provided the original work is properly cited.

Chapter 3

Photos: Chapter Opener: Stocktrek Images/Getty; Figure 3.5 (left to right): NASA, Aflo Co., Ltd./Alamy Stock Photo, Martin Shields/Alamy Stock Photo, jfmdesign/Getty Images, and YuryKara/Shutterstock; Figure 3.14B: Google Earth; Figure 3.13B (inset left): pavel891/Shutterstock; Figure 3.13B (inset right): Jim Feliciano/Shutterstock; Figure 3.15B (left): yhelfman/Deposit Photos; Figure 3.15B (right): AP Photo/The Wichita Eagle, Travis Heying; Figure 3.16: mauritius images GmbH/Alamy Stock Photo; Figure 3.17: imageBROKER/Alamy Stock Photo; Figure 3.18: Paulo Oliveira/Alamy Stock Photo; Figure 3.20A (left to right): blickwinkel/Alamy Stock Photo, Petlin Dmitry/Shutterstock, and Sergey Uryadnikov/Alamy Stock Photo; Figure 3.20D: Panama Landscapes by Oyvind Martinsen/Alamy Stock Photo; Figure 3.20E: imageBROKER/Alamy Stock Photo; Figure 3.21A (left to right): Octavio Campos Salles/Alamy Stock Photo, Chris Mattison/Alamy Stock Photo, and Ariadne Van Zandbergen/Alamy Stock Photo; Figure 3.21D: Courtesy of Michael P. Marchetti; Figure 3.21E: Tim Plowden/Alamy Stock Photo; Figure 3.22A (left to right): Gabbro/Alamy Stock Photo, Life on White/Alamy Stock Photo, and Johan Swanepoel/Alamy Stock Photo; Figure 3.22D: Ryan Faas/Alamy Stock Photo; Figure 3.22E: Cultura Creative Ltd/Alamy Stock Photo; Figure 3.23A (left to right): HAWK Photography NAMIBIA/Shutterstock, fivespots/Shutterstock, and PhotoStock-Israel/Alamy Stock Photo; Figure 3.23D (left): Courtesy of Michael P. Marchetti; Figure 3.23D (right): Matthew J. Kirsch/Alamy Stock Photo; 3.23E: Timothy Mulholland/Alamy Stock Photo; Figure 3.24A (left to right): PureStock/Alamy Stock Photo, Zoonar GmbH/Alamy Stock Photo, and blickwinkel/Alamy Stock Photo; Figure 3.24D: Garey Lennox/Alamy Stock Photo; Figure 3.24E: Courtesy of Michael P. Marchetti; Figure 3.25A (left to right): Dominique Braud/Dembinsky Photo Associates/Alamy Stock Photo, Life on White/Alamy Stock Photo, and Prisma by Dukas Presseagentur GmbH/Alamy Stock Photo; Figure 3.25D: Clint Farlinger/Alamy Stock Photo; Figure 3.25E: All Canada Photos/Alamy Stock Photo; Figure 3.26A (left to right): Stone Nature Photography/Alamy Stock Photo, nikkytok/Shutterstock, and RudiErnst/Shutterstock; Figure 3.26D (left): Prisma by Dukas Presseagentur GmbH/Alamy Stock Photo; Figure 3.26D (right): agefotostock/Alamy Stock Photo; Figure 3.26E (left): Imagebroker/Alamy Stock Photo; Figure 3.26E (right): All Canada Photos/Alamy Stock Photo; Figure 3.27A (left to right): Jim Cumming/Alamy Stock Photo, FotoRequest/Shutterstock, and Ryan Hagerty, U.S. Fish and Wildlife Service; Figure 3.27D: MShieldsPhotos/Alamy Stock Photo; Figure 3.27E: Harry Collins/Alamy Stock Photo; Figure 3.28A (left to right): Arterra Picture Library/Alamy Stock Photo, FotoRequest/Shutterstock, and Arterra Picture Library/Alamy Stock Photo; Figure 3.28D: Mint Images Limited/Alamy Stock Photo; Figure 3.28E: Erkki Makkonen/Alamy Stock Photo; Figure 3.32: Used with permission of Encyclopædia Britannica Online, © Encyclopædia Britannica, Inc.; Figure 3.33: Edwin Remsberg/Alamy Stock Photo; Figure 3.34: Images & Stories/Alamy Stock Photo; Figure 3.35: BIOSPHOTO/Alamy Stock Photo; Figure 3.36: Mark Conlin/Alamy Stock Photo; Figure 3.37: Kim Karpeles/Alamy Stock Photo; Figure 3.38: Kevin Griffin/Alamy Stock Photo; Figure 3.39 (left): Alena Vikhareva/Alamy Stock Photo; Figure 3.39 (right): George Ostertag/Alamy Stock Photo; Figure 3.40: Bo Valentino/Alamy Stock Photo; Figure 3.41: Jacques Descloitres, MODIS Rapid Response Team, NASA/GSFC; Figure 3.43: Woods Hole Oceanographic Institution and P. Caiger; Figure 3.48: robertharding/Alamy Stock Photo; Figure 3.50: Brandi Mueller/Getty Images; Figure 3.53: Dorling Kindersley Ltd/Alamy Stock Photo.

Other: Figure 3.9: Courtesy of L. David Rooper; Figure 3.12: Adapted from Michael Pidwirny, "Surface and Subsurface Ocean Currents: Ocean Current Map," *Understanding Physical Geography*. Reprinted with permission; Figure 3.15B: Annual climatology of Wichita and San Francisco from "Climographs," drought.unl.edu. Reused with permission of the National Drought Mitigation Center; Figures 3.20–3.28: Part C from ClimateCharts.net. Licensed under a Creative Commons Attribution 4.0 International license (https://creativecommons.org/licenses/by/4.0/). See Laura Zepner, Pierre Karrasch, Felix Wiemann, and Lars Bernard (2020). ClimateCharts.net—an interactive climate analysis web platform, International Journal of Digital Earth, DOI: 10.1080/17538947.2020.189112; Figure 3.42: Adapted from xkcd.com. Used with permission; Figure 3.52: Adapted from IPCC, 2007: *Climate Change 2007: The Physical Science Basis. Contribution of Working Group I to the Fourth Assessment Report of the Intergovernmental Panel on Climate Change* [Solomon, S., D. Qin, M. Manning, Z. Chen, M. Marquis, K.B. Averyt, M. Tignor, and H.L. Miller (eds.)]. Cambridge University Press, Cambridge, United Kingdom and New York, NY, USA, 996 pp. Reused with permission from the IPCC.

Chapter 4

Photos: Chapter Opener: Frank Hecker/Alamy Stock Photo; Figure 4.1A: Chris Bosworth/Alamy Stock Photo; Figure 4.1B: Holmes Garden Photos/Alamy Stock Photo; Figure 4.3A: Cannasue/Getty Images; Figure 4.3B: Imagebroker/Alamy Stock Photo; Figure 4.5A: GFC Collection/Alamy Stock Photo; Figure 4.5B: Darlene Murawski; Figure 4.6A: Natural Visions/Alamy Stock Photo; Figure 4.6B: Panther Media GmbH/Alamy Stock Photo; Figure

4.6C: blickwinkel/Alamy Stock Photo; Figure 4.8: Arterra Picture Library/Alamy Stock Photo; Figure 4.9: Biophoto Associates/Science Source; Figure 4.10 (left): GlobalP/Getty Images; Figure 4.10 (right): Yves Lanceau/Nature Picture Library; Figure 4.11 (top to bottom): GlobalP/Getty Images and Yves Lanceau/Nature Picture Library; Figure 4.17: All Canada Photos/Alamy Stock Photo; Figure 4.22: Ger Bosma/Alamy Stock Photo; Figure 4.23: All Canada Photos/Alamy Stock Photo; Figure 4.26: Rudh, A., Rogell, B., and Höglund, J. (2007), "Non-gradual variation in colour morphs of the strawberry poison frog *Dendrobates pumilio*: genetic and geographical isolation suggest a role for selection in maintaining polymorphism," *Molecular Ecology* 16: 4284–4294. With permission from John Wiley & Sons; Figure 4.27: ANT Photo Library/Science Source.

Other: Figure 4.19: Data from Boose E, Gould E. (2022). Harvard Forest Climate Data since 1964. Harvard Forest Data Archive: HF300 (v.5). Environmental Data Initiative (https://doi.org/10.6073/pasta/03dc1f107ca816675b4983daf7a97dbe); Figure 4.26: Republished with permission of John Wiley & Sons, from Andreas Rudh, Björn Rogell, and Jacob Höglund, "Non-gradual variation in colour morphs of the strawberry poison frog *Dendrobates pumilio*: genetic and geographical isolation suggest a role for selection in maintaining polymorphism," *Molecular Ecology* 16(20): 4284–4294 (2007). Permission conveyed through Copyright Clearance Center, Inc.; Figure 4.28: Adapted by permission from Springer Nature: *Nature* 429, Tigga Kingston and Stephen J. Rossiter, "Harmonic-hopping in Wallacea's bats," pp. 654–657. Copyright © 2004, Macmillan Magazines Ltd; Figure 4.29: Adapted from Gabriel Pigeon, Marco Festa-Bianchet, David W. Coltman, et al., "Intense selective hunting leads to artificial evolution in horn size," *Evolutionary Applications* 9(4): 521–530 (2016). https://doi.org/10.1111/eva.12358. This is an open access article distributed under the terms of the Creative Commons Attribution License (https://creativecommons.org/licenses/by/4.0/).

Chapter 5

Photos: Chapter Opener: imageBROKER/Alamy Stock Photo; Figure 5.1: Marie Lemerle/Alamy Stock Photo; Figure 5.2A: Greatstock/Alamy Stock Photo; Figure 5.2B: Wolfgang Kaehler/Alamy Stock Photo; Figure 5.3: Neil Lucas/Nature Picture Library; Figure 5.4: Sjo/Getty Images; Figure 5.5: Design Pics Inc/Alamy Stock Photo; Figure 5.6 (left): Suradech Prapairat/Shutterstock; Figure 5.6 (right): Kyryl Gorlov/Alamy Stock Photo; Figure 5.7 (left to right): William Mullins/Alamy Stock Photo, wonderful-Earth.net/Alamy Stock Photo, and Rick & Nora Bowers/Alamy Stock Photo; Figure 5.9: Zoonar GmbH/Alamy Stock Photo; Figure 5.10: Courtesy of Michael P. Marchetti; Figure 5.11: Rick & Nora Bowers/Alamy Stock Photo; Figure 5.14: Alfio Scisetti/Alamy Stock Photo; Figure 5.15: InfoFlowersPlants/Shutterstock; Figure 5.16: vaeenma/Deposit Photos; Figure 5.17: john t. fowler/Alamy Stock Photo; Figure 5.21 (left): Wichai Prasomsri1/Shutterstock, Arterra Picture Library/Alamy Stock Photo, YAY Media AS/Alamy Stock Photo, Panther Media GmbH/Alamy Stock Photo, Nature Picture Library/Alamy Stock Photo, and Eric Isselee/Shutterstock; Figure 5.21 (right): James D Coppinger/Dembinsky Photo Associates/Alamy Stock Photo, Chris Mattison/Alamy Stock Photo, Kletr/Shutterstock, Charles Brutlag/Shutterstock, Anan Kaewkhammul/Alamy Stock Photo, and LJSphotography/Alamy Stock Photo; Figure 5.23: Stephanie Jackson—Australian wildlife collection/

Alamy Stock Photo; 5.24: Matthijs Kuijpers/Alamy Stock Photo; Figure 5.25: dpa picture alliance/Alamy Stock Photo; Figure 5.26A: Stone Nature Photography/Alamy Stock Photo; Figure 5.26B: Stephen Osman/Los Angeles Times via Getty Images.

Other: Figure 5.8A: From Glimcher, Paul W., *Decisions, Uncertainty, and the Brain*, p. 218, © 2003 Massachusetts Institute of Technology, by permission of The MIT Press; Figure 5.8B: Reprinted from *Animal Behaviour* 25(9), John R. Krebs, Jonathan T. Erichsen, Michael I. Webber, et al., "Optimal prey selection in the great tit (*Parus major*)," pp. 30–38. © 1997, with permission from the Association for the Study of Animal Behaviour; Figure 5.17: Republished with permission of John Wiley & Sons, from Patricia T. Tomlinson and Paul D. Anderson, "Ontogeny affects response of northern red oak seedlings to elevated CO_2 and water stress. II. Recent photosynthate distribution and growth," *New Phytologist* 140(3): 493–504 (2008). Permission conveyed through Copyright Clearance Center, Inc.; Figure 5.20A: Republished with permission of John Wiley & Sons, from Christopher J. Still, Joseph A. Berry, G. James Collatz, et al., "Global distribution of C3 and C4 vegetation: Carbon cycle implications," *Global Biogeochemical Cycles* 17(1): 6-1–6-14, 2003. Permission conveyed through Copyright Clearance Center, Inc.; Figures 5.22 and 5.23: Republished with permission of University of Chicago Press, from Raymond B. Huey and Montgomery Slatkin, "Costs and Benefits of Thermoregulation," *The Quarterly Review of Biology* 51(3): 363–384 (1976). Permission conveyed through Copyright Clearance Center, Inc.; Figure 5.24: Reprinted from *Journal of Thermal Biology* 48, Sara Valenzuela-Ceballos, Gamaliel Castañeda, Tamara Rioja-Paradela, et al., "Variation in the thermal ecology of an endemic iguana from Mexico reduces its vulnerability to global warming," pp. 56–64. © 2015, with permission from Elsevier; Figure 5.25B,C: Adapted by permission from Springer Nature: Springer, *Oecologia* 36, J. R. Ehleringer and O. Björkman, "Pubescence and leaf spectral characteristics in a desert shrub, Encelia farinose," pp. 151–162. © 1978, Springer-Verlag; Figure 5.31: Adapted by permission from Springer Nature: *Nature* 268, Richard J. Cowie, "Optimal foraging in great tits (*Parus major*)," pp. 137–139. © 1977, Nature Publishing Group.

Chapter 6

Photos: Chapter Opener: Cindy Hopkins/Alamy Stock Photo; Figure 6.1: Daniel Dempster Photography/Alamy Stock Photo; Figure 6.2 (top): Arterra Picture Library/Alamy Stock Photo; Figure 6.2 (middle): Georgette Douwma/Getty Images; Figure 6.2 (bottom): Papilio/Alamy Stock Photo; Figure 6.3 (left): Courtesy of Vulcan Inc.; Figure 6.3 (right): Vadim Petrakov/Shutterstock; Figure 6.4A: Will Rayment/NZ Whale and Dolphin Trust; Figure 6.4B: ArteSub/Alamy Stock Photo; Figure 6.4C: Natural History Library/Alamy Stock Photo; Figure 6.5: Wyoming Game and Fish Department/Public Domain; Figure 6.6: agefotostock/Alamy Stock Photo; Figure 6.7: vuttichai chaiya/Shutterstock; Figure 6.8A: Robert McGouey/Wildlife/Alamy Stock Photo; Figure 6.8B: heckepics/Getty Images; Figure 6.8C: agefotostock/Alamy Stock Photo; Figure 6.13: Christian Musat/Alamy Stock Photo; Figure 6.18: Christian Musat/Alamy Stock Photo; Figure 6.19A: Christian Musat/Alamy Stock Photo; Figure 6.19B: CLS Digital Arts/Shutterstock; Figure 6.19C: Nature Photographers Ltd/Alamy Stock Photo; Figure 6.19D: Zoonar GmbH/Alamy Stock Photo; Figure 6.20A: Colleen Gara/Getty Images; Figure 6.20B: Drake Fleege/Alamy Stock Photo; Figure 6.24: blickwinkel/Alamy Stock

Photo; Figure 6.25 (left to right): Gilbert S. Grant/Science Source, Carol Dembinsky/Dembinsky Photo Associates/Alamy Stock Photo, Carol Dembinsky/Dembinsky Photo Associates/Alamy Stock Photo, and Pat Canova/Alamy Stock Photo; Figure 6.31A: J. Griffis Smith/TxDOT; Figure 6.31B: RGB Ventures/SuperStock/Alamy Stock Photo.

Other: Figure 6.21: Adapted by permission from Springer Nature: Springer, *Oecologia* 143, Kelley M. Stewart, R. Terry Bowyer, Brian L. Dick, et al., "Density-dependent effects on physical condition and reproduction in North American elk: an experimental test," pp. 85–93. © 2004, Springer-Verlag; Figure 6.24A: Used by permission of the State of Queensland (Department of Agriculture and Fisheries); Figure 6.24B: Data from Boose, E., Gould, E. (2022). Harvard Forest Climate Data since 1964. Harvard Forest Data Archive: HF300 (v.5). Environmental Data Initiative (https://doi.org/10.6073/pasta/03dc1f107ca816675b4983daf7a97dbe); Table 6.4: Republished with permission from John Wiley & Sons, from Larry B. Crowder, Deborah T. Crouse, Selina S. Heppell, et al., "Predicting the Impact of Turtle Excluder Devices on Loggerhead Sea Turtle Populations," *Ecological Applications* 4(3): 437–445 (1994); permission conveyed through Copyright Clearance Center, Inc.

Chapter 7

Photos: Chapter Opener: agefotostock/Alamy Stock Photo; Figure 7.1A: NHPA/SuperStock; Figure 7.1B: Science History Images/Alamy Stock Photo; Figure 7.4A: imageBROKER/Alamy Stock Photo; Figure 7.4C: Ivan Kmit/Alamy Stock Photo; Figure 7.17: Terry Donnelly/Alamy Stock Photo; Figure 7.19: Tomasz Klejdysz/Shutterstock; Figure 7.21A: blickwinkel/Alamy Stock Photo; Figure 7.21B: ardea.com/Mary Evans/David/Pantheon/SuperStock; Figure 7.29A: Janet Horton/Alamy Stock Photo; Figure 7.29B: Michael Rose/Alamy Stock Photo; Figure 7.30A (left): yamaoyaji/Shutterstock; Figure 7.30A (right): Jeff Mondragon/Alamy Stock Photo; Figure 7.35: Pfennig, D.W., Wund, M.A., Snell-Rood, E.C., Cruickshank, T., Schlichting, C.D., and Moczek, A.P. (2010). "Phenotypic plasticity's impacts on diversification and speciation," *Trends in Ecology & Evolution* 25(8): 459–467. Used with permission from Elsevier.

Other: Figure 7.4B: Cougar or mountain lion distribution map, California NatureMapping Program, naturemappingfoundation.org. Courtesy of K. Dvornich; Figure 7.20: Republished with permission of John Wiley & Sons, from Thomas Park, "Interspecies competition in populations of *Trilobium confusum* Duval and *Trilobium castaneum* Herbst," *Ecological Monographs* 18: 265–306 (1948). Permission conveyed through Copyright Clearance Center, Inc.; Figure 7.30B: Republished with permission of Springer Nature, from K.D. Fausch, S. Nakano, and K. Ishigaki, "Distribution of two congeneric charrs in streams of Hokkaido Island, Japan: considering multiple factors across scales," *Oecologia* 100(1-2): 1–12 (1994). Permission conveyed through Copyright Clearance Center, Inc.; Figures 7.31 and 7.32: Republished with permission of John Wiley & Sons, from Yoshinori Taniguchi and Shigeru Nakano, "Condition-specific competition: implications for the altitudinal distribution of stream fishes," *Ecology* 81(7): 2027–2039 (2000). Permission conveyed through Copyright Clearance Center, Inc.; Figure 7.34: Republished with permission of the American Association for the Advancement of Science, from Dolph Schluter, Trevor D. Price, and Peter R. Grant, "Ecological Character Displacement in Darwin's Finches," *Science* 227(4690): 1056–1059 (1985). Permission conveyed through Copyright Clearance Center, Inc.

Chapter 8

Photos: Chapter Opener: Avalon.red/Alamy Stock Photo; Figure 8.1A: Mark Conlin/VWPics/Alamy Stock Photo; Figure 8.1B: Courtesy of Michael P. Marchetti; Figure 8.1C: Courtesy of Michael P. Marchetti; Figure 8.2: Perry van Munster/Alamy Stock Photo; Figure 8.7: Jeffrey Lepore/Science Source; Figure 8.8A: Nature Picture Library/Alamy Stock Photo; Figure 8.8B: Paul Zenk; Figure 8.9A: gerard lacz/Alamy Stock Photo; Figure 8.9B: Richard Wayne Collens/Shutterstock; Figure 8.19: Courtesy of Michael P. Marchetti; Figure 8.20 (left): AfriPics.com/Alamy Stock Photo; Figure 8.20 (right): Seyms Brugger/Shutterstock; Figure 8.21: Bill Gorum/Alamy Stock Photo; Figure 8.23: Sundry Photography/Alamy Stock Photo, BIOSPHOTO/Alamy Stock Photo, Yuval Helfman/Alamy Stock Photo, Richard Mittleman/Gon2Foto/Alamy Stock Photo, Andrew Kandel/Alamy Stock Photo, Clarence Holmes Wildlife/Alamy Stock Photo, Charles Melton/Alamy Stock Photo, Joyce Gross, Clarence Holmes Wildlife/Alamy Stock Photo, Martin Shields/Alamy Stock Photo, and courtesy of Michael P. Marchetti; Figure 8.25: Dicrocoelium dendriticum (YPM IZ 094296). Digital Image: Yale Peabody Museum of Natural History. Photo by Daniel J. Drew (2017), Juan Cuadros/CDC, Eric Isselee/Shutterstock, JAH/Alamy Stock Photo, Marcel Derweduwen/Alamy Stock Photo, and Juniors Bildarchiv GmbH/Alamy Stock Photo; Figure 8.26: FLPA/Alamy Stock Photo; Figure 8.27A: Courtesy of Michael P. Marchetti; Figure 8.27B: Courtesy of Michael P. Marchetti; Figure 8.28A: Paul Lund/Alamy Stock Photo; Figure 8.28B: Gerry Bishop/Alamy Stock Photo; Figure 8.29: Rolf Nussbaumer Photography/Alamy Stock Photo; Figure 8.30 (top left): Robert Hamilton/Alamy Stock Photo; Figure 8.30 (top right): Bryan Reynolds/Alamy Stock Photo; Figure 8.30 (bottom): Bryan Reynolds/Alamy Stock Photo; Figure 8.31A: agefotostock/Alamy Stock Photo; Figure 8.31B: Christian Musat/Alamy Stock Photo; Figure 8.31C: EyeEm/Alamy Stock Photo; Figure 8.31D: Beth Swanson/Alamy Stock Photo; Figure 8.32 (top left): Paul Starosta/Getty Images; Figure 8.32 (top right): Dani Purnomo/Alamy Stock Photo; Figure 8.32 (bottom left): Sandra Standbridge/Alamy Stock Photo; Figure 8.32 (bottom right): Robert Shantz/Alamy Stock Photo; Figure 8.33: Greene, H.W. and McDiarmid, R.W., "Coral snake mimicry: Does it occur?" *Science* 213(4513): 1207–1212 (1981). With permission from AAAS; Figure 8.35: Seth Golub; Figure 8.36: Courtesy of Michael P. Marchetti; Figure 8.37: blickwinkel/Alamy Stock Photo; Figure 8.38: Bazzano Photography/Alamy Stock Photo.

Other: Figure 8.4: Republished with permission of John Wiley & Sons, from Ronald A. Knapp, Kathleen R. Matthews, and Orlando Sarnelle, "Resistance and resilience of alpine lake fauna to fish introductions," *Ecological Monographs* 71(3): 401–421 (2001). Permission conveyed through Copyright Clearance Center, Inc.; Figure 8.6B: Republished with permission of the American Association for the Advancement of Science, from G.F. Gause, "Experimental Analysis of Vito Volterra's Mathematical Theory of the Struggle for Existence," *Science* 79(2036): 16–17 (1934). Permission conveyed through Copyright Clearance Center, Inc.; Figure 8.33: From Harry W. Greene and Roy W. McDiarmid, "Coral snake mimicry: Does it occur?" *Science* 213(4513): 1207–1212 (1981). Reprinted with permission from AAAS; Figure 8.34: Adapted by permission from Springer Nature: *Nature* 410, David W. Pfennig, William R. Harcombe, and Karin S. Pfennig, "Frequency-dependent Batesian mimicry," p. 323. © 2001, Nature Publishing Group.

Chapter 9

Photos: Chapter Opener: Geoff Smith/Alamy Stock Photo; Figure 9.1: adrian hepworth/Alamy Stock Photo, Redmond Durrell/Alamy Stock Photo, Michael Siluk/Alamy Stock Photo, D. Kucharski K. Kucharska/Shutterstock, and Nature Picture Library/Alamy Stock Photo; Figure 9.2A: Lipatova Maryna/Shutterstock; Figure 9.2B: Daniel Murphy; Figure 9.2C: blickwinkel/Alamy Stock Photo; Figure 9.3 (top to bottom): Kevin Ebi/Alamy Stock Photo, Michael Rose/Alamy Stock Photo, and Rick & Nora Bowers/Alamy Stock Photo; Figure 9.4: Brent Durand/Getty Images; Figure 9.8: Dr. Jeremy Burgess/Science Source; Figure 9.9: Kevin Schafer/Alamy Stock Photo; Figure 9.10: imageBROKER/Alamy Stock Photo; Figure 9.13A: Penny Frith; Figure 9.13B: Martin Lindsay/Alamy Stock Photo; Figure 9.14: imageBROKER/Alamy Stock Photo; Figure 9.20: Courtesy of Michael P. Marchetti; Figure 9.21: Science Photo Library/Alamy Stock Photo; Figure 9.22: Gary K Smith/Alamy Stock Photo; Figure 9.23: Mark Conlin/Alamy Stock Photo; Figure 9.24A: Peter Yeeles/Alamy Stock Photo; Figure 9.24B: FLPA/Alamy Stock Photo; 9.25: Courtesy of Peter Green.

Other: Figure 9.6: From Jeremy E. Koenig, et al., "Succession of microbial consortia in the developing infant gut microbiome," *PNAS* 108(Supplement 1): 4578–4585 (2010). Reprinted with permission of the National Academy of Science; Figure 9.7: Reprinted from *Gastroenterology* 145(5), Loek P. Smits, Kristien E.C. Bouter, Willem M. de Vos, et al., "Therapeutic Potential of Fecal Microbiota Transplantation," pp. 946–953. © 2013, with permission from the American Gastroenterological Association.

Chapter 10

Photos: Chapter Opener: Greg Vaughn/Alamy Stock Photo; Figure 10.2: Michele Burgess/SuperStock, AlessandraRCstock/Alamy Stock Photo, Joyce Gross, Bob Gibbons/Alamy Stock Photo, and VIKVAD/Alamy Stock Photo; Figure 10.3: Courtesy of Michael P. Marchetti; Figure 10.5: Life on White/Alamy Stock Photo; Figure 10.12 (top): Courtesy of Michael Marchetti; Figure 10.12 (bottom): Cavan Images/Alamy Stock Photo; Figure 10.14: Michele Burgess/SuperStock, AlessandraRCstock/Alamy Stock Photo, Joyce Gross, Bob Gibbons/Alamy Stock Photo, and VIKVAD/Alamy Stock Photo; Figure 10.15: Michele Burgess/SuperStock, AlessandraRCstock/Alamy Stock Photo, Joyce Gross, Bob Gibbons/Alamy Stock Photo, Tomas Calle/Alamy Stock Photo, Mc Photo/Alamy Stock Photo, Patrick Barron/Alamy Stock Photo, blickwinkel/Alamy Stock Photo, Don Johnston MA/Alamy Stock Photo, and VIKVAD/Alamy Stock Photo; Figure 10.17: D.M. Lavigne, "Ecological interactions between marine mammals, commercial fisheries, and their prey: unravelling the tangled web," *Oceanographic Literature Review* 44(3): 229 (1997). With permission from Elsevier; Figure 10.18: Courtesy of Jennifer Dunne; Figure 10.19: Courtesy of Jennifer Dunne; Figure 10.20: W. Leggett/National Park Service; Figure 10.24: Courtesy of Bill Ripple; Figure 10.26: K.D. Leperi/Alamy Stock Photo; Figure 10.27B: Terrence Walters/USDA; Figure 10.28: Courtesy of Michael P. Marchetti; Figure 10.29: Tony Watson/Alamy Stock Photo; Figure 10.30A: Michael S. Nolan/AGE Fotostock; Figure 10.30B: imageBROKER/Alamy Stock Photo; Figure 10.30C: moose henderson/Alamy Stock Photo; Figure 10.31: All Canada Photos/Alamy Stock Photo; Figure 10.33: wonderful-Earth.net/Alamy Stock Photo, William Mullins/Alamy Stock Photo, and Rick & Nora Bowers/Alamy Stock Photo; Figure 10.34: FLPA/Malcolm Schuyl/AGE Fotostock.

Other: Figures 10.3, 10.9, and 10.10: Courtesy of Sudeep Chandra. Figure 10.13: Republished with permission from John Wiley & Sons, from Lev R. Ginzburg and H. Resit Akcakaya, "Consequences of ratio-dependent predation for steady-state properties of ecosystems," *Ecology* 73(5): 1536–1543 (1992). Permission conveyed through Copyright Clearance Center, Inc.; Figure 10.16: Republished with permission of University of Chicago Press, from Gary A. Polis, "Complex trophic interactions in deserts: An empirical critique of food-web theory," *The American Naturalist* 138(1): 123–155 (1991). Permission conveyed through Copyright Clearance Center, Inc.; Figure 10.22: Republished with permission of CSIRO Publishing, from Megan Layhee, Michael Marchetti, Sudeep Chandra, et al., "Impacts of aquatic invasive species and land use on stream food webs in Kaua'i, Hawai'i," *Pacific Conservation Biology* 20(3): 252–271 (2014). Permission conveyed through Copyright Clearance Center, Inc.; Figure 10.23: Reprinted from *Biological Conservation* 145(1), William J. Ripple and Robert L. Beschta, "Trophic cascades in Yellowstone: The first 15 years after wolf reintroduction," pp. 205–213, © 2012, with permission from Elsevier; Figure 10.25: Reprinted from *Biological Conservation* 198, Robert L. Beschta and William J. Ripple, "Riparian vegetation recover in Yellowstone: The first two decades after wolf reintroduction," pp. 93–103, © 2016, with permission from Elsevier; Figure 10.26: Map of Yellowstone National Park courtesy of Karl Musser; Figure 10.27A: Adapted from Gary A. Polis and Stephen D. Hurd, "Extraordinarily high spider densities on islands: Flow of energy from the marine to terrestrial food webs and the absence of predation," *PNAS* 92: 4382–4386 (1995). © 1995 National Academy of Sciences, U.S.A.; Figure 10.27C: Republished with permission of University of Chicago Press, from Gary A. Polis and Stephen D. Hurd, "Linking marine and terrestrial food webs: Allochthonous input from the ocean supports high secondary productivity on small islands and coastal land communities," *The American Naturalist* 147(3): 396–423 (1996). Permission conveyed through Copyright Clearance Center, Inc.

Chapter 11

Photos: Chapter Opener: Alberto Carrera/Alamy Stock Photo; Figure 11.2: Jochen Tack/Alamy Stock Photo; Figure 11.3: Hebert, P.D.N., et al. (2004), "Ten species in one: DNA barcoding reveals cryptic species in the Neotropical skipper butterfly Astraptes fulgerator." *Proc. Natl. Acad. Sci. U.S.A.* 101: 14812–14817. © 2004 National Academy of Sciences, U.S.A.; Figure 11.4: Courtesy of Mike Baird; Figure 11.5: dpa picture alliance archive/Alamy Stock Photo; Figure 11.7: Maurice Leponce; Figure 11.9: Simia Attentive/Shutterstock, Oleg_Z/Shutterstock, Anastasy Yarmolovich/Alamy Stock Photo, directphoto.bz/Alamy Stock Photo, and Vitolga/Shutterstock; Figure 11.13: Courtesy of Michael P. Marchetti; Figure 11.14A: KenWiedemann/Getty Images; Figure 11.14B: David R. Frazier Photolibrary, Inc./Alamy Stock Photo; Figure 11.15A: ephotocorp/Alamy Stock Photo; Figure 11.20A: IreneuszB/Alamy Stock Photo; Figure

11.20B: Danny Ye/Alamy Stock Photo; Figure 11.30: Morley Read/Alamy Stock Photo; Figure 11.33: Nature Picture Library/Alamy Stock Photo; Figure 11.35A: Imagebroker/Alamy Stock Photo; Figure 11.35B: Whitworth Images/Getty Images.

Other: Figure 11.12, 11.4: Adapted by permission from Springer Nature: Nature, *Nature Education Knowledge* 3(10), Wilco C.E.P. Verberk, "Explaining General Patterns in Species Abundance and Distributions," p. 38. © 2011 Wilco C.E.P. Verberk, under exclusive license to Springer Nature Limited; Figure 11.16: Republished with permission from John Wiley & Sons, from F.W. Preston, "The Canonical distribution of commonness and rarity: Part I," *Ecology* 43(2): 182–215 (1962). Permission conveyed through Copyright Clearance Center, Inc.; Figure 11.18: Adapted from Richard Hoffmann, Melanie Tietje, and Martin Aberhan, "Diversity partitioning in Phanerozoic benthic marine communities," *PNAS* 116(1): 79–83 (2019). https://doi.org/10.1073/pnas.1814487116. © 2019 The Authors. Published by PNAS. This open access article is distributed under Creative Commons Attribution License 4.0 (CC BY). http://creativecommons.org/licenses/by/4.0/; Figure 11.19: Reprinted from *Biological Conservation* 203, Agnieszka Nobis, Michal Zmihorski, and Dorota Kotowska, "Linking the diversity of native flora to land cover heterogeneity and plant invasions in a river valley," pp. 17–24. © 2016, with permission from Elsevier; Figure 11.23: Reprinted from *Trends in Ecology & Evolution* 11, Kevin J. Gasaton, "Species range-size distributions: Patterns, mechanisms, and implications," pp. 679–688. © 1996, with permission from Elsevier; Figures 11.25 and 11.26: Republished with permission from John Wiley & Sons, from Alison G. Boyer and Walter Jetz, "Extinctions and the loss of ecological function in island bird communities," *Global Ecology and Biogeography* 23(6): 679–688 (2014). Permission conveyed through Copyright Clearance Center, Inc.; Figure 11.27: Reprinted from *Trends in Ecology & Evolution* 28, Marten Winter, Vincent Devictor, and Oliver Schweiger, "Phylogenetic diversity and nature conservation: Where are we?" pp. 199–204. © 2013, with permission from Elsevier; Figure 11.28: Adapted from Jacqueline Howard, "An Exhaustive Lemur Family Tree Sheds Light on These Rare, Threatened Primates," Huffpost.com, April 25, 2016. Reused with permission of Stephen Nash; Figure 11.29: Adapted from Matt Davis, Soren Faurby, and Jens-Christian Svenning, "Mammal diversity will take millions of years to recover from the current biodiversity crisis," *PNAS* 115(44): 11262–11267 (2018). Reprinted with permission of National Academy of Sciences; Figure 11.31A: Adapted from Wilhelm Barthlott, M. Daud Rafiqpoor, and Walter R. Erdelen, "Bionics and biodiversity—bio-inspired technical innovation for a sustainable future," in J. Knippers, et al. (eds.), *Biomimetic Research for Architecture and Building Construction, Biologically-Inspired Systems* 9 (2016). Reused with permission; Figures 11.31B and 11.34: Republished with permission from The Royal Society, from T. Jonathan Davies and Lauren B. Buckley, "Phylogenetic diversity as a window into the evolutionary and biogeographic histories of present-day richness gradients for mammals," *Philosophical Transactions of The Royal Society B* 366: 2414–2425 (2011). Permission conveyed through Copyright Clearance Center, Inc.; Figure 11.31C: Adapted from IUCN, Conservation International, and NatureServe (2008). An Analysis of Amphibians on the 2008 IUCN Red List (www.iucn.redlist.org/amphibians). Reused with permission; Figure 11.32: Republished with permission from John Wiley & Sons, from Viola Clausnitzer, Klaas-Douwe B. Dijkstra, Robert Koch, et al., "Focus on African freshwaters: Hotspots of dragonfly diversity and conservation concern," *Frontiers in Ecology and the Environment* 10(3): 129–134 (2012). Permission conveyed through Copyright Clearance Center, Inc.; Figure 11.36: Adapted with permission of Springer Nature: *Nature* 501, Rick D. Stuart-Smith, Amanda E. Bates, Jonathan S. Lefcheck, et al., "Integrating abundance and functional traits reveals new global hotspots of fish diversity," pp. 539–542. © 2013, Nature Publishing group, a division of Macmillan Publishers Limited. All rights reserved; Figure 11.37: Adapted with permission of Springer Nature: *Nature* 403, Norman Myers, Russell A. Mittermeier, Cristina G. Mittermeier, et al., "Biodiversity hotspots for conservation priorities," pp. 853–858. © 2000, Macmillan Magazines Ltd.

Chapter 12

Photos: Chapter Opener: USFWS Photo/Alamy Stock Photo; Figure 12.2A: Courtesy of Michael P. Marchetti; Figure 12.3A: Lynn Amaral/Alamy Stock Photo; Figure 12.3B: WaterFrame/Alamy Stock Photo; Figure 12.3C: All Canada Photos/Alamy Stock Photo; Figure 12.4: Nature Picture Library/Alamy Stock Photo; Figure 12.5A: All Canada Photos/Alamy Stock Photo; Figure 12.6: Richard Ellis/Alamy Stock Photo; Figure 12.7: Avalon.red/Alamy Stock Photo; Figure 12.8: All Canada Photos/Alamy Stock Photo; Figure 12.9: STUDIO75/Alamy Stock Photo; Figure 12.13: Steve Apps/Alamy Stock Photo; Figure 12.14: Pally/Alamy Stock Photo; Figure 12.15: All Canada Photos/Alamy Stock Photo; Figure 12.17A: Hans Blossey/Alamy Stock Photo; Figure 12.17B: Anthony Dunn/Alamy Stock Photo; Figure 12.20: Worldspec/NASA/Alamy Stock Photo; Figure 12.21: Johanna Turner; Figure 12.22A: 24BY36/Alamy Stock Photo; Figure 12.22B: Design Pics Inc/Alamy Stock Photo; Figure 12.27: Michael Rosebrock/Alamy Stock Photo; Figure 12.29: Michael Wheatley/Alamy Stock Photo.

Other: Figure 12.3C: From Fink. D., T. Auer, A. Johnston, et al. eBird Status and Trends. Data Version: 2020, Released: 2021. Cornell Lab of Ornithology, Ithaca, New York. Reused with permission of Cornell Lab of Ornithology. This material uses data from the eBird Status and Trends Project at the Cornell Lab of Ornithology, eBird.org. Any opinions, findings, and conclusions or recommendations expressed in this material are those of the authors and do not necessarily reflect the views of the Cornell Lab of Ornithology; Figure 12.5B: Republished with permission from John Wiley & Sons, from Terry D. Beacham, Richard J. Beamish, John R. Candy, et al., "Stock-specific migration pathways of juvenile Sockeye Salmon in British Columbia waters and the Gulf of Alaska," *Transactions of the American Fisheries Society* 14(6): 1386–1403 (2014). Permission conveyed through Copyright Clearance Center, Inc.; Figure 12.6: Based on Monarch Butterfly: Fall & Spring Migration map by Paul Mirocha. Reproduced with permission of Monarch Watch; Figure 12.8B: Republished with permission from John Wiley & Sons, from Jeffrey P. Hoover, "Decision rules for site fidelity in a migratory bird, the Prothonotary Warbler," *Ecology* 84(2): 416–430 (2003). Permission conveyed through Copyright Clearance Center, Inc.; Figure 12.9B: Adapted from Ingo Kowarik and Moritz von der Lippe, "Secondary wind dispersal enhances long-distance dispersal of an invasive species in urban road corridors," *NeoBiota* 9: 49–70 (2011). Reprinted with permission of the authors; Figure 12.23: Adapted from Jared M. Diamond, "Biogeographic kinetics: estimation of relaxation times for avifaunas of southwest Pacific islands," *PNAS* 69(11): 3199–3203 (1972). Reprinted with permission of the author; Figure 12.28: Repub-

lished with permission from John Wiley & Sons, from Daniel S. Simberloff and Edward O. Wilson, "Experimental zoogeography of islands. A two-year record of colonization," *Ecology* 51(5): 934–937 (1970). Permission conveyed through Copyright Clearance Center, Inc.

Chapter 13

Photos: Chapter Opener: Doug Perrine/Alamy Stock Photo; Figure 13.1: Google Earth; Figure 13.2A (left): john lambing/Alamy Stock Photo; Figure 13.2A (center): agefotostock/Alamy Stock Photo; Figure 13.2A (right): William Mullins/Alamy Stock Photo; Figure 13.3 (left to right): Galyna Andrushko/Alamy Stock Photo, Dennis Frates/Alamy Stock Photo, Alexey Kamenskiy/Alamy Stock Photo, and courtesy of Michael P. Marchetti; Figure 13.4A: Purestock/Alamy Stock Photo; Figure 13.4B: imageBROKER/Alamy Stock Photo; Figure 13.5A: Kraig Lieb/Alamy Stock Photo; Figure 13.5B: Francisco Blanco/Alamy Stock Photo; Figure 13.5C: UPI/Alamy Stock Photo; Figure 13.5D: Tom Uhlman/Alamy Stock Photo; Figure 13.6A: Gail Wagner; Figure 13.6B: Tim Gainey/Alamy Stock Photo; Figure 13.7 (left): Jennifer Booher/Alamy Stock Photo, Wichai Prasomsri1/Shutterstock, CLS Digital Arts/Shutterstock, Eric Isselee/Shutterstock, AGAMI Photo Agency/Alamy Stock Photo, AGAMI Photo Agency/Alamy Stock Photo, www.pqpictures.co.uk/AlamyStock Photo, Jim Lane/Alamy Stock Photo; Figure 13.7 (right): Nature Photographers Ltd/Alamy Stock Photo, Jakub Krechowicz/Shutterstock, akepong srichaichana/Shutterstock, 3DMI/Shutterstock, Billion Photos/Shutterstock, and kostasgr/Shutterstock; Figure 13.8A: World History Archive/Alamy Stock Photo; Figure 13.8B: Jim Corwin/Alamy Stock Photo; Figure 13.9: Courtesy of Roger del Moral; Figure 13.11: Spring Images/Alamy Stock Photo; Figure 13.14: HHelene/Alamy Stock Photo; Figure 13.15: Tomas Kaspar/Alamy Stock Photo; Figure 13.16: Tom Uhlman/Alamy Stock Photo; Figure 13.17: M.Svetlana/Shutterstock, Mongkon N. Thongsai/Shutterstock, and Aonprom Photo/Shutterstock; Figure 13.18: M.Svetlana/Shutterstock and Mongkon N. Thongsai/Shutterstock; Figure 13.19: M.Svetlana/Shutterstock, Mongkon N. Thongsai/Shutterstock, Aonprom Photo/Shutterstock, and ANote/Alamy Stock Photo; Figure 13.22 (left): Courtesy of Paul van Giersbergen; Figure 13.22 (right): Lars Petersson; Figure 13.23: Malcolm Storey/www.bioimages.org.uk, Science History Images/Alamy Stock Photo, Michael Abbey/Science Source, Sinclair Stammers/Science Source, Ernie Cooper/Shutterstock, Frank Fox/Science Source, and blickwinkel/Alamy Stock Photo; Figure 13.25: Rich Reiner; Figure 13.26A: Courtesy of Jason P. Andras and Kate Ballantine; Figure 13.27: Mike Hill/Alamy Stock Photo; Figure 13.28: Carver Mostardi/Alamy Stock Photo; Figure 13.29A: Hulton-Deutsch Collection/CORBIS/Corbis via Getty Images; Figure 13.29B: The New York Times/Redux; Figure 13.30: ITAR-TASS News Agency/Alamy Stock Photo; Figure 13.33A: FLPA/Alamy Stock Photo; Figure 13.33B: ZUMA Press, Inc./Alamy Stock Photo.

Other: Figure 13.10: Republished with permission from Cengage, from Eugene Pleasants Odum and Roger L. Kroodsma, *Fundamentals of Ecology*, 3rd Edition (1971). Permission conveyed through Copyright Clearance Center, Inc.; Figure 13.12: Adapted from Robert H. MacArthur, and Joseph H. Connell, *The Biology of Populations*. © 1966 by John Wiley & Sons, Inc. Reprinted with permission of John Wiley & Sons, Inc.; Figure 13.13: Courtesy of Roger del Moral; Figure 13.24: Republished with permission of University of Chicago Press, from James A.

Drake, "Community-assembly mechanics and the structure of an experimental species ensemble," *The American Naturalist* 137(1): 1–26 (1991). Permission conveyed through Copyright Clearance Center, Inc.; Figure 13.26BC: Adapted from Jason P. Andras, William G. Rodriguez-Reillo, Alexander Truchon, et al., "Rewilding the small stuff: The effect of ecological restoration on prokaryotic communities of peatland soils," *FEMS Microbiology Ecology* 96(10): fiaa144 (2020). By permission of the Federation of European Microbiological Societies.

Chapter 14

Photos: Chapter Opener: Gaja Snover/Alamy Stock Photo; Figure 14.1: D. Finn/American Museum of Natural History; Figure 14.2: BIOSPHOTO/Alamy Stock Photo, Eric Isselee/Shutterstock, Wichai Prasomsri1/Shutterstock, and Manfred Ruckszio/Alamy Stock Photo; Figure 14.3A: Nature Picture Library/Alamy Stock Photo; Figure 14.3B: Seaphotoart/Alamy Stock Photo; Figure 14.4: Valentyna Chukhlyebova/Alamy Stock Photo; Figure 14.10: Nina Bednarsek, National Oceanic and Atmospheric Administration Pacific Marine Environmental Laboratory; Figure 14.11: Fabricius, K., Langdon, C., Uthicke, S., et al. "Losers and winners in coral reefs acclimatized to elevated carbon dioxide concentrations," *Nature Clim Change* 1: 165–169 (2011). Published 2011 by Nature Publishing Group. Reproduced with permission of SNCSC. Photos courtesy of Katharina Fabricius and Australian Institute of Marine Science; Figure 14.14: Courtesy of Neil Entwistle; Figure 14.15A: Design Pics Inc/Alamy Stock Photo; Figure 14.15B: Zoonar GmbH/Alamy Stock Photo; Figure 14.18: Ashley Cooper/Alamy Stock Photo; Figure 14.20: NASA/Landsat/Alamy Stock Photo; Figure 14.24: Courtesy of John Aber/Harvard Forest; Figure 14.29B: Courtesy of Rod Fensham; Figure 14.30A: Yon Marsh/Alamy Stock Photo; Figure 14.30B: Custom Life Science Images/Alamy Stock Photo; Figure 14.30C: Custom Life Science Images/Alamy Stock Photo.

Other: Figure 14.4: Republished with permission of Annual Reviews, from Daniel M. Alongi, "Carbon cycling and storage in mangrove forests," *Annual Review of Marine Science* 6: 195–219 (2004). Permission conveyed through Copyright Clearance Center, Inc.; Figure 14.5: Adapted with permission of John Wiley & Sons, from James Hutchison, Andrea Manica, Ruth Swetnam, et al., "Predicting global patterns in mangrove forest biomass," *Conservation Letters* 7(3): 233–240 (2013). Permission conveyed through Copyright Clearance Center, Inc.; Figure 14.9: Courtesy of Andrew Yool; Figures 14.11D and 14.12: Adapted with permission of Springer Nature: *Nature Climate Change* 1, Katharina E. Fabricius, Chris Langdon, Sven Uthicke, et al., "Losers and winners in coral reefs acclimatized to elevated carbon dioxide concentrations," pp. 165–169. © 2011, Nature Publishing Group; Figure 14.16: Republished with permission of John Wiley & Sons, from James M. Helfield and Robert J. Naiman, "Effects of salmon-derived nitrogen on riparian forest growth and implications for stream productivity," *Ecology* 82(9): 2403–2409 (2001). Permission conveyed through Copyright Clearance Center, Inc.; Figure 14.17: Republished with permission of John Wiley & Sons, from Joseph E. Merz and Peter B. Moyle, "Salmon, wildlife, and wine: Marine-derived nutrients in human-dominated ecosystems of Central California," *Ecological Applications* 16(3): 999–1009 (2006). Permission conveyed through Copyright Clearance Center, Inc.; Figure 14.19: Adapted from James N. Galloway, Wilfried Winiwarter, Adrian Leip, et al., "Nitrogen footprints: past, present and future," *Environmental Research Letters* 9(11): 115003 (2014). © IOP Publishing. Reproduced with permission. https://doi.org/10.1088/1748-9326/9/11/115003; Figures 14.22 and 14.23: Republished with permission of John Wiley & Sons, from Colin Averill, Michael C. Dietze, and Jennifer Bhatnagar, "Continental-scale nitrogen pollution is shifting forest mycorrhizal associations and soil carbon stocks," *Global Change Biology* 24(10): 4544–4553 (2018). Permission conveyed through Copyright Clearance Center, Inc.; Figure 14.25: Republished with permission of the American Association for the Advancement of Science, from Carly J. Stevens, Nancy B. Dise, J. Owen Mountford, et al., "Impact of nitrogen deposition on the species richness of grasslands," *Science* 303(5665): 1876–1879 (2004). Permission conveyed through Copyright Clearance Center, Inc.; Figure 14.28: Adapted from IPCC (2013): *Climate Change 2013: The Physical Science Basis. Contribution of Working Group I to the Fifth Assessment Report of the Intergovernmental Panel on Climate Change* [Stocker, T.F., D. Qin, G.-K. Plattner, M. Tignor, S.K. Allen, J. Boschung, A. Nauels, Y. Xia, V. Bex, and P.M. Midgley (eds.)]. Cambridge University Press, Cambridge, United Kingdom and New York, NY, USA, 1585pp. Reused with permission from the IPCC. Original IPCC caption: Precipitation change patterns derived from transient simulations from the CMIP5 ensembles, scaled to 1°C of global mean surface temperature change. The patterns have been calculated by computing 20-year averages at the end of the 21st century over the period 1986–2005 for the available simulations under all RCPs, taking their percentage difference and normalizing it, grid-point by grid-point, by the corresponding value of global average temperature change for each model and scenario. The normalized patterns have then been averaged across models and scenarios. The colour scale represents percent per 1°C of global average temperature change. Stippling indicates where the mean change averaged over all realizations is larger than the 95% percentile of the distribution of models. Figures 14.31B and 14.32: Adapted from Cristhiana P. Röpke, Sidinéia Amadio, Jansen Zuanon, et al., "Simultaneous abrupt shifts in hydrology and fish assemblage structure in a floodplain lake in the central Amazon," *Scientific Reports* 7: 40170 (2017). https://doi.org/10.1038/srep40170. This work is licensed under a Creative Commons Attribution 4.0 International License. http://creativecommons.org/licenses/by/4.0/; Figure 14.33: Republished with permission of the American Association for the Advancement of Science, from Paul J. Durack, Susan E. Wijffels, and Richard J. Matear, "Ocean salinities reveal strong global water cycle intensification during 1950 to 2000," *Science* 336(6080): 455–458 (2012). Permission conveyed through Copyright Clearance Center, Inc.; Figure 14.34: Adapted with permission of Springer Nature: *Nature* 486, Bradley J. Cardinale, J. Emmett Duffy, Andrew Gonzalez, et al., "Biodiversity loss and its impact on humanity," pp. 59–67. © 2012, Nature Publishing Group, a division of Macmillan Publishers Limited. All rights reserved; Table 14.1: Adapted from William H. Schlesinger, "On the fate of anthropogenic nitrogen," *PNAS* 106(1): 203–208 (2009). Reprinted with permission of National Academy of Sciences.

Chapter 15

Photos: Chapter Opener: Album/Alamy Stock Photo; Figure 15.2: J.P. Briner; Figure 15.5: Caner CIFTCI/Alamy Stock Photo; Figure 15.6 (left to right): Nuwat Chanthachanthuek/Alamy Stock Photo, Library of Congress, ITAR-TASS News Agency/Alamy Stock Photo, Lee Rentz/Alamy Stock Photo, Todd Bannor/Alamy Stock Photo, WaterFrame/Alamy Stock Photo, and Peter Yeeles/Alamy Stock Photo; Figure 15.8: Courtesy of Matthew Sileo and Rosyid A Azhar/Shutterstock; Figure 15.9 (left): Arterra Picture Library/Alamy Stock Photo; Figure 15.9 (right): imageBROKER/Alamy Stock Photo; Figure 15.13: Kelvin Aitken/VWPics/Alamy Stock Photo; Figure 15.16: Lucia Terui/Getty Images; Figure 15.19: Nature Picture Library/Alamy Stock Photo; Figure 15.21: Reinhard Dirscherl/Alamy Stock Photo, Joao Ponces/Alamy Stock Photo, Stocktrek Images, Inc./Alamy Stock Photo, Sam Hodge/Alamy Stock Photo, Helmut Corneli/Alamy Stock Photo, Gabriele Rohde/Shutterstock, and RDW-Underwater/Alamy Stock Photo; Figure 15.23: Zoonar GmbH/Alamy Stock Photo; Figure 15.24: Natural History Archive/Alamy Stock Photo; Figure 15.34: Juan Vilata/Alamy Stock Photo.

Other: Figure 15.3: Republished with permission of the American Association for the Advancement of Science, from Colin N. Waters, Jan Zalasiewicz, Colin Summerhayes, et al., "The Anthropocene is functionally and stratigraphically distinct from the Holocene," *Science* 351(6269): aad2622 (2016). Permission conveyed through Copyright Clearance Center, Inc.; Figure 15.4: Republished with permission of John Wiley & Sons, from Fernando Valladares, Silvia Matesanz, François Guilhaumon, et al., "The effects of phenotypic plasticity and local adaptation on forecasts of species range shifts under climate change," *Ecology Letters* 17: 1351–1364 (2014). Permission conveyed through Copyright Clearance Center, Inc.; Figure 15.7B: Republished with permission of the American Association for the Advancement of Science, from M.C. Urban, G. Bocedi, A.P. Hendry, et al., "Improving the forecast for biodiversity under climate change," *Science* 353(6304): aad8466 (2016). Permission conveyed through Copyright Clearance Center, Inc.; Figure 15.9: Courtesy of The Woodland Trust and Jean Combes. Nature's Calendar data is freely available to students upon request at https://naturescalendar.woodlandtrust.org.uk/; Figure 15.10: Republished with permission of the American Association for the Advancement of Science, from Gerardo Ceballos, Paul R. Erlich, Anthony D. Baronsky, et al., "Accelerated modern human-induced species losses: Entering the sixth mass extinction," *Science Advances* 1(4): e1400253 (2015). © The Authors, some rights reserved; exclusive licensee AAAS. Distributed under a CC BY-NC 4.0 License (http://creativecommons.org/licenses/by-nc/4.0/). Figures 15.13 and 15.14: Republished with permission of John Wiley & Sons, from Francesco Ferretti, Boris Worm, Gregory L. Britten, et al., "Patterns and ecosystem consequences of shark declines in the ocean," *Ecology Letters* 13: 1055–1071 (2010). Permission conveyed through Copyright Clearance Center, Inc.; Figure 15.17A: From Philip W. Lambdon, Peter Pyšek, Corina Basnou, et al., "Alien flora of Europe: species diversity, temporal trends, geographical patterns and research needs," *Preslia* 80: 101–149 (2008). Reused with permission of Czech Botanical Society; Figures 15.17B and 15.18AB: Republished with permission of John Wiley & Sons, from Philip E. Hulme, "Trade, transport and trouble: Managing invasive species pathways in an era of globalization," *Journal of Applied Ecology* 46: 10–18 (2009). Permission conveyed through Copyright Clearance Center, Inc.; Figure 15.20: Reprinted from *Environmental Research* 111, Jesús Ernesto Arias-González, Carlos González-Gándara, José Luis Cabrera, et al., "Predicted impact of the lionfish *Pterois volitans* on the food web of a Caribbean coral reef," pp. 917–925. © 2011, with permission from Elsevier; Figure 15.21: Adapted with permission

INDEX

high atmospheric pressure, 62, *62*

Hilborn, Ray, 14

Hoeksema, Jason, 370–71

Hofmann, Richard, 453, *453*

Holocene, 624, *625*, 650

homeotherms, 191, *192*, 196

homologous chromosomes, *131*

homozygous genotype, 132, *133*

Hoopes, Martha, 141, 144, *144*, 446, *447*, 448

Hoover, Jeffrey, 493, *493*

horseweed (*Erigeron canadensis*), 538, 543, 546

hosts, 308, 333–34, 335

hotspots of biodiversity, 471, 474, 476, *477*

house, ecological, 1–2, *2*, 4, 202

HSS (Hairston, Smith, and Slobodkin), 402–3, 414

Huey, Raymond, *193*, 193–94

humans
    energy to feed population, 418–19
    energy transfer up food chain to, 393–94, *394*
    gut microbiota, 360–61, *361*
    health improved by nature, 660
    history of impact on Earth, 623
    landscape modification by, *654*, 654–55, *655*
    movement of, 483–84, *484*, 493
    population, *3*, 4, 209, 650, *650*
    water loss in body, 181–82, *182*
    See also anthropogenic effects

humpback whale (*Megaptera novaeangliae*), 487, *488*, 489–91

hunting
    human selective factors, 157, *157*
    population counts based on, 214, *214*

hurricanes
    flooding reduced by wetlands, *657*, 657–58
    mangrove damage, *533*, 575

Hutchinson, G. Evelyn, 197

Hutchinsonian niche, 197–98, 199, 200, *200*

hydrologic cycle. *See* water cycle

hydrosphere, 56, *56*

hyperparasitoids, 333

hypolimnion, *99*

hypotheses, 8, 9

hypothesis testing
    best-fit comparisons, 13–17, *15*, *16*
    experimental, 8–11, *9*, *10*, 20

I

ice, density of, 98

ice cores, 624, *625*

icefish, white-blooded, 75, *76*

ice sheets, melting of, *110*

immigration, 217

incidence function models, 511

indirect effects
    common types of, 421–27
    concept of, 388–89

facilitative, 377, 380

of keystone species, 422–24

in recovery of ecosystem function, 659

in resource competition, 260

inducible defenses, 344

Industrial Revolution, 623

industrial waste sites, revegetation of, 532

infauna, 104, *104*

ingested energy, 167–68, 170–71

inhibition model of succession, 543, *544–45*, 546
    disturbances and, 559
    evidence for, 547
    priority effects and, 549

inorganic carbon, 574–75, 582

inorganic nitrogen, 591
    AM fungi and, *601*, 601–2

insectivores, 463, 465

insects
    counting of populations, 214
    declining populations, 635, *636*
    global species richness, 439–40
    as herbivores, 330–32, *331*, *332*
    increase in non-native introductions, 647, *648*
    migratory behavior, 491, *492*, 493
    parasitoid predators of, 333
    See also ants; beetles; butterflies; dragonflies; moths, defenses of

instantaneous per capita rate of population growth (*r*), 226

interference competition, 260

intermediate disturbance hypothesis, *562*, 562–63, 564

intermediate host, 333–34

intertidal systems
    intermediate disturbance hypothesis and, 562
    See also rocky intertidal food webs; tide pools

intrinsic rate of increase (r_{max}), 229
    r-selected species and, 535

invasional meltdown, *379*, 379–80, *381*

invasive species
    Australian spotted jellyfish, 639
    capacity of countries to address, 647, *649*
    changing trophic architecture, 412–13, *413*
    definition of, 640
    extinctions caused by, 290, 346
    future of, 647, *648*, *649*
    global trade and, 647, *648*
    homogenization of habitats and, 4
    lionfish, *642*, 642–43, *643*, *644*
    spatial diversity patterns and, 453–54, *454*
    See also non-native species

inverse density dependence, 327, *328*

Io moth (*Automeris io*), *338*, 338

island biogeography
    equilibrium theory, 514–18, *515*, *516*
    experimental evidence, 516–17, *517*, 518t

Pacific island bird species and, 514, *514*

species-area relationship and, 450, *513*, 513–14

isoclines
    in competition phase plane, *271*, 271–72, *272*
    in economic model of trade, 371, *371*, 373, *374*
    in exploitation phase plane, 316–19, *317*
    zero net growth (ZNGI), 280–82, *281*, *282*

isolation acquisition isocline (IAI), 371, *371*

iteration of model, *34*, 35
    in adaptive modeling, 630, *631*

iterative prisoner's dilemma (IPD), 369–70

iteroparous organisms, 169

J

jack-o'-lantern mushroom (Omphalotus olearius), 308, *309*

jellyfish
    alternative stable state and, *639*, 639–40
    asexual reproduction in, *128*, 128

Jetz, Walter, *464*, 464–65, *465*

just-so stories, 8, *8*

K

kangaroo rat (*Dipodomys merriami*), 180–82, *181*, *182*

kapok tree (*Ceiba pentandra*), 344

Karner blue butterfly (*Lycaeides melissa samuelensis*), 503, *503*

Kaye, Jason, 655

Keeling, Charles David, 584–85

Keeling curve, 585, *585*

Keever, Catherine, 538, 540, 541, 549

kelp forests, 103, *103*
    protected by otters, 424

keystone metaphor, 422, *422*

keystone mutualists, 422–23, *424*

keystone predators, 422, *423*, 424–25

keystone species, 422–24

kidneys, and water balance, 182, *182*

Kie, John, 236–37, *237*

killer whales (orcas), 490, *490*

kingdom, *125*

Kipling, Rudyard, 8, *8*

Knapp, Roland, 309–10

Koenig, Jeremy, 360

Krebs, Charles, 12, *13*

Krebs, John, 176–77, *177*

K-selected species, 535, 535–36, 543, 546, 554, 562

L

lace monitor (*Varanus varius*), *193*, 193–94

lady beetles, *127*, 326, 342

lakes
    anthropogenic runoff to, 396, *396*, 400
    circulation between layers, 98–99, *99*
    trophic status of, *399*, 399–400

source-sink dynamics and, 508–10, *509*, *510*

See also habitat patches; Levins metapopulation model

methane

from fossil fuels, 584

in Great Acceleration, *3*, *4*

as greenhouse gas, 584

in ice cores, *625*

in initial Earth atmosphere, 622

mangrove carbon sink and, 580

from microorganisms in sediment, 576, 578

methanotrophy, in restored wetlands, *555*, 556

microparasites, 333

symbiotic, 358

migration, 487–92

definition of, 485–86

vs. dispersal, 485–87, *486*, 494

mimicry, *340*, 340–42, *341*, *342*

mines, succession on sites of, 532, *533*, 554

mites (*Demodex folliculorum*), 376, *376*

models, ecological, 17–20, 30–38

categorical variables in, 40

conceptual, 17–18, *18*, 31–34, *33*, 45–46, *46*

deterministic, 37, 44

mathematical, 18–19, *34*, 34–38, *36*, *38*

parameter variation in, 36–38, *38*

reality and, 19, *19*, 30

simulation models, 18–19, *19*, 37–38, *38*

stochastic, 37–38, *38*, 44, 564

mollusks

giant African snail, 380, *381*

increase in non-native introductions, 647, *648*

periwinkle snail, 11, *12*

monarch butterfly (*Danaus plexippus*)

life stages, 210, *210*

migration, 491, *492*

monotonic damping, 234, *235*

monsoons, 609

in the Amazon, 611–12

Moore, Thomas, 126, *126*

morphology

cryptic species and, 436, *436*, *437*

definition of species based on, 122–28, *123*

moths, defenses of, 337, *337*, 338, *338*

mountain lion (*Puma concolor*)

defense against, 343, *343*

dispersal distance, 512, *512*

fundamental niche, 263–64, *264*

as keystone species, 422, *423*

in New England, 437–38

mountains

biological zones, 96, *96*

rain shadows and, 71, *72*

mountain yellow-legged frogs (*Rana muscosa* and *R. sierrae*), 309–10, *310*

Mount St. Helens succession, *536*, 536–37, *537*

early colonizers, 542, *542*, 543

long-term trends, *540*, 540–42

seed dispersal, 541, *541*

spatial mosaic, 541, *542*

Tilman model and, 548

movement

essential questions, 485

multispecies models, 513–19

See also dispersal; migration

Moyle, Peter, 596–97

Müller, Fritz, 340–41

Müllerian mimicry, *340*, 340–41

multiple hypothesis testing, 13–17, *15*, *16*

Mumby, Peter, 659, *659*

mutations, *135*, 135–36, 135*t*, 137, 146

mutualism

benefits of, 358–62

cheating in, *368*, 368–69, *369*, 376

conceptual model, 363

conclusions about, 375–76

dispersal and, 504

economic model, 370–75, *371*, *372*, *373*, *374*

game theory models, 366–70, *367*

keystone mutualists, 422–23, *424*

Lotka-Volterra approaches, 363–66, *365*

between non-native species, 379–80

(+, +) notation for, 354

obligate and facultative, 354–58

partner specificity in, 355–56

of plants with diazotrophs, 592

in pollination, 355, 356–57, *357*, 358–59, *359*, 422–23, *424*

in restoration projects, 556

in seed dispersal, 355, *357*, 357, 359–60, 422, 541

See also mycorrhizae

mycorrhizae

mutualism with plants, 361, *361*

mutualism with trees, 370–75, *372*, *373*, *374*

nitrogen and, 361, *601*, 601–2

two types of, 600–602, *601*

nature-based solutions, 658

nature reserves, 518

n-dimensional hypervolume, 199, *200*

nearshore biological zones, 100–104. *See also* coral reefs; estuaries; mangrove forests

nectivores, 463

nematodes, 437, 440

neritic zone, 97

net primary production (NPP)

humans' appropriation of, 650

in mangrove forests, 578

by phytoplankton, 586

neutral alleles, 136, 137

neutral mutations, 136

neutral stability, 319, *319*, 320, *320*, 325, 326

Allee effects and, *328*, 329

niche

basic concepts, 197–99

climate change and, 625–28, 627*t*

of Devil's Hole pupfish, 198–200, *199*, *200*, *201*

as position in ecosystem, 201, *202*

range size and, 461, 641

resource utilization curves and, 200, *200*, *201*

niche flexibility, 626–28, 627*t*

niche overlap, 263, *263*, 264–65, *265*

character displacement and, *294*, 294–95, *295*, 296

in stable coexistence, 275

niche shift, 298

by non-native fish species, 645, *645*

niche space, 198–99, *199*

functional traits and, 463

Nile perch (*Lates niloticus*), 561, *561*

nitrate (NO_3^-), 591, 592, 593

denitrification and, *594*, 595

in fertilizers, 592, 593, 595, 599

in ice cores, 625

nitrogen

for anemone from clownfish, 362

atmospheric, 591–92, 623

in food webs, 593, 595–97, *596*, *597*

forms of, 591–92

mycorrhizae and, 361, *601*, 601–2

in soil, 542, *542*

nitrogen, anthropogenic, 597–98, *598*

ecological effects of, 598–603

excess affecting plants and communities, 602, 602–3, *603*

forests and, 600–603, *601*, *602*

in nitrogen budget, 598–600, 599*t*

removed in wetland soils, *555*, 556

runoff into coastal waters, *3*, *4*, 600, *600*

runoff into lakes, 396, *396*, 400

nitrogenase, 592

nitrogen assimilation, 593

nitrogen budget, 598–600, 599*t*

nitrogen cycle, 593, *594*, 595

altered by eastern white pine, 646, *646*

nitrogen fixation
  biological, 592, 593, 606, 646
  industrial, 597–98, *598*, 599, 600
  by invasive plants, 646
  phosphorus and, 606
nitrogen isotopes, 411–13, *412*
  marine-derived, *597*
  of native and non-native fish, 645, *645*
nitrogenous wastes, 593
nitrogen oxides
  combustion engines and, 598, 599
  lightning and, 592
Nobis, Agnieszka, 453–54, *454*
nodes of phylogeny, 465, *466*
nominal categorical variables, 40, *40*
Nomura's jellyfish (*Nemopilema nomurai*), 639
nonequilibrium dynamics, 564, 565
non-native species
  Colorado River fish, 645, *645*
  definition of, 640
  emerging infectious diseases and, 346
  global distribution, *640*, 640–41, *641*
  in haphazard solutions, 658
  homogenization of habitats and, 4
  in Lake Tahoe, 391, 396
  as predators, 560, 561, *561*
  rise in number of, 346
  slowing rate of introductions, 647, *648*
  trait divergence of, 157
  See also invasive species
normal distribution, 48–49, *49*
North Carolina Piedmont, *538*, 538–39, *539*, 540, 541, 543, 546
nuclear fallout, 624, *625*
null hypothesis, 9, 46–48
numerical data, 40–42, *41*
nutrient budgets, 572–74, *573*
nutrient cycles, 572–73
  ecosystem engineers and, 645
nutrients
  acquired in mutualism, 360–61, *361*, 362
  anthropogenic runoff into lakes, 396, *396*, 400
  traded in mutualism, 370–75

O

oak trees, change in budburst, 632, *633*
obligate mutualism, 354–56
observation, 5–8
  hypothesis testing and, 9, 10, 11
  large-scale and quantitative, *7*, 7–8, 13–14
ocean acidification
  coral reefs and, *588*, 588–89, *589*
  Great Acceleration in, *3*, 4
  pH change over two centuries, 587, *587*
  weakening carbonate parts, 587, *587*, 588–89
oceanic carbon, 585–89, *586*, *587*
oceanic circulation, *69*, 69–70
  thermally-driven layers and, 98–99
oceanic zone, 104–5, *105*

ocean salinity, 612–13, *614*
Oculina patagonica, 358, *358*
Olden, Julian, 645
oligotrophic lakes, *399*, 399–400
omnivores, 403, 407
  squirrels as example, 409–10, *410*
ontogenetic diet shifts, 405, 407
open ocean, 104–5, *105*
open populations, 211
optimal foraging, 170–79
  basic theory, 172–73
  comparing prey choices, 174–76, *175*
  maximizing fitness and, 170–71, *172*
  with mobile vs. immobile prey, 179
  simple model, 173–74
  tests of, 176–79, *177*, *178*
optimization, and trade-offs, 166–67
orcas (*Orcinus orca*), 490, *490*
order, *125*
ordinal categorical variables, 40, *40*, 42
organically grown produce, 345–46
organic carbon, 574–75
  dissolved (DOC), 578–79, *580*
  EM fungi and, 601, 602
  particulate (POC), 579, *580*
organic nitrogen, 591
  EM fungi and, 601
organism, definition of, 1
oxygen
  of early atmosphere, 622–23
  phytoplankton blooms and, 400

P

Paine, Robert T., 422, 424
Paramecium aurelia, *261*, 261–62, *262*, 264, 266, 267–68, 274
Paramecium caudatum
  as competitor, *261*, 261–62, *262*, 264, 266, 267–68, 274
  as prey, 310–12, *311*, 321
parameters, 35–36, *36*
  in adaptive modeling, 630, *631*
  birth and death rates used as, 219
  provided by data, 44
  variation in, 36–38, *38*
parameter space, 36
parasites, 308, 332–35
  in food webs, 408–9, *409*
parasitoids, 333
Park, Thomas, 277, *277*, 281
parrotfishes (family Scaridae), 659, *659*
particulate organic carbon (POC), 579, *580*
patches. *See* habitat patches
pathogens, 308, 333
  symbiotic, 358
peacocks, 148, *148*
peat wetlands
  as carbon sinks, 581, 582, 656
  prokaryotic soil communities, *555*, 555–56
pelagic zone, 97, 104–6, *105*
Pendleton, Linwood, 582

per capita rate of population growth
  finite (λ), 219
  instantaneous (r), 226
  intrinsic (r_{max}), 229
  See also lambda (λ)
per capita rates, 219
per capita rates of birth and death, 219
  density dependence and, 228
per capita survival rate, 219
periphyton, 411, 412, *412*
  of headwater streams, 421
periwinkle snail (*Littorina littorea*), 11, *12*
permafrost
  in boreal forest, 93
  carbon stores in, 114
  in tundra, 95
Petchey, Owen, 464, *464*
Pfennig, David, 296–97, 342, *342*
phase-plane graphs
  for Lotka-Volterra competition model, 270, *270*
  for Lotka-Volterra exploiter-prey model, 316, *317*
phenology, 140–43, *143*
  climate change and, 632–34, *633*
phenotype, 131–32, *133*
phosphate
  diatom competition for, 280, 281–83, 288, *288*
  mining of, 604, 606
  from mycorrhizae, 361, *361*, 370–75, *372*, *373*, *374*, 604
phosphorus
  anthropogenic runoff into lakes, 396, *396*, 400
  biomass in lake food chains and, 400, *401*
phosphorus cycle, 603–6, *605*
photic zone, 97, 97–98, 105, *106*
photorespiration, 184
photosynthesis
  in aquatic zones, 97–98
  carbon dioxide capture for, 183–86, *185*, *186*
  chemical reaction of, 574–75
  climate change and, 189, *189*, *190*
  as intake of energy, 169
  oxygen-rich atmosphere and, 622–23
  phosphorus and nitrogen affecting, 606
  primary productivity by means of, 76, 389
  rate of, 392, *392*, 606
  types of, 184–87, *185*, *186*, *187t*
phylogenetic distance, 466, *466*
phylogenetic diversity (PD), 465–68, *468*
  global pattern, *473*, 473–74, *474*
  lost in extinctions, 467–68, *468*, 476
phylogenies, 465–66, *466*
phylum, *125*
physiology, 166
  climate change and, 628, *629*
phytoplankton, 105
  anthropogenic coastal runoff and, 600, *600*, *601*

tiger shark, 422, *423*, 636

    trophic cascades and, 637, *638*, 640

side-blotched lizard (*Uta stansburiana*), 17–18, *18*

signal crayfish (*Pacifastacus leniusculus*), *32*

    conceptual population model, 32, *33*, 45–46

    mathematical population model, 34–35, *36*

    parameters in model, *35*, 35–37

    relationship between variables, 45–46, *46*

    sampling of population, 43

    statistical significance and, 46–47

silica dioxide, as plant defense, 344

silicate, diatom competition for, 280, 281–83, 288, *288*

Simberloff, Daniel, 379, 516–17, *517*

simulation models, 18–19, *19*, 37–38, *38*

sink ecosystems, 579

    turnover rate of, 581

sink patch, 509, *509*

sister species, 467

Sitka spruce (*Picea sitchensis*), 596, *596*

Slatkin, Montgomery, *193*, 193–94

Slatyer, Ralph, 542, 546–47, 549

Slobodkin, Lawrence, 400, 402–3

slow carbon cycle, 582, 584

Smith, Frederick, 400, 402–3

snow, 63, 608

snowshoe hare (*Lepus americanus*), 312, *312*, 329, 332

socioecological systems, 656

sockeye salmon (*Oncorhynchus nerka*), 491, *491*, *596*

soil microbes

    bacteria in carbon budget, 590

    bacteria in mangrove forests, 576

    carbon cycle and, 590

    nitrogen cycle and, 595

    in secondary succession, 534

    in wetlands, *555*, 555–56

soil respiration, 578

soils

    agriculture and, 114

    biomes and, 77

    carbon in, 590, 656

    of floodplains, 595

    formation of, 533

    fungal networks in tropical rainforests and, 79

    ground-foraging birds and, 465

    nitrogen in, 542, *542*, 606

    phosphorus in, 606

    in primary succession, 543

    in secondary succession, 533–34

    See also sediments; soil microbes

solar radiation, 57–59, *58*, *59*

    rainfall near equator and, 61, *62*

    See also sunlight

solstices, *58*, *59*

sound transects, 213, *213*

source ecosystem, 579

source patch, 509, *509*

source-sink dynamics, 508–10, *509*, *510*

spadefoot toad tadpoles, *296*, 296–97, *297*

sparse species, 460

spatial dynamics. *See* movement

spatial heterogeneity

    local extinction and, 16, *16*

    See also habitat heterogeneity

spatially explicit models, 510–11, 519

spatially implicit models, 510, 519

spatial mosaic, of plant communities, 541, *542*

Spea bombifrons and *S. multiplicata*, *296*, 296–97, *297*, 298

speciation, 150–56

    allopatric, 154

    anthropogenic habitat fragmentation and, 157

    gene flow and, 150–51

    genetic exchange and, 129, *129*

    geographic isolation and, 151–54, *152*, *153*

    hierarchy leading to, 129, *130*

    reproductive isolation and, 150–51, *152*, *153*, 154

    sympatric, *154*, 154–56, *155*

species

    biological species concept, 127–29

    classification of, 123, *124–25*, 156

    formally described, *434*, 434–39

    made up of populations, 129, *130*, 132

    modern definitions, 128–29

    morphological definition, 122–28, *123*

    undescribed, 435–37, 438–39

species-abundance distribution, *442*, 442–45

    log-normal, *443*, 443–44

    for wetland plants, 444–45, *445*

species-area relationship, 445–50

    island size and, *513*, 513–14

    land-use decisions and, 476

    power function, 448–50, *449*

    sampling design, 446, *447*, 448

species diversity

    herbivory as driver of, 332

    intermediate levels of disturbance and, *562*, 562–63

    See also Shannon-Wiener diversity index (H'); species richness (S)

species evenness, *441*, 441–42

    diversity indexes and, 456

    of fish in water cycle changes, 612, *612*

    hierarchical sampling design, 446

    Shannon-Wiener diversity index and, 456–58, 456*t*, 457*t*, 540

    succession and, 540

species richness (S)

    definition of, 433

    diversity indexes and, 456

    hierarchical sampling design, 446

    intermediate disturbance hypothesis and, *562*, 562–63

    island biogeography and, 513–18

land conversions and, 476

    as limited metric, 455

    nitrogen on grassland plants and, 603, *603*

    ocean acidification and, 589

    Shannon-Wiener diversity index and, 456–58, 456*t*, 457*t*

    spatial diversity metrics, 450–54, *452*, *453*, *454*

    successional increase in, 540, *540*

    of trees in tropical rainforests, 332

    See also biodiversity; global species richness

specific epithet, 122

spiny-tailed iguana (*Ctenosaura oaxacana*), *194*, 194–95

Spix's macaw (*Cyanopsitta spixii*), 455, *455*

spotted hyena (*Crocuta crocuta*), 166

spotted owl (*Strix occidentalis*), 505, *505*

spotted St. John's wort (Hypericum punctatum), *144*, 144–46

stabilizing factor, 323

stabilizing selection, *145*, 147

stable coexistence

    Allee effects and, *328*, 329

    in Lotka-Volterra competition model, 274–75, *275*, 279

    in Lotka-Volterra exploitation model, 323, 326, *328*, 329

    in Tilman model, 285, *286*, 287, 288, *288*

stable isotopes, 411–13, *412*, *413*, 597

    of native and non-native fish, 645, *645*

stable limit cycles, 234, *235*

stable stage distribution, 249

stable states, alternative, 639, *639*

stage structure, 240–42, *241*

stage-structured matrix model, 242, 244–50, *245*, 245*t*

    calculations in, 246–50, *247*, *248*, *249*

standard deviation, 48–49, *49*

Standish, Rachel, 658

statistical significance, 46, 47–48

steady state, 525, 563–64

Steffen, Will, 3–4

Stevens, Carly, 603

stinking cedar (*Torreya taxifolia*), 627, *627*

St. John's wort (Hypericum perforatum), 139–46, *142*, *143*

    native species *H. punctatum* and, *144*, 144–46

stochastic models, 37–38, *38*, 44, 564

Stoermer, Eugene F., 2–3, 624

stoma (*plural*, stomata), 183, *183*

strawberry poison dart frog (*Oophaga pumilio*), 153

streams

    nitrogen in, 599

    phosphorus in, 604, 606

    See also lotic ecosystems; rivers

strongly interacting species, 413–14, 422–24, 426–27. *See also* ecosystem engineers